ALGEBRA

Quadratic Formula

The solutions of the quadratic equation
$ax^2 + bx + c = 0$ are given by

$$x = \frac{-b \pm \sqrt{b^2 - 4ac}}{2a}.$$

Factorial notation

For each positive integer n,
$$n! = n(n-1)(n-2) \cdots 3 \cdot 2 \cdot 1;$$
by definition, $0! = 1$.

Radicals

$$\sqrt[n]{x^m} = \left(\sqrt[n]{x}\right)^m = x^{m/n}$$

Exponents

$$(ab)^r = a^r b^r \qquad a^r a^s = a^{r+s} \qquad x^{-n} = \frac{1}{x^n}$$

$$(a^r)^s = a^{rs} \qquad \frac{a^r}{a^s} = a^{r-s}$$

Binomial Formula

$$(x + y)^2 = x^2 + 2xy + y^2$$
$$(x + y)^3 = x^3 + 3x^2 y + 3xy^2 + y^3$$
$$(x + y)^4 = x^4 + 4x^3 y + 6x^2 y^2 + 4xy^3 + y^4$$

In general, $(x + y)^n = x^n + \binom{n}{1}x^{n-1}y + \binom{n}{2}x^{n-2}y^2$

$$+ \cdots + \binom{n}{k}x^{n-k}y^k + \cdots + \binom{n}{n-1}xy^{n-1} + y^n,$$

where the binomial coefficient $\binom{n}{m}$ is the integer $\dfrac{n!}{m!(n-m)!}$.

Factoring

If n is a positive integer, then
$$x^n - y^n = (x - y)(x^{n-1} + x^{n-2}y + x^{n-3}y^2 + \cdots$$
$$+ x^{n-k-1}y^k + \cdots + xy^{n-2} + y^{n-1}).$$

If n is an *odd* positive integer, then
$$x^n + y^n = (x + y)(x^{n-1} - x^{n-2}y + x^{n-3}y^2 - \cdots$$
$$\pm x^{n-k-1}y^k \mp \cdots - xy^{n-2} + y^{n-1}).$$

GEOMETRY

Distance Formulas

Distance on the real number line:
$$d = |a - b|$$

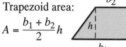

Distance in the coordinate plane:

$$d = \sqrt{(x_1 - x_2)^2 + (y_1 - y_2)^2}$$

Equations of Lines and Circles

Slope-intercept equation:
$$y = mx + b$$

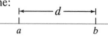

Point-slope equation:
$$y - y_1 = m(x - x_1)$$

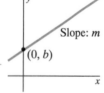

Circle with center (h, k) and
radius r:
$$(x - h)^2 + (y - k)^2 = r^2$$

Triangle area:
$$A = \tfrac{1}{2}bh$$

Rectangle area:
$$A = bh$$

Trapezoid area:
$$A = \frac{b_1 + b_2}{2}h$$

Circle area:
$$A = \pi r^2$$
Circumference:
$$C = 2\pi r$$

Sphere volume:
$$V = \tfrac{4}{3}\pi r^3$$
Surface area:
$$A = 4\pi r^2$$

Cylinder volume:
$$V = \pi r^2 h$$
Curved
surface area:
$$A = 2\pi rh$$

Cone volume:
$$V = \tfrac{1}{3}\pi r^2 h$$
Curved surface area:
$$A = \pi r \sqrt{r^2 + h^2}$$

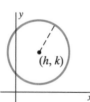

TRIGONOMETRY

$$\sin^2 A + \cos^2 A = 1 \qquad \text{(the *fundamental identity*)}$$
$$\tan^2 A + 1 = \sec^2 A$$

$$\cos 2A = \cos^2 A - \sin^2 A = 1 - 2\sin^2 A = 2\cos^2 A - 1$$
$$\sin 2A = 2 \sin A \cos A$$

$$\cos(A + B) = \cos A \cos B - \sin A \sin B$$
$$\cos(A - B) = \cos A \cos B + \sin A \sin B$$
$$\sin(A + B) = \sin A \cos B + \cos A \sin B$$
$$\sin(A - B) = \sin A \cos B - \cos A \sin B$$

$$\cos^2 A = \frac{1 + \cos 2A}{2} \qquad \sin^2 A = \frac{1 - \cos 2A}{2}$$

See the Appendices for more reference formulas.

TABLE OF INTEGRALS

ELEMENTARY FORMS

$$1 \quad \int u \, dv = uv - \int v \, du$$

$$2 \quad \int u^n \, du = \frac{1}{n+1} u^{n+1} + C \quad \text{if } n \neq -1$$

$$3 \quad \int \frac{du}{u} = \ln|u| + C$$

$$4 \quad \int e^u \, du = e^u + C$$

$$5 \quad \int a^u \, du = \frac{a^u}{\ln a} + C$$

$$6 \quad \int \sin u \, du = -\cos u + C$$

$$7 \quad \int \cos u \, du = \sin u + C$$

$$8 \quad \int \sec^2 u \, du = \tan u + C$$

$$9 \quad \int \csc^2 u \, du = -\cot u + C$$

$$10 \quad \int \sec u \tan u \, du = \sec u + C$$

$$11 \quad \int \csc u \cot u \, du = -\csc u + C$$

$$12 \quad \int \tan u \, du = \ln|\sec u| + C$$

$$13 \quad \int \cot u \, du = \ln|\sin u| + C$$

$$14 \quad \int \sec u \, du = \ln|\sec u + \tan u| + C$$

$$15 \quad \int \csc u \, du = \ln|\csc u - \cot u| + C$$

$$16 \quad \int \frac{du}{\sqrt{a^2 - u^2}} = \sin^{-1}\frac{u}{a} + C$$

$$17 \quad \int \frac{du}{a^2 + u^2} = \frac{1}{a} \tan^{-1}\frac{u}{a} + C$$

$$18 \quad \int \frac{du}{a^2 - u^2} = \frac{1}{2a} \ln\left|\frac{u+a}{u-a}\right| + C$$

$$19 \quad \int \frac{du}{u\sqrt{u^2 - a^2}} = \frac{1}{a} \sec^{-1}\left|\frac{u}{a}\right| + C$$

TRIGONOMETRIC FORMS

$$20 \quad \int \sin^2 u \, du = \frac{1}{2}u - \frac{1}{4}\sin 2u + C$$

$$21 \quad \int \cos^2 u \, du = \frac{1}{2}u + \frac{1}{4}\sin 2u + C$$

$$22 \quad \int \tan^2 u \, du = \tan u - u + C$$

$$23 \quad \int \cot^2 u \, du = -\cot u - u + C$$

$$24 \quad \int \sin^3 u \, du = -\frac{1}{3}(2 + \sin^2 u)\cos u + C$$

$$25 \quad \int \cos^3 u \, du = \frac{1}{3}(2 + \cos^2 u)\sin u + C$$

$$26 \quad \int \tan^3 u \, du = \frac{1}{2}\tan^2 u + \ln|\cos u| + C$$

$$27 \quad \int \cot^3 u \, du = -\frac{1}{2}\cot^2 u - \ln|\sin u| + C$$

$$28 \quad \int \sec^3 u \, du = \frac{1}{2}\sec u \tan u + \frac{1}{2}\ln|\sec u + \tan u| + C$$

$$29 \quad \int \csc^3 u \, du = -\frac{1}{2}\csc u \cot u + \frac{1}{2}\ln|\csc u - \cot u| + C$$

$$30 \quad \int \sin au \sin bu \, du = \frac{\sin(a-b)u}{2(a-b)} - \frac{\sin(a+b)u}{2(a+b)} + C \quad \text{if } a^2 \neq b^2$$

$$31 \quad \int \cos au \cos bu \, du = \frac{\sin(a-b)u}{2(a-b)} + \frac{\sin(a+b)u}{2(a+b)} + C \quad \text{if } a^2 \neq b^2$$

$$32 \quad \int \sin au \cos bu \, du = -\frac{\cos(a-b)u}{2(a-b)} - \frac{\cos(a+b)u}{2(a+b)} + C \quad \text{if } a^2 \neq b^2$$

$$33 \quad \int \sin^n u \, du = -\frac{1}{n}\sin^{n-1} u \cos u + \frac{n-1}{n}\int \sin^{n-2} u \, du$$

$$34 \quad \int \cos^n u \, du = \frac{1}{n}\cos^{n-1} u \sin u + \frac{n-1}{n}\int \cos^{n-2} u \, du$$

$$35 \quad \int \tan^n u \, du = \frac{1}{n-1}\tan^{n-1} u - \int \tan^{n-2} u \, du \quad \text{if } n \neq 1$$

$$36 \quad \int \cot^n u \, du = -\frac{1}{n-1}\cot^{n-1} u - \int \cot^{n-2} u \, du \quad \text{if } n \neq 1$$

$$37 \quad \int \sec^n u \, du = \frac{1}{n-1}\sec^{n-2} u \tan u + \frac{n-2}{n-1}\int \sec^{n-2} u \, du \quad \text{if } n \neq 1$$

$$38 \quad \int \csc^n u \, du = -\frac{1}{n-1}\csc^{n-2} u \cot u + \frac{n-2}{n-1}\int \csc^{n-2} u \, du \quad \text{if } n \neq 1$$

$$39a \quad \int \sin^n u \cos^m u \, du = -\frac{\sin^{n-1} u \cos^{m+1} u}{n+m} + \frac{n-1}{n+m}\int \sin^{n-2} u \cos^m u \, du \quad \text{if } n \neq -m$$

$$39b \quad \int \sin^n u \cos^m u \, du = -\frac{\sin^{n+1} u \cos^{m-1} u}{n+m} + \frac{m-1}{n+m}\int \sin^n u \cos^{m-2} u \, du \quad \text{if } m \neq -n$$

$$40 \quad \int u \sin u \, du = \sin u - u \cos u + C$$

$$41 \quad \int u \cos u \, du = \cos u + u \sin u + C$$

$$42 \quad \int u^n \sin u \, du = -u^n \cos u + n \int u^{n-1} \cos u \, du$$

$$43 \quad \int u^n \cos u \, du = u^n \sin u - n \int u^{n-1} \sin u \, du$$

FOURTH
EDITION

Multivariable
Calculus
with Analytic Geometry

Multivariable
Calculus
with Analytic Geometry

C. H. Edwards, Jr.
The University of Georgia, Athens

David E. Penney
The University of Georgia, Athens

PRENTICE HALL Inc.
Englewood Cliffs, New Jersey 07632

Acquisitions Editor: Jacqueline Wood
Developmental Editor: Maurice Esses
Production Editor: Kelly Dickson
Cover Design: Carole Gigùere
Cover Image: Michael Portman
Page Layout: Andrew Zutis, Karen Noferi
Text Composition: Interactive Composition Corporation

MATLAB® is a registered trademark of:
The MathWorks/24 Prime Park Way/ Natick MA 01760

ISBN 0-13-300583-6

Prentice-Hall Inc., Englewood Cliffs, New Jersey
Prentice-Hall International (UK) Limited, London
Prentice-Hall of Australia Pty. Limited, Sydney
Prentice-Hall Hispanoamericana, S.A., Mexico City
Prentice-Hall of India Private Limited, New Delhi
Prentice-Hall of Japan, Inc., Tokyo
Simon & Schuster Asia Private Limited., Singapore
Editora Prentice-Hall do Brasil, Ltda., Rio de Janeiro

Contents

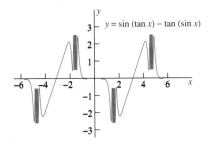
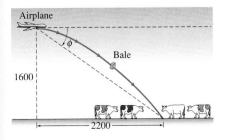

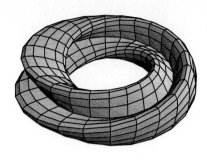

About the Authors

C. Henry Edwards, University of Georgia, received his Ph.D. from the University of Tennessee in 1960. He then taught at the University of Wisconsin for three years and spent a year at the Institute for Advanced Study (Princeton) as an Alfred P. Sloan Research Fellow. Professor Edwards has just completed his thirty-fifth year of teaching (including teaching calculus almost every year) and has received numerous university-wide teaching awards. His scholarly career has ranged from research and the direction of dissertations in topology to the history of mathematics to applied mathematics to computers and technology in mathematics (his focus in recent years). In addition to his calculus, advanced calculus, linear algebra, and differential equations textbooks, he is well known to calculus instructors as the author of *The Historical Development of the Calculus* (Springer-Verlag, 1979). He has served as a principal investigator on three recent NSF-supported projects: (1) A project to introduce technology throughout the mathematics curricula in two northeastern Georgia public school systems (including *Maple* for beginning algebra students); (2) a *Calculus with Mathematica* pilot program at the University of Georgia; and (3) a *MATLAB*-based computer lab project for upper-division numerical analysis and applied mathematics students.

David E. Penney, University of Georgia, completed his Ph.D. at Tulane University in 1965 while teaching at the University of New Orleans. Earlier he had worked in experimental biophysics at both Tulane University and the Veteran's Administration Hospital in New Orleans. He actually began teaching calculus in 1957 and has taught the course almost every term since then. He joined the mathematics department at Georgia in 1966 and has since received numerous university-wide teaching awards. He is the author of several research papers in number theory and topology and is author or co-author of books on linear algebra, differential equations, and calculus.

Preface

The role and practice of mathematics in the world at large is now undergoing a revolution that is driven largely by computational technology. Calculators and computer systems provide students and teachers with mathematical power that no previous generation could have imagined. We read even in daily newspapers of stunning current events like the recently announced proof of Fermat's last theorem. Surely *today* is the most exciting time in all history to be mathematically alive! So in preparing this new edition of **CALCULUS with Analytic Geometry**, we wanted first of all to bring at least some sense of this excitement to the students who will use it.

We also realize that the calculus course is a principal gateway to technical and professional careers for a still increasing number of students in an ever widening range of curricula. Wherever we look—in business and government, in science and technology—almost every aspect of professional work in the world involves mathematics. We therefore have re-thought once again the goal of providing calculus students the solid foundation for their subsequent work that they deserve to get from their calculus textbook.

For the first time since the original version was published in 1982, the text for this fourth edition has been reworked from start to finish. Discussions and explanations have been rewritten throughout in language that (we hope) today's students will find more lively and accessible. Seldom-covered topics have been trimmed to accommodate a leaner calculus course. Historical and biographical notes have been added to show students the human face of calculus. Graphics calculator and computer lab projects (with Derive, Maple, and *Mathematica* options) for key sections throughout the text have been added. Indeed, a new spirit and flavor reflecting the prevalent interest in graphics calculators and computer systems will be discernible throughout this edition. Consistent with the graphical emphasis of the current calculus reform movement, the number of figures in the text has been almost doubled, with must new computer-generated artwork added. Many of these additional figures serve to illustrate a more deliberative and exploratory approach to problem-solving. Our own teaching experience suggests that use of contemporary technology can make calculus more concrete and accessible to many students.

Fourth Edition Features

In preparing this edition, we have benefited from many valuable comments and suggestions from users of the first three editions. This revision was so pervasive that the individual changes are too numerous to be detailed in a preface, but the following paragraphs summarize those that may be of widest interest.

ADDITIONAL PROBLEMS   The number of problems has steadily increased since the first edition, and now totals well over 6000. In the third edition we inserted many additional practice exercises near the beginnings of problem sets to insure that students gain sufficient confidence and computational skill before moving on to the more conceptual problems that constitute the real goal of calculus. In this edition we have added graphics-based problems that emphasize conceptual understanding and accommodate student use of graphics calculators.

NEW EXAMPLES AND COMPUTATIONAL DETAILS   In many sections throughout this edition, we have inserted a simpler first example or have replaced existing examples with ones that are computationally simpler. Moreover, we have inserted an additional line or two of computational detail in many of the worked-out examples to make them easier for student readers to follow. The purpose of these computational changes is to make the computations themselves less of a barrier to conceptual understanding.

PROJECT MATERIAL   Several supplementary projects have been inserted in each chapter—a total of four dozen in all. Each project employs some aspect of modern computational technology to illustrate the principal ideas of the preceding section, and typically contains additional problems intended for solution with the use of a graphics calculator or computer. Figures and data illustrate the use of graphics calculators and computer systems such as Derive, Maple, and *Mathematica*. This project material is suitable for use in a computer/calculator lab that is conducted in association with a standard calculus course, perhaps meeting weekly. It can also be used as a basis for graphics calculator or computer assignments that students will complete outside of class, or for individual study.

COMPUTER GRAPHICS   Now that graphics calculators and computers are here to stay, an increased emphasis on graphical visualization along with numeric and symbolic work is possible as well as desirable. About 300 new MATLAB-generated figures illustrate the kind of figures students using graphics calculators can produce for themselves. Many of these are included with new graphical problem material. *Mathematica*-generated color graphics are included to highlight all sections involving 3-dimensional material.

HISTORICAL MATERIAL   Historical and biographical chapter openings have been inserted to give students a sense of the development of our subject by real, live human beings. Both authors are fond of the history of mathematics, and believe that it can favorably influence our teaching of mathematics. For this reason, numerous historical comments appear in the text itself.

INTRODUCTORY CHAPTERS   Chapters 1 and 2 have been streamlined for a leaner and quicker start on calculus. Chapter 1 concentrates on functions and graphs. It now includes a section cataloging the elementary functions of calculus, and provides a foundation for an earlier emphasis on transcendental functions. Chapter 1 now concludes with a section addressing the question "What *is* calculus?" Chapter 2 on limits begins with a section on tangent lines to motivate the official introduction of limits in Section 2.2. In contrast with the third edition, trigonometric limits now are treated throughout Chapter 2, in order to encourage a richer and more visual introduction to the limit concept.

DIFFERENTIATION CHAPTERS   The sequence of topics in Chapters 3 and 4 varies a bit from the most traditional order. We attempt to build student confidence by introducing topics more nearly in order of increasing difficulty. The chain rule appears quite early (in Section 3.3) and we cover the basic techniques for differentiating algebraic functions before discussing maxima and minima in Sections 3.5 and 3.6. The appearance of inverse functions is now delayed until Chapter 7. Section 3.7 now treats the derivatives of all six trigonometric functions. Implicit differentiation and related rates are combined in a single section (Section 3.8). The mean value theorem and its applications are deferred to Chapter 4. Sections 4.4 on the first derivative test and 4.6 on higher derivatives and concavity have been simplified and streamlined. A great deal of new graphic material has been added in the curve-sketching sections that conclude Chapter 4.

INTEGRATION CHAPTERS   New and simpler examples have been inserted throughout Chapters 5 and 6. Antiderivatives (formerly at the end of Chapter 4) now begin Chapter 5. Section 5.4 (Riemann sums) has been simplified greatly, with upper and lower sums eliminated and endpoint and midpoint sums emphasized instead. Many instructors now believe that first applications of integration ought not be confined to the standard area and volume computations; Section 6.5 is an optional section that introduces separable differential equations. To eliminate redundancy, the material on centroids and the theorems of Pappus is delayed to Chapter 15 (Multiple Integrals) where it can be treated in a more natural context.

EARLY TRANSCENDENTAL FUNCTIONS OPTIONS   In the "regular version" of this book the appearance of exponential and logarithmic functions is delayed until Chapter 7 (after integration). In the present "early transcendental version" these functions are introduced at the earliest practical opportunity in differential calculus—in Section 3.8, immediately following the differentiation of trigonometric functions in Section 3.7. Section 3.8 begins with an intuitive approach to exponential functions regarded as variable powers of a constant base, followed by the elementary idea of a logarithm as "the power to which the base a must be raised to get the number x". On this basis, the section includes a low-key review of the laws of exponents and of logarithms, and investigates somewhat informally the differentiation of exponential and logarithmic functions. Consequently, a diverse collection of transcendental functions is available for use in examples and problems throughout the balance of differential calculus (Chapters 3 and 4) and in the study of integral calculus (Chapters 5 and 6). Chapter 7 returns to exponential and logarithmic functions, offering a more complete and rigorous treatment plus further applications. But Section 7.2—based on the formal definition of the logarithm as an integral—actually can be covered anytime after the integral has been defined early in Chapter 5 (along with as much of the remainder of Chapter 7 as the instructor desires). Thus this version of the text is designed to support a course syllabus that includes exponential functions early in differential calculus, as well as logarithmic functions (defined as integrals) early in integral calculus.

The remaining transcendental functions—inverse trigonometric and hyperbolic—are now treated in Chapter 8. This newly organized chapter now includes also indeterminate forms and l'Hopital's rule (much earlier than in the third edition).

STREAMLINING TECHNIQUES OF INTEGRATION   Chapter 9 is organized to accommodate those instructors who feel that methods of formal integration now require less emphasis, in view of modern techniques for both numerical and symbolic integration. Presumably everyone will want to cover the first four sections of the chapter (through integration by parts in Section 9.4.). The method of partial fractions appears in Section 9.5, and trigonometric substitutions and integrals involving quadratic polynomials follow in Section 9.6 and 9.7. Improper integrals now appear in Section 9.8, and the more specialized rationalizing substitutions have been relegated to the Chapter 9 miscellaneous problems. This rearrangement of Chapter 9 makes it more convenient to stop wherever the instructor desires.

INFINITE SERIES   After the usual introduction to convergence of infinite sequences and series in Section 11.2 and 11.3, a combined treatment of Taylor polynomials and Taylor series appears in Section 11.4. This makes it possible for the instructor to experiment with a much briefer treatment of infinite series, but still offer exposure to the Taylor series that are so important for applications.

DIFFERENTIAL EQUATIONS   Many calculus instructors now believe that differential equations should be seen as early and as often as possible. The very simplest differential equations (of the form $y' = f(x)$) appear in a subsection at the end of Section 5.2 (Antiderivatives). Section 6.5 illustrates applications of integration to the solution of separable differential equations. Section 9.5 includes applications of the method of partial fractions to population problems and the logistic equation. In such ways we have distributed enough of the spirit and flavor of differential equations throughout the text that it seemed expeditious to eliminate the (former) final chapter devoted solely to differential equations. However, those who so desire can arrange with the publisher to obtain for supplemental use appropriate sections of Edwards and Penney, *Elementary Differential Applications with Boundary Value Problems,* third edition (Englewood Cliffs, N.J.: Prentice-Hall, 1993).

Maintaining Traditional Strengths

While many new features have been added, five related objectives remained in constant view· concreteness, readability, motivation, applicability, and accuracy.

CONCRETENESS

The power of calculus is impressive in its precise answers to realistic questions and problems. In the necessary conceptual development of the subject, we keep in sight the central question: How does one actually *compute* it? We place special emphasis on concrete examples, applications, and problems that serve both to highlight the development of the theory and to demonstrate the remarkable versatility of calculus in the investigation of important scientific questions.

READABILITY

Difficulties in learning mathematics often are complicated by language difficulties. Our writing style stems from the belief that crisp exposition, both intuitive and precise, makes mathematics more accessible—and hence more readily learned—with no loss of rigor. We hope our language is clear and attractive to students and that they can and actually will read it, thereby enabling the instructor to concentrate class time on the less routine aspects of teaching calculus.

MOTIVATION

Our exposition is centered around examples of the use of calculus to solve real problems of interest to real people. In selecting such problems for examples and exercises, we took the view that stimulating interest and motivating effective study go hand in hand. We attempt to make it clear to students how the knowledge gained with each new concept or technique will be worth the effort expended. In theoretical discussions, especially, we try to provide an intuitive picture of the goal before we set off in pursuit of it.

APPLICATIONS

Its diverse applications are what attract many students to calculus, and realistic applications provide valuable motivation and reinforcement for all students. This book is well-known for the broad range of applications that we include, but it is neither necessary nor desirable that the course cover all the applications in the book. Each section or subsection that may be omitted without loss of continuity is marked with an asterisk. This provides flexibility for each instructor to determine his or her own flavor and emphasis.

ACCURACY

Our coverage of calculus is complete (though we hope it is somewhat less than encyclopedic). Still more than its predecessors, this edition was subjected to a comprehensive reviewing process to help ensure accuracy. For example, essentially every problem answer appearing in the Answers Section at the back of the book in this edition has been verified using *Mathematica*. With regard to the selection and sequence of mathematical topics, our approach is traditional. However, close examination of the treatment of standard topics may betray our own participation in the current movement to revitalize the teaching of calculus. We continue to favor an intuitive approach that emphasizes both conceptual understanding and care in the formulation of definitions and key concepts of calculus. Some proofs that may be omitted at the discretion of the instructor are placed at the ends of sections, and others are deferred to the book's appendices. In this way we leave ample room for variation in seeking the proper balance between rigor and intuition.

Supplementary Material

Answers to most of the odd-numbered problems appear in the back of the book. Solutions to most problems (other than those odd-numbered ones for which an answer alone is sufficient) are available in the *Instructor's Solutions Manual*. A subset of this manual, containing solutions to problems numbered 1, 4, 7, 10, · · · is availabe as a *Student's Solutions Manual*. A collection of some 1700 additional problems suitable for use as test questions, the *Calculus Test Item File,* is available (in both electronic and hard copy form) for use by instructors. Finally, an *Instructor's Edition* including section-by-section teaching outlines and suggestions is available to those who are using this book to teach calculus.

STUDENT LAB MANUALS

A variety of additional supplements are provided by the publisher, including author-written project manuals keyed to specific calculators and computer systems. These include

> Calculus Projects Using Derive
> Calculus Projects Using HP Graphics Calculators
> Calculus Projects Using Maple
> Calculus Projects Using *Mathematica*
> Calculus Projects Using MATLAB
> Calculus Projects Using TI Graphics Calculators
> Calculus Projects Using X(PLORE)

Each of these manuals consists of versions of the text's 48 projects that have been expanded where necessary to take advantage of specific computational technology in the teaching of calculus. Each project is designed to provide the basis for an outside class assignment that will engage students for a period of several days (or perhaps longer). The Derive, Maple, *Mathematica,* and MATLAB manuals are accompanied by command-specific diskettes that will relieve students of much of the burden of typing (by providing "templates" for the principal commands used in each project). In many cases, the diskettes also contain additional discussion and examples.

Acknowledgments

All experienced textbook authors know the value of critical reviewing during the preparation and revision of a manuscript. In our work on various editions of this book, we have benefited greatly from the advice (and frequently the consent) of the following exceptionally able reviewers:

Leon E. Arnold, Delaware County Community College
H. L. Bentley, University of Toledo
Michael L. Berry, West Virginia Wesleyan College
William Blair, Northern Illinois University
George Cain, Georgia Institute of Technology
Wil Clarke, Atlantic Union College
Peter Colwell, Iowa State University
James W. Daniel, University of Texas at Austin
Robert Devaney, Boston University
Dan Drucker, Wayne State University
William B. Francis, Michigan Technological University
Dianne H. Haber, Westfield State College
John C. Higgins, Brigham Young University
W. Cary Huffman, Loyola University of Chicago
Calvin Jongsma, Dordt College
Louise E. Knouse, Le Tourneau College
Morris Kalka, Tulane University
Catherine Lilly, Westfield State College
Joyce Longman, Villanova University
E. D. McCune, Stephen F. Austin State University
Arthur L. Moser, Illinois Central College
Barbara Moses, Bowling Green University
Barbara L. Osofsky, Rutgers University at New Brunswick
John Petro, Western Michigan University
Wayne B. Powell, Oklahoma State University
James P. Qualey, Jr., University of Colorado
Thomas Roe, South Dakota State University
Lawrence Runyan, Shoreline Community College
William L. Siegmann, Rensselaer Polytechnic Institute
John Spellman, Southwest Texas State University
Virginia Taylor, University of Lowell
Samuel A. Truitt, Jr., Middle Tennessee State University
Robert Urbanski, Middlesex County College
Robert Whiting, Villanova University
Cathleen M. Zucco, LeMoyne College

Many of the best improvements that have been made must be credited to colleagues and users of the first three editions throughout the United States, Canada, and abroad. We are grateful to all those, especially students, who have written to us, and hope they will continue to do so. We thank Terri Bittner of Laurel Tutoring (San Carlos, CA) who with her staff checked the accuracy of every example solution and odd-numbered answer. We also believe that the quality of the finished book itself is adequate testimony of the skill, diligence, and talent of an exceptional staff at Prentice-Hall; we owe special thanks to George Lobell, mathematics editor; Karen Karlin, development editor, Ed Thomas, production editor; Andrew Zutis, designer; and Network Graphics who did the illustrations. Finally, we again are unable to thank Alice Fitzgerald Edwards and Carol Wilson Penney adequately for their continued assistance, encouragement, support, and patience.

C. H. E., Jr.
D. E. P.

Athens, Georgia

Acknowledgments

PROJECTS

The following projects employ various technologies and provide the basis for individual study or for laboratory assignments.

Infinite Series

Srinivasa Ramanujan (1887–1920)

❑ On a January day in 1913, the eminent Cambridge mathematics professor G. H. Hardy received a lettter from an unknown 25-year-old clerk in the accounting department of a government office in Madras, India. Its author, Srinivasa Ramanujan, had no university education, he admitted—he had flunked out—but "after leaving school I have employed the spare time at my disposal to work at Mathematics. . . . I have not trodden through the conventional regular course . . . but am striking out a new path for myself." The 10 pages that followed listed in neat handwritten script approximately 50 formulas, most dealing with integrals and infinite series that Ramanujan had discovered, and asked Hardy's advice whether they contained anything of value. The formulas were of such exotic and unlikely appearance that Hardy at first suspected a hoax,

but he and his colleague J. E. Littlewood soon realized that they were looking at the work of an extraordinary mathematical genius.

❑ Thus began one of the most romantic episodes in the history of mathematics. In April 1914 Ramanujan arrived in England a poor, self-taught Indian mathematical amateur called to collaborate as an equal with the most sophisticated professional mathematicians of the day. For the next three years a steady stream of remarkable discoveries poured forth from his pen. But in 1917 he fell seriously ill, apparently with tuberculosis. The following year he returned to India to attempt to regain his health but never recovered, and he died in 1920 at the age of 32. Up to the very end he worked feverishly to record his final discoveries. He left behind notebooks outlining work whose completion has occupied prominent mathematicians throughout the twentieth century.

❑ With the possible exception of Euler, no one before or since has exhibited Ramanujan's virtuosity with infinite series. An example of his discoveries is the infinite series

$$\frac{1}{\pi} = \frac{\sqrt{8}}{9801} \sum_{n=0}^{\infty} \frac{(4n)!}{(n!)^4} \frac{(1103 + 26{,}390n)}{396^{4n}},$$

whose first term yields the familiar approximation $\pi \approx 3.14159$, with each additional term giving π to roughly eight more decimal places of accuracy. For instance, just four terms of Ramanujan's

series are needed to calculate the 30-place approximation

$$\pi \approx 3.14159\ 26535\ 89793$$
$$23846\ 26433\ 83279$$

that suffices for virtually any imaginable "practical" application—if the universe were a sphere with a radius of 10 billion light years, then this value of π would give its circumference accurate to the nearest hundredth of an inch. But in recent years Ramanujan's ideas have been used to calculate the value of π accurate to a *billion* decimal places! Indeed, such gargantuan computations of π are commonly used to check the accuracy of new supercomputers.

A typical page of Ramanujan's letter to Hardy, listing formulas Ramanujan had discovered, but with no hint of proof or derivation.

11.1

Introduction

Fig. 11.1.1 Subdivision of an interval to illustrate Zeno's paradox

In the fifth century B.C., the Greek philosopher Zeno proposed the following paradox: In order for a runner to travel a given distance, the runner must first travel halfway, then half the remaining distance, then half the distance that yet remains, and so on ad infinitum. But, Zeno argued, it is clearly impossible for a runner to take infinitely many steps in a finite period of time, so motion from one point to another is impossible.

Zeno's paradox suggests the infinite subdivision of $[0, 1]$ indicated in Fig. 11.1.1. There is one subinterval of length $1/2^n$ for each integer $n = 1, 2, 3, \ldots$. If the length of the interval is the sum of the lengths of the subintervals into which it is divided, then it would appear that

$$1 = \frac{1}{2} + \frac{1}{4} + \frac{1}{8} + \frac{1}{16} + \cdots + \frac{1}{2^n} + \cdots,$$

with infinitely many terms somehow adding up to 1. But the formal infinite sum

$$1 + 2 + 3 + \cdots + n + \cdots$$

of all the positive integers seems meaningless—it does not appear to add up to *any* (finite) value.

The question is this: What, if anything, do we mean by the sum of an *infinite* collection of numbers? This chapter explores conditions under which an *infinite* sum

$$a_1 + a_2 + a_3 + \cdots + a_n + \cdots,$$

known as an *infinite series,* is meaningful. We discuss methods for computing the sum of an infinite series and applications of the algebra and calculus of infinite series. Infinite series are important in science and mathematics because many functions either arise most naturally in the form of infinite series or have infinite series representations (such as the Taylor series of Section 11.4) that are useful for numerical computations.

11.2

Infinite Sequences

An **infinite sequence** of real numbers is a function whose domain of definition is the set of all positive integers. Thus if s is a sequence, then to each positive integer n there corresponds a real number $s(n)$. Ordinarily, a sequence is most conveniently described by listing its values in order, beginning with $s(1)$:

$$s(1), s(2), s(3), \ldots, s(n), \ldots.$$

With subscript notation rather than function notation, we usually write

$$s_1, s_2, s_3, \ldots, s_n, \ldots \qquad (1)$$

for this list of values. The values in this list are the **terms** of the sequence; s_1 is the first term, s_2 the second term, s_n the **nth term.**

We use the notation $\{s_n\}_{n=1}^{\infty}$, or simply $\{s_n\}$, as an abbreviation for the **ordered** list in Eq. (1), and we may refer to the sequence by saying simply "the sequence $\{s_n\}$." When a particular sequence is so described, the nth term s_n is generally (though not always) given by a formula in terms of its subscript n. In this case, listing the first few terms of the sequence often helps us to see it more concretely.

EXAMPLE 1 The following table lists explicitly the first four terms of several sequences.

$\{s_n\}_1^\infty$	$s_1, s_2, s_3, s_4, \ldots$
$\left\{\dfrac{1}{n}\right\}_1^\infty$	$1, \dfrac{1}{2}, \dfrac{1}{3}, \dfrac{1}{4}, \ldots$
$\left\{\dfrac{1}{10^n}\right\}_1^\infty$	$0.1, 0.01, 0.001, 0.0001, \ldots$
$\left\{\dfrac{1}{n!}\right\}_1^\infty$	$1, \dfrac{1}{2}, \dfrac{1}{6}, \dfrac{1}{24}, \ldots$
$\left\{\sin\dfrac{n\pi}{2}\right\}_1^\infty$	$1, 0, -1, 0, \ldots$
$\{1 + (-1)^n\}_1^\infty$	$0, 2, 0, 2, \ldots$

EXAMPLE 2 The **Fibonacci sequence** $\{F_n\}$ may be defined as follows:

$$F_1 = 1, \qquad F_2 = 1, \quad \text{and} \quad F_{n+1} = F_n + F_{n-1} \qquad \text{for } n \geqq 2.$$

The first ten terms of the Fibonacci sequence are

$$1, 1, 2, 3, 5, 8, 13, 21, 34, 55.$$

This is a *recursively defined sequence* —after the initial terms are given, each term is defined in terms of its predecessors. Such sequences are particularly important in computer science; a computer's state at each tick of its internal clock typically depends on its state at the previous tick.

The limit of a sequence is defined in much the same way as the limit of an ordinary function (Section 2.2).

Definition *Limit of a Sequence*
We say that the sequence $\{s_n\}$ **converges** to the real number L, or has the **limit** L, and we write

$$\lim_{n \to \infty} s_n = L, \tag{2}$$

provided that s_n can be made as close to L as we please merely by choosing n to be sufficiently large. That is, given any number $\epsilon > 0$, there exists an integer N such that

$$|s_n - L| < \epsilon \qquad \text{for all } n \geqq N. \tag{3}$$

If the sequence $\{s_n\}$ does *not* converge, then we say that $\{s_n\}$ **diverges.**

Figure 11.2.1 illustrates geometrically the definition of the limit of a sequence. Because

$$|s_n - L| < \epsilon \quad \text{means that} \quad L - \epsilon < s_n < L + \epsilon,$$

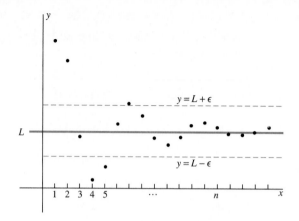

Fig. 11.2.1 The point (n, s_n) approaches the line $y = L$ as $n \to \infty$.

the condition in (3) means that if $n \geq N$, then the point (n, s_n) lies between the horizontal lines $y = L - \epsilon$ and $y = L + \epsilon$.

EXAMPLE 3 Prove that $\lim\limits_{n \to \infty} \dfrac{1}{n} = 0$.

Proof We need to show this: To each positive real number ϵ, there corresponds an integer N such that, for all $n \geq N$,

$$\left| \frac{1}{n} - 0 \right| = \frac{1}{n} < \epsilon.$$

It suffices to choose any fixed integer $N > 1/\epsilon$. For example, let N denote the *smallest* integer that is greater than the real number $1/\epsilon$ (Fig. 11.2.2). Then $n \geq N$ implies that

$$\frac{1}{n} \leq \frac{1}{N} < \epsilon,$$

as desired. ❑

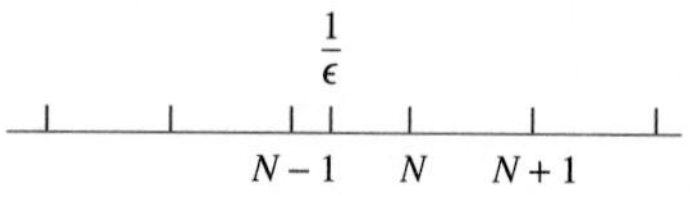

Fig. 11.2.2 The integer N of Example 3

EXAMPLE 4 (a) The sequence $\{(-1)^n\}$ diverges because its successive terms "oscillate" between the two values $+1$ and -1. Hence $(-1)^n$ cannot approach any single value as $n \to \infty$. (b) The terms of the sequence $\{n^2\}$ increase without bound as $n \to \infty$. Thus $\{n^2\}$ diverges. In this case, we might also say that $\{n^2\}$ diverges to *infinity*.

USING LIMIT LAWS

The limit laws in Section 2.2 for limits of functions have natural analogues for limits of sequences. Their proofs are based on techniques similar to those used in Appendix B.

Theorem 1   Limit Laws for Sequences

If the limits

$$\lim_{n \to \infty} a_n = A \quad \text{and} \quad \lim_{n \to \infty} b_n = B$$

exist (so A and B are real numbers), then

1. $\lim_{n \to \infty} ca_n = cA \qquad$ (c any real number);

2. $\lim_{n \to \infty} (a_n + b_n) = A + B$;

3. $\lim_{n \to \infty} a_n b_n = AB$;

4. $\lim_{n \to \infty} \dfrac{a_n}{b_n} = \dfrac{A}{B}$.

In part 4 we must assume that $B \neq 0$ and that $b_n \neq 0$ for all sufficiently large values of n.

Theorem 2   Substitution Law for Sequences

If $\lim_{n \to \infty} a_n = A$ and the function f is continuous at $x = A$, then

$$\lim_{n \to \infty} f(a_n) = f(A).$$

Theorem 3   Squeeze Law for Sequences

If $a_n \leqq b_n \leqq c_n$ for all n and

$$\lim_{n \to \infty} a_n = L = \lim_{n \to \infty} c_n,$$

then $\lim_{n \to \infty} b_n = L$ as well.

These theorems can be used to compute limits of many sequences formally, without recourse to the definition. For example, if k is a positive integer and c is a constant, then Example 3 and the product law (Theorem 1, part 3) give

$$\lim_{n \to \infty} \frac{c}{n^k} = c \cdot 0 \cdot 0 \cdots 0 = 0.$$

EXAMPLE 5   Show that $\lim_{n \to \infty} \dfrac{(-1)^n \cos n}{n^2} = 0$.

Solution   This result follows from the squeeze law and from the fact that $1/n^2 \to 0$ as $n \to \infty$, because

$$-\frac{1}{n^2} \leqq \frac{(-1)^n \cos n}{n^2} \leqq \frac{1}{n^2}.$$

EXAMPLE 6   Show that if $a > 0$, then $\lim_{n \to \infty} \sqrt[n]{a} = 1$.

Solution   We apply the substitution law with $f(x) = a^x$ and $A = 0$. Because $1/n \to 0$ as $n \to \infty$ and f is continuous at $x = 0$, this gives

$$\lim_{n \to \infty} a^{1/n} = a^0 = 1.$$

EXAMPLE 7 The limit laws and the continuity of $f(x) = \sqrt{x}$ at $x = 4$ yield

$$\lim_{n \to \infty} \sqrt{\frac{4n - 1}{n + 1}} = \sqrt{\lim_{n \to \infty} \frac{4 - \dfrac{1}{n}}{1 + \dfrac{1}{n}}} = \sqrt{4} = 2.$$

EXAMPLE 8 Show that if $|r| < 1$, then $\lim\limits_{n \to \infty} r^n = 0$.

Solution   Because $|r^n| = |(-r)^n|$, we may assume that $0 < r < 1$. Then $1/r = 1 + a$ for some number $a > 0$, so the binomial formula yields

$$\frac{1}{r^n} = (1 + a)^n = 1 + na + \{\text{positive terms}\} > 1 + na;$$

$$0 < r^n < \frac{1}{1 + na}.$$

Now $1/(1 + na) \to 0$ as $n \to \infty$. Therefore, the squeeze law implies that $r^n \to 0$ as $n \to \infty$.

Figure 11.2.3 shows the graph of a function f such that $\lim_{x \to \infty} f(x) = L$. If the sequence $\{a_n\}$ is defined by the formula $a_n = f(n)$ for each positive integer n, then all the points $(n, f(n))$ lie on the graph of $y = f(x)$. It therefore follows from the definition of the limit of a function that $\lim\limits_{n \to \infty} a_n = L$ as well.

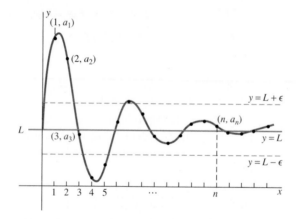

Fig. 11.2.3 The limit of the sequence is the limit of the function.

Theorem 4   Limits of Functions and Sequences
If $a_n = f(n)$ for each positive integer n, then

$$\lim_{x \to \infty} f(x) = L \quad \text{implies} \quad \lim_{n \to \infty} a_n = L. \tag{4}$$

The converse of the statement in (4) is generally false. For example, take $f(x) = \sin \pi x$ and, for each positive integer n, let $a_n = f(n) = \sin n\pi$. Then

$$\lim_{n \to \infty} a_n = \lim_{n \to \infty} \sin n\pi = 0, \quad \text{but}$$

$$\lim_{n \to \infty} f(x) = \lim_{n \to \infty} \sin nx \quad \text{does not exist.}$$

Because of (4) we can use **l'Hôpital's rule for sequences:** If $a_n = f(n)$, $b_n = g(n)$, and $f(x)/g(x)$ has the indeterminate form ∞/∞ as $x \to \infty$, then

$$\lim_{n \to \infty} \frac{a_n}{b_n} = \lim_{x \to \infty} \frac{f(x)}{g(x)} = \lim_{x \to \infty} \frac{f'(x)}{g'(x)}, \tag{5}$$

provided that f and g satisfy the other hypotheses of l'Hôpital's rule, including the assumption that the right-hand limit exists.

EXAMPLE 9 Show that $\displaystyle\lim_{n \to \infty} \frac{\ln n}{n} = 0$.

Solution The function $(\ln x)/x$ is defined for all $x \geqq 1$ and agrees with the given sequence $\{(\ln n)/n\}$ when $x = n$, a positive integer. Because $(\ln x)/x$ has the indeterminate form ∞/∞ as $x \to \infty$, l'Hôpital's rule gives

$$\lim_{n \to \infty} \frac{\ln n}{n} = \lim_{x \to \infty} \frac{\ln x}{x} = \lim_{x \to \infty} \frac{\dfrac{1}{x}}{1} = 0.$$

EXAMPLE 10 Show that $\displaystyle\lim_{n \to \infty} \sqrt[n]{n} = 1$.

Solution First we note that

$$\ln \sqrt[n]{n} = \ln n^{1/n} = \frac{\ln n}{n} \to 0 \qquad \text{as } n \to \infty,$$

by Example 9. By the substitution law with $f(x) = e^x$, this gives

$$\lim_{n \to \infty} n^{1/n} = \lim_{n \to \infty} \exp(\ln n^{1/n}) = e^0 = 1.$$

EXAMPLE 11 Find $\displaystyle\lim_{n \to \infty} \frac{3n^3}{e^{2n}}$.

Solution We apply l'Hôpital's rule repeatedly, although we must be careful at each intermediate step to verify that we still have an indeterminate form. Thus we find that

$$\lim_{n \to \infty} \frac{3n^3}{e^{2n}} = \lim_{x \to \infty} \frac{3x^3}{e^{2x}} = \lim_{x \to \infty} \frac{9x^2}{2e^{2x}} = \lim_{x \to \infty} \frac{18x}{4e^{2x}} = \lim_{x \to \infty} \frac{18}{8e^{2x}} = 0.$$

BOUNDED MONOTONIC SEQUENCES

The set of all *rational* numbers has by itself all the most familiar elementary algebraic properties of the entire real number system. To guarantee the existence of irrational numbers, we must assume in addition a "completeness property" of the real numbers. Otherwise, the real line might have "holes" where the irrational numbers ought to be. One way of stating this completeness property is in terms of the convergence of an important type of sequence, a bounded monotonic sequence.

The sequence $\{a_n\}_1^\infty$ is said to be **increasing** if

$$a_1 \leqq a_2 \leqq a_3 \leqq \cdots \leqq a_n \leqq \cdots$$

and **decreasing** if

$$a_1 \geqq a_2 \geqq a_3 \geqq \cdots \geqq a_n \geqq \cdots.$$

The sequence $\{a_n\}$ is **monotonic** if it is *either* increasing *or* decreasing. The sequence $\{a_n\}$ is **bounded** if there is a number M such that $|a_n| \leq M$ for all n. The following assertion may be taken to be an axiom for the real number system.

> **Bounded Monotonic Sequence Property**
> Every bounded monotonic infinite sequence converges—that is, has a finite limit.

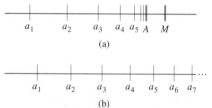

(a)

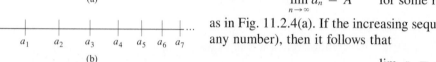

(b)

Fig. 11.2.4 (a) A bounded increasing sequence; (b) an increasing sequence that is not bounded above

Suppose, for example, that the increasing sequence $\{a_n\}_1^\infty$ is bounded above by a number M, meaning that $a_n \leq M$ for all n. Because it is also bounded below (by a_1, for instance), the bounded monotonic sequence property implies that

$$\lim_{n \to \infty} a_n = A \qquad \text{for some real number } A \leq M,$$

as in Fig. 11.2.4(a). If the increasing sequence $\{a_n\}$ is *not* bounded above (by any number), then it follows that

$$\lim_{n \to \infty} a_n = +\infty,$$

as in Fig. 11.2.4(b) (see Problem 38).

EXAMPLE 12 Investigate the sequence $\{a_n\}$ that is defined recursively by

$$a_1 = \sqrt{2}, \qquad a_{n+1} = \sqrt{2 + a_n} \qquad \text{for } n \geq 1. \tag{6}$$

Solution The first four terms of $\{a_n\}$ are

$$\sqrt{2}, \quad \sqrt{2 + \sqrt{2}}, \quad \sqrt{2 + \sqrt{2 + \sqrt{2}}}, \quad \sqrt{2 + \sqrt{2 + \sqrt{2 + \sqrt{2}}}}.$$

If the sequence $\{a_n\}$ has a limit A, then A would seem to be the natural interpretation of the value of the infinite expression

$$\sqrt{2 + \sqrt{2 + \sqrt{2 + \sqrt{2 + \cdots}}}}.$$

The sequence $\{a_n\}$ defined in (6) is an increasing sequence with $a_n < 2$ for all n, as Problem 43 asks you to verify. Therefore, the bounded monotonic sequence property implies that the sequence $\{a_n\}$ has a limit A. It does not tell us what the number A is. But now that we know that the limit A of the sequence $\{a_n\}$ exists, we can write

$$A = \lim_{n \to \infty} a_{n+1} = \lim_{n \to \infty} \sqrt{2 + a_n} = \sqrt{2 + A},$$

and thus

$$A^2 = 2 + A.$$

The roots of this equation are -1 and 2. It is clear that $A > 0$, so we conclude that

$$\lim_{n \to \infty} a_n = 2.$$

Thus

$$\sqrt{2 + \sqrt{2 + \sqrt{2 + \sqrt{2 + \cdots}}}} = 2.$$

To indicate what the bounded monotonic sequence property has to do with the "completeness property" of the real numbers, in Problem 42 we outline a proof, using this property, of the existence of the number $\sqrt{2}$. In Problems 45 and 46, we outline a proof of the equivalence of the bounded monotonic sequence property and another common statement of the completeness of the real numbers—the *least upper bound property*.

11.2 Problems

In Problems 1 through 35, determine whether or not the sequence $\{a_n\}$ converges, and find its limit if it does converge.

1. $a_n = \dfrac{2n}{5n - 3}$

2. $a_n = \dfrac{1 - n^2}{2 + 3n^2}$

3. $a_n = \dfrac{n^2 - n + 7}{2n^3 + n^2}$

4. $a_n = \dfrac{n^3}{10n^2 + 1}$

5. $a_n = 1 + \left(\frac{9}{10}\right)^n$

6. $a_n = 2 - \left(-\frac{1}{2}\right)^n$

7. $a_n = 1 + (-1)^n$

8. $a_n = \dfrac{1 + (-1)^n}{\sqrt{n}}$

9. $a_n = \dfrac{1 + (-1)^n\sqrt{n}}{\left(\frac{3}{2}\right)^n}$

10. $a_n = \dfrac{\sin n}{3^n}$

11. $a_n = \dfrac{\sin^2 n}{\sqrt{n}}$

12. $a_n = \sqrt{\dfrac{2 + \cos n}{n}}$

13. $a_n = n \sin \pi n$

14. $a_n = n \cos \pi n$

15. $a_n = \pi^{-(\sin n)/n}$

16. $a_n = 2^{\cos n\pi}$

17. $a_n = \dfrac{\ln n}{\sqrt{n}}$

18. $a_n = \dfrac{\ln 2n}{\ln 3n}$

19. $a_n = \dfrac{(\ln n)^2}{n}$

20. $a_n = n \sin\left(\dfrac{1}{n}\right)$

21. $a_n = \dfrac{\tan^{-1} n}{n}$

22. $a_n = \dfrac{n^3}{e^{n/10}}$

23. $a_n = \dfrac{2^n + 1}{e^n}$

24. $a_n = \dfrac{\sinh n}{\cosh n}$

25. $a_n = \left(1 + \dfrac{1}{n}\right)^n$

26. $a_n = (2n + 5)^{1/n}$

27. $a_n = \left(\dfrac{n - 1}{n + 1}\right)^n$

28. $a_n = (0.001)^{-1/n}$

29. $a_n = \sqrt[n]{2^{n+1}}$

30. $a_n = \left(1 - \dfrac{2}{n^2}\right)^n$

31. $a_n = \left(\dfrac{2}{n}\right)^{3/n}$

32. $a_n = (-1)^n(n^2 + 1)^{1/n}$

33. $a_n = \left(\dfrac{2 - n^2}{3 + n^2}\right)^n$

34. $a_n = \dfrac{\left(\frac{2}{3}\right)^n}{1 - \sqrt[n]{n}}$

35. $a_n = \dfrac{\left(\frac{2}{3}\right)^n}{\left(\frac{1}{2}\right)^n + \left(\frac{9}{10}\right)^n}$

36. Suppose that $\lim\limits_{n \to \infty} a_n = A$. Prove that $\lim\limits_{n \to \infty} |a_n| = |A|$.

37. Prove that $\{(-1)^n a_n\}$ diverges if $\lim\limits_{n \to \infty} a_n = A \neq 0$.

38. Suppose that $\{a_n\}$ is an increasing sequence that is not bounded. Prove that $\lim\limits_{n \to \infty} a_n = +\infty$.

39. Suppose that $A > 0$. Given x_1 arbitrary, define the sequence $\{x_n\}$ recursively as follows:

$$x_{n+1} = \frac{1}{2}\left(x_n + \frac{A}{x_n}\right) \qquad \text{if } n \geq 1.$$

Prove that if $L = \lim\limits_{n \to \infty} x_n$ exists, then $L = \pm\sqrt{A}$.

40. Let $\{F_n\}$ be the Fibonacci sequence of Example 2. Assume that

$$\tau = \lim_{n \to \infty} \frac{F_{n+1}}{F_n}$$

exists, and prove that $\tau = \frac{1}{2}(1 + \sqrt{5})$.

41. Let the sequence $\{a_n\}$ be defined recursively by

$$a_1 = 2; \qquad a_{n+1} = \frac{1}{2}(a_n + 4) \qquad \text{for } n \geq 1.$$

(a) Prove by induction on n that $a_n < 4$ for each n and that $\{a_n\}$ is an increasing sequence. (b) Find the limit of this sequence.

42. For each positive integer n, let a_n be the largest integral multiple of $1/10^n$ such that $a_n^2 \leq 2$. (a) Prove that $\{a_n\}$ is a bounded increasing sequence, so $A = \lim_{n \to \infty} a_n$ exists. (b) Prove that if $A^2 > 2$, then $a_n^2 > 2$ for n sufficiently large. (c) Prove that if $A^2 < 2$, then $a_n^2 < B$ for some number $B < 2$ and all sufficiently large n. (d) Conclude that $A^2 = 2$.

43. (a) Square both sides of the defining equation $a_{n+1} = \sqrt{2 + a_n}$ of the sequence in Example 12 to show that if $a_n < 2$, then it follows that $a_{n+1} < 2$. Why does this fact imply that $a_n < 2$ for all n? (b) Compute $(a_{n+1})^2 - a_n^2$ to show that $a_{n+1} > a_n$ for all n.

44. Use the method of Example 12 to show that

(a) $\sqrt{6 + \sqrt{6 + \sqrt{6 + \cdots}}} = 3$;

(b) $\sqrt{20 + \sqrt{20 + \sqrt{20 + \cdots}}} = 5$.

Problems 45 and 46 deal with the least upper bound *property of the real numbers: If the nonempty set S of real numbers has an upper bound, then S has a least upper bound. The number M is an* **upper bound** *for the set S if $x \leqq M$ for all x in S. The upper bound L of S is a* **least upper bound** *for S if no number smaller than L is a least upper bound for S. You can show easily that if the set S has least upper bounds L_1 and L_2, then $L_1 = L_2$; in other words, if a least upper bound for a set exists, then it is unique.*

45. Prove that the least upper bound property implies the bounded monotonic sequence property. [*Suggestion:* If $\{a_n\}$ is a bounded increasing sequence and A is the least upper bound of the set $\{a_n : n \geqq 1\}$ of terms of the sequence, you can prove that $A = \lim\limits_{n \to \infty} a_n$.]

46. Prove that the bounded monotonic sequence property implies the least upper bound property. [*Suggestion:* For each positive integer n, let a_n be the least integral multiple of $1/10^n$ that is an upper bound of the set S. Prove that $\{a_n\}$ is a bounded decreasing sequence and then that $A = \lim\limits_{n \to \infty} a_n$ is a least upper bound for S.]

11.3

Infinite Series and Convergence

An **infinite series** is an expression of the form

$$\sum_{n=1}^{\infty} a_n = a_1 + a_2 + a_3 + \cdots + a_n + \cdots, \qquad (1)$$

where $\{a_n\}$ is an infinite sequence of real numbers. The number a_n is called the **nth term** of the series. The symbol $\sum_{n=1}^{\infty} a_n$ is simply an abbreviation for the right-hand side of Eq. (1). An example of an infinite series is the series

$$\sum_{n=1}^{\infty} \frac{1}{2^n} = \frac{1}{2} + \frac{1}{4} + \frac{1}{8} + \frac{1}{16} + \cdots + \frac{1}{2^n} + \cdots$$

that we mentioned in Section 11.1. The nth term of this particular infinite series is $a_n = 1/2^n$.

To say what such an infinite sum means, we introduce the *partial sums* of the infinite series in Eq. (1). The **nth partial sum S_n** of the series is the sum of its first n terms:

$$S_n = a_1 + a_2 + a_3 + \cdots + a_n. \qquad (2)$$

Thus each infinite series is associated with an infinite **sequence of partial sums**

$$S_1, S_2, S_3, \ldots, S_n, \ldots.$$

We define the *sum* of the infinite series to be the limit of its sequence of partial sums, provided that this limit exists.

Definition *The Sum of an Infinite Series*

We say that the infinite series

$$\sum_{n=1}^{\infty} a_n \quad \textbf{converges} \text{ (or is } \textbf{convergent}\text{)}$$

with **sum** S provided that the limit of its sequence of partial sums,

$$S = \lim_{n \to \infty} S_n, \qquad (3)$$

exists (and is finite). Otherwise we say that the series **diverges** (or is **divergent**). If a series diverges, then it has no sum.

Thus an infinite series is a limit of finite sums,

$$S = \sum_{n=1}^{\infty} a_n = \lim_{N \to \infty} \sum_{n=1}^{N} a_n,$$

provided that this limit exists.

EXAMPLE 1 Show that the series

$$\sum_{n=1}^{\infty} \left(\tfrac{1}{2}\right)^n = \tfrac{1}{2} + \tfrac{1}{4} + \tfrac{1}{8} + \tfrac{1}{16} + \cdots$$

converges, and find its sum.

Solution The first four partial sums are

$$S_1 = \tfrac{1}{2}, \qquad S_2 = \tfrac{3}{4}, \qquad S_3 = \tfrac{7}{8}, \quad \text{and} \quad S_4 = \tfrac{15}{16}.$$

It seems likely that $S_n = (2^n - 1)/2^n$, and indeed this follows easily by induction on n, because

$$S_{n+1} = S_n + \frac{1}{2^{n+1}} = \frac{2^n - 1}{2^n} + \frac{1}{2^{n+1}} = \frac{2^{n+1} - 2 + 1}{2^{n+1}} = \frac{2^{n+1} - 1}{2^{n+1}}.$$

Hence the sum of the given series is

$$S = \lim_{n \to \infty} S_n = \lim_{n \to \infty} \frac{2^n - 1}{2^n} = \lim_{n \to \infty} \left(1 - \frac{1}{2^n}\right) = 1.$$

EXAMPLE 2 Show that the series

$$\sum_{n=1}^{\infty} (-1)^{n+1} = 1 - 1 + 1 - 1 + \cdots$$

diverges.

Solution The sequence of partial sums of this series is

$$1, 0, 1, 0, 1, \ldots,$$

which has no limit. Therefore the series diverges.

EXAMPLE 3 Show that the infinite series

$$\sum_{n=1}^{\infty} \frac{1}{n(n + 1)}$$

converges, and find its sum.

Solution We need a formula for the nth partial sum S_n so that we can evaluate its limit as $n \to \infty$. To find such a formula, we begin with the observation that the nth term of the series is

$$a_n = \frac{1}{n(n + 1)} = \frac{1}{n} - \frac{1}{n + 1}.$$

(In more complicated cases, such as those in Problems 36 through 40, such a decomposition can be obtained by the method of partial fractions.) It follows that the sum of the first n terms of the given series is

$$S_n = \left(1 - \frac{1}{2}\right) + \left(\frac{1}{2} - \frac{1}{3}\right) + \left(\frac{1}{3} - \frac{1}{4}\right)$$

$$+ \left(\frac{1}{4} - \frac{1}{5}\right) + \cdots + \left(\frac{1}{n} - \frac{1}{n+1}\right)$$

$$= 1 - \frac{1}{n+1} = \frac{n}{n+1}.$$

Hence

$$\sum_{n=1}^{\infty} \frac{1}{n(n+1)} = \lim_{n \to \infty} \frac{n}{n+1} = 1.$$

The sum for S_n in Example 3, called a *telescoping* sum, provides us with a way to find the sums of certain series. The series in Examples 1 and 2 are examples of a more common and more important type of series, the *geometric series*.

Definition   *Geometric Series*

The series $\sum_{n=0}^{\infty} a_n$ is said to be a **geometric series** if each term after the first is a fixed multiple of the term immediately before it. That is, there is a number r, called the **ratio** of the series, such that

$$a_{n+1} = ra_n \qquad \text{for all } n \geqq 0.$$

Thus every geometric series takes the form

$$a_0 + ra_0 + r^2 a_0 + r^3 a_0 + \cdots = \sum_{n=0}^{\infty} r^n a_0. \qquad (4)$$

It is convenient to begin the summation at $n = 0$, and thus we regard the sum

$$S_n = a_0(1 + r + r^2 + \cdots + r^n)$$

of the first $n + 1$ terms to be the nth partial sum of the series.

EXAMPLE 4   The infinite series

$$\sum_{n=0}^{\infty} \frac{2}{3^n} = 2 + \frac{2}{3} + \frac{2}{9} + \cdots + \frac{2}{3^n} + \cdots$$

is a geometric series whose first term is $a_0 = 2$ and whose ratio is $r = \frac{1}{3}$.

Theorem 1   *Sum of a Geometric Series*

If $|r| < 1$, then the geometric series in Eq. (4) converges, and its sum is

$$S = \sum_{n=0}^{\infty} r^n a_0 = \frac{a_0}{1 - r}. \qquad (5)$$

If $|r| \geqq 1$ and $a_0 \neq 0$, then the geometric series diverges.

Proof If $r = 1$, then $S_n = (n + 1)a_0$, so the series certainly diverges if $a_0 \neq 0$. If $r = -1$ and $a_0 \neq 0$, then the series diverges by an argument like the one in Example 2. So we may suppose that $|r| \neq 1$. Then the elementary identity

$$1 + r + r^2 + \cdots + r^n = \frac{1 - r^{n+1}}{1 - r}$$

follows if we multiply each side by $1 - r$. Hence the nth partial sum of the geometric series is

$$S_n = a_0(1 + r + r^2 + \cdots + r^n) = a_0\left(\frac{1}{1 - r} - \frac{r^{n+1}}{1 - r}\right).$$

If $|r| < 1$, then $r^{n+1} \to 0$ as $n \to \infty$, by Example 7 in Section 11.2. So in this case the geometric series converges to

$$S = \lim_{n \to \infty} a_0\left(\frac{1}{1 - r} - \frac{r^{n+1}}{1 - r}\right) = \frac{a_0}{1 - r}.$$

But if $|r| > 1$, then $\lim_{n \to \infty} r^{n+1}$ does not exist, so $\lim_{n \to \infty} S_n$ does not exist. This establishes the theorem. ❑

EXAMPLE 5 With $a_0 = 1$ and $r = -\frac{1}{2}$, we find that

$$1 - \frac{1}{2} + \frac{1}{4} - \frac{1}{8} + \cdots = \sum_{n=0}^{\infty} \left(-\frac{1}{2}\right)^n = \frac{1}{1 - (-\frac{1}{2})} = \frac{2}{3}.$$

Theorem 2 implies that the operations of addition and of multiplication by a constant can be carried out term by term in the case of *convergent* series. Because the sum of an infinite series is the limit of its sequence of partial sums, this theorem follows immediately from the limit laws for sequences (Theorem 1 of Section 11.2).

Theorem 2 *Termwise Addition and Multiplication*
If the series $A = \Sigma a_n$ and $B = \Sigma b_n$ converge to the indicated sums and c is a constant, then the series $\Sigma (a_n + b_n)$ and Σca_n also converge, with sums

1. $\displaystyle\sum (a_n + b_n) = A + B$;
2. $\displaystyle\sum ca_n = cA$.

The geometric series in Eq. (5) may be used to find the rational number represented by a given infinite repeating decimal.

EXAMPLE 6

$$0.55555\ldots = \frac{5}{10} + \frac{5}{100} + \frac{5}{1000} + \cdots$$

$$= \sum_{n=0}^{\infty} \frac{5}{10}\left(\frac{1}{10}\right)^n = \frac{\frac{5}{10}}{1 - \frac{1}{10}} = \frac{5}{10} \cdot \frac{10}{9} = \frac{5}{9}.$$

In a more complicated situation, we may need to use the termwise algebra of Theorem 2:

$$0.72828\,28\ldots = \frac{7}{10} + \frac{28}{10^3} + \frac{28}{10^5} + \frac{28}{10^7} + \cdots$$

$$= \frac{7}{10} + \frac{28}{10^3}\left(1 + \frac{1}{10^2} + \frac{1}{10^4} + \cdots\right)$$

$$= \frac{7}{10} + \frac{28}{1000}\sum_{n=0}^{\infty}\left(\frac{1}{100}\right)^n = \frac{7}{10} + \frac{28}{1000}\left(\frac{1}{1 - \frac{1}{100}}\right)$$

$$= \frac{7}{10} + \frac{28}{1000}\cdot\frac{100}{99} = \frac{7}{10} + \frac{28}{990} = \frac{721}{990}.$$

This technique can be used to show that every repeating infinite decimal represents a rational number. Consequently, the decimal expansions of irrational numbers such as π, e, and $\sqrt{2}$ must be nonrepeating as well as infinite. Conversely, if p and q are integers with $q \neq 0$, then long division of q into p yields a repeating decimal expansion for the rational number p/q, because such a division can yield at each stage only q possible different remainders.

EXAMPLE 7   Suppose that Paul and Mary toss a fair six-sided die in turn until one of them wins by getting the first "six." If Paul tosses first, calculate the probability that he will win the game.

Solution   Because the die is fair, the probability that Paul gets a "six" on the first round is $\frac{1}{6}$. The probability that he gets the game's first "six" on the second round is $\left(\frac{5}{6}\right)^2\left(\frac{1}{6}\right)$—the product of the probability $\left(\frac{5}{6}\right)^2$ that neither Paul nor Mary rolls a "six" in the first round and the probability $\frac{1}{6}$ that Paul rolls a "six" in the second round. Paul's probability p of getting the first "six" in the game is the *sum* of his probabilities of getting it in the first round, in the second round, in the third round, and so on. Hence

$$p = \frac{1}{6} + \left(\frac{5}{6}\right)^2\left(\frac{1}{6}\right) + \left(\frac{5}{6}\right)^2\left(\frac{5}{6}\right)^2\left(\frac{1}{6}\right) + \cdots = \frac{1}{6}\left[1 + \left(\frac{5}{6}\right)^2 + \left(\frac{5}{6}\right)^4 + \cdots\right]$$

$$= \frac{1}{6}\cdot\frac{1}{1 - \left(\frac{5}{6}\right)^2} = \frac{1}{6}\cdot\frac{36}{11} = \frac{6}{11}.$$

Because he has the advantage of tossing first, Paul has more than the fair probability $\frac{1}{2}$ of getting the first "six" and thus winning the game.

Theorem 3 is often useful in showing that a given series does *not* converge.

> **Theorem 3   The *n*th-Term Test for Divergence**
> If either
>
> $$\lim_{n \to \infty} a_n \neq 0$$
>
> or this limit does not exist, then the infinite series $\sum a_n$ diverges.

Proof We want to show under the stated hypothesis that the series $\Sigma\, a_n$ diverges. It suffices to show that *if* the series $\Sigma\, a_n$ does converge, then $\lim\limits_{n\to\infty} a_n = 0$. So suppose that $\Sigma\, a_n$ converges with sum $S = \lim\limits_{n\to\infty} S_n$, where

$$S_n = a_1 + a_2 + a_3 + \cdots + a_n$$

is the *n*th partial sum of the series. Because $a_n = S_n - S_{n-1}$,

$$\lim_{n\to\infty} a_n = \lim_{n\to\infty} (S_n - S_{n-1}) = \lim_{n\to\infty} S_n - \lim_{n\to\infty} S_{n-1} = S - S = 0.$$

Consequently, if $\lim\limits_{n\to\infty} a_n \neq 0$, then the series $\Sigma\, a_n$ diverges. ❑

EXAMPLE 8 The series

$$\sum_{n=1}^{\infty} (-1)^{n-1} n^2 = 1 - 4 + 9 - 16 + 25 - \cdots$$

diverges because $\lim\limits_{n\to\infty} a_n$ does not exist, whereas the series

$$\sum_{n=1}^{\infty} \frac{n}{3n + 1} = \frac{1}{4} + \frac{2}{7} + \frac{3}{10} + \frac{4}{13} + \cdots$$

diverges because

$$\lim_{n\to\infty} \frac{n}{3n + 1} = \frac{1}{3} \neq 0.$$

WARNING   The converse of Theorem 3 is *false*! The condition

$$\lim_{n\to\infty} a_n = 0$$

is necessary *but not sufficient* to guarantee convergence of the series

$$\sum_{n=1}^{\infty} a_n.$$

That is, a series may satisfy the condition $a_n \to 0$ as $n \to \infty$ and yet diverge. An important example of a divergent series with terms that approach zero is the **harmonic series**

$$\sum_{n=1}^{\infty} \frac{1}{n} = 1 + \frac{1}{2} + \frac{1}{3} + \frac{1}{4} + \frac{1}{5} + \cdots. \qquad (6)$$

Theorem 4
The harmonic series diverges.

Proof Because each term of the harmonic series is positive, its sequence of partial sums $\{S_n\}$ is increasing. We shall prove that

$$\lim_{n\to\infty} S_n = +\infty,$$

and thus that the harmonic series diverges, by showing that there are arbitrarily large partial sums. Consider the closed interval $[0, k]$ on the *x*-axis, where k is a positive integer. On its subinterval $[0, 1]$, imagine a rectangle of height

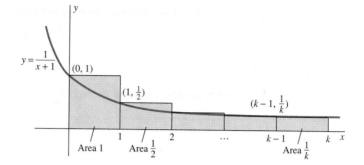

Fig. 11.3.1 Idea of the proof of Theorem 4

1 with that subinterval as its base. On the subinterval $[1, 2]$, imagine a rectangle of height $\frac{1}{2}$ whose base is that subinterval. Build a rectangle of height $\frac{1}{3}$ over the subinterval $[2, 3]$, a rectangle of height $\frac{1}{4}$ over the subinterval $[3, 4]$, and so on. The last rectangle will have the subinterval $[k - 1, k]$ as its base and will have height $1/k$. These rectangles are shown in Fig. 11.3.1.

Next, note that the total area of all k of these rectangles is the kth partial sum of the harmonic series:

$$S_k = 1 + \frac{1}{2} + \frac{1}{3} + \frac{1}{4} + \cdots + \frac{1}{k}.$$

Finally, the graph of the function

$$f(x) = \frac{1}{x + 1}$$

—also shown in Fig. 11.3.1—passes through the upper left-hand corner point of each rectangle. Because f is decreasing for all $x \geq 0$, the graph of f is never above the top of any rectangle on the corresponding subinterval. So the sum of the areas of the k rectangles is greater than the area under the graph of f over $[0, k]$. That is,

$$S_k \geq \int_0^k \frac{1}{x + 1}\,dx = \left[\ln(x + 1)\right]_0^k = \ln(k + 1).$$

But

$$\lim_{k \to \infty} \ln(k + 1) = +\infty,$$

so $\ln(k + 1)$ takes on arbitrarily large positive values with increasing k. Consequently, because $S_k \geq \ln(k + 1)$ for all $k \geq 1$, S_k also takes on arbitrarily large positive values. Therefore $\lim_{k \to \infty} S_k = +\infty$, and hence the harmonic series diverges. ❑

If the sequence of partial sums of the series Σa_n diverges to infinity, then we say that the series **diverges to infinity,** and we write

$$\sum_{n=1}^{\infty} a_n = \infty.$$

The series $\Sigma(-1)^{n+1}$ of Example 2 is a series that diverges but does not diverge to infinity. In the nineteenth century it was common to say that such a series was divergent by oscillation; today we say merely that it diverges.

Our proof of Theorem 4 shows that

$$\sum_{n=1}^{\infty} \frac{1}{n} = \infty.$$

But the partial sums of the harmonic series diverge to infinity very slowly. If N_A denotes the smallest integer such that

$$\sum_{n=1}^{N_A} \frac{1}{n} \geqq A,$$

then it is known that

$$N_5 = 83,$$

$$N_{10} = 12{,}367,$$

$$N_{20} = 272{,}400{,}600,$$

$$N_{100} \approx 1.5 \times 10^{43}, \quad \text{and}$$

$$N_{1000} \approx 1.1 \times 10^{434}.$$

(This can be verified with the aid of a programmable calculator.)

Thus you would need to add more than a quarter of a billion terms of the harmonic series to get a partial sum that exceeds 20. At this point each of the next few terms would be approximately $0.000000004 = 4 \times 10^{-9}$. The number of terms you'd have to add to reach 1000 is far larger than the estimated number of elementary particles in the entire universe (10^{80}).[†]

Theorem 5 says that if two infinite series have the same terms from some point on, then either both series converge or both series diverge. The proof is left for Problem 43.

Theorem 5   *Series that Are Eventually the Same*
If there exists a positive integer k such that $a_n = b_n$ for all $n > k$, then the series $\Sigma\, a_n$ and $\Sigma\, b_n$ either both converge or both diverge.

It follows that a *finite* number of terms can be changed, deleted from, or adjoined to an infinite series without altering its convergence or divergence (although the *sum* of a convergent series will generally be changed by such alterations). In particular, taking $b_n = 0$ for $n \leqq k$ and $b_n = a_n$ for $n > k$, we see that the series

$$\sum_{n=1}^{\infty} a_n$$

and the series

$$\sum_{n=k+1}^{\infty} a_n$$

that is obtained by deleting its first k terms either both converge or both diverge.

[†] If you enjoy such large numbers, see the article "Partial sums of inifinite series, and how they grow," by R. P. Boas, Jr., in *American Mathematical Monthly* **84** (1977): 237–248.

11.3 Problems

In Problems 1 through 29, determine whether the given infinite series converges or diverges. If it converges, find its sum.

1. $1 + \dfrac{1}{3} + \dfrac{1}{9} + \cdots + \dfrac{1}{3^n} + \cdots$

2. $1 + e^{-1} + e^{-2} + \cdots + e^{-n} + \cdots$

3. $1 + 3 + 5 + 7 + \cdots + (2n - 1) + \cdots$

4. $\dfrac{1}{2} + \dfrac{1}{\sqrt{2}} + \dfrac{1}{\sqrt[3]{2}} + \cdots + \dfrac{1}{\sqrt[n]{2}} + \cdots$

5. $1 - 2 + 4 - 8 + \cdots + (-2)^n + \cdots$

6. $1 - \tfrac{1}{4} + \tfrac{1}{16} - \cdots + (-\tfrac{1}{4})^n + \cdots$

7. $4 + \dfrac{4}{3} + \dfrac{4}{9} + \dfrac{4}{27} + \cdots + \dfrac{4}{3^n} + \cdots$

8. $\dfrac{1}{3} + \dfrac{2}{9} + \dfrac{4}{27} + \cdots + \dfrac{2^{n-1}}{3^n} + \cdots$

9. $1 + (1.01) + (1.01)^2 + (1.01)^3 + \cdots + (1.01)^n + \cdots$

10. $1 + \dfrac{1}{\sqrt{2}} + \dfrac{1}{\sqrt[3]{3}} + \cdots + \dfrac{1}{\sqrt[n]{n}} + \cdots$

11. $\displaystyle\sum_{n=0}^{\infty} \dfrac{(-1)^n n}{n + 1}$ **12.** $\displaystyle\sum_{n=1}^{\infty} \left(\dfrac{e}{10}\right)^n$

13. $\displaystyle\sum_{n=0}^{\infty} (-1)^n \left(\dfrac{3}{e}\right)^n$ **14.** $\displaystyle\sum_{n=0}^{\infty} \dfrac{3^n - 2^n}{4^n}$

15. $\displaystyle\sum_{n=1}^{\infty} (\sqrt{2})^{1-n}$ **16.** $\displaystyle\sum_{n=1}^{\infty} \left(\dfrac{2}{n} - \dfrac{1}{2^n}\right)$

17. $\displaystyle\sum_{n=1}^{\infty} \dfrac{n}{10n + 17}$ **18.** $\displaystyle\sum_{n=1}^{\infty} \dfrac{\sqrt{n}}{\ln(n + 1)}$

19. $\displaystyle\sum_{n=1}^{\infty} (5^{-n} - 7^{-n})$ **20.** $\displaystyle\sum_{n=0}^{\infty} \dfrac{1}{1 + (\tfrac{9}{10})^n}$

21. $\displaystyle\sum_{n=1}^{\infty} \left(\dfrac{e}{\pi}\right)^n$ **22.** $\displaystyle\sum_{n=1}^{\infty} \left(\dfrac{\pi}{e}\right)^n$

23. $\displaystyle\sum_{n=0}^{\infty} (\tfrac{100}{99})^n$ **24.** $\displaystyle\sum_{n=0}^{\infty} (\tfrac{99}{100})^n$

25. $\displaystyle\sum_{n=0}^{\infty} \dfrac{1 + 2^n + 3^n}{5^n}$ **26.** $\displaystyle\sum_{n=0}^{\infty} \dfrac{1 + 2^n + 5^n}{3^n}$

27. $\displaystyle\sum_{n=0}^{\infty} \dfrac{7 \cdot 5^n + 3 \cdot 11^n}{13^n}$ **28.** $\displaystyle\sum_{n=1}^{\infty} \sqrt[n]{2}$

29. $\displaystyle\sum_{n=1}^{\infty} [(\tfrac{7}{11})^n - (\tfrac{3}{5})^n]$

30. Use the method of Example 6 to verify that

(a) $0.666\,666\,666 \ldots = \tfrac{2}{3}$; (b) $0.111\,111\,111 \ldots = \tfrac{1}{9}$;

(c) $0.249\,999\,999 \ldots = \tfrac{1}{4}$; (d) $0.999\,999\,999 \ldots = 1$.

In Problems 31 through 35, find the rational number represented by the given repeating decimal.

31. $0.4747\,4747 \ldots$ **32.** $0.2525\,2525 \ldots$

33. $0.123\,123\,123 \ldots$ **34.** $0.3377\,3377\,3377 \ldots$

35. $3.14159\,14159\,14159 \ldots$

In Problems 36 through 40, use the method of Example 3 to find a formula for the nth partial sum S_n, and then compute the sum of the infinite series.

36. $\displaystyle\sum_{n=1}^{\infty} \dfrac{1}{4n^2 - 1}$ **37.** $\displaystyle\sum_{n=1}^{\infty} \ln\left(\dfrac{n + 1}{n}\right)$

38. $\displaystyle\sum_{n=0}^{\infty} \dfrac{4}{16n^2 - 8n - 3}$ **39.** $\displaystyle\sum_{n=2}^{\infty} \dfrac{2}{n^2 - 1}$

40. $\dfrac{1}{1 \cdot 3} + \dfrac{1}{2 \cdot 4} + \dfrac{1}{3 \cdot 5} + \dfrac{1}{4 \cdot 6} + \cdots$

41. Prove: If Σa_n diverges and c is a nonzero constant, then Σca_n diverges.

42. Suppose that Σa_n converges and that Σb_n diverges. Prove that $\Sigma (a_n + b_n)$ diverges.

43. Let S_n and T_n denote the nth partial sums of Σa_n and Σb_n, respectively. Suppose that $a_n = b_n$ for all $n > k$. Show that $S_n - T_n = S_k - T_k$ if $n > k$. Hence prove Theorem 5.

44. Suppose that $0 < x \leqq 1$. Integrate both sides of the identity

$$\dfrac{1}{1 + t} = 1 - t + t^2 - t^3 + \cdots$$

$$+ (-1)^n t^n + \dfrac{(-1)^{n+1} t^{n+1}}{1 + t}$$

from $t = 0$ to $t = x$ to show that

$$\ln(1 + x) = x - \dfrac{x^2}{2} + \dfrac{x^3}{3} - \cdots + (-1)^n \dfrac{x^{n+1}}{n + 1} + R_n,$$

where $\displaystyle\lim_{n \to \infty} R_n = 0$. Hence conclude that

$$\ln(1 + x) = \sum_{n=1}^{\infty} (-1)^{n+1} \dfrac{x^n}{n}$$

if $0 < x \leqq 1$.

45. Criticize the following "proof" that $2 = 1$. Substitution of $x = 1$ into the result of Problem 44 gives the fact that

$$\ln 2 = 1 - \tfrac{1}{2} + \tfrac{1}{3} - \tfrac{1}{4} + \cdots.$$

If

$$S = 1 + \tfrac{1}{2} + \tfrac{1}{3} + \tfrac{1}{4} + \cdots,$$

then

$$\ln 2 = S - 2 \cdot (\tfrac{1}{2} + \tfrac{1}{4} + \tfrac{1}{6} + \tfrac{1}{8} + \cdots) = S - S = 0.$$

Hence $2 = e^{\ln 2} = e^0 = 1$.

46. A ball has *bounce coefficient* $r < 1$ if, when it is dropped from a height h, it bounces back to a height of rh (Fig. 11.3.2). Suppose that such a ball is dropped from the initial height a and subsequently bounces infinitely many times. Use a geometric series to show that the total up-and-down distance it travels in all its bouncing is

$$D = a\frac{1 + r}{1 - r}.$$

Note that D is *finite*.

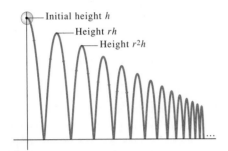

Fig. 11.3.2 Successive bounces of the ball of Problems 46 and 47

47. A ball with bounce coefficient $r = 0.64$ (see Problem 46) is dropped from an initial height of $a = 4$ ft. Use a geometric series to compute the total time required for it to complete its infinitely many bounces. The time required for a ball to drop h feet (from rest) is $\sqrt{2h/g}$ seconds where $g = 32$ ft/s^2.

48. Suppose that the government spends \$1 billion and that each recipient of a fraction of this wealth spends 90% of the dollars that he or she receives. In turn, the secondary recipients spend 90% of the dollars they receive, and so on. How much total spending results from the original injection of \$1 billion into the economy?

49. A tank initially contains a mass M_0 of air. Each stroke of a vacuum pump removes 5% of the air in the container.

Compute (a) the mass M_n of air remaining in the tank after n strokes of the pump; (b) $\lim_{n \to \infty} M_n$.

50. Paul and Mary toss a fair coin in turn until one of them wins the game by getting the first "head." Calculate for each the probability that he or she wins the game.

51. Peter, Paul, and Mary toss a fair coin in turn until one of them wins by getting the first "head." Calculate for each the probability that he or she wins the game. Check your answer by verifying that the sum of the three probabilities is 1.

52. Peter, Paul, and Mary roll a fair die in turn until one of them wins the game by getting the first "six." Calculate for each the probability that he or she wins the game. Check your answer by verifying that the sum of the three probabilities is 1.

53. A pane of a certain type of glass reflects half the incident light, absorbs one-fourth, and transmits one-fourth. A window is made of two panes of this glass separated by a small space (Fig. 11.3.3). What fraction of the incident light I is transmitted by the double window?

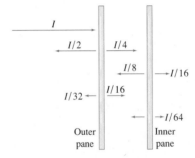

Fig. 11.3.3 The double-pane window of Problem 53

54. Criticize the following evaluation of the sum of an infinite series:

Let $x = 1 - 2 + 4 - 8 + 16 - 32 + 64 - \cdots$.
Then $2x = 2 - 4 + 8 - 16 + 32 - 64 + \cdots$.
Add the equations to obtain $3x = 1$. Thus $x = \tfrac{1}{3}$, and "therefore"

$$1 - 2 + 4 - 8 + 16 - 32 + 64 - \cdots = \tfrac{1}{3}.$$

11.3 Projects

These projects require the use of a programmable calculator or computer. It is not otherwise practical to calculate partial sums with large numbers of terms.

The TI-81/85 and BASIC programs listed in Fig. 11.3.4 can be used to approximate the sum of the convergent infinite series

$$\sum_{n=1}^{\infty} a_n = a_1 + a_2 + a_3 + \cdots$$

TI-81/85	BASIC	Comments
`PRGM1:SERIES`	`DEF FN A(N) = (1/2)^N`	Define Y1 $= a(n)$
`:Disp "K"`		
`:Input K`	`INPUT "K"; K`	Number of terms to sum
`:0->S`	`S = 0`	Initialize sum
`:1->N`		Initialize index
`:Lbl 1`	`FOR N = 1 TO K`	Begin loop
`:Y1->T`	`T = FN A(N)`	Compute next term
`:S+T->S`	`S = S + T`	Add next term
`:N+1->N`	`NEXT`	Next N
`:If N ≤ K`		End loop
`:Goto 1`		
`:Disp "SUM="`	`PRINT "Sum = "; S`	Display sum
`:Disp S`		
`:End`	`END`	

Fig. 11.3.4 TI calculator and BASIC programs for calculating partial sums of Σa_n

by calculating the kth partial sum

$$S_k = \sum_{n=1}^{k} a_n = a_1 + a_2 + a_3 + \cdots + a_k.$$

In the case of an infinite series $\Sigma_{n=0}^{\infty} a_n$ beginning at $n = 0$, the initial term a_0 must be added manually or the program altered appropriately.

Even if you don't program in BASIC or in the TI calculator language, you may find it helpful to compare these alternative descriptions of the same computational algorithm. For the TI calculator program, the formula $a_n = a(n)$ that gives the nth term as a function of n must be defined (as Y1) in the Y = menu before the program is executed. (For instance, enter `Y1=(1/2)∧N` to sum the geometric series $\Sigma_{n=1}^{\infty} (1/2)^n$.) In the BASIC version the nth term function is defined in the first line of the program itself. The desired number k of terms to sum is entered while the program is running.

Most computer algebra systems include a SUM function that can be used to calculate partial sums directly. Suppose that the nth term function $a(n)$ (expressing a_n as a a function of n) has already been defined appropriately. Then all the single-line commands listed in Fig. 11.3.5 calculate (in various systems) the kth partial sum $\Sigma_{n=1}^{k} a_n$. The similarities among these commands in different systems are striking.

Program	Command
Derive	`SUM( a(n), n, 1, k )`
HP-48SX	`'Σ (n=1,k,a(n))'`
Maple	`sum( a(n), n=1..k )`
Mathematica	`Sum[ a[n], {n, 1, k} ]`
TI-85	`sum seq( a(n), n, 1, k, 1 )`
(X)PLORE	`sum( a(n), n=1 to k )`

Fig. 11.3.5 Commands for calculating the kth partial sum of the infinite series Σa_n

PROJECT A Calculate partial sums of the geometric series $\Sigma_{n=0}^{\infty} r^n$ with $r = 0.2, 0.5, 0.75, 0.9, 0.95,$ and 0.99. For each value of r, calculate k-term partial sums with $k = 10, 20, 30, \ldots$, continuing until two successive results agree to four or five decimal places. (For $r = 0.95$ and 0.99, you may decide to use $k = 100, 200, 300, \ldots$.) How does the apparent rate of convergence—as measured by the number of terms required for the desired accuracy—depend on the value of r?

PROJECT B Calculate partial sums of the harmonic series $\Sigma\, 1/n$ with $k = 100, 200, 300, \ldots$ terms (or with $k = 1000, 2000, 3000, \ldots$ if you have a powerful microcomputer). Interpret your results in light of our discussion of the harmonic series.

PROJECT C In Section 11.4 we will see that the famous number e is given by

$$e = 1 + \sum_{n=1}^{\infty} \frac{1}{n!} = 1 + \frac{1}{1!} + \frac{1}{2!} + \frac{1}{3!} + \cdots,$$

where the *factorial* $n! = 1 \cdot 2 \cdot 3 \cdots n$ denotes the product of the first n positive integers. Sum enough terms of this rapidly convergent infinite series to convince yourself that $e = 2.7\,1828\,1828$ (accurate to nine decimal places).

11.4
Taylor Series and Taylor Polynomials

The infinite series we studied in Section 11.3 have *constant* terms, and the sum of such a series (assuming it converges) is a *number*. In contrast, much of the practical importance of infinite series derives from the fact that many functions have useful representations as infinite series with *variable* terms.

EXAMPLE 1 If we write $r = x$ for the ratio in a geometric series, Theorem 1 in Section 11.3 gives the infinite series representation

$$\frac{1}{1 - x} = \sum_{n=0}^{\infty} x^n = 1 + x + x^2 + x^3 + \cdots \tag{1}$$

of the function $f(x) = 1/(1 - x)$. The infinite series in Eq. (1) converges to $1/(1 - x)$ for every number x in the interval $(-1, 1)$. Each partial sum

$$1 + x + x^2 + x^3 + \cdots + x^n \tag{2}$$

of the geometric series in Eq. (1) is a polynomial approximation to the function $f(x) = 1/(1 - x)$. If $|x| < 1$, then the convergence of the series in Eq. (1) implies that the approximation

$$\frac{1}{1 - x} \approx 1 + x + x^2 + x^3 + \cdots + x^n \tag{3}$$

is accurate if n is sufficiently large.

REMARK The approximation in (3) could be used to calculate numerical quotients with a calculator that has only $+$, $-$, $\times$ keys (but no $\div$ key). For instance,

$$\frac{329}{73} = \frac{3.29}{0.73} = 3.29 \times \frac{1}{1 - 0.27}$$

$$\approx (3.29)[1 + (0.27) + (0.27)^2 + \cdots + (0.27)^{10}]$$

$$\approx (3.29)(1.36986); \quad \text{thus}$$

$$\frac{329}{73} \approx 4.5068,$$

accurate to four decimal places. This is a simple illustration of the use of polynomial approximation for numerical computation.

The definitions of the various elementary transcendental functions leave it unclear how to compute their values precisely, except at a few isolated points. For example,

$$\ln x = \int_1^x \frac{1}{t}\, dt \qquad (x > 0)$$

by definition, so obviously $\ln 1 = 0$, but no other value of $\ln x$ is obvious. The natural exponential function is the inverse of $\ln x$, so it is clear that $e^0 = 1$, but it is not at all clear how to compute e^x for $x \neq 0$. Indeed, even such an innocent-looking expression as $\sqrt{x}$ is not computable (precisely and in a finite number of steps) unless x happens to be the square of a rational number.

But *any* value of a polynomial

$$P(x) = c_0 + c_1 x + c_2 x^2 + \cdots + c_n x^n$$

with known coefficients $c_0, c_1, c_2, \ldots, c_n$ is easy to calculate—as in the preceding remark, only addition and multiplication are required. One goal of this section is to use the fact that polynomial values are so readily computable to help us calculate approximate values of functions such as $\ln x$ and e^x.

POLYNOMIAL APPROXIMATIONS AND TAYLOR'S FORMULA

Suppose that we want to calculate (or, at least, closely approximate) a specific value $f(x_0)$ of a given function f. It would suffice to find a polynomial $P(x)$ with a graph that is very close to that of f on some interval containing x_0. For then we could use the value $P(x_0)$ as an approximation to the actual value of $f(x_0)$. Once we know how to find such an approximating polynomial $P(x)$, the next question would be how accurately $P(x_0)$ approximates the desired value $f(x_0)$.

The simplest example of polynomial approximation is the linear approximation

$$f(x) \approx f(a) + f'(a)(x - a)$$

obtained by writing $\Delta x = x - a$ in the linear approximation formula, Eq. (3) of Section 4.2. The graph of the first-degree polynomial

$$P_1(x) = f(a) + f'(a)(x - a) \tag{4}$$

is the line tangent to the curve $y = f(x)$ at the point $(a, f(a))$; see Fig. 11.4.1. This first-degree polynomial agrees with f and with its first derivative at $x = a$. That is,

$$P_1(a) = f(a) \quad \text{and} \quad P_1'(a) = f'(a).$$

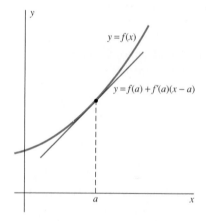

Fig. 11.4.1 The tangent line at $(a, f(a))$ is the best linear approximation to $y = f(x)$ near a.

EXAMPLE 2 Suppose that $f(x) = \ln x$ and that $a = 1$. Then $f(1) = 0$ and $f'(1) = 1$, so $P_1(x) = x - 1$. Hence we expect that $\ln x \approx x - 1$ for x near 1. With $x = 1.1$, we find that

$$P_1(1.1) = 0.1000, \quad \text{whereas} \quad \ln(1.1) \approx 0.0953.$$

The error in this approximation is about 5%.

To better approximate $\ln x$ near $x = 1$, let us look for a second-degree polynomial

$$P_2(x) = c_0 + c_1 x + c_2 x^2$$

that has not only the same value and the same first derivative as does f at $x = 1$, but has also the same second derivative there: $P_2''(1) = f''(1) = -1$. To satisfy these conditions, we must have

$$P_2(1) = c_2 + c_1 + c_0 = 0,$$
$$P_2'(1) = 2c_2 + c_1 = 1,$$
$$P_2''(1) = 2c_2 = -1.$$

When we solve these equations, we find that $c_0 = -\frac{3}{2}$, $c_1 = 2$, and $c_2 = -\frac{1}{2}$, so

$$P_2(x) = -\tfrac{1}{2}x^2 + 2x - \tfrac{3}{2}.$$

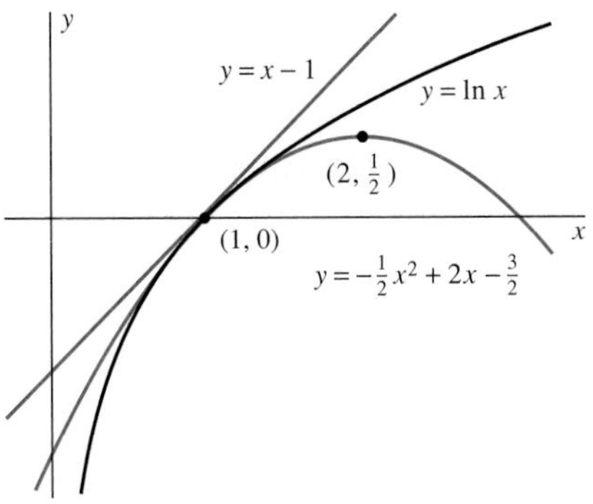

With $x = 1.1$ we find that $P_2(1.1) = 0.0950$, which is accurate to three decimal places because $\ln(1.1) \approx 0.0953$. The graph of $y = -\frac{1}{2}x^2 + 2x - \frac{3}{2}$ is a parabola through $(1, 0)$ with the same value, slope, *and curvature* there as $y = \ln x$ (Fig. 11.4.2).

Fig. 11.4.2 The linear and parabolic approximations to $y = \ln x$ near the point $(1, 0)$ (Example 2)

The tangent line and the parabola used in the computations of Example 2 illustrate one general approach to polynomial approximation. To approximate the function $f(x)$ near $x = a$, we look for an nth-degree polynomial

$$P_n(x) = c_0 + c_1 x + c_2 x^2 + \cdots + c_n x^n$$

such that its value at a and the values of its first n derivatives at a agree with the corresponding values of f. That is, we require that

$$
\begin{aligned}
P_n(a) &= f(a), \\
P_n'(a) &= f'(a), \\
P_n''(a) &= f''(a), \\
&\;\;\vdots \\
P_n^{(n)}(a) &= f^{(n)}(a).
\end{aligned}
\tag{5}
$$

We can use these $n + 1$ conditions to evaluate the values of the $n + 1$ coefficients $c_0, c_1, \ldots, c_n$.

The algebra involved is much simpler, however, if we begin with $P_n(x)$ expressed as an nth-degree polynomial in powers of $x - a$ rather than in powers of x:

$$P_n(x) = b_0 + b_1(x - a) + b_2(x - a)^2 + \cdots + b_n(x - a)^n. \tag{6}$$

Then substitution of $x = a$ into Eq. (3) yields

$$b_0 = P_n(a) = f(a)$$

by the first condition in Eq. (5). Substitution of $x = a$ into

$$P_n'(x) = b_1 + 2b_2(x - a) + 3b_3(x - a)^2 + \cdots + nb_n(x - a)^{n-1}$$

yields

$$b_1 = P_n'(a) = f'(a)$$

by the second condition in Eq. (5). Next, substitution of $x = a$ into

$$P_n''(x) = 2b_2 + 3 \cdot 2b_3(x - a) + \cdots + n(n - 1)b_n(x - a)^{n-2}$$

yields $2b_2 = P_n''(a) = f''(a)$, so

$$b_2 = \tfrac{1}{2} f''(a).$$

We continue this process to find $b_3, b_4, \ldots, b_n$. In general, the constant term in the kth derivative $P_n^{(k)}(x)$ is $k!b_k$, because it is the kth derivative of the kth-degree term $b_k(x - a)^k$ in $P_n(x)$:

$$P_n^{(k)}(x) = k!b_k + \{\text{powers of } x - a\}.$$

(Recall that $k! = 1 \cdot 2 \cdot 3 \cdots (k - 1) \cdot k$ denotes the *factorial* of the positive integer k, read "k factorial.") So when we substitute $x = a$ into $P_n^{(k)}(x)$, we find that

$$k!b_k = P_n^{(k)}(a) = f^{(k)}(a)$$

and thus that

$$b_k = \frac{f^{(k)}(a)}{k!} \tag{7}$$

for $k = 1, 2, 3, \ldots, n$.

Indeed, Eq. (7) holds also for $k = 0$ if we use the universal convention that $0! = 1$ and agree that the zeroth derivative $g^{(0)}$ of the function g is just g itself. With such conventions, our computations establish the following theorem.

Theorem 1   *The nth-Degree Taylor Polynomial*

Suppose that the first n derivatives of the function $f(x)$ exist at $x = a$. Let $P_n(x)$ be the nth-degree polynomial

$$P_n(x) = \sum_{k=0}^{n} \frac{f^{(k)}(a)}{k!}(x - a)^k$$

$$= f(a) + f'(a)(x - a) + \frac{f''(a)}{2!}(x - a)^2$$

$$+ \cdots + \frac{f^{(n)}(a)}{n!}(x - a)^n. \tag{8}$$

Then the values of $P_n(x)$ and its first n derivatives agree, at $x = a$, with the values of f and its first n derivatives there. That is, the equations in (5) all hold.

The polynomial in Eq. (8) is called the **nth-degree Taylor polynomial of the function f at the point $x = a$.** Note that $P_n(x)$ is a polynomial in powers of $x - a$ rather than in powers of x. To use $P_n(x)$ effectively for the approximation of $f(x)$ near a, we must be able to compute the value $f(a)$ and the values of its derivatives $f'(a), f''(a)$, and so on, all the way to $f^{(n)}(a)$.

The line $y = P_1(x)$ is simply the line tangent to the curve $y = f(x)$ at the point $(a, f(a))$. Thus $y = f(x)$ and $y = P_1(x)$ have the same slope at this point. Now recall from Section 4.6 that the second derivative measures the way the curve $y = f(x)$ is bending as it passes through $(a, f(a))$. Therefore, let us call $f''(a)$ the "concavity" of $y = f(x)$ at $(a, f(a))$. Then, because $P_2''(a) = f''(a)$, it follows that $y = P_2(x)$ has the same value, the same slope, and the same concavity at $(a, f(a))$ as does $y = f(x)$. Moreover, $P_3(x)$ and $f(x)$ will also have the same rate of change of concavity at $(a, f(a))$. Such observations suggest that the larger n is, the more closely the nth-degree Taylor polynomial will approximate $f(x)$ for x near a.

EXAMPLE 3 Find the nth-degree Taylor polynomial of $f(x) = \ln x$ at $a = 1$.

Solution The first few derivatives of $f(x) = \ln x$ are

$$f'(x) = \frac{1}{x}, \quad f''(x) = -\frac{1}{x^2}, \quad f^{(3)}(x) = \frac{2}{x^3}, \quad f^{(4)}(x) = -\frac{3!}{x^4}, \quad f^{(5)}(x) = \frac{4!}{x^5}.$$

The pattern is clear:

$$f^{(k)}(x) = (-1)^{k-1}\frac{(k-1)!}{x^k} \quad \text{for } k \geq 1.$$

Hence $f^{(k)}(1) = (-1)^{k-1}(k-1)!$, so Eq. (8) gives

$$P_n(x) = (x - 1) - \frac{1}{2}(x - 1)^2 + \frac{1}{3}(x - 1)^3$$
$$- \frac{1}{4}(x - 1)^4 + \cdots + \frac{(-1)^{n-1}}{n}(x - 1)^n.$$

With $n = 2$ we obtain the quadratic polynomial

$$P_2(x) = (x - 1) - \tfrac{1}{2}(x - 1)^2 = -\tfrac{1}{2}x^2 + 2x - \tfrac{3}{2},$$

the same as in Example 2. With the third-degree Taylor polynomial

$$P_3(x) = (x - 1) - \tfrac{1}{2}(x - 1)^2 + \tfrac{1}{3}(x - 1)^3,$$

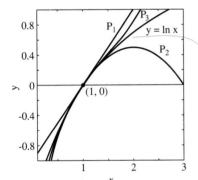

Fig. 11.4.3 The graphs of $f(x) = \ln x$ and its first three Taylor polynomial approximations at $x = 1$ (Example 3)

we can go a step further in approximating $\ln(1.1)$: The value $P_3(1.1) = 0.095333\ldots$ is correct to four decimal places. See Fig. 11.4.3 for the graphs of $\ln x$, $P_1(x)$, $P_2(x)$, and $P_3(x)$.

In the common case $a = 0$, the nth-degree Taylor polynomial in Eq. (8) reduces to

$$P_n(x) = f(0) + f'(0) \cdot x + \frac{f''(0)}{2!}x^2 + \cdots + \frac{f^{(n)}(0)}{n!}x^n. \tag{9}$$

EXAMPLE 4 Find the nth-degree Taylor polynomial for $f(x) = e^x$ at $a = 0$.

Solution This is the easiest of all Taylor polynomials to compute, because $f^{(k)}(x) = e^x$ for all $k \geq 0$. Hence $f^{(k)}(0) = 1$ for all $k \geq 0$, so Eq. (9) yields

$$P_n(x) = 1 + x + \frac{x^2}{2!} + \frac{x^3}{3!} + \cdots + \frac{x^n}{n!}.$$

The first few Taylor polynomials of the natural exponential function at $a = 0$ are, therefore,

$$P_0(x) = 1,$$

$$P_1(x) = 1 + x,$$

$$P_2(x) = 1 + x + \tfrac{1}{2}x^2,$$

$$P_3(x) = 1 + x + \tfrac{1}{2}x^2 + \tfrac{1}{6}x^3,$$

$$P_4(x) = 1 + x + \tfrac{1}{2}x^2 + \tfrac{1}{6}x^3 + \tfrac{1}{24}x^4,$$

$$P_5(x) = 1 + x + \tfrac{1}{2}x^2 + \tfrac{1}{6}x^3 + \tfrac{1}{24}x^4 + \tfrac{1}{120}x^5.$$

The table in Fig. 11.4.4 shows how these polynomials approximate $f(x) = e^x$ for $x = 0.1$ and for $x = 0.5$. At least for these two values of x, the closer x is to $a = 0$, the more rapidly $P_n(x)$ appears to approach $f(x)$ as n increases.

$x = 0.1$

n	$P_n(X)$	e^x	$e^x - P_n(x)$
0	1.00000	1.10517	0.10517
1	1.10000	1.10517	0.00517
2	1.10500	1.10517	0.00017
3	1.10517	1.10517	0.00000
4	1.10517	1.10517	0.00000

$x = 0.5$

n	$P_n(X)$	e^x	$e^x - P_n(x)$
0	1.00000	1.64872	0.64872
1	1.50000	1.64872	0.14872
2	1.62500	1.64872	0.02372
3	1.64583	1.64872	0.00289
4	1.64844	1.64872	0.00028
5	1.64879	1.64872	0.00002

Fig. 11.4.4 Approximating $y = e^x$ with Taylor polynomials at $a = 0$

The closeness with which $P_n(x)$ approximates $f(x)$ is measured by the difference

$$R_n(x) = f(x) - P_n(x),$$

for which

$$f(x) = P_n(x) + R_n(x). \tag{10}$$

This difference $R_n(x)$ is called the **nth-degree remainder for $f(x)$ at $x = a$**. It is the *error* made if the value $f(x)$ is replaced by the approximation $P_n(x)$.

The theorem that lets us estimate the error, or remainder, $R_n(x)$ is called **Taylor's formula**, after Brook Taylor (1685–1731), a follower of Newton who introduced Taylor polynomials in an article published in 1715. The particular expression for $R_n(x)$ that we give next is called the *Lagrange form* for the remainder because it first appeared in 1797 in a book written by the French mathematician Joseph Louis Lagrange (1736–1813).

> **Theorem 2** *Taylor's Formula*
>
> Suppose that the $(n + 1)$th derivative of the function f exists on an interval containing the points a and b. Then
>
> $$f(b) = f(a) + f'(a)(b - a) + \frac{f''(a)}{2!}(b - a)^2$$
>
> $$+ \frac{f^{(3)}(a)}{3!}(b - a)^3 + \cdots + \frac{f^{(n)}(a)}{n!}(b - a)^n$$
>
> $$+ \frac{f^{(n+1)}(z)}{(n + 1)!}(b - a)^{n+1} \tag{11}$$
>
> for some number z between a and b.

REMARK   With $n = 0$, Eq. (11) reduces to the equation

$$f(b) = f(a) + f'(z) \cdot (b - a)$$

in the conclusion of the mean value theorem (Section 4.3). Thus Taylor's formula is a far-reaching generalization of the mean value theorem of differential calculus.

The proof of Taylor's formula is given in Appendix H. If we replace b by x in Eq. (11), we get the **nth-degree Taylor formula with remainder at $x = a$**,

$$f(x) = f(a) + f'(a)(x - a) + \frac{f''(a)}{2!}(x - a)^2 + \frac{f^{(3)}(a)}{3!}(x - a)^3$$

$$+ \cdots + \frac{f^{(n)}(a)}{n!}(x - a)^n + \frac{f^{(n+1)}(z)}{(n + 1)!}(x - a)^{n+1}, \tag{12}$$

where z is some number between a and x. Thus the nth-degree remainder term is

$$R_n(x) = \frac{f^{(n+1)}(z)}{(n + 1)!}(x - a)^{n+1}, \tag{13}$$

which is easy to remember—it's the same as the *last* term of $P_{n+1}(x)$, except that $f^{(n+1)}(a)$ is replaced by $f^{(n+1)}(z)$.

EXAMPLE 3 continued   To estimate the accuracy of the approximation

$$\ln(1.1) \approx 0.095333,$$

we substitute $x = 1$ into the formula

$$f^{(k)}(x) = (-1)^{k-1}\frac{(k-1)!}{x^k}$$

for the kth derivative of $f(x) = \ln x$ and get

$$f^{(k)}(1) = (-1)^{k-1}(k-1)!.$$

Hence the third-degree Taylor formula *with remainder* at $a = 1$ is

$$\ln x = (x - 1) - \frac{1}{2}(x - 1)^2 + \frac{1}{3}(x - 1)^3 - \frac{3!}{4! \, z^4}(x - 1)^4$$

with z between $a = 1$ and x. With $x = 1.1$ this gives

$$\ln(1.1) = 0.095333 \ldots - \frac{(0.1)^4}{4z^4}$$

with $1 < z < 1.1$. The value $z = 1$ gives the largest possible value $(0.1)^4/4 = 0.000025$ of the remainder term. It follows that

$$0.0953083 < \ln(1.1) < 0.0953334,$$

so we can conclude that $\ln(1.1) = 0.0953$ to four-place accuracy.

TAYLOR SERIES

If the function f has derivatives of all orders, then we can write Taylor's formula [Eq. (11)] with any degree n that we please. Ordinarily, the exact value of z in the Taylor remainder term in Eq. (13) is unknown. Nevertheless, we can sometimes use Eq. (13) to show that the remainder approaches zero as $n \to \infty$:

$$\lim_{n \to \infty} R_n(x) = 0 \qquad (14)$$

for some particular *fixed* value of x. Then Eq. (10) gives

$$f(x) = \lim_{n \to \infty} [P_n(x) + R_n(x)] = \lim_{n \to \infty} P_n(x) = \lim_{n \to \infty} \sum_{k=0}^{n} \frac{f^{(k)}(a)}{k!}(x - a)^k;$$

that is,

$$f(x) = \sum_{k=0}^{\infty} \frac{f^{(k)}(a)}{k!}(x - a)^k. \qquad (15)$$

The infinite series

$$\sum_{n=0}^{\infty} \frac{f^{(n)}(a)}{n!}(x - a)^n = f(a) + f'(a)(x - a) + \frac{f''(a)}{2}(x - a)^2$$

$$+ \cdots + \frac{f^{(n)}(a)}{n!}(x - a)^n + \cdots \qquad (16)$$

is called the **Taylor series** of the function f at $x = a$. Its partial sums are the successive Taylor polynomials of f at $x = a$.

We can write the Taylor series of a function f without knowing that it converges. But if the limit in Eq. (14) can be established, then it follows as in Eq. (15) that the Taylor series in Eq. (16) actually converges to $f(x)$. If so, then we can approximate the value of $f(x)$ accurately by calculating the value of a Taylor polynomial of f of sufficiently high degree.

EXAMPLE 5   In Example 4 we noted that if $f(x) = e^x$, then $f^{(k)}(x) = e^x$ for all $k \geq 0$. Hence the Taylor formula

$$f(x) = f(0) + f'(0) \cdot x + \frac{f''(0)}{2!}x^2 + \cdots + \frac{f^{(n)}(0)}{n!}x^n + \frac{f^{(n+1)}(z)}{(n + 1)!}x^{n+1}$$

at $a = 0$ gives

$$e^x = 1 + x + \frac{x^2}{2!} + \frac{x^3}{3!} + \cdots + \frac{x^n}{n!} + \frac{e^z x^{n+1}}{(n+1)!} \tag{17}$$

for some z between 0 and x. Thus the remainder term $R_n(x)$ satisfies the inequalities

$$0 < |R_n(x)| < \frac{|x|^{n+1}}{(n+1)!} \qquad \text{if } x < 0,$$

$$0 < |R_n(x)| < \frac{e^x x^{n+1}}{(n+1)!} \qquad \text{if } x > 0.$$

Therefore, the fact that

$$\lim_{n \to \infty} \frac{x^n}{n!} = 0 \tag{18}$$

for all x (see Problem 49) implies that $\lim_{n \to \infty} R_n(x) = 0$ for all x. This means that the Taylor series for e^x converges to e^x for all x, and we may write

$$e^x = \sum_{n=0}^{\infty} \frac{x^n}{n!} = 1 + x + \frac{x^2}{2!} + \frac{x^3}{3!} + \frac{x^4}{4!} + \cdots. \tag{19}$$

The series in Eq. (19) is the most famous and most important of all Taylor series.

With $x = 1$, Eq. (19) yields a numerical series

$$e = \sum_{n=0}^{\infty} \frac{1}{n!} = 1 + \frac{1}{1!} + \frac{1}{2!} + \frac{1}{3!} + \frac{1}{4!} + \cdots \tag{20}$$

for the number e itself. The 10th and 20th partial sums of this series gives the approximations

$$e \approx 1 + \frac{1}{1!} + \frac{1}{2!} + \cdots + \frac{1}{10!} \approx 2.71828\,18$$

and

$$e \approx 1 + \frac{1}{1!} + \frac{1}{2!} + \cdots + \frac{1}{20!} \approx 2.71828\,18284\,59045\,235,$$

both of which are accurate to the number of decimal places shown.

EXAMPLE 6 To find the Taylor series at $a = 0$ for $f(x) = \cos x$, we first calculate the derivatives

$$f(x) = \cos x, \qquad\qquad f'(x) = -\sin x,$$
$$f''(x) = -\cos x, \qquad\qquad f^{(3)}(x) = \sin x,$$
$$f^{(4)}(x) = \cos x, \qquad\qquad f^{(5)}(x) = -\sin x,$$
$$\vdots \qquad\qquad\qquad \vdots$$
$$f^{(2n)}(x) = (-1)^n \cos x, \qquad f^{(2n+1)}(x) = (-1)^{n+1} \sin x.$$

It follows that

$$f^{(2n)}(0) = (-1)^n \quad \text{but} \quad f^{(2n+1)}(0) = 0,$$

so the Taylor polynomials and Taylor series of $f(x) = \cos x$ include only terms of *even* degree. The Taylor formula of degree $2n$ for $\cos x$ at $a = 0$ is

$$\cos x = 1 - \frac{x^2}{2!} + \frac{x^4}{4!} - \cdots + (-1)^n \frac{x^{2n}}{(2n)!} + (-1)^{n+1} \frac{\cos z}{(2n+2)!} x^{2n+2},$$

where z is between 0 and x. Because $|\cos z| \leq 1$ for all z, it follows from Eq. (18) that the remainder term approaches zero as $n \to \infty$ *for all* x. Hence the desired Taylor series of $f(x) = \cos x$ at $a = 0$ is

$$\cos x = \sum_{n=0}^{\infty} \frac{(-1)^n x^{2n}}{(2n)!} = 1 - \frac{x^2}{2!} + \frac{x^4}{4!} - \frac{x^6}{6!} + \cdots. \qquad (21)$$

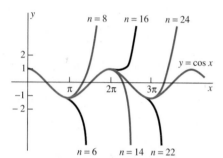

Fig. 11.4.5 Approximating $\cos x$ with nth-degree Taylor polynomials

In Problem 43 we ask you to show similarly that the Taylor series at $a = 0$ of $f(x) = \sin x$ is

$$\sin x = \sum_{n=0}^{\infty} \frac{(-1)^n x^{2n+1}}{(2n+1)!} = x - \frac{x^3}{3!} + \frac{x^5}{5!} - \frac{x^7}{7!} + \cdots. \qquad (22)$$

Figures 11.4.5 and 11.4.6 illustrate the increasingly better approximations to $\cos x$ and $\sin x$ that we get by using more and more terms of the series in Eqs. (21) and (22).

The case $a = 0$ of Taylor's series is called the **Maclaurin series** of the function $f(x)$,

$$\sum_{n=0}^{\infty} \frac{f^{(n)}(0)}{n!} x^n = f(0) + f'(0) \cdot x + \frac{f''(0)}{2!} x^2 + \frac{f^{(3)}(0)}{3!} x^3 + \cdots. \qquad (23)$$

Colin Maclaurin (1698–1746) was a Scottish mathematician who used this series as a basic tool in a calculus book he published in 1742. The three Maclaurin series

$$e^x = \sum_{n=0}^{\infty} \frac{x^n}{n!} = 1 + x + \frac{x^2}{2!} + \frac{x^3}{3!} + \frac{x^4}{4!} + \cdots, \qquad (19)$$

$$\cos x = \sum_{n=0}^{\infty} \frac{(-1)^n x^{2n}}{(2n)!} = 1 - \frac{x^2}{2!} + \frac{x^4}{4!} - \frac{x^6}{6!} + \cdots, \quad \text{and} \qquad (21)$$

$$\sin x = \sum_{n=0}^{\infty} \frac{(-1)^n x^{2n+1}}{(2n+1)!} = x - \frac{x^3}{3!} + \frac{x^5}{5!} - \frac{x^7}{7!} + \cdots \qquad (22)$$

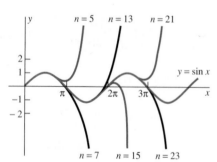

Fig. 11.4.6 Approximating $\sin x$ with nth-degree Taylor polynomials

(which actually were discovered by Newton) bear careful examination and comparison. Observe that

❏ The terms in the *even* cosine series are the *even*-degree terms in the exponential series but with alternating signs.

❏ The terms in the *odd* sine series are the *odd*-degree terms in the exponential series but with alternating signs.

These series are *identities* that hold for all values of x. Consequently, new series can be derived by substitution, as in Examples 7 and 8.

EXAMPLE 7 The substitution $x = -t^2$ into Eq. (19) yields

$$e^{-t^2} = 1 - t^2 + \frac{t^4}{2!} - \frac{t^6}{3!} + \cdots + (-1)^n \frac{t^{2n}}{n!} + \cdots.$$

EXAMPLE 8 The substitution $x = 2t$ into Eq. (22) gives

$$\sin 2t = 2t - \frac{4}{3}t^3 + \frac{4}{15}t^5 - \frac{8}{315}t^7 + \cdots.$$

*THE NUMBER π

In Section 5.3 we described how Archimedes used polygons inscribed in and circumscribed about the unit circle to show that $3\frac{10}{71} < \pi < 3\frac{1}{7}$. With the aid of electronic computers, π has been calculated to well over a *billion* decimal places. We describe now some of the methods that have been used for such computations.[†]

We begin with the elementary algebraic identity

$$\frac{1}{1+x} = 1 - x + x^2 - x^3 + \cdots + (-1)^{k-1}x^{k-1} + \frac{(-1)^k x^k}{1+x}, \quad (24)$$

which can be verified by multiplying both sides by $1 + x$. We substitute t^2 for x and $n + 1$ for k and thus find that

$$\frac{1}{1+t^2} = 1 - t^2 + t^4 - t^6 + \cdots + (-1)^n t^{2n} + \frac{(-1)^{n+1} t^{2n+2}}{1+t^2}.$$

Because $D_t \tan^{-1} t = 1/(1 + t^2)$, integration of both sides of this last equation from $t = 0$ to $t = x$ gives

$$\tan^{-1} x = x - \frac{x^3}{3} + \frac{x^5}{5} - \frac{x^7}{7} + \cdots + (-1)^n \frac{x^{2n+1}}{2n+1} + R_{2n+1}, \quad (25)$$

where

$$|R_{2n+1}| = \left| \int_0^x \frac{t^{2n+2}}{1+t^2}\, dt \right| \leq \left| \int_0^x t^{2n+2}\, dt \right| = \frac{|x|^{2n+3}}{2n+3}. \quad (26)$$

This estimate of the error makes it clear that

$$\lim_{n \to \infty} R_n = 0$$

if $|x| \leq 1$. Hence we obtain the Taylor series for the inverse tangent function:

$$\tan^{-1} x = \sum_{n=0}^{\infty} (-1)^n \frac{x^{2n+1}}{2n+1} = x - \frac{x^3}{3} + \frac{x^5}{5} - \frac{x^7}{7} + \cdots, \quad (27)$$

valid for $-1 \leq x \leq 1$.

If we substitute $x = 1$ into Eq. (27), we obtain *Leibniz's series*

$$\frac{\pi}{4} = 1 - \frac{1}{3} + \frac{1}{5} - \frac{1}{7} + \cdots.$$

Although this is a beautiful series, it is not an effective way to compute π. But the error estimate in Eq. (26) shows that we can use Eq. (25) to calculate $\tan^{-1} x$ if $|x|$ is small. For example, if $x = \frac{1}{5}$, then the fact that

$$\frac{1}{9 \cdot 5^9} \approx 0.000000057 < 0.0000001$$

[†] For a chronicle of humanity's perennial fascination with the number π, see Howard Eves, *An Introduction to the History of Mathematics*, (Boston: Allyn and Bacon, 4th ed., 1976), pp. 96–102.

implies that the approximation

$$\tan^{-1}\left(\tfrac{1}{5}\right) \approx \tfrac{1}{5} - \tfrac{1}{3}\left(\tfrac{1}{5}\right)^3 + \tfrac{1}{5}\left(\tfrac{1}{5}\right)^5 - \tfrac{1}{7}\left(\tfrac{1}{5}\right)^7$$

is accurate to six decimal places.

Accurate inverse tangent calculations lead to accurate computations of the number π. For example, we can use the addition formula for the tangent function to show (Problem 46) that

$$\frac{\pi}{4} = 4\,\tan^{-1}\left(\frac{1}{5}\right) - \tan^{-1}\left(\frac{1}{239}\right). \tag{28}$$

In 1706, John Machin (?–1751) used Eq. (28) to calculate the first 100 decimal places of π. (In Problem 48 we ask you to use it to show that $\pi = 3.14159$ to five decimal places.) In 1844 the lightning-fast mental calculator Zacharias Dase (1824–1861) of Germany computed the first 200 decimal places of π, using the related formula

$$\frac{\pi}{4} = \tan^{-1}\left(\frac{1}{2}\right) + \tan^{-1}\left(\frac{1}{5}\right) + \tan^{-1}\left(\frac{1}{8}\right). \tag{29}$$

You might enjoy verifying this formula (see Problem 47). A recent computation of 1 million decimal places of π used the formula

$$\frac{\pi}{4} = 12\,\tan^{-1}\left(\frac{1}{18}\right) + 8\,\tan^{-1}\left(\frac{1}{57}\right) - 5\,\tan^{-1}\left(\frac{1}{239}\right).$$

For derivations of this formula and others like it, with further discussion of the computations of the number π, see the article "An algorithm for the calculation of π" by George Miel in the *American Mathematical Monthly* **86** (1979), pp. 694–697. Although no practical application is likely to require more than ten or twelve decimal places of π, these computations provide dramatic evidence of the power of Taylor's formula. Moreover, the number π continues to serve as a challenge both to human ingenuity and to the accuracy and efficiency of modern electronic computers. For an account of how investigations of the Indian mathematical genius Srinivasa Ramanujan (1887–1920) have led recently to the computation of over a billion decimal places of π, see the article "Ramanujan and pi," Jonathan M. Borwein and Peter B. Borwein, *Scientific American* (Feb. 1988), pp. 112–117.

11.4 Problems

In Problems 1 through 10, find Taylor's formula for the given function f at a = 0. Find both the Taylor polynomial $P_n(x)$ of the indicated degree n and the remainder term $R_n(x)$.

1. $f(x) = e^{-x}, \quad n = 5$

2. $f(x) = \sin x, \quad n = 4$

3. $f(x) = \cos x, \quad n = 4$

4. $f(x) = \dfrac{1}{1-x}, \quad n = 4$

5. $f(x) = \sqrt{1+x}, \quad n = 3$

6. $f(x) = \ln(1+x), \quad n = 4$

7. $f(x) = \tan x, \quad n = 3$

8. $f(x) = \arctan x, \quad n = 2$

9. $f(x) = \sin^{-1}x, \quad n = 2$

10. $f(x) = x^3 - 3x^2 + 5x - 7, \quad n = 4$

In Problems 11 through 20, find the Taylor polynomial with remainder by using the given values of a and n.

11. $f(x) = e^x; \quad a = 1, n = 4$

12. $f(x) = \cos x; \quad a = \pi/4, n = 3$

13. $f(x) = \sin x$; $a = \pi/6$, $n = 3$

14. $f(x) = \sqrt{x}$; $a = 100$, $n = 3$

15. $f(x) = \dfrac{1}{(x-4)^2}$; $a = 5$, $n = 5$

16. $f(x) = \tan x$; $a = \pi/4$, $n = 4$

17. $f(x) = \cos x$; $a = \pi$, $n = 4$

18. $f(x) = \sin x$; $a = \pi/2$, $n = 4$

19. $f(x) = x^{3/2}$; $a = 1$, $n = 4$

20. $f(x) = \dfrac{1}{\sqrt{1-x}}$; $a = 0$, $n = 4$

In Problems 21 through 28, find the Maclaurin series of the given function f by substitution in one of the known series in Eqs. (19), (21), and (22).

21. $f(x) = e^{-x}$ **22.** $f(x) = e^{2x}$

23. $f(x) = e^{-3x}$ **24.** $f(x) = \exp(x^3)$

25. $f(x) = \cos 2x$ **26.** $f(x) = \sin \dfrac{x}{2}$

27. $f(x) = \sin x^2$ **28.** $f(x) = \cos \sqrt{x}$

In Problems 29 through 42, find the Taylor series [Eq. (16)] of the given function at the indicated point a.

29. $f(x) = \ln(1 + x)$, $a = 0$

30. $f(x) = \dfrac{1}{1-x}$, $a = 0$

31. $f(x) = e^{-x}$, $a = 0$

32. $f(x) = \cosh x$, $a = 0$

33. $f(x) = \ln x$, $a = 1$

34. $f(x) = \sin x$, $a = \pi/2$

35. $f(x) = \cos x$, $a = \pi/4$

36. $f(x) = e^{2x}$, $a = 0$

37. $f(x) = \sinh x$, $a = 0$

38. $f(x) = \dfrac{1}{(1-x)^2}$, $a = 0$

39. $f(x) = \dfrac{1}{x}$, $a = 1$

40. $f(x) = \cos x$, $a = \pi/2$

41. $f(x) = \sin x$, $a = \pi/4$

42. $f(x) = \sqrt{1 + x}$, $a = 0$

43. Derive, as in Example 5, the Taylor series in Eq. (22) of $f(x) = \sin x$ at $a = 0$.

44. Granted that it is valid to differentiate the sine and cosine Taylor series in a term-by-term manner, use these series to verify that $D_x \cos x = -\sin x$ and $D_x \sin x = \cos x$.

45. Use the Taylor series for the cosine and sine functions to verify that $\cos(-x) = \cos x$ and $\sin(-x) = -\sin x$ for all x.

46. Beginning with $\alpha = \tan^{-1}(\frac{1}{5})$, use the addition formula

$$\tan(A + B) = \frac{\tan A + \tan B}{1 - \tan A \tan B}$$

to show in turn that: (a) $\tan 2\alpha = \frac{5}{12}$; (b) $\tan 4\alpha = \frac{120}{119}$; (c) $\tan(\pi/4 - 4\alpha) = -\frac{1}{239}$. Finally, show that (c) implies Eq. (28).

47. Apply the addition formula for the tangent function to verify Eq. (29).

48. Use Eqs. (25), (26), and (28) to show that $\pi = 3.14159$ to five decimal places. [*Suggestion:* Compute $\arctan(\frac{1}{5})$ and $\arctan(\frac{1}{239})$, both with error less than 5×10^{-8}. You can do this by applying Eq. (25) with $n = 4$ and with $n = 1$. Carry your computations to seven decimal places, and keep track of errors.]

49. Prove that

$$\lim_{n \to \infty} \frac{x^n}{n!} = 0$$

if x is a real number. [*Suggestion:* Choose an integer k such that $k > |2x|$, and let $L = |x|^k/k!$. Then show that

$$\frac{|x|^n}{n!} < \frac{L}{2^{n-k}}$$

if $n > k$.]

50. (a) Substitute $-x$ for x in Eq. (24). (b) Now suppose that t is a positive real number but is otherwise arbitrary. Integrate both sides in the result of part (a) from $x = 0$ to $x = t$ to show that

$$\ln(1 + t) = t - \frac{t^2}{2} + \frac{t^3}{3} - \cdots + (-1)^{k-1}\frac{t^k}{k} + R_k,$$

where

$$|R_k| \leq \frac{t^{k+1}}{k + 1}.$$

(c) Conclude that

$$\ln(1 + t) = t - \frac{t^2}{2} + \frac{t^3}{3} - \cdots = \sum_{k=1}^{\infty} (-1)^{k-1}\frac{t^k}{k}$$

if $0 \leq t \leq 1$.

51. Use the formula $\ln 2 = \ln(\frac{5}{4}) + 2 \ln(\frac{6}{5}) + \ln(\frac{10}{9})$ to calculate $\ln 2$ accurate to three decimal places. See part (b) of Problem 50 for the Taylor formula for $\ln(1 + x)$.

11.4 Project

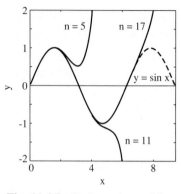

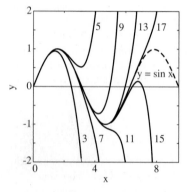

Fig. 11.4.7 Taylor polynomial approximations to sin x

Fig. 11.4.8 Taylor polynomial approximations to sin x of degrees $n = 3, 5, 7, \ldots, 17$

This project involves the use of a graphics calculator or computer graphing program to plot Taylor polynomials. By plotting several successive Taylor polynomials on the same set of coordinate axes, we get a visual sense of the way in which a function is approximated by partial sums of its Taylor series.

Figure 11.4.7 shows the graph of sin x and its Taylor polynomials of degrees $n = 5, 11$, and 17. Using a graphics calculator with a Y = menu, you can generate this figure by plotting the functions defined by

```
Y1=X − X^3/3! + X^5/5!
Y2=Y1 − X^7/7! + X^9/9! − X^11/11!
Y3=Y2 + X^13/13! − X^15/15! + X^17/17!
Y4=sin X
```

With a calculator or computer system that has a $\boxed{\text{sum}}$ function (Fig. 11.3.5), we can define high-degree Taylor polynomials by commands such as the *X(PLORE)* command

```
SUM( (x^n)/n!, n=0 to k )
```

which defines the kth partial sum of the exponential series in Eq. (19) of this section. With a calculator or computer that can plot a number of graphs simultaneously, we can generate a picture like Fig. 11.4.8, which shows the graphs of the Taylor polynomials of sin x of degrees $n = 3, 5, 7, \ldots, 17$.

For every function given in Problems 1 through 7, generate several pictures, each of which shows the function's graph and several Taylor polynomial approximations.

1. $f(x) = e^{-x}$ **2.** $f(x) = \sin x$

3. $f(x) = \cos x$ **4.** $f(x) = \dfrac{1}{1 + x}$

5. $f(x) = \ln(1 + x)$ **6.** $f(x) = \dfrac{1}{1 + x^2}$

7. $f(x) = \tan^{-1} x$

11.5
The Integral Test

A Taylor series (as in Section 11.4) is a special type of infinite series with *variable* terms. We saw that Taylor's formula can sometimes be used—as in the case of the exponential, sine, and cosine series—to establish the convergence of such a series.

But given an infinite series $\Sigma\, a_n$ with *constant* terms, it is the exception rather than the rule when a simple formula for the nth partial sum of that series can be found and used directly to determine whether the series converges or diverges. There are, however, several *convergence tests* that use the *terms* of an infinite series rather than its partial sums. Such a test, when successful, will tell us whether or not the series converges. Once we know that the series $\Sigma\, a_n$ does converge, it is then a separate matter to *find* its sum S. It may be necessary to approximate S by adding sufficiently many terms; in this case we shall need to know how many terms are required for the desired accuracy.

612

Here and in Section 11.6, we concentrate our attention on **positive-term series**—that is, series with terms that are all positive. If $a_n > 0$ for all n, then

$$S_1 < S_2 < S_3 < \cdots < S_n < \cdots,$$

so the sequence $\{S_n\}$ of partial sums of the series is increasing. Hence there are just two possibilities. If the sequence $\{S_n\}$ is *bounded*—there exists a number M such that $S_n \leqq M$ for all n—then the bounded monotonic sequence property (Section 11.2) implies that $S = \lim_{n\to\infty} S_n$ exists, so the series $\Sigma\, a_n$ *converges*. Otherwise, it diverges to infinity (by Problem 38 in Section 11.2).

A similar alternative holds for improper integrals. Suppose that the function f is continuous and positive-valued for $x \geqq 1$. Then it follows (from Problem 35) that the improper integral

$$\int_1^\infty f(x)\, dx = \lim_{b\to\infty} \int_1^b f(x)\, dx \tag{1}$$

either converges (the limit is a real number) or diverges to infinity (the limit is $+\infty$). This analogy between positive-term series and improper integrals of positive functions is the key to the **integral test**. We compare the behavior of the series $\Sigma\, a_n$ with that of the improper integral in Eq. (1), where f is an appropriately chosen function. [Among other things, we require that $f(n) = a_n$ for all n.]

Theorem 1   *The Integral Test*

Suppose that $\Sigma\, a_n$ is a positive-term series and that f is a positive-valued, decreasing, continuous function for $x \geqq 1$. If $f(n) = a_n$ for all integers $n \geqq 1$, then the series and the improper integral

$$\sum_{n=1}^\infty a_n \quad \text{and} \quad \int_1^\infty f(x)\, dx$$

either both converge or both diverge.

Proof   Because f is a decreasing function, the rectangular polygon with area

$$S_n = a_1 + a_2 + a_3 + \cdots + a_n$$

shown in Fig. 11.5.1 contains the region under $y = f(x)$ from $x = 1$ to $x = n + 1$. Hence

$$\int_1^{n+1} f(x)\, dx \leqq S_n. \tag{2}$$

Similarly, the rectangular polygon with area

$$S_n - a_1 = a_2 + a_3 + a_4 + \cdots + a_n$$

shown in Fig. 11.5.2 is contained in the region under $y = f(x)$ from $x = 1$ to $x = n$. Hence

$$S_n - a_1 \leqq \int_1^n f(x)\, dx. \tag{3}$$

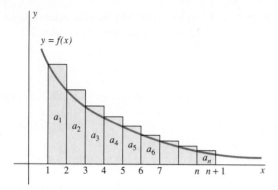

Fig. 11.5.1 Underestimating the partial sums with an integral

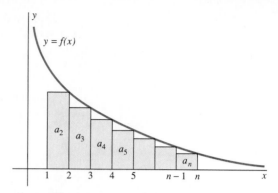

Fig. 11.5.2 Overestimating the partial sums with an integral

Suppose first that the improper integral $\int_1^\infty f(x)\,dx$ diverges (necessarily to $+\infty$). Then

$$\lim_{n \to \infty} \int_1^{n+1} f(x)\,dx = +\infty,$$

so it follows from Eq. (2) that $\lim_{n \to \infty} S_n = +\infty$ as well, and hence the infinite series $\Sigma\, a_n$ likewise diverges.

Now suppose instead that the improper integral $\int_1^\infty f(x)\,dx$ converges and has the (finite) value I. Then (3) implies that

$$S_n \leqq a_1 + \int_1^n f(x)\,dx \leqq a_1 + I,$$

so the increasing sequence $\{S_n\}$ is bounded. Thus the infinite series

$$\sum_{n=1}^\infty a_n = \lim_{n \to \infty} S_n$$

converges as well. Hence we have shown that the infinite series and the improper integral either both converge or both diverge. ❑

EXAMPLE 1 We used a version of the integral test to prove in Section 11.3 that the harmonic series

$$\sum_{n=1}^\infty \frac{1}{n} = 1 + \frac{1}{2} + \frac{1}{3} + \frac{1}{4} + \cdots$$

diverges. Using the test as stated in Theorem 1 is a little simpler: We note that $f(x) = 1/x$ is positive, continuous, and decreasing for $x \geqq 1$ and that $f(n) = 1/n$ for each positive integer n. Now

$$\int_1^\infty \frac{1}{x}\,dx = \lim_{b \to \infty} \int_1^b \frac{1}{x}\,dx = \lim_{b \to \infty} \left[\ln x\right]_1^b = \lim_{b \to \infty} (\ln b - \ln 1) = +\infty.$$

Thus the improper integral diverges and, therefore, so does the harmonic series.

The harmonic series is the case $p = 1$ of the **p-series**

$$\sum_{n=1}^\infty \frac{1}{n^p} = 1 + \frac{1}{2^p} + \frac{1}{3^p} + \cdots + \frac{1}{n^p} + \cdots. \tag{4}$$

Whether the p-series converges or diverges depends on the value of p.

614

EXAMPLE 2 Show that the *p*-series converges if $p > 1$ but diverges if $0 < p \leq 1$.

Solution The case $p = 1$ has already been settled in Example 1. If $p > 0$ but $p \neq 1$, then the function $f(x) = 1/x^p$ satisfies the conditions of the integral test, and

$$\int_1^\infty \frac{1}{x^p}\, dx = \lim_{b \to \infty} \int_1^b \frac{1}{x^p}\, dx = \lim_{b \to \infty}\left[-\frac{1}{(p-1)x^{p-1}} \right]_1^b$$

$$= \lim_{b \to \infty} \frac{1}{p-1}\left(1 - \frac{1}{b^{p-1}} \right).$$

If $p > 1$, then

$$\int_1^\infty \frac{1}{x^p}\, dx = \frac{1}{p-1} < \infty,$$

so the integral and the series both converge. But if $0 < p < 1$, then

$$\int_1^\infty \frac{1}{x^p}\, dx = \lim_{b \to \infty} \frac{1}{1-p}(b^{1-p} - 1) = \infty,$$

ar d in this case the integral and the series both diverge.

As specific examples, the series

$$\sum_{n=1}^\infty \frac{1}{n^2} = 1 + \frac{1}{2^2} + \frac{1}{3^2} + \cdots + \frac{1}{n^2} + \cdots$$

converges ($p = 2$), whereas the series

$$\sum_{n=1}^\infty \frac{1}{\sqrt{n}} = 1 + \frac{1}{\sqrt{2}} + \frac{1}{\sqrt{3}} + \cdots + \frac{1}{\sqrt{n}} + \cdots$$

diverges ($p = \frac{1}{2}$).

Now suppose that the positive-term series $\Sigma\, a_n$ converges by the integral test and that we wish to approximate its sum by adding sufficiently many of its initial terms. The difference between the sum S and the *n*th partial sum S_n is the **remainder**

$$R_n = S - S_n = a_{n+1} + a_{n+2} + a_{n+3} + \cdots. \tag{5}$$

This remainder is the error made when the sum S is estimated by using in its place the partial sum S_n.

Theorem 2 *The Integral Test Remainder Estimate*

Suppose that the infinite series and improper integral

$$\sum_{n=1}^\infty a_n \quad \text{and} \quad \int_1^\infty f(x)\, dx$$

satisfy the hypotheses of the integral test, and suppose in addition that both converge. Then

$$\int_{n+1}^\infty f(x)\, dx \leq R_n \leq \int_n^\infty f(x)\, dx, \tag{6}$$

where R_n is the remainder given in Eq. (5).

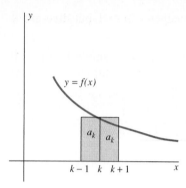

Fig. 11.5.3 Establishing the integral test remainder estimate

Proof We see from Figure 11.5.3 that

$$\int_k^{k+1} f(x) \, dx \leqq a_k \leqq \int_{k-1}^k f(x) \, dx$$

for $k = n + 1, n + 2, \ldots$. We add these inequalities for all such values of k, and the result is the inequality in (6), because

$$\sum_{k=n+1}^{\infty} \int_k^{k+1} f(x) \, dx = \int_{n+1}^{\infty} f(x) \, dx$$

and

$$\sum_{k=n+1}^{\infty} \int_{k-1}^k f(x) \, dx = \int_n^{\infty} f(x) \, dx. \qquad \square$$

If we substitute $R_n = S - S_n$, then it follows from (6) that the sum S of the series satisfies the inequality

$$S_n + \int_{n+1}^{\infty} f(x) \, dx \leqq S \leqq S_n + \int_n^{\infty} f(x) \, dx. \tag{7}$$

If the nth partial sum is known and the difference

$$\int_n^{n+1} f(x) \, dx$$

between the two integrals is small, then (7) provides an accurate estimate of the sum S.

EXAMPLE 3 We will see in Section 11.8 that the exact sum of the p-series with $p = 2$ is $\pi^2/6$, thus giving the beautiful formula

$$\frac{\pi^2}{6} = 1 + \frac{1}{2^2} + \frac{1}{3^2} + \frac{1}{4^2} + \cdots . \tag{8}$$

Approximate the number π by applying to the series in Eq. (8) the integral test remainder estimate with $n = 50$.

Solution The sum of the first 50 terms in Eq. (8) is, to the accuracy shown,

$$\sum_{n=1}^{50} \frac{1}{n^2} = 1.62513\,273.$$

You can add these 50 terms one by one on a pocket calculator in a few minutes, but this is precisely the sort of thing for which a computer or programmable calculator is effective.

Because

$$\int_a^{\infty} \frac{1}{x^2} \, dx = \lim_{b \to \infty} \left[-\frac{1}{x} \right]_a^b = \frac{1}{a},$$

the integral test remainder estimate with $f(x) = 1/x^2$ gives $\frac{1}{51} \leqq R_{50} \leqq \frac{1}{50}$, so (7) yields

$$\frac{1}{51} + 1.6251327 \leqq \frac{\pi^2}{6} \leqq \frac{1}{50} + 1.62513\,28.$$

We multiply by 6, extract the square root, and round to four-place accuracy. The result is that

$$3.1414 < \pi < 3.1418.$$

With $n = 200$, a programmable calculator or computer gives

$$\sum_{n=1}^{200} \frac{1}{n^2} = 1.63994\,655,$$

so

$$\frac{1}{201} + 1.63994\,65 \leqq \frac{\pi^2}{6} \leqq \frac{1}{200} + 1.63994\,66.$$

This leads to the inequality $3.14158 < \pi < 3.14161$, and it follows that $\pi = 3.1416$ rounded to four decimal places.

EXAMPLE 4 Show that the series

$$\sum_{n=2}^{\infty} \frac{1}{n(\ln n)^2}$$

converges, and find how many terms you would need to add to find its sum accurate to within 0.01.

Solution We begin the sum at $n = 2$ because $\ln 1 = 0$. Let $f(x) = x(\ln x)^2$. Then

$$\int_{a}^{\infty} \frac{1}{x(\ln x)^2} \, dx = \lim_{b \to \infty} \left[-\frac{1}{\ln x} \right]_{a}^{b} = \frac{1}{\ln a}.$$

With $a = 2$, the series converges by the integral test. With $a = n$, the right-hand inequality in (6) gives $R_n \leqq 1/(\ln n)$, so we need

$$\frac{1}{\ln n} \leqq 0.01, \qquad \ln n \geqq 100, \qquad n \geqq e^{100} \approx 2.7 \times 10^{43}.$$

This is a far larger number of terms than any conceivable computer could add within the expected lifetime of the universe. But accuracy to within 0.05 would require only that $n \geqq 4.85 \times 10^8$, fewer than half a billion terms—well within the range of a modern computer.

11.5 Problems

In Problems 1 through 28, use the integral test to test the given series for convergence.

1. $\displaystyle\sum_{n=1}^{\infty} \frac{n}{n^2 + 1}$

2. $\displaystyle\sum_{n=1}^{\infty} \frac{n}{e^{n^2}}$

3. $\displaystyle\sum_{n=1}^{\infty} \frac{1}{\sqrt{n + 1}}$

4. $\displaystyle\sum_{n=1}^{\infty} \frac{1}{(n + 1)^{4/3}}$

5. $\displaystyle\sum_{n=1}^{\infty} \frac{1}{n^2 + 1}$

6. $\displaystyle\sum_{n=1}^{\infty} \frac{1}{n(n + 1)}$

7. $\displaystyle\sum_{n=2}^{\infty} \frac{1}{n \ln n}$

8. $\displaystyle\sum_{n=1}^{\infty} \frac{\ln n}{n}$

9. $\displaystyle\sum_{n=1}^{\infty} \frac{1}{2^n}$

10. $\displaystyle\sum_{n=1}^{\infty} \frac{n}{e^n}$

11. $\displaystyle\sum_{n=1}^{\infty} \frac{n^2}{e^n}$

12. $\displaystyle\sum_{n=1}^{\infty} \frac{1}{17n - 13}$

13. $\displaystyle\sum_{n=1}^{\infty} \frac{\ln n}{n^2}$

14. $\displaystyle\sum_{n=1}^{\infty} \frac{n + 1}{n^2}$

15. $\displaystyle\sum_{n=1}^{\infty} \frac{n}{n^4 + 1}$

16. $\displaystyle\sum_{n=1}^{\infty} \frac{1}{n^3 + n}$

17. $\displaystyle\sum_{n=1}^{\infty} \frac{2n + 5}{n^2 + 5n + 7}$

18. $\displaystyle\sum_{n=1}^{\infty} \ln\left(\frac{n+1}{n}\right)$

19. $\displaystyle\sum_{n=1}^{\infty} \ln\left(1 + \frac{1}{n^2}\right)$

20. $\displaystyle\sum_{n=1}^{\infty} \frac{2^{1/n}}{n^2}$

21. $\displaystyle\sum_{n=1}^{\infty} \frac{n}{4n^2 + 5}$

22. $\displaystyle\sum_{n=1}^{\infty} \frac{n}{(4n^2 + 5)^{3/2}}$

23. $\displaystyle\sum_{n=2}^{\infty} \frac{1}{n\sqrt{\ln n}}$

24. $\displaystyle\sum_{n=2}^{\infty} \frac{1}{n(\ln n)^3}$

25. $\displaystyle\sum_{n=1}^{\infty} \frac{1}{4n^2 + 9}$

26. $\displaystyle\sum_{n=1}^{\infty} \frac{n + 1}{n + 100}$

27. $\displaystyle\sum_{n=1}^{\infty} \frac{n}{n^4 + 2n^2 + 1}$

28. $\displaystyle\sum_{n=1}^{\infty} \frac{1}{(n + 1)^3}$

In Problems 29 through 31, tell why the integral test does not apply to the given infinite series.

29. $\displaystyle\sum_{n=1}^{\infty} \frac{(-1)^n}{n}$

30. $\displaystyle\sum_{n=1}^{\infty} e^{-n}\sin n$

31. $\displaystyle\sum_{n=1}^{\infty} \frac{2 + \sin n}{n^2}$

In Problems 32 through 34, find the least positive integer n such that the remainder R_n in Theorem 2 is less than E.

32. $\displaystyle\sum_{n=1}^{\infty} \frac{1}{n^2}$; $E = 0.0001$

33. $\displaystyle\sum_{n=1}^{\infty} \frac{1}{n^2}$; $E = 0.00005$

34. $\displaystyle\sum_{n=1}^{\infty} \frac{1}{n^6}$; $E = 2 \times 10^{-11}$

35. Suppose that the function f is continuous and positive-valued for $x \geq 1$. Let $b_n = \int_1^n f(x)\,dx$ for $n = 1, 2, 3,$ .... (a) Suppose that the increasing sequence $\{b_n\}$ is bounded, so $B = \lim_{n\to\infty} b_n$ exists. Prove that

$$\int_1^{\infty} f(x)\,dx = B.$$

(b) Prove that if the sequence $\{b_n\}$ is not bounded, then

$$\int_1^{\infty} f(x)\,dx = +\infty.$$

36. Show that the series $\displaystyle\sum_{n=2}^{\infty} \frac{1}{n(\ln n)^p}$ converges if $p > 1$ and diverges if $p \leq 1$.

37. Use the integral test remainder estimate to find

$$\sum_{n=1}^{\infty} \frac{1}{n^5}$$

with three digits to the right of the decimal correct or correctly rounded.

38. Using the integral test remainder estimate, how many terms are needed to approximate

$$\sum_{n=1}^{\infty} \frac{1}{n^{3/2}}$$

with two digits to the right of the decimal correct?

39. Deduce from the inequalities in (2) and (3) with the function $f(x) = 1/x$ that

$$\ln n \leq 1 + \frac{1}{2} + \frac{1}{3} + \cdots + \frac{1}{n} \leq 1 + \ln n$$

for $n = 1, 2, 3,$ .... If a computer adds 1 million terms of the harmonic series per second, how long will it take for the partial sum to reach 50?

40. (a) Let

$$c_n = 1 + \frac{1}{2} + \frac{1}{3} + \cdots + \frac{1}{n} - \ln n$$

for $n = 1, 2, 3,$ .... Deduce from Problem 39 that $0 \leq c_n \leq 1$ for all n. (b) Note that

$$\int_n^{n+1} \frac{1}{x}\,dx \geq \frac{1}{n + 1}.$$

Conclude that the sequence $\{c_n\}$ is decreasing. Therefore, the sequence $\{c_n\}$ converges. The number

$$\gamma = \lim_{n\to\infty}\left(1 + \frac{1}{2} + \frac{1}{3} + \cdots + \frac{1}{n} - \ln n\right) \approx 0.57722$$

is known as **Euler's constant**.

41. It is known that

$$\sum_{n=1}^{\infty} \frac{1}{n^4} = \frac{\pi^4}{90}.$$

Use the integral test remainder estimate and the first 10 terms of this series to show that $\pi = 3.1416$ rounded to four decimal places.

11.5 Project

You can readily use the programs listed in Fig. 11.3.4 to calculate partial sums of the p-series

$$\zeta(p) = \sum_{n=1}^{\infty} \frac{1}{n^p} = 1 + \frac{1}{2^p} + \frac{1}{3^p} + \cdots, \qquad (9)$$

618

where the notation $\zeta(p)$—read "zeta of p."—is a standard abbreviation for this p-series. Given the value of p to be used, you need only define `Y1=1/N^P` before you execute the TI calculator program, or define `A(N)=1/N^P` in the first line of the BASIC program.

Alternatively, with an appropriate calculator or computer system you can use a SUM function such as the *Maple* command

```
sum( 1/n^2, n=1..1000 )
```

to calculate the sum of the first thousand terms of $\zeta(2) = \Sigma\, 1/n^2$.

In Problems 1 through 5, use the listed known value of $\zeta(p)$ to see how accurately you can approximate the value

$$\pi = 3.14159\,26535\,89793\,\ldots.$$

For instance, in Problem 2 you might calculate the kth partial sum s_k of $\Sigma 1/n^4$ for $k = 10, 20, 40, 80, \ldots$ and observe how accurately $\sqrt[4]{90 s_k}$ approximates π.

1. $\zeta(2) = \dfrac{\pi^2}{6}$

2. $\zeta(4) = \dfrac{\pi^4}{90}$

3. $\zeta(6) = \dfrac{\pi^6}{945}$

4. $\zeta(8) = \dfrac{\pi^8}{9450}$

5. $\zeta(10) = \dfrac{\pi^{10}}{93555}$

6. Calculate partial sums of the harmonic series $\Sigma\, 1/n$ to approximate Euler's constant

$$\gamma = \lim_{n \to \infty} \left(1 + \frac{1}{2} + \frac{1}{3} + \cdots + \frac{1}{n} - \ln n \right) \approx 0.57722$$

(see Problem 40 of this section). Unless you use a very powerful microcomputer, you will probably have to be content with accuracy to two or three decimal places only.

11.6

Comparison Tests for Positive-Term Series

With the integral test we attempt to determine whether or not an infinite series converges by comparing it with an improper integral. The methods of this section involve comparing the terms of the *positive-term* series Σa_n with those of another positive-term series Σb_n whose convergence or divergence is known. We have already developed two families of *reference series* for the role of the known series Σb_n; these are the geometric series of Section 11.3 and the p-series of Section 11.5. They are well adapted for our new purposes because their convergence or divergence is quite easy to determine. Recall that the geometric series Σr^n converges if $|r| < 1$ and diverges if $|r| \geq 1$, and the p-series $\Sigma 1/n^p$ converges if $p > 1$ and diverges if $0 < p \leq 1$.

Let Σa_n and Σb_n be positive-term series. Then we say that the series Σb_n **dominates** the series Σa_n provided that $a_n \leq b_n$ for all n. Theorem 1 says that the positive-term series Σa_n converges if it is dominated by a convergent series and diverges if it dominates a positive-term divergent series.

> **Theorem 1   *Comparison Test***
>
> Suppose that $\Sigma\, a_n$ and $\Sigma\, b_n$ are positive-term series. Then
>
> **1.** $\Sigma\, a_n$ converges if $\Sigma\, b_n$ converges and $a_n \leqq b_n$ for all n;
> **2.** $\Sigma\, a_n$ diverges if $\Sigma\, b_n$ diverges and $a_n \geqq b_n$ for all n.

Proof   Denote the nth partial sums of the series $\Sigma\, a_n$ and $\Sigma\, b_n$ by S_n and T_n, respectively. Then $\{S_n\}$ and $\{T_n\}$ are increasing sequences. To prove part (1), suppose that $\Sigma\, b_n$ converges, so $T = \lim_{n \to \infty} T_n$ exists (T is a real number). Then the fact that $a_n \leqq b_n$ for all n implies that $S_n \leqq T_n \leqq T$ for all n. Thus the sequence $\{S_n\}$ of partial sums of $\Sigma\, a_n$ is bounded and increasing and therefore converges. Thus $\Sigma\, a_n$ converges.

Part (2) is merely a restatement of part (1). If the series $\Sigma\, a_n$ converged, then the fact that $\Sigma\, a_n$ dominates $\Sigma\, b_n$ would imply—by part 1, with a_n and b_n interchanged—that $\Sigma\, b_n$ converged. But $\Sigma\, b_n$ diverges, so it follows that $\Sigma\, a_n$ must also diverge.   ❑

We know by Theorem 5 of Section 11.3 that the convergence or divergence of an infinite series is not affected by the addition or deletion of a finite number of terms. Consequently, the conditions $a_n \leqq b_n$ and $a_n \geqq b_n$ in the two parts of the comparison test really need to hold only for all $n \geqq k$, where k is some fixed positive integer. Thus we can say that the positive-term series $\Sigma\, a_n$ converges if it is "eventually dominated" by the convergent positive-term series $\Sigma\, b_n$.

EXAMPLE 1   Because

$$\frac{1}{n(n + 1)(n + 2)} < \frac{1}{n^3}$$

for all $n \geqq 1$, the series

$$\sum_{n=1}^{\infty} \frac{1}{n(n + 1)(n + 2)} = \frac{1}{1 \cdot 2 \cdot 3} + \frac{1}{2 \cdot 3 \cdot 4} + \frac{1}{3 \cdot 4 \cdot 5} + \cdots$$

is dominated by the series $\Sigma\, 1/n^3$, which is a convergent p-series with $p = 3$. Both are positive-term series, and hence the series $\Sigma\, 1/[n(n + 1)(n + 2)]$ converges by part (1) of the comparison test.

EXAMPLE 2   Because

$$\frac{1}{\sqrt{2n - 1}} > \frac{1}{\sqrt{2n}}$$

for all $n \geqq 1$, the positive-term series

$$\sum_{n=1}^{\infty} \frac{1}{\sqrt{2n - 1}} = 1 + \frac{1}{\sqrt{3}} + \frac{1}{\sqrt{5}} + \frac{1}{\sqrt{7}} + \cdots$$

dominates the series

$$\sum_{n=1}^{\infty} \frac{1}{\sqrt{2n}} = \frac{1}{\sqrt{2}} \sum_{n=1}^{\infty} \frac{1}{n^{1/2}}.$$

But $\Sigma 1/n^{1/2}$ is a divergent p-series with $p = \frac{1}{2}$, and a constant nonzero multiple of a divergent series diverges. So part (2) of the comparison test shows that the series $\Sigma 1/\sqrt{2n-1}$ also diverges.

EXAMPLE 3 Test the series

$$\sum_{n=0}^{\infty} \frac{1}{n!} = 1 + \frac{1}{1!} + \frac{1}{2!} + \frac{1}{3!} + \cdots$$

for convergence.

Solution We note first that if $n \geq 1$, then

$$n! = n(n-1)(n-2) \cdots 3 \cdot 2 \cdot 1$$
$$\geq 2 \cdot 2 \cdot 2 \cdots 2 \cdot 2 \cdot 1; \qquad \text{(the same number of factors)}$$

that is, $n! \geq 2^{n-1}$ for $n \geq 1$. Thus

$$\frac{1}{n!} \leq \frac{1}{2^{n-1}} \text{ for } n \geq 1,$$

so the series

$$\sum_{n=0}^{\infty} \frac{1}{n!} \text{ is dominated by the series } 1 + \sum_{n=1}^{\infty} \frac{1}{2^{n-1}} = 1 + \sum_{n=0}^{\infty} \frac{1}{2^n},$$

which is a convergent geometric series (after the first term). Both are positive-term series, and so by the comparison test the series converges. We saw in Section 11.4 that the sum of this series is the number e, so

$$e = 1 + \frac{1}{1!} + \frac{1}{2!} + \frac{1}{3!} + \cdots + \frac{1}{n!} + \cdots.$$

Indeed, this series provides perhaps the simplest way of showing that

$$e \approx 2.71828\,18284\,59045\,23536.$$

Suppose that Σa_n is a positive-term series such that $a_n \to 0$ as $n \to \infty$. Then, in connection with the nth-term divergence test of Section 11.3, the series Σa_n has at least a *chance* of converging. How do we choose an appropriate positive-term series Σb_n with which to compare it? A good idea is to pick b_n as a *simple* function of n, simpler than a_n but such that a_n and b_n approach zero at the same rate as $n \to \infty$. If the formula for a_n is a fraction, we can try discarding all but the terms of largest magnitude in its numerator and denominator to form b_n. For example, if

$$a_n = \frac{3n^2 + n}{n^4 + \sqrt{n}},$$

then we reason that n is small in comparison with $3n^2$ and $\sqrt{n}$ is small in comparison with n^4 when n is quite large. This suggests that we choose $b_n = 3n^2/n^4 = 3/n^2$. The series $\Sigma 3/n^2$ converges ($p = 2$), but when we attempt to compare Σa_n and Σb_n, we find that $a_n \geq b_n$ (rather than $a_n \leq b_n$). Consequently, the comparison test does not apply immediately—the fact that Σa_n dominates a convergent series does *not* imply that Σa_n itself converges. Theorem 2 provides a convenient way of handling such a situation.

> **Theorem 2** *Limit Comparison Test*
>
> Suppose that $\Sigma\, a_n$ and $\Sigma\, b_n$ are positive-term series. If the limit
>
> $$L = \lim_{n \to \infty} \frac{a_n}{b_n}$$
>
> exists and $0 < L < +\infty$, then either both series converge or both series diverge.

Proof   Choose two fixed positive numbers P and Q such that $P < L < Q$. Then $P < a_n/b_n < Q$ for n sufficiently large, and so

$$P b_n < a_n < Q b_n$$

for all sufficiently large values of n. If $\Sigma\, b_n$ converges, then $\Sigma\, a_n$ is eventually dominated by the convergent series $\Sigma\, Q b_n = Q \Sigma\, b_n$, so part (1) of the comparison test implies that $\Sigma\, a_n$ also converges. If $\Sigma\, b_n$ diverges, then $\Sigma\, a_n$ eventually dominates the divergent series $\Sigma\, P b_n = P \Sigma\, b_n$, so part (2) of the comparison test implies that $\Sigma\, a_n$ also diverges. Thus the convergence of either series implies the convergence of the other. $\square$

EXAMPLE 4   With

$$a_n = \frac{3n^2 + n}{n^4 + \sqrt{n}} \quad \text{and} \quad b_n = \frac{1}{n^2}$$

(motivated by the discussion preceding Theorem 2), we find that

$$\lim_{n \to \infty} \frac{a_n}{b_n} = \lim_{n \to \infty} \frac{3n^4 + n^3}{n^4 + \sqrt{n}} = \lim_{n \to \infty} \frac{3 + \dfrac{1}{n}}{1 + \dfrac{1}{n^{7/2}}} = 3.$$

Because $\Sigma\, 1/n^2$ is a convergent p-series ($p = 2$), the limit comparison test tells us that the series

$$\sum_{n=1}^{\infty} \frac{3n^2 + n}{n^4 + \sqrt{n}}$$

also converges.

EXAMPLE 5   Test for convergence: $\displaystyle\sum_{n=1}^{\infty} \frac{1}{2n + \ln n}$.

Solution   Because $\lim_{n \to \infty} (\ln n)/n = 0$ by l'Hôpital's rule, $\ln n$ is very small in comparison with $2n$ when n is large. We therefore take $a_n = 1/(2n + \ln n)$ and, ignoring the constant coefficient 2, we take $b_n = 1/n$. Then we find that

$$\lim_{n \to \infty} \frac{a_n}{b_n} = \lim_{n \to \infty} \frac{n}{2n + \ln n} = \lim_{n \to \infty} \frac{1}{2 + \dfrac{\ln n}{n}} = \frac{1}{2}.$$

Because the harmonic series $\Sigma\, 1/n = \Sigma\, b_n$ diverges, it follows that the given series $\Sigma\, a_n$ also diverges.

It is important to realize that if $L = \lim(a_n/b_n)$ is either zero or ∞, then the limit comparison test does not apply. (See Problem 42 for a discussion of what conclusions may sometimes be drawn in these cases.) Note, for example, that if $a_n = 1/n^2$ and $b_n = 1/n$, then $\lim(a_n/b_n) = 0$. But in this case $\Sigma\, a_n$ converges, whereas $\Sigma\, b_n$ diverges.

We close our discussion of positive-term series with the observation that the sum of a convergent *positive*-term series is not altered by grouping or rearranging its terms. For example, let $\Sigma\, a_n$ be a convergent positive-term series and consider

$$\sum_{n=1}^{\infty} b_n = (a_1 + a_2 + a_3) + a_4 + (a_5 + a_6) + \cdots.$$

That is, the new series has terms $b_1 = a_1 + a_2 + a_3, b_2 = a_4, b_3 = a_5 + a_6,$ and so on. Then every partial sum T_n of $\Sigma\, b_n$ is equal to some partial sum $S_{n'}$ of $\Sigma\, a_n$. Because $\{S_n\}$ is an increasing sequence with limit $S = \Sigma\, a_n$, it follows easily that $\{T_n\}$ is an increasing sequence with the same limit. Thus $\Sigma\, b_n = S$ as well. The argument is more subtle if terms of $\Sigma\, a_n$ are moved "out of place," as in

$$\sum_{n=1}^{\infty} b_n = a_2 + a_1 + a_4 + a_3 + a_6 + a_5 + \cdots,$$

but the same conclusion holds: Any rearrangement of a convergent *positive*-term series also converges, and it converges to the same sum.

Similarly, it is easy to prove that any grouping or rearrangement of a divergent positive-term series also diverges. But these observations all fail in the case of an infinite series with both positive and negative terms. For example, the series $\Sigma\, (-1)^n$ diverges, but it has the convergent grouping

$$(-1 + 1) + (-1 + 1) + (-1 + 1) + \cdots = 0 + 0 + 0 + \cdots = 0.$$

It follows from Problem 44 of Section 11.3 that

$$\ln 2 = 1 - \frac{1}{2} + \frac{1}{3} - \frac{1}{4} + \frac{1}{5} - \cdots,$$

but the rearrangement

$$1 + \frac{1}{3} - \frac{1}{2} + \frac{1}{5} + \frac{1}{7} - \frac{1}{4} + \frac{1}{9} + \frac{1}{11} - \frac{1}{6} + \cdots$$

converges instead to $\frac{3}{2} \ln 2$. This series for $\ln 2$ even has rearrangements that converge to zero and others that diverge to $+\infty$. One could write a book on this fascinating aspect of infinite series.

11.6 Problems

Use comparison tests to determine whether the infinite series in Problems 1 through 35 converge or diverge.

1. $\displaystyle\sum_{n=1}^{\infty} \frac{1}{n^2 + n + 1}$

2. $\displaystyle\sum_{n=1}^{\infty} \frac{n^3 + 1}{n^4 + 2}$

3. $\displaystyle\sum_{n=1}^{\infty} \frac{1}{n + \sqrt{n}}$

4. $\displaystyle\sum_{n=1}^{\infty} \frac{1}{n + n^{3/2}}$

5. $\displaystyle\sum_{n=1}^{\infty} \frac{1}{1 + 3^n}$

6. $\displaystyle\sum_{n=1}^{\infty} \frac{10n^2}{n^4 + 1}$

7. $\displaystyle\sum_{n=2}^{\infty} \frac{10n^2}{n^3 - 1}$

8. $\displaystyle\sum_{n=1}^{\infty} \frac{n^2 - n}{n^4 + 2}$

9. $\displaystyle\sum_{n=1}^{\infty} \frac{1}{\sqrt{37n^3 + 3}}$

10. $\displaystyle\sum_{n=1}^{\infty} \frac{1}{\sqrt{n^2 + 1}}$

11. $\displaystyle\sum_{n=1}^{\infty} \frac{\sqrt{n}}{n^2 + n}$

12. $\displaystyle\sum_{n=1}^{\infty} \frac{1}{3 + 5^n}$

13. $\displaystyle\sum_{n=2}^{\infty} \frac{1}{\ln n}$

14. $\displaystyle\sum_{n=1}^{\infty} \frac{1}{n - \ln n}$

15. $\displaystyle\sum_{n=1}^{\infty} \frac{\sin^2 n}{n^2 + 1}$

16. $\displaystyle\sum_{n=1}^{\infty} \frac{\cos^2 n}{3^n}$

17. $\displaystyle\sum_{n=1}^{\infty} \frac{n + 2^n}{n + 3^n}$

18. $\displaystyle\sum_{n=1}^{\infty} \frac{1}{2^n + 3^n}$

19. $\displaystyle\sum_{n=2}^{\infty} \frac{1}{n^2 \ln n}$

20. $\displaystyle\sum_{n=1}^{\infty} \frac{1}{n^{1 + \sqrt{n}}}$

21. $\displaystyle\sum_{n=1}^{\infty} \frac{\ln n}{n^2}$

22. $\displaystyle\sum_{n=1}^{\infty} \frac{\arctan n}{n}$

23. $\displaystyle\sum_{n=1}^{\infty} \frac{\sin^2(1/n)}{n^2}$

24. $\displaystyle\sum_{n=1}^{\infty} \frac{e^{1/n}}{n}$

25. $\displaystyle\sum_{n=1}^{\infty} \frac{\ln n}{e^n}$

26. $\displaystyle\sum_{n=1}^{\infty} \frac{n^2 + 2}{n^3 + 3n}$

27. $\displaystyle\sum_{n=1}^{\infty} \frac{n^{3/2}}{n^2 + 4}$

28. $\displaystyle\sum_{n=1}^{\infty} \frac{1}{n \cdot 2^n}$

29. $\displaystyle\sum_{n=1}^{\infty} \frac{3}{4 + \sqrt{n}}$

30. $\displaystyle\sum_{n=1}^{\infty} \frac{n^2 + 1}{e^n(n + 1)^2}$

31. $\displaystyle\sum_{n=1}^{\infty} \frac{2n^2 - 1}{n^2 \cdot 3^n}$

32. $\displaystyle\sum_{n=1}^{\infty} \frac{1}{\sqrt[3]{2n^4 + 1}}$

33. $\displaystyle\sum_{n=1}^{\infty} \frac{2 + \sin n}{n^2}$

34. $\displaystyle\sum_{n=1}^{\infty} \frac{\ln n}{n^3}$

35. $\displaystyle\sum_{n=1}^{\infty} \frac{(n + 1)^n}{n^{n+1}}$ [*Suggestion:* $\displaystyle\lim_{n \to \infty} \left(1 + \frac{1}{n}\right)^n = e.$]

36. (a) Prove that $\ln n < n^{1/8}$ for all sufficiently large values of n. (b) Explain why part (a) shows that the series $\Sigma\, 1/(\ln n)^8$ diverges.

37. Prove that if $\Sigma\, a_n$ is a convergent positive-term series, then $\Sigma\, (a_n/n)$ converges.

38. Suppose that $\Sigma\, a_n$ is a convergent positive-term series and that $\{c_n\}$ is a sequence of positive numbers with limit zero. Prove that $\Sigma\, a_n c_n$ converges.

39. Use the result of Problem 38 to prove that if $\Sigma\, a_n$ and $\Sigma\, b_n$ are convergent positive-term series, then $\Sigma\, a_n b_n$ converges.

40. Prove that the series

$$\sum_{n=1}^{\infty} \frac{1}{1 + 2 + 3 + \cdots + n}$$

converges.

41. Use the result of Problem 40 in Section 11.5 to prove that the series

$$\sum_{n=1}^{\infty} \frac{1}{1 + \dfrac{1}{2} + \dfrac{1}{3} + \cdots + \dfrac{1}{n}}$$

diverges.

42. Adapt the proof of the limit comparison test to prove the following two results. (a) Suppose that $\Sigma\, a_n$ and $\Sigma\, b_n$ are positive-term series and that $\Sigma\, b_n$ converges. If

$$L = \lim_{n \to \infty} \frac{a_n}{b_n} = 0,$$

then $\Sigma\, a_n$ converges. (b) Suppose that $\Sigma\, a_n$ and $\Sigma\, b_n$ are positive-term series and that $\Sigma\, b_n$ diverges. If

$$L = \lim_{n \to \infty} \frac{a_n}{b_n} = +\infty,$$

then $\Sigma\, a_n$ diverges.

11.7
Alternating Series and Absolute Convergence

In Sections 11.5 and 11.6 we concentrated on positive-term series. Now we discuss infinite series that have both positive terms and negative terms. An important example is a series with terms that are alternately positive and negative. An **alternating series** is an infinite series of the form

$$\sum_{n=1}^{\infty} (-1)^{n+1} a_n = a_1 - a_2 + a_3 - a_4 + a_5 - \cdots \tag{1}$$

or of the form $\displaystyle\sum_{n=1}^{\infty} (-1)^n a_n$, where $a_n > 0$ for all n. For example, the series

$$\sum_{n=1}^{\infty} \frac{(-1)^{n+1}}{n} = 1 - \frac{1}{2} + \frac{1}{3} - \frac{1}{4} + \frac{1}{5} - \cdots$$

is an alternating series. Theorem 1 shows that this series converges because the sequence of absolute values of its terms is decreasing and has limit zero.

> **Theorem 1 Alternating Series Test**
> If $a_n > a_{n+1} > 0$ for all n and $\lim\limits_{n\to\infty} a_n = 0$, then the alternating series in Eq. (1) converges.

Proof We first consider the even-numbered partial sums $S_2, S_4, S_6, \ldots, S_{2n}$. We may write

$$S_{2n} = (a_1 - a_2) + (a_3 - a_4) + \cdots + (a_{2n-1} - a_{2n}).$$

Because $a_k - a_{k+1} \geqq 0$ for all k, the sequence $\{S_{2n}\}$ is increasing. Also, because

$$S_{2n} = a_1 - (a_2 - a_3) - \cdots - (a_{2n-2} - a_{2n-1}) - a_{2n},$$

$S_{2n} \leqq a_1$ for all n, so the increasing sequence $\{S_{2n}\}$ is bounded above. Hence the limit

$$S = \lim_{n\to\infty} S_{2n}$$

exists by the bounded monotonic sequence property of Section 11.2. It remains only for us to verify that the odd-numbered partial sums $S_1, S_3, S_5, \ldots$ also converge to S. But $S_{2n+1} = S_{2n} + a_{2n+1}$ and $\lim\limits_{n\to\infty} a_{2n+1} = 0$, so

$$\lim_{n\to\infty} S_{2n+1} = \lim_{n\to\infty} S_{2n} + \lim_{n\to\infty} a_{2n+1} = S.$$

Thus $\lim\limits_{n\to\infty} S_n = S$, and therefore the series converges. ◻

EXAMPLE 1 The series

$$\sum_{n=1}^{\infty} \frac{(-1)^{n+1}}{2n-1} = 1 - \frac{1}{3} + \frac{1}{5} - \frac{1}{7} + \frac{1}{9} - \cdots$$

satisfies the conditions of Theorem 1 and therefore converges. The alternating series test does not tell us the sum of this series, but we saw in Section 11.4 that its sum is $\pi/4$.

EXAMPLE 2 The series

$$\sum_{n=1}^{\infty} \frac{(-1)^{n+1}n}{2n-1} = 1 - \frac{2}{3} + \frac{3}{5} - \frac{4}{7} + \frac{5}{9} - \cdots$$

is an alternating series, and it is easy to verify that

$$a_n = \frac{n}{2n-1} > \frac{n+1}{2n+1} = a_{n+1}$$

for all $n \geqq 1$. But

$$\lim_{n\to\infty} a_n = \frac{1}{2} \neq 0,$$

so the alternating series test *does not apply*. (This fact alone does not imply that the series in question diverges—many series in Sections 11.5 and 11.6 converge even though the alternating series test does not apply. But the series of this example diverges by the nth-term divergence test.)

If a series converges by the alternating series test, then Theorem 2 shows how to approximate its sum with any desired degree of accuracy—*if* you have a computer fast enough to add a large number of its terms.

Theorem 2   *Alternating Series Error Estimate*

Suppose that the series $\Sigma(-1)^{n+1}a_n$ satisfies the conditions of the alternating series test and therefore converges. Let S denote the sum of the series. Denote by $R_n = S - S_n$ the error made in replacing S by the nth partial sum S_n of the series. Then R_n has the same sign as the next term $(-1)^{n+2}a_{n+1}$ of the series, and

$$0 < |R_n| < a_{n+1}. \tag{2}$$

In particular, the *sum S of a convergent alternating series lies between any two consecutive partial sums*. This follows from the proof of Theorem 1, where we saw that $\{S_{2n}\}$ is an increasing sequence, whereas $\{S_{2n+1}\}$ is a decreasing sequence, each converging to S. The resulting inequalities

$$S_{2n-1} > S > S_{2n} = S_{2n-1} - a_{2n}$$

and

$$S_{2n} < S < S_{2n+1} = S_{2n} + a_{2n+1}$$

imply the inequality in (2).

EXAMPLE 3   We saw in Section 11.4 that

$$e^x = \sum_{n=0}^{\infty} \frac{x^n}{n!}$$

for all x and thus that

$$\frac{1}{e} = e^{-1} = 1 - 1 + \frac{1}{2!} - \frac{1}{3!} + \frac{1}{4!} - \cdots.$$

Use this alternating series to compute e^{-1} accurate to four decimal places.

Solution   We want $|R_n| < 1/(n + 1)! \leqq 0.00005$. The least value of n for which this inequality holds is $n = 7$. Then

$$e^{-1} = 1 - 1 + \frac{1}{2!} - \frac{1}{3!} + \frac{1}{4!} - \frac{1}{5!} + \frac{1}{6!} - \frac{1}{7!} + R_7 = 0.367857 + R_7.$$

(We are carrying six decimal places because we want four-place accuracy in the final answer.) Now the inequality in (2) gives

$$0 < R_7 < \frac{1}{8!} < 0.000025.$$

Thus

$$0.367857 < e^{-1} < 0.367882.$$

Thus $e^{-1} = 0.3679$ rounded to four places, and $e = 2.718$ rounded to three places.

ABSOLUTE CONVERGENCE

The series

$$\sum_{n=1}^{\infty} \frac{(-1)^{n+1}}{n} = 1 - \frac{1}{2} + \frac{1}{3} - \frac{1}{4} + \frac{1}{5} - \cdots$$

converges, but if we simply replace each term with its absolute value, we get the *divergent* series

$$1 + \frac{1}{2} + \frac{1}{3} + \frac{1}{4} + \frac{1}{5} + \cdots .$$

In contrast, the *convergent* series

$$\sum_{n=1}^{\infty} \frac{(-1)^n}{2^n} = 1 - \frac{1}{2} + \frac{1}{4} - \frac{1}{8} + \cdots = \frac{2}{3}$$

has the property that the associated positive-term series

$$1 + \frac{1}{2} + \frac{1}{4} + \frac{1}{8} + \cdots = 2$$

also converges. Theorem 3 tells us that if a series of *positive* terms converges, then we may insert minus signs in front of any of the terms—every other one, for instance—and the resulting series will also converge.

Theorem 3 *Absolute Convergence Implies Convergence*
If the series $\sum |a_n|$ converges, then so does the series $\sum a_n$.

Proof   Suppose that the series $\sum |a_n|$ converges. Note that

$$0 \leqq a_n + |a_n| \leqq 2|a_n|$$

for all n. Let $b_n = a_n + |a_n|$. It then follows from the comparison test that the positive-term series $\sum b_n$ converges, because it is dominated by the convergent series $\sum 2|a_n|$. It is easy to verify, too, that the termwise difference of two convergent series also converges. Hence we now see that the series

$$\sum a_n = \sum (b_n - |a_n|) = \sum b_n - \sum |a_n|$$

converges. ❏

   Thus we have another convergence test, one not limited to positive-term or to alternating series: Given the series $\sum a_n$, test the series $\sum |a_n|$ for convergence. If the latter converges, then so does the former. (But the converse is *not* true!) This phenomenon motivates us to make the following definition.

Definition *Absolute Convergence*
The series $\sum a_n$ is said to **converge absolutely** (and is called **absolutely convergent**) provided that the series

$$\sum |a_n| = |a_1| + |a_2| + |a_3| + \cdots + |a_n| + \cdots$$

converges.

Thus we have explained the title of Theorem 3, and we can rephrase the theorem as follows: *If a series converges absolutely, then it converges.* The two examples preceding Theorem 3 show that a convergent series may either converge absolutely or fail to do so:

$$1 - \tfrac{1}{2} + \tfrac{1}{4} - \tfrac{1}{8} + \cdots$$

is an absolutely convergent series because

$$1 + \tfrac{1}{2} + \tfrac{1}{4} + \tfrac{1}{8} + \cdots$$

converges, whereas

$$1 - \tfrac{1}{2} + \tfrac{1}{3} - \tfrac{1}{4} + \tfrac{1}{5} - \cdots$$

is a series that, though convergent, is *not* absolutely convergent. A series that converges but does not converge absolutely is said to be **conditionally convergent**. Consequently, the terms *absolutely convergent, conditionally convergent,* and *divergent* are simultaneously all-inclusive and mutually exclusive. (There's no such thing as "absolute divergence.")

There is some advantage in the application of Theorem 3, because to apply it we test the *positive*-term series $\Sigma |a_n|$ for convergence—and we have a variety of tests, such as comparison tests or the integral test, designed for use on positive-term series.

Note also that absolute convergence of the series Σa_n means that *another* series $\Sigma |a_n|$ converges, and the two sums will generally differ. For example, with $a_n = (-\tfrac{1}{3})^n$, the formula for the sum of a geometric series gives

$$\sum_{n=0}^{\infty} a_n = \sum_{n=0}^{\infty} \left(-\frac{1}{3}\right)^n = \frac{1}{1 - (-\tfrac{1}{3})} = \frac{3}{4},$$

whereas

$$\sum_{n=0}^{\infty} |a_n| = \sum_{n=0}^{\infty} \left(\frac{1}{3}\right)^n = \frac{1}{1 - \tfrac{1}{3}} = \frac{3}{2}.$$

EXAMPLE 4   Discuss the convergence of the series

$$\sum_{n=1}^{\infty} \frac{\cos n}{n^2} = \cos 1 + \frac{\cos 2}{4} + \frac{\cos 3}{9} + \cdots.$$

Solution   Let $a_n = (\cos n)/n^2$. Then

$$|a_n| = \frac{|\cos n|}{n^2} \leqq \frac{1}{n^2}$$

for all $n \geqq 1$. Hence the positive-term series $\Sigma |a_n|$ converges by the comparison test, because it is dominated by the convergent p-series $\Sigma (1/n^2)$. Thus the given series is absolutely convergent, and it therefore converges by Theorem 3.

One reason for the importance of absolute convergence is the fact (proved in advanced calculus) that the terms of an absolutely convergent series may be regrouped or rearranged without changing the sum of the series. As we suggested at the end of Section 11.6, this is *not* true of conditionally convergent series.

THE RATIO TEST AND THE ROOT TEST

Our next two convergence tests involve a way of measuring the rate of growth or decrease of the sequence $\{a_n\}$ of terms of a series to determine whether $\Sigma\, a_n$ converges absolutely or diverges.

Theorem 4 *The Ratio Test*

Suppose that the limit

$$\rho = \lim_{n \to \infty} \left| \frac{a_{n+1}}{a_n} \right| \tag{3}$$

either exists or is infinite. Then the infinite series $\Sigma\, a_n$ of nonzero terms

1. Converges absolutely if $\rho < 1$;
2. Diverges if $\rho > 1$.

If $\rho = 1$, the ratio test is inconclusive.

Proof If $\rho < 1$, choose a (fixed) number r with $\rho < r < 1$. Then Eq. (3) implies that there exists an integer N such that $|a_{n+1}| \leqq r|a_n|$ for all $n \geqq N$. It follows that

$$|a_{N+1}| \leqq r|a_N|,$$

$$|a_{N+2}| \leqq r|a_{N+1}| \leqq r^2|a_N|,$$

$$|a_{N+3}| \leqq r|a_{N+2}| \leqq r^3|a_N|,$$

and in general that

$$|a_{N+k}| \leqq r^k|a_N| \qquad \text{for } k \geqq 0.$$

Hence the series

$$|a_N| + |a_{N+1}| + |a_{N+2}| + \cdots$$

is dominated by the geometric series

$$|a_N|(1 + r + r^2 + r^3 + \cdots),$$

and the latter converges because $r < 1$. Thus the series $\Sigma\, |a_n|$ converges, so the series $\Sigma\, a_n$ converges absolutely.

If $\rho > 1$, then Eq. (3) implies that there exists a positive integer N such that $|a_{n+1}| > |a_n|$ for all $n \geqq N$. It follows that $|a_n| > |a_N| > 0$ for all $n > N$. Thus the sequence $\{a_n\}$ cannot approach zero as $n \to \infty$, and consequently, by the nth-term divergence test, the series $\Sigma\, a_n$ diverges. $\quad\square$

To see that $\Sigma\, a_n$ may either converge or diverge if $\rho = 1$, consider the divergent series $\Sigma\,(1/n)$ and the convergent series $\Sigma\,(1/n^2)$. You should verify that, for both series, the value of the ratio ρ is 1.

EXAMPLE 5 Consider the series

$$\sum_{n=1}^{\infty} \frac{(-1)^n 2^n}{n!} = -2 + \frac{4}{2!} - \frac{8}{3!} + \frac{16}{4!} - \cdots.$$

Then

$$\rho = \lim_{n \to \infty} \left| \frac{a_{n+1}}{a_n} \right| = \lim_{n \to \infty} \left| \frac{\dfrac{(-1)^{n+1}2^{n+1}}{(n+1)!}}{\dfrac{(-1)^n 2^n}{n!}} \right| = \lim_{n \to \infty} \frac{2}{n+1} = 0.$$

Because $\rho < 1$, the series converges absolutely.

EXAMPLE 6 Test for convergence: $\displaystyle\sum_{n=1}^{\infty} \frac{n}{2^n}$.

Solution We have

$$\rho = \lim_{n \to \infty} \left| \frac{a_{n+1}}{a_n} \right| = \lim_{n \to \infty} \frac{\dfrac{n+1}{2^{n+1}}}{\dfrac{n}{2^n}} = \lim_{n \to \infty} \frac{n+1}{2n} = \frac{1}{2}.$$

Because $\rho < 1$, this series converges (absolutely).

EXAMPLE 7 Test for convergence: $\displaystyle\sum_{n=1}^{\infty} \frac{3^n}{n^2}$.

Solution Here we have

$$\rho = \lim_{n \to \infty} \left| \frac{a_{n+1}}{a_n} \right| = \lim_{n \to \infty} \frac{\dfrac{3^{n+1}}{(n+1)^2}}{\dfrac{3^n}{n^2}} = \lim_{n \to \infty} \frac{3n^2}{(n+1)^2} = 3.$$

In this case $\rho > 1$, so the given series diverges.

Theorem 5 *The Root Test*

Suppose that the limit

$$\rho = \lim_{n \to \infty} \sqrt[n]{|a_n|} \qquad (4)$$

exists or is infinite. Then the infinite series $\Sigma\, a_n$

1. Converges absolutely if $\rho < 1$;
2. Diverges if $\rho > 1$.

If $\rho = 1$, the test is inconclusive.

Proof If $\rho < 1$, choose a (fixed) number r such that $\rho < r < 1$. Then $|a_n|^{1/n} < r$, and hence $|a_n| < r^n$, for n sufficiently large. Thus the series $\Sigma |a_n|$ is eventually dominated by the convergent geometric series Σr^n. Therefore $\Sigma |a_n|$ converges, and so the series $\Sigma\, a_n$ converges absolutely.

If $\rho > 1$, then $|a_n|^{1/n} > 1$, and hence $|a_n| > 1$, for n sufficiently large. Therefore the nth-term test for divergence implies that the series $\Sigma\, a_n$ diverges. ❑

The ratio test is generally simpler to apply than the root test, and therefore it is ordinarily the one to try first. But there are certain series for which the root test succeeds and the ratio test fails, as in Example 8.

EXAMPLE 8 Consider the series

$$\sum_{n=1}^{\infty} \frac{1}{2^{n+(-1)^n}} = \frac{1}{2} + \frac{1}{1} + \frac{1}{8} + \frac{1}{4} + \frac{1}{32} + \frac{1}{16} + \cdots.$$

Then $a_{n+1}/a_n = 2$ if n is even, whereas $a_{n+1}/a_n = \frac{1}{8}$ if n is odd. So the limit required for the ratio test does not exist. But

$$\lim_{n\to\infty} |a_n|^{1/n} = \lim_{n\to\infty} \left| \frac{1}{2^{n+(-1)^n}} \right|^{1/n} = \lim_{n\to\infty} \frac{1}{2} \left| \frac{1}{2^{(-1)^n/n}} \right| = \frac{1}{2},$$

so the given series converges by the root test. (Its convergence also follows from the fact that it is a rearrangement of the positive-term convergent geometric series $\Sigma\, 1/2^n$.)

11.7 Problems

Determine whether or not the alternating series in Problems 1 through 10 converge.

1. $\displaystyle\sum_{n=1}^{\infty} \frac{(-1)^{n+1}}{n^2}$

2. $\displaystyle\sum_{n=1}^{\infty} \frac{(-1)^{n+1} n}{3n + 2}$

3. $\displaystyle\sum_{n=1}^{\infty} \frac{(-1)^{n+1} n}{n^2 + 1}$

4. $\displaystyle\sum_{n=2}^{\infty} \frac{(-1)^n n}{\ln n}$

5. $\displaystyle\sum_{n=2}^{\infty} \frac{(-1)^n}{\ln n}$

6. $\displaystyle\sum_{n=1}^{\infty} \frac{(-1)^{n+1}}{\sqrt{2n + 1}}$

7. $\displaystyle\sum_{n=0}^{\infty} \frac{(-1)^n}{n!}$

8. $\displaystyle\sum_{n=1}^{\infty} \frac{(-1)^{n+1}(1.01)^n}{n^4 + 1}$

9. $\displaystyle\sum_{n=1}^{\infty} \frac{(-1)^{n+1}}{\sqrt[n]{n}}$

10. $\displaystyle\sum_{n=1}^{\infty} \frac{(-1)^{n+1} n!}{(2n)!}$

Determine whether the series in Problems 11 through 32 converge absolutely, converge conditionally, or diverge.

11. $\displaystyle\sum_{n=1}^{\infty} \frac{(-1)^{n+1}}{2^n}$

12. $\displaystyle\sum_{n=1}^{\infty} \frac{1}{n^2 + 1}$

13. $\displaystyle\sum_{n=1}^{\infty} \frac{(-1)^{n+1} \ln n}{n}$

14. $\displaystyle\sum_{n=1}^{\infty} \frac{1}{n^n}$

15. $\displaystyle\sum_{n=1}^{\infty} \left(\frac{10}{n}\right)^n$

16. $\displaystyle\sum_{n=1}^{\infty} \frac{3^n}{n!n}$

17. $\displaystyle\sum_{n=0}^{\infty} \frac{(-10)^n}{n!}$

18. $\displaystyle\sum_{n=1}^{\infty} \frac{(-1)^{n+1} n!}{n^n}$

19. $\displaystyle\sum_{n=1}^{\infty} (-1)^{n+1} \left(\frac{n}{n + 1}\right)^n$

20. $\displaystyle\sum_{n=1}^{\infty} \frac{n! \, n^2}{(2n)!}$

21. $\displaystyle\sum_{n=1}^{\infty} \left(\frac{\ln n}{n}\right)^n$

22. $\displaystyle\sum_{n=0}^{\infty} \frac{(-1)^n 2^{3n}}{7^n}$

23. $\displaystyle\sum_{n=0}^{\infty} (-1)^n (\sqrt{n + 1} - \sqrt{n})$

24. $\displaystyle\sum_{n=1}^{\infty} n(\tfrac{3}{4})^n$

25. $\displaystyle\sum_{n=1}^{\infty} \left[\ln\!\left(\frac{1}{n}\right) \right]^n$

26. $\displaystyle\sum_{n=0}^{\infty} \frac{(n!)^2}{(2n)!}$

27. $\displaystyle\sum_{n=1}^{\infty} \frac{(-1)^{n+1} 3^n}{n(2^n + 1)}$

28. $\displaystyle\sum_{n=1}^{\infty} \frac{(-1)^{n+1} \arctan n}{n}$

29. $\displaystyle\sum_{n=1}^{\infty} \frac{(-1)^{n+1} n!}{1 \cdot 3 \cdot 5 \cdots (2n - 1)}$

30. $\displaystyle\sum_{n=1}^{\infty} (-1)^{n+1} \frac{1 \cdot 3 \cdot 5 \cdots (2n - 1)}{1 \cdot 4 \cdot 7 \cdots (3n - 2)}$

31. $\displaystyle\sum_{n=1}^{\infty} \frac{(n + 2)!}{3^n (n!)^2}$

32. $\displaystyle\sum_{n=1}^{\infty} \frac{(-1)^{n+1} n^n}{3^{n^2}}$

In Problems 33 through 36, find the least positive integer n such that $|R_n| = |S - S_n| < 0.0005$, so the nth partial sum S_n of the given alternating series approximates its sum S accurate to three decimal places.

33. $\displaystyle\sum_{n=1}^{\infty} \frac{(-1)^{n+1}}{n}$

34. $\displaystyle\sum_{n=1}^{\infty} \frac{(-1)^{n+1}}{n^2}$

35. $\displaystyle\sum_{n=0}^{\infty} \frac{(-1)^n}{3^n}$

36. $\displaystyle\sum_{n=1}^{\infty} \frac{(-1)^{n+1}}{n^n}$

In Problems 37 through 40, approximate the sum of the given series accurate to three decimal places.

37. $\displaystyle e^{-1/2} = \sum_{n=0}^{\infty} \frac{(-1)^n}{n! 2^n}$

38. $\displaystyle \cos 1 = \sum_{n=0}^{\infty} \frac{(-1)^n}{(2n)!}$

39. $\displaystyle \ln(1.1) = \sum_{n=1}^{\infty} \frac{(-1)^{n+1}(0.1)^n}{n}$

40. $\displaystyle\sum_{n=1}^{\infty} \frac{(-1)^{n+1}}{n^5}$

41. Approximate the sum of the series

$$\frac{\pi^2}{12} = \sum_{n=1}^{\infty} \frac{(-1)^{n+1}}{n^2}$$

with error less than 0.01. Use the corresponding partial sum and error estimate to verify that $3.13 < \pi < 3.15$.

42. Prove that $\Sigma |a_n|$ diverges if the series Σa_n diverges.

43. Give an example of a pair of convergent series Σa_n and Σb_n such that $\Sigma a_n b_n$ diverges.

44. (a) Suppose that r is a (fixed) number such that $|r| < 1$. Use the ratio test to prove that the series

$$\sum_{n=0}^{\infty} nr^n$$

converges. Let S denote its sum. (b) Show that

$$(1 - r)S = \sum_{n=1}^{\infty} r^n.$$

Show how to conclude that

$$\sum_{n=0}^{\infty} nr^n = \frac{r}{(1 - r)^2}.$$

11.8
Power Series

The most important infinite series representations of functions are those whose terms are constant multiples of (successive) integral powers of the independent variable x—that is, series that resemble "infinite polynomials." For example, we discussed in Section 11.4 the geometric series

$$\frac{1}{1 - x} = 1 + x + x^2 + x^3 + \cdots \qquad (|x| < 1) \qquad (1)$$

and the Taylor series

$$e^x = \sum_{n=0}^{\infty} \frac{x^n}{n!} = 1 + x + \frac{x^2}{2!} + \frac{x^3}{3!} + \frac{x^4}{4!} + \cdots, \qquad (2)$$

$$\cos x = \sum_{n=0}^{\infty} (-1)^n \frac{x^{2n}}{(2n)!} = 1 - \frac{x^2}{2!} + \frac{x^4}{4!} - \frac{x^6}{6!} + \cdots, \quad \text{and} \qquad (3)$$

$$\sin x = \sum_{n=0}^{\infty} (-1)^n \frac{x^{2n+1}}{(2n + 1)!} = x - \frac{x^3}{3!} + \frac{x^5}{5!} - \frac{x^7}{7!} + \cdots. \qquad (4)$$

There we used Taylor's formula to show that the series in Eqs. (2) through (4) converge, for all x, to the functions e^x, $\cos x$, and $\sin x$, respectively. Here we investigate the convergence of a "power series" without knowing in advance the function (if any) to which it converges.

All the infinite series in Eqs. (1) through (4) have the form

$$\sum_{n=0}^{\infty} a_n x^n = a_0 + a_1 x + a_2 x^2 + \cdots + a_n x^n + \cdots \qquad (5)$$

with the constant *coefficients* $a_0, a_1, a_2, \ldots$. An infinite series of this form is called a **power series** in (powers of) x. In order that the initial terms of the two sides of Eq. (5) agree, we adopt here the convention that $x^0 = 1$ even if $x = 0$.

CONVERGENCE OF POWER SERIES

The power series in Eq. (5) obviously converges when $x = 0$. In general, it will converge for some nonzero values of x and diverge for others. Because of the way in which powers of x are involved, the ratio test is particularly effective in determining the values of x for which a power series converges.

Assume that the limit

$$\rho = \lim_{n \to \infty} \left| \frac{a_{n+1}}{a_n} \right| \tag{6}$$

exists. This is the limit that we need if we want to apply the ratio test to the series $\Sigma \, a_n$ of constants. To apply the ratio test to the power series in Eq. (5), we write $u_n = a_n x^n$ and compute the limit

$$\lim_{n \to \infty} \left| \frac{u_{n+1}}{u_n} \right| = \lim_{n \to \infty} \left| \frac{a_{n+1} x^{n+1}}{a_n x^n} \right| = \rho |x|. \tag{7}$$

If $\rho = 0$, then $\Sigma \, a_n x^n$ converges absolutely for all x. If $\rho = +\infty$, then $\Sigma \, a_n x^n$ diverges for all $x \neq 0$. If ρ is a positive real number, we see from Eq. (7) that $\Sigma \, a_n x^n$ converges absolutely for all x such that $\rho |x| < 1$—that is, when

$$|x| < R = \frac{1}{\rho} = \lim_{n \to \infty} \left| \frac{a_n}{a_{n+1}} \right|. \tag{8}$$

In this case the ratio test also implies that $\Sigma \, a_n x^n$ diverges if $|x| > R$ but is inconclusive when $x = \pm R$. We have therefore proved Theorem 1, under the additional hypothesis that the limit in Eq. (6) exists. In Problems 43 and 44 we outline a proof that does not require this additional hypothesis.

Theorem 1   Convergence of Power Series
If $\Sigma \, a_n x^n$ is a power series, then either

1. The series converges absolutely for all x, or
2. The series converges only when $x = 0$, or
3. There exists a number $R > 0$ such that $\Sigma \, a_n x^n$ converges absolutely if $|x| < R$ and diverges if $|x| > R$.

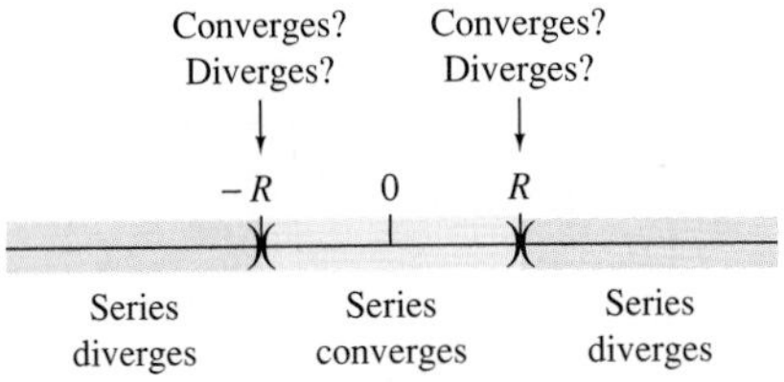

Converges?
Diverges?   Converges?
                   Diverges?

$-R$ 0 R

Series   Series   Series
diverges   converges   diverges

Fig. 11.8.1 The interval of convergence if

$$0 < R = \lim_{n \to \infty} \left| \frac{a_n}{a_{n+1}} \right| < \infty$$

The number R of case 3 is called the **radius of convergence** of the power series $\Sigma \, a_n x^n$. We shall write $R = \infty$ in case 1 and $R = 0$ in case 2. The set of all real numbers x for which the series converges is called its **interval of convergence** (Fig. 11.8.1); note that this set *is* an interval. If $0 < R < \infty$, then the interval of convergence is one of the intervals

$$(-R, R), \qquad (-R, R], \qquad [-R, R), \quad \text{or} \quad [-R, R].$$

When we substitute either of the endpoints $x = \pm R$ into the series $\Sigma \, a_n x^n$, we obtain an infinite series with constant terms whose convergence must be determined separately. Because these will be numerical series, the earlier tests of this chapter are appropriate.

EXAMPLE 1 Find the interval of convergence of the series

$$\sum_{n=1}^{\infty} \frac{x^n}{n \cdot 3^n}.$$

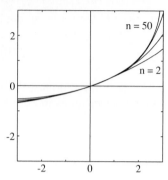

Fig. 11.8.2 Graphs of partial sums of the series of Example 1, with degrees $n = 2, 4, 10, 50$

Solution With $u_n = x_n/(n \cdot 3^n)$ we find that

$$\lim_{n \to \infty} \left| \frac{u_{n+1}}{u_n} \right| = \lim_{n \to \infty} \left| \frac{\dfrac{x^{n+1}}{(n+1) \cdot 3^{n+1}}}{\dfrac{x^n}{n \cdot 3^n}} \right| = \lim_{n \to \infty} \frac{n|x|}{3(n+1)} = \frac{|x|}{3}.$$

Now $|x|/3 < 1$ provided that $|x| < 3$, so the ratio test implies that the given series converges absolutely if $|x| < 3$ and diverges if $|x| > 3$. When $x = 3$, we have the divergent harmonic series $\Sigma\,(1/n)$, and when $x = -3$ we have the convergent alternating series $\Sigma\,(-1)^n/n$. Thus the interval of convergence of the given power series is $[-3, 3)$. We see graphically in Fig. 11.8.2 the difference between convergence at $x = -3$ and divergence at $x = +3$.

EXAMPLE 2 Find the interval of convergence of $\displaystyle\sum_{n=0}^{\infty} \frac{2^n x^n}{n!}$.

Solution With $u_n = 2^n x^n/n!$, we find that

$$\lim_{n \to \infty} \left| \frac{u_{n+1}}{u_n} \right| = \lim_{n \to \infty} \left| \frac{\dfrac{2^{n+1} x^{n+1}}{(n+1)!}}{\dfrac{2^n x^n}{n!}} \right| = \lim_{n \to \infty} \frac{2|x|}{n+1} = 0$$

for all x. Hence the ratio test implies that the power series converges for all x, and its interval of convergence is $(-\infty, \infty)$, the entire real line.

EXAMPLE 3 Find the interval of convergence of the series $\displaystyle\sum_{n=0}^{\infty} n^n x^n$.

Solution With $u_n = n^n x^n$ we find that

$$\lim_{n \to \infty} \left| \frac{u_{n+1}}{u_n} \right| = \lim_{n \to \infty} \left| \frac{(n+1)^{n+1} x^{n+1}}{n^n x^n} \right| = \lim_{n \to \infty} (n+1) \left(1 + \frac{1}{n} \right)^n |x| = +\infty$$

for all $x \neq 0$, because

$$\lim_{n \to \infty} \left(1 + \frac{1}{n} \right)^n = e.$$

Thus the given series diverges for all $x \neq 0$, and its interval of convergence consists of the single point $x = 0$.

EXAMPLE 4 Use the ratio test to verify that the Taylor series for $\cos x$ in Eq. (3) converges for all x.

Solution With $u_n = (-1)^n x^{2n}/(2n)!$ we find that

$$\lim_{n \to \infty} \left| \frac{u_{n+1}}{u_n} \right| = \lim_{n \to \infty} \left| \frac{\dfrac{(-1)^{n+1} x^{2n+2}}{(2n+2)!}}{\dfrac{(-1)^n x^{2n}}{(2n)!}} \right| = \lim_{n \to \infty} \frac{x^2}{(2n+1)(2n+2)} = 0$$

for all x, so the series converges for all x.

IMPORTANT   In Example 4, the ratio test tells us only that the series for cos *x* converges to *some* number, *not* necessarily the particular number cos *x*. The argument of Section 11.4, using Taylor's formula with remainder, is required to establish that the sum of the series is actually cos *x*.

An infinite series of the form

$$\sum_{n=0}^{\infty} a_n(x - c)^n = a_0 + a_1(x - c) + a_2(x - c)^2 + \cdots, \tag{9}$$

where *c* is a constant, is called a **power series in** (powers of) *x* − *c*. By the same reasoning that led us to Theorem 1, with x^n replaced by $(x - c)^n$ throughout, we conclude that either

1. The series in Eq. (9) converges absolutely for all *x*, or
2. That series converges only when *x* − *c* = 0—that is, when *x* = *c*—or
3. There exists a number *R* > 0 such that the series in Eq. (9) converges absolutely if $|x - c| < R$ and diverges if $|x - c| > R$.

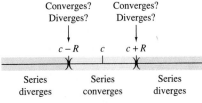

Converges?   Converges?
Diverges?   Diverges?

c − *R*   *c*   *c* + *R*

Series   Series   Series
diverges   converges   diverges

Fig. 11.8.3 The interval of convergence of

$$\sum_{n=0}^{\infty} a_n(x - c)^n$$

As in the case of a power series with *c* = 0, the number *R* is called the **radius of convergence** of the series, and the **interval of convergence** of the series $\sum a_n(x - c)^n$ is the set of all numbers *x* for which it converges (Fig. 11.8.3). As before, when $0 < R < \infty$, the convergence of the series at the endpoints *x* = *c* − *R* and *x* = *c* + *R* of its interval of convergence must be checked separately.

EXAMPLE 5   Determine the interval of convergence of the series

$$\sum_{n=1}^{\infty} \frac{(-1)^n(x - 2)^n}{n \cdot 4^n}.$$

Solution   We let $u_n = (-1)^n(x - 2)^n/(n \cdot 4^n)$. Then

$$\lim_{n \to \infty} \left| \frac{u_{n+1}}{u_n} \right| = \lim_{n \to \infty} \left| \frac{\dfrac{(-1)^{n+1}(x - 2)^{n+1}}{(n + 1) \cdot 4^{n+1}}}{\dfrac{(-1)^n(x - 2)^n}{n \cdot 4^n}} \right|$$

$$= \lim_{n \to \infty} \frac{|x - 2|}{4} \cdot \frac{n}{n + 1} = \frac{|x - 2|}{4}.$$

Hence the given series converges when $|x - 2|/4 < 1$—that is, when $|x - 2| < 4$—so the radius of convergence is *R* = 4. Because *c* = 2, the series converges when −2 < *x* < 6 and diverges if either *x* < −2 or *x* > 6. When *x* = −2, the series reduces to the divergent harmonic series, and when *x* = 6, it reduces to the convergent alternating series $\sum (-1)^n/n$. Thus the interval of convergence of the given power series is (−2, 6].

POWER SERIES REPRESENTATIONS OF FUNCTIONS

Power series are important tools for computing (or approximating) values of functions. Suppose that the series $\sum a_n x^n$ converges to the value *f*(*x*); that is,

$$f(x) = a_0 + a_1 x + a_2 x^2 + \cdots + a_n x^n + \cdots$$

for each x in the interval of convergence of the power series. Then we call $\Sigma a_n x^n$ a **power series representation** of $f(x)$. For example, the geometric series Σx^n in Eq. (1) is a power series representation of the function $f(x) = 1/(1 - x)$ on the interval $(-1, 1)$.

We saw in Section 11.4 how Taylor's formula with remainder can often be used to find a power series representation of a given function. Recall that the nth-degree Taylor's formula for $f(x)$ at $x = a$ is

$$f(x) = f(a) + f'(a)(x - a) + \frac{f''(a)}{2!}(x - a)^2 + \frac{f^{(3)}(a)}{3!}(x - a)^3$$

$$+ \cdots + \frac{f^{(n)}(a)}{n!}(x - a)^n + R_n(x). \tag{10}$$

The remainder $R_n(x)$ is given by

$$R_n(x) = \frac{f^{(n+1)}(z)}{(n + 1)!}(x - a)^{n+1},$$

where z is some number between a and x. If we let $n \to \infty$ in Eq. (10), we obtain Theorem 2.

Theorem 2   Taylor Series Representations

Suppose that the function f has derivatives of all orders on some interval containing a and also that

$$\lim_{n \to \infty} R_n(x) = 0 \tag{11}$$

for each x in that interval. Then

$$f(x) = \sum_{n=0}^{\infty} \frac{f^{(n)}(a)}{n!}(x - a)^n \tag{12}$$

for each x in the interval.

The power series in Eq. (12) is the **Taylor series** of the function f **at** $x = a$ (or *in powers of $x - a$*, or *with center a*). If $a = 0$, we obtain the power series

$$f(x) = \sum_{n=0}^{\infty} \frac{f^{(n)}(0)}{n!}x^n = f(0) + f'(0)x + \frac{f''(0)}{2!}x^2 + \cdots, \tag{13}$$

commonly called the **Maclaurin series** of f. Thus the power series in Eqs. (2) through (4) are the Maclaurin series of the functions e^x, $\cos x$, and $\sin x$, respectively.

Upon replacing x by $-x$ in the Maclaurin series for e^x, we obtain

$$e^{-x} = 1 - x + \frac{x^2}{2!} - \frac{x^3}{3!} + \cdots + (-1)^n \frac{x^n}{n!} + \cdots.$$

Let us add the series for e^x and e^{-x} and divide by 2. This gives

$$\cosh x = \frac{e^x + e^{-x}}{2} = \frac{1}{2}\left(1 + x + \frac{x^2}{2!} + \frac{x^3}{3!} + \frac{x^4}{4!} + \cdots\right)$$

$$+ \frac{1}{2}\left(1 - x + \frac{x^2}{2!} - \frac{x^3}{3!} + \frac{x^4}{4!} - \cdots\right),$$

so

$$\cosh x = 1 + \frac{x^2}{2!} + \frac{x^4}{4!} + \frac{x^6}{6!} + \cdots.$$

Similarly,

$$\sinh x = x + \frac{x^3}{3!} + \frac{x^5}{5!} + \frac{x^7}{7!} + \cdots.$$

Note the strong resemblance to Eqs. (3) and (4), the series for cos x and sin x, respectively.

Upon replacing x by $-x^2$ in the series for e^x, we obtain

$$e^{-x^2} = \sum_{n=0}^{\infty} (-1)^n \frac{x^{2n}}{n!} = 1 - x^2 + \frac{x^4}{2!} - \frac{x^6}{3!} + \cdots.$$

Because this power series converges to $\exp(-x^2)$ for all x, it must be the Maclaurin series for $\exp(-x^2)$ (see Problem 40). Think how tedious it would be to compute the derivatives of $\exp(-x^2)$ needed to write its Maclaurin series directly from Eq. (13).

THE BINOMIAL SERIES

Example 6 gives one of the most famous and useful of all series, the *binomial series,* which was discovered by Newton in the 1660s. It is the infinite series generalization of the (finite) binomial formula of elementary algebra.

EXAMPLE 6 Suppose that α is a nonzero real number. Show that the Maclaurin series of $f(x) = (1 + x)^\alpha$ is

$$(1 + x)^\alpha = 1 + \sum_{n=1}^{\infty} \frac{\alpha(\alpha - 1)(\alpha - 2) \cdots (\alpha - n + 1)}{n!} x^n$$

$$= 1 + \alpha x + \frac{\alpha(\alpha - 1)}{2!} x^2 + \frac{\alpha(\alpha - 1)(\alpha - 2)}{3!} x^3 + \cdots. \quad (14)$$

Also determine the interval of convergence of this **binomial series**.

Solution To derive the series itself, we simply list all the derivatives of $f(x) = (1 + x)^\alpha$, including its "zeroth" derivative:

$$f(x) = (1 + x)^\alpha,$$

$$f'(x) = \alpha(1 + x)^{\alpha-1},$$

$$f''(x) = \alpha(\alpha - 1)(1 + x)^{\alpha-2},$$

$$f^{(3)}(x) = \alpha(\alpha - 1)(\alpha - 2)(1 + x)^{\alpha-3},$$

$$\vdots$$

$$f^{(n)}(x) = \alpha(\alpha - 1)(\alpha - 2) \cdots (\alpha - n + 1)(1 + x)^{\alpha-n}.$$

Thus

$$f^{(n)}(0) = \alpha(\alpha - 1)(\alpha - 2) \cdots (\alpha - n + 1).$$

If we substitute this value of $f^{(n)}(0)$ into the Maclaurin series formula in Eq. (13), we get the binomial series in Eq. (14).

To determine the interval of convergence of the binomial series, we let

$$u_n = \frac{\alpha(\alpha - 1)(\alpha - 2) \cdots (\alpha - n + 1)}{n!} x^n.$$

We find that

$$\lim_{n \to \infty} \left| \frac{u_{n+1}}{u_n} \right| = \lim_{n \to \infty} \left| \frac{\dfrac{\alpha(\alpha - 1)(\alpha - 2) \cdots (\alpha - n)x^{n+1}}{(n + 1)!}}{\dfrac{\alpha(\alpha - 1)(\alpha - 2) \cdots (\alpha - n + 1)x^n}{n!}} \right|$$

$$= \lim_{n \to \infty} \left| \frac{(\alpha - n)x}{n + 1} \right| = |x|.$$

Hence the ratio test shows that the binomial series converges absolutely if $|x| < 1$ and diverges if $|x| > 1$. Its convergence at the endpoints $x = \pm 1$ depends on the value of α; we shall not pursue this problem. Problem 41 outlines a proof that the sum of the binomial series actually is $(1 + x)^\alpha$ if $|x| < 1$.

If $\alpha = k$, a positive integer, then the coefficient of x^n is zero for $n > k$, and the binomial series reduces to the binomial formula

$$(1 + x)^k = \sum_{n=0}^{k} \frac{k!}{n!(k - n)!} x^n.$$

Otherwise Eq. (14) is an infinite series. For example, with $\alpha = \frac{1}{2}$, we obtain

$$\sqrt{1 + x} = 1 + \frac{\frac{1}{2}}{1!}x + \frac{(\frac{1}{2})(-\frac{1}{2})}{2!}x^2 + \frac{(\frac{1}{2})(-\frac{1}{2})(-\frac{3}{2})}{3!}x^3$$

$$+ \frac{(\frac{1}{2})(-\frac{1}{2})(-\frac{3}{2})(-\frac{5}{2})}{4!}x^4 + \cdots$$

$$= 1 + \frac{1}{2}x - \frac{1}{8}x^2 + \frac{1}{16}x^3 - \frac{5}{128}x^4 + \cdots. \tag{15}$$

If we replace x by $-x$ and take $\alpha = -\frac{1}{2}$, we get the series

$$\frac{1}{\sqrt{1 - x}} = 1 + \frac{-\frac{1}{2}}{1!}(-x) + \frac{(-\frac{1}{2})(-\frac{3}{2})}{2!}(-x)^2$$

$$+ \cdots + \frac{1 \cdot 3 \cdot 5 \cdots (2n - 1)}{n! \cdot 2^n}x^n + \cdots,$$

which in summation notation takes the form

$$\frac{1}{\sqrt{1 - x}} = 1 + \sum_{n=1}^{\infty} \frac{1 \cdot 3 \cdot 5 \cdots (2n - 1)}{2 \cdot 4 \cdot 6 \cdots (2n)}x^n. \tag{16}$$

We will find this series quite useful in Example 10 and in Problem 42.

DIFFERENTIATION AND INTEGRATION OF POWER SERIES

Sometimes it is inconvenient to compute the repeated derivatives of a function in order to find its Taylor series. An alternative method of finding new power series is by the differentiation and integration of known power series.

Suppose that a power series representation of the function $f(x)$ is known. Then Theorem 3 (we leave its proof to advanced calculus) implies that the function $f(x)$ may be differentiated by separately differentiating the individual terms in its power series. That is, the power series obtained by termwise differentiation converges to the derivative $f'(x)$. Similarly, a function can be integrated by termwise integration of its power series.

Theorem 3 *Termwise Differentiation and Integration*

Suppose that the function f has a power series representation

$$f(x) = \sum_{n=0}^{\infty} a_n x^n = a_0 + a_1 x + a_2 x^2 + a_3 x^3 + \cdots$$

with nonzero radius of convergence R. Then f is differentiable on $(-R, R)$ and

$$f'(x) = \sum_{n=1}^{\infty} n a_n x^{n-1} = a_1 + 2a_2 x + 3a_3 x^2 + 4a_4 x^3 + \cdots. \quad (17)$$

Also,

$$\int_0^x f(t)\, dt = \sum_{n=0}^{\infty} \frac{a_n x^{n+1}}{n+1} = a_0 x + \tfrac{1}{2} a_1 x^2 + \tfrac{1}{3} a_2 x^3 + \cdots \quad (18)$$

for each x in $(-R, R)$. Moreover, the power series in Eqs. (17) and (18) have the same radius of convergence R.

REMARK   Although we omit the proof of Theorem 3, we observe that the radius of convergence of the series in Eq. (17) is

$$R = \lim_{n \to \infty} \left| \frac{n a_n}{(n+1) a_{n+1}} \right| = \left(\lim_{n \to \infty} \frac{n}{n+1} \right) \left(\lim_{n \to \infty} \left| \frac{a_n}{a_{n+1}} \right| \right) = \lim_{n \to \infty} \left| \frac{a_n}{a_{n+1}} \right|.$$

Thus, by Eq. (8), the power series for $f(x)$ and the power series for $f'(x)$ have the same radius of convergence (under the assumption that the preceding limit exists).

EXAMPLE 7   Termwise differentiation of the geometric series for

$$f(x) = \frac{1}{1-x}$$

yields

$$\frac{1}{(1-x)^2} = D_x\left(\frac{1}{1-x} \right) = D_x(1 + x + x^2 + x^3 + \cdots)$$

$$= 1 + 2x + 3x^2 + 4x^3 + \cdots.$$

Thus

$$\frac{1}{(1-x)^2} = \sum_{n=1}^{\infty} n x^{n-1} = \sum_{n=0}^{\infty} (n+1) x^n.$$

The series converges to $1/(1-x)^2$ if $-1 < x < 1$.

EXAMPLE 8 Replacement of x by $-t$ in the geometric series of Example 7 gives

$$\frac{1}{1+t} = 1 - t + t^2 - t^3 + \cdots + (-1)^n t^n + \cdots.$$

Because $D_t \ln(1 + t) = 1/(1 + t)$, termwise integration from $t = 0$ to $t = x$ now gives

$$\ln(1 + x) = \int_0^x \frac{1}{1+t}\, dt$$

$$= \int_0^x (1 - t + t^2 - \cdots + (-1)^n t^n + \cdots)\, dt;$$

$$\ln(1 + x) = x - \frac{1}{2}x^2 + \frac{1}{3}x^3 - \frac{1}{4}x^4 + \cdots + \frac{(-1)^{n-1}}{n}x^n + \cdots \quad (19)$$

if $|x| < 1$.

EXAMPLE 9 Find a power series representation for the arctangent function.

Solution Because $D_t \tan^{-1} t = 1/(1 + t^2)$, termwise integration of the series

$$\frac{1}{1+t^2} = 1 - t^2 + t^4 - t^6 + t^8 - \cdots$$

gives

$$\tan^{-1} x = \int_0^x \frac{1}{1+t^2}\, dt = \int_0^x (1 - t^2 + t^4 - t^6 + t^8 - \cdots)\, dt.$$

Therefore,

$$\tan^{-1} x = x - \tfrac{1}{3}x^3 + \tfrac{1}{5}x^5 - \tfrac{1}{7}x^7 + \tfrac{1}{9}x^9 - \cdots \quad (20)$$

if $-1 < x < 1$.

EXAMPLE 10 Find a power series representation for the arcsine function.

Solution First we substitute t^2 for x in Eq. (16). This yields

$$\frac{1}{\sqrt{1-t^2}} = 1 + \sum_{n=1}^{\infty} \frac{1 \cdot 3 \cdot 5 \cdots (2n-1)}{2 \cdot 4 \cdot 6 \cdots (2n)} t^{2n}$$

if $|t| < 1$. Because $D_t \sin^{-1} t = 1/\sqrt{1-t^2}$, termwise integration of this series from $t = 0$ to $t = x$ gives

$$\sin^{-1} x = \int_0^x \frac{1}{\sqrt{1-t^2}}\, dt = x + \sum_{n=1}^{\infty} \frac{1 \cdot 3 \cdot 5 \cdots (2n-1)}{2 \cdot 4 \cdot 6 \cdots (2n)} \cdot \frac{x^{2n+1}}{2n+1} \quad (21)$$

if $|x| < 1$. Problem 42 shows how to use this series to derive the series

$$\frac{\pi^2}{6} = 1 + \frac{1}{2^2} + \frac{1}{3^2} + \frac{1}{4^2} + \cdots + \frac{1}{n^2} + \cdots,$$

which we used in Example 3 of Section 11.5 to approximate the number π.

Theorem 3 has this important consequence: If both power series $\Sigma\, a_n x^n$ and $\Sigma\, b_n x^n$ converge and, for all x with $|x| < R$ $(R > 0)$, $\Sigma\, a_n x^n = \Sigma\, b_n x^n$, then $a_n = b_n$ for all n. In particular, the Taylor series of a function is its unique power series representation (if any). See Problem 40.

11.8 Problems

Find the interval of convergence of the power series in Problems 1 through 20.

1. $\displaystyle\sum_{n=1}^{\infty} \frac{1}{n} x^n$

2. $\displaystyle\sum_{n=0}^{\infty} \frac{(-1)^n}{n^2 + 1} x^n$

3. $\displaystyle\sum_{n=1}^{\infty} (-1)^{n+1} n^2 x^n$

4. $\displaystyle\sum_{n=1}^{\infty} n! x^n$

5. $\displaystyle\sum_{n=1}^{\infty} \frac{(-1)^{n+1} x^{2n}}{2n - 1}$

6. $\displaystyle\sum_{n=1}^{\infty} \frac{n x^n}{5^n}$

7. $\displaystyle\sum_{n=0}^{\infty} (5x - 3)^n$

8. $\displaystyle\sum_{n=1}^{\infty} \frac{(2x - 1)^n}{n^4 + 16}$

9. $\displaystyle\sum_{n=1}^{\infty} \frac{2^n (x - 3)^n}{n^2}$

10. $\displaystyle\sum_{n=1}^{\infty} \frac{n!}{n^n} x^n$ (Do not test the endpoints; the series diverges at each.)

11. $\displaystyle\sum_{n=1}^{\infty} \frac{(2n)!}{n!} x^n$

12. $\displaystyle\sum_{n=1}^{\infty} \frac{1 \cdot 3 \cdot 5 \cdots (2n + 1)}{n!} x^n$ (Do not test the endpoints; the series diverges at each.)

13. $\displaystyle\sum_{n=1}^{\infty} \frac{n^3 (x + 1)^n}{3^n}$

14. $\displaystyle\sum_{n=1}^{\infty} \frac{(-1)^{n+1} (x - 2)^n}{n^2}$

15. $\displaystyle\sum_{n=1}^{\infty} \frac{(3 - x)^n}{n^3}$

16. $\displaystyle\sum_{n=1}^{\infty} \frac{(-1)^{n+1} 10^n}{n!} (x - 10)^n$

17. $\displaystyle\sum_{n=1}^{\infty} \frac{n!}{2^n} (x - 5)^n$

18. $\displaystyle\sum_{n=1}^{\infty} \frac{(-1)^{n+1}}{n \cdot 10^n} (x - 2)^n$

19. $\displaystyle\sum_{n=0}^{\infty} x^{(2^n)}$

20. $\displaystyle\sum_{n=0}^{\infty} \left(\frac{x^2 + 1}{5} \right)^n$

In Problems 21 through 30, use power series established in this section to find a power series representation of the given function. Then determine the radius of convergence of the resulting series.

21. $f(x) = x^2 e^{-3x}$

22. $f(x) = \dfrac{1}{10 + x}$

23. $f(x) = \sin x^2$

24. $f(x) = \cos^2 x$ [*Suggestion:* $\cos^2 x = \frac{1}{2}(1 + \cos 2x)$.]

25. $f(x) = \sqrt[3]{1 - x}$

26. $f(x) = (1 + x^2)^{3/2}$

27. $f(x) = (1 + x)^{-3}$

28. $f(x) = \dfrac{1}{\sqrt{9 + x^3}}$

29. $f(x) = \dfrac{\ln(1 + x)}{x}$

30. $f(x) = \dfrac{x - \arctan x}{x^3}$

In Problems 31 through 36, find a power series representation for the given function $f(x)$ by using termwise integration.

31. $f(x) = \displaystyle\int_0^x \sin t^3 \, dt$

32. $f(x) = \displaystyle\int_0^x \frac{\sin t}{t} \, dt$

33. $f(x) = \displaystyle\int_0^x \exp(-t^3) \, dt$

34. $f(x) = \displaystyle\int_0^x \frac{\arctan t}{t} \, dt$

35. $f(x) = \displaystyle\int_0^x \frac{1 - \exp(-t^2)}{t^2} \, dt$

36. $\tanh^{-1} x = \displaystyle\int_0^x \frac{1}{1 - t^2} \, dt$

37. Deduce from the arctangent series (Example 9) that

$$\pi = \frac{6}{\sqrt{3}} \sum_{n=0}^{\infty} \frac{(-1)^n}{2n + 1} \left(\frac{1}{3} \right)^n .$$

Then use this alternating series to show that $\pi = 3.14$ accurate to two decimal places.

38. Substitute the Maclaurin series for $\sin x$, and then assume the validity of termwise integration of the resulting series, to derive the formula

$$\int_0^\infty e^{-t} \sin xt \, dt = \frac{x}{1 + x^2} \qquad (|x| < 1).$$

Use the fact that

$$\int_0^\infty t^n e^{-t} \, dt = \Gamma(n + 1) = n!$$

(Section 9.8).

39. (a) Deduce from the Maclaurin series for e^t that

$$\frac{1}{x^x} = \sum_{n=0}^{\infty} \frac{(-1)^n}{n!} (x \ln x)^n .$$

(b) Assuming the validity of termwise integration of the series in part (a), use the integral formula of Problem 31 in Section 9.8 to conclude that

$$\int_0^1 \frac{1}{x^x} \, dx = \sum_{n=1}^{\infty} \frac{1}{n^n} .$$

40. Suppose that $f(x)$ is represented by the power series $\sum_{n=0}^{\infty} a_n x^n$ for all x in some open interval centered at $x = 0$. Show by repeated termwise differentiation of the series, substituting $x = 0$ each time, that $a_n = f^{(n)}(0)/n!$ for all n. Thus the only power series in x that represents a function at and near $x = 0$ is its Maclaurin series.

41. (a) Consider the binomial series

$$f(x) = \sum_{n=0}^{\infty} \frac{\alpha(\alpha - 1)(\alpha - 2) \cdots (\alpha - n + 1)}{n!} x^n,$$

which converges (to *something*) if $|x| < 1$. Compute the derivative $f'(x)$ by termwise differentiation, and show that it satisfies the differential equation $(1 + x)f'(x) = \alpha f(x)$.
(b) Solve the differential equation in part (a) to obtain $f(x) = C(1 + x)^\alpha$ for some constant C. Finally, show that $C = 1$. Thus the binomial series converges to $(1 + x)^\alpha$ if $|x| < 1$.

42. (a) Show by direct integration that

$$\int_0^1 \frac{\arcsin x}{\sqrt{1 - x^2}} dx = \frac{\pi^2}{8}.$$

(b) Use the result of Problem 50 in Section 9.4 to show that

$$\int_0^1 \frac{x^{2n+1}}{\sqrt{1 - x^2}} dx = \frac{2 \cdot 4 \cdot 6 \cdots (2n)}{1 \cdot 3 \cdot 5 \cdots (2n + 1)}.$$

(c) Substitute the series of Example 10 for $\arcsin x$ into the integral of part (a); then use the integral of part (b) to integrate termwise. Conclude that

$$\int_0^1 \frac{\arcsin x}{\sqrt{1 - x^2}} dx = 1 + \frac{1}{3^2} + \frac{1}{5^2} + \cdots.$$

(d) Note that

$$\sum_{n=1}^{\infty} \frac{1}{n^2} = \sum_{n=1}^{\infty} \frac{1}{(2n - 1)^2} + \sum_{n=1}^{\infty} \frac{1}{(2n)^2}.$$

Use this information and parts (a) and (c) to show that

$$\sum_{n=1}^{\infty} \frac{1}{n^2} = \frac{\pi^2}{6}.$$

43. Prove that if the power series $\sum a_n x^n$ converges for some $x = x_0 \neq 0$, then it converges absolutely for all x such that $|x| < |x_0|$. [*Suggestion:* Conclude from the fact that $\lim_{n \to \infty} a_n x_0^n = 0$ that $|a_n x^n| \leq |x/x_0|^n$ for n sufficiently large. Thus the series $\sum |a_n x^n|$ is eventually dominated by the geometric series $\sum |x/x_0|^n$, which converges if $|x| < |x_0|$.]

44. Suppose that the power series $\sum a_n x^n$ converges for some but not all nonzero values of x. Let S be the set of real numbers for which the series converges absolutely. (a) Conclude from Problem 43 that the set S is bounded above. (b) Let λ be the least upper bound of the set S (see Problem 46 of Section 11.2). Then show that $\sum a_n x^n$ converges absolutely if $|x| < \lambda$ and diverges if $|x| > \lambda$. Explain why this proves Theorem 1 without the additional hypothesis that $\lim_{n \to \infty} |a_{n+1}/a_n|$ exists.

11.8 Project

The dashed curve in Fig. 11.8.4 is the graph of the function

$$\frac{\sin x}{x} = 1 - \frac{x^2}{3!} + \frac{x^4}{5!} - \frac{x^6}{7!} + \cdots, \tag{22}$$

and the other graphs are those of eight of the curve's Taylor polynomial approximations. The dashed curve in Fig. 11.8.5 is the graph of the antiderivative

$$f(x) = \int_0^x \frac{\sin t}{t} dt = x - \frac{x^3}{3! \, 3} + \frac{x^5}{5! \, 5} - \frac{x^7}{7! \, 7} + \cdots \tag{23}$$

of the function in Eq. (22); the other graphs are those of eight of the curve's Taylor polynomial approximations.

First verify the Taylor series in Eq. (22) and (23). Then use them to produce your own versions of Figs. 11.8.4 and 11.8.5. You may want to plot a smaller or large number of Taylor polynomials, depending on the type of calculator or computer graphing program you have available. You may find useful the series summation techniques described in the projects of Sections 11.3 and 11.4.

Then do the same with one of the other function-antiderivative pairs of Problems 31 through 36 of this section.

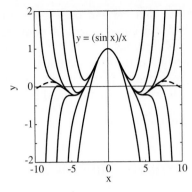

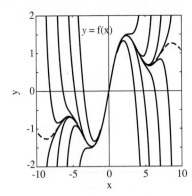

Fig. 11.8.4 Approximating the function $(\sin x)/x$ with its Taylor polynomials of degrees $n = 2, 4, 6, 8, \ldots, 16$

Fig. 11.8.5 Approximating the function $f(x) = \int_0^x (\sin t)/t \, dt$ with its Taylor polynomials of degrees $n = 3, 5, 7, \ldots, 17$

11.9
Power Series Computations

Power series often are used to approximate numerical values of functions and integrals. *Alternating* power series are especially useful. Recall the alternating series remainder (or "error") estimate of Section 11.7 (Theorem 2). It may be applied to a convergent alternating series $\Sigma(-1)^{n+1}a_n$ whose terms are decreasing in magnitude:

$$a_1 > a_2 > a_3 > \cdots > a_n > a_{n+1} > \cdots.$$

The remainder estimate says that when such a convergent alternating series is *truncated*—the terms that follow a_n are simply chopped off and discarded—the error E made has the same sign as the first term a_k omitted and $|E| < |a_k|$: The error is less in magnitude than the first omitted term in the truncation.

EXAMPLE 1 Use the binomial series

$$\sqrt{1 + x} = 1 + \tfrac{1}{2}x - \tfrac{1}{8}x^2 + \tfrac{1}{16}x^3 - \tfrac{5}{128}x^4 + \cdots$$

to approximate $\sqrt{105}$ and estimate the accuracy.

Solution   The binomial series shown here is, after the first term, an alternating series. Hence

$$\sqrt{105} = \sqrt{100 + 5} = 10\sqrt{1 + 0.05}$$
$$= 10[1 + \tfrac{1}{2}(0.05) - \tfrac{1}{8}(0.05)^2 + \tfrac{1}{16}(0.05)^3 + E]$$
$$\approx 10 \cdot (1.02469\,531 + E) = 10.2469\,531 + 10E,$$

where the error E is negative and

$$|10E| < \tfrac{50}{128}(0.05)^4 < 0.000003.$$

Consequently,

$$10.246950 < \sqrt{105} < 10.246953,$$

so $\sqrt{105} = 10.24695$ to five decimal places.

Suppose that we had been asked in advance to approximate $\sqrt{105}$ accurate to five decimal places. A convenient way to do this is to continue writing terms of the series until it is clear that they have become too small in magnitude to affect the fifth decimal place. A good rule of thumb is to use two more decimal places in the computations than are required in the final answer. Thus we use seven decimal places in this case and get

$$\sqrt{105} = 10(1 + 0.05)^{1/2}$$

$$\approx 10(1 + 0.025 - 0.0003125 + 0.0000078 - 0.0000002 + \cdots)$$

$$\approx 10.246951 \approx 10.24695.$$

EXAMPLE 2   Approximate

$$\int_0^1 \frac{1 - \cos x}{x^2} dx$$

accurate to five places.

Solution   We replace $\cos x$ by its Maclaurin series and get

$$\int_0^1 \frac{1 - \cos x}{x^2} dx = \int_0^1 \frac{1}{x^2} \left(\frac{x^2}{2!} - \frac{x^4}{4!} + \frac{x^6}{6!} - \cdots \right) dx$$

$$= \int_0^1 \left(\frac{1}{2!} - \frac{x^2}{4!} + \frac{x^4}{6!} - \frac{x^6}{8!} + \cdots \right) dx$$

$$= \frac{1}{2!} - \frac{1}{4!\,3} + \frac{1}{6!\,5} - \frac{1}{8!\,7} + \cdots.$$

Because this last series is a convergent alternating series, it follows that

$$\int_0^1 \frac{1 - \cos x}{x^2} dx = \frac{1}{2!} - \frac{1}{4!\,3} + \frac{1}{6!\,5} + E \approx 0.4863889 + E,$$

where E is negative and

$$|E| < \frac{1}{8!\,7} < 0.0000036.$$

Therefore,

$$\int_0^1 \frac{1 - \cos x}{x^2} dx \approx 0.48639$$

rounded to five decimal places.

EXAMPLE 3   The binomial series with $\alpha = \frac{1}{3}$ gives

$$(1 + x^2)^{1/3} = 1 + \tfrac{1}{3}x^2 - \tfrac{1}{9}x^4 + \tfrac{5}{81}x^6 - \tfrac{10}{243}x^8 + \cdots,$$

which alternates after its first term. Use the first five terms of this series to estimate the value of

$$\int_0^{1/2} \sqrt[3]{1 + x^2}\, dx.$$

Solution   Termwise integration of the binomial series gives

$$\int_0^{1/2} \sqrt[3]{1 + x^2} \, dx = \int_0^{1/2} (1 + \tfrac{1}{3}x^2 - \tfrac{1}{9}x^4 + \tfrac{5}{81}x^6 - \tfrac{10}{243}x^8 + \cdots) \, dx$$

$$= \left[x + \tfrac{1}{9}x^3 - \tfrac{1}{45}x^5 + \tfrac{5}{567}x^7 - \tfrac{10}{2187}x^9 + \cdots \right]_0^{1/2}$$

$$= \tfrac{1}{2} + \tfrac{1}{9}\left(\tfrac{1}{2}\right)^3 - \tfrac{1}{45}\left(\tfrac{1}{2}\right)^5 + \tfrac{5}{567}\left(\tfrac{1}{2}\right)^7 - \tfrac{10}{2187}\left(\tfrac{1}{2}\right)^9 + \cdots.$$

Because this last series is a convergent alternating series, it follows that

$$\int_0^{1/2} \sqrt[3]{1 + x^2} \, dx = \tfrac{1}{2} + \tfrac{1}{9}\left(\tfrac{1}{2}\right)^3 - \tfrac{1}{45}\left(\tfrac{1}{2}\right)^5 + \tfrac{5}{567}\left(\tfrac{1}{2}\right)^7 + E \approx 0.513263 + E,$$

where E is negative and

$$|E| < \tfrac{10}{2187}\left(\tfrac{1}{2}\right)^9 < 0.000009.$$

Therefore,

$$0.513254 < \int_0^{1/2} \sqrt[3]{1 + x^2} \, dx < 0.513264,$$

so

$$\int_0^{1/2} \sqrt[3]{1 + x^2} \, dx = 0.51326 \pm 0.00001.$$

THE ALGEBRA OF POWER SERIES

Theorem 1, which we state without proof, implies that power series may be added and multiplied much like polynomials. The guiding principle is that of collecting coefficients of like powers of x.

Theorem 1   *Adding and Multiplying Power Series*
Let $\sum a_n x^n$ and $\sum b_n x^n$ be power series with nonzero radii of convergence. Then

$$\sum_{n=0}^{\infty} a_n x^n + \sum_{n=0}^{\infty} b_n x^n = \sum_{n=0}^{\infty} (a_n + b_n)x^n \qquad (1)$$

and

$$\left(\sum_{n=0}^{\infty} a_n x^n \right)\left(\sum_{n=0}^{\infty} b_n x^n \right) = \sum_{n=0}^{\infty} c_n x^n$$

$$= a_0 b_0 + (a_0 b_1 + a_1 b_0)x + (a_0 b_2 + a_1 b_1 + a_2 b_0)x^2 + \cdots, \qquad (2)$$

where

$$c_n = a_0 b_n + a_1 b_{n-1} + \cdots a_{n-1} b_1 + a_n b_0. \qquad (3)$$

The series in Eqs. (1) and (2) converge for any x that lies interior to the intervals of convergence of both $\sum a_n x^n$ and $\sum b_n x^n$.

Thus if $\Sigma\, a_n x^n$ and $\Sigma\, b_n x^n$ are power series representations of the functions $f(x)$ and $g(x)$, respectively, then the product power series $\Sigma\, c_n x^n$ found by "ordinary multiplication" and collection of terms is a power series representation of the product function $f(x)g(x)$. This fact can also be used to divide one power series by another, *provided* that the quotient is known to have a power series representation.

EXAMPLE 4   Assume that the tangent function has a power series representation $\tan x = \Sigma\, a_n x^n$. Use the Maclaurin series for $\sin x$ and $\cos x$ to find a_0, a_1, a_2, and a_3.

Solution   We multiply series to obtain

$$\sin x = \tan x \cos x$$

$$= (a_0 + a_1 x + a_2 x^2 + a_3 x^3 + \cdots)\left(1 - \frac{x^2}{2} + \frac{x^4}{24} - \cdots\right)$$

$$= a_0 + a_1 x + \left(a_2 - \frac{1}{2}a_0\right)x^2 + \left(a_3 - \frac{1}{2}a_1\right)x^3 + \cdots.$$

But because

$$\sin x = x - \tfrac{1}{6}x^3 + \tfrac{1}{120}x^5 - \cdots,$$

comparison of coefficients gives the equations

$$
\begin{array}{rcl}
a_0 & = & 0, \\
a_1 & = & 1, \\
-\tfrac{1}{2}a_0 + a_2 & = & 0, \\
-\tfrac{1}{2}a_1 + a_3 & = & -\tfrac{1}{6}.
\end{array}
$$

Then we find that $a_0 = 0$, $a_1 = 1$, $a_2 = 0$, and $a_3 = \tfrac{1}{3}$. So

$$\tan x = x + \tfrac{1}{3}x^3 + \cdots.$$

Things are not always as they first appear. The continuation of the tangent series is

$$\tan x = x + \tfrac{1}{3}x^3 + \tfrac{2}{15}x^5 + \tfrac{17}{315}x^7 + \cdots.$$

For the general form of the nth coefficient, see K. Knopp's *Theory and Application of Infinite Series* (New York: Hafner Press, 1971), p. 204. You may check that the first few terms agree with the result of ordinary division:

$$
1 - \tfrac{1}{2}x^2 + \tfrac{1}{24}x^4 - \cdots \overline{)x - \tfrac{1}{3}x^3 + \tfrac{1}{120}x^5 + \cdots.}
$$

$$x + \tfrac{1}{3}x^3 + \tfrac{2}{15}x^5 + \cdots$$

POWER SERIES AND INDETERMINATE FORMS

According to Theorem 3 of Section 11.8, a power series is differentiable and therefore continuous within its interval of convergence. It follows that

$$\lim_{x \to c} \sum_{n=0}^{\infty} a_n(x - c)^n = a_0. \tag{4}$$

Examples 5 and 6 illustrate the use of this simple observation to find the limit of the indeterminate form $f(x)/g(x)$. The technique is first to substitute power series representations for $f(x)$ and $g(x)$.

EXAMPLE 5 Find $\lim\limits_{x \to 0} \dfrac{\sin x - \arctan x}{x^2 \ln(1 + x)}$.

Solution The power series of Eqs. (4), (19), and (20) in Section 11.8 give

$$\sin x - \arctan x = (x - \tfrac{1}{6}x^3 + \tfrac{1}{120}x^5 - \cdots) - (x - \tfrac{1}{3}x^3 + \tfrac{1}{5}x^5 - \cdots)$$

$$= \tfrac{1}{6}x^3 - \tfrac{23}{120}x^5 + \cdots$$

and

$$x^2 \ln(1 + x) = x^2(x - \tfrac{1}{2}x^2 + \tfrac{1}{3}x^3 - \cdots) = x^3 - \tfrac{1}{2}x^4 + \tfrac{1}{3}x^5 - \cdots.$$

Hence

$$\lim_{x \to 0} \frac{\sin x - \arctan x}{x^2 \ln(1 + x)} = \lim_{x \to 0} \frac{\tfrac{1}{6}x^3 - \tfrac{23}{120}x^5 + \cdots}{x^3 - \tfrac{1}{2}x^4 + \cdots}$$

$$= \lim_{x \to 0} \frac{\tfrac{1}{6} - \tfrac{23}{120}x^2 + \cdots}{1 - \tfrac{1}{2}x + \cdots} = \frac{1}{6}.$$

EXAMPLE 6 Find $\lim\limits_{x \to 1} \dfrac{\ln x}{x - 1}$.

Solution We first replace x by $x - 1$ in the power series for $\ln(1 + x)$ used in Example 5. This gives us

$$\ln x = (x - 1) - \tfrac{1}{2}(x - 1)^2 + \tfrac{1}{3}(x - 1)^3 - \cdots.$$

Hence

$$\lim_{x \to 1} \frac{\ln x}{x - 1} = \lim_{x \to 1} \frac{(x - 1) - \tfrac{1}{2}(x - 1)^2 + \tfrac{1}{3}(x - 1)^3 - \cdots}{x - 1}$$

$$= \lim_{x \to 1} \left[1 - \frac{1}{2}(x - 1) + \frac{1}{3}(x - 1)^2 - \cdots \right] = 1.$$

The method of Examples 5 and 6 provides a useful alternative to l'Hôpital's rule, especially when repeated differentiation of numerator and denominator is inconvenient or too time-consuming.

11.9 Problems

In Problems 1 through 10, use an infinite series to approximate the indicated number accurate to three decimal places.

1. $\sqrt[3]{65}$ **2.** $\sqrt[4]{630}$ **3.** $\sin(0.5)$ **4.** $e^{-0.2}$

5. $\tan^{-1}(0.5)$ **6.** $\ln(1.1)$ **7.** $\sin\left(\dfrac{\pi}{10}\right)$

8. $\cos\left(\dfrac{\pi}{20}\right)$ **9.** $\sin 10°$ **10.** $\cos 5°$

In Problems 11 through 20, use an infinite series to approximate the value of the given integral accurate to three decimal places.

11. $\displaystyle\int_0^1 \frac{\sin x}{x}\, dx$ **12.** $\displaystyle\int_0^1 \frac{\sin x}{\sqrt{x}}\, dx$

13. $\displaystyle\int_0^{0.5} \frac{\arctan x}{x}\, dx$ **14.** $\displaystyle\int_0^1 \sin x^2\, dx$

15. $\displaystyle\int_0^{0.1} \frac{\ln(1 + x)}{x}\, dx$ **16.** $\displaystyle\int_0^{0.5} \frac{1}{\sqrt{1 + x^4}}\, dx$

17. $\displaystyle\int_0^{0.5} \frac{1 - e^{-x}}{x}\, dx$ **18.** $\displaystyle\int_0^{0.5} \sqrt{1 + x^3}\, dx$

19. $\displaystyle\int_0^1 e^{-x^2}\, dx$ **20.** $\displaystyle\int_0^{0.5} \frac{1}{1 + x^5}\, dx$

In Problems 21 through 26, use power series rather than l'Hôpital's rule to evaluate the given limit.

21. $\lim\limits_{x \to 0} \dfrac{1 + x - e^x}{x^2}$ **22.** $\lim\limits_{x \to 0} \dfrac{x - \sin x}{x^3 \cos x}$

23. $\lim\limits_{x \to 0} \dfrac{1 - \cos x}{x(e^x - 1)}$

24. $\lim\limits_{x \to 0} \dfrac{e^x - e^{-x} - 2x}{x - \arctan x}$

25. $\lim\limits_{x \to 0} \left(\dfrac{1}{x} - \dfrac{1}{\sin x} \right)$

26. $\lim\limits_{x \to 1} \dfrac{\ln(x^2)}{x - 1}$

27. Derive the geometric series by long division of $1 - x$ into 1.

28. Derive the series for $\tan x$ listed in Example 4 by long division of the Taylor series of $\cos x$ into the Taylor series of $\sin x$.

29. Derive the geometric series representation of $1/(1 - x)$ by finding $a_0, a_1, a_2, \ldots$ such that

$$(1 - x)(a_0 + a_1 x + a_2 x^2 + a_3 x^3 + \cdots) = 1.$$

30. Derive the first five coefficients in the binomial series for $\sqrt{1 + x}$ by finding $a_0, a_1, a_2, a_3,$ and a_4 such that

$$(a_0 + a_1 x + a_2 x^2 + a_3 x^3 + a_4 x^4 + \cdots)^2 = 1 + x.$$

31. Use the method of Example 4 to find the coefficients $a_0, a_1, a_2, a_3,$ and a_4 in the series

$$\sec x = \frac{1}{\cos x} = \sum_{n=0}^{\infty} a_n x^n.$$

32. Multiply the geometric series for $1/(1 - x)$ and the series for $\ln(1 - x)$ to show that if $|x| < 1$, then

$$\frac{1}{1 - x} \ln(1 - x) = x + (1 + \tfrac{1}{2})x^2 + (1 + \tfrac{1}{2} + \tfrac{1}{3})x^3$$
$$+ (1 + \tfrac{1}{2} + \tfrac{1}{3} + \tfrac{1}{4})x^4 + \cdots.$$

33. Take as known the logarithmic series

$$\ln(1 + x) = x - \tfrac{1}{2}x^2 + \tfrac{1}{3}x^3 - \tfrac{1}{4}x^4 + \cdots.$$

Find the first four coefficients in the series for e^x by finding $a_0, a_1, a_2,$ and a_3 such that

$$1 + x = e^{\ln(1 + x)} = \sum_{n=0}^{\infty} a_n (x - \tfrac{1}{2}x^2 + \tfrac{1}{3}x^3 - \tfrac{1}{4}x^4 + \cdots)^n.$$

This is exactly how the power series for e^x was first discovered (by Newton)!

34. Use the method of Example 4 to show that

$$\frac{x}{\sin x} = 1 + \frac{1}{6}x^2 + \frac{7}{360}x^4 + \cdots.$$

35. Show that long division of power series gives

$$\frac{2 + x}{1 + x + x^2} = 2 - x - x^2 + 2x^3 - x^4 - x^5 + 2x^6$$
$$- x^7 - x^8 + 2x^9 - x^{10} - x^{11} + \cdots.$$

Show also that the radius of convergence of this series is $R = 1$.

11.9 Project

This project explores the use of known power series such as

$$\sin x = x - \frac{x^3}{6} + \frac{x^5}{120} - \frac{x^7}{5040} + \cdots, \tag{5}$$

$$\tan x = x + \frac{x^3}{3} + \frac{2x^5}{15} + \frac{17x^7}{315} + \cdots, \tag{6}$$

$$\sin^{-1} x = x + \frac{x^3}{6} + \frac{3x^5}{40} + \frac{5x^7}{112} + \cdots, \tag{7}$$

$$\tan^{-1} x = x - \frac{x^3}{3} + \frac{x^5}{5} - \frac{x^7}{7} + \cdots \tag{8}$$

to evaluate certain indeterminate forms for which the use of l'Hôpital's rule would be inconvenient or impractical.

1. Evaluate

$$\lim_{x \to 0} \frac{\sin x - \tan x}{\sin^{-1} x - \tan^{-1} x}. \tag{9}$$

When you substitute Eqs. (5) and (6) into the numerator and Eqs. (7) and (8) into the denominator, you will find that both the numerator series and the denominator series have leading terms that are multiples of x^3. Hence division of each term by x^3 leads quickly to the desired value of the indeterminate form. Can you explain why this implies that l'Hôpital's rule would have to be applied three times in succession? Would you like to calculate (by hand) the third derivative of the denominator in (9)?

2. Evaluate

$$\lim_{x \to 0} \frac{\sin(\tan x) - \tan(\sin x)}{\sin^{-1}(\tan^{-1} x) - \tan^{-1}(\sin^{-1} x)}. \tag{10}$$

This is a much more substantial problem, and the use of a computer algebra system is highly desirable. The table in Fig. 11.9.1 lists commands in common systems that generate the Taylor series

$$\sin(\tan x) = x + \frac{x^3}{6} - \frac{x^5}{40} - \frac{55x^7}{1008} + \cdots. \tag{11}$$

Program	Command
Derive	`taylor( sin(tan(x)), x,0,7 )`
Maple	`taylor( sin(tan(x)), x=0,8 )`
Mathematica	`Series[ Sin[Tan[x]], {x,0,7} ]`

Fig. 11.9.1 Computer algebra system commands that generate the series in Eq. (11)

If one of these systems is available to you, generate similarly the Taylor series

$$\tan(\sin x) = x + \frac{x^3}{6} - \frac{x^5}{40} - \frac{107x^7}{5040} + \cdots, \tag{12}$$

$$\sin^{-1}(\tan^{-1} x) = x - \frac{x^3}{6} + \frac{13x^5}{120} - \frac{341x^7}{5040} + \cdots, \quad \text{and} \tag{13}$$

$$\tan^{-1}(\sin^{-1} x) = x - \frac{x^3}{6} + \frac{13x^5}{120} - \frac{173x^7}{5040} + \cdots. \tag{14}$$

Then show that when you substitute Eqs. (11) and (12) into the numerator of Eq. (10) and Eqs. (13) and (14) into the denominator, both resulting series have leading terms that are multiples of x^7. Hence division of each series by x^7 leads quickly to the desired evaluation. It follows that seven successive applications of l'Hôpital's rule would be required to evaluate the indeterminate form in Eq. (10). (Why?) Can you even conceive of calculating the seventh derivative of either the numerator or denominator in (10)? The seventh derivative of $\sin(\tan x)$ is a sum of sixteen terms, a typical one of which is $3696 \cos(\tan x) \sec^2 x \tan^2 x$.

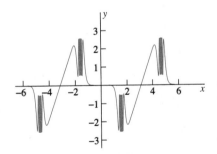

Fig. 11.9.2 The graph $y = \sin(\tan x) - \tan(\sin x)$

The fact that the power series for both

$$f(x) = \sin(\tan x) - \tan(\sin x) \tag{15}$$

and

$$g(x) = \sin^{-1}(\tan^{-1} x) - \tan^{-1}(\sin^{-1} x) \tag{16}$$

begin with the term involving x^7 means, geometrically, that the graphs of both are quite "flat" near the origin (Figs. 11.9.2 and 11.9.3).

The graph of $y = f(x)$ in Fig. 11.9.2 is especially exotic. Can you explain the conspicuous oscillations that appear in certain sections of the graph?

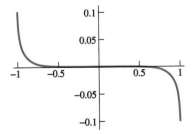

Fig. 11.9.3 The graph $y = \sin^{-1}(\tan^{-1} x) - \tan^{-1}(\sin^{-1} x)$

Chapter 11 Review: DEFINITIONS, CONCEPTS, RESULTS

Use the following list as a guide to concepts that you may need to review.

1. Definition of the limit of a sequence
2. The limit laws for sequences
3. The bounded monotonic sequence property
4. Definition of the sum of an infinite series
5. Formula for the sum of a geometric series
6. The nth-term test for divergence
7. Divergence of the harmonic series
8. The nth-degree Taylor polynomial of the function f at the point $x = a$
9. Taylor's formula with remainder
10. The Taylor series of the elementary transcendental functions
11. The integral test
12. Convergence of p-series
13. The comparison and limit comparison tests
14. The alternating series test
15. Absolute convergence: definition *and* the fact that it implies convergence
16. The ratio test
17. The root test
18. Power series; the radius and interval of convergence
19. The binomial series
20. Termwise differentiation and integration of power series
21. The use of power series to approximate values of functions and integrals
22. The product of two power series
23. The use of power series to evaluate indeterminate forms

Chapter 11 Miscellaneous Problems

In Problems 1 through 15, determine whether or not the sequence $\{a_n\}$ converges, and find its limit if it does converge.

1. $a_n = \dfrac{n^2 + 1}{n^2 + 4}$

2. $a_n = \dfrac{8n - 7}{7n - 8}$

3. $a_n = 10 - (0.99)^n$

4. $a_n = n \sin \pi n$

5. $a_n = \dfrac{1 + (-1)^n \sqrt{n}}{n + 1}$

6. $a_n = \sqrt{\dfrac{1 + (-0.5)^n}{n + 1}}$

7. $a_n = \dfrac{\sin 2n}{n}$

8. $a_n = 2^{-(\ln n)/n}$

9. $a_n = (-1)^{\sin(n\pi/2)}$

10. $a_n = \dfrac{(\ln n)^3}{n^2}$

11. $a_n = \dfrac{1}{n} \sin \dfrac{1}{n}$

12. $a_n = \dfrac{n - e^n}{n + e^n}$

13. $a_n = \dfrac{\sinh n}{n}$

14. $a_n = \left(1 + \dfrac{2}{n}\right)^{2n}$

15. $a_n = (2n^2 + 1)^{1/n}$

Determine whether each infinite series in Problems 16 through 30 converges or diverges.

16. $\displaystyle\sum_{n=1}^{\infty} \dfrac{(n^2)!}{n^n}$

17. $\displaystyle\sum_{n=1}^{\infty} \dfrac{(-1)^{n+1} \ln n}{n}$

18. $\displaystyle\sum_{n=0}^{\infty} \dfrac{3^n}{2^n + 4^n}$

19. $\displaystyle\sum_{n=0}^{\infty} \dfrac{n!}{e^{n^2}}$

20. $\displaystyle\sum_{n=1}^{\infty} \dfrac{1}{n^{3/2}} \sin \dfrac{1}{n}$

21. $\displaystyle\sum_{n=0}^{\infty} \dfrac{(-2)^n}{3^n + 1}$

22. $\displaystyle\sum_{n=1}^{\infty} 2^{-(2/n^2)}$

23. $\displaystyle\sum_{n=2}^{\infty} \dfrac{(-1)^n n}{(\ln n)^3}$

24. $\displaystyle\sum_{n=1}^{\infty} \dfrac{(-1)^n}{10^{1/n}}$

25. $\displaystyle\sum_{n=1}^{\infty} \dfrac{\sqrt{n} + \sqrt[3]{n}}{n^2 + n^3}$

26. $\displaystyle\sum_{n=1}^{\infty} \dfrac{(-1)^{n+1}}{n^{[1+(1/n)]}}$

27. $\displaystyle\sum_{n=1}^{\infty} \dfrac{(-1)^{n+1} \arctan n}{\sqrt{n}}$

28. $\displaystyle\sum_{n=1}^{\infty} n \sin \dfrac{1}{n}$

29. $\displaystyle\sum_{n=3}^{\infty} \dfrac{1}{n(\ln n)(\ln \ln n)}$

30. $\displaystyle\sum_{n=3}^{\infty} \dfrac{1}{n(\ln n)(\ln \ln n)^2}$

Find the interval of convergence of the power series in Problems 31 through 40.

31. $\displaystyle\sum_{n=0}^{\infty} \dfrac{2^n x^n}{n!}$

32. $\displaystyle\sum_{n=0}^{\infty} \dfrac{(3x)^n}{2^{n+1}}$

33. $\displaystyle\sum_{n=1}^{\infty} \dfrac{(x - 1)^n}{n \cdot 3^n}$

34. $\displaystyle\sum_{n=0}^{\infty} \dfrac{(2x - 3)^n}{4^n}$

35. $\displaystyle\sum_{n=1}^{\infty} \frac{(-1)^n x^n}{4n^2 - 1}$ **36.** $\displaystyle\sum_{n=0}^{\infty} \frac{(2x - 1)^n}{n^2 + 1}$

37. $\displaystyle\sum_{n=0}^{\infty} \frac{n!\, x^{2n}}{10^n}$ **38.** $\displaystyle\sum_{n=2}^{\infty} \frac{x^n}{\ln n}$

39. $\displaystyle\sum_{n=0}^{\infty} \frac{1 + (-1)^n}{2(n!)} x^n$ **40.** $\displaystyle\sum_{n=1}^{\infty} \left(1 + \frac{1}{n}\right)^n (x - 1)^n$

Find the set of values of x for which each series in Problems 41 through 43 converges.

41. $\displaystyle\sum_{n=1}^{\infty} (x - n)^n$ **42.** $\displaystyle\sum_{n=1}^{\infty} (\ln x)^n$ **43.** $\displaystyle\sum_{n=0}^{\infty} \frac{e^{nx}}{n!}$

44. Find the rational number that has repeated decimal expansion $2.7\,1828\,1828\,1828\ldots$.

45. Give an example of two convergent numerical series $\Sigma\, a_n$ and $\Sigma\, b_n$ such that the series $\Sigma\, a_n b_n$ diverges.

46. Prove that if $\Sigma\, a_n$ is a convergent positive-term series, then $\Sigma\, a_n^2$ converges.

47. Let the sequence $\{a_n\}$ be defined recursively by

$$a_1 = 1, \qquad a_{n+1} = 1 + \frac{1}{1 + a_n} \qquad \text{if } n \geq 1.$$

The limit of the sequence $\{a_n\}$ is the value of the *continued fraction*

$$1 + \cfrac{1}{2 + \cfrac{1}{2 + \cfrac{1}{2 + \cfrac{1}{2 + \cdots}}}}.$$

Assuming that $A = \lim_{n \to \infty} a_n$ exists, prove that $A = \sqrt{2}$.

48. Let $\{F_n\}_1^{\infty}$ be the Fibonacci sequence of Example 2 in Section 11.2. (a) Prove that $0 < F_n \leq 2^n$ for all $n \geq 1$, and hence conclude that the power series $F(x) = \sum_{n=1}^{\infty} F_n x^n$ converges if $|x| < \frac{1}{2}$. (b) Show that $(1 - x - x^2)F(x) = x$, so

$$F(x) = \frac{x}{1 - x - x^2}.$$

49. We say that the *infinite product* indicated by

$$\prod_{n=1}^{\infty} (1 + a_n) = (1 + a_1)(1 + a_2)(1 + a_3) \cdots$$

converges provided that the infinite series

$$S = \sum_{n=1}^{\infty} \ln(1 + a_n)$$

converges, in which case the value of the infinite product is e^S. Use the integral test to prove that

$$\prod_{n=1}^{\infty} \left(1 + \frac{1}{n}\right)$$

diverges.

50. Prove that the infinite product (see Problem 49)

$$\prod_{n=1}^{\infty} \left(1 + \frac{1}{n^2}\right)$$

converges, and use the integral test remainder estimate to approximate its value. The actual value of this infinite product is known to be $(\sinh \pi)/\pi \approx 3.67607\,791$.

In Problems 51 through 55, use infinite series to approximate the indicated number accurate to three decimal places.

51. $\sqrt[5]{1.5}$ **52.** $\ln(1.2)$ **53.** $\displaystyle\int_0^{0.5} e^{-x^2}\, dx$

54. $\displaystyle\int_0^{0.5} \sqrt[3]{1 + x^4}\, dx$ **55.** $\displaystyle\int_0^1 \frac{1 - e^{-x}}{x}\, dx$

56. Substitute the Maclaurin series for $\sin x$ into that for e^x to obtain

$$e^{\sin x} = 1 + x + \tfrac{1}{2}x^2 - \tfrac{1}{8}x^4 + \cdots.$$

57. Substitute the Maclaurin series for the cosine and then integrate termwise to derive the formula

$$\int_0^{\infty} e^{-t^2} \cos 2xt\, dt = \frac{\sqrt{\pi}}{2} e^{-x^2}.$$

Use the reduction formula

$$\int_0^{\infty} t^{2n} e^{-t^2}\, dt = \frac{2n - 1}{2} \int_0^{\infty} t^{2n-2} e^{-t^2}\, dt$$

derived in Problem 42 of Section 9.4. The validity of this improper termwise integration is subject to verification.

58. Prove that

$$\tanh^{-1} x = \int_0^x \frac{1}{1 - t^2}\, dt = \sum_{n=0}^{\infty} \frac{x^{2n+1}}{2n + 1}$$

if $|x| < 1$.

59. Prove that

$$\sinh^{-1} x = \int_0^x \frac{1}{\sqrt{1 + t^2}}\, dt$$

$$= \sum_{n=0}^{\infty} (-1)^n \frac{1 \cdot 3 \cdot 5 \cdots (2n - 1)}{2 \cdot 4 \cdot 6 \cdots (2n)} \cdot \frac{x^{2n+1}}{2n + 1}$$

if $|x| < 1$.

60. Suppose that $\tan y = \Sigma\, a_n y^n$. Determine a_0, a_1, a_2, and a_3 by substituting the inverse tangent series [Eq.(27) of Section 11.4] into the equation

$$x = \tan(\tan^{-1} x) = \sum_{n=0}^{\infty} a_n (\tan^{-1} x)^n.$$

61. According to *Stirling's series,* the value of $n!$ for large n is given to a close approximation by

$$n! = \sqrt{2\pi n} \left(\frac{n}{e}\right)^n e^{\mu(n)},$$

where

$$\mu(n) = \frac{1}{12n} - \frac{1}{360n^3} + \frac{1}{1260n^5}.$$

Substitute $\mu(n)$ into Maclaurin's series for e^x to show that

$$e^{\mu(n)} = 1 + \frac{1}{12n} + \frac{1}{288n^2} - \frac{139}{51,840n^3} + \cdots.$$

Can you show that the next term in the last series is $-571/(2,488,320n^4)$?

62. Define

$$T(n) = \int_0^{\pi/4} \tan^n x \, dx$$

for $n \geqq 0$. (a) Show by "reduction" of the integral that

$$T(n + 2) = \frac{1}{n + 1} - T(n)$$

for $n \geqq 0$. (b) Conclude that $T(n) \to 0$ as $n \to \infty$. (c) Show that $T_0 = \pi/4$ and that $T_1 = \frac{1}{2}\ln 2$. (d) Prove by induction on n that

$$T(2n) = (-1)^{n+1}\left(1 - \frac{1}{3} + \frac{1}{5} - \cdots \pm \frac{1}{2n - 1} - \frac{\pi}{4}\right).$$

(e) Conclude from parts (b) and (d) that

$$1 - \frac{1}{3} + \frac{1}{5} - \frac{1}{7} + \cdots = \frac{\pi}{4}.$$

(f) Prove by induction on n that

$$T(2n + 1) = \frac{1}{2}(-1)^n\left(1 - \frac{1}{2} + \frac{1}{3} - \cdots \pm \frac{1}{n} - \ln 2\right).$$

(g) Conclude from parts (b) and (f) that

$$1 - \frac{1}{2} + \frac{1}{3} - \frac{1}{4} + \cdots = \ln 2.$$

63. Prove as follows that the number e is irrational. First suppose to the contrary that $e = p/q$, where p and q are positive integers. Note that $q > 1$. Write

$$\frac{p}{q} = e = \frac{1}{0!} + \frac{1}{1!} + \frac{1}{2!} + \frac{1}{3!} + \cdots + \frac{1}{q!} + R_q,$$

where $0 < R_q < 3/(q + 1)!$. (Why?) Then show that multiplication of both sides of this equation by $q!$ would lead to the contradiction that one side of the result is an integer but the other side is not.

64. Evaluate the infinite product (see Problem 49)

$$\prod_{n=2}^{\infty} \frac{n^2}{n^2 - 1}$$

by finding an explicit formula for

$$\prod_{n=2}^{k} \frac{n^2}{n^2 - 1} \qquad (k \geqq 2)$$

and then taking the limit as $k \to \infty$.

Parametric Curves and Vectors in the Plane

Johannes Kepler (1571–1630)

❑ Johannes Kepler has been called the watershed between ancient and modern, between the middle ages and the age of reason. Philosophically and spiritually he stood with one foot firmly placed in both eras. As the "Mathematicus of the Province" of Styria (in Austria), his responsibilities included both teaching mathematics at the university in Gratz and the preparation of an annual calendar of astrological forecasts. He realized that astrology is founded on superstition but attempted (unsuccessfully) by careful study of astronomy to discover some rational basis for the subject.

❑ Early in his career Kepler convinced himself that the spacing of the orbits of the six planets then known—Mercury, Venus, Earth, Mars, Jupiter, and Saturn—was determined by the five

perfect solids of Plato: the tetrahedron, the cube, the octahedron (bounded by 8 triangles), the dodecahedron (bounded by 12 pentagons), and the icosahedron (bounded by 20 triangles). To model the solar system, he started

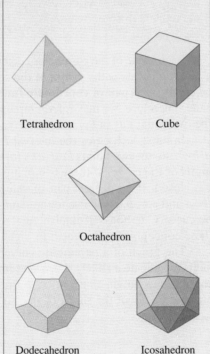

The five perfect solids of Plato

with a sphere centered at the Sun and containing the orbit of Saturn. In this sphere he inscribed a cube. A sphere inscribed in this cube contained the orbit of Jupiter. Inscribed in *this* sphere was a tetrahedron containing the sphere of Mars. Between the spheres of Mars and Earth was the dodecahedron, between Earth and Venus the icosahe-

dron, and between Venus and Mercury the octahedron. Indeed, Kepler argued, there *were* only six planets (false!) because there are only five perfect solids (true!) to separate them in this manner.

❑ For 20 years Kepler analyzed mathematically the observational data of the Danish astronomer Tycho Brahe in search of proof of Kepler's arcane theory. Instead, Kepler discovered the truth—that the orbit of each planet is an ellipse with the Sun at one focus. Moreover, the planets' periods of revolution and their speeds in their orbits obey "Kepler's laws" (Section 12.5). Kepler's laws led in turn to Newton's theory of gravitation, which, more than any single other scientific discovery, shaped the modern vision of a rational universe governed by science rather than by superstition.

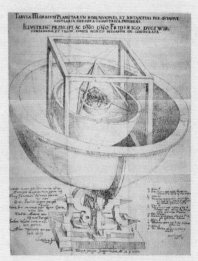

Kepler's model of the solar system

12.1
Parametric Curves

Until now we have encountered *curves* mainly as graphs of equations. An equation of the form $y = f(x)$ or of the form $x = g(y)$ determines a curve by giving one of the coordinate variables explicitly as a function of the other. An equation of the form $F(x, y) = 0$ may also determine a curve, but then each variable is given implicitly as a function of the other.

Another important type of curve is the trajectory of a point moving in the coordinate plane. The motion of the point can be described by giving its position $(x(t), y(t))$ at time t. Such a description involves expressing both the rectangular-coordinates variables x and y as functions of a third variable, or *parameter*, t, rather than as functions of one another. In this context a **parameter** is an independent variable (not a constant, as is sometimes meant in popular usage). This approach motivates the following definition.

Definition   *Parametric Curve*
A **parametric curve** C in the plane is a pair of functions

$$x = f(t), \qquad y = g(t), \tag{1}$$

that give x and y as continuous functions of the real number t (the parameter) in some interval I.

Each value of the parameter t determines a point $(f(t), g(t))$, and the set of all such points is the **graph** of the curve C. Often the distinction between the curve—the pair of **coordinate functions** f and g — and the graph is not made. Therefore, we may refer interchangeably to the curve and to its graph. The two equations in (1) are called the **parametric equations** of the curve.

In most such cases the interval I will be a closed interval of the form $[a, b]$; if so, the two points $P_a(f(a), g(a))$ and $P_b(f(b), g(b))$ are called the **endpoints** of the curve. If these two points coincide, then we say that the curve C is **closed**. If no distinct pair of values of t, except possibly for the values $t = a$ and $t = b$, gives rise to the same point on the graph of the curve, then the curve C is non-self-intersecting, and we say that the curve is **simple**. These concepts are illustrated in Fig. 12.1.1.

The graph of a parametric curve may be sketched by plotting enough points to indicate its likely shape. In some cases we can eliminate the parameter t and thus obtain an equation in x and y. This equation may give us more information about the shape of the curve.

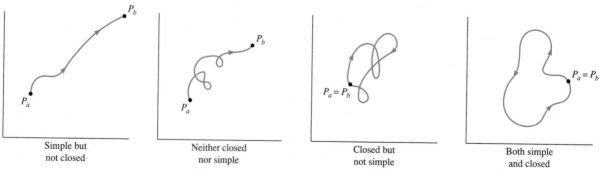

Simple but not closed   Neither closed nor simple   Closed but not simple   Both simple and closed

Fig. 12.1.1   Parametric curves may be simple or not, closed or not.

t	x	y
0	1	0
$\pi/4$	$1/\sqrt{2}$	$1/\sqrt{2}$
$\pi/2$	0	1
$3\pi/4$	$-1/\sqrt{2}$	$1/\sqrt{2}$
π	-1	0
$5\pi/4$	$-1/\sqrt{2}$	$-1/\sqrt{2}$
$3\pi/2$	0	-1
$7\pi/4$	$1/\sqrt{2}$	$-1/\sqrt{2}$
2π	1	0

(a)

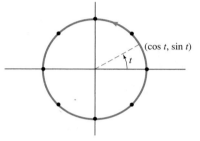

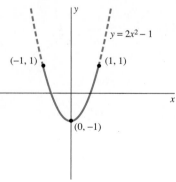

Fig. 12.1.2 (a) Table of values for Example 1; (b) the corresponding graph

EXAMPLE 1 Determine the graph of the curve

$$x = \cos t, \qquad y = \sin t, \qquad 0 \leqq t \leqq 2\pi. \qquad (2)$$

Solution Figure 12.1.2(a) shows a table of values of x and y that correspond to multiples of $\pi/4$ for the parameter t. These values give the eight points highlighted in Fig. 12.1.2(b), all of which lie on the unit circle. This suggests that the graph is, in fact, the unit circle. To verify this, we note that the fundamental identity of trigonometry gives

$$x^2 + y^2 = \cos^2 t + \sin^2 t = 1,$$

so every point of the graph lies on the circle with equation $x^2 + y^2 = 1$. Conversely, the point of the circle with angular (polar) coordinate t is the point $(\cos t, \sin t)$ of the graph. Thus the graph is precisely the unit circle.

What is lost in the process in Example 1 is the information about how the graph is produced as t goes from 0 to 2π. But this is easy to determine by inspection. As t travels from 0 to 2π, the point $(\cos t, \sin t)$ begins at $(1, 0)$ and travels counterclockwise around the circle, ending at $(1, 0)$ when $t = 2\pi$.

A given figure in the plane may be the graph of different curves. To speak more loosely, a given curve may have different **parametrizations**. For example, the graph of the curve

$$x = \frac{1 - t^2}{1 + t^2}, \qquad y = \frac{2t}{1 + t^2}, \qquad -\infty < t < +\infty$$

also lies on the unit circle, because we find that $x^2 + y^2 = 1$ here as well. If t begins at 0 and increases, then the point $P(x(t), y(t))$ begins at $(1, 0)$ and travels along the upper half of the circle. If t begins at 0 and decreases, then the point $P(x(t), y(t))$ travels along the lower half. As t approaches either $+\infty$ or $-\infty$, the point P approaches the point $(-1, 0)$. Thus the graph consists of the unit circle with the single point $(-1, 0)$ deleted. A slight modification of the curve of Example 1,

$$x = \cos t, \qquad y = \sin t, \qquad -\pi < t < \pi,$$

is a different parametrization of this graph.

EXAMPLE 2 Eliminate the parameter to determine the graph of the parametric curve

$$x = t - 1, \qquad y = 2t^2 - 4t + 1, \qquad 0 \leqq t \leqq 2.$$

Solution We substitute $t = x + 1$ (from the equation for x) into the equation for y. This yields

$$y = 2(x + 1)^2 - 4(x + 1) + 1 = 2x^2 - 1$$

for $-1 \leqq x \leqq 1$. Thus the graph of the given curve is a portion of the parabola $y = 2x^2 - 1$ (Fig 12.1.3). As t increases from 0 to 2, the point $(t - 1, 2t^2 - 4t + 1)$ travels along the parabola from $(-1, 1)$ to $(1, 1)$.

Fig. 12.1.3 The curve of Example 2 is part of a parabola.

The parabolic arc of Example 2 can be reparametrized with

$$x = \sin t, \qquad y = 2 \sin^2 t - 1.$$

Now, as t increases, the point $(\sin t, 2\sin^2 t - 1)$ travels back and forth along the parabola between the two points $(-1, 1)$ and $(1, 1)$, rather like the bob of a pendulum.

Examples 1 and 2 involve parametric curves in which we can eliminate the parameter and thus obtain an explicit equation $y = f(x)$. Conversely, any explicitly presented curve $y = f(x)$ can be viewed as a parametric curve by writing

$$x = t, \qquad y = f(t),$$

with t running through the values in the original domain of f. For example, the straight line

$$y - y_0 = m(x - x_0)$$

that has slope m and passes through the point (x_0, y_0) may be parametrized by

$$x = t, \qquad y = y_0 + m(t - x_0).$$

An even simpler parametrization is obtained with $x - x_0 = t$. This gives

$$x = x_0 + t, \qquad y = y_0 + mt.$$

The use of parametric equations $x = x(t)$, $y = y(t)$ is most advantageous when elimination of the parameter is either impossible or would lead to an equation $y = f(x)$ that is considerably more complicated than the original parametric equations. This often happens when the curve is a geometric locus or the path of a point moving under specified conditions.

EXAMPLE 3   The curve traced by a point P on the edge of a rolling circle is called a **cycloid**. The circle rolls along a straight line without slipping or

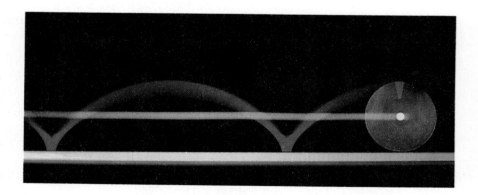

Cycloid generated by a point
of light on a rolling circle

stopping. (You will see a cycloid if you watch a patch of bright paint on the tire of a bicycle that crosses your path from left to right.) Find parametric equations for the cycloid if the line along which the circle rolls is the x-axis, the circle is above the x-axis but always tangent to it, and the point P begins at the origin.

Solution   Evidently the cycloid consists of a series of arches. We take as parameter t the angle (in radians) through which the circle has turned since it began with P at the origin. This is the angle TCP in Fig. 12.1.4.

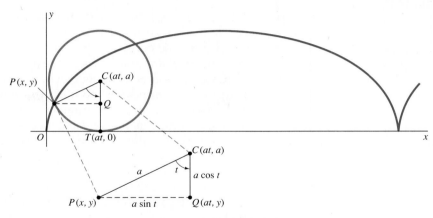

Fig. 12.1.4 The cycloid and the right triangle *CPQ* (Example 3)

The distance the circle has rolled is $|OT|$, so this is also the length of the circumference subtended by the angle *TCP*. Thus $|OT| = at$ if a is the radius of the circle, so the center C of the rolling circle has coordinates (at, a) when the angle *TCP* is t. The right triangle *CPQ* of Fig. 12.1.4 provides us with the relations

$$at - x = a \sin t \quad \text{and} \quad a - y = a \cos t.$$

Therefore the cycloid—the path of the moving point P—has parametric equations

$$x = a(t - \sin t), \quad y = a(1 - \cos t). \tag{3}$$

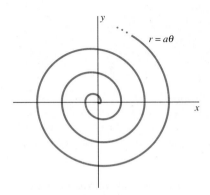

Fig. 12.1.5 A bead sliding down a wire—the brachistochrone problem

NOTE Figure 12.1.5 shows a bead sliding down a frictionless wire from point P to point Q. The *brachistochrone problem* asks what shape the wire should be to minimize the bead's time of descent from P to Q. In June of 1696, John Bernoulli proposed the brachistochrone problem as a public challenge, with a 6-month deadline (later extended to Easter 1697 at Leibniz's request). Isaac Newton, then retired from academic life and serving as Warden of the Mint in London, received Bernoulli's challenge on January 29, 1697. The very next day he communicated his own solution—the curve of minimal descent time is an arc of an inverted cycloid—to the Royal Society of London.

POLAR CURVES AS PARAMETRIC CURVES

A curve given in polar coordinates by the equation $r = f(\theta)$ can be regarded as a parametric curve with parameter θ. To see this, we recall that the equations $x = r \cos \theta$ and $y = r \sin \theta$ allow us to change from polar to rectangular coordinates. We replace r with $f(\theta)$, and this gives the parametric equations

$$x = f(\theta) \cos \theta, \qquad y = f(\theta) \sin \theta, \tag{4}$$

which express x and y in terms of the parameter θ.

Fig. 12.1.6 The spiral of Archimedes (Example 4)

EXAMPLE 4 The *spiral of Archimedes* has the polar-coordinates equation $r = a\theta$ (Fig. 12.1.6). The equations in (4) give the spiral the parametrization

$$x = a\theta \cos \theta, \qquad y = a\theta \sin \theta.$$

LINES TANGENT TO PARAMETRIC CURVES

The parametric curve $x = f(t)$, $y = g(t)$ is called **smooth** if the derivatives $f'(t)$ and $g'(t)$ are continuous and never simultaneously zero. In some neighborhood of each point of its graph, a smooth parametric curve can be described in one or possibly both of the forms $y = F(x)$ and $x = G(y)$. To see why this is so, suppose (for example) that $f'(t) > 0$ on the interval I. Then $f(t)$ is an increasing function on I and therefore has an inverse function $t = \phi(x)$ there. If we substitute $t = \phi(x)$ into the equation $y = g(t)$, then we get

$$y = g(\phi(x)) = F(x).$$

We can use the chain rule to compute the slope dy/dx of the line tangent to a smooth parametric curve. Differentiation of $y = F(x)$ with respect to t yields

$$\frac{dy}{dt} = \frac{dy}{dx} \cdot \frac{dx}{dt},$$

so

$$\frac{dy}{dx} = \frac{dy/dt}{dx/dt} = \frac{g'(t)}{f'(t)} \tag{5}$$

at any point where $f'(t) \neq 0$. The tangent line is vertical at any point where $f'(t) = 0$ but $g'(t) \neq 0$.

Equation (5) gives $y' = dy/dx$ as a function of t. Another differentiation with respect to t, again by using the chain rule, results in the formula

$$\frac{dy'}{dt} = \frac{dy'}{dx} \cdot \frac{dx}{dt},$$

so

$$\frac{d^2y}{dx^2} = \frac{dy'}{dx} = \frac{dy'/dt}{dx/dt}. \tag{6}$$

EXAMPLE 5   Calculate dy/dx and d^2y/dx^2 for the cycloid with the parametric equations in (3).

Solution   We begin with

$$x = a(t - \sin t), \quad y = a(1 - \cos t). \tag{3}$$

Then Eq. (5) gives

$$\frac{dy}{dx} = \frac{dy/dt}{dx/dt} = \frac{a \sin t}{a(1 - \cos t)} = \frac{\sin t}{1 - \cos t}. \tag{7}$$

This derivative is zero when t is an odd multiple of π, so the tangent line is horizontal at the midpoint of each arch of the cycloid. The endpoints of the arches correspond to even multiples of π, where both the numerator and the denominator in Eq. (7) are zero. These are isolated points (called *cusps*) at which the cycloid fails to be a smooth curve.

Next, Eq. (6) yields

$$\frac{d^2y}{dx^2} = \frac{(\cos t)(1 - \cos t) - (\sin t)(\sin t)}{(1 - \cos t)^2 \cdot a(1 - \cos t)} = -\frac{1}{a(1 - \cos t)^2}.$$

Because $d^2y/dx^2 < 0$ for all t (except for the isolated even multiples of π), this shows that each arch of the cycloid is concave downward (Fig. 12.1.4).

The slope dy/dx can be computed in terms of polar coordinates as well as rectangular coordinates. Given a polar-coordinates curve $r = f(\theta)$, we use the parametrization

$$x = f(\theta) \cos \theta, \qquad y = f(\theta) \sin \theta$$

shown in (4). Then Eq. (5), with θ in place of t, gives

$$\frac{dy}{dx} = \frac{dy/d\theta}{dx/d\theta} = \frac{f'(\theta) \sin \theta + f(\theta) \cos \theta}{f'(\theta) \cos \theta - f(\theta) \sin \theta}, \tag{8}$$

or, alternatively, denoting $f'(\theta)$ by r',

$$\frac{dy}{dx} = \frac{r' \sin \theta + r \cos \theta}{r' \cos \theta - r \sin \theta}. \tag{9}$$

Equation (9) has the following useful consequence. Let ψ denote the angle between the tangent line at P and the radius OP (extended) from the origin (Fig. 12.1.7). Then

$$\cot \psi = \frac{1}{r} \cdot \frac{dr}{d\theta} \qquad (0 \le \psi \le \pi). \tag{10}$$

In Problem 26 we indicate how Eq. (10) can be derived from Eq. (9).

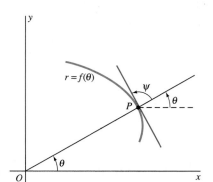

Fig. 12.1.7 The interpretation of the angle ψ [see Eq. (10)]

EXAMPLE 6 Consider the *logarithmic spiral* with polar equation $r = e^{\theta}$. Show that $\psi = \pi/4$ at every point of the spiral, and write the equation of its tangent line at the point $(e^{\pi/2}, \pi/2)$.

Solution Because $dr/d\theta = e^{\theta}$, Eq. (10) tells us that $\cot \psi = e^{\theta}/e^{\theta} = 1$. Thus $\psi = \pi/4$. When $\theta = \pi/2$, Eq. (9) gives

$$\frac{dy}{dx} = \frac{e^{\pi/2} \sin(\pi/2) + e^{\pi/2} \cos(\pi/2)}{e^{\pi/2} \cos(\pi/2) - e^{\pi/2} \sin(\pi/2)} = -1.$$

But when $\theta = \pi/2$, we have $x = 0$ and $y = e^{\pi/2}$. It follows that an equation of the desired tangent line is

$$y - e^{\pi/2} = -x; \quad \text{that is,} \quad x + y = e^{\pi/2}.$$

The line and the spiral appear in Fig. 12.1.8.

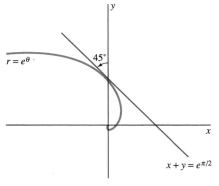
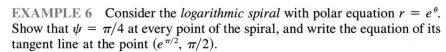

Fig. 12.1.8 The angle ψ is always 45° for the logarithmic spiral (Example 6)

12.1 Problems

In Problems 1 through 12, eliminate the parameter and then sketch the curve.

1. $x = t + 1, \quad y = 2t - 1$
2. $x = t^2 + 1, \quad y = 2t^2 - 1$
3. $x = t^2, \quad y = t^3$
4. $x = \sqrt{t}, \quad y = 3t - 2$

5. $x = t + 1, \quad y = 2t^2 - t - 1$
6. $x = t^2 + 3t, \quad y = t - 2$
7. $x = e^t, \quad y = 4e^{2t}$
8. $x = 2e^t, \quad y = 2e^{-t}$
9. $x = 5 \cos t, \quad y = 3 \sin t$
10. $x = 2 \cosh t, \quad y = 3 \sinh t$

11. $x = \sec t, \quad y = \tan t$

12. $x = \cos 2t, \quad y = \sin t$

In Problems 13 through 17, (a) first write the equation of the line tangent to the given parametric curve at the point that corresponds to the given value of t, and (b) then calculate d^2y/dx^2 to determine whether the curve is concave upward or concave downward at this point.

13. $x = 2t^2 + 1, \quad y = 3t^3 + 2; \quad t = 1$

14. $x = \cos^3 t, \quad y = \sin^3 t; \quad t = \pi/4$

15. $x = t \sin t, \quad y = t \cos t; \quad t = \pi/2$

16. $x = e^t, \quad y = e^{-t}; \quad t = 0$

17. $x = \dfrac{3t}{1 + t^3}, \quad y = \dfrac{3t^2}{1 + t^3}; \quad t = 1$

In Problems 18 through 21, find the angle ψ between the radius OP and the tangent line at the point P that corresponds to the given value of θ.

18. $r = 1/\theta, \quad \theta = 1$ **19.** $r = \exp(\theta\sqrt{3}), \quad \theta = \pi/2$

20. $r = 1 - \cos\theta, \; \theta = \pi/3$ **21.** $r = \sin 3\theta, \; \theta = \pi/6$

22. Eliminate t to determine the graph of the parametric curve $x = t^2, \; y = t^3 - 3t, \; -2 \leqq t \leqq 2$. Note that $t = \sqrt{x}$ if $t > 0$, whereas $t = -\sqrt{x}$ if $t < 0$. Your graph should contain a loop and should be symmetric about the x-axis.

23. The curve C is determined by the parametric equations $x = e^{-t}, y = e^{2t}$. Calculate dy/dx and d^2y/dx^2 directly from these parametric equations. Conclude that C is concave upward at every point. Then sketch C.

24. The graph of the folium of Descartes with rectangular equation $x^3 + y^3 = 3xy$ appears in Fig. 12.1.9. Parametrize its loop as follows: Let P be the point of intersection of the line $y = tx$ with the loop; then solve for the coordinates x and y of P in terms of t.

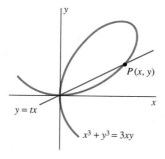

Fig. 12.1.9 The loop of the folium of Descartes (Problem 24)

25. Parametrize the parabola $y^2 = 4px$ by expressing x and y as functions of the slope m of the tangent line at the point $P(x, y)$ of the parabola.

26. Let P be a point of the curve with polar equation $r = f(\theta)$, and let ψ be the angle between the extended radius OP and the tangent line at P. Let α be the angle of inclination of this tangent line, measured counterclockwise from the horizontal. Then $\psi = \alpha - \theta$. Verify Eq. (10) by substituting $\tan \alpha = dy/dx$ from Eq. (9) and $\tan \theta = y/x = (\sin \theta)/(\cos \theta)$ into the identity

$$\cot \psi = \frac{1}{\tan(\alpha - \theta)} = \frac{1 + \tan \alpha \tan \theta}{\tan \alpha - \tan \theta}.$$

27. Let P_0 be the highest point of the circle of Fig. 12.1.4(a)—the circle that generates the cycloid of Example 3. Show that the line through P_0 and the point P of the cycloid [the point P is shown in Fig. 12.1.4(a)] is tangent to the cycloid at P. This fact gives a geometric construction of the line tangent to the cycloid.

28. A circle of radius b rolls without slipping inside a circle of radius $a > b$. The path of a point P fixed on the circumference of the rolling circle is called a *hypocycloid* (Fig. 12.1.10). Let P begin its journey at $A(a, 0)$, and let t be the angle AOC, where O is the center of the large circle and C is the center of the rolling circle. Show that the coordinates of P are given by the parametric equations

$$x = (a - b) \cos t + b \cos\left(\frac{a - b}{b}t\right),$$

$$y = (a - b) \sin t - b \sin\left(\frac{a - b}{b}t\right).$$

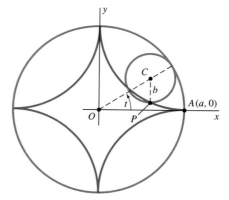

Fig. 12.1.10 The hypocycloid of Problem 28

29. If $b = a/4$ in Problem 28, show that the parametric equations of the hypocycloid reduce to

$$x = a \cos^3 t, \qquad y = a \sin^3 t.$$

30. (a) Prove that the hypocycloid of Problem 29 is the graph of the equation $x^{2/3} + y^{2/3} = a^{2/3}$. (b) Find all points of this hypocycloid where its tangent line is either horizontal or vertical, and find the intervals on which it is concave upward and those on which it is concave downward. (c) Sketch this hypocycloid.

31. Consider a point P on the spiral of Archimedes, the curve shown in Fig. 12.1.11 with polar equation $r = a\theta$. Archimedes viewed the path of P as compounded of two motions, one with speed a directly away from the origin O and another a circular motion with unit angular speed around O. This suggests Archimedes' result that the line PQ in the figure is tangent to the spiral at P. Prove that this is indeed true.

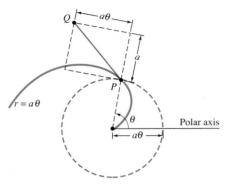

Fig. 12.1.11 The segment PQ is tangent to the spiral (Archimedes' result; Problem 31)

32. (a) Deduce from Eq. (7) that if t is not an integral multiple of 2π, then the slope of the tangent line at the corresponding point of the cycloid is $\cot(t/2)$. (b) Conclude that at the cusp of the cycloid where t is an integral multiple of 2π, the cycloid has a vertical tangent line.

33. A *loxodrome* is a curve $r = f(\theta)$ such that the tangent line at P and the radius OP in Fig. 12.1.7 make a constant angle. Use Eq. (10) to prove that every loxodrome is of the form $r = Ae^{k\theta}$, where A and k are constants. Thus every loxodrome is a logarithmic spiral similar to the one considered in Example 6.

34. Let a curve be described in polar coordinates by $r = f(\theta)$, where f is continuous. If $f(\alpha) = 0$, then the origin is the point of the curve corresponding to $\theta = \alpha$. Deduce from the parametrization $x = f(\theta) \cos \theta$, $y = f(\theta) \sin \theta$ that the line tangent to the curve at this point makes the angle α with the positive x-axis. For example, the *cardioid* $r = f(\theta) = 1 - \sin \theta$ shown in Fig. 12.1.12 is tangent to the y-axis at the origin. And, indeed, $f(\pi/2) = 0$: The y-axis is the line $\theta = \alpha = \pi/2$.

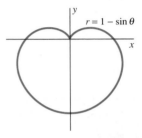

Fig. 12.1.12 The cardioid of Problem 34

12.1 Projects

The most fun in plotting parametric curves comes from trying your own hand—especially if you have a graphics calculator or computer with graphing utility to do the real work. Try various values of the constants $a, b, p, \ldots$ in the projects. When sines and cosines are involved, the interval $0 \leq t \leq 2\pi$ is a reasonable one to try first. You will need to experiment with different viewing windows to find one that shows the whole curve (or the most interesting part of it).

PROJECT A Given: $x = at - b \sin t$, $y = a - b \cos t$. This *trochoid* is traced by a point P on a solid wheel of radius a as it rolls along the x-axis; the distance of P from the center of the wheel is $b > 0$. (The graph is a cycloid if $a = b$.) Try both cases $a > b$ and $a < b$.

PROJECT B Given:

$$x = (a - b) \cos t + b \cos\left(\frac{a - b}{b} t\right),$$

$$y = (a - b) \sin t - b \sin\left(\frac{a - b}{b} t\right).$$

This is a *hypocycloid*—the path of a point P on a circle of radius b that rolls along the inside of a circle of radius $a > b$. Figure 12.1.10 shows the case $b = a/4$.

PROJECT C Given:

$$x = (a + b) \cos t - b \cos\left(\frac{a + b}{b} t\right),$$

$$y = (a + b) \sin t - b \sin\left(\frac{a + b}{b} t\right).$$

This is an *epicycloid* traced by a point P on a circle of radius b that rolls along the outside of a circle of radius $a > b$.

PROJECT D Given:

$$x = a \cos t - b \cos\frac{at}{2}, \qquad y = a \sin t - b \sin\frac{at}{2}.$$

This is an *epitrochoid*—it is to an epicycloid what a trochoid is to a cycloid. With $a = 8$ and $b = 5$, you should get the curve shown in Fig. 12.1.13.

PROJECT E Given:

$$x = a \cos t + b \cos\frac{at}{2}, \qquad y = a \sin t - b \sin\frac{at}{2}.$$

With $a = 8$ and $b = 5$, this *hypotrochoid* looks like the curve shown in Fig. 12.1.14.

PROJECT F Given: $x = \cos at$, $y = \sin bt$. These are the *Lissajous curves* that typically appear on oscilloscopes in physics or electronics laboratories. The Lissajous curve with $a = 3$ and $b = 5$ is shown in Fig. 12.1.15.

PROJECT G Consider the parametric equations

$$x = a \cos t - b \cos pt, \qquad y = c \sin t - d \sin qt.$$

The values $a = 16$, $b = 5$, $c = 12$, $d = 3$, $p = \frac{47}{3}$, and $q = \frac{44}{3}$ yield the "slinky curve" shown in Fig. 12.1.16. Experiment with various combinations of constants to see if you can produce a prettier picture.

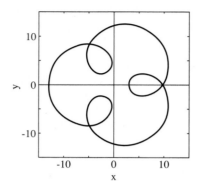

Fig. 12.1.13 The epitrochoid with $a = 8$, $b = 5$ (Project D)

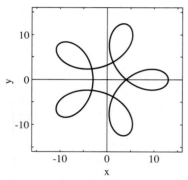

Fig. 12.1.14 The hypotrochoid with $a = 8$, $b = 5$ (Project E)

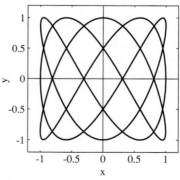

Fig. 12.1.15 The Lissajous curve with $a = 3$, $b = 5$ (Project F)

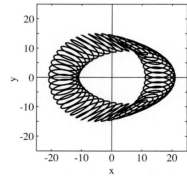

Fig. 12.1.16 The slinky curve of Project G

12.2
Integral Computations with Parametric Curves

In Chapter 6 we discussed the computation of a variety of geometric quantities associated with the graph $y = f(x)$ of a nonnegative function on the interval $[a, b]$. These included:

Area under the curve:	$A = \displaystyle\int_a^b y \, dx.$	(1)

Volume of revolution around the x-axis:	$V_x = \displaystyle\int_a^b \pi y^2 \, dx.$	(2a)

Volume of revolution around the y-axis:	$V_y = \displaystyle\int_a^b 2\pi x y \, dx.$	(2b)

Arc length of the curve:	$s = \displaystyle\int_0^s ds = \int_a^b \sqrt{1 + (y')^2} \, dx.$	(3)

Area of surface of revolution around the x-axis:	$A_x = \displaystyle\int_{x=a}^b 2\pi y \, ds.$	(4a)

Area of surface of revolution around the y-axis:	$A_y = \displaystyle\int_{x=a}^b 2\pi x \, ds.$	(4b)

We substitute $y = f(x)$ into each of these integrals before we integrate from $x = a$ to $x = b$.

We now want to compute these same quantities for a smooth parametric curve

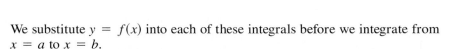

$$x = f(t), \qquad y = g(t), \qquad \alpha \leq t \leq \beta.$$

To ensure that this parametric curve looks like the graph of a function defined on some interval $[a, b]$ of the x-axis, we assume that $f(t)$ *is either an increasing or a decreasing function of t on $[\alpha, \beta]$ and that $g(t) \geq 0$*. If $a = f(\alpha)$ and $b = f(\beta)$, then the curve will be traced from left to right as t increases, whereas if $a = f(\beta)$ and $b = f(\alpha)$, it is traced from right to left (Fig. 12.2.1).

Any one of the quantities A, V_x, V_y, s, A_x, and A_y may then be computed by making the substitutions

$$x = f(t), \qquad\qquad y = g(t),$$
$$dx = f'(t) \, dt, \qquad dy = g'(t) \, dt, \quad \text{and} \tag{5}$$
$$ds = \sqrt{[f'(t)]^2 + [g'(t)]^2} \, dt$$

in the appropriate integral formula in Eqs. (1) through (4). In the case of Eqs. (1) and (2), which contain dx, we then integrate from $t = \alpha$ to $t = \beta$ in part (a) of Fig. 12.2.1 or from $t = \beta$ to $t = \alpha$ in part (b). Thus the proper choice of limits on t corresponds to traversing the curve *from left to right*. For example,

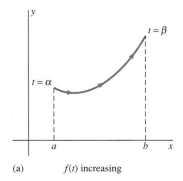

(a) $f(t)$ increasing

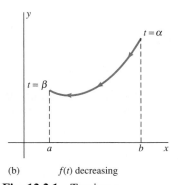

(b) $f(t)$ decreasing

Fig. 12.2.1 Tracing a parametrized curve: (a) $f(t)$ increasing; (b) $f(t)$ decreasing

$$A = \int_{\alpha}^{\beta} g(t) f'(t) \, dt \qquad \text{if } f(\alpha) < f(\beta),$$

whereas

$$A = \int_{\beta}^{\alpha} g(t) f'(t) \, dt \qquad \text{if } f(\beta) < f(\alpha).$$

The validity of this method of evaluating the integrals in Eqs. (1) and (2) follows from Theorem 1 of Section 5.7, on integration by substitution.

In the case of Eqs. (3) and (4), which contain ds, we integrate from $t = \alpha$ to $t = \beta$ in either case. To see why this is so, recall from Eq. (5) of Section 12.1 that $dy/dx = g'(t)/f'(t)$ if $f'(t) \neq 0$ on $[\alpha, \beta]$. Hence

$$s = \int_{a}^{b} \sqrt{1 + \left(\frac{dy}{dx}\right)^2} \, dx = \int_{f^{-1}(a)}^{f^{-1}(b)} \sqrt{1 + \left[\frac{g'(t)}{f'(t)}\right]^2} \, f'(t) \, dt.$$

Because $f'(t) > 0$ if $f(\alpha) = a$ and $f(\beta) = b$, whereas $f'(t) < 0$ if $f(\alpha) = b$ and $f(\beta) = a$, it follows that

$$s = \int_{\alpha}^{\beta} \sqrt{1 + \left[\frac{g'(t)}{f'(t)}\right]^2} \, |f'(t)| \, dt,$$

and so

$$s = \int_{\alpha}^{\beta} \sqrt{[f'(t)]^2 + [g'(t)]^2} \, dt = \int_{\alpha}^{\beta} \sqrt{\left(\frac{dx}{dt}\right)^2 + \left(\frac{dy}{dt}\right)^2} \, dt. \qquad (6)$$

This formula, derived under the assumption that $f'(t) \neq 0$ on $[\alpha, \beta]$, may be taken to be the *definition* of arc length for an arbitrary smooth parametric curve. Similarly, the area of a surface of revolution is defined for smooth parametric curves as the result of first making the substitution of Eq. (5) into Eq. (4a) or (4b) and then integrating from $t = \alpha$ to $t = \beta$.

The infinitesimal "triangle" in Fig. 12.2.2 serves as a convenient device for remembering the substitution

$$ds = \sqrt{[f'(t)]^2 + [g'(t)]^2} \, dt.$$

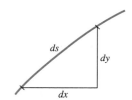

Fig. 12.2.2 Nearly a right triangle for dx and dy close to zero

When we apply the Pythagorean theorem to this right triangle, it leads to the symbolic manipulation

$$ds = \sqrt{(dx)^2 + (dy)^2} = \sqrt{\left(\frac{dx}{dt}\right)^2 + \left(\frac{dy}{dt}\right)^2} \, dt = \sqrt{[f'(t)]^2 + [g'(t)]^2} \, dt. \quad (7)$$

EXAMPLE 1 Use the parametrization $x = a \cos t$, $y = a \sin t$ of the circle with center $O(0, 0)$ and radius a to find the volume V and surface area A of the sphere obtained by revolving this circle around the x-axis.

Solution Half the sphere is obtained by revolving the first quadrant of the circle (Fig. 12.2.3). The left-to-right direction along the curve is from $t = \pi/2$ to $t = 0$, so Eq. (2a) gives

$$V = 2 \int_{\pi/2}^{0} \pi(a \sin t)^2 (-a \sin t \, dt) = 2\pi a^3 \int_{0}^{\pi/2} (1 - \cos^2 t) \sin t \, dt$$

$$= 2\pi a^3 \left[-\cos t + \tfrac{1}{3}\cos^3 t \right]_{0}^{\pi/2} = \tfrac{4}{3}\pi a^3.$$

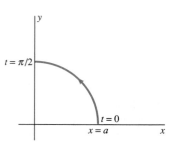

Fig. 12.2.3 The quarter-circle of Example 1

The arc-length differential for the parametrized curve is

$$ds = \sqrt{(-a \sin t)^2 + (a \cos t)^2} \, dt = a \, dt.$$

Hence Eq. (4a) gives

$$A = 2 \int_0^{\pi/2} 2\pi(a \sin t)(a \, dt) = 4\pi a^2 \int_0^{\pi/2} \sin t \, dt = 4\pi a^2 \left[-\cos t \right]_0^{\pi/2} = 4\pi a^2.$$

Of course, the results of Example 1 are familiar. In contrast, Example 2 requires the methods of this section.

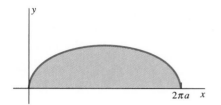

Fig. 12.2.4 The cycloidal arch of Example 2

EXAMPLE 2 Find the area under, and the arc length of, the cycloidal arch of Fig. 12.2.4. Its parametric equations are

$$x = a(t - \sin t), \qquad y = a(1 - \cos t), \qquad 0 \leq t \leq 2\pi.$$

Solution We use the identity $1 - \cos t = 2 \sin^2(t/2)$ and a consequence of Problem 50 in Section 9.4:

$$\int_0^\pi \sin^{2n} u \, du = \pi \cdot \frac{1}{2} \cdot \frac{3}{4} \cdot \frac{5}{6} \cdots \frac{2n-1}{2n}.$$

Because $dx = a(1 - \cos t) \, dt$ and the left-to-right direction along the curve is from $t = 0$ to $t = 2\pi$, Eq. (1) gives

$$A = \int_0^{2\pi} a(1 - \cos t) \cdot a(1 - \cos t) \, dt = a^2 \int_0^{2\pi} (1 - \cos t)^2 \, dt$$

$$= 4a^2 \int_0^{2\pi} \sin^4 \frac{t}{2} \, dt = 8a^2 \int_0^\pi \sin^4 u \, du \qquad \left(u = \frac{t}{2} \right)$$

$$= 8a^2 \cdot \pi \cdot \tfrac{1}{2} \cdot \tfrac{3}{4} = 3\pi a^2$$

for the area under one arch of the cycloid. The arc-length differential is

$$ds = \sqrt{a^2(1 - \cos t)^2 + (a \sin t)^2} \, dt$$

$$= a\sqrt{2(1 - \cos t)} \, dt = 2a \sin \left(\frac{t}{2} \right) dt,$$

so Eq. (3) gives

$$s = \int_0^{2\pi} 2a \sin \frac{t}{2} \, dt = \left[-4a \cos \frac{t}{2} \right]_0^{2\pi} = 8a$$

for the length of one arch of the cycloid.

PARAMETRIC POLAR COORDINATES

Suppose that a parametric curve is determined by giving its polar coordinates

$$r = r(t), \qquad \theta = \theta(t), \qquad \alpha \leq t \leq \beta$$

as functions of the parameter t. Then this curve is described in rectangular coordinates by the parametric equations

$$x(t) = r(t) \cos \theta(t), \qquad y(t) = r(t) \sin \theta(t), \qquad \alpha \leq t \leq \beta,$$

giving x and y as functions of t. The latter parametric equations may then be used in the integral formulas in Eqs. (1) through (4).

To compute ds, we first calculate the derivatives

$$\frac{dx}{dt} = (\cos\theta)\frac{dr}{dt} - (r\sin\theta)\frac{d\theta}{dt}, \qquad \frac{dy}{dt} = (\sin\theta)\frac{dr}{dt} + (r\cos\theta)\frac{d\theta}{dt}.$$

Upon substituting these expressions for dx/dt and dy/dt into Eq. (7) and making algebraic simplifications, we find that the arc-length differential in parametric polar coordinates is

$$ds = \sqrt{\left(\frac{dr}{dt}\right)^2 + \left(r\frac{d\theta}{dt}\right)^2}\ dt. \tag{8}$$

In the case of a curve with the explicit polar-coordinates equation $r = f(\theta)$, we may use θ itself as the parameter. Then Eq. (8) takes the simpler form

$$ds = \sqrt{\left(\frac{dr}{d\theta}\right)^2 + r^2}\ d\theta. \tag{9}$$

This formula is easy to remember with the aid of the tiny "almost-triangle" shown in Fig. 12.2.5.

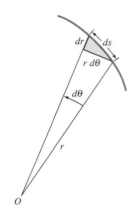

Fig. 12.2.5 The differential triangle in polar coordinates

EXAMPLE 3 Find the perimeter s of the cardioid with polar equation $r = 1 + \cos\theta$. Find also the surface area A generated by revolving the cardioid around the x-axis.

Solution   The cardioid is shown in Fig. 12.2.6. Equation (9) gives

$$ds = \sqrt{(-\sin\theta)^2 + (1+\cos\theta)^2}\ d\theta = \sqrt{2(1+\cos\theta)}\ d\theta$$

$$= \sqrt{4\cos^2\left(\frac{\theta}{2}\right)}\ d\theta = \left|\,2\cos\left(\frac{\theta}{2}\right)\,\right|\ d\theta.$$

Hence $ds = 2\cos(\theta/2)\ d\theta$ on the upper half of the cardioid, where $0 \leq \theta \leq \pi$, so $\cos(\theta/2) \geq 0$. Therefore,

$$s = 2\int_0^\pi 2\cos\frac{\theta}{2}\ d\theta = 8\left[\sin\frac{\theta}{2}\right]_0^\pi = 8.$$

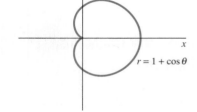

Fig. 12.2.6 The cardioid of Example 3

The surface area of revolution is

$$A = \int_*^{**} 2\pi y\ ds = \int_{\theta=0}^\pi 2\pi(r\sin\theta)\ ds = \int_0^\pi 2\pi(1+\cos\theta)(\sin\theta)\cdot 2\cos\frac{\theta}{2}\ d\theta$$

$$= 16\pi\int_0^\pi \cos^4\frac{\theta}{2}\sin\frac{\theta}{2}\ d\theta = 16\pi\left[-\frac{2}{5}\cos^5\frac{\theta}{2}\right]_0^\pi = \frac{32\pi}{5}.$$

12.2 Problems

In Problems 1 through 6, find the area of the region that lies between the given parametric curve and the x-axis.

1. $x = t^3$, $\quad y = 2t^2 + 1$; $\quad -1 \leq t \leq 1$
2. $x = e^{3t}$, $\quad y = e^{-t}$; $\quad 0 \leq t \leq \ln 2$
3. $x = \cos t$, $\quad y = \sin^2 t$; $\quad 0 \leq t \leq \pi$
4. $x = 2 - 3t$, $\quad y = e^{2t}$; $\quad 0 \leq t \leq 1$

5. $x = \cos t$, $\quad y = e^t$; $\quad 0 \leq t \leq \pi$
6. $x = 1 - e^t$, $\quad y = 2t + 1$; $\quad 0 \leq t \leq 1$

In Problems 7 through 10, find the volume obtained by revolving around the x-axis the region described in the given problem.

7. Problem 1   8. Problem 2

9. Problem 3   10. Problem 5

In Problems 11 through 16, find the arc length of the given curve.

11. $x = 2t,\quad y = \frac{2}{3}t^{3/2};\quad 5 \le t \le 12$

12. $x = \frac{1}{2}t^2;\quad y = \frac{1}{3}t^3;\quad 0 \le t \le 1$

13. $x = \sin t - \cos t,\quad y = \sin t + \cos t;$
    $\pi \le 4t \le 2\pi$

14. $x = e^t \sin t,\quad y = e^t \cos t;\quad 0 \le t \le \pi$

15. $r = e^{\theta/2};\quad 0 \le \theta \le 4\pi$

16. $r = \theta;\quad 2\pi \le \theta \le 4\pi$

In Problems 17 through 22, find the area of the surface of revolution generated by revolving the given curve around the indicated axis.

17. $x = 1 - t,\quad y = 2t^{1/2},\quad 1 \le t \le 4;$ the x-axis

18. $x = 2t^2 + t^{-1},\quad y = 8t^{1/2},\quad 1 \le t \le 2;$ the x-axis

19. $x = t^3,\quad y = 2t + 3,\quad -1 \le t \le 1;$ the y-axis

20. $x = 2t + 1,\quad y = t^2 + t,\quad 0 \le t \le 3;$ the y-axis

21. $r = 4 \sin \theta;$ the x-axis

22. $r = e^\theta,\quad 0 \le \theta \le \pi/2;$ the y-axis

23. Find the volume generated by revolving around the x-axis the region under the cycloidal arch of Example 2.

24. Find the area of the surface generated by revolving around the x-axis the cycloidal arch of Example 2.

25. Use the parametrization $x = a \cos t,\ y = b \sin t$ to find (a) the area bounded by the ellipse $x^2/a^2 + y^2/b^2 = 1$; (b) the volume of the ellipsoid generated by revolving this ellipse around the x-axis.

26. Find the area bounded by the loop of the parametric curve $x = t^2,\ y = t^3 - 3t$ of Problem 22 in Section 12.1.

27. Use the parametrization $x = t \cos t,\ y = t \sin t$ of the Archimedean spiral to find the arc length of the first full turn of this spiral (corresponding to $0 \le t \le 2\pi$).

28. The circle $(x - b)^2 + y^2 = a^2$ with radius $a < b$ and center $(b, 0)$ can be parametrized by

$$x = b + a \cos t,\qquad y = a \sin t,\qquad 0 \le t \le 2\pi.$$

Find the surface area of the torus obtained by revolving this circle around the y-axis (Fig. 12.2.7).

29. The *astroid* (four-cusped hypocycloid) has equation $x^{2/3} + y^{2/3} = a^{2/3}$ (Fig. 12.1.10) and the parametrization

$$x = a \cos^3 t,\quad y = a \sin^3 t,\quad 0 \le t \le 2\pi.$$

Find the area of the region bounded by the astroid.

30. Find the total length of the astroid of Problem 29.

31. Find the area of the surface obtained by revolving the astroid of Problem 29 around the x-axis.

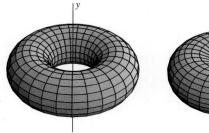
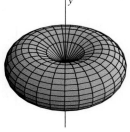

Fig. 12.2.7 The torus of Problem 28

Fig. 12.2.8 The rotated lemniscate of Problem 32

32. Find the area of the surface generated by revolving the lemniscate $r^2 = 2a^2 \cos 2\theta$ around the y-axis (Fig. 12.2.8). [*Suggestion:* Use Eq. (9) and note that $r\, dr = -2a^2 \sin 2\theta\, d\theta$.]

33. Figure 12.2.9 shows the graph of the parametric curve

$$x = t^2\sqrt{3},\quad y = 3t - \frac{1}{3}t^3.$$

The shaded region is bounded by the part of the curve for which $-3 \le t \le 3$. Find its area.

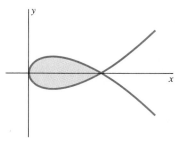

Fig. 12.2.9 The parametric curve of Problems 33 through 36

34. Find the arc length of the loop in Problem 33.

35. Find the volume of the solid obtained by revolving around the x-axis the shaded region in Fig. 12.2.9.

36. Find the surface area of revolution generated by revolving around the x-axis the loop of Fig. 12.2.9.

37. (a) With reference to Problem 24 and Fig. 12.1.9 in Section 12.1, show that the arc length of the first-quadrant loop of the folium of Descartes is

$$s = 6 \int_0^1 \frac{\sqrt{1 + 4t^2 - 4t^3 - 4t^5 + 4t^6 + t^8}}{(1 + t^3)^2}\, dt.$$

(b) If you have a programmable calculator or computer, apply Simpson's approximation to obtain $s \approx 4.9175$.

38. Find the surface area generated by rotating around the y-axis the cycloidal arch of Example 2. [*Suggestion:* $\sqrt{x^2} = x$ only if $x \ge 0$.]

39. Find the volume generated by rotating around the y-axis the region under the cycloidal arch of Example 2.

These projects illustrate the use of numerical integration techniques (as described in the projects of Sections 5.4 and 5.9) to approximate the parametric arc-length integral

$$s = \int_a^b \sqrt{[x'(t)]^2 + [y'(t)]^2}\, dt. \tag{10}$$

Consider the ellipse with equation

$$\frac{x^2}{a^2} + \frac{y^2}{b^2} = 1 \qquad (a > b) \tag{11}$$

and eccentricity $\epsilon = (1/a)\sqrt{a^2 - b^2}$. Substitute the parametrization

$$x = a \cos t, \qquad y = b \sin t \tag{12}$$

into Eq. (10) to show that the perimeter of the ellipse is given by the *elliptic integral*

$$p = 4a \int_0^{\pi/2} \sqrt{1 - \epsilon^2 \cos^2 t}\, dt. \tag{13}$$

This integral is known to be nonelementary if $0 < \epsilon < 1$. A common simple approximation to it is

$$p \approx \pi(A + R), \tag{14}$$

where

$$A = \frac{1}{2}(a + b) \quad \text{and} \quad R = \sqrt{\frac{a^2 + b^2}{2}}$$

denote the arithmetic mean and root-square mean, respectively, of the semiaxes a and b of the ellipse.

PROJECT A Choose several different ellipses by selecting values of a and b with $a > b > 0$. For each ellipse, estimate the perimeter given in Eq. (13) by using the right-endpoint and left-endpoint approximations R_n and L_n, the trapezoidal approximation $T_n = (R_n + L_n)/2$, the midpoint approximation M_n, and/or Simpson's approximation S_n. In each case compare these approximations for $n = 10, 20, 40, \ldots$ subintervals. Construct a table showing your results along with the simple estimate of Eq. (14).

PROJECT B If we ignore the perturbing effects of the sun and the planets other than the earth, the orbit of the moon is an almost perfect ellipse with the earth at one focus. Assume that this ellipse has major semiaxis $a = 384{,}403$ km (exactly) and eccentricity $\epsilon = 0.0549$ (exactly). Approximate the perimeter p of this ellipse [Eq. (13)] to the nearest meter.

12.3
Vectors in the Plane

A physical quantity such as length, temperature, or mass can be specified in terms of a single real number, its *magnitude*. Such a quantity is called a **scalar**. Other physical quantities, such as force and velocity, possess both

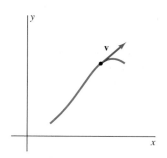

Fig. 12.3.1 A velocity vector may be represented by an arrow.

magnitude and *direction*; these entities are called **vector quantities**, or simply **vectors**.

For example, to specify the velocity of a moving point in the coordinate plane, we must give both the rate at which it moves (its speed) and the direction of that motion. The combination is the **velocity vector** of the moving point. It is convenient to represent this velocity vector by an arrow, located at the current position of the moving point on its trajectory (Fig. 12.3.1).

Although the arrow (represented by a directed line segment) carries the desired information—both magnitude (the segment's length) and direction—it is a pictorial representation rather than a quantitative object. The following formal definition of a vector captures the essence of magnitude in combination with direction.

Definition of Vector

A **vector v** in the Cartesian plane is an ordered pair of real numbers that has the form $\langle a, b \rangle$. We write $\mathbf{v} = \langle a, b \rangle$ and call a and b the **components** of the vector **v**.

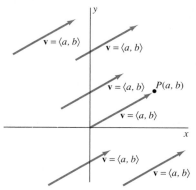

Fig. 12.3.2 All these arrows represent the same vector $\mathbf{v} = \langle a, b \rangle$.

The directed line segment $\overrightarrow{OP}$ from the origin O to the point $P(a, b)$ is one geometric representation of the vector **v**. For this reason, the vector $\mathbf{v} = \langle a, b \rangle$ is called the **position vector** of the point $P(a, b)$. In fact, the relationship between $\mathbf{v} = \langle a, b \rangle$ and $P(a, b)$ is so close that, in certain contexts, it is convenient to confuse the two deliberately—to regard **v** and P as the same mathematical object. In other contexts the distinction between the two is important; in particular, *any* directed line segment with the same direction and magnitude (length) as $\overrightarrow{OP}$ is a representation of **v**, as Fig. 12.3.2 suggests. What is important about the vector **v** is not usually *where* it is, but how long it is and which way it points.

The magnitude associated with the vector $\mathbf{v} = \langle a, b \rangle$, called its **length**, is denoted by $v = |\mathbf{v}|$ and is defined to be

$$v = |\mathbf{v}| = |\langle a, b \rangle| = \sqrt{a^2 + b^2}. \tag{1}$$

For example, the length of the vector $\mathbf{v} = \langle 1, -2 \rangle$ is

$$|\mathbf{v}| = |\langle 1, -2 \rangle| = \sqrt{(1)^2 + (-2)^2} = \sqrt{5}.$$

The notation $v = |\mathbf{v}|$ is used because the length of a vector is in many ways analogous to the absolute value of a real number.

The only vector with length zero is the **zero vector** with both components zero, denoted by $\mathbf{0} = \langle 0, 0 \rangle$. The zero vector is unique in that it has no specific direction.

In printing the name of a vector, we use **boldface** type to distinguish vectors from other mathematical objects, such as the scalars a and b that constitute the components of the vector $\mathbf{v} = \langle a, b \rangle$. Because the use of boldface is impractical in handwritten or typewritten work, a suitable alternative is to place a *right harpoon* over every symbol that denotes a vector; thus in type you will see (and should use in handwritten work) $\vec{v} = \langle a, b \rangle$. There is no need for a harpoon over a vector $\langle a, b \rangle$ already identified by angle brackets, so none should be used there.

ALGEBRAIC OPERATIONS WITH VECTORS

The operations of addition and multiplication of real numbers have analogues for vectors. We shall define each of these operations of *vector algebra* in terms of components of vectors and then give a geometric interpretation in terms of arrows.

Definition   *Equality of Vectors*
The two vectors $\mathbf{u} = \langle u_1, u_2 \rangle$ and $\mathbf{v} = \langle v_1, v_2 \rangle$ are **equal** provided that $u_1 = v_1$ and $u_2 = v_2$.

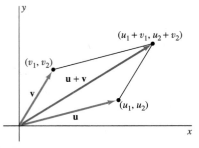

Fig. 12.3.3   Parallel directed segments representing equal vectors

In other words, two vectors are equal if and only if corresponding components are the same. Moreover, two directed line segments $\overrightarrow{PQ}$ and $\overrightarrow{RS}$ that have the same length and direction represent equal vectors. Because if $P = (p_1, p_2)$, $Q = (q_1, q_2)$, $R = (r_1, r_2)$, and $S = (s_1, s_2)$, then $\overrightarrow{PQ}$ represents the vector $\langle q_1 - p_1, q_2 - p_2 \rangle$, whereas $\overrightarrow{RS}$ represents the vector $\langle s_1 - r_1, s_2 - r_2 \rangle$. But $PQRS$ is a parallelogram because $\mathbf{u} = \overrightarrow{PQ}$ and $\mathbf{v} = \overrightarrow{RS}$ have the same length and direction—they represent opposite sides of a parallelogram (Fig. 12.3.3). Consequently, it follows that $q_1 - p_1 = s_1 - r_1$ and $q_2 - p_2 = s_2 - r_2$. Hence the vectors $\mathbf{u}$ and $\mathbf{v}$ are equal.

Definition   *Addition of Vectors*
The **sum** $\mathbf{u} + \mathbf{v}$ of the two vectors $\mathbf{u} = \langle u_1, u_2 \rangle$ and $\mathbf{v} = \langle v_1, v_2 \rangle$ is the vector

$$\mathbf{u} + \mathbf{v} = \langle u_1 + v_1, u_2 + v_2 \rangle. \tag{2}$$

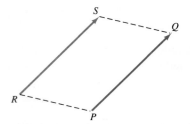

Fig. 12.3.4   The triangle law is a geometric interpretation of vector addition.

For example, the sum of the vectors $\mathbf{u} = \langle 4, 3 \rangle$ and $\mathbf{v} = \langle -5, 2 \rangle$ is

$$\langle 4, 3 \rangle + \langle -5, 2 \rangle = \langle 4 + (-5), 3 + 2 \rangle = \langle -1, 5 \rangle.$$

Thus we add vectors by adding corresponding components—that is, by *componentwise addition*. The geometric interpretation of vector addition is the **triangle law of addition**, illustrated in Fig. 12.3.4, where the labeled lengths indicate why this interpretation is valid. An equivalent interpretation is the **parallelogram law of addition**, illustrated in Fig. 12.3.5.

It is natural to write $2\mathbf{u} = \mathbf{u} + \mathbf{u}$. But if $\mathbf{u} = \langle u_1, u_2 \rangle$, then

$$2\mathbf{u} = \mathbf{u} + \mathbf{u} = \langle u_1, u_2 \rangle + \langle u_1, u_2 \rangle = \langle 2u_1, 2u_2 \rangle.$$

This suggests that multiplication of a vector by a scalar (real number) also is defined in a componentwise manner.

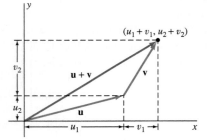

Fig. 12.3.5   The parallelogram law for vector addition

Definition   *Multiplication of a Vector by a Scalar*
If $\mathbf{u} = \langle u_1, u_2 \rangle$ and c is a real number, then the **scalar multiple** $c\mathbf{u}$ is the vector

$$c\mathbf{u} = \langle cu_1, cu_2 \rangle. \tag{3}$$

Ch. 12 / *Parametric Curves and Vectors in the Plane*

Note that

$$|c\mathbf{u}| = \sqrt{(cu_1)^2 + (cu_2)^2} = |c|\sqrt{(u_1)^2 + (u_2)^2} = |c| \cdot |\mathbf{u}|.$$

Thus the length of $c\mathbf{u}$ is $|c|$ times that of $\mathbf{u}$. The **negative** of the vector $\mathbf{u}$ is the vector

$$-\mathbf{u} = (-1)\mathbf{u} = \langle -u_1, -u_2 \rangle,$$

with the same length as $\mathbf{u}$ but the opposite direction. We say that the two nonzero vectors $\mathbf{u}$ and $\mathbf{v}$ have

❑ The **same direction** if $\mathbf{u} = c\mathbf{v}$ for some $c > 0$;

❑ **Opposite directions** if $\mathbf{u} = c\mathbf{v}$ for some $c < 0$.

The geometric interpretation of scalar multiplication is that $c\mathbf{u}$ is the vector with length $|c| \cdot |\mathbf{u}|$, with the same direction as $\mathbf{u}$ if $c > 0$ but the opposite direction if $c < 0$ (Fig. 12.3.6).

The **difference** $\mathbf{u} - \mathbf{v}$ of the vectors $\mathbf{u} = \langle u_1, u_2 \rangle$ and $\mathbf{v} = \langle v_1, v_2 \rangle$ is defined to be

$$\mathbf{u} - \mathbf{v} = \mathbf{u} + (-\mathbf{v}) = \langle u_1 - v_1, u_2 - v_2 \rangle. \tag{4}$$

If we think of $\langle u_1, u_2 \rangle$ and $\langle v_1, v_2 \rangle$ as position vectors of the points P and Q, respectively, then $\mathbf{u} - \mathbf{v}$ may be represented by the arrow $\overrightarrow{QP}$ from Q to P. We may therefore write

$$\mathbf{u} - \mathbf{v} = \overrightarrow{OP} - \overrightarrow{OQ} = \overrightarrow{QP},$$

as illustrated in Fig. 12.3.7.

EXAMPLE 1 Suppose that $\mathbf{u} = \langle 4, -3 \rangle$ and $\mathbf{v} = \langle -2, 3 \rangle$. Find $|\mathbf{u}|$ and the vectors $\mathbf{u} + \mathbf{v}$, $\mathbf{u} - \mathbf{v}$, $3\mathbf{u} - 2\mathbf{v}$, and $2\mathbf{u} + 4\mathbf{v}$.

Solution

$$|\mathbf{u}| = \sqrt{4^2 + (-3)^2} = \sqrt{25} = 5.$$

$$\mathbf{u} + \mathbf{v} = \langle 4 + (-2), -3 + 3 \rangle = \langle 2, 0 \rangle.$$

$$\mathbf{u} - \mathbf{v} = \langle 4 - (-2), -3 - 3 \rangle = \langle 6, -6 \rangle.$$

$$3\mathbf{u} = \langle 3 \cdot 4, 3 \cdot (-3) \rangle = \langle 12, -9 \rangle$$

$$-2\mathbf{v} = \langle -2 \cdot (-2), -2 \cdot 3 \rangle = \langle 4, -6 \rangle.$$

$$2\mathbf{u} + 4\mathbf{v} = \langle 2 \cdot 4 + 4 \cdot (-2), 2 \cdot (-3) + 4 \cdot 3 \rangle = \langle 0, 6 \rangle.$$

The familiar algebraic properties of real numbers carry over to the following analogous properties of vector addition and scalar multiplication. Let $\mathbf{a}$, $\mathbf{b}$, and $\mathbf{c}$ be vectors and r and s real numbers. Then

1. $\mathbf{a} + \mathbf{b} = \mathbf{b} + \mathbf{a}$,
2. $\mathbf{a} + (\mathbf{b} + \mathbf{c}) = (\mathbf{a} + \mathbf{b}) + \mathbf{c}$,
3. $r(\mathbf{a} + \mathbf{b}) = r\mathbf{a} + r\mathbf{b}$, $\qquad (5)$
4. $(r + s)\mathbf{a} = r\mathbf{a} + s\mathbf{a}$,
5. $(rs)\mathbf{a} = r(s\mathbf{a}) = s(r\mathbf{a})$.

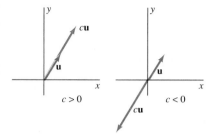

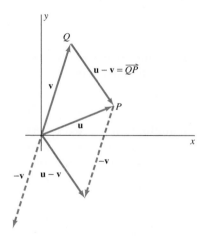

Fig. 12.3.6 The vector $c\mathbf{u}$ may have the same direction as $\mathbf{u}$ or the opposite direction, depending on the sign of c.

Fig. 12.3.7 Geometric interpretation of the difference $\mathbf{u} - \mathbf{v}$

You can easily verify these identities by working with components. For example, if $\mathbf{a} = \langle a_1, a_2 \rangle$ and $\mathbf{b} = \langle b_1, b_2 \rangle$, then

$$r(\mathbf{a} + \mathbf{b}) = r\langle a_1 + b_1, a_2 + b_2 \rangle = \langle r(a_1 + b_1), r(a_2 + b_2) \rangle$$
$$= \langle ra_1 + rb_1, ra_2 + rb_2 \rangle = \langle ra_1, ra_2 \rangle + \langle rb_1, rb_2 \rangle = r\mathbf{a} + r\mathbf{b}.$$

The proofs of the other four identities in (5) are left as exercises.

THE UNIT VECTORS i AND j

A **unit** vector is a vector of length 1. If $\mathbf{a} = \langle a_1, a_2 \rangle \neq \mathbf{0}$, then

$$\mathbf{u} = \frac{\mathbf{a}}{|\mathbf{a}|} \tag{6}$$

is the unit vector with the same direction as $\mathbf{a}$, because

$$|\mathbf{u}| = \sqrt{\left(\frac{a_1}{|\mathbf{a}|}\right)^2 + \left(\frac{a_2}{|\mathbf{a}|}\right)^2} = \frac{1}{|\mathbf{a}|}\sqrt{a_1{}^2 + a_2{}^2} = 1.$$

For example, if $\mathbf{a} = \langle 3, -4 \rangle$, then $|\mathbf{a}| = 5$. Thus $\langle \frac{3}{5}, -\frac{4}{5} \rangle$ is a unit vector that has the same direction as $\mathbf{a}$.

Two particular unit vectors play a special role, the vectors

$$\mathbf{i} = \langle 1, 0 \rangle \quad \text{and} \quad \mathbf{j} = \langle 0, 1 \rangle.$$

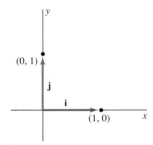

The first points in the positive x-direction; the second, in the positive y-direction (Fig. 12.3.8). Together they provide a useful alternative notation for vectors. If $\mathbf{a} = \langle a_1, a_2 \rangle$, then

$$\mathbf{a} = \langle a_1, 0 \rangle + \langle 0, a_2 \rangle = a_1\langle 1, 0 \rangle + a_2\langle 0, 1 \rangle = a_1\mathbf{i} + a_2\mathbf{j}. \tag{7}$$

Fig. 12.3.8 The vectors **i** and **j**

Thus every vector in the plane is a **linear combination** of **i** and **j**. The usefulness of this notation is based on the fact that such linear combinations of **i** and **j** may be manipulated as if they were ordinary sums. For example, if

$$\mathbf{a} = a_1\mathbf{i} + a_2\mathbf{j} \quad \text{and} \quad \mathbf{b} = b_1\mathbf{i} + b_2\mathbf{j},$$

then

$$\mathbf{a} + \mathbf{b} = (a_1\mathbf{i} + a_2\mathbf{j}) + (b_1\mathbf{i} + b_2\mathbf{j}) = (a_1 + b_1)\mathbf{i} + (a_2 + b_2)\mathbf{j}.$$

Also,

$$c\mathbf{a} = c(a_1\mathbf{i} + a_2\mathbf{j}) = (ca_1)\mathbf{i} + (ca_2)\mathbf{j}.$$

EXAMPLE 2 Suppose that $\mathbf{a} = 2\mathbf{i} - 3\mathbf{j}$ and $\mathbf{b} = 3\mathbf{i} + 4\mathbf{j}$. Express $5\mathbf{a} - 3\mathbf{b}$ in terms of **i** and **j**.

Solution

$$5\mathbf{a} - 3\mathbf{b} = 5 \cdot (2\mathbf{i} - 3\mathbf{j}) - 3 \cdot (3\mathbf{i} + 4\mathbf{j})$$
$$= (10 - 9)\mathbf{i} + (-15 - 12)\mathbf{j} = \mathbf{i} - 27\mathbf{j}.$$

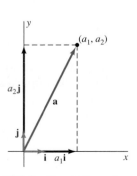

Fig. 12.3.9 Resolution of $\mathbf{a} = \langle a_1, a_2 \rangle$ into its horizontal and vertical components

Equation (7) expresses the vector $\mathbf{a} = \langle a_1, a_2 \rangle$ as the sum of a horizontal vector $a_1\mathbf{i}$ and a vertical vector $a_2\mathbf{j}$, as Fig. 12.3.9 shows. The decomposition, or "resolution," of a vector into its horizontal and vertical components is an important technique in the study of vector quantities. For example, a force $\mathbf{F}$ may be decomposed into its horizontal and vertical components $F_1\mathbf{i}$ and $F_2\mathbf{j}$, respectively. The physical effect of the single force $\mathbf{F}$ is the same as

Ch. 12 / Parametric Curves and Vectors in the Plane

the combined effect of the separate forces $F_1\mathbf{i}$ and $F_2\mathbf{j}$. (This is an instance of the empirically verifiable parallelogram law of addition of forces.) Because of this decomposition, many two-dimensional problems can be reduced to one-dimensional problems, the latter solved, and the two results combined (again by vector methods) to give the solution of the original problem.

PERPENDICULAR VECTORS AND THE DOT PRODUCT

Two nonzero vectors $\mathbf{a}$ and $\mathbf{b}$ are said to be **perpendicular** if, when they are represented as position vectors $\mathbf{a} = \overrightarrow{OP}$ and $\mathbf{b} = \overrightarrow{OQ}$, the line segments $\overrightarrow{OP}$ and $\overrightarrow{OQ}$ are perpendicular. We need a simple way to determine whether or not two given vectors are perpendicular.

Figure 12.3.10 shows the two nonzero vectors

$$\mathbf{a} = \langle a_1, a_2 \rangle \quad \text{and} \quad \mathbf{b} = \langle b_1, b_2 \rangle. \tag{8}$$

Their difference

$$\mathbf{a} - \mathbf{b} = \langle a_1 - b_1, a_2 - b_2 \rangle \tag{9}$$

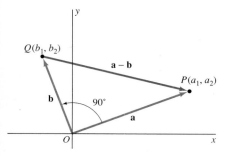

Fig. 12.3.10 Two perpendicular vectors

is represented by the third side $\overrightarrow{QP}$ of the triangle OPQ. By the Pythagorean theorem and its converse, this triangle has a right angle at O if and only if

$$|\mathbf{a}|^2 + |\mathbf{b}|^2 = |\mathbf{a} - \mathbf{b}|^2;$$

$$(a_1^2 + a_2^2) + (b_1^2 + b_2^2) = (a_1 - b_1)^2 + (a_2 - b_2)^2$$

$$= (a_1^2 - 2a_1b_1 + b_1^2) + (a_2^2 - 2a_2b_2 + b_2^2);$$

the latter is true if and only if

$$a_1b_1 + a_2b_2 = 0. \tag{10}$$

This last condition is the condition for perpendicularity of the vectors $\mathbf{a}$ and $\mathbf{b}$, and we have proved the following theorem.

Theorem *Test for Perpendicular Vectors*

The two nonzero vectors $\mathbf{a} = \langle a_1, a_2 \rangle$ and $\mathbf{b} = \langle b_1, b_2 \rangle$ are perpendicular if and only if

$$a_1b_1 + a_2b_2 = 0. \tag{10}$$

NOTE It is consistent with Eq. (10) to regard the zero vector $\mathbf{0} = \langle 0, 0 \rangle$ as perpendicular to every vector.

The particular combination of the components of $\mathbf{a}$ and $\mathbf{b}$ in Eq. (10) has important applications and is called the *dot product* of $\mathbf{a}$ and $\mathbf{b}$.

Definition *Dot Product of Two Vectors*

The **dot product** $\mathbf{a} \cdot \mathbf{b}$ of two vectors $\mathbf{a} = \langle a_1, a_2 \rangle$ and $\mathbf{b} = \langle b_1, b_2 \rangle$ is the real number

$$\mathbf{a} \cdot \mathbf{b} = a_1b_1 + a_2b_2. \tag{11}$$

The dot product of two vectors is not a vector, but a *scalar*. For this reason it is sometimes called the *scalar product* of vectors. For example,

$$\langle 2, -3 \rangle \cdot \langle -4, 2 \rangle = 2 \cdot (-4) + (-3) \cdot 2 = -14.$$

The following algebraic properties of the dot product are easy to verify by componentwise calculations. For all vectors $\mathbf{a}$, $\mathbf{b}$, and $\mathbf{c}$ and all scalars r,

1. $\mathbf{a} \cdot \mathbf{a} = |\mathbf{a}|^2$,
2. $\mathbf{a} \cdot \mathbf{b} = \mathbf{b} \cdot \mathbf{a}$,
3. $\mathbf{a} \cdot (\mathbf{b} + \mathbf{c}) = \mathbf{a} \cdot \mathbf{b} + \mathbf{a} \cdot \mathbf{c}$,
4. $(r\mathbf{a}) \cdot \mathbf{b} = r(\mathbf{a} \cdot \mathbf{b}) = \mathbf{a} \cdot (r\mathbf{b})$. (12)

For examples of how to prove such statements,

$$\mathbf{a} \cdot \mathbf{a} = (a_1)^2 + (a_2)^2 = |\mathbf{a}|^2$$

and

$$\mathbf{a} \cdot (\mathbf{b} + \mathbf{c}) = \langle a_1, a_2 \rangle \cdot \langle b_1 + c_1, b_2 + c_2 \rangle = a_1(b_1 + c_1) + a_2(b_2 + c_2)$$
$$= (a_1 b_1 + a_2 b_2) + (a_1 c_1 + a_2 c_2) = \mathbf{a} \cdot \mathbf{b} + \mathbf{a} \cdot \mathbf{c}.$$

Because $\mathbf{i} \cdot \mathbf{i} = \mathbf{j} \cdot \mathbf{j} = 1$ and $\mathbf{i} \cdot \mathbf{j} = \mathbf{j} \cdot \mathbf{i} = 0$, application of the properties in (12) gives

$$\mathbf{a} \cdot \mathbf{b} = (a_1 \mathbf{i} + a_2 \mathbf{j}) \cdot (b_1 \mathbf{i} + b_2 \mathbf{j})$$
$$= (a_1 b_1)\mathbf{i} \cdot \mathbf{i} + (a_1 b_2 + a_2 b_1)\mathbf{i} \cdot \mathbf{j} + (a_2 b_2)\mathbf{j} \cdot \mathbf{j}$$
$$= a_1 b_1 + a_2 b_2,$$

in accord with the definition of $\mathbf{a} \cdot \mathbf{b}$. Thus we can compute dot products of vectors expressed in terms of $\mathbf{i}$ and $\mathbf{j}$ by "ordinary" algebra.

The test for perpendicularity is best remembered in terms of the dot product.

Corollary *Test for Perpendicular Vectors*
The two nonzero vectors $\mathbf{a}$ and $\mathbf{b}$ are perpendicular if and only if $\mathbf{a} \cdot \mathbf{b} = 0$.

EXAMPLE 3 Determine whether or not the following pairs of vectors are perpendicular: (a) $\langle 3, 4 \rangle$ and $\langle 8, -6 \rangle$; (b) $\langle 2, 3 \rangle$ and $\langle 4, -3 \rangle$.

Solution   (a) $\langle 3, 4 \rangle \cdot \langle 8, -6 \rangle = 24 - 24 = 0$, so these two vectors are perpendicular. (b) $\langle 2, 3 \rangle \cdot \langle 4, -3 \rangle = 8 - 9 = -1 \neq 0$, so $\langle 2, 3 \rangle$ and $\langle 4, -3 \rangle$ are *not* perpendicular.

In Section 13.1 we shall see that $\mathbf{a} \cdot \mathbf{b} = |\mathbf{a}||\mathbf{b}| \cos \theta$, where θ is the angle between the vectors $\mathbf{a}$ and $\mathbf{b}$. The case $\theta = \pi/2$ gives the preceding corollary as a special case.

12.3 Problems

In Problems 1 through 8, find $|\mathbf{a}|$, $|-2\mathbf{b}|$, $|\mathbf{a} - \mathbf{b}|$, $\mathbf{a} + \mathbf{b}$, *and* $3\mathbf{a} - 2\mathbf{b}$. *Also determine whether or not* $\mathbf{a}$ *and* $\mathbf{b}$ *are perpendicular.*

1. $\mathbf{a} = \langle 1, -2 \rangle$, $\mathbf{b} = \langle -3, 2 \rangle$
2. $\mathbf{a} = \langle 3, 4 \rangle$, $\mathbf{b} = \langle -4, 3 \rangle$
3. $\mathbf{a} = \langle -2, -2 \rangle$, $\mathbf{b} = \langle -3, -4 \rangle$
4. $\mathbf{a} = -2\langle 4, 7 \rangle$, $\mathbf{b} = -3\langle -4, -2 \rangle$
5. $\mathbf{a} = \mathbf{i} + 3\mathbf{j}$, $\mathbf{b} = 2\mathbf{i} - 5\mathbf{j}$
6. $\mathbf{a} = 2\mathbf{i} - 5\mathbf{j}$, $\mathbf{b} = \mathbf{i} - 6\mathbf{j}$
7. $\mathbf{a} = 4\mathbf{i}$, $\mathbf{b} = -7\mathbf{j}$
8. $\mathbf{a} = -\mathbf{i} - \mathbf{j}$, $\mathbf{b} = 2\mathbf{i} + 2\mathbf{j}$

In Problems 9 through 12, find a unit vector $\mathbf{u}$ *with the same direction as the given vector* $\mathbf{a}$. *Express* $\mathbf{u}$ *in terms of* $\mathbf{i}$ *and* $\mathbf{j}$. *Also find a unit vector* $\mathbf{v}$ *with the direction opposite that of* $\mathbf{a}$.

9. $\mathbf{a} = \langle -3, -4 \rangle$
10. $\mathbf{a} = \langle 5, -12 \rangle$
11. $\mathbf{a} = 8\mathbf{i} + 15\mathbf{j}$
12. $\mathbf{a} = 7\mathbf{i} - 24\mathbf{j}$

In Problems 13 through 16, find the vector $\mathbf{a}$, *expressed in terms of* $\mathbf{i}$ *and* $\mathbf{j}$, *that is represented by the arrow* $\overrightarrow{PQ}$ *in the plane.*

13. $P = (3, 2)$, $Q = (3, -2)$
14. $P = (-3, 5)$, $Q = (-3, 6)$
15. $P = (-4, 7)$, $Q = (4, -7)$
16. $P = (1, -1)$, $Q = (-4, -1)$

In Problems 17 through 20, determine whether or not the given vectors $\mathbf{a}$ *and* $\mathbf{b}$ *are perpendicular.*

17. $\mathbf{a} = \langle 6, 0 \rangle$, $\mathbf{b} = \langle 0, -7 \rangle$
18. $\mathbf{a} = 3\mathbf{j}$, $\mathbf{b} = 3\mathbf{i} - \mathbf{j}$
19. $\mathbf{a} = 2\mathbf{i} - \mathbf{j}$, $\mathbf{b} = 4\mathbf{j} + 8\mathbf{i}$
20. $\mathbf{a} = 8\mathbf{i} + 10\mathbf{j}$, $\mathbf{b} = 15\mathbf{i} - 12\mathbf{j}$

21. Find a vector that has the same direction as $5\mathbf{i} - 7\mathbf{j}$ and is (a) three times its length; (b) one-third its length.

22. Find a vector that has the opposite direction from $-3\mathbf{i} + 5\mathbf{j}$ and is (a) four times its length; (b) one-fourth its length.

23. Find a vector of length 5 with (a) the same direction as $7\mathbf{i} - 3\mathbf{j}$; (b) the opposite direction from $8\mathbf{i} + 5\mathbf{j}$.

24. For what numbers c are the vectors $\langle c, 2 \rangle$ and $\langle c, -8 \rangle$ perpendicular?

25. For what numbers c are the vectors $2c\mathbf{i} - 4\mathbf{j}$ and $3\mathbf{i} + c\mathbf{j}$ perpendicular?

26. Given the three points $A(2, 3)$, $B(-5, 7)$, and $C(1, -5)$,

verify by direct computation of the vectors and their sum that $\overrightarrow{AB} + \overrightarrow{BC} + \overrightarrow{CA} = \mathbf{0}$.

In Problems 27 through 32, give a componentwise proof of the indicated property of vector algebra. Take $\mathbf{a} = \langle a_1, a_2 \rangle$ *and* $\mathbf{b} = \langle b_1, b_2 \rangle$ *throughout.*

27. $\mathbf{a} + (\mathbf{b} + \mathbf{c}) = (\mathbf{a} + \mathbf{b}) + \mathbf{c}$
28. $(r + s)\mathbf{a} = r\mathbf{a} + s\mathbf{a}$
29. $(rs)\mathbf{a} = r(s\mathbf{a})$
30. $\mathbf{a} \cdot \mathbf{b} = \mathbf{b} \cdot \mathbf{a}$
31. $(r\mathbf{a}) \cdot \mathbf{b} = r(\mathbf{a} \cdot \mathbf{b})$
32. If $\mathbf{a} + \mathbf{b} = \mathbf{a}$, then $\mathbf{b} = \mathbf{0}$.

In Problems 33 through 35, assume the following fact: If an airplane flies with velocity vector $\mathbf{v}_a$ *relative to the air and the velocity of the wind is* $\mathbf{w}$, *then the velocity vector of the plane relative to the ground is* $\mathbf{v}_g = \mathbf{v}_a + \mathbf{w}$ *(Fig. 12.3.11). The vector* $\mathbf{v}_a$ *is called the* **apparent velocity vector**, *whereas* $\mathbf{v}_g$ *is called the* **true velocity vector**.

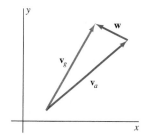

Fig. 12.3.11 The vectors of Problems 33 through 35
Apparent velocity: $\mathbf{v}_a$ Wind velocity: $\mathbf{w}$
True velocity: $\mathbf{v}_g = \mathbf{v}_a + \mathbf{w}$

33. Suppose that the wind is blowing from the northeast at 50 mi/h and that the pilot wishes to fly due east at 500 mi/h. What should the plane's apparent velocity vector be?

34. Repeat Problem 33 with the phrase *due east* replaced by *due west*.

35. Repeat Problem 33 in the case that the pilot wishes to fly northwest at 500 mi/h.

36. In the triangle ABC, let M and N be the midpoints of AB and AC, respectively. Show that $\overrightarrow{MN} = \frac{1}{2}\overrightarrow{BC}$. Conclude that the line segment joining the midpoints of two sides of a triangle is parallel to the third side. How are their lengths related?

37. Prove that the diagonals of a parallelogram $ABCD$ bisect each other. [*Suggestion:* If M and N are the mid-

points of the diagonals AC and BD, respectively, and O is the origin, show that $\overrightarrow{OM} = \overrightarrow{ON}$.]

38. Use vectors to prove that the midpoints of the four sides of an arbitrary quadrilateral are the vertices of a parallelogram.

39. Prove that the vector $\mathbf{n} = a\mathbf{i} + b\mathbf{j}$ is perpendicular to the line with equation $ax + by + c = 0$ [*Suggestion:* If $P(x_1, y_1)$ and $Q(x_2, y_2)$ are two points on the line, show that $\mathbf{n} \cdot \overrightarrow{PQ} = 0$.]

40. Figure 12.3.12 shows the vector $\mathbf{a}_\perp$ obtained by rotating the vector $\mathbf{a} = a_1\mathbf{i} + a_2\mathbf{j}$ through a counterclockwise angle of $90°$. Show that

$$\mathbf{a}_\perp = -a_2\mathbf{i} + a_1\mathbf{j}.$$

[*Suggestion:* Begin by writing $\mathbf{a} = (r \cos \theta)\mathbf{i} + (r \sin \theta)\mathbf{j}$.]

41. Is the dot product associative? That is, if $\mathbf{a}$, $\mathbf{b}$, and $\mathbf{c}$ are vectors, does the equation $\mathbf{a} \cdot (\mathbf{b} \cdot \mathbf{c}) = (\mathbf{a} \cdot \mathbf{b}) \cdot \mathbf{c}$ always hold? Does this question make any sense?

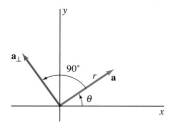

Fig. 12.3.12 Rotate $\mathbf{a}$ counterclockwise $90°$ to obtain $\mathbf{a}_\perp$ (Problem 40).

12.4

Motion and Vector-Valued Functions

We now use vectors to study the motion of a point in the coordinate plane. If the coordinates of the moving point at time t are given by the parametric equations $x = f(t)$, $y = g(t)$, then the vector

$$\mathbf{r}(t) = f(t)\mathbf{i} + g(t)\mathbf{j}, \quad \text{or} \quad \mathbf{r} = x\mathbf{i} + y\mathbf{j}, \tag{1}$$

is called the **position vector** of the point. Equation (1) determines a **vector-valued function** $\mathbf{r} = \mathbf{r}(t)$ that associates with the number t the position vector $\mathbf{r}(t)$ of the moving point. In the bracket notation for vectors, a vector-valued function is an ordered pair of real-valued functions: $\mathbf{r}(t) = \langle f(t), g(t) \rangle$.

Much of the calculus of (ordinary) real-valued functions applies to vector-valued functions. To begin with, the **limit** of a vector-valued function $\mathbf{r} = \langle f, g \rangle$ is defined as follows:

$$\lim_{t \to a} \mathbf{r}(t) = \left\langle \lim_{t \to a} f(t), \lim_{t \to a} g(t) \right\rangle = \mathbf{i}\left(\lim_{t \to a} f(t) \right) + \mathbf{j}\left(\lim_{t \to a} g(t) \right), \tag{2}$$

provided that the limits in the latter two expressions exist. Thus we take limits of vector-valued functions by taking limits of their component functions.

We say that $\mathbf{r} = \mathbf{r}(t)$ is **continuous** at the number a provided that

$$\lim_{t \to a} \mathbf{r}(t) = \mathbf{r}(a).$$

This amounts to saying that $\mathbf{r}$ is continuous at a if and only if its component functions f and g are continuous at a.

The derivative $\mathbf{r}'(t)$ of the vector-valued function $\mathbf{r}(t)$ is defined in almost exactly the same way as the derivative of a real-valued function. Specifically,

$$\mathbf{r}'(t) = \lim_{h \to 0} \frac{\mathbf{r}(t + h) - \mathbf{r}(t)}{h}, \tag{3}$$

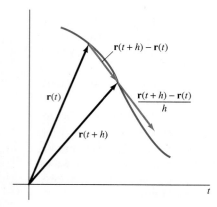

Fig. 12.4.1 Geometry of the derivative of a vector-valued function

provided that this limit exists. A geometric note: Fig. 12.4.1 suggests that $\mathbf{r}'(t)$ will be tangent to the curve swept out by $\mathbf{r}$ if $\mathbf{r}'(t)$ is attached to the curve at the point of evaluation. Note that because $| \mathbf{r}(t + h) - \mathbf{r}(t) |$ is the distance from the point with position vector $\mathbf{r}(t)$ to the point with position vector $\mathbf{r}(t + h)$, the quotient

$$\frac{|\mathbf{r}(t + h) - \mathbf{r}(t)|}{h}$$

equals the average speed of a particle that travels from $\mathbf{r}(t)$ to $\mathbf{r}(t + h)$ in time h. Consequently, the limit in Eq. (3) yields both the direction of motion and the instantaneous speed of a particle moving with position vector $\mathbf{r}(t)$ along a curve.

Our next result implies the simple *but important* fact that $\mathbf{r}'(t)$ can be calculated by componentwise differentiation. We shall also denote derivatives by

$$\mathbf{r}'(t) = D_t\,\mathbf{r}(t) = \frac{d\mathbf{r}}{dt}.$$

Theorem 1 *Componentwise Differentiation*
Suppose that

$$\mathbf{r}(t) = \langle f(t),\, g(t) \rangle = f(t)\mathbf{i} + g(t)\mathbf{j},$$

where both f and g are differentiable functions. Then

$$\mathbf{r}'(t) = \langle f'(t),\, g'(t) \rangle = f'(t)\mathbf{i} + g'(t)\mathbf{j}. \tag{4}$$

That is, if $\mathbf{r} = x\mathbf{i} + y\mathbf{j}$, then

$$\frac{d\mathbf{r}}{dt} = \frac{dx}{dt}\mathbf{i} + \frac{dy}{dt}\mathbf{j}.$$

Proof We take the limit in Eq. (3) simply by taking limits of components. We find that

$$
\begin{aligned}
\mathbf{r}'(t) &= \lim_{h \to 0} \frac{\mathbf{r}(t + h) - \mathbf{r}(t)}{h} \\
&= \lim_{h \to 0} \frac{f(t + h)\mathbf{i} + g(t + h)\mathbf{j} - f(t)\mathbf{i} - g(t)\mathbf{j}}{h} \\
&= \left(\lim_{h \to 0} \frac{f(t + h) - f(t)}{h} \right)\mathbf{i} + \left(\lim_{h \to 0} \frac{g(t + h) - g(t)}{h} \right)\mathbf{j} \\
&= f'(t)\mathbf{i} + g'(t)\mathbf{j}. \quad \square
\end{aligned}
$$

In Eq. (5) of Section 12.1, we saw that the line tangent to the parametric curve $x = f(t),\ y = g(t)$ has slope $g'(t)/f'(t)$ and therefore is parallel to the line through $(0, 0)$ and $(f'(t), g'(t))$. Hence, by Theorem 1, we may visualize the derivative vector $\mathbf{r}'(t)$ as a tangent to the curve at the point $(f(t), g(t))$ (Fig. 12.4.2).

Theorem 2 tells us that the formulas for computing derivatives of sums and products of vector-valued functions are formally similar to those for real-valued functions.

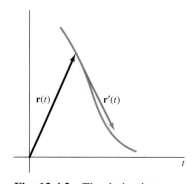

Fig. 12.4.2 The derivative vector is tangent to the curve at the point of evaluation.

Theorem 2 *Differentiation Formulas*

Let $\mathbf{u}(t)$ and $\mathbf{v}(t)$ be differentiable vector-valued functions. Let $h(t)$ be a differentiable real-valued function and let c be a (constant) scalar. Then

1. $D_t[\mathbf{u}(t) + \mathbf{v}(t)] = \mathbf{u}'(t) + \mathbf{v}'(t),$

2. $D_t[c\mathbf{u}(t)] = c\mathbf{u}'(t),$

3. $D_t[h(t)\mathbf{u}(t)] = h'(t)\mathbf{u}(t) + h(t)\mathbf{u}'(t),$

4. $D_t[\mathbf{u}(t)\cdot\mathbf{v}(t)] = \mathbf{u}'(t)\cdot\mathbf{v}(t) + \mathbf{u}(t)\cdot\mathbf{v}'(t).$

Proof We shall prove part (4) and leave the other parts as exercises. If

$$\mathbf{u}(t) = \langle f_1(t), f_2(t)\rangle \quad \text{and} \quad \mathbf{v}(t) = \langle g_1(t), g_2(t)\rangle,$$

then

$$\mathbf{u}(t)\cdot\mathbf{v}(t) = f_1(t)g_1(t) + f_2(t)g_2(t).$$

Hence the product rule for ordinary real-valued functions gives

$$
\begin{aligned}
D_t[\mathbf{u}(t)\cdot\mathbf{v}(t)] &= D_t[f_1(t)g_1(t) + f_2(t)g_2(t)] \\
&= [f_1'(t)g_1(t) + f_2'(t)g_2(t)] + [f_1(t)g_1'(t) + f_2(t)g_2'(t)] \\
&= \mathbf{u}'(t)\cdot\mathbf{v}(t) + \mathbf{u}(t)\cdot\mathbf{v}'(t). \quad \square
\end{aligned}
$$

Now we can discuss the motion of a point with position vector $\mathbf{r}(t) = f(t)\mathbf{i} + g(t)\mathbf{j}$. Its **velocity vector** $\mathbf{v}(t)$ and **acceleration vector** $\mathbf{a}(t)$ are defined as follows:

$$\mathbf{v}(t) = \mathbf{r}'(t) = f'(t)\mathbf{i} + g'(t)\mathbf{j}, \tag{5}$$

$$\mathbf{a}(t) = \mathbf{v}'(t) = f''(t)\mathbf{i} + g''(t)\mathbf{j}. \tag{6}$$

We also write

$$\mathbf{v} = \frac{d\mathbf{r}}{dt} = \frac{dx}{dt}\mathbf{i} + \frac{dy}{dt}\mathbf{j}$$

and

$$\mathbf{a} = \frac{d^2\mathbf{r}}{dt^2} = \frac{d^2x}{dt^2}\mathbf{i} + \frac{d^2y}{dt^2}\mathbf{j}.$$

The moving point has **speed** $v(t)$ and **scalar acceleration** $a(t)$, scalar-valued functions that are merely the lengths of the corresponding velocity and acceleration vectors. Thus

$$v(t) = |\mathbf{v}(t)| = \sqrt{\left(\frac{dx}{dt}\right)^2 + \left(\frac{dy}{dt}\right)^2} \tag{7}$$

and

$$a(t) = |\mathbf{a}(t)| = \sqrt{\left(\frac{d^2x}{dt^2}\right)^2 + \left(\frac{d^2y}{dt^2}\right)^2}. \tag{8}$$

Do *not* assume that the scalar acceleration a and the derivative dv/dt of the speed are equal. Although this sometimes holds for motion in a straight line, it is generally false for two-dimensional motion.

EXAMPLE 1 A moving particle has position vector

$$\mathbf{r}(t) = t\mathbf{i} + t^2\mathbf{j}.$$

Find the velocity and acceleration vectors and the speed and the scalar acceleration when $t = 2$.

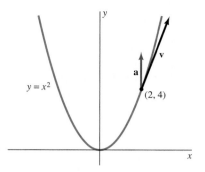

Fig. 12.4.3 The velocity and acceleration vectors at $t = 2$ (Example 1)

Solution Because $\mathbf{r}(2) = 2\mathbf{i} + 4\mathbf{j}$, the particle is at the point $(2, 4)$ on the parabola $y = x^2$ when $t = 2$. Its velocity vector is

$$\mathbf{v}(t) = \mathbf{i} + 2t\mathbf{j}.$$

So $\mathbf{v}(2) = \mathbf{i} + 4\mathbf{j}$ and $v(2) = |\mathbf{v}(2)| = \sqrt{17}$. The *velocity* $\mathbf{v}(2)$ is a *vector*, whereas the *speed* $v(2)$ is a *scalar*.

The acceleration vector is $\mathbf{a} = 2\mathbf{j}$, so the acceleration of the particle is the (constant) vector $2\mathbf{j}$. In particular, when $t = 2$, its acceleration is $2\mathbf{j}$ and its scalar acceleration is $a = 2$. Figure 12.4.3 shows the trajectory of the particle with the vectors $\mathbf{v}(2)$ and $\mathbf{a}(2)$ attached at the point corresponding to $t = 2$.

INTEGRATION OF VECTOR-VALUED FUNCTIONS

Integrals of vector-valued functions are defined by analogy with the definition of an integral of a real-valued function:

$$\int_a^b \mathbf{r}(t)\, dt = \lim_{\Delta t \to 0} \sum_{i=1}^{n} \mathbf{r}(t_i^*)\, \Delta t, \tag{9}$$

where t_i^* is a point of the ith subinterval of a division of $[a, b]$ into n subintervals, all with the same length $\Delta t = (b - a)/n$.

If $\mathbf{r}(t) = f(t)\mathbf{i} + g(t)\mathbf{j}$ is continuous on $[a, b]$, then—by taking limits componentwise—we get

$$\int_a^b \mathbf{r}(t)\, dt = \lim_{\Delta t \to 0} \sum_{i=1}^{n} \mathbf{r}(t_i^*)\, \Delta t$$

$$= \mathbf{i}\left(\lim_{\Delta t \to 0} \sum_{i=1}^{n} f(t_i^*)\, \Delta t \right) + \mathbf{j}\left(\lim_{\Delta t \to 0} \sum_{i=1}^{n} g(t_i^*)\Delta t \right).$$

This gives the result that

$$\int_a^b \mathbf{r}(t)\, dt = \mathbf{i}\left(\int_a^b f(t)\, dt \right) + \mathbf{j}\left(\int_a^b g(t)\, dt \right). \tag{10}$$

Thus *a vector-valued function may be integrated componentwise.*

Now suppose that $\mathbf{R}(t)$ is an *antiderivative* of $\mathbf{r}(t)$, meaning that $\mathbf{R}'(t) = \mathbf{r}(t)$. That is, if $\mathbf{R}(t) = F(t)\mathbf{i} + G(t)\mathbf{j}$, then

$$\mathbf{R}'(t) = F'(t)\mathbf{i} + G'(t)\mathbf{j} = f(t)\mathbf{i} + g(t)\mathbf{j} = \mathbf{r}(t).$$

Then componentwise integration yields

$$\int_a^b \mathbf{r}(t)\, dt = \mathbf{i}\left(\int_a^b f(t)\, dt \right) + \mathbf{j}\left(\int_a^b g(t)\, dt \right) = \mathbf{i}\left[F(t) \right]_a^b + \mathbf{j}\left[G(t) \right]_a^b$$

$$= [F(b)\mathbf{i} + G(b)\mathbf{j}] - [F(a)\mathbf{i} + G(a)\mathbf{j}].$$

Thus the *fundamental theorem of calculus* for vector-valued functions takes the form

$$\int_a^b \mathbf{r}(t)\,dt = \left[\mathbf{R}(t)\right]_a^b = \mathbf{R}(b) - \mathbf{R}(a), \tag{11}$$

where $\mathbf{R}'(t) = \mathbf{r}(t)$.

Vector integration is the basis for at least one method of navigation. If a submarine is cruising beneath the icecap at the North Pole, as in Fig. 12.4.4,

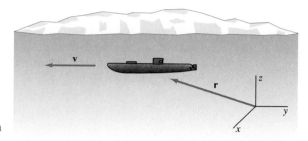

Fig. 12.4.4 A submarine beneath the polar icecap

and thus can use neither visual nor radio methods to determine its position, there is an alternative. Build a sensitive gyroscope-accelerometer combination and install it in the submarine. The device continuously measures the sub's acceleration vector, beginning at the time $t = 0$ when its position $\mathbf{r}(0)$ and velocity $\mathbf{v}(0)$ are known. Because $\mathbf{v}'(t) = \mathbf{a}(t)$, Eq. (10) gives

$$\int_0^t \mathbf{a}(t)\,dt = \left[\mathbf{v}(t)\right]_0^t = \mathbf{v}(t) - \mathbf{v}(0),$$

so

$$\mathbf{v}(t) = \mathbf{v}(0) + \int_0^t \mathbf{a}(t)\,dt.$$

Thus the velocity at every time $t \geqq 0$ is known. Similarly, because $\mathbf{r}'(t) = \mathbf{v}(t)$, a second integration gives

$$\mathbf{r}(t) = \mathbf{r}(0) + \int_0^t \mathbf{v}(t)\,dt$$

for the position of the sub at every time t. On-board computers can be programmed to carry out these integrations (perhaps by using Simpson's approximation) and continuously provide captain and crew with the submarine's (almost) exact position and velocity.

Indefinite integrals of vector-valued functions may be computed also. If $\mathbf{R}'(t) = \mathbf{r}(t)$, then every antiderivative of $\mathbf{r}(t)$ is of the form $\mathbf{R}(t) + \mathbf{C}$ for some constant vector $\mathbf{C}$. We therefore write

$$\int \mathbf{r}(t)\,dt = \mathbf{R}(t) + \mathbf{C} \quad \text{if} \quad \mathbf{R}'(t) = \mathbf{r}(t), \tag{12}$$

on the basis of a componentwise computation similar to the one leading to Eq. (11).

Ch. 12 / Parametric Curves and Vectors in the Plane

EXAMPLE 2 Suppose that a moving point has initial position $\mathbf{r}(0) = 2\mathbf{i}$, initial velocity $\mathbf{v}(0) = \mathbf{i} - \mathbf{j}$, and acceleration $\mathbf{a}(t) = 2\mathbf{i} + 6t\mathbf{j}$. Find its position and velocity at time t.

Solution Because $\mathbf{a}(t) = \mathbf{v}'(t)$, Eq. (12) gives

$$\mathbf{v}(t) = \int \mathbf{a}(t)\, dt + \mathbf{C}_1 = \int (2\mathbf{i} + 6t\mathbf{j})\, dt + \mathbf{C}_1 = 2t\mathbf{i} + 3t^2\mathbf{j} + \mathbf{C}_1.$$

To evaluate $\mathbf{C}_1$, we use the fact that $\mathbf{v}(0)$ is known. We substitute $t = 0$ into both sides of the last equation and find that $\mathbf{C}_1 = \mathbf{v}(0) = \mathbf{i} - \mathbf{j}$. So

$$\mathbf{v}(t) = (2t\mathbf{i} + 3t^2\mathbf{j}) + (\mathbf{i} - \mathbf{j}) = (2t + 1)\mathbf{i} + (3t^2 - 1)\mathbf{j}.$$

Next,

$$\mathbf{r}(t) = \int \mathbf{v}(t)\, dt + \mathbf{C}_2 = (t^2 + t)\mathbf{i} + (t^3 - t)\mathbf{j} + \mathbf{C}_2.$$

Again we substitute $t = 0$, and we find that $\mathbf{C}_2 = \mathbf{r}(0) = 2\mathbf{i}$. Hence

$$\mathbf{r}(t) = (t^2 + t + 2)\mathbf{i} + (t^3 - t)\mathbf{j}$$

is the position vector of the moving point.

MOTION OF PROJECTILES

Suppose that a projectile is launched from the point (x_0, y_0), with y_0 denoting its initial height above the surface of the earth. Let α be the angle of inclination from the horizontal of its initial velocity vector $\mathbf{v}_0$ (Fig. 12.4.5). Then its initial position vector is

$$\mathbf{r}_0 = x_0\mathbf{i} + y_0\mathbf{j}, \tag{13a}$$

and from Fig. 12.4.5 we see that

$$\mathbf{v}_0 = (v_0 \cos \alpha)\mathbf{i} + (v_0 \sin \alpha)\mathbf{j}, \tag{13b}$$

where $v_0 = |\mathbf{v}_0|$ is the initial speed of the projectile.

We suppose that the motion takes place sufficiently close to the surface that we may assume the earth is flat and that gravity is perfectly uniform. Then, if we also ignore air resistance, the acceleration of the projectile is

$$\mathbf{a} = \frac{d\mathbf{v}}{dt} = -g\mathbf{j},$$

where $g \approx 32$ ft/s². Antidifferentiation gives

$$\mathbf{v}(t) = -gt\mathbf{j} + \mathbf{C}_1.$$

Put $t = 0$ in both sides of this last equation. This shows that $\mathbf{C}_1 = \mathbf{v}_0$ (as expected!) and thus that

$$\mathbf{v}(t) = \frac{d\mathbf{r}}{dt} = -gt\mathbf{j} + \mathbf{v}_0.$$

Another antidifferentiation gives

$$\mathbf{r}(t) = -\tfrac{1}{2}gt^2\mathbf{j} + \mathbf{v}_0 t + \mathbf{C}_2.$$

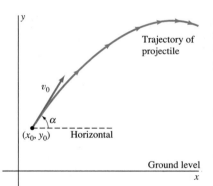

Fig. 12.4.5 Trajectory of a projectile launched at the angle α

Now substitution of $t = 0$ yields $\mathbf{C}_2 = \mathbf{r}_0$, so the position vector of the projectile at time t is

$$\mathbf{r}(t) = -\tfrac{1}{2}gt^2\mathbf{j} + \mathbf{v}_0 t + \mathbf{r}_0. \tag{14}$$

Equations (13a) and (13b) now give

$$\mathbf{r}(t) = [(v_0 \cos \alpha)t + x_0]\mathbf{i} + [-\tfrac{1}{2}gt^2 + (v_0 \sin \alpha)t + y_0]\mathbf{j},$$

so the parametric equations of the particle's trajectory are

$$x = (v_0 \cos \alpha)t + x_0, \tag{15}$$

$$y = -\tfrac{1}{2}gt^2 + (v_0 \sin \alpha)t + y_0. \tag{16}$$

EXAMPLE 3 An airplane is flying horizontally at an altitude of 1600 ft to pass directly over snowbound cattle on the ground and release hay there. The plane's speed is a constant 150 mi/h (220 ft/s). At what angle of sight ϕ (between the horizontal and the direct line to the target) should a bale of hay be released in order to hit the target?

Solution See Fig. 12.4.6. We take $x_0 = 0$ where the bale of hay is released at time $t = 0$. Then $y_0 = 1600$ (ft), $v_0 = 220$ (ft/s), and $\alpha = 0$. Then Eqs. (15) and (16) take the forms

$$x = 220t, \qquad y = -16t^2 + 1600.$$

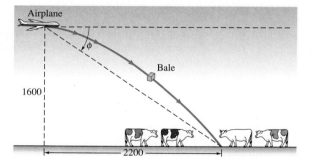

Fig. 12.4.6 Trajectory of the hay bale of Example 3

From the second of these equations we find that $t = 10$ (s) when the bale of hay hits the ground ($y = 0$). It has then traveled a horizontal distance

$$x(10) = 220 \cdot 10 = 2200 \quad \text{(ft)}.$$

Hence the required angle of sight is

$$\phi = \tan^{-1}\left(\frac{1600}{2200}\right) \approx 36°.$$

12.4 Problems

In Problems 1 through 8, find the values of $\mathbf{r}'(t)$ and $\mathbf{r}''(t)$ for the indicated value of t.

1. $\mathbf{r}(t) = 3\mathbf{i} - 2\mathbf{j};\quad t = 1$
2. $\mathbf{r}(t) = t^2\mathbf{i} - t^3\mathbf{j};\quad t = 2$
3. $\mathbf{r}(t) = e^{2t}\mathbf{i} + e^{-t}\mathbf{j};\quad t = 0$

4. $\mathbf{r}(t) = \mathbf{i} \cos t + \mathbf{j} \sin t;\quad t = \pi/4$
5. $\mathbf{r}(t) = 3\mathbf{i} \cos 2\pi t + 3\mathbf{j} \sin 2\pi t;\quad t = 3/4$
6. $\mathbf{r}(t) = 5\mathbf{i} \cos t + 4\mathbf{j} \sin t;\quad t = \pi$
7. $\mathbf{r}(t) = \mathbf{i} \sec t + \mathbf{j} \tan t;\quad t = 0$
8. $\mathbf{r}(t) = (2t + 3)\mathbf{i} + (6t - 5)\mathbf{j};\quad t = 10$

Calculate the integrals in Problems 9 through 12.

9. $\displaystyle\int_0^{\pi/4} (\mathbf{i} \sin t + 2\mathbf{j} \cos t)\, dt$ **10.** $\displaystyle\int_1^e \left(\frac{1}{t}\mathbf{i} - \mathbf{j}\right) dt$

11. $\displaystyle\int_0^2 t^2(1 + t^3)^{3/2}\mathbf{i}\, dt$ **12.** $\displaystyle\int_0^1 (\mathbf{i}e^t - \mathbf{j}te^{-t^2})\, dt$

In Problems 13 through 15, apply Theorem 2 to compute the derivative $D_t[\mathbf{u}(t) \cdot \mathbf{v}(t)]$.

13. $\mathbf{u}(t) = 3t\mathbf{i} - \mathbf{j}, \quad \mathbf{v}(t) = 2\mathbf{i} - 5t\mathbf{j}$

14. $\mathbf{u}(t) = t\mathbf{i} + t^2\mathbf{j}, \quad \mathbf{v}(t) = t^2\mathbf{i} - t\mathbf{j}$

15. $\mathbf{u}(t) = \langle \cos t, \sin t \rangle, \quad \mathbf{v}(t) = \langle \sin t, -\cos t \rangle$

In Problems 16 through 20, find the velocity $\mathbf{v}(t)$ and position $\mathbf{r}(t)$ corresponding to the given values of the acceleration $\mathbf{a}(t)$, initial velocity $\mathbf{v}(0)$, and initial position $\mathbf{r}(0)$.

16. $\mathbf{a} = 0, \quad \mathbf{r}_0 = 2\mathbf{i} + 3\mathbf{j}, \quad \mathbf{v}_0 = -2\mathbf{j}$

17. $\mathbf{a} = 2\mathbf{j}, \quad \mathbf{r}_0 = 0, \quad \mathbf{v}_0 = \mathbf{i}$

18. $\mathbf{a} = \mathbf{i} - \mathbf{j}, \quad \mathbf{r}_0 = \mathbf{j}, \quad \mathbf{v}_0 = \mathbf{i} + \mathbf{j}$

19. $\mathbf{a} = t\mathbf{i} + t^2\mathbf{j}, \quad \mathbf{r}_0 = \mathbf{i}, \quad \mathbf{v}_0 = 0$

20. $\mathbf{a} = -\mathbf{i} \sin t + \mathbf{j} \cos t, \quad \mathbf{r}_0 = \mathbf{i}, \quad \mathbf{v}_0 = \mathbf{j}$

Problems 21 through 26 deal with a projectile fired from the origin (so $x_0 = y_0 = 0$) with initial speed v_0 and initial angle of inclination α. The range of the projectile is the horizontal distance it travels before it returns to the ground.

21. If $\alpha = 45°$, what value of v_0 gives a range of 1 mi?

22. If $\alpha = 60°$ and the range is $R = 1$ mi, what is the maximum height attained by the projectile?

23. Deduce from Eqs. (15) and (16) the fact that the range is

$$R = \tfrac{1}{16}v_0^2 \sin \alpha \cos \alpha.$$

24. Given the initial speed v_0, find the angle α that maximizes the range. [*Suggestion:* Use the result of Problem 23.]

25. Suppose that $v_0 = 160$ (ft/s). Find the maximum height y_{max} and the range R of the projectile if (a) $\alpha = 30°$; (b) $\alpha = 45°$; (c) $\alpha = 60°$.

26. The projectile of Problem 25 is to be fired at a target 600 ft away, and there is a hill 300 ft high midway between the gun site and this target. At what initial angle of inclination should the projectile be fired?

27. A projectile is to be fired horizontally from the top of a 100-m cliff at a target 1 km from the base of the cliff. What should be the initial velocity of the projectile? (Use $g = 9.8$ m/s².)

28. A bomb is dropped (initial speed zero) from a helicopter hovering at a height of 800 m. A projectile is fired

from a gun located on the ground 800 m west of the point directly beneath the helicopter. The projectile is supposed to intercept the bomb at a height of exactly 400 m. If the projectile is fired at the same instant that the bomb is dropped, what should be its initial velocity and angle of inclination?

29. Suppose, more realistically, that the projectile of Problem 28 is fired 1 s after the bomb is dropped. What should be its initial velocity and angle of inclination?

30. An artillery gun with a muzzle velocity of 1000 ft/s is located atop a seaside cliff 500 ft high. At what initial inclination angle (or angles) should it fire a projectile in order to hit a ship at sea 20,000 ft from the base of the cliff?

31. Suppose that the vector-valued functions $\mathbf{u}(t)$ and $\mathbf{v}(t)$ both have limits as $t \to a$. Prove:

(a) $\displaystyle\lim_{t \to a} (\mathbf{u}(t) + \mathbf{v}(t)) = \lim_{t \to a} \mathbf{u}(t) + \lim_{t \to a} \mathbf{v}(t)$;

(b) $\displaystyle\lim_{t \to a} (\mathbf{u}(t) \cdot \mathbf{v}(t)) = \left(\lim_{t \to a} \mathbf{u}(t)\right) \cdot \left(\lim_{t \to a} \mathbf{v}(t)\right)$.

32. Prove part (1) of Theorem 2.

33. Prove part (2) of Theorem 2.

34. Suppose that both the vector-valued function $\mathbf{r}(t)$ and the real-valued function $h(t)$ are differentiable. Deduce the chain rule for the vector-valued functions,

$$D_t[\mathbf{r}(h(t))] = h'(t)\mathbf{r}'(h(t)),$$

in componentwise fashion from the ordinary chain rule.

35. A point moves with constant speed, so its velocity vector $\mathbf{v}$ satisfies the condition

$$|\mathbf{v}|^2 = \mathbf{v} \cdot \mathbf{v} = C \quad \text{(a constant)}.$$

Prove that the velocity and acceleration vectors of the point are always perpendicular to each other.

36. A point moves on a circle whose center is at the origin. Use the dot product to show that the position and velocity vectors of the moving points are always perpendicular.

37. A point moves on the hyperbola $x^2 - y^2 = 1$ with position vector

$$\mathbf{r}(t) = \mathbf{i} \cosh \omega t + \mathbf{j} \sinh \omega t$$

(the number ω is a constant). Prove that the acceleration vector $\mathbf{a}(t)$ satisfies the equation $\mathbf{a}(t) = c\mathbf{r}(t)$, where c is a positive constant. What sort of external force would produce this kind of motion?

38. Suppose that a point moves on the ellipse

$$\frac{x^2}{a^2} + \frac{y^2}{b^2} = 1$$

with position vector $\mathbf{r}(t) = \mathbf{i}a \cos \omega t + \mathbf{j}b \sin \omega t$ (ω is a constant). Prove that the acceleration vector $\mathbf{a}$ satisfies the equation $\mathbf{a}(t) = c\mathbf{r}(t)$, where c is a negative constant. To what sort of external force $\mathbf{F}(t)$ does this motion correspond?

39. A point moves in the plane with constant acceleration vector $\mathbf{a} = a\mathbf{j}$. Prove that its path is a parabola or a straight line.

40. Suppose that a particle is subject to no force, so its acceleration vector $\mathbf{a}(t)$ is identically zero. Prove that the particle travels along a straight line at constant speed (Newton's first law of motion).

41. *Uniform Circular Motion* Consider a particle that moves counterclockwise around the circle with center $(0, 0)$ and radius r at a constant angular speed of ω radians per second (Fig. 12.4.7). If its initial position is $(r, 0)$, then its position vector is

$$\mathbf{r}(t) = \mathbf{i}r \cos \omega t + \mathbf{j}r \sin \omega t.$$

(a) Show that the velocity vector of the particle is tangent to the circle and that the speed of the particle is

$$v(t) = |\mathbf{v}(t)| = r\omega.$$

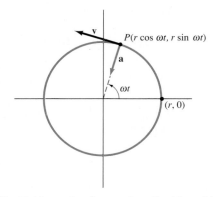

Fig. 12.4.7 Uniform circular motion (Problem 41)

(b) Show that the acceleration vector $\mathbf{a}$ of the particle is directed opposite to $\mathbf{r}$ and that

$$a(t) = |\mathbf{a}(t)| = r\omega^2.$$

*12.5

Orbits of Planets and Satellites

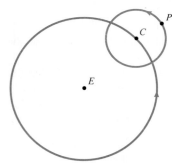

Fig. 12.5.1 The small circle is the epicycle.

Ancient Greek mathematicians and astronomers developed an elaborate mathematical model to account for the complicated motions of the sun, moon, and six planets then known as viewed from the earth. A combination of uniform circular motions was used to describe the motion of each body about the earth—if the earth is placed at the origin, then each body *does* orbit the earth.

In this system, it was typical for a planet P to travel uniformly around a small circle (the *epicycle*) with center C, which in turn traveled uniformly around a circle centered at the earth, labeled E in Fig. 12.5.1. The radii of the circles and the angular speeds of P and C around them were chosen to match the observed motion of the planet as closely as possible. For greater accuracy, the ancient Greeks could use secondary circles. In fact, several circles were required for each body in the solar system. The theory of epicycles reached its definitive form in Ptolemy's *Almagest* of the second century A.D.

In 1543, Copernicus altered Ptolemy's approach by placing the center of each primary circle at the sun rather than at the earth. This change was of much greater philosophical than mathematical importance. For, contrary to popular belief, this *heliocentric system* was *not* simpler that Ptolemy's geocentric system. Indeed, Copernicus's system actually required more circles.

It was Johannes Kepler (1571–1630) who finally got rid of all these circles. On the basis of a detailed analysis of a lifetime of planetary observations by the Danish astronomer Tycho Brahe, Kepler stated the following three propositions, now known as **Kepler's laws of planetary motion**:

1. The orbit of each plant is an ellipse with the sun at one focus.

2. The radius vector from the sun to a planet sweeps out area at a constant rate.

3. The *square* of the period of revolution of a planet is proportional to the *cube* of the major semiaxis of its elliptical orbit.

In his *Principia Mathematica* (1687) Newton showed that Kepler's laws follow from the basic principles of mechanics ($F = ma$, and so on) and the inverse-square law of gravitational attraction. His success in using mathematics to explain natural phenomena ("I now demonstrate the frame of the System of the World") inspired confidence that the universe could be understood and perhaps even mastered. This new confidence permanently altered humanity's perception of itself and of its place in the scheme of things.

In this section we show how Kepler's laws can be derived as indicated previously. To begin, we set up a coordinate system in which the sun is located at the origin in the plane of motion of a planet. Let $r = r(t)$ and $\theta = \theta(t)$ be the polar coordinates at time t of the planet as it orbits the sun. We want first to split the planet's position, velocity, and acceleration vectors $\mathbf{r}$, $\mathbf{v}$, and $\mathbf{a}$ into *radial* and *transverse* components. To do so, we introduce at each point (r, θ) of the plane (the origin excepted) the *unit* vectors

$$\mathbf{u}_r = \mathbf{i} \cos \theta + \mathbf{j} \sin \theta, \qquad \mathbf{u}_\theta = -\mathbf{i} \sin \theta + \mathbf{j} \cos \theta. \tag{1}$$

If we substitute $\theta = \theta(t)$, then $\mathbf{u}_r$ and $\mathbf{u}_\theta$ become functions of t. The **radial** unit vector $\mathbf{u}_r$ always points directly away from the origin; the **transverse** unit vector $\mathbf{u}_\theta$ is obtained from $\mathbf{u}_r$ by a 90° counterclockwise rotation (Fig. 12.5.2).

In Problem 6 we ask you to verify, by componentwise differentiation of the equations in (1), that

$$\frac{d\mathbf{u}_r}{dt} = \mathbf{u}_\theta \frac{d\theta}{dt} \quad \text{and} \quad \frac{d\mathbf{u}_\theta}{dt} = -\mathbf{u}_r \frac{d\theta}{dt}. \tag{2}$$

The position vector $\mathbf{r}$ points directly away from the origin and has length $|\mathbf{r}| = r$, so

$$\mathbf{r} = r\mathbf{u}_r. \tag{3}$$

Differentiation of both sides of Eq. (3) with respect to t gives

$$\mathbf{v} = \frac{d\mathbf{r}}{dt} = \mathbf{u}_r \frac{dr}{dt} + r \frac{d\mathbf{u}_r}{dt}.$$

We use the first equation in (2) and find that the planet's velocity vector is

$$\mathbf{v} = \mathbf{u}_r \frac{dr}{dt} + r \frac{d\theta}{dt} \mathbf{u}_\theta. \tag{4}$$

Thus we have expressed the velocity $\mathbf{v}$ in terms of the radial vector $\mathbf{u}_r$ and the transverse vector $\mathbf{u}_\theta$.

We differentiate both sides of Eq. (4) and thereby find that

$$\mathbf{a} = \frac{d\mathbf{v}}{dt} = \left(\mathbf{u}_r \frac{d^2 r}{dt^2} + \frac{dr}{dt} \frac{d\mathbf{u}_r}{dt} \right) + \left(\frac{dr}{dt} \frac{d\theta}{dt} \mathbf{u}_\theta + r \frac{d^2 \theta}{dt^2} \mathbf{u}_\theta + r \frac{d\theta}{dt} \frac{d\mathbf{u}_\theta}{dt} \right).$$

Then, by using the equations in (2) and collecting the coefficients of $\mathbf{u}_r$ and $\mathbf{u}_\theta$ (Problem 7), we obtain the decomposition

$$\mathbf{a} = \left[\frac{d^2 r}{dt^2} - r \left(\frac{d\theta}{dt} \right)^2 \right] \mathbf{u}_r + \left[\frac{1}{r} \frac{d}{dt} \left(r^2 \frac{d\theta}{dt} \right) \right] \mathbf{u}_\theta \tag{5}$$

of the acceleration vector into its radial and transverse components.

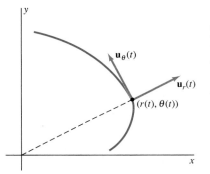

Fig. 12.5.2 The radial and transverse unit vectors $\mathbf{u}_r$ and $\mathbf{u}_\theta$.

KEPLER'S THREE LAWS OF PLANETARY MOTION

Let M denote the mass of the sun and m the mass of an orbiting planet. The inverse-square law of gravitation in vector form is

$$\mathbf{F} = m\mathbf{a} = -\frac{GMm}{r^2}\mathbf{u}_r,$$

so the acceleration of the planet is given *also* by

$$\mathbf{a} = -\frac{\mu}{r^2}\mathbf{u}_r, \tag{6}$$

where $\mu = GM$. We equate the transverse components in Eqs. (5) and (6) and thus obtain

$$\frac{1}{r}\cdot\frac{d}{dt}\left(r^2\frac{d\theta}{dt}\right) = 0.$$

We drop the factor $1/r$, then antidifferentiate both sides. We find that

$$r^2\frac{d\theta}{dt} = h \qquad (h \text{ a constant}). \tag{7}$$

We know from Section 10.3 that if $A(t)$ denotes the area swept out by the planet's radius vector from time 0 to time t (Fig. 12.5.3), then

$$A(t) = \int_{*}^{**}\frac{1}{2}r^2\,d\theta = \int_{0}^{t}\frac{1}{2}r^2\frac{d\theta}{dt}\,dt.$$

Now we apply the fundamental theorem of calculus, which yields

$$\frac{dA}{dt} = \frac{1}{2}r^2\frac{d\theta}{dt}. \tag{8}$$

When we compare Eqs. (7) and (8), we see that

$$\frac{dA}{dt} = \frac{h}{2}. \tag{9}$$

Because $h/2$ is a constant, we have derived Kepler's second law: The radius vector from sun to planet sweeps out area at a constant rate.

Now we derive Kepler's first law. Choose coordinate axes so that at time $t = 0$ the planet is on the polar axis and is at its closest point of approach to the sun, with initial position vector $\mathbf{r}_0$ and initial vector $\mathbf{v}_0$ (Fig. 12.5.4). By Eq. (4),

$$\mathbf{v}_0 = r_0\,\theta'(0)\mathbf{u}_\theta = v_0\mathbf{j} \tag{10}$$

because, when $t = 0$, $\mathbf{u}_\theta = \mathbf{j}$ and $dr/dt = 0$ (as r is minimal then).

From Eq. (6) we have

$$\frac{d\mathbf{v}}{dt} = -\frac{\mu}{r^2}\mathbf{u}_r = -\frac{\mu}{r^2}\left(-\frac{1}{d\theta/dt}\cdot\frac{d\mathbf{u}_\theta}{dt}\right)$$

$$= \frac{\mu}{r^2(d\theta/dt)}\cdot\frac{d\mathbf{u}_\theta}{dt} = \frac{\mu}{h}\cdot\frac{d\mathbf{u}_\theta}{dt},$$

by using the second equation in (2) in the first step and then Eq. (7).

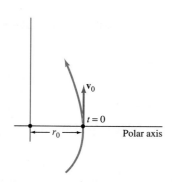

Fig. 12.5.3 Area swept out by the radius vector

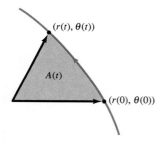

Fig. 12.5.4 Setup of the coordinate system for the derivation of Kepler's first law

Ch. 12 / *Parametric Curves and Vectors in the Plane*

Next we antidifferentiate and find that

$$\mathbf{v} = \frac{\mu}{h}\mathbf{u}_\theta + \mathbf{C}.$$

To find $\mathbf{C}$, we substitute $t = 0$. This yields $\mathbf{u}_\theta = \mathbf{j}$ and $\mathbf{v} = v_0\mathbf{j}$. We get

$$\mathbf{C} = v_0\mathbf{j} - \frac{\mu}{h}\mathbf{j} = \left(v_0 - \frac{\mu}{h}\right)\mathbf{j}.$$

Consequently,

$$\mathbf{v} = \frac{\mu}{h}\mathbf{u}_\theta + \left(v_0 - \frac{\mu}{h}\right)\mathbf{j}. \tag{11}$$

Now we take the dot product of each side in Eq. (11) with the unit vector $\mathbf{u}_\theta$. Remember that

$$\mathbf{u}_\theta \cdot \mathbf{v} = r\frac{d\theta}{dt} \quad \text{and} \quad \mathbf{u}_\theta \cdot \mathbf{j} = \cos\theta,$$

consequences of Eqs. (4) and (1), respectively. The result is

$$r\frac{d\theta}{dt} = \frac{\mu}{h} + \left(v_0 - \frac{\mu}{h}\right)\cos\theta.$$

But $r(d\theta/dt) = h/r$ by Eq. (7). It follows that

$$\frac{h}{r} = \frac{\mu}{h} + \left(v_0 - \frac{\mu}{h}\right)\cos\theta.$$

We solve for r and find that

$$r = \frac{h^2/\mu}{1 + \left(\dfrac{v_0 h}{\mu} - 1\right)\cos\theta}.$$

Next, we see from Eqs. (7) and (10) that

$$h = r_0 v_0, \tag{12}$$

so finally we may write the equation of the planet's orbit:

$$r = \frac{(r_0 v_0)^2/\mu}{1 + \left(\dfrac{r_0 v_0{}^2}{\mu} - 1\right)\cos\theta} = \frac{pe}{1 + e\cos\theta}. \tag{13}$$

Equation (13) is the polar-coordinates equation of a conic section whose focus is at the origin and with eccentricity

$$e = \frac{r_0 v_0{}^2}{GM} - 1 \tag{14}$$

(see Problem 17). Because the nature of a conic section is determined by its eccentricity, we see that the planet's orbit is:

❑ A circle if $r_0 v_0{}^2 = GM$,
❑ An ellipse if $GM < r_0 v_0{}^2 < 2GM$,
❑ A parabola if $r_0 v_0{}^2 = 2GM$,
❑ A hyperbola if $r_0 v_0{}^2 > 2GM$.

$$(15)$$

This comprehensive description is a generalization of Kepler's first law. Of course, the elliptical case is the one that holds for planets in the solar system.

Let us consider further the case in which the orbit is an ellipse with major semiaxis a and minor semiaxis b. (The case in which the ellipse is actually a circle, with $a = b$, will be a special case of this discussion.) The constant

$$pe = \frac{h^2}{GM}$$

used in Eq. (13) satisfies the equations

$$pe = a(1 - e^2) = a\left(1 - \frac{a^2 - b^2}{a^2}\right) = \frac{b^2}{a} \tag{16}$$

(see Problem 16). We evaluate these two expressions for pe and find that

$$h^2 = GM\frac{b^2}{a}.$$

Now let T denote the period of revolution of the planet in its elliptical orbit. Then we see from Eq. (9) that the area of the ellipse is $A = \frac{1}{2}hT = \pi ab$ and thus that

$$T^2 = \frac{4\pi^2 a^2 b^2}{h^2} = \frac{4\pi^2 a^2 b^2}{GMb^2/a},$$

so

$$T^2 = \gamma a^3, \tag{17}$$

where $\gamma = 4\pi^2/GM$ is a constant. This is Kepler's third law.

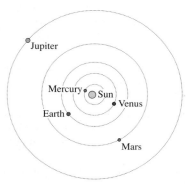

Fig. 12.5.5 The inner planets of the solar system (Example 1)

EXAMPLE 1 The period of revolution of Mercury in its elliptical orbit about the sun is 87.97 days, whereas that of earth is 365.26 days. Compute the major semiaxis (in astronomical units) of the orbit of Mercury. See Fig. 12.5.5.

Solution The major semiaxis of the earth's orbit is, by definition, 1 AU. So Eq. (17) gives the value of the constant $\gamma = (365.26)^2$ (day^2/AU^3). Hence the major semiaxis of the orbit of Mercury is

$$a = \left(\frac{T^2}{\gamma}\right)^{1/3} = \left(\frac{(87.97)^2}{(365.26)^2}\right)^{1/3} \approx 0.387 \text{ (AU)}.$$

As yet we have considered only planets in orbits about the sun. But Kepler's laws and the equations of this section apply to bodies in orbit about any common central mass, so long as they move solely under the influence of *its* gravitational attraction. Examples include satellites (artificial or natural) orbiting the earth or the moons of Jupiter. For earth, we can compute μ by beginning with the values

$$g = \frac{GM}{R^2} \approx 32.16$$

ft/s^2 (approximately 9.8 m/s^2) for surface gravitational acceleration and the radius 3960 (mi) of the earth—which we assume to be spherical. Then

$$\mu = GM = R^2 g = (3960)^2\left(\frac{32.16}{5280}\right) \approx 95,500 \quad \text{(mi}^3\text{/s}^2\text{)}.$$

Fig. 12.5.6 A communications satellite in orbit around the earth (Example 2)

EXAMPLE 2 A communications relay satellite is to be placed in a circular orbit about the earth and is to have a period of revolution of 24 h. This is a *geosynchronous* orbit in which the satellite appears to be stationary in the sky. Assume that the earth's natural moon has a period of 27.32 days in a circular orbit of radius 238,850 mi. What should be the radius of the satellite's orbit? See Fig. 12.5.6.

Solution Equation (17), when applied to the moon, yields

$$(27.32)^2 = k(238,850)^2.$$

For the stationary satellite that has period $T = 1$ (day), it yields $1^2 = kr^3$, where r is the radius of the geosynchronous orbit. To eliminate k, we divide the second of these equations by the first and find that

$$r^3 = \frac{(238,850)^3}{(27.32)^2}.$$

Thus r is approximately 26,330 mi. The radius of the earth is about 3960 mi, so the satellite will be 22,370 mi above the surface.

12.5 Problems

For each parametric curve in Problems 1 through 5, express its velocity and acceleration vectors in terms of $\mathbf{u}_r$ and $\mathbf{u}_\theta$—that is, in the form $A(t)\mathbf{u}_r + B(t)\mathbf{u}_\theta$.

1. $r = a$, $\theta = t$ **2.** $r = 2\cos t$, $\theta = t$

3. $r = t$, $\theta = t$ **4.** $r = e^t$, $\theta = 2t$

5. $r = 3\sin 4t$, $\theta = 2t$

6. Derive both equations in (2) by differentiation of the equations in (1).

7. Derive Eq. (5) by differentiating Eq. (4).

8. Consider a body in an elliptical orbit with major and minor semiaxes a and b and period of revolution T. (a) Deduce from Eq. (4) that $v = r(d\theta/dt)$ when the body is nearest to and farthest from its foci. (b) Then apply Kepler's second law to conclude that $v = 2\pi ab/(rT)$ at the body's nearest and farthest points.

In Problems 9 through 12, apply the equation of part (b) of Problem 8 to compute the speed (in miles per second) of the given body at the nearest and farthest points of its orbit. Convert 1 AU, the major semiaxis of the earth's orbit, into 92,956,000 mi.

9. Mercury: $a = 0.387$ AU, $e = 0.206$, $T = 87.97$ days

10. The earth: $e = 0.0167$, $T = 365.26$ days

11. The moon: $a = 238,900$ mi, $e = 0.055$, $T = 27.32$ days

12. An earth satellite: $a = 10,000$ mi, $e = 0.5$

13. Assuming the earth to be a sphere with radius 3960 mi, find the altitude above the earth's surface of a satellite in a circular orbit that has a period of revolution of 1 h.

14. Given the fact that Jupiter's period of (almost) circular revolution about the sun is 11.86 yr, calculate the distance of Jupiter from the sun.

15. Suppose that an earth satellite in elliptical orbit varies in altitude from 100 to 1000 mi above the earth's surface (presumed spherical). Find this satellite's period of revolution.

16. Substitute $\theta = 0$ and $\theta = \pi$ into Eq. (13) to deduce that $pe = a(1 - e^2)$.

17. Figure 12.5.7 shows a conic section with eccentricity e and focus at the origin. Derive Eq. (13) from the defining relation $|OP| = e|PQ|$ of the conic section.

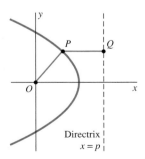

Fig. 12.5.7 A conic section with eccentricity e: $|OP| = e|PQ|$ (Problem 17).

Chapter 12 Review: DEFINITIONS AND CONCEPTS

Use the following list as a guide to concepts that you may need to review.

1. Definition of a parametric curve and of a smooth parametric curve

2. The slope of the line tangent to a smooth parametric curve (both in rectangular and in polar coordinates)

3. Integral computations with parametric curves [Eqs. (1) through (4) of Section 12.2]

4. Arc length of a parametric curve

5. Vectors: their definition, length, equality, addition, multiplication by scalars, and dot product

6. Test for perpendicular vectors

7. Vector-valued functions, velocity vectors, and acceleration vectors

8. Componentwise differentiation and integration of vector-valued functions

9. The equations of motion of a projectile

10. Kepler's three laws of planetary motion

11. The radial and transverse unit vectors

12. Polar decomposition of velocity and acceleration vectors

13. Outline of the derivation of Kepler's laws from Newton's law of gravitation

Chapter 12 Miscellaneous Problems

In Problems 1 through 5, eliminate the parameter and sketch the curve.

1. $x = 2t^3 - 1, \quad y = 2t^3 + 1$

2. $x = \cosh t, \quad y = \sinh t$

3. $x = 2 + \cos t, \quad y = 1 - \sin t$

4. $x = \cos^4 t, \quad y = \sin^4 t$

5. $x = 1 + t^2, \quad y = t^3$

In Problems 6 through 10, write an equation of the line tangent to the given curve at the indicated point.

6. $x = t^2, \quad y = t^3; \quad t = 1$

7. $x = 3 \sin t, \quad y = 4 \cos t; \quad t = \pi/4$

8. $x = e^t, \quad y = e^{-t}; \quad t = 0$

9. $r = \theta; \quad \theta = \pi/2$

10. $r = 1 + \sin \theta; \quad \theta = \pi/3$

In Problems 11 through 14, find the area of the region between the given curve and the x-axis.

11. $x = 2t + 1, \quad y = t^2 + 3; \quad -1 \le t \le 2$

12. $x = e^t, \quad y = e^{-t}; \quad 0 \le t \le 10$

13. $x = 3 \sin t, \quad y = 4 \cos t, \quad 0 \le t \le \pi/2$

14. $x = \cosh t, \quad y = \sinh t; \quad 0 \le t \le 1$

In Problems 15 through 19, find the arc length of the given curve.

15. $x = t^2, \quad y = t^3; \quad 0 \le t \le 1$

16. $x = \ln(\cos t), \quad y = t; \quad 0 \le t \le \pi/4$

17. $x = 2t, \quad y = t^3 + 1/(3t); \quad 1 \le t \le 2$

18. $r = \sin \theta; \quad 0 \le \theta \le \pi$

19. $r = \sin^3(\theta/3); \quad 0 \le \theta \le \pi$

In Problems 20 through 24, find the area of the surface generated by revolving the given curve around the x-axis.

20. $x = t^2 + 1, \quad y = 3t; \quad 0 \le t \le 2$

21. $x = 4t^{1/2}, \quad y = \frac{1}{3}t^3 + \frac{1}{2}t^{-2}; \quad 1 \le t \le 4$

22. $r = 4 \cos \theta$

23. $r = e^{\theta/2}; \quad 0 \le \theta \le \pi$

24. $x = e^t \cos t, \quad y = e^t \sin t; \quad 0 \le t \le \pi/2$

25. Consider the rolling circle of radius a that was used to generate the cycloid in Example 3 of Section 12.1. Suppose that this circle is the rim of a disk, and let Q be a point of this disk at distance $b < a$ from its center. Find parametric equations for the curve traced by Q as the circle rolls along the x-axis. Assume that Q begins at the point $(0, a - b)$. Sketch this curve, which is called a **trochoid**.

26. If the smaller circle of Problem 28 in Section 12.1 rolls around the *outside* of the larger circle, the path of the point P is called an **epicycloid**. Show that it has parametric equations

$$x = (a + b) \cos t - b \cos\left(\frac{a + b}{b} t\right),$$

$$y = (a + b) \sin t - b \sin\left(\frac{a + b}{b} t\right).$$

27. Suppose that $b = a$ in Problem 26. Show that the epicycloid is then the cardioid $r = 2a(1 - \cos \theta)$ translated a units to the right.

28. Find the area of the surface generated by revolving the lemniscate $r^2 = 2a^2 \cos 2\theta$ around the x-axis.

29. Find the volume generated by revolving around the y-axis the area under the cycloid

$$x = a(t - \sin t), \qquad y = a(1 - \cos t), \qquad 0 \leq t \leq 2\pi.$$

30. Show that the length of one arch of the hypocycloid of Problem 28 in Section 12.1 is $s = 8b(a - b)/a$.

31. Let ABC be an isosceles triangle with $|AB| = |AC|$. Let M be the midpoint of BC. Use the dot product to show that AM and BC are perpendicular.

32. Use the dot product to show that the diagonals of a rhombus (a parallelogram with all four sides equal) are perpendicular to each other.

33. The acceleration of a certain particle is

$$\mathbf{a} = \mathbf{i} \sin t - \mathbf{j} \cos t.$$

Assume that the particle begins at time $t = 0$ at the point $(0, 1)$ and has initial velocity $\mathbf{v}_0 = -\mathbf{i}$. Show that its path is a circle.

34. A particle moves in an attracting central force field with force proportional to the distance from the origin. This implies that the particle's acceleration vector is $\mathbf{a} = -\omega^2 \mathbf{r}$, where $\mathbf{r}$ is the position vector of the particle. Assume that the particle's initial position is $\mathbf{r}_0 = p\mathbf{i}$ and that its initial velocity is $\mathbf{v}_0 = q\omega\mathbf{j}$. Show that the trajectory of the particle is the ellipse with equation $x^2/p^2 + y^2/q^2 = 1$. [*Suggestion:* If $x''(t) = -k^2x(t)$

(where k is constant), then $x(t) = A \cos kt + B \sin kt$ for some constants A and B.]

35. At time $t = 0$, a ground target is 160 ft from a gun and is moving directly away from it with a constant speed of 80 ft/s. If the muzzle velocity of the gun is 320 ft/s, at what angle of elevation α should it be fired in order to strike the moving target?

36. Suppose that a gun with muzzle velocity v_0 is located at the foot of a hill with a 30° slope. At what angle of elevation (from the horizontal) should the gun be fired in order to maximize its range, as measured up the hill?

37. Assume that a body has an elliptical orbit

$$r = \frac{pe}{1 + e \cos \theta}$$

and satisfies Kepler's second law in the form

$$r^2 \frac{d\theta}{dt} = h \qquad \text{(a constant)}.$$

Then deduce from Eq. (5) in Section 12.5 that its acceleration vector $\mathbf{a}$ satisfies the equation

$$\mathbf{a} = \frac{k}{r^2}\mathbf{u}_r \qquad (k \text{ a constant}).$$

This shows that Newton's inverse-square law of gravitation follows from Kepler's first and second laws.

Vectors, Curves, and Surfaces in Space

J. W. Gibbs (1839–1903)

❑ The study of vector quantities with both direction and magnitude (such as force and velocity) dates back at least to Newton's *Principia Mathematica*. But the vector notation and terminology used in science and mathematics today was largely "born in the U.S.A." The modern system of vector analysis was created independently (and almost simultaneously) in the 1880s by the American mathematical physicist Josiah Willard Gibbs and the British electrical engineer Oliver Heaviside (1850–1925). Gibbs' first vector publication—*Elements of Vector Analysis* (1881)—appeared slightly earlier than Heaviside's (an 1885 paper) and was more systematic and

complete in its exposition of the foundations of the subject.

❑ Gibbs, the son of a Yale University professor, grew up in New Haven and attended Yale as an undergraduate. He studied both Latin and mathematics and remained at Yale for graduate work in engineering. In 1863 he received one of the first Ph.D. degrees awarded in the United States (and apparently the very first in engineering). After several years of postdoctoral study of mathematics and physics in France and Germany, he returned to Yale, where he served for more than three decades as professor of mathematical physics.

❑ The educational careers of Gibbs and Heaviside were as different as their vectors were similar. At 16 Heaviside quit school, and at 18 he began work as a telegraph operator. Starting with this very practical introduction to electricity, he published in 1872 the first of a series of electrical papers leading to his famous three-volume treatise *Electromagnetic Theory*. The first volume of this treatise appeared in 1893 and included Heaviside's own systematic presentation of modern vector analysis.

❑ Gibbs introduced the **i**, **j**, **k** notation now standard for three-

dimensional vectors, adapting it from the algebra of "quaternions," in which the Irish mathematician William Rowan Hamilton (1805–1865) earlier had used i, j, and k to denote three distinct square roots of -1. Gibbs was the first to define clearly both the scalar (dot) product $\mathbf{a} \cdot \mathbf{b}$ and the vector (cross) product $\mathbf{a} \times \mathbf{b}$ of the vectors **a** and **b**. He observed that $\mathbf{b} \times \mathbf{a} = -\mathbf{a} \times \mathbf{b}$, in contrast with the commutativity of multiplication of ordinary numbers. In the Section 13.4 Project, you will see that the vector product is a key to the analysis of the "curve" of a baseball pitch.

Does a curveball really curve?

13.1
Rectangular Coordinates and Three-Dimensional Vectors

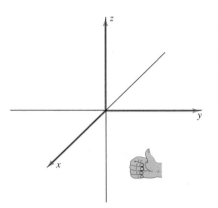

Fig. 13.1.1 The right-handed coordinate system

In the first twelve chapters we discussed many aspects of the calculus of functions of a *single* variable. The geometry of such functions is two-dimensional, because the graph of a function of a single variable is a curve in the coordinate plane. Most of the remaining chapters deal with the calculus of functions of *several* (two or more) independent variables. The geometry of functions of two variables is three-dimensional, because the graphs of such functions are generally surfaces in space.

Rectangular coordinates in the plane may be generalized to rectangular coordinates in space. A point in space is determined by giving its location relative to three mutually perpendicular **coordinate axes** that pass through the origin O. We shall always draw the x-, y-, and z-axes as shown in Fig. 13.1.1, with arrows indicating the positive direction along each axis. With this configuration of axes, our rectangular coordinate system is said to be **right-handed:** If you curl the fingers of your right hand in the direction of a $90°$ rotation from the positive x-axis to the positive y-axis, then your thumb points in the direction of the positive z-axis. If the x- and y-axes were interchanged, then the coordinate system would be left-handed. These two coordinate systems are different in that it is impossible to bring one into coincidence with the other by means of rotations and translations. Similarly, the L- and D-alanine molecules shown in Fig. 13.1.2 are different; you can digest the left-handed ("levo") version but not the right-handed ("dextro") version. In this book we shall discuss right-handed coordinate systems exclusively and always draw the x-, y-, and z-axes with the orientation shown in Fig. 13.1.1.

The three coordinate axes taken in pairs determine the three **coordinate planes**:

❑ The (horizontal) xy-plane, where $z = 0$;
❑ The (vertical) yz-plane, where $x = 0$; and
❑ The (vertical) xz-plane, where $y = 0$.

The point P in space is said to have **rectangular coordinates** (x, y, z) if

❑ x is its signed distance from the yz-plane,
❑ y is its signed distance from the xz-plane, and
❑ z is its signed distance from the xy-plane

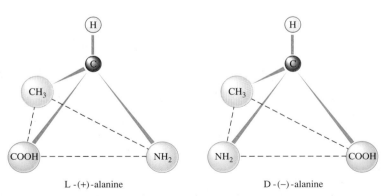

Fig. 13.1.2 The stereoisomers of the amino acid alanine are physically and biologically different even though they have the same molecular formula.

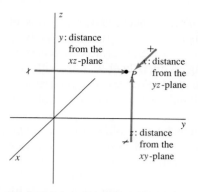

Fig. 13.1.3 Locating the point P by using rectangular coordinates

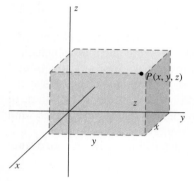

Fig. 13.1.4 "Completing the box" to show P with the illusion of the third dimension

(Fig. 13.1.3). In this case we may describe the location of the point P simply by calling it "the point $P(x, y, z)$." There is a natural one-to-one correspondence between ordered triples (x, y, z) of real numbers and points P in space; this correspondence is called a **rectangular coordinate system** in space. In Fig. 13.1.4 the point P is located in the **first octant**—the eighth of space in which all three rectangular coordinates are positive.

If we apply the Pythagorean theorem to the right triangles $P_1 QR$ and $P_1 RP_2$ in Fig. 13.1.5, we get

$$|P_1 P_2|^2 = |RP_2|^2 + |P_1 R|^2 = |RP_2|^2 + |QR|^2 + |P_1 Q|^2$$
$$= (x_1 - x_2)^2 + (y_1 - y_2)^2 + (z_1 - z_2)^2.$$

Thus the **distance formula** for the **distance** $|P_1 P_2|$ between the points P_1 and P_2 is

$$|P_1 P_2| = \sqrt{(x_1 - x_2)^2 + (y_1 - y_2)^2 + (z_1 - z_2)^2}. \tag{1}$$

For example, the distance between the points $P_1(1, 3, -2)$ and $P_2(4, -3, 1)$ is

$$|P_1 P_2| = \sqrt{(4 - 1)^2 + (-3 - 3)^2 + (1 + 2)^2} = \sqrt{54} \approx 7.34847.$$

You can apply the distance formula in Eq. (1) to show that the **midpoint** M of the line segment joining $P_1(x_1, y_1, z_1)$ and $P_2(x_2, y_2, z_2)$ is

$$M\left(\frac{x_1 + x_2}{2}, \frac{y_1 + y_2}{2}, \frac{z_1 + z_2}{2}\right) \tag{2}$$

(see Problem 45).

The **graph** of an equation in three variables x, y, and z is the set of all points in space with rectangular coordinates that satisfy that equation. In general, the graph of an equation in three variables is a *two-dimensional surface* in $\mathbf{R}^3$ (three-dimensional space with rectangular coordinates). For example, let $C(h, k, l)$ be a fixed point. Then the graph of the equation

$$(x - h)^2 + (y - k)^2 + (z - l)^2 = r^2 \tag{3}$$

is the set of all points $P(x, y, z)$ at distance $r > 0$ from the fixed point C. This

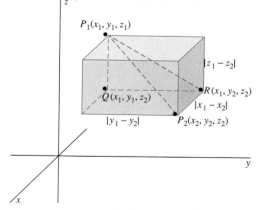

Fig. 13.1.5 The distance between P_1 and P_2 is the length of the long diagonal of the box.

Ch. 13 / Vectors, Curves, and Surfaces in Space

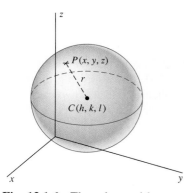

Fig. 13.1.6 The sphere with center (h, k, l) and radius r

means that Eq. (3) is the equation of the **sphere with radius r and center $C(h, k, l)$** shown in Fig. 13.1.6. Moreover, given an equation of the form

$$x^2 + y^2 + z^2 + Ax + By + Cz + D = 0,$$

we can attempt—by completing the square in each variable—to write it in the form of Eq. (3) and thereby show that its graph is a sphere.

EXAMPLE 1   Determine the graph of the equation

$$x^2 + y^2 + z^2 + 4x + 2y - 6z - 2 = 0.$$

Solution   We complete the square in each variable. The equation then takes the form

$$(x^2 + 4x + 4) + (y^2 + 2y + 1) + (z^2 - 6z + 9) = 16.$$

That is,

$$(x + 2)^2 + (y + 1)^2 + (z - 3)^2 = 4^2.$$

Thus the graph of the given equation is a sphere with radius 4 and center $(-2, -1, 3)$.

VECTORS IN SPACE

The discussion of vectors in the plane in Section 12.3 may be repeated almost verbatim for vectors in space. The principal difference is that a vector in space has three components rather than two. The vector determined by the point $P(x, y, z)$ is its **position vector** $\mathbf{v} = \overrightarrow{OP} = \langle x, y, z \rangle$, which is represented (Fig. 13.1.7) by the arrow from the origin O to P or by any parallel translate of this arrow. The distance formula in Eq. (1) gives

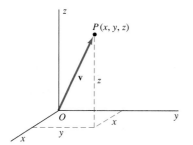

Fig. 13.1.7   The segment OP is a realization of the vector $\mathbf{v} = \overrightarrow{OP}$.

$$|\mathbf{v}| = \sqrt{x^2 + y^2 + z^2} \tag{4}$$

for the **length** (or **magnitude**) of the vector $\mathbf{v} = \langle x, y, z \rangle$.

The vector $\overrightarrow{AB}$ represented in Fig. 13.1.8 by the arrow from $A(a_1, a_2, a_3)$ to $B(b_1, b_2, b_3)$ is defined to be

$$\overrightarrow{AB} = \langle b_1 - a_1, b_2 - a_2, b_3 - a_3 \rangle.$$

Its length is simply the distance between the two points A and B.

What it means for two vectors in space to be equal is essentially the same as in the case of two-dimensional vectors: The vectors $\mathbf{a} = \langle a_1, a_2, a_3 \rangle$ and $\mathbf{b} = \langle b_1, b_2, b_3 \rangle$ are **equal** provided that $a_1 = b_1$, $a_2 = b_2$, and $a_3 = b_3$. That is, two vectors are equal exactly when corresponding components are equal.

We define addition and scalar multiplication of vectors exactly as we did in Section 12.3, taking into account that the vectors now have three components rather than two: The **sum** of the vectors $\mathbf{a} = \langle a_1, a_2, a_3 \rangle$ and $\mathbf{b} = \langle b_1, b_2, b_3 \rangle$ is the vector

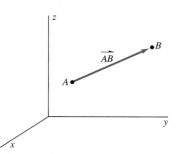

Fig. 13.1.8   The segment AB is an instance of the vector $\overrightarrow{AB}$.

$$\mathbf{a} + \mathbf{b} = \langle a_1 + b_1, a_2 + b_2, a_3 + b_3 \rangle. \tag{5}$$

Because $\mathbf{a}$ and $\mathbf{b}$ lie in a plane (although not necessarily the xy-plane) if their initial points coincide, addition of vectors obeys the same **parallelogram law** as in the two-dimensional case (Fig. 13.1.9).

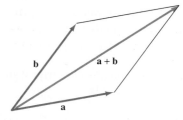

Fig. 13.1.9 The parallelogram law for addition of vectors

If c is a real number, then the **scalar multiple** $c\mathbf{a}$ is the vector

$$c\mathbf{a} = \langle ca_1, ca_2, ca_3 \rangle. \tag{6}$$

The length of $c\mathbf{a}$ is $|c|$ times the length of $\mathbf{a}$, and $c\mathbf{a}$ has the same direction as $\mathbf{a}$ if $c > 0$ but the opposite direction if $c < 0$. The following algebraic properties of vector addition and scalar multiplication are easy to establish; they follow from computations with components, exactly as in Section 12.3:

$$\mathbf{a} + \mathbf{b} = \mathbf{b} + \mathbf{a},$$
$$\mathbf{a} + (\mathbf{b} + \mathbf{c}) = (\mathbf{a} + \mathbf{b}) + \mathbf{c},$$
$$r(\mathbf{a} + \mathbf{b}) = r\mathbf{a} + r\mathbf{b}, \tag{7}$$
$$(r + s)\mathbf{a} = r\mathbf{a} + s\mathbf{a},$$
$$(rs)\mathbf{a} = r(s\mathbf{a}) = s(r\mathbf{a}).$$

EXAMPLE 2 If $\mathbf{a} = \langle 3, 4, 12 \rangle$ and $\mathbf{b} = \langle -4, 3, 0 \rangle$, then

$$\mathbf{a} + \mathbf{b} = \langle 3 - 4, 4 + 3, 12 + 0 \rangle = \langle -1, 7, 12 \rangle,$$
$$|\mathbf{a}| = \sqrt{3^2 + 4^2 + 12^2} = \sqrt{169} = 13,$$
$$2\mathbf{a} = \langle 2 \cdot 3, 2 \cdot 4, 2 \cdot 12 \rangle = \langle 6, 8, 24 \rangle,$$
$$2\mathbf{a} - 3\mathbf{b} = \langle 6 + 12, 8 - 9, 24 - 0 \rangle = \langle 18, -1, 24 \rangle.$$

A **unit vector** is a vector of length 1. We can express any vector in space (or space vector) in terms of the three **basic unit vectors**

$$\mathbf{i} = \langle 1, 0, 0 \rangle, \qquad \mathbf{j} = \langle 0, 1, 0 \rangle, \qquad \mathbf{k} = \langle 0, 0, 1 \rangle.$$

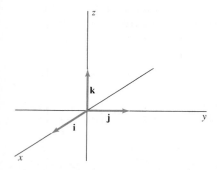

Fig. 13.1.10 The basic unit vectors $\mathbf{i}$, $\mathbf{j}$, and $\mathbf{k}$

When located with their initial points at the origin, these basic unit vectors form a right-handed triple of vectors pointing in the positive directions along the three coordinate axes (Fig. 13.1.10).

The space vector $\mathbf{a} = \langle a_1, a_2, a_3 \rangle$ can be written as

$$\mathbf{a} = a_1\mathbf{i} + a_2\mathbf{j} + a_3\mathbf{k},$$

a linear combination of the basic unit vectors. As in the two-dimensional case, the usefulness of this representation is that algebraic operations involving vectors may be carried out simply by collecting coefficients of $\mathbf{i}$, $\mathbf{j}$, and $\mathbf{k}$. For example, if $\mathbf{a} = \langle a_1, a_2, a_3 \rangle$ and $\mathbf{b} = \langle b_1, b_2, b_3 \rangle$, then

$$\mathbf{a} + \mathbf{b} = (a_1\mathbf{i} + a_2\mathbf{j} + a_3\mathbf{k}) + (b_1\mathbf{i} + b_2\mathbf{j} + b_3\mathbf{k})$$
$$= (a_1 + b_1)\mathbf{i} + (a_2 + b_2)\mathbf{j} + (a_3 + b_3)\mathbf{k}.$$

THE DOT PRODUCT IN SPACE

The **dot product** of the two vectors

$$\mathbf{a} = a_1\mathbf{i} + a_2\mathbf{j} + a_3\mathbf{k} \quad \text{and} \quad \mathbf{b} = b_1\mathbf{i} + b_2\mathbf{j} + b_3\mathbf{k}$$

is defined almost exactly as before: Multiply corresponding components and then add the results. Thus

$$\mathbf{a} \cdot \mathbf{b} = a_1 b_1 + a_2 b_2 + a_3 b_3. \tag{8}$$

If $a_3 = b_3 = 0$, then we may think of **a** and **b** as vectors in the xy-plane; indeed, any arrows representing them have parallel translates that lie in the xy-plane. Then the definition in Eq. (8) reduces to the one given in Section 12.3 for the dot product of two vectors in the plane. The three-dimensional dot product has the same list of properties as the two-dimensional dot product. We can routinely establish Eqs. (9) by working with components:

$$\mathbf{a} \cdot \mathbf{a} = |\mathbf{a}|^2,$$

$$\mathbf{a} \cdot \mathbf{b} = \mathbf{b} \cdot \mathbf{a},$$

$$\mathbf{a} \cdot (\mathbf{b} + \mathbf{c}) = \mathbf{a} \cdot \mathbf{b} + \mathbf{a} \cdot \mathbf{c}, \tag{9}$$

$$(r\mathbf{a}) \cdot \mathbf{b} = r(\mathbf{a} \cdot \mathbf{b}) = \mathbf{a} \cdot (r\mathbf{b}).$$

EXAMPLE 3 If $\mathbf{a} = \langle 3, 4, 12 \rangle$ and $\mathbf{b} = \langle -4, 3, 0 \rangle$, then

$$\mathbf{a} \cdot \mathbf{b} = 3 \cdot (-4) + 4 \cdot 3 + 12 \cdot 0 = -12 + 12 + 0 = 0.$$

If $\mathbf{c} = \langle 4, 5, -3 \rangle$, then

$$\mathbf{a} \cdot \mathbf{c} = 3 \cdot 4 + 4 \cdot 5 + 12 \cdot (-3) = 12 + 20 - 36 = -4.$$

IMPORTANT   The dot product of two *vectors* is a *scalar*—that is, a real number. For this reason, the dot product is often called the *scalar product*.

The significance of the dot product lies in its geometric interpretation. Let the vectors **a** and **b** be represented by the position vectors $\overrightarrow{OP}$ and $\overrightarrow{OQ}$, respectively. Then the angle θ between **a** and **b** is the angle at O in triangle OPQ of Fig. 13.1.11. We say that **a** and **b** are **parallel** if $\theta = 0$ or if $\theta = \pi$ and that **a** and **b** are **perpendicular** if $\theta = \pi/2$. For convenience, we regard the zero vector $\mathbf{0} = \langle 0, 0, 0 \rangle$ as parallel to *and* perpendicular to *every* vector.

> **Theorem 1**   *Interpretation of the Dot Product*
> If θ is the angle between the vectors **a** and **b**, then
> $$\mathbf{a} \cdot \mathbf{b} = |\mathbf{a}||\mathbf{b}| \cos \theta. \tag{10}$$

Proof   If either $\mathbf{a} = \mathbf{0}$ or $\mathbf{b} = \mathbf{0}$, then Eq. (10) follows immediately. If the vectors **a** and **b** are parallel, then $\mathbf{b} = t\mathbf{a}$ with either $t > 0$ and $\theta = 0$ or $t < 0$ and $\theta = \pi$. In either case, both sides in Eq. (10) reduce to $t|\mathbf{a}|^2$, so again the conclusion of Theorem 1 follows.

We turn to the general case in which the vectors $\mathbf{a} = \overrightarrow{OP}$ and $\mathbf{b} = \overrightarrow{OQ}$ are nonzero and nonparallel. Then

$$|\overrightarrow{QP}| = |\mathbf{a} - \mathbf{b}|^2 = (\mathbf{a} - \mathbf{b}) \cdot (\mathbf{a} - \mathbf{b}) = \mathbf{a} \cdot \mathbf{a} - \mathbf{a} \cdot \mathbf{b} - \mathbf{b} \cdot \mathbf{a} + \mathbf{b} \cdot \mathbf{b}$$

$$= |\mathbf{a}|^2 + |\mathbf{b}|^2 - 2\mathbf{a} \cdot \mathbf{b}.$$

But $c = |\overrightarrow{QP}|$ is the side of triangle OPQ (Fig. 13.1.11) that is opposite the angle θ included between the sides $a = |\mathbf{a}|$ and $b = |\mathbf{b}|$. Hence the law of cosines gives

$$|\overrightarrow{QP}|^2 = c^2 = a^2 + b^2 - 2ab \cos \theta = |\mathbf{a}|^2 + |\mathbf{b}|^2 - 2|\mathbf{a}||\mathbf{b}| \cos \theta.$$

Finally, comparison of these two expressions for $|\overrightarrow{QP}|^2$ yields Eq. (10). ❑

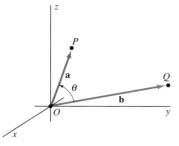

Fig. 13.1.11   The angle θ between the vectors **a** and **b**

This theorem tells us that the angle θ between the nonzero vectors **a** and **b** can be found by using the equation

$$\cos \theta = \frac{\mathbf{a} \cdot \mathbf{b}}{|\mathbf{a}||\mathbf{b}|}. \tag{11}$$

This immediately implies the perpendicularity test of Section 12.3: *The two vectors* **a** *and* **b** *are perpendicular* ($\theta = \pi/2$) *if and only if* $\mathbf{a} \cdot \mathbf{b} = 0$. For instance, the vectors **a** and **b** of Example 3 are perpendicular, because we found there that $\mathbf{a} \cdot \mathbf{b} = 0$.

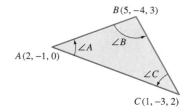

$B(5, -4, 3)$

$A(2, -1, 0)$ ∠A ∠B ∠C $C(1, -3, 2)$

Fig. 13.1.12 The triangle of Example 4

EXAMPLE 4 Find the angles shown in the triangle of Fig. 13.1.12 with vertices at $A(2, -1, 0)$, $B(5, -4, 3)$, and $C(1, -3, 2)$.

Solution We apply Eq. (10) with $\theta = \angle A$, $\mathbf{a} = \overrightarrow{AB} = \langle 3, -3, 3 \rangle$, and $\mathbf{b} = \overrightarrow{AC} = \langle -1, -2, 2 \rangle$. This yields

$$\angle A = \cos^{-1}\left(\frac{\overrightarrow{AB} \cdot \overrightarrow{AC}}{|\overrightarrow{AB}||\overrightarrow{AC}|}\right) = \cos^{-1}\left(\frac{\langle 3, -3, 3\rangle \cdot \langle -1, -2, 2\rangle}{\sqrt{27}\,\sqrt{9}}\right)$$

$$= \cos^{-1}\left(\frac{9}{\sqrt{27}\,\sqrt{9}}\right) \approx 54.74°.$$

Similarly,

$$\angle B = \cos^{-1}\left(\frac{\overrightarrow{BA} \cdot \overrightarrow{BC}}{|\overrightarrow{BA}||\overrightarrow{BC}|}\right) = \cos^{-1}\left(\frac{\langle -3, 3, -3\rangle \cdot \langle -4, 1, -1\rangle}{\sqrt{27}\,\sqrt{18}}\right)$$

$$= \cos^{-1}\left(\frac{18}{\sqrt{27}\,\sqrt{18}}\right) \approx 35.26°.$$

Then $\angle C = 180° - \angle A - \angle B = 90°$. As a check, note that

$$\overrightarrow{CA} \cdot \overrightarrow{CB} = \langle 1, 2, -2\rangle \cdot \langle 4, -1, 1\rangle = 0.$$

So the angle at C is, indeed, a right angle.

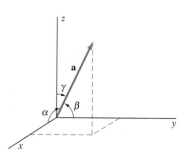

Fig. 13.1.13 The direction angles of the vector **a**

The **direction angles** of the nonzero vector $\mathbf{a} = \langle a_1, a_2, a_3 \rangle$ are the angles α, β, and γ that it makes with the vectors **i**, **j**, and **k**, respectively (Fig. 13.1.13). The cosines of these angles, $\cos \alpha$, $\cos \beta$, and $\cos \gamma$, are called the **direction cosines** of the vector **a**. When we replace **b** in Eq. (11) by **i**, **j**, and **k** in turn, we find that

$$\cos \alpha = \frac{\mathbf{a} \cdot \mathbf{i}}{|\mathbf{a}||\mathbf{i}|} = \frac{a_1}{|\mathbf{a}|},$$

$$\cos \beta = \frac{\mathbf{a} \cdot \mathbf{j}}{|\mathbf{a}||\mathbf{j}|} = \frac{a_2}{|\mathbf{a}|}, \quad \text{and} \tag{12}$$

$$\cos \gamma = \frac{\mathbf{a} \cdot \mathbf{k}}{|\mathbf{a}||\mathbf{k}|} = \frac{a_3}{|\mathbf{a}|}.$$

That is, the direction cosines of **a** are the components of the *unit vector* $\mathbf{a}/|\mathbf{a}|$ with the same direction as **a**. Consequently

$$\cos^2 \alpha + \cos^2 \beta + \cos^2 \gamma = 1. \tag{13}$$

EXAMPLE 5 Find the direction angles of the vector $\mathbf{a} = 2\mathbf{i} + 3\mathbf{j} - \mathbf{k}$.

Solution Because $|\mathbf{a}| = \sqrt{14}$, the equations in (12) give

$$\alpha = \cos^{-1}\left(\frac{2}{\sqrt{14}}\right) \approx 57.69°, \qquad \beta = \cos^{-1}\left(\frac{3}{\sqrt{14}}\right) \approx 36.70°,$$

$$\text{and} \quad \gamma = \cos^{-1}\left(\frac{-1}{\sqrt{14}}\right) \approx 105.50°.$$

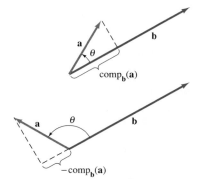

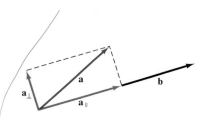

Fig. 13.1.14 The component of **a** along **b**

Sometimes we need to find the component of one vector **a** in the direction of another *nonzero* vector **b**. Think of the two vectors located with the same initial point (Fig. 13.1.14). Then the (scalar) **component of a along b**, denoted by $\text{comp}_b\, \mathbf{a}$, is numerically the length of the perpendicular projection of **a** onto the straight line determined by **b**. The number $\text{comp}_b\, \mathbf{a}$ is positive if the angle θ between **a** and **b** is acute (so **a** and **b** point in the same general direction) and negative if $\theta > \pi/2$. Thus $\text{comp}_a\, \mathbf{b} = |\mathbf{a}|\cos\theta$ in either case. Equation (10) then gives

$$\text{comp}_b\, \mathbf{a} = \frac{|\mathbf{a}||\mathbf{b}|\cos\theta}{|\mathbf{b}|} = \frac{\mathbf{a}\cdot\mathbf{b}}{|\mathbf{b}|}. \tag{14}$$

There is no need to memorize this formula, for—in practice—we can always read $\text{comp}_b\, \mathbf{a} = |\mathbf{a}|\cos\theta$ from the figure and then apply Eq. (10) to eliminate $\cos\theta$. Note that $\text{comp}_b\, \mathbf{a}$ is a scalar, not a vector.

EXAMPLE 6 Given $\mathbf{a} = \langle 4, -5, 3\rangle$ and $\mathbf{b} = \langle 2, 1, -2\rangle$, express **a** as the sum of a vector $\mathbf{a}_\|$ parallel to **b** and a vector $\mathbf{a}_\perp$ perpendicular to **b**.

Solution Our method of solution is motivated by the diagram in Fig. 13.1.15. We take

$$\mathbf{a}_\| = (\text{comp}_b\, \mathbf{a})\frac{\mathbf{b}}{|\mathbf{b}|} = \frac{\mathbf{a}\cdot\mathbf{b}}{|\mathbf{b}|^2}\mathbf{b} = \frac{8 - 5 - 6}{9}\mathbf{b}$$

$$= -\frac{1}{3}\langle 2, 1, -2\rangle = \left\langle -\frac{2}{3}, -\frac{1}{3}, \frac{2}{3}\right\rangle,$$

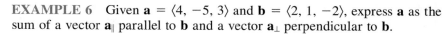

and

$$\mathbf{a}_\perp = \mathbf{a} - \mathbf{a}_\| = \langle 4, -5, 3\rangle - \left\langle -\frac{2}{3}, -\frac{1}{3}, \frac{2}{3}\right\rangle = \left\langle \frac{14}{3}, -\frac{14}{3}, \frac{7}{3}\right\rangle.$$

Fig. 13.1.15 Construction of $\mathbf{a}_\|$ and $\mathbf{a}_\perp$ (Example 6)

The diagram makes our choice of $\mathbf{a}_\|$ plausible, and we have deliberately chosen $\mathbf{a}_\perp$ so that $\mathbf{a} = \mathbf{a}_\| + \mathbf{a}_\perp$. To verify that the vector $\mathbf{a}_\|$ is indeed parallel to **b**, we simply note that it is a scalar multiple of **b**. To verify that $\mathbf{a}_\perp$ is perpendicular to **b**, we compute the dot product

$$\mathbf{a}_\perp \cdot \mathbf{b} = \tfrac{28}{3} - \tfrac{14}{3} - \tfrac{14}{3} = 0.$$

Thus $\mathbf{a}_\|$ and $\mathbf{a}_\perp$ have the required properties.

One important application of vector components is to the definition and computation of work. Recall that the work W done by a constant force F exerted along the line of motion in moving a particle a distance d is given by $W = Fd$. But what if the force is a constant vector **F** pointing in some

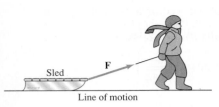

Sled
F
Line of motion

Fig. 13.1.16 The vector force **F** is constant but acts at an angle to the line of motion (Example 7, Problem 36).

direction other than the line of motion, as when a child pulls a sled against the resistance of friction (Fig. 13.1.16)? Suppose that **F** moves a particle along the line segment from P to Q, and let $\mathbf{u} = \overrightarrow{PQ}$. Then the **work** W done by **F** in moving the particle along the line from P to Q is, *by definition*, the product of the component of **F** along **u** and the distance moved:

$$W = (\text{comp}_\mathbf{u}\, \mathbf{F})\,|\mathbf{u}|. \tag{15}$$

EXAMPLE 7 Show that the work done by a constant force **F** in moving a particle along the line segment from P to Q is

$$W = \mathbf{F} \cdot \mathbf{D}, \tag{16}$$

where $\mathbf{D} = \overrightarrow{PQ}$ is the displacement vector. This formula is the vector generalization of the scalar formula $W = Fd$.

Solution We combine Eqs. (14) and (15). This gives

$$W = (\text{comp}_\mathbf{D}\,\mathbf{F})|\mathbf{D}| = \frac{\mathbf{F}\cdot\mathbf{D}}{|\mathbf{D}|}|\mathbf{D}| = \mathbf{F}\cdot\mathbf{D}.$$

13.1 Problems

In Problems 1 through 5, find (a) $2\mathbf{a} + \mathbf{b}$, (b) $3\mathbf{a} - 4\mathbf{b}$, (c) $\mathbf{a} \cdot \mathbf{b}$, (d) $|\mathbf{a} - \mathbf{b}|$, and (e) $\mathbf{a}/|\mathbf{a}|$.

1. $\mathbf{a} = \langle 2, 5, -4\rangle$, $\mathbf{b} = \langle 1, -2, -3\rangle$
2. $\mathbf{a} = \langle -1, 0, 2\rangle$, $\mathbf{b} = \langle 3, 4, -5\rangle$
3. $\mathbf{a} = \mathbf{i} + \mathbf{j} + \mathbf{k}$, $\mathbf{b} = \mathbf{j} - \mathbf{k}$
4. $\mathbf{a} = 2\mathbf{i} - 3\mathbf{j} + 5\mathbf{k}$, $\mathbf{b} = 5\mathbf{i} + 3\mathbf{j} - 7\mathbf{k}$
5. $\mathbf{a} = 2\mathbf{i} - \mathbf{j}$, $\mathbf{b} = \mathbf{j} - 3\mathbf{k}$

6 through 10. Find the angle between the vectors **a** and **b** in Problems 1 through 5.

11 through 15. Find $\text{comp}_\mathbf{a}\,\mathbf{b}$ and $\text{comp}_\mathbf{b}\,\mathbf{a}$ for the vectors **a** and **b** given in Problems 1 through 5.

In Problems 16 through 20, write the equation of the indicated sphere.

16. Center $(3, 1, 2)$, radius 5
17. Center $(-2, 1, -5)$, radius $\sqrt{7}$
18. One diameter: the segment joining $(3, 5, -3)$ and $(7, 3, 1)$
19. Center $(4, 5, -2)$, passing through the point $(1, 0, 0)$
20. Center $(3, -4, 3)$, tangent to the xz-plane

In Problems 21 through 23, find the center and radius of the sphere that has the given equation.

21. $x^2 + y^2 + z^2 + 4x - 6y = 0$
22. $x^2 + y^2 + z^2 - 8x - 9y + 10z + 40 = 0$
23. $3x^2 + 3y^2 + 3z^2 - 18z - 48 = 0$

In Problems 24 through 30, describe the graph of the given equation in geometric terms, using plain, clear language.

24. $x = 0$ 25. $z = 10$ 26. $xy = 0$
27. $xyz = 0$ 28. $x^2 + y^2 + z^2 + 7 = 0$
29. $x^2 + y^2 + z^2 - 2x + 1 = 0$
30. $x^2 + y^2 + z^2 - 6x + 8y + 25 = 0$

In Problems 31 through 33, find the direction angles of the vector $\overrightarrow{PQ}$.

31. $P(1, -1, 0)$, $Q(3, 4, 5)$
32. $P(2, -3, 5)$, $Q(1, 0, -1)$
33. $P(-1, -2, -3)$, $Q(5, 6, 7)$

*In Problems 34 and 35, find the work W done by the force **F** in moving a particle in a straight line from P to Q.*

34. $\mathbf{F} = \mathbf{i} - \mathbf{k}$; $P(0, 0, 0), Q(3, 1, 0)$
35. $\mathbf{F} = 2\mathbf{i} - 3\mathbf{j} + 5\mathbf{k}$; $P(5, 3, -4), Q(-1, -2, 5)$
36. Suppose that the force vector in Fig. 13.1.16 is inclined at an angle of 30° to the ground. If the child exerts a constant force of 20 lb, how much work (in foot-pounds) is done by pulling the sled a distance of 100 ft along the ground?
37. Suppose that the horizontal and vertical components of the vectors shown in Fig. 13.1.17 balance (the algebraic sum of the horizontal components is zero, as is the sum of the vertical components). How much work is done by the constant force **F** (parallel to the inclined plane) in pulling the weight mg up the inclined plane a vertical height h?

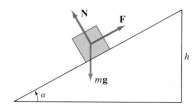

Fig. 13.1.17   The inclined plane of Problem 37

38. Prove the **Cauchy-Schwarz inequality**:

$$|\mathbf{a} \cdot \mathbf{b}| \leq |\mathbf{a}||\mathbf{b}|$$

for all pairs of vectors **a** and **b**.

39. Given two arbitrary vectors **a** and **b**, prove that they satisfy the **triangle inequality**

$$|\mathbf{a} + \mathbf{b}| \leq |\mathbf{a}| + |\mathbf{b}|.$$

[*Suggestion:* Square both sides.]

40. Prove that if **a** and **b** are arbitrary vectors, then

$$|\mathbf{a} - \mathbf{b}| \geq |\mathbf{a}| - |\mathbf{b}|.$$

[*Suggestion:* Write $\mathbf{a} = (\mathbf{a} - \mathbf{b}) + \mathbf{b}$; then apply the triangle inequality of Problem 39.]

41. Find the area of the triangle with vertices $A(1, 1, 1)$, $B(3, -2, 3)$, and $C(3, 4, 6)$.

42. Find the three angles of the triangle of Problem 41.

43. Find the angle between any longest diagonal of a cube and any edge that diagonal meets.

44. Prove that the three points $P(0, -2, 4)$, $Q(1, -3, 5)$, and $R(4, -6, 8)$ lie on a single straight line.

45. Prove that the point M given in Eq. (2) is indeed the midpoint of the segment $P_1 P_2$. [*Note:* You must prove *both* that M is equally distant from P_1 and P_2 *and* that M lies on the segment $P_1 P_2$.]

46. Given vectors **a** and **b**, let $a = |\mathbf{a}|$ and $b = |\mathbf{b}|$. Prove that the vector $\mathbf{c} = (b\mathbf{a} + a\mathbf{b})/(a + b)$ bisects the angle between **a** and **b**.

47. Let **a**, **b**, and **c** be three vectors in the xy-plane with **a** and **b** nonzero and not parallel. Show that there exist scalars α and β such that $\mathbf{c} = \alpha\mathbf{a} + \beta\mathbf{b}$. [*Suggestion:* Begin by expressing **a**, **b**, and **c** in terms of **i**, **j**, and **k**.]

48. Let $ax + by + c = 0$ be the equation of the line L in the xy-plane with normal vector **n**. Let $P_0(x_0, y_0)$ be a point on this line and $P_1(x_1, y_1)$ be a point not on L. Prove that the perpendicular distance from P_1 to L is

$$d = \frac{|\mathbf{n} \cdot \overrightarrow{P_0 P_1}|}{|\mathbf{n}|} = \frac{|ax_1 + by_1 + c|}{\sqrt{a^2 + b^2}}.$$

49. Given the two points $A(3, -2, 4)$ and $B(5, 7, -1)$, write an equation in x, y, and z that says that the point $P(x, y, z)$ is equally distant from the points A and B. Then simplify this equation, and give a geometric description of the set of all such points $P(x, y, z)$.

50. Given the fixed point $A(1, 3, 5)$, the point $P(x, y, z)$, and the vector $\mathbf{n} = \mathbf{i} - \mathbf{j} + 2\mathbf{k}$, use the dot product to help you write an equation in x, y, and z that says this: **n** and $\overrightarrow{AP}$ are perpendicular. Then simplify this equation, and give a geometric description of the set of all such points $P(x, y, z)$.

51. Prove that the points $(0, 0, 0)$, $(1, 1, 0)$, $(1, 0, 1)$, and $(0, 1, 1)$ are the vertices of a regular tetrahedron by showing that each of the six edges has length $\sqrt{2}$. Then use the dot product to find the angle between any two edges of the tetrahedron.

52. The methane molecule, CH_4, is arranged with the four hydrogen atoms at the vertices of a regular tetrahedron and with the carbon atom at its center (Fig. 13.1.18). Suppose that the axes and scale are chosen so that the tetrahedron is that of Problem 51, with its center at $(\frac{1}{2}, \frac{1}{2}, \frac{1}{2})$. Find the *bond angle* α between the lines from the carbon atom to two of the hydrogen atoms.

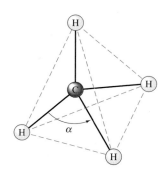

Fig. 13.1.18   The methane bond angle α of Problem 52

13.2
The Vector Product of Two Vectors

We often need to find a vector that is perpendicular to each of two vectors **a** and **b** in space. A routine way of doing this is provided by the **vector product**, or **cross product**, $\mathbf{a} \times \mathbf{b}$ of the vectors **a** and **b**. This vector product is quite unlike the dot product $\mathbf{a} \cdot \mathbf{b}$ in that $\mathbf{a} \cdot \mathbf{b}$ is a scalar, whereas $\mathbf{a} \times \mathbf{b}$ is a vector.

The vector product of the vectors $\mathbf{a} = \langle a_1, a_2, a_3 \rangle$ and $\mathbf{b} = \langle b_1, b_2, b_3 \rangle$ can be defined by the formula

$$\mathbf{a} \times \mathbf{b} = \langle a_2 b_3 - a_3 b_2, \ a_3 b_1 - a_1 b_3, \ a_1 b_2 - a_2 b_1 \rangle. \tag{1}$$

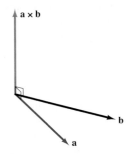

Fig. 13.2.1 The vector product $\mathbf{a} \times \mathbf{b}$ is perpendicular to both $\mathbf{a}$ and $\mathbf{b}$.

Although this formula seems unmotivated, it has a redeeming feature: the product $\mathbf{a} \times \mathbf{b}$ is perpendicular both to $\mathbf{a}$ and to $\mathbf{b}$, as suggested in Fig. 13.2.1.

Theorem 1 *Perpendicularity of the Vector Product*
The vector product $\mathbf{a} \times \mathbf{b}$ is perpendicular both to $\mathbf{a}$ and to $\mathbf{b}$.

Proof We show that $\mathbf{a} \times \mathbf{b}$ is perpendicular to $\mathbf{a}$ by showing that the dot product of $\mathbf{a}$ and $\mathbf{a} \times \mathbf{b}$ is zero. With the components as in Eq. (1), we find that

$$\mathbf{a} \cdot (\mathbf{a} \times \mathbf{b}) = a_1(a_2b_3 - a_3b_2) + a_2(a_3b_1 - a_1b_3) + a_3(a_1b_2 - a_2b_1)$$

$$= a_1a_2b_3 - a_1a_3b_2 + a_2a_3b_1 - a_2a_1b_3 + a_3a_1b_2 - a_3a_2b_1$$

$$= 0.$$

A similar computation shows that $\mathbf{b} \cdot (\mathbf{a} \times \mathbf{b}) = 0$ as well, so $\mathbf{a} \times \mathbf{b}$ is also perpendicular to the vector $\mathbf{b}$. ❏

You need not memorize Eq. (1), because there is an alternative version involving determinants that is easy to remember. Recall that a *determinant* of order 2 is defined as follows:

$$\begin{vmatrix} a_1 & a_2 \\ b_1 & b_2 \end{vmatrix} = a_1b_2 - a_2b_1. \tag{2}$$

EXAMPLE 1

$$\begin{vmatrix} 2 & -1 \\ 3 & 4 \end{vmatrix} = 2 \cdot 4 - (-1) \cdot 3 = 11.$$

A determinant of order 3 can be defined in terms of determinants of order 2:

$$\begin{vmatrix} a_1 & a_2 & a_3 \\ b_1 & b_2 & b_3 \\ c_1 & c_2 & c_3 \end{vmatrix} = +a_1 \begin{vmatrix} b_2 & b_3 \\ c_2 & c_3 \end{vmatrix} - a_2 \begin{vmatrix} b_1 & b_3 \\ c_1 & c_3 \end{vmatrix} + a_3 \begin{vmatrix} b_1 & b_2 \\ c_1 & c_2 \end{vmatrix}. \tag{3}$$

Each element a_i of the first row is multiplied by the 2-by-2 "subdeterminant" obtained by deleting the row *and* column that contain a_i. Note in Eq. (3) that signs are attached to the a_i in accord with the checkerboard pattern

$$\begin{vmatrix} + & - & + \\ - & + & - \\ + & - & + \end{vmatrix}.$$

Equation (3) is an expansion of the 3-by-3 determinant along its first row. It can be expanded along any other row or column as well. For example, its expansion along its second column is

$$\begin{vmatrix} a_1 & a_2 & a_3 \\ b_1 & b_2 & b_3 \\ c_1 & c_2 & c_3 \end{vmatrix} = -a_2 \begin{vmatrix} b_1 & b_3 \\ c_1 & c_3 \end{vmatrix} + b_2 \begin{vmatrix} a_1 & a_3 \\ c_1 & c_3 \end{vmatrix} - c_2 \begin{vmatrix} a_1 & a_3 \\ b_1 & b_3 \end{vmatrix}.$$

In linear algebra we prove that all such expansions yield the same value for the determinant.

Although we can expand a determinant of order 3 along any row or column, we shall use only expansions along the first row, as in Eq. (3).

EXAMPLE 2

$$\begin{vmatrix} 1 & 3 & -2 \\ 2 & -1 & 4 \\ -3 & 7 & 5 \end{vmatrix} = 1 \cdot \begin{vmatrix} -1 & 4 \\ 7 & 5 \end{vmatrix} - 3 \cdot \begin{vmatrix} 2 & 4 \\ -3 & 5 \end{vmatrix} + (-2) \cdot \begin{vmatrix} 2 & -1 \\ -3 & 7 \end{vmatrix}$$

$$= 1 \cdot (-5 - 28) + (-3) \cdot (10 + 12) + (-2) \cdot (14 - 3)$$

$$= -33 - 66 - 22 = -121.$$

Equation (1) for the vector product of the vectors $\mathbf{a} = a_1\mathbf{i} + a_2\mathbf{j} + a_3\mathbf{k}$ and $\mathbf{b} = b_1\mathbf{i} + b_2\mathbf{j} + b_3\mathbf{k}$ is equivalent to

$$\mathbf{a} \times \mathbf{b} = \begin{vmatrix} a_2 & a_3 \\ b_2 & b_3 \end{vmatrix}\mathbf{i} - \begin{vmatrix} a_1 & a_3 \\ b_1 & b_3 \end{vmatrix}\mathbf{j} + \begin{vmatrix} a_1 & a_2 \\ b_1 & b_2 \end{vmatrix}\mathbf{k}. \qquad (4)$$

This is easy to verify by expanding the 2-by-2 determinants on the right-hand side and noting that the three components of the right-hand side of Eq. (1) result. Motivated by Eq. (4), we write

$$\mathbf{a} \times \mathbf{b} = \begin{vmatrix} \mathbf{i} & \mathbf{j} & \mathbf{k} \\ a_1 & a_2 & a_3 \\ b_1 & b_2 & b_3 \end{vmatrix}. \qquad (5)$$

The "symbolic determinant" in this equation is to be evaluated by expanding along its first row, just as in Eq. (3) and just as though it were an ordinary determinant with real number entries. The result of this expansion is the right-hand side of Eq. (4). The components of the *first* vector $\mathbf{a}$ in $\mathbf{a} \times \mathbf{b}$ form the *second* row of the 3-by-3 determinant, and the components of the *second* vector $\mathbf{b}$ form the *third* row. The order of the vectors $\mathbf{a}$ and $\mathbf{b}$ is important because, as we soon shall see, $\mathbf{a} \times \mathbf{b}$ is generally *not* equal to $\mathbf{b} \times \mathbf{a}$: The vector product is *not commutative*.

Equation (5) for the vector product is the form most convenient for computational purposes.

EXAMPLE 3 If $\mathbf{a} = 3\mathbf{i} - \mathbf{j} + 2\mathbf{k}$ and $\mathbf{b} = 2\mathbf{i} + 2\mathbf{j} - \mathbf{k}$, then

$$\mathbf{a} \times \mathbf{b} = \begin{vmatrix} \mathbf{i} & \mathbf{j} & \mathbf{k} \\ 3 & -1 & 2 \\ 2 & 2 & -1 \end{vmatrix} = \begin{vmatrix} -1 & 2 \\ 2 & -1 \end{vmatrix}\mathbf{i} - \begin{vmatrix} 3 & 2 \\ 2 & -1 \end{vmatrix}\mathbf{j} + \begin{vmatrix} 3 & -1 \\ 2 & 2 \end{vmatrix}\mathbf{k}$$

$$= (1 - 4)\mathbf{i} - (-3 - 4)\mathbf{j} + (6 - (-2))\mathbf{k}.$$

Thus

$$\mathbf{a} \times \mathbf{b} = -3\mathbf{i} + 7\mathbf{j} + 8\mathbf{k}.$$

You might now pause to verify (by using the dot product) that the vector $-3\mathbf{i} + 7\mathbf{j} + 8\mathbf{k}$ is perpendicular both to $\mathbf{a}$ and to $\mathbf{b}$.

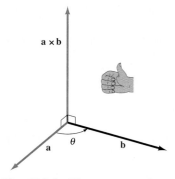

Fig. 13.2.2 The vectors **a**, **b**, and **a** × **b**—in that order—form a right-handed triple.

If the vectors **a** and **b** share the same initial point, then Theorem 1 implies that **a** × **b** is normal to the plane determined by **a** and **b** (Fig. 13.2.2). There are still two possible directions for **a** × **b**, but if **a** × **b** ≠ **0**, then the triple **a**, **b**, **a** × **b** is a *right-handed* triple in exactly the same sense as the triple **i**, **j**, **k**. Thus if the thumb of your right hand points in the direction of **a** × **b**, then your fingers curl in the direction of rotation (less than 180°) from **a** to **b**.

Once we have established the direction of **a** × **b**, we can describe the vector product in completely geometric terms by telling what the length $|\mathbf{a} \times \mathbf{b}|$ of the vector **a** × **b** is. This is given by the formula

$$|\mathbf{a} \times \mathbf{b}|^2 = |\mathbf{a}|^2|\mathbf{b}|^2 - (\mathbf{a} \cdot \mathbf{b})^2. \qquad (6)$$

We can verify this vector identity routinely (though tediously) by writing $\mathbf{a} = \langle a_1, a_2, a_3 \rangle$ and $\mathbf{b} = \langle b_1, b_2, b_3 \rangle$, computing both sides of Eq. (6), and then noting that the results are equal (Problem 28).

GEOMETRIC SIGNIFICANCE OF THE VECTOR PRODUCT

Equation (6) tells us what $|\mathbf{a} \times \mathbf{b}|$ is, but Theorem 2 reveals the geometric significance of the vector product.

Theorem 2 *Length of the Vector Product*
Let θ be the angle between the nonzero vectors **a** and **b** (measured so that $0 \le \theta \le \pi$). Then

$$|\mathbf{a} \times \mathbf{b}| = |\mathbf{a}||\mathbf{b}|\sin\theta. \qquad (7)$$

Proof We begin with Eq. (6) and use the fact that $\mathbf{a} \cdot \mathbf{b} = |\mathbf{a}||\mathbf{b}|\cos\theta$. Thus

$$|\mathbf{a} \times \mathbf{b}|^2 = |\mathbf{a}|^2|\mathbf{b}|^2 - (\mathbf{a} \cdot \mathbf{b})^2 = |\mathbf{a}|^2|\mathbf{b}|^2 - (|\mathbf{a}||\mathbf{b}|\cos\theta)^2$$

$$= |\mathbf{a}|^2|\mathbf{b}|^2(1 - \cos^2\theta) = |\mathbf{a}|^2|\mathbf{b}|^2\sin^2\theta.$$

Equation (7) now follows after we take the positive square root of both sides. (This is the correct root on the right-hand side, because $\sin\theta \ge 0$ for $0 \le \theta \le \pi$.) ❑

Corollary *Parallel Vectors*
Two nonzero vectors **a** and **b** are parallel ($\theta = 0$ or $\theta = \pi$) if and only if **a** × **b** = **0**.

In particular, the vector product of any vector with itself is the zero vector. Also, Eq. (1) shows immediately that the vector product of any vector with the zero vector is the zero vector itself. Thus

$$\mathbf{a} \times \mathbf{a} = \mathbf{a} \times \mathbf{0} = \mathbf{0} \times \mathbf{a} = \mathbf{0} \qquad (8)$$

for every vector **a**.

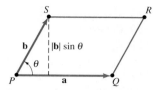

Fig. 13.2.3 The area of the parallelogram *PQRS* is $|\mathbf{a} \times \mathbf{b}|$.

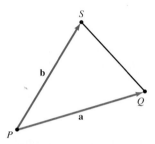

Fig. 13.2.4 The area of $\triangle PQS$ is $\frac{1}{2}|\mathbf{a} \times \mathbf{b}|$.

Equation (7) has an important geometric interpretation. Suppose that $\mathbf{a}$ and $\mathbf{b}$ are represented by adjacent sides of a parallelogram *PQRS*, with $\mathbf{a} = \overrightarrow{PQ}$ and $\mathbf{b} = \overrightarrow{PS}$ (Fig. 13.2.3). The parallelogram then has base of length $|\mathbf{a}|$ and height $|\mathbf{b}| \sin \theta$, so its area is

$$A = |\mathbf{a}||\mathbf{b}| \sin \theta = |\mathbf{a} \times \mathbf{b}|. \qquad (9)$$

Thus *the length of the vector product* $\mathbf{a} \times \mathbf{b}$ *is numerically the same as the area of the parallelogram determined by* $\mathbf{a}$ *and* $\mathbf{b}$. It follows that the area of the triangle *PQS* in Fig. 13.2.4, whose area is half that of the parallelogram, is

$$\tfrac{1}{2}A = \tfrac{1}{2}|\mathbf{a} \times \mathbf{b}| = \tfrac{1}{2}|\overrightarrow{PQ} \times \overrightarrow{PS}|. \qquad (10)$$

Equation (10) gives a quick way to compute the area of a triangle—even one in space—without the need of finding any of its angles.

EXAMPLE 4 Find the area of the triangle with vertices $A(3, 0, -1)$, $B(4, 2, 5)$, and $C(7, -2, 4)$.

Solution $\overrightarrow{AB} = \langle 1, 2, 6 \rangle$ and $\overrightarrow{AC} = \langle 4, -2, 5 \rangle$, so

$$\overrightarrow{AB} \times \overrightarrow{AC} = \begin{vmatrix} \mathbf{i} & \mathbf{j} & \mathbf{k} \\ 1 & 2 & 6 \\ 4 & -2 & 5 \end{vmatrix} = 22\mathbf{i} + 19\mathbf{j} - 10\mathbf{k}.$$

Therefore, by Eq. (10), the area of triangle *ABC* is

$$\tfrac{1}{2}\sqrt{22^2 + 19^2 + (-10)^2} = \tfrac{1}{2}\sqrt{945} \approx 15.37.$$

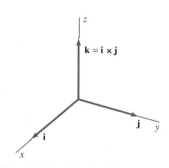

Fig. 13.2.5 The basic unit vectors in space

Now let $\mathbf{u}$, $\mathbf{v}$, $\mathbf{w}$ be a right-handed triple of mutually perpendicular *unit* vectors. The angle between any two of these is $\theta = \pi/2$, and $|\mathbf{u}| = |\mathbf{v}| = |\mathbf{w}| = 1$. Thus it follows from Eq. (7) that $\mathbf{u} \times \mathbf{v} = \mathbf{w}$. When we apply this observation to the basic unit vectors $\mathbf{i}$, $\mathbf{j}$, and $\mathbf{k}$ (Fig. 13.2.5), we see that

$$\mathbf{i} \times \mathbf{j} = \mathbf{k}, \quad \mathbf{j} \times \mathbf{k} = \mathbf{i}, \quad \text{and} \quad \mathbf{k} \times \mathbf{i} = \mathbf{j}. \qquad (11a)$$

But

$$\mathbf{j} \times \mathbf{i} = -\mathbf{k}, \quad \mathbf{k} \times \mathbf{j} = -\mathbf{i}, \quad \text{and} \quad \mathbf{i} \times \mathbf{k} = -\mathbf{j}. \qquad (11b)$$

These observations, together with the fact that

$$\mathbf{i} \times \mathbf{i} = \mathbf{j} \times \mathbf{j} = \mathbf{k} \times \mathbf{k} = \mathbf{0}, \qquad (11c)$$

also follow directly from the original definition of the vector product [in the form of Eq. (5)]. The products in Eq. (11a) are easily remembered in terms of the sequence

$$\mathbf{i}, \quad \mathbf{j}, \quad \mathbf{k}, \quad \mathbf{i}, \quad \mathbf{j}, \quad \mathbf{k}, \dots.$$

The product of any two consecutive unit vectors, in the order in which they appear in this sequence, is the next in the sequence.

NOTE *The vector product is not commutative:* $\mathbf{i} \times \mathbf{j} \neq \mathbf{j} \times \mathbf{i}$. Instead, it is **anticommutative**: For any two vectors $\mathbf{a}$ and $\mathbf{b}$, $\mathbf{a} \times \mathbf{b} = -(\mathbf{b} \times \mathbf{a})$. This is the first part of Theorem 3.

> **Theorem 3** *Algebraic Properties of the Vector Product*
> If **a**, **b**, and **c** are vectors and k is a real number, then
>
> 1. $\mathbf{a} \times \mathbf{b} = -(\mathbf{b} \times \mathbf{a})$;　　　　　　　　　　　　　　(12)
> 2. $(k\mathbf{a}) \times \mathbf{b} = \mathbf{a} \times (k\mathbf{b}) = k(\mathbf{a} \times \mathbf{b})$;　　　　　　(13)
> 3. $\mathbf{a} \times (\mathbf{b} + \mathbf{c}) = (\mathbf{a} \times \mathbf{b}) + (\mathbf{a} \times \mathbf{c})$;　　　　(14)
> 4. $\mathbf{a} \cdot (\mathbf{b} \times \mathbf{c}) = (\mathbf{a} \times \mathbf{b}) \cdot \mathbf{c}$;　　　　　　　(15)
> 5. $\mathbf{a} \times (\mathbf{b} \times \mathbf{c}) = (\mathbf{a} \cdot \mathbf{c})\mathbf{b} - (\mathbf{a} \cdot \mathbf{b})\mathbf{c}$.　　　(16)

The proofs of Eqs. (12) through (15) are straightforward applications of the definition of the vector product in terms of components. See Problem 25 for an outline of the proof of Eq. (16).

We can find vector products of vectors expressed in terms of the basic unit vectors **i**, **j**, and **k** by means of computations that closely resemble those of ordinary algebra. We simply apply the algebraic properties summarized in Theorem 3 together with the relations in Eq. (11) giving the various products of the three unit vectors. We must be careful to preserve the order of factors, because vector multiplication is not commutative—although, of course, we should not hesitate to use Eq. (12). For example,

$$(\mathbf{i} - 2\mathbf{j} + 3\mathbf{k}) \times (3\mathbf{i} + 2\mathbf{j} - 4\mathbf{k})$$

$$= 3(\mathbf{i} \times \mathbf{i}) + 2(\mathbf{i} \times \mathbf{j}) - 4(\mathbf{i} \times \mathbf{k}) - 6(\mathbf{j} \times \mathbf{i}) - 4(\mathbf{j} \times \mathbf{j})$$

$$+ 8(\mathbf{j} \times \mathbf{k}) + 9(\mathbf{k} \times \mathbf{i}) + 6(\mathbf{k} \times \mathbf{j}) - 12(\mathbf{k} \times \mathbf{k})$$

$$= 3 \cdot \mathbf{0} + 2\mathbf{k} - 4 \cdot (-\mathbf{j}) - 6 \cdot (-\mathbf{k}) - 4 \cdot \mathbf{0}$$

$$+ 8\mathbf{i} + 9\mathbf{j} + 6 \cdot (-\mathbf{i}) - 12 \cdot \mathbf{0}$$

$$= 2\mathbf{i} + 13\mathbf{j} + 8\mathbf{k}.$$

SCALAR TRIPLE PRODUCTS

Let us examine the product $\mathbf{a} \cdot (\mathbf{b} \times \mathbf{c})$ that appears in Eq. (15). This expression would not make sense were the parentheses around $\mathbf{a} \cdot \mathbf{b}$, because $\mathbf{a} \cdot \mathbf{b}$ is a scalar, and thus we could not form the vector product of $\mathbf{a} \cdot \mathbf{b}$ with the vector **c**. This means that we may omit the parentheses—the expression $\mathbf{a} \cdot \mathbf{b} \times \mathbf{c}$ is not ambiguous—but we keep them for simplicity. The dot product of the vectors **a** and $\mathbf{b} \times \mathbf{c}$ is a real number, called the **scalar triple product** of the vectors **a**, **b**, and **c**. Equation (15) implies the curious fact that we can interchange the operations · (dot) and × (cross) without affecting the value of the expression:

$$\mathbf{a} \cdot (\mathbf{b} \times \mathbf{c}) = (\mathbf{a} \times \mathbf{b}) \cdot \mathbf{c}$$

for all vectors **a**, **b**, and **c**.

To compute the scalar triple product in terms of components, write $\mathbf{a} = \langle a_1\ a_2\ a_3 \rangle$, $\mathbf{b} = \langle b_1\ b_2\ b_3 \rangle$, and $\mathbf{c} = \langle c_1,\ c_2,\ c_3 \rangle$. Then

$$\mathbf{b} \times \mathbf{c} = (b_2 c_3 - b_3 c_2)\mathbf{i} - (b_1 c_3 - b_3 c_1)\mathbf{j} + (b_1 c_2 - b_2 c_1)\mathbf{k},$$

so

$$\mathbf{a} \cdot (\mathbf{b} \times \mathbf{c}) = a_1(b_2c_3 - b_3c_2) - a_2(b_1c_3 - b_3c_1) + a_3(b_1c_2 - b_2c_1).$$

But the expression on the right is the value of the 3-by-3 determinant

$$\mathbf{a} \cdot (\mathbf{b} \times \mathbf{c}) = \begin{vmatrix} a_1 & a_2 & a_3 \\ b_1 & b_2 & b_3 \\ c_1 & c_2 & c_3 \end{vmatrix}. \tag{17}$$

This is the quickest way to compute the scalar triple product.

EXAMPLE 5 If $\mathbf{a} = 2\mathbf{i} - 3\mathbf{k}$, $\mathbf{b} = \mathbf{i} + \mathbf{j} + \mathbf{k}$, and $\mathbf{c} = 4\mathbf{j} - \mathbf{k}$, then

$$\mathbf{a} \cdot (\mathbf{b} \times \mathbf{c}) = \begin{vmatrix} 2 & 0 & -3 \\ 1 & 1 & 1 \\ 0 & 4 & -1 \end{vmatrix}$$

$$= +2 \cdot \begin{vmatrix} 1 & 1 \\ 4 & -1 \end{vmatrix} - 0 \cdot \begin{vmatrix} 1 & 1 \\ 0 & -1 \end{vmatrix} + (-3) \cdot \begin{vmatrix} 1 & 1 \\ 0 & 4 \end{vmatrix}$$

$$= 2 \cdot (-5) + (-3) \cdot 4 = -22.$$

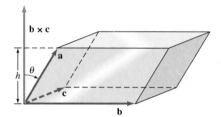

Fig. 13.2.6 The volume of the parallelepiped is $|\mathbf{a} \cdot (\mathbf{b} \times \mathbf{c})|$.

The importance of the scalar triple product for applications depends on the following geometric interpretation. Let $\mathbf{a}$, $\mathbf{b}$, and $\mathbf{c}$ be three vectors with the same initial point. Figure 13.2.6 shows the parallelepiped determined by these vectors—that is, with arrows representing these vectors as adjacent edges. If the vectors $\mathbf{a}$, $\mathbf{b}$, and $\mathbf{c}$ are coplanar (lie in a single plane), then the parallelepiped is *degenerate* and its volume is zero. Theorem 4 holds whether or not the three vectors are coplanar, but it is most useful when they are not.

Theorem 4 *Scalar Triple Products and Volume*
The volume V of the parallelepiped determined by the vectors $\mathbf{a}$, $\mathbf{b}$, and $\mathbf{c}$ is the absolute value of the scalar triple product $\mathbf{a} \cdot (\mathbf{b} \times \mathbf{c})$; that is,

$$V = |\mathbf{a} \cdot (\mathbf{b} \times \mathbf{c})|. \tag{18}$$

Proof If the three vectors are coplanar, then $\mathbf{a}$ and $\mathbf{b} \times \mathbf{c}$ are perpendicular, so $V = |\mathbf{a} \cdot (\mathbf{b} \times \mathbf{c})| = 0$. Assume that they are not coplanar. By Eq. (9) the area of the base (determined by $\mathbf{b}$ and $\mathbf{c}$) of the parallelepiped is $A = |\mathbf{b} \times \mathbf{c}|$.

Now let α be the *acute* angle between $\mathbf{a}$ and the line through $\mathbf{b} \times \mathbf{c}$ that is perpendicular to the base. Then the height of the parallelepiped is $h = |\mathbf{a}| \cos \alpha$. If θ is the angle between the vectors $\mathbf{a}$ and $\mathbf{b} \times \mathbf{c}$, then either $\theta = \alpha$ or $\theta = \pi - \alpha$. Hence $\cos \alpha = |\cos \theta|$, so

$$V = Ah = |\mathbf{b} \times \mathbf{c}||\mathbf{a}| \cos \alpha = |\mathbf{a}||\mathbf{b} \times \mathbf{c}||\cos \theta| = |\mathbf{a} \cdot (\mathbf{b} \times \mathbf{c})|.$$

Thus we have verified Eq. (18). □

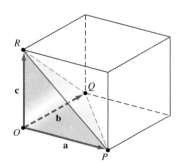

Fig. 13.2.7 The pyramid (and parallelepiped) of Example 6

EXAMPLE 6 Figure 13.2.7 shows the pyramid $OPQR$ and the parallelepiped both determined by the vectors $\mathbf{a} = \overrightarrow{OP} = \langle 3, 2, -1 \rangle$, $\mathbf{b} = \overrightarrow{OQ} = \langle -2, 5, 1 \rangle$, and $\mathbf{c} = \overrightarrow{OR} = \langle 2, 1, 5 \rangle$. The volume of the pyramid is

$V = \frac{1}{3}Ah$, where h is its height and the area of its base OPQ is *half* the area of the corresponding base of the parallelepiped. It therefore follows from Eqs. (17) and (18) that V is one-sixth the volume of the parallelepiped:

$$V = \frac{1}{6}|\mathbf{a} \cdot (\mathbf{b} \times \mathbf{c})| = \frac{1}{6}\begin{vmatrix} 3 & 2 & -1 \\ -2 & 5 & 1 \\ 2 & 1 & 5 \end{vmatrix} = \frac{108}{6} = 18.$$

EXAMPLE 7 Use the scalar triple product to show that the points $A(1, -1, 2)$, $B(2, 0, 1)$, $C(3, 2, 0)$, and $D(5, 4, -2)$ are coplanar.

Solution It's enough to show that the vectors $\overrightarrow{AB} = \langle 1, 1, -1 \rangle$, $\overrightarrow{AC} = \langle 2, 3, -2 \rangle$, and $\overrightarrow{AD} = \langle 4, 5, -4 \rangle$ are coplanar. But their scalar triple product is

$$\begin{vmatrix} 1 & 1 & -1 \\ 2 & 3 & -2 \\ 4 & 5 & -4 \end{vmatrix} = 1 \cdot (-2) - 1 \cdot 0 + (-1) \cdot (-2) = 0,$$

so Theorem 4 guarantees that the parallelepiped determined by these three vectors has volume zero. Hence the four given points are coplanar.

The vector product occurs quite often in scientific applications. For example, suppose that a body in space is free to rotate about the fixed point O. If a force $\mathbf{F}$ acts at a point P of the body, that force causes the body to rotate. This effect is measured by the **torque vector $\boldsymbol{\tau}$** defined by the relation

$$\boldsymbol{\tau} = \mathbf{r} \times \mathbf{F},$$

where $\mathbf{r} = \overrightarrow{OP}$. The straight line through O determined by $\boldsymbol{\tau}$ is the axis of rotation, and the length

$$|\boldsymbol{\tau}| = |\mathbf{r}||\mathbf{F}| \sin \theta$$

is the **moment** of the force $\mathbf{F}$ about this axis (Fig. 13.2.8).

Another application of the vector product involves the force exerted on a moving charged particle by a magnetic field. This force is important in particle accelerators and in television picture tubes; controlling the paths of the ions is accomplished through the interplay of electric and magnetic fields. In such circumstances, the force $\mathbf{F}$ on the particle due to a magnetic field depends on three things: the charge q of the particle, its velocity vector $\mathbf{v}$, and the magnetic field vector $\mathbf{B}$ at the instantaneous location of the particle. And it turns out that

$$\mathbf{F} = (q\mathbf{v}) \times \mathbf{B}.$$

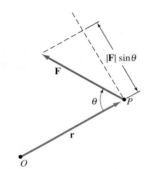

Fig. 13.2.8 The torque vector $\boldsymbol{\tau}$ is normal to both $\mathbf{r}$ and $\mathbf{F}$.

13.2 Problems

Find $\mathbf{a} \times \mathbf{b}$ in Problems 1 through 4.

1. $\mathbf{a} = \langle 5, -1, -2 \rangle$, $\mathbf{b} = \langle -3, 2, 4 \rangle$
2. $\mathbf{a} = \langle 3, -2, 0 \rangle$, $\mathbf{b} = \langle 0, 3, -2 \rangle$
3. $\mathbf{a} = \mathbf{i} - \mathbf{j} + 3\mathbf{k}$, $\mathbf{b} = -2\mathbf{i} + 3\mathbf{j} + \mathbf{k}$
4. $\mathbf{a} = 4\mathbf{i} + 2\mathbf{j} - 2\mathbf{k}$, $\mathbf{b} = 2\mathbf{i} - 5\mathbf{j} + 5\mathbf{k}$
5. Apply Eq. (5) to verify Eqs. (11a).

6. Apply Eq. (5) to verify Eqs. (11b).

7. Prove that the vector product is not associative by calculating and comparing $\mathbf{a} \times (\mathbf{b} \times \mathbf{c})$ and $(\mathbf{a} \times \mathbf{b}) \times \mathbf{c}$ with $\mathbf{a} = \mathbf{i}$, $\mathbf{b} = \mathbf{i} + \mathbf{j}$, and $\mathbf{c} = \mathbf{i} + \mathbf{j} + \mathbf{k}$.

8. Find nonzero vectors $\mathbf{a}$, $\mathbf{b}$, and $\mathbf{c}$ such that $\mathbf{a} \times \mathbf{b} = \mathbf{a} \times \mathbf{c}$ but $\mathbf{b} \neq \mathbf{c}$.

9. Suppose that the three vectors **a**, **b**, and **c** are mutually perpendicular. Prove that $\mathbf{a} \times (\mathbf{b} \times \mathbf{c}) = \mathbf{0}$.

10. Find the area of the triangle with vertices $P(1, 1, 0)$, $Q(1, 0, 1)$, and $R(0, 1, 1)$.

11. Find the area of the triangle with vertices $P(1, 3, -2)$, $Q(2, 4, 5)$, and $R(-3, -2, 2)$.

12. Find the volume of the parallelepiped with adjacent edges $\overrightarrow{OP}$, $\overrightarrow{OQ}$, and $\overrightarrow{OR}$, where P, Q, and R are the points given in Problem 10.

13. (a) Find the volume of the parallelepiped with adjacent edges $\overrightarrow{OP}$, $\overrightarrow{OQ}$, and $\overrightarrow{OR}$, where P, Q, and R are the points given in Problem 11. (b) Find the volume of the pyramid with vertices O, P, Q, and R.

14. Find a unit vector **n** perpendicular to the plane through the three points P, Q, and R of Problem 11. Then find the distances from the origin to this plane by computing $\mathbf{n} \cdot \overrightarrow{OP}$.

15. Figure 13.2.9 shows a polygonal plot of land, with angles and lengths measured by a surveyor. First find the coordinates of each vertex and then use the vector product [as in Eq. (10)] to calculate the area of the plot.

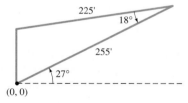

Fig. 13.2.9 Problem 15

16. Repeat Problem 15 with the plot shown in Fig. 13.2.10.

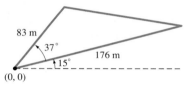

Fig. 13.2.10 Problem 16

17. Repeat Problem 15 with the plot shown in Fig. 13.2.11. [*Suggestion:* First divide the plot into two triangles.]

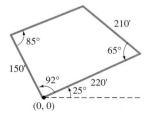

Fig. 13.2.11 Problem 17

18. Repeat Problem 15 with the plot shown in Fig. 13.2.12.

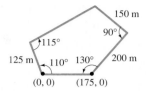

Fig. 13.2.12 Problem 18

19. Apply Eq. (5) to verify Eq. (12), the anticommutativity of the vector product.

20. Apply Eq. (17) to verify the identity for scalar triple products found in Eq. (15).

21. Suppose that P and Q are points on a line L in space. Let A be a point not on L (Fig. 13.2.13). (a) Calculate in two ways the area of the triangle APQ to show that the perpendicular distance from A to the line L is $d = |\overrightarrow{AP} \times \overrightarrow{AQ}| / |\overrightarrow{PQ}|$. (b) Use this formula to compute the distance from the point $A(1, 0, 1)$ to the line through the two points $P(2, 3, 1)$ and $Q(-3, 1, 4)$.

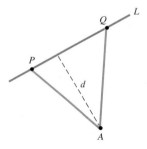

Fig. 13.2.13 Problem 21

22. Suppose that A is a point not on the plane determined by the three points P, Q, and R. Calculate in two ways the volume of the pyramid $APQR$ to show that the perpendicular distance from A to this plane is

$$d = \frac{|\overrightarrow{AP} \cdot (\overrightarrow{AQ} \times \overrightarrow{AR})|}{|\overrightarrow{PQ} \times \overrightarrow{PR}|}.$$

Use this formula to compute the distance from the point $A(1, 0, 1)$ to the plane through the points $P(2, 3, 1)$, $Q(3, -1, 4)$, and $R(0, 0, 2)$.

23. Suppose that P_1 and Q_1 are two points on the line L_1 and that P_2 and Q_2 are two points on the line L_2. If the lines L_1 and L_2 are not parallel, then the perpendicular distance d between them is the projection of $\overrightarrow{P_1 P_2}$ onto a vector **n** that is perpendicular both to $\overrightarrow{P_1 Q_1}$ and $\overrightarrow{P_2 Q_2}$. Prove that

$$d = \frac{|\overrightarrow{P_1 P_2} \cdot (\overrightarrow{P_1 Q_1} \times \overrightarrow{P_2 Q_2})|}{|\overrightarrow{P_1 Q_1} \times \overrightarrow{P_2 Q_2}|}.$$

24. Use the following method to establish that the **vector triple product** $(\mathbf{a} \times \mathbf{b}) \times \mathbf{c}$ is equal to $(\mathbf{a} \cdot \mathbf{c})\mathbf{b} - (\mathbf{b} \cdot \mathbf{c})\mathbf{a}$. (a) Let $\mathbf{I}$ be a unit vector in the direction of $\mathbf{a}$, and let $\mathbf{J}$ be a unit vector perpendicular to $\mathbf{I}$ and parallel to the plane of $\mathbf{a}$ and $\mathbf{b}$. Let $\mathbf{K} = \mathbf{I} \times \mathbf{J}$. Explain why there are scalars a_1, b_1, b_2, c_1, c_2, and c_3 such that

$$\mathbf{a} = a_1 \mathbf{I}, \quad \mathbf{b} = b_1 \mathbf{I} + b_2 \mathbf{J}, \quad \mathbf{c} = c_1 \mathbf{I} + c_2 \mathbf{J} + c_3 \mathbf{K}.$$

(b) Now show that

$$(\mathbf{a} \times \mathbf{b}) \times \mathbf{c} = -a_1 b_2 c_2 \mathbf{I} + a_1 b_2 c_1 \mathbf{J}.$$

(c) Finally, substitute for $\mathbf{I}$ and $\mathbf{J}$ in terms of $\mathbf{a}$ and $\mathbf{b}$.

25. By permutation of the vectors $\mathbf{a}$, $\mathbf{b}$, and $\mathbf{c}$, deduce from Problem 24 that

$$\mathbf{a} \times (\mathbf{b} \times \mathbf{c}) = (\mathbf{a} \cdot \mathbf{c})\mathbf{b} - (\mathbf{a} \cdot \mathbf{b})\mathbf{c}$$

[this is Eq. (16)].

26. Deduce from the orthogonality properties of the vector product that the vector $(\mathbf{a} \times \mathbf{b}) \times (\mathbf{c} \times \mathbf{d})$ can be written in the form $r_1 \mathbf{a} + r_2 \mathbf{b}$ and in the form $s_1 \mathbf{c} + s_2 \mathbf{d}$.

27. Consider the triangle in the xy-plane that has vertices $(x_1, y_1, 0)$, $(x_2, y_2, 0)$, and $(x_3, y_3, 0)$. Use the vector product to prove that the area of this triangle is *half* the *absolute value* of the determinant

$$\begin{vmatrix} 1 & x_1 & y_1 \\ 1 & x_2 & y_2 \\ 1 & x_3 & y_3 \end{vmatrix}.$$

28. Given the vectors $\mathbf{a} = \langle a_1, a_2, a_3 \rangle$ and $\mathbf{b} = \langle b_1, b_2, b_3 \rangle$, verify Eq. (6),

$$|\mathbf{a} \times \mathbf{b}|^2 = |\mathbf{a}|^2 |\mathbf{b}|^2 - (\mathbf{a} \cdot \mathbf{b})^2,$$

by computing each side in terms of the components of $\mathbf{a}$ and $\mathbf{b}$.

13.3

Lines and Planes in Space

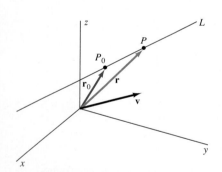

Fig. 13.3.1 Finding the equation of the line L that passes through the point P_0 and is parallel to the vector $\mathbf{v}$

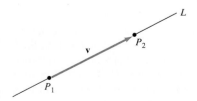

Fig. 13.3.2 The line L of Example 1

A straight line in space is determined by any two points P_0 and P_1 on it. Alternatively, a line in space can be specified by giving a point P_0 on it *and* a vector, such as $\overrightarrow{P_0 P_1}$, that determines the direction of the line.

To investigate equations that describe lines in space, let us begin with a straight line L that passes through the point $P_0(x_0, y_0, z_0)$ and is parallel to the vector $\mathbf{v} = a\mathbf{i} + b\mathbf{j} + c\mathbf{k}$ (Fig. 13.3.1). Then another point $P(x, y, z)$ lies on the line L if and only if the vectors $\mathbf{v}$ and $\overrightarrow{P_0 P}$ are parallel, in which case

$$\overrightarrow{P_0 P} = t\mathbf{v} \tag{1}$$

for some real number t. If $\mathbf{r}_0 = \overrightarrow{OP_0}$ and $\mathbf{r} = \overrightarrow{OP}$ are the position vectors of the points P_0 and P, respectively, then $\overrightarrow{P_0 P} = \mathbf{r} - \mathbf{r}_0$. Hence Eq. (1) gives the **vector equation**

$$\mathbf{r} = \mathbf{r}_0 + t\mathbf{v} \tag{2}$$

describing the line L. As indicated in Fig. 13.3.1, $\mathbf{r}$ is the position vector of an *arbitrary* point P on the line L, and Eq. (2) gives $\mathbf{r}$ in terms of the parameter t, the position vector $\mathbf{r}_0$ of a *fixed* point P_0 on L, and the fixed vector $\mathbf{v}$ that determines the direction of L.

The left- and right-hand sides of Eq. (2) are equal, and each side is a vector. So corresponding components are also equal. When we write the resulting equations, we get a scalar description of the line L. Because $\mathbf{r}_0 = \langle x_0, y_0, z_0 \rangle$ and $\mathbf{r} = \langle x, y, z \rangle$, Eq. (2) thereby yields the three scalar equations

$$x = x_0 + at, \qquad y = y_0 + bt, \qquad z = z_0 + ct. \tag{3}$$

These are **parametric equations** of the line L that passes through the point (x_0, y_0, z_0) and is parallel to the vector $\mathbf{v} = \langle a, b, c \rangle$.

EXAMPLE 1 Write parametric equations of the line L that passes through the points $P_1(1, 2, 2)$ and $P_2(3, -1, 3)$ of Fig. 13.3.2.

710

Solution The line L is parallel to the vector

$$\mathbf{v} = \overrightarrow{P_1 P_2} = (3\mathbf{i} - \mathbf{j} + 3\mathbf{k}) - (\mathbf{i} + 2\mathbf{j} + 2\mathbf{k}) = 2\mathbf{i} - 3\mathbf{j} + \mathbf{k},$$

so we take $a = 2$, $b = -3$, and $c = 1$. With P_1 as the fixed point, the equations in (3) give

$$x = 1 + 2t, \qquad y = 2 - 3t, \qquad z = 2 + t$$

as parametric equations of L. In contrast, with P_2 as the fixed point and with

$$-2\mathbf{v} = -4\mathbf{i} + 6\mathbf{j} - 2\mathbf{k}$$

as the direction vector, the equations in (3) yield the parametric equations

$$x = 3 - 4t, \qquad y = -1 + 6t, \qquad z = 3 - 2t.$$

Thus the parametric equations of a line are not unique.

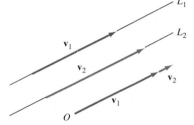

Fig. 13.3.3 Parallel lines

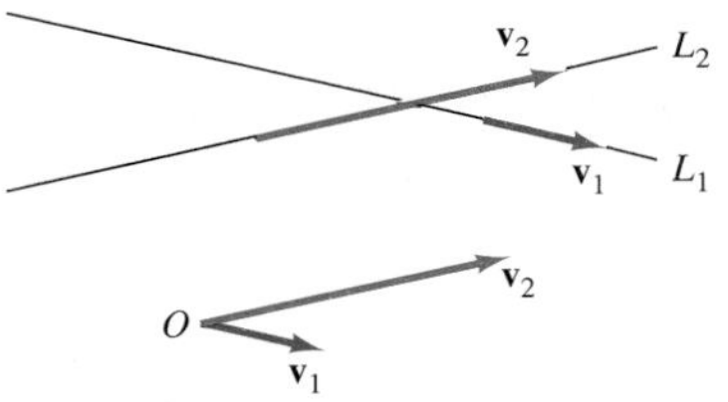

Given two straight lines L_1 and L_2 with parametric equations

$$x = x_1 + a_1 t, \qquad y = y_1 + b_1 t, \qquad z = z_1 + c_1 t \tag{4}$$

and

$$x = x_2 + a_2 s, \qquad y = y_2 + b_2 s, \qquad z = z_2 + c_2 s, \tag{5}$$

respectively, we can see at a glance whether or not L_1 and L_2 are parallel. Because L_1 is parallel to $\mathbf{v}_1 = \langle a_1, b_1, c_1 \rangle$ and L_2 is parallel to $\mathbf{v}_2 = \langle a_2, b_2, c_2 \rangle$, it follows that the lines L_1 and L_2 are parallel if and only if the vectors $\mathbf{v}_1$ and $\mathbf{v}_2$ are scalar multiples of each other (Fig. 13.3.3). If the two lines are not parallel, we can attempt to find a point of intersection by solving the equations

$$x_1 + a_1 t = x_2 + a_2 s \quad \text{and} \quad y_1 + b_1 t = y_2 + b_2 s$$

Fig. 13.3.4 Skew lines

simultaneously for s and t. If these values of s and t satisfy also the equation $z_1 + c_1 t = z_2 + c_2 s$, then we have found a point of intersection. Its rectangular coordinates can be found by substituting the resulting value of t into Eq. (4) [or the resulting value of s into Eq. (5)]. Otherwise, the lines L_1 and L_2 do not intersect. Two nonparallel and nonintersecting lines in space are called **skew lines** (Fig. 13.3.4).

EXAMPLE 2 The line L_1 with parametric equations

$$x = 1 + 2t, \qquad y = 2 - 3t, \qquad z = 2 + t$$

passes through the point $P_1(1, 2, 2)$ (discovered by substitution of $t = 0$) and is parallel to the vector $\mathbf{v}_1 = \langle 2, -3, 1 \rangle$. The line L_2 with parametric equations

$$x = 3 + 4t, \qquad y = 1 - 6t, \qquad z = 5 + 2t$$

passes through the point $P_2(3, 1, 5)$ and is parallel to the vector $\mathbf{v}_2 = \langle 4, -6, 2 \rangle$. Because $\mathbf{v}_2 = 2\mathbf{v}_1$, we see that L_1 and L_2 are parallel.

But are L_1 and L_2 actually different lines, or are we perhaps dealing with two different parametrizations of the same line? To answer this question, we note that $\overrightarrow{P_1 P_2} = \langle 2, -1, 3 \rangle$ is not parallel to $\mathbf{v}_1 = \langle 2, -3, 1 \rangle$. Thus the point P_2 does not lie on the line L_1, and hence the lines L_1 and L_2 are indeed distinct.

If all the coefficients a, b, and c in (3) are nonzero, then we can eliminate the parameter t. Simply solve each equation for t and then set the resulting expressions equal to each other. This gives

$$\frac{x - x_0}{a} = \frac{y - y_0}{b} = \frac{z - z_0}{c}. \tag{6}$$

These are called the **symmetric equations** of the line L. If one or more of a or b or c is zero, this means that L lies in a plane parallel to one of the coordinate planes, and in this case the line does not have symmetric equations. For example, if $c = 0$, then L lies in the horizontal plane $z = z_0$. Of course, it is still possible for us to write equations for L that don't include the parameter t; if $c = 0$, for instance, but a and b are nonzero, then we could describe the line L as the simultaneous solution of the equations

$$\frac{x - x_0}{a} = \frac{y - y_0}{b}, \qquad z = z_0.$$

EXAMPLE 3 Find both parametric and symmetric equations of the line L through the points $P_0(3, 1, -2)$ and $P_1(4, -1, 1)$. Find also the points at which L intersects the three coordinate planes.

Solution The line L is parallel to the vector $\mathbf{v} = \overrightarrow{P_0 P_1} = \langle 1, -2, 3 \rangle$, so we take $a = 1$, $b = -2$, and $c = 3$. The equations in (3) then give the parametric equations

$$x = 3 + t, \qquad y = 1 - 2t, \qquad z = -2 + 3t$$

of L, whereas the equations in (6) give the symmetric equations

$$\frac{x - 3}{1} = \frac{y - 1}{-2} = \frac{z + 2}{3}.$$

To find the point at which L intersects the xy-plane, we set $z = 0$ in the symmetric equations. This gives

$$\frac{x - 3}{1} = \frac{y - 1}{-2} = \frac{2}{3},$$

and so $x = \frac{11}{3}$ and $y = -\frac{1}{3}$. Thus L meets the xy-plane at the point $(\frac{11}{3}, -\frac{1}{3}, 0)$. Similarly, $x = 0$ gives $(0, 7, -11)$ for the point where L meets the yz-plane, and $y = 0$ gives $(\frac{7}{2}, 0, -\frac{1}{2})$ for its intersection with the xz-plane.

PLANES IN SPACE

A plane $\mathcal{P}$ in space is determined by a point $P_0(x_0, y_0, z_0)$ through which $\mathcal{P}$ passes and a line through P_0 that is normal to $\mathcal{P}$. Alternatively, we may be given P_0 on $\mathcal{P}$ and a normal vector $\mathbf{n} = \langle a, b, c \rangle$ to the plane $\mathcal{P}$. The point $P(x, y, z)$ lies on the plane $\mathcal{P}$ if and only if the vectors $\mathbf{n}$ and $\overrightarrow{P_0 P}$ are perpendicular (Fig. 13.3.5), in which case $\mathbf{n} \cdot \overrightarrow{P_0 P} = 0$. We write $\overrightarrow{P_0 P} = \mathbf{r} - \mathbf{r}_0$, where $\mathbf{r}$ and $\mathbf{r}_0$ are the position vectors $\mathbf{r} = \overrightarrow{OP}$ and $\mathbf{r}_0 = \overrightarrow{OP_0}$ of the points P and P_0, respectively. Thus we obtain a **vector equation**

$$\mathbf{n} \cdot (\mathbf{r} - \mathbf{r}_0) = 0 \tag{7}$$

of the plane $\mathcal{P}$.

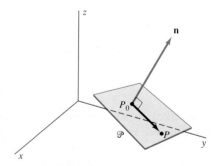

Fig. 13.3.5 Because $\mathbf{n}$ is normal to $\mathcal{P}$, it follows that $\mathbf{n}$ is normal to $\overrightarrow{P_0 P}$ for all points P in $\mathcal{P}$.

712

If we substitute $\mathbf{n} = \langle a, b, c \rangle$, $\mathbf{r} = \langle x, y, z \rangle$, and $\mathbf{r}_0 = \langle x_0, y_0, z_0 \rangle$ into Eq. (7), we thereby obtain a **scalar equation**

$$a(x - x_0) + b(y - y_0) + c(z - z_0) = 0 \qquad (8)$$

of the plane through $P_0(x_0, y_0, z_0)$ with normal vector $\mathbf{n} = \langle a, b, c \rangle$.

EXAMPLE 4 An equation of the plane through $P_0(-1, 5, 2)$ with normal vector $\mathbf{n} = \langle 1, -3, 2 \rangle$ is

$$1 \cdot (x + 1) + (-3) \cdot (y - 5) + 2 \cdot (z - 2) = 0;$$

that is, $x - 3y + 2z = -12$.

IMPORTANT The coefficients of x, y, and z in the last equation are the components of the normal vector. This is always the case, because we can write Eq. (8) in the form

$$ax + by + cz = d, \qquad (9)$$

where $d = ax_0 + by_0 + cz_0$. Conversely, every *linear equation* in x, y, and z of the form in Eq. (9) represents a plane in space provided that the coefficients a, b, and c are not all zero. The reason is that if $c \neq 0$, for instance, then we can choose x_0 and y_0 arbitrarily and solve the equation $ax_0 + by_0 + cz_0 = d$ for z_0. With these values, Eq. (9) takes the form

$$ax + by + cz = ax_0 + by_0 + cz_0;$$

that is,

$$a(x - x_0) + b(y - y_0) + c(z - z_0) = 0,$$

so this equation represents the plane through (x_0, y_0, z_0) with normal vector $\langle a, b, c \rangle$.

EXAMPLE 5 Find an equation for the plane through the three points $P(2, 4, -3)$, $Q(3, 7, -1)$ and $R(4, 3, 0)$.

Solution We want to use Eq. (8), so we first need a vector $\mathbf{n}$ that is normal to the plane in question. One easy way to obtain such a normal vector is by the vector product. Let

$$\mathbf{n} = \overrightarrow{PQ} \times \overrightarrow{PR} = \begin{vmatrix} \mathbf{i} & \mathbf{j} & \mathbf{k} \\ 1 & 3 & 2 \\ 2 & -1 & 3 \end{vmatrix} = 11\mathbf{i} + \mathbf{j} - 7\mathbf{k}.$$

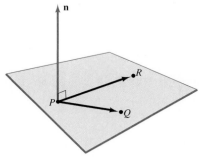

Fig. 13.3.6 The normal vector $\mathbf{n}$ as a vector product (Example 5)

Because $\overrightarrow{PQ}$ and $\overrightarrow{PR}$ are in the plane, their vector product $\mathbf{n}$ is normal to the plane (Fig. 13.3.6). Hence the plane has equation

$$11(x - 2) + (y - 4) - 7(z + 3) = 0.$$

After simplifying, we write the equation as

$$11x + y - 7z = 47.$$

Two planes with normal vectors $\mathbf{n}$ and $\mathbf{m}$ are said to be **parallel** provided that $\mathbf{n}$ and $\mathbf{m}$ are parallel. Otherwise, the two planes meet in a straight

Fig. 13.3.7 The intersection of two nonparallel planes is a straight line.

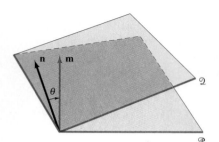

Fig. 13.3.8 Vectors **m** and **n** normal to the planes $\mathcal{P}$ and $\mathcal{Q}$, respectively.

line (Fig. 13.3.7). We define the angle between the two planes to be the angle between their normal vectors **n** and **m**, as in Fig. 13.3.8.

EXAMPLE 6   Find the angle θ between the planes with equations

$$2x + 3y - z = -3 \quad \text{and} \quad 4x + 5y + z = 1.$$

Then write symmetric equations of their line of intersection L.

Solution   The vectors $\mathbf{n} = \langle 2, 3, -1 \rangle$ and $\mathbf{m} = \langle 4, 5, 1 \rangle$ are normal to the two planes, so

$$\cos\theta = \frac{\mathbf{n} \cdot \mathbf{m}}{|\mathbf{n}||\mathbf{m}|} = \frac{22}{\sqrt{14}\,\sqrt{42}}.$$

Hence $\theta = \cos^{-1}\left(\frac{11}{21}\sqrt{3}\right) \approx 24.87°$.

To determine the line of intersection of the two planes, we need first to find a point P_0 that lies on L. We can do this by substituting an arbitrarily chosen value of x into the equations of the given planes and then solving the resulting equations for y and z. With $x = 1$ we get the equations

$$2 + 3y - z = -3,$$
$$4 + 5y + z = 1.$$

The common solution is $y = -1$, $z = 2$. Thus the point $P_0(1, -1, 2)$ lies on the line L.

Next we need a vector **v** parallel to L. The vectors **n** and **m** normal to the two planes are both perpendicular to L, so their vector product is parallel to L. Alternatively, we can find a second point P_1 on L by substituting a second value of x into the equations of the given planes and solving for y and z, as before. With $x = 5$ we obtain the equations

$$10 + 3y - z = -3,$$
$$20 + 5y + z = 1,$$

with common solution $y = -4$, $z = 1$. Thus we obtain a second point $P_1(5, -4, 1)$ and thereby the vector

$$\mathbf{v} = \overrightarrow{P_0 P_1} = \langle 4, -3, -1 \rangle$$

parallel to L. From Eq. (6) we now find symmetric equations

$$\frac{x-1}{4} = \frac{y+1}{-3} = \frac{z-2}{-1}$$

of the line of intersection of the two given planes.

In conclusion, we note that the symmetric equations of a line L present the line as an intersection of planes: We can rewrite the equations in (6) in the form

$$b(x - x_0) - a(y - y_0) = 0,$$
$$c(x - x_0) - a(z - z_0) = 0, \tag{10}$$
$$c(y - y_0) - b(z - z_0) = 0.$$

These are the equations of three planes that intersect in the line L. The first has normal vector $\langle b, -a, 0 \rangle$, a vector parallel to the xy-plane. So the first plane is perpendicular to the xy-plane. Similarly, the second plane is perpendicular to the xz-plane, and the third is perpendicular to the yz-plane.

The equations in (10) are symmetric equations of the line that passes through $P_0(x_0, y_0, z_0)$ and is parallel to $\mathbf{v} = \langle a, b, c \rangle$. Unlike the equations in (6), these equations are meaningful whether or not all the components a, b, and c of $\mathbf{v}$ are nonzero. They have a special form, though, if one of the two components is zero. If, say, $a = 0$, then the first two equations in (10) take the form $x = x_0$. The line is then the intersection of the two planes $x = x_0$ and $c(y - y_0) = b(z - z_0)$.

13.3 Problems

In Problems 1 through 3, write parametric equations of the straight line that passes through the point P and is parallel to the vector $\mathbf{v}$.

1. $P(0, 0, 0)$, $\quad \mathbf{v} = \mathbf{i} + 2\mathbf{j} + 3\mathbf{k}$

2. $P(3, -4, 5)$, $\quad \mathbf{v} = -2\mathbf{i} + 7\mathbf{j} + 3\mathbf{k}$

3. $P(4, 13, -3)$, $\quad \mathbf{v} = 2\mathbf{i} - 3\mathbf{k}$

In Problems 4 through 6, write parametric equations of the straight line that passes through the points P_1 *and* P_2.

4. $P_1(0, 0, 0)$, $\quad P_2(-6, 3, 5)$

5. $P_1(3, 5, 7)$, $\quad P_2(6, -8, 10)$

6. $P_1(3, 5, 7)$, $\quad P_2(6, 5, 4)$

In Problems 7 through 10, write an equation of the plane with normal vector $\mathbf{n}$ *that passes through the point P.*

7. $P(0, 0, 0)$, $\mathbf{n} = \langle 1, 2, 3 \rangle$

8. $P(3, -4, 5)$, $\mathbf{n} = \langle -2, 7, 3 \rangle$

9. $P(5, 12, 13)$, $\mathbf{n} = \mathbf{i} - \mathbf{k}$ $\quad$ **10.** $P(5, 12, 13)$, $\mathbf{n} = \mathbf{j}$

In Problems 11 through 16, write both parametric and symmetric equations for the indicated straight line.

11. Through $P(2, 3, -4)$ and parallel to $\mathbf{v} = \langle 1, -1, -2 \rangle$

12. Through $P(2, 5, -7)$ and $Q(4, 3, 8)$

13. Through $P(1, 1, 1)$ and perpendicular to the xy-plane

14. Through the origin and perpendicular to the plane with equation $x + y + z = 1$

15. Through $P(2, -3, 4)$ and perpendicular to the plane with equation $2x - y + 3z = 4$

16. Through $P(2, -1, 5)$ and parallel to the line with parametric equations $x = 3t$, $y = 2 + t$, $z = 2 - t$

In Problems 17 through 24, write an equation of the indicated plane.

17. Through $P(5, 7, -6)$ and parallel to the xz-plane

18. Through $P(1, 0, -1)$ with normal vector $\mathbf{n} = \langle 2, 2, -1 \rangle$

19. Through $P(10, 4, -3)$ with normal vector $\mathbf{n} = \langle 7, 11, 0 \rangle$

20. Through $P(1, -3, 2)$ with normal vector $\mathbf{n} = \overrightarrow{OP}$

21. Through the origin and parallel to the plane with equation $3x + 4y = z + 10$

22. Through $P(5, 1, 4)$ and parallel to the plane with equation $x + y - 2z = 0$

23. Through the origin and the points $P(1, 1, 1)$ and $Q(1, -1, 3)$

24. Through the points $A(1, 0, -1)$, $B(3, 3, 2)$, and $C(4, 5, -1)$.

In Problems 25 through 27, find the angle between the planes with the given equations.

25. $x = 10$ and $x + y + z = 0$

26. $2x - y + z = 5$ and $x + y - z = 1$

27. $x - y - 2z = 1$ and $x - y - 2z = 5$

28. Find parametric equations of the line of intersection of the planes $2x + y + z = 4$ and $3x - y + z = 3$.

29. Write symmetric equations for the line through $P(3, 3, 1)$ that is parallel to the line of Problem 28.

30. Find an equation of the plane through $P(3, 3, 1)$ that is perpendicular to the planes $x + y = 2z$ and $2x + z = 10$.

31. Find an equation of the plane through $(1, 1, 1)$ that intersects the xy-plane in the same line as does the plane $3x + 2y - z = 6$.

32. Find an equation for the plane that passes through the point $P(1, 3, -2)$ and contains the line of intersection of the planes $x - y + z = 1$ and $x + y - z = 1$.

33. Find an equation of the plane that passes through the points $P(1, 0, -1)$ and $Q(2, 1, 0)$ and is parallel to the line of intersection of the planes $x + y + z = 5$ and $3x - y = 4$.

34. Prove that the lines $x - 1 = \frac{1}{2}(y + 1) = z - 2$ and $x - 2 = \frac{1}{3}(y - 2) = \frac{1}{2}(z - 4)$ intersect. Find an equation of the (only) plane that contains them both.

35. Prove that the line of intersection of the planes $x + 2y - z = 2$ and $3x + 2y + 2z = 7$ is parallel to the line $x = 1 + 6t$, $y = 3 - 5t$, $z = 2 - 4t$. Find an equation of the plane determined by these two lines.

36. Show that the perpendicular distance D from the point $P_0(x_0, y_0, z_0)$ to the plane $ax + by + cz = d$ is

$$D = \frac{|ax_0 + by_0 + cz_0 - d|}{\sqrt{a^2 + b^2 + c^2}}.$$

[*Suggestion:* The line that passes through P_0 and is perpendicular to the given plane has parametric equations $x = x_0 + at$, $y = y_0 + bt$, $z = z_0 + ct$. Let $P_1(x_1, y_1, z_1)$ be the point of this line, corresponding to $t = t_1$, at which it intersects the given plane. Solve for t_1, and then compute $D = |P_0P_1|$.]

In Problems 37 and 38, use the formula of Problem 36 to find the distance between the given point and the given plane.

37. The origin and the plane $x + y + z = 10$

38. The point $P(5, 12, -13)$ and the plane with equation $3x + 4y + 5z = 12$

39. Prove that any two skew lines lie in parallel planes.

40. Use the formula of Problem 36 to show that the perpendicular distance D between the two parallel planes $ax + by + cz + d_1 = 0$ and $ax + by + cz + d_2 = 0$ is

$$D = \frac{|d_1 - d_2|}{\sqrt{a^2 + b^2 + c^2}}.$$

41. The line L_1 is described by the equations

$$x - 1 = 2y + 2, \qquad z = 4.$$

The line L_2 passes through the points $P(2, 1, -3)$ and $Q(0, 8, 4)$. (a) Show that L_1 and L_2 are skew lines. (b) Use the results of Problems 39 and 40 to find the perpendicular distance between L_1 and L_2.

13.4
Curves and Motion in Space

In Section 12.4 we used vector-valued functions to discuss curves and motion in the coordinate plane. Much of that discussion applies, with only minor changes, to curves and motion in three-dimensional space. The principal difference is that the vectors now have three components rather than two.

Think of a point that moves along a curve in space. We can describe its position at time t by *parametric equations*

$$x = f(t), \qquad y = g(t), \quad z = h(t)$$

that give its coordinates at time t. Alternatively, we can give the location of the point by its **position vector**

$$\mathbf{r} = f(t)\mathbf{i} + g(t)\mathbf{j} + h(t)\mathbf{k} = x\mathbf{i} + y\mathbf{j} + z\mathbf{k}, \tag{1}$$

shown in Fig. 13.4.1.

We can differentiate a three-dimensional vector-valued function such as $\mathbf{r}(t)$ in a componentwise manner, just like a two-dimensional vector-valued function (see Theorem 1 in Section 12.4). Thus the **velocity vector** $\mathbf{v} = \mathbf{v}(t)$ of the moving point at time t is given by

$$\mathbf{v}(t) = \mathbf{r}'(t) = f'(t)\mathbf{i} + g'(t)\mathbf{j} + h'(t)\mathbf{k}; \tag{2a}$$

in differential notation,

$$\mathbf{v} = \frac{d\mathbf{r}}{dt} = \frac{dx}{dt}\mathbf{i} + \frac{dy}{dt}\mathbf{j} + \frac{dz}{dt}\mathbf{k}. \tag{2b}$$

Its **acceleration vector** $\mathbf{a} = \mathbf{a}(t)$ is given by

$$\mathbf{a}(t) = \mathbf{v}'(t) = f''(t)\mathbf{i} + g''(t)\mathbf{j} + h''(t)\mathbf{k}; \tag{3a}$$

alternatively,

$$\mathbf{a} = \frac{d\mathbf{v}}{dt} = \frac{d^2x}{dt^2}\mathbf{i} + \frac{d^2y}{dt^2}\mathbf{j} + \frac{d^2z}{dt^2}\mathbf{k}. \tag{3b}$$

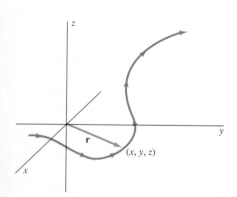

Fig. 13.4.1 The position vector $\mathbf{r} = \langle x, y, z \rangle$ of a moving particle in space

The **speed** $v(t)$ and **scalar acceleration** $a(t)$ of the moving point are the lengths of its velocity and acceleration vectors, respectively:

$$v(t) = |\mathbf{v}(t)| = \sqrt{\left(\frac{dx}{dt}\right)^2 + \left(\frac{dy}{dt}\right)^2 + \left(\frac{dz}{dt}\right)^2} \tag{4}$$

and

$$a(t) = |\mathbf{a}(t)| = \sqrt{\left(\frac{d^2x}{dt^2}\right)^2 + \left(\frac{d^2y}{dt^2}\right)^2 + \left(\frac{d^2z}{dt^2}\right)^2}. \tag{5}$$

EXAMPLE 1 The parametric equations of a moving point are

$$x(t) = a \cos \omega t, \qquad y(t) = a \sin \omega t, \qquad z = bt,$$

where a, b, and ω are all positive. Describe the path of the point in geometric terms. Compute its velocity, speed, and acceleration at time t.

Solution Because

$$x^2 + y^2 = a^2 \cos^2 \omega t + a^2 \sin^2 \omega t = a^2,$$

the path of the moving point lies on the **cylinder** that stands above and below the circle $x^2 + y^2 = a^2$ in the xy-plane, extending infinitely far in both directions (Fig. 13.4.2). The given parametric equations for x and y tell us that the projection of the point into the xy-plane moves counterclockwise around the circle $x^2 + y^2 = a^2$ with angular speed ω. Meanwhile, because $z = bt$, the point is itself rising with vertical speed b. Its path on the cylinder is a spiral called a **helix** (also shown in Fig. 13.4.2).

The derivative of the position vector

$$\mathbf{r}(t) = (a \cos \omega t)\mathbf{i} + (a \sin \omega t)\mathbf{j} + (bt)\mathbf{k}$$

of the moving point is its velocity vector

$$\mathbf{v}(t) = (-a\omega \sin \omega t)\mathbf{i} + (a\omega \cos \omega t)\mathbf{j} + b\mathbf{k}. \tag{6}$$

Another differentiation gives its acceleration vector

$$\begin{aligned}
\mathbf{a}(t) &= (-a\omega^2 \cos \omega t)\mathbf{i} + (-a\omega^2 \sin \omega t)\mathbf{j} \\
&= -a\omega^2(\mathbf{i} \cos \omega t + \mathbf{j} \sin \omega t).
\end{aligned} \tag{7}$$

The speed of the moving point is a constant, because

$$v(t) = |\mathbf{v}(t)| = \sqrt{a^2\omega^2 + b^2}.$$

Note that the acceleration vector is a horizontal vector of length $a\omega^2$. Moreover, if we think of $\mathbf{a}(t)$ as attached to the moving point at the time t of evaluation—so that the initial point of $\mathbf{a}(t)$ is the terminal point of $\mathbf{r}(t)$—then $\mathbf{a}(t)$ points directly toward the point $(0, 0, bt)$ on the *z-axis*.

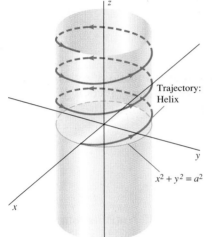

Trajectory: Helix

$x^2 + y^2 = a^2$

Fig. 13.4.2 The point of Example 1 moves in a helical path.

REMARK The helix of Example 1 is a typical trajectory of a charged particle in a constant magnetic field. Such a particle must satisfy both Newton's law $\mathbf{F} = m\mathbf{a}$ and the magnetic force law $\mathbf{F} = (q\mathbf{v}) \times \mathbf{B}$ mentioned in Section 13.2. Hence its velocity and acceleration vectors must satisfy the equation

$$(q\mathbf{v}) \times \mathbf{B} = m\mathbf{a}. \tag{8}$$

If the constant magnetic field is vertical, $\mathbf{B} = B\mathbf{k}$, then with the velocity vector of Eq. (6) we find that

$$(q\mathbf{v}) \times \mathbf{B} = q \begin{vmatrix} \mathbf{i} & \mathbf{j} & \mathbf{k} \\ -a\omega \sin \omega t & a\omega \cos \omega t & b \\ 0 & 0 & B \end{vmatrix} = qa\omega B(\mathbf{i} \cos \omega t + \mathbf{j} \sin \omega t).$$

The acceleration vector of Eq. (7) gives

$$m\mathbf{a} = -ma\omega^2(\mathbf{i} \cos \omega t + \mathbf{j} \sin \omega t).$$

When we compare the last two results, we see that the helix of Example 1 satisfies Eq. (8) provided that

$$qa\omega B = -ma\omega^2; \quad \text{that is,} \quad \omega = -\frac{qB}{m}.$$

For example, this equation would determine the angular speed ω for the helical trajectory of electrons ($q < 0$) in a cathode-ray tube placed in a constant magnetic field parallel to the axis of the tube (Fig. 13.4.3).

Fig. 13.4.3 A spiraling electron in a cathode-ray tube

PROPERTIES OF DIFFERENTIATION

Differentiation of three-dimensional vector-valued functions satisfies the same formal properties that we listed in Theorem 2 of Section 12.4 for two-dimensional vector-valued functions:

1. $D_t[\mathbf{u}(t) + \mathbf{v}(t)] = \mathbf{u}'(t) + \mathbf{v}'(t)$,
2. $D_t[c\mathbf{u}(t)] = c\mathbf{u}'(t)$,
3. $D_t[h(t)\mathbf{u}(t)] = h'(t)\mathbf{u}(t) + h(t)\mathbf{u}'(t)$,
4. $D_t[\mathbf{u}(t) \cdot \mathbf{v}(t)] = \mathbf{u}'(t) \cdot \mathbf{v}(t) + \mathbf{u}(t) \cdot \mathbf{v}'(t)$.

(9)

Here, $\mathbf{u}(t)$ and $\mathbf{v}(t)$ are vector-valued functions with differentiable component functions, and $h(t)$ is a differentiable real-valued function. In addition, the expected product rule for the derivative of a vector product holds:

5. $D_t[\mathbf{u}(t) \times \mathbf{v}(t)] = \mathbf{u}'(t) \times \mathbf{v}(t) + \mathbf{u}(t) \times \mathbf{v}'(t)$

(10)

The order of the factors in Eq. (10) *must* be preserved because the vector product is not commutative. We can verify each of the properties in (9) and (10) routinely by componentwise differentiation, as in Section 12.4.

EXAMPLE 2   If $\mathbf{u}(t) = 3\mathbf{j} + 4t\mathbf{k}$ and $\mathbf{v}(t) = 5t\mathbf{i} - 4\mathbf{k}$, then

$$\mathbf{u}'(t) = 4\mathbf{k} \quad \text{and} \quad \mathbf{v}'(t) = 5\mathbf{i}.$$

Hence Eq. (10) yields

$$\begin{aligned} D_t[\mathbf{u}(t) \times \mathbf{v}(t)] &= 4\mathbf{k} \times (5t\mathbf{i} - 4\mathbf{k}) + (3\mathbf{j} + 4t\mathbf{k}) \times 5\mathbf{i} \\ &= 20t(\mathbf{k} \times \mathbf{i}) - 16(\mathbf{k} \times \mathbf{k}) + 15(\mathbf{j} \times \mathbf{i}) + 20t(\mathbf{k} \times \mathbf{i}) \\ &= 20t\mathbf{j} - 16 \cdot \mathbf{0} + 15 \cdot (-\mathbf{k}) + 20t\mathbf{j} \\ &= 40t\mathbf{j} - 15\mathbf{k}, \end{aligned}$$

without our having to calculate the vector product $\mathbf{u}(t) \times \mathbf{v}(t)$.

Ch. 13 / Vectors, Curves, and Surfaces in Space

INTEGRATION OF VECTOR-VALUED FUNCTIONS

Just as in Section 12.4, vector-valued functions are integrated in *component-wise* fashion:

$$\int [f(t)\mathbf{i} + g(t)\mathbf{j} + h(t)\mathbf{k}]\, dt$$

$$= \left(\int f(t)\, dt\right)\mathbf{i} + \left(\int g(t)\, dt\right)\mathbf{j} + \left(\int h(t)\, dt\right)\mathbf{k}. \qquad (11)$$

If $\mathbf{r}(t)$, $\mathbf{v}(t)$, and $\mathbf{a}(t)$ are the position, velocity, and acceleration vectors, respectively, of a particle moving in space, then the facts that

$$\frac{d\mathbf{r}}{dt} = \mathbf{v} \quad \text{and} \quad \frac{d\mathbf{v}}{dt} = \mathbf{a}$$

[Eqs. (2) and (3)] imply that

$$\mathbf{v}(t) = \int \mathbf{a}(t)\, dt \qquad (12)$$

and

$$\mathbf{r}(t) = \int \mathbf{v}(t)\, dt. \qquad (13)$$

Each of these indefinite integrals involves a *vector* constant of integration.

EXAMPLE 3 A ball is thrown northward into the air from the origin in *xyz*-space (the *xy*-plane represents the ground, and the positive *y*-axis points north). The initial velocity (vector) of the ball is

$$\mathbf{v}_0 = \mathbf{v}(0) = 80\mathbf{j} + 80\mathbf{k}.$$

The spin of the ball causes an eastward acceleration of 2 ft/s² in addition to gravitational acceleration. Thus the acceleration vector produced by the combination of gravity and spin is

$$\mathbf{a}(t) = 2\mathbf{i} - 32\mathbf{k}.$$

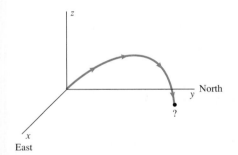

Fig. 13.4.4 The trajectory of the ball of Example 3

First find the velocity vector $\mathbf{v}(t)$ of the ball and its position vector $\mathbf{r}(t)$. Then determine where and with what speed the ball hits the ground (Fig. 13.4.4).

Solution When we antidifferentiate $\mathbf{a}(t)$ we get

$$\mathbf{v}(t) = \int \mathbf{a}(t)\, dt = \int (2\mathbf{i} - 32\mathbf{k})\, dt = 2t\mathbf{i} - 32t\mathbf{k} + \mathbf{c}_1.$$

We substitute $t = 0$ to find that $\mathbf{c}_1 = \mathbf{v}_0 = 80\mathbf{j} + 80\mathbf{k}$, so

$$\mathbf{v}(t) = 2t\mathbf{i} + 80\mathbf{j} + (80 - 32t)\mathbf{k}.$$

Another antidifferentiation yields

$$\mathbf{r}(t) = \int \mathbf{v}(t)\, dt = \int [2t\mathbf{i} + 80\mathbf{j} + (80 - 32t)\mathbf{k}]\, dt$$

$$= t^2\mathbf{i} + 80t\mathbf{j} + (80t - 16t^2)\mathbf{k} + \mathbf{c}_2,$$

and substitution of $t = 0$ gives $c_2 = r(0) = 0$. Hence the position vector of the ball is

$$\mathbf{r}(t) = t^2\mathbf{i} + 80t\mathbf{j} + (80t - 16t^2)\mathbf{k}.$$

The ball hits the ground when $z = 80t - 16t^2 = 0$; that is, when $t = 5$. Its position vector then is

$$\mathbf{r}(5) = 5^2\mathbf{i} + 80 \cdot 5\mathbf{j} = 25\mathbf{i} + 400\mathbf{j},$$

so the ball has traveled 25 ft eastward and 400 ft northward. Its velocity vector at impact is

$$\mathbf{v}(5) = 2 \cdot 5\mathbf{i} + 80\mathbf{j} + (80 - 32 \cdot 5)\mathbf{k} = 10\mathbf{i} + 80\mathbf{j} - 80\mathbf{k},$$

so its speed when it hits the ground is

$$v(5) = |\mathbf{v}(5)| = \sqrt{10^2 + 80^2 + (-80)^2},$$

approximately 113.58 ft/s. Because the ball started with initial speed $v_0 = (80^2 + 80^2)^{1/2} \approx 113.14$ ft/s, its eastward acceleration has slightly increased its terminal speed.

13.4 Problems

In Problems 1 through 8, the position vector $\mathbf{r}(t)$ of a particle moving in space is given. Find its velocity and acceleration vectors and its speed at time t.

1. $\mathbf{r}(t) = 3\mathbf{i} - 4\mathbf{j} + 5t\mathbf{k}$

2. $\mathbf{r}(t) = 3\mathbf{i} - 4t\mathbf{j} + 5t^2\mathbf{k}$

3. $\mathbf{r}(t) = t\mathbf{i} + t^2\mathbf{j} + t^3\mathbf{k}$

4. $\mathbf{r}(t) = t^2(3\mathbf{i} + 4\mathbf{j} - 12\mathbf{k})$

5. $\mathbf{r}(t) = t\mathbf{i} + 3e^t\mathbf{j} + 4e^t\mathbf{k}$

6. $\mathbf{r}(t) = e^t\mathbf{i} + e^{2t}\mathbf{j} + e^{3t}\mathbf{k}$

7. $\mathbf{r}(t) = (3\cos t)\mathbf{i} + (3\sin t)\mathbf{j} - 4t\mathbf{k}$

8. $\mathbf{r}(t) = 12t\mathbf{i} + (5\sin 2t)\mathbf{j} - (5\cos 2t)\mathbf{k}$

In Problems 9 through 16, the acceleration vector $\mathbf{a}(t)$, the initial position $\mathbf{r}_0 = \mathbf{r}(0)$, and the initial velocity $\mathbf{v}_0 = \mathbf{v}(0)$ of a particle moving in xyz-space are given. Find its position vector $\mathbf{r}(t)$ at time t.

9. $\mathbf{a}(t) = 2\mathbf{i} - 4\mathbf{k}$; $\mathbf{r}_0 = \mathbf{0}$; $\mathbf{v}_0 = 10\mathbf{j}$

10. $\mathbf{a}(t) = \mathbf{i} - \mathbf{j} + 3\mathbf{k}$; $\mathbf{r}_0 = 5\mathbf{i}$; $\mathbf{v}_0 = 7\mathbf{j}$

11. $\mathbf{a}(t) = 2\mathbf{j} - 6t\mathbf{k}$; $\mathbf{r}_0 = 2\mathbf{i}$; $\mathbf{v}_0 = 5\mathbf{k}$

12. $\mathbf{a}(t) = 6t\mathbf{i} - 5\mathbf{j} + 12t^2\mathbf{k}$; $\mathbf{r}_0 = 3\mathbf{i} + 4\mathbf{j}$; $\mathbf{v}_0 = 4\mathbf{j} - 5\mathbf{k}$

13. $\mathbf{a}(t) = t\mathbf{i} + t^2\mathbf{j} + t^3\mathbf{k}$; $\mathbf{r}_0 = 10\mathbf{i}$; $\mathbf{v}_0 = 10\mathbf{j}$

14. $\mathbf{a}(t) = t\mathbf{i} + e^{-t}\mathbf{j}$; $\mathbf{r}_0 = 3\mathbf{i} + 4\mathbf{j}$; $\mathbf{v}_0 = 5\mathbf{k}$

15. $\mathbf{a}(t) = \mathbf{i}\cos t + \mathbf{j}\sin t$; $\mathbf{r}_0 = \mathbf{j}$; $\mathbf{v}_0 = -\mathbf{i} + 5\mathbf{k}$

16. $\mathbf{a}(t) = -9(\mathbf{i}\sin 3t + \mathbf{j}\cos 3t) + 4\mathbf{k}$; $\mathbf{r}_0 = 3\mathbf{i} + 4\mathbf{j}$; $\mathbf{v}_0 = 2\mathbf{i} - 7\mathbf{k}$

17. The parametric equations of a moving point are

$$x(t) = 3\cos 2t, \qquad y(t) = 3\sin 2t, \qquad z(t) = 8t.$$

Find its velocity, speed, and acceleration at time $t = 7\pi/8$.

18. Use the equations in (9) and (10) to calculate

$$D_t[\mathbf{u}(t) \cdot \mathbf{v}(t)] \quad \text{and} \quad D_t[\mathbf{u}(t) \times \mathbf{v}(t)]$$

if $\mathbf{u}(t) = \langle t, t^2, t^3 \rangle$ and $\mathbf{v}(t) = \langle e^t, \cos t, \sin t \rangle$.

19. Verify the result obtained in Example 2 by first calculating $\mathbf{u}(t) \times \mathbf{v}(t)$ and then differentiating.

20. Verify that

$$D_t[\mathbf{u}(t) \times \mathbf{v}(t)] = \mathbf{u}'(t) \times \mathbf{v}(t) + \mathbf{u}(t) \times \mathbf{v}'(t).$$

21. A point moves on a sphere centered at the origin. Show that its velocity vector is always tangent to the sphere.

22. A particle moves with constant speed along a curve in space. Show that its velocity and acceleration vectors are always perpendicular.

23. Find the maximum height reached by the ball in Example 3 and also its speed at that height.

24. The **angular momentum** $\mathbf{L}(t)$ and **torque** $\boldsymbol{\tau}(t)$ of a moving particle of mass m with position vector $\mathbf{r}(t)$ are defined to be

$$\mathbf{L}(t) = m\mathbf{r}(t) \times \mathbf{v}(t), \qquad \boldsymbol{\tau}(t) = m\mathbf{r}(t) \times \mathbf{a}(t).$$

Prove that $\mathbf{L}'(t) = \boldsymbol{\tau}(t)$. It follows that $\mathbf{L}(t)$ must be constant if $\boldsymbol{\tau} \equiv \mathbf{0}$; this is the law of conservation of angular momentum.

25. Suppose that a particle is moving under the influence of a *central* force field $\mathbf{R} = k\,\mathbf{r}$, where k is a scalar function of x, y, and z. Conclude that the trajectory of the particle lies in a *fixed* plane through the origin.

26. A baseball is thrown with an initial velocity of 160 ft/s straight upward from the ground. It experiences a downward gravitational acceleration of 32 ft/s². Because of spin, it experiences also a (horizontal) northward acceleration of 0.1 ft/s²; otherwise, the air has no effect on its motion. How far north of the throwing point will the ball land?

27. A baseball is hit with an initial velocity of 96 ft/s and an initial inclination angle of 15° from ground level straight down a foul line. Because of spin it experiences a horizontal acceleration of 2 ft/s² perpendicular to the foul line; otherwise, the air has no effect on its motion. When the ball hits the ground, how far is it from the foul line?

28. A gun fires a shell with a muzzle velocity of 150 m/s. While the shell is in the air, it experiences a downward (vertical) gravitational acceleration of 9.8 m/s² and an eastward (horizontal) Coriolis acceleration of 5 cm/s²; ignore air resistance. The target is 1500 m due north of the gun, and both the gun and the target are on level ground. Halfway between them is a hill 600 m high. Tell precisely how to aim the gun—both compass heading and inclination from the horizontal—so that the shell will clear the hill and hit the target.

13.4 Project

Have you ever wondered whether a baseball pitch really curves or whether it's some sort of optical illusion? In this project you'll use calculus to settle the matter.

Suppose that a pitcher throws a ball toward home plate (60 ft away, as in Fig. 13.4.5) and gives it a spin of S revolutions per second counterclockwise (as viewed from above) about a vertical axis through the center of the ball. This spin is described by the *spin vector* $\mathbf{S}$ that points along the axis of revolution in the right-handed direction and has length S (Fig. 13.4.6).

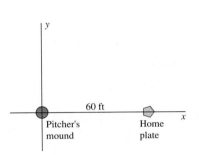

Fig. 13.4.5 The x-axis points toward home plate.

We know from studies of aerodynamics that this spin causes a difference in air pressure on the sides of the ball toward and away from this spin. Studies also show that this pressure difference results in a *spin acceleration*

$$\mathbf{a}_s = c\mathbf{S} \times \mathbf{v} \tag{14}$$

of the ball (where c is an empirical constant). The total acceleration of the ball is then

$$\mathbf{a} = (c\mathbf{S} \times \mathbf{v}) - g\mathbf{k}, \tag{15}$$

where $g \approx 32$ ft/s² is the gravitational acceleration. Here we will ignore any other effects of air resistance.

With the spin vector $\mathbf{S} = S\mathbf{k}$ pointing upward, as in Fig. 13.4.6, show first that

$$\mathbf{S} \times \mathbf{v} = -Sv_y\mathbf{i} + Sv_x\mathbf{j}, \tag{16}$$

where v_x is the component of $\mathbf{v}$ in the x-direction, v_y is the component of $\mathbf{v}$ in the y-direction.

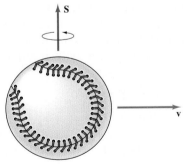

Fig. 13.4.6 The spin and velocity vectors

For a ball pitched along the x-axis, v_x is much larger than v_y, and so the approximation $\mathbf{S} \times \mathbf{v} = Sv_x\mathbf{j}$ is sufficiently accurate for our purposes. We may then take the acceleration vector of the ball to be

$$\mathbf{a} = cSv_x\mathbf{j} - g\mathbf{k}. \tag{17}$$

Now suppose that the pitcher throws the ball from the initial position $x_0 = y_0 = 0$, $z_0 = 5$ (ft), with initial velocity vector

$$\mathbf{v}_0 = 120\mathbf{i} - 2\mathbf{j} + 4\mathbf{k} \tag{18}$$

(with components in feet per second, so $v_0 \approx 120$ ft/s, about 82 mi/h) and with a spin of $S = \frac{80}{3}$ rev/s. A reasonable value of c is

$$c = 0.005 \text{ ft/s}^2 \quad \text{per ft/s of velocity and rev/s of spin,}$$

although the precise value depends on whether the pitcher has (accidentally, of course) scuffed the ball or administered some foreign substance to it.

Show first that these values of the parameters yield

$$\mathbf{a} = 16\mathbf{j} - 32\mathbf{k}$$

for the ball's acceleration vector. Then integrate twice in succession to find the ball's position vector

$$\mathbf{r}(t) = x(t)\mathbf{i} + y(t)\mathbf{j} + z(t)\mathbf{k}.$$

Use your results to fill in the following table, giving the pitched ball's horizontal deflection y and height z (above the ground) at quarter-second intervals.

t (s)	x (ft)	y (ft)	z (ft)
0.0	0	0	5
0.25	30	?	?
0.50	60	?	?

Suppose that the batter gets a "fix" on the pitch by observing the ball during the first quarter-second and prepares to swing. After 0.25 s does the pitch still appear to be straight on target toward home plate at a height of 5 ft?

What happens to the ball during the final quarter-second of its approach to home plate—*after* the batter has begun to swing the bat? What were the ball's horizontal and vertical deflections during this brief period? What is your conclusion? Does the pitched ball really "curve" or not?

13.5

Curvature and Acceleration

The speed of a moving point is closely related to the arc length of its trajectory. The arc-length formula for parametric curves in space (or space curves) is a natural generalization of the formula for parametric plane curves [Eq. (6) of Section 12.2]. The **arc length** s along the smooth curve with position vector

$$\mathbf{r}(t) = f(t)\mathbf{i} + g(t)\mathbf{j} + h(t)\mathbf{k} = x\mathbf{i} + y\mathbf{j} + z\mathbf{k} \tag{1}$$

from the point $\mathbf{r}(a)$ to the point $\mathbf{r}(b)$ is, by definition,

$$s = \int_a^b \sqrt{[x'(t)]^2 + [y'(t)]^2 + [z'(t)]^2}\, dt$$

$$= \int_a^b \sqrt{\left(\frac{dx}{dt}\right)^2 + \left(\frac{dy}{dt}\right)^2 + \left(\frac{dz}{dt}\right)^2}\, dt. \tag{2}$$

We see from Eq. (4) in Section 13.4 that the integrand is the speed $v(t) = |\mathbf{r}'(t)|$ of the moving point with position vector $\mathbf{r}(t)$, so

$$s = \int_a^b v(t)\, dt. \tag{3}$$

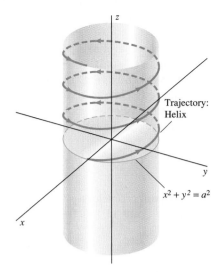

Fig. 13.5.1 The helix of Example 1

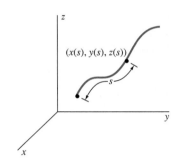

Fig. 13.5.2 A curve parametrized by arc length s

EXAMPLE 1 Find the arc length of one turn (from $t = 0$ to $t = 2\pi/\omega$) of the helix shown in Fig. 13.5.1. This helix has the parametric equations

$$x(t) = a \cos \omega t, \qquad y(t) = a \sin \omega t, \qquad z(t) = bt.$$

Solution We found in Example 1 of Section 13.4 that

$$v(t) = \sqrt{a^2\omega^2 + b^2}.$$

Hence Eq. (3) gives

$$s = \int_0^{2\pi/\omega} \sqrt{a^2\omega^2 + b^2} \, dt = \frac{2\pi}{\omega} \sqrt{a^2\omega^2 + b^2}.$$

For instance, if $a = b = \omega = 1$, then $s = 2\pi\sqrt{2}$, which is $\sqrt{2}$ times the circumference of the circle in the xy-plane over which the helix lies.

Let $s(t)$ denote the arc length along a smooth curve from its initial point $\mathbf{r}(a)$ to the variable point $\mathbf{r}(t)$ (Fig. 13.5.2). Then, from Eq. (3), we obtain the **arc-length function** $s(t)$ of the curve:

$$s(t) = \int_a^t v(\tau) \, d\tau. \tag{4}$$

The fundamental theorem of calculus then gives

$$\frac{ds}{dt} = v. \tag{5}$$

Thus *the speed of the moving point is the time rate of change of its arc-length function.* If $v(t) > 0$ for all t, then it follows that $s(t)$ is an increasing function of t and therefore has an inverse function $t(s)$. When we replace t by $t(s)$ in the curve's original parametric equations, we obtain the **arc-length parametrization**

$$x = x(s), \qquad y = y(s), \qquad z = z(s).$$

This gives the position of the moving point as a function of arc length measured along the curve from its initial point (see Fig. 13.5.2).

EXAMPLE 2 If we take $a = 5$, $b = 12$, and $\omega = 1$ for the helix of Example 1, the velocity formula $v = (a^2\omega^2 + b^2)^{1/2}$ yields

$$v = \sqrt{5^2 \cdot 1^2 + 12^2} = \sqrt{169} = 13.$$

Hence Eq. (5) gives $ds/dt = 13$, so

$$s = 13t,$$

taking $s = 0$ when $t = 0$ and thereby measuring arc length from the natural starting point $(5, 0, 0)$. When we substitute $t = s/13$ and the numerical values of a, b, and ω into the original parametric equations of the helix, we get the arc-length parametrization

$$x(s) = 5 \cos \frac{s}{13}, \qquad y(s) = 5 \sin \frac{s}{13}, \qquad z(s) = \frac{12s}{13}$$

of the helix.

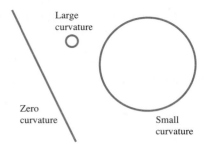

Large
curvature

Zero
curvature

Small
curvature

Fig. 13.5.3 The intuitive idea
of curvature

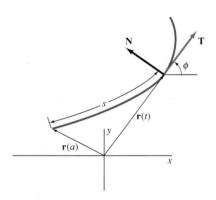

N

T

ϕ

s

$\mathbf{r}(t)$

y

$\mathbf{r}(a)$

x

Fig. 13.5.4 The unit tangent
vector **T**

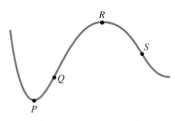

R

S

Q

P

Fig. 13.5.5 The curvature is
large at P and R, small at Q
and S.

CURVATURE OF PLANE CURVES

The word *curvature* has an intuitive meaning that we need to make precise. Most people would agree that a straight line does not curve at all, whereas a circle of small radius is more curved than a circle of large radius (Fig. 13.5.3). This judgment may be based on a feeling that curvature is "rate of change of direction." The direction of a curve is determined by its velocity vector, so you would expect the idea of curvature to have something to do with the rate at which the velocity vector is turning.

Let

$$\mathbf{r}(t) = x(t)\mathbf{i} + y(t)\mathbf{j}, \qquad a \leqq t \leqq b, \tag{6}$$

be the position vector of a smooth *plane* curve with nonzero velocity vector $\mathbf{v}(t) = \mathbf{r}'(t)$. The curve's **unit tangent vector** at the point $\mathbf{r}(t)$ is the unit vector

$$\mathbf{T}(t) = \frac{\mathbf{v}(t)}{|\mathbf{v}(t)|} = \frac{\mathbf{v}(t)}{v(t)}, \tag{7}$$

where $v(t) = |\mathbf{v}(t)|$ is the speed. Now denote by ϕ the angle of inclination of T, measured counterclockwise from the positive x-axis (Fig. 13.5.4). Then

$$\mathbf{T} = \mathbf{i} \cos \phi + \mathbf{j} \sin \phi. \tag{8}$$

We can express the unit tangent vector **T** of Eq. (8) as a function of the arc-length parameter s indicated in Fig. 13.5.4. Then the rate at which **T** is turning is measured by the derivative

$$\frac{d\mathbf{T}}{ds} = \frac{d\mathbf{T}}{d\phi} \cdot \frac{d\phi}{ds} = (-\mathbf{i} \sin \phi + \mathbf{j} \cos \phi) \frac{d\phi}{ds}. \tag{9}$$

Note that

$$\left| \frac{d\mathbf{T}}{ds} \right| = \left| \frac{d\phi}{ds} \right| \tag{10}$$

because the vector on the right-hand side of Eq. (9) is a unit vector.

The **curvature** at a point of a plane curve, denoted by κ (lowercase Greek kappa), is therefore defined to be

$$\kappa = \left| \frac{d\phi}{ds} \right|, \tag{11}$$

the absolute value of the rate of change of the angle ϕ with respect to arc length s. We define the curvature κ in terms of $d\phi/ds$ rather than $d\phi/dt$ because the latter depends not only on the shape of the curve, but also on the speed of the moving point $\mathbf{r}(t)$. For a straight line the angle ϕ is a constant, so the curvature given by Eq. (11) is zero. If you imagine a point that is moving with constant speed along a curve, the curvature is greatest at points where ϕ changes the most rapidly, such as the points P and R on the curve of Fig. 13.5.5. The curvature is least at points such as Q and S, where ϕ is changing the least rapidly.

We need to derive a formula that is effective in computing the curvature of a smooth parametric plane curve $x = x(t)$, $y = y(t)$. First we note that

$$\phi = \tan^{-1}\left(\frac{dy}{dx}\right) = \tan^{-1}\left(\frac{y'(t)}{x'(t)}\right)$$

provided $x'(t) \neq 0$. Hence

$$\frac{d\phi}{dt} = \frac{y''x' - y'x''}{(x')^2} \div \left(1 + \left(\frac{y'}{x'}\right)^2\right) = \frac{x'y'' - x''y'}{(x')^2 + (y')^2},$$

where primes denote derivatives with respect to t. Because $v = ds/dt > 0$, Eq. (11) gives

$$\kappa = \left|\frac{d\phi}{ds}\right| = \left|\frac{d\phi}{dt} \cdot \frac{dt}{ds}\right| = \frac{1}{v}\left|\frac{d\phi}{dt}\right|;$$

thus

$$\kappa = \frac{|x'y'' - x''y'|}{[(x')^2 + (y')^2]^{3/2}} = \frac{|x'y'' - x''y'|}{v^3}. \tag{12}$$

At a point where $x'(t) = 0$, we know that $y'(t) \neq 0$, because the curve is smooth. Thus we will obtain the same result if we begin with the equation $\phi = \cot^{-1}(x'/y')$.

An explicitly described curve $y = f(x)$ may be regarded as a parametric curve $x = x$, $y = f(x)$. Then $x' = 1$ and $x'' = 0$, so Eq. (12)—with x in place of t as the parameter—becomes

$$\kappa = \frac{|y''|}{[1 + (y')^2]^{3/2}} = \frac{|d^2y/dx^2|}{[1 + (dy/dx)^2]^{3/2}}. \tag{13}$$

EXAMPLE 3 Show that the curvature at each point of a circle of radius a is $\kappa = 1/a$.

Solution With the familiar parametrization $x = a \cos t$, $y = a \sin t$ of such a circle centered at the origin, we let primes denote derivatives with respect to t and obtain

$$x' = -a \sin t, \qquad y' = a \cos t,$$
$$x'' = -a \cos t, \qquad y'' = -a \sin t.$$

Hence Eq. (12) gives

$$\kappa = \frac{|(-a \sin t)(-a \sin t) - (-a \cos t)(a \cos t)|}{[(-a \sin t)^2 + (a \cos t)^2]^{3/2}} = \frac{a^2}{a^3} = \frac{1}{a}.$$

Alternatively, we could have used Eq. (13). Our point of departure would then be the equation $x^2 + y^2 = a^2$ of the same circle, and we would compute y' and y'' by implicit differentiation (see Problem 27).

It follows immediately from Eqs. (8) and (9) that

$$\mathbf{T} \cdot \frac{d\mathbf{T}}{ds} = 0,$$

so the unit tangent vector $\mathbf{T}$ and its derivative vector $d\mathbf{T}/ds$ are perpendicular. The *unit* vector $\mathbf{N}$ that points in the direction of $d\mathbf{T}/ds$ is called the **principal unit normal vector** to the curve. Because $\kappa = |d\phi/ds| = |d\mathbf{T}/ds|$ by Eq. (10), it follows that

$$\frac{d\mathbf{T}}{ds} = \kappa \mathbf{N}. \tag{14}$$

Intuitively, **N** *is the unit normal vector to the curve that points in the direction in which the curve is bending.*

Suppose that P is a point on a parametrized curve where $\kappa \neq 0$. Consider the circle that is tangent to the curve at P and has the same curvature there. The center of the circle is to lie on the concave side of the curve—that is, on the side toward which the normal vector **N** points. This circle is called the **osculating circle** (or **circle of curvature**) of the curve at the given point because it touches the curve so closely there. (*Osculum* is the Latin word for *kiss*.) Let ρ be the radius of the osculating circle, and let γ be the position vector of its center. Thus $\gamma = \overrightarrow{OC}$, where C is the center of the osculating circle (Fig. 13.5.6). Then ρ is called the **radius of curvature** of the curve at the point P, and γ is called the (vector) **center of curvature** of the curve at P.

Example 3 implies that the radius of curvature is

$$\rho = \frac{1}{\kappa}, \tag{15}$$

and the fact that $|\mathbf{N}| = 1$ implies that the position vector of the center of curvature is

$$\gamma = \mathbf{r} + \rho\mathbf{N} \qquad (\mathbf{r} = \overrightarrow{OP}). \tag{16}$$

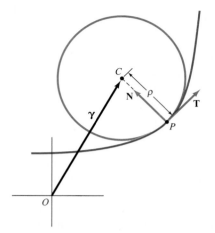

Fig. 13.5.6 Osculating circle, radius of curvature, and center of curvature

EXAMPLE 4 Determine the vectors **T** and **N**, the curvature κ, and the center of curvature of the parabola $y = x^2$ at the point $(1, 1)$.

Solution If the parabola is parametrized by $x = t$, $y = t^2$, then its position vector is $\mathbf{r}(t) = t\mathbf{i} + t^2\mathbf{j}$, so $\mathbf{v}(t) = \mathbf{i} + 2t\mathbf{j}$. The speed is $v(t) = (1 + 4t^2)^{1/2}$, so Eq. (7) yields

$$\mathbf{T}(t) = \frac{\mathbf{v}(t)}{v(t)} = \frac{\mathbf{i} + 2t\mathbf{j}}{\sqrt{1 + 4t^2}}.$$

By substituting $t = 1$, we find that the unit tangent vector at $(1, 1)$ is

$$\mathbf{T} = \frac{1}{\sqrt{5}}\mathbf{i} + \frac{2}{\sqrt{5}}\mathbf{j}.$$

Because the parabola is concave upward at $(1, 1)$, the principal unit normal vector is the upward-pointing unit vector

$$\mathbf{N} = -\frac{2}{\sqrt{5}}\mathbf{i} + \frac{1}{\sqrt{5}}\mathbf{j}$$

that is perpendicular to **T**. (Note that $\mathbf{T} \cdot \mathbf{N} = 0$.) If $y = x^2$, then $dy/dx = 2x$ and $d^2y/dx^2 = 2$, so Eq. (13) yields

$$\kappa = \frac{|y''|}{[1 + (y')^2]^{3/2}} = \frac{2}{(1 + 4x^2)^{3/2}}.$$

So at the point $(1, 1)$ we find the curvature and radius of curvature to be

$$\kappa = \frac{2}{5\sqrt{5}} \quad \text{and} \quad \rho = \frac{5\sqrt{5}}{2},$$

respectively.

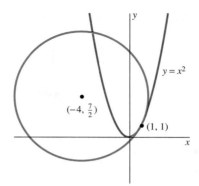

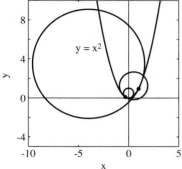

Fig. 13.5.8 The parabola $y = x^2$ and three of its osculating circles (Example 4)

Fig. 13.5.7 The osculating circle of Example 4

Next, Eq. (16) gives the center of curvature as

$$\boldsymbol{\gamma} = \langle 1, 1 \rangle + \frac{5\sqrt{5}}{2} \left\langle -\frac{2}{\sqrt{5}}, \frac{1}{\sqrt{5}} \right\rangle = \left\langle -4, \frac{7}{2} \right\rangle.$$

The equation of the osculating circle to the parabola is, therefore,

$$(x + 4)^2 + (y - \tfrac{7}{2})^2 = \rho^2 = \tfrac{125}{4}.$$

The parabola and this osculating circle are shown in Fig. 13.5.7. Figure 13.5.8 shows this large osculating circle at $(1, 1)$ as well as the small osculating circle at $(0, 0)$ and the medium-sized osculating circle at $(-\frac{1}{2}, \frac{1}{4})$.

CURVATURE OF CURVES IN SPACE

Consider now a moving particle in space with twice-differentiable position vector $\mathbf{r}(t)$. Suppose also that the velocity vector $\mathbf{v}(t)$ is never zero. The **unit tangent vector** at time t is defined, as before, to be

$$\mathbf{T}(t) = \frac{\mathbf{v}(t)}{|\mathbf{v}(t)|} = \frac{\mathbf{v}(t)}{v(t)}, \tag{17}$$

so

$$\mathbf{v} = v\mathbf{T}. \tag{18}$$

We defined the curvature of a plane curve to be $\kappa = |d\phi/dt|$, where ϕ is the angle of inclination of $\mathbf{T}$ from the positive x-axis. For a space curve, there is no single angle that determines the direction of $\mathbf{T}$, so we adopt the following approach (which leads to the same value for curvature when applied to a space curve that happens to lie in the xy-plane). Differentiation of the identity $\mathbf{T} \cdot \mathbf{T} = 1$ with respect to arc length gives

$$\mathbf{T} \cdot \frac{d\mathbf{T}}{ds} = 0.$$

It follows that the vectors $\mathbf{T}$ and $d\mathbf{T}/ds$ are always perpendicular.

Then we define the **curvature** κ of the curve at the point $\mathbf{r}(t)$ to be

$$\kappa = \left| \frac{d\mathbf{T}}{ds} \right| = \left| \frac{d\mathbf{T}}{dt} \frac{dt}{ds} \right| = \frac{1}{v} \left| \frac{d\mathbf{T}}{dt} \right|. \tag{19}$$

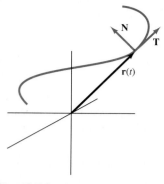

Fig. 13.5.9 The principal unit normal vector **N** points in the direction in which the curve is turning.

At a point where $\kappa \neq 0$, we define the **principal unit normal vector N** to be

$$\mathbf{N} = \frac{d\mathbf{T}/ds}{|d\mathbf{T}/ds|} = \frac{1}{\kappa}\frac{d\mathbf{T}}{ds}, \qquad (20)$$

so

$$\frac{d\mathbf{T}}{ds} = \kappa \mathbf{N}. \qquad (21)$$

Equation (21) shows that **N** has the same direction as $d\mathbf{T}/ds$ (Fig. 13.5.9), and Eq. (20) shows that **N** is a unit vector. Because Eq. (21) is the same as Eq. (14), we see that the present definitions of κ and **N** agree with those given earlier in the two-dimensional case.

EXAMPLE 5 Compute the curvature κ of the helix of Example 1, the helix with parametric equations

$$x(t) = a \cos \omega t, \qquad y(t) = a \sin \omega t, \qquad z(t) = bt.$$

Solution In Example 1 of Section 13.4, we computed the velocity vector

$$\mathbf{v} = \mathbf{i}(-a\omega \sin \omega t) + \mathbf{j}(a\omega \cos \omega t) + b\mathbf{k}$$

and speed

$$v = |\mathbf{v}| = \sqrt{a^2\omega^2 + b^2}.$$

Hence Eq. (17) gives the unit tangent vector

$$\mathbf{T} = \frac{\mathbf{v}}{v} = \frac{\mathbf{i}(-a\omega \sin \omega t) + \mathbf{j}(a\omega \cos \omega t) + b\mathbf{k}}{\sqrt{a^2\omega^2 + b^2}}.$$

Then

$$\frac{d\mathbf{T}}{dt} = \frac{\mathbf{i}(-a\omega^2 \cos \omega t) + \mathbf{j}(-a\omega^2 \sin \omega t)}{\sqrt{a^2\omega^2 + b^2}},$$

so Eq. (19) gives

$$\kappa = \frac{1}{v}\left|\frac{d\mathbf{T}}{dt}\right| = \frac{a\omega^2}{a^2\omega^2 + b^2}$$

for the curvature of the helix of Example 1 (Section 13.4). Note that the helix has constant curvature. Also note that, if $b = 0$ (so that the helix reduces to a circle of radius a in the xy-plane), our result reduces to $\kappa = 1/a$, in agreement with our computation of the curvature of a circle in Example 3.

NORMAL AND TANGENTIAL COMPONENTS OF ACCELERATION

We may apply Eq. (21) to analyze the meaning of the acceleration vector of a moving particle with velocity vector **v** and speed v. Then Eq. (17) gives $\mathbf{v} = v\mathbf{T}$, so the acceleration vector of the particle is

$$\mathbf{a} = \frac{d\mathbf{v}}{dt} = \frac{dv}{dt}\mathbf{T} + v\frac{d\mathbf{T}}{dt} = \frac{dv}{dt}\mathbf{T} + v\frac{d\mathbf{T}}{ds}\frac{ds}{dt}.$$

But $ds/dt = v$, so Eq. (21) gives

$$\mathbf{a} = \frac{dv}{dt}\mathbf{T} + \kappa v^2 \mathbf{N}. \tag{22}$$

Because $\mathbf{T}$ and $\mathbf{N}$ are unit vectors tangent and normal to the curve, respectively, Eq. (22) provides a *decomposition of the acceleration vector* into its components tangent and normal to the trajectory. The **tangential component**

$$a_T = \frac{dv}{dt} \tag{23}$$

is the rate of change of speed of the particle, whereas the **normal component**

$$a_N = \kappa v^2 = \frac{v^2}{\rho} \tag{24}$$

measures the rate of change of its direction of motion. The decomposition

$$\mathbf{a} = a_T \mathbf{T} + a_N \mathbf{N} \tag{25}$$

is illustrated in Fig. 13.5.10.

As an application of Eq. (22), think of a train moving along a straight track with constant speed v, so that $a_T = 0 = a_N$ (the latter because $\kappa = 0$ for a straight line). Suppose that at time $t = 0$, the train enters a circular curve of radius ρ. At that instant, it will *suddenly* be subjected to a normal acceleration of magnitude v^2/ρ, proportional to the *square* of the speed of the train. A passenger in the train will experience a sudden jerk to the side. If v is large, the stresses may be great enough to damage the track or derail the train. It is for exactly this reason that railroads are built not with curves shaped like arcs of circles but with *approach curves* in which the curvature, and hence the normal acceleration, build up smoothly.

EXAMPLE 6   A particle moves in the xy-plane with parametric equations

$$x(t) = \tfrac{3}{2}t^2, \qquad y(t) = \tfrac{4}{3}t^3.$$

Find the tangential and normal components of its acceleration vector when $t = 1$.

Solution   The trajectory and the vectors $\mathbf{N}$ and $\mathbf{T}$ appear in Fig. 13.5.11. There $\mathbf{N}$ and $\mathbf{T}$ are shown attached at the point of evaluation, at which $t = 1$. The particle has position vector

$$\mathbf{r}(t) = \tfrac{3}{2}t^2\mathbf{i} + \tfrac{4}{3}t^3\mathbf{j}$$

and thus velocity

$$\mathbf{v}(t) = 3t\mathbf{i} + 4t^2\mathbf{j}.$$

Hence its speed is

$$v(t) = \sqrt{9t^2 + 16t^4},$$

from which we calculate

$$a_T = \frac{dv}{dt} = \frac{9t + 32t^3}{\sqrt{9t^2 + 16t^4}}.$$

Thus $v = 5$ and $a_T = \tfrac{41}{5}$ when $t = 1$.

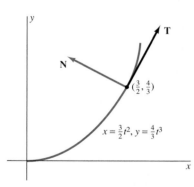

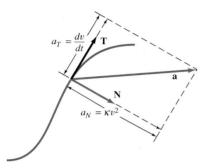

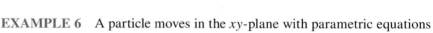

Fig. 13.5.10   Resolution of the acceleration vector $\mathbf{a}$ into its tangential and normal components

Fig. 13.5.11   The moving particle of Example 6

To use Eq. (12) to compute the curvature at $t = 1$, we compute $dx/dt = 3t$, $dy/dt = 4t^2$, $d^2x/dt^2 = 3$, and $d^2y/dt^2 = 8t$. Thus at $t = 1$ we have

$$\kappa = \frac{|x'y'' - x''y'|}{v^3} = \frac{|3 \cdot 8 - 3 \cdot 4|}{5^3} = \frac{12}{125}.$$

Hence

$$a_N = \kappa v^2 = \tfrac{12}{125} \cdot 5^2 = \tfrac{12}{5}$$

when $t = 1$. As a check (Problem 28), you might compute $\mathbf{T}$ and $\mathbf{N}$ when $t = 1$ and verify that

$$\tfrac{41}{5}\mathbf{T} + \tfrac{12}{5}\mathbf{N} = \mathbf{a} = 3\mathbf{i} + 8\mathbf{j}.$$

It remains for us to see how to compute a_T, a_N, and $\mathbf{N}$ effectively in the case of a space curve. We would prefer to have formulas that explicitly contain only the vectors $\mathbf{r}$, $\mathbf{v}$, and $\mathbf{a}$.

If we compute the dot product of $\mathbf{v} = v\mathbf{T}$ with the acceleration $\mathbf{a}$ as given in Eq. (22) and use the facts that $\mathbf{T} \cdot \mathbf{T} = 1$ and $\mathbf{T} \cdot \mathbf{N} = 0$, we get

$$\mathbf{v} \cdot \mathbf{a} = v\mathbf{T} \cdot \left(\frac{dv}{dt}\mathbf{T}\right) + (v\mathbf{T}) \cdot (\kappa v^2 \mathbf{N}) = v\,\frac{dv}{dt}.$$

It follows that

$$a_T = \frac{dv}{dt} = \frac{\mathbf{v} \cdot \mathbf{a}}{v} = \frac{\mathbf{r}'(t) \cdot \mathbf{r}''(t)}{|\mathbf{r}'(t)|}. \qquad (26)$$

Similarly, when we compute the vector product of $\mathbf{v} = v\mathbf{T}$ with each side of Eq. (22), we find that

$$\mathbf{v} \times \mathbf{a} = \left(v\mathbf{T} \times \frac{dv}{dt}\mathbf{T}\right) + (v\mathbf{T} \times \kappa v^2 \mathbf{N}) = \kappa v^3 \mathbf{T} \times \mathbf{N}.$$

Because κ and v are nonnegative and because $\mathbf{T} \times \mathbf{N}$ is a unit vector, we may conclude that

$$\kappa = \frac{|\mathbf{v} \times \mathbf{a}|}{v^3} = \frac{|\mathbf{r}'(t) \times \mathbf{r}''(t)|}{|\mathbf{r}'(t)|^3}. \qquad (27)$$

It now follows from Eq. (24) that

$$a_N = \frac{|\mathbf{r}'(t) \times \mathbf{r}''(t)|}{|\mathbf{r}'(t)|}. \qquad (28)$$

The curvature of a space curve often is not as easy to compute directly from the definition as we found in the case of the helix of Example 5. It is generally more convenient to use Eq. (27). Once $\mathbf{a}$, $\mathbf{T}$, a_T, and a_N have been computed, we can rewrite Eq. (25) as

$$\mathbf{N} = \frac{\mathbf{a} - a_T\mathbf{T}}{a_N} \qquad (29)$$

to find the principal unit normal vector.

EXAMPLE 7 Compute $\mathbf{T}$, $\mathbf{N}$, κ, a_T, and a_N at the point $(1, \frac{1}{2}, \frac{1}{3})$ of the twisted cubic with parametric equations

$$x(t) = t, \qquad y(t) = \tfrac{1}{2}t^2, \qquad z(t) = \tfrac{1}{3}t^3.$$

Solution Differentiation of the position vector

$$\mathbf{r}(t) = \langle t, \tfrac{1}{2}t^2, \tfrac{1}{3}t^3 \rangle$$

gives

$$\mathbf{r}'(t) = \langle 1, t, t^2 \rangle \quad \text{and} \quad \mathbf{r}''(t) = \langle 0, 1, 2t \rangle.$$

When we substitute $t = 1$, we obtain

$$\mathbf{v}(1) = \langle 1, 1, 1 \rangle \qquad \text{(velocity)},$$
$$v(1) = |\mathbf{v}(1)| = \sqrt{3} \qquad \text{(speed), and}$$
$$\mathbf{a}(1) = \langle 0, 1, 2 \rangle \qquad \text{(acceleration)}$$

at the point $(1, \frac{1}{2}, \frac{1}{3})$. Then Eq. (26) gives the tangential component of acceleration:

$$a_T = \frac{\mathbf{v} \cdot \mathbf{a}}{v} = \frac{3}{\sqrt{3}} = \sqrt{3}.$$

Because

$$\mathbf{v} \times \mathbf{a} = \begin{vmatrix} \mathbf{i} & \mathbf{j} & \mathbf{k} \\ 1 & 1 & 1 \\ 0 & 1 & 2 \end{vmatrix} = \langle 1, -2, 1 \rangle,$$

Eq. (27) gives the curvature:

$$\kappa = \frac{|\mathbf{v} \times \mathbf{a}|}{v^3} = \frac{\sqrt{6}}{(\sqrt{3})^3} = \frac{\sqrt{2}}{3}.$$

The normal component of acceleration is $a_N = \kappa v^2 = \sqrt{2}$. The unit tangent vector is

$$\mathbf{T} = \frac{\mathbf{v}}{v} = \frac{1}{\sqrt{3}} \langle 1, 1, 1 \rangle = \frac{\mathbf{i} + \mathbf{j} + \mathbf{k}}{\sqrt{3}}.$$

Finally, Eq. (29) gives

$$\mathbf{N} = \frac{\mathbf{a} - a_T \mathbf{T}}{a_N} = \frac{1}{\sqrt{2}} (\langle 0, 1, 2 \rangle - \langle 1, 1, 1 \rangle) = \frac{1}{\sqrt{2}} \langle -1, 0, 1 \rangle = \frac{-\mathbf{i} + \mathbf{k}}{\sqrt{2}}.$$

13.5 Problems

Find the arc length of the curves described in Problems 1 through 6.

1. $x = 3 \sin 2t$, $y = 3 \cos 2t$, $z = 8t$; from $t = 0$ to $t = \pi$

2. $x = t$, $y = t^2/\sqrt{2}$, $z = t^3/3$; from $t = 0$ to $t = 1$

3. $x = 6e^t \cos t$, $y = 6e^t \sin t$, $z = 17e^t$; from $t = 0$ to $t = 1$

4. $x = t^2/2$, $y = \ln t$, $z = t\sqrt{2}$; from $t = 1$ to $t = 2$

5. $x = 3t \sin t$, $y = 3t \cos t$, $z = 2t^2$; from $t = 0$ to $t = \frac{4}{5}$

6. $x = 2e^t$, $y = e^{-t}$, $z = 2t$; from $t = 0$ to $t = 1$

In Problems 7 through 12, find the curvature of the given plane curve at the indicated point.

7. $y = x^3$ at $(0, 0)$ **8.** $y = x^3$ at $(-1, -1)$

9. $y = \cos x$ at $(0, 1)$

10. $x = t - 1$, $y = t^2 + 3t + 2$, where $t = 2$

11. $x = 5 \cos t$, $y = 4 \sin t$, where $t = \pi/4$

12. $x = 5 \cosh t$, $y = 3 \sinh t$, where $t = 0$

In Problems 13 through 16, find the point or points of the given curve at which the curvature is a maximum.

13. $y = e^x$ **14.** $y = \ln x$

15. $x = 5 \cos t$, $y = 3 \sin t$ **16.** $xy = 1$

For the plane curves in Problem 17 through 21, find the unit tangent and normal vectors at the indicated point.

17. $y = x^3$ at $(-1, -1)$

18. $x = t^3$, $y = t^2$ at $(-1, 1)$

19. $x = 3 \sin 2t$, $y = 4 \cos 2t$, where $t = \pi/6$

20. $x = t - \sin t$, $y = 1 - \cos t$, where $t = \pi/2$

21. $x = \cos^3 t$, $y = \sin^3 t$, where $t = 3\pi/4$

The position vector of a particle moving in the plane is given in Problems 22 through 26. Find the tangential and normal components of the acceleration vector.

22. $\mathbf{r}(t) = 3\mathbf{i} \sin \pi t + 3\mathbf{j} \cos \pi t$

23. $\mathbf{r}(t) = (2t + 1)\mathbf{i} + (3t^2 - 1)\mathbf{j}$

24. $\mathbf{r}(t) = \mathbf{i} \cosh 3t + \mathbf{j} \sinh 3t$

25. $\mathbf{r}(t) = \mathbf{i}t \cos t + \mathbf{j}t \sin t$

26. $\mathbf{r}(t) = \langle e^t \sin t, e^t \cos t \rangle$

27. Use Eq. (13) to compute the curvature of the circle with equation $x^2 + y^2 = a^2$.

28. Verify the equation $\frac{41}{5}\mathbf{T} + \frac{12}{5}\mathbf{N} = 3\mathbf{i} + 8\mathbf{j}$ given at the end of Example 6.

In Problems 29 through 31, find the equation of the osculating circle for the given plane curve at the indicated point.

29. $y = 1 - x^2$ at $(0, 1)$

30. $y = e^x$ at $(0, 1)$

31. $xy = 1$ at $(1, 1)$

Find the curvature κ of the space curves with position vectors given in Problems 32 through 36.

32. $\mathbf{r}(t) = t\mathbf{i} + (2t - 1)\mathbf{j} + (3t + 5)\mathbf{k}$

33. $\mathbf{r}(t) = t\mathbf{i} + \mathbf{j} \sin t + \mathbf{k} \cos t$

34. $\mathbf{r}(t) = \langle t, t^2, t^3 \rangle$

35. $\mathbf{r}(t) = \langle e^t \cos t, e^t \sin t, e^t \rangle$

36. $\mathbf{r}(t) = \mathbf{i}t \sin t + \mathbf{j}t \cos t + \mathbf{k}t$

37 through **41.** Find the tangential and normal components of acceleration a_T and a_N for the curves of Problems 32 through 36, respectively.

In Problems 42 through 45, find the unit vectors $\mathbf{T}$ and $\mathbf{N}$ for the given curve at the indicated point.

42. The curve of Problem 34 at $(1, 1, 1)$

43. The curve of Problem 33 at $(0, 0, 1)$

44. The curve of Problem 3 at $(6, 0, 17)$

45. The curve of Problem 35 at $(1, 0, 1)$

46. Find $\mathbf{T}$, $\mathbf{N}$, a_T, and a_N as functions of t for the helix of Example 1.

47. Find the arc-length parametrization of the line

$$x(t) = 2 + 4t, \qquad y(t) = 1 - 12t, \qquad z(t) = 3 + 3t$$

in terms of the arc length s measured from the initial point $(2, 1, 3)$.

48. Find the arc-length parametrization of the circle

$$x(t) = 2 \cos t, \qquad y(t) = 2 \sin t, \qquad z = 0$$

in terms of the arc length s measured from the initial point $(2, 0, 0)$.

49. Find the arc-length parametrization of the helix

$$x(t) = 3 \cos t, \qquad y(t) = 3 \sin t, \qquad z(t) = 4t$$

in terms of the arc length s measured from the initial point $(3, 0, 0)$.

50. Substitute $x = t$, $y = f(t)$, and $z = 0$ into Eq. (27) to verify that the curvature of the plane curve $y = f(x)$ is

$$\kappa = \frac{|f''(x)|}{[1 + (f'(x))^2]^{3/2}}.$$

51. A particle moves under the influence of a force that is always perpendicular to its direction of motion. Show that the speed of the particle must be constant.

52. Deduce from Eq. (20) that

$$\kappa = \frac{\sqrt{a^2 - a_T^2}}{v^2} = \frac{[(x'')^2 + (y'')^2 - (v')^2]^{1/2}}{(x')^2 + (y')^2},$$

where primes denote differentiation with respect to t.

53. Apply the formula of Problem 52 to calculate the curvature of the curve

$$x(t) = \cos t + t \sin t, \qquad y(t) = \sin t - t \cos t.$$

54. The folium of Descartes with equation $x^3 + y^3 = 3xy$ is shown in Fig. 13.5.12. Find the curvature and center

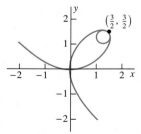

Fig. 13.5.12 The folium of Descartes (Problem 54)

of curvature of this folium at the point $(\frac{3}{2}, \frac{3}{2})$. Begin by calculating dy/dx and d^2y/dx^2 by implicit differentiation.

55. Determine the constants A, B, C, D, E, and F so that the curve

$$y = Ax^5 + Bx^4 + Cx^3 + Dx^2 + Ex + F$$

does, simultaneously, all of the following;

❑ Joins the two points $(0, 0)$ and $(1, 1)$;
❑ Has slope 0 at $(0, 0)$ and slope 1 at $(1, 1)$;
❑ Has curvature 0 at both $(0, 0)$ and $(1, 1)$.

The curve in question is shown in color in Fig. 13.5.13.

Why would this be a good curve to join the railroad tracks, shown in black in the figure?

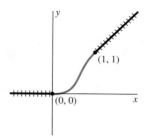

Fig. 13.5.13 Connecting railroad tracks (Problem 55)

13.6
Cylinders and Quadric Surfaces

Just as the graph of an equation $f(x, y) = 0$ is generally a curve in the xy-plane, the graph of an equation in three variables is generally a surface in space. A function F of three variables associates a real number $F(x, y, z)$ with each ordered triple (x, y, z) of real numbers. The **graph** of the equation

$$F(x, y, z) = 0 \qquad (1)$$

is the set of all points whose coordinates (x, y, z) satisfy this equation. We shall refer to the graph of such an equation as a **surface**. For instance, the graph of the equation

$$x^2 + y^2 + z^2 - 1 = 0$$

is a familiar surface, the unit sphere centered at the origin. But note that the graph of Eq. (1) does not always agree with our intuitive notion of a surface. For example, the graph of the equation

$$(x^2 + y^2)(y^2 + z^2)(z^2 + x^2) = 0$$

consists of the points lying on the three coordinate axes in space, because

❑ $x^2 + y^2 = 0$ implies that $x = y = 0$ (the z-axis);
❑ $y^2 + z^2 = 0$ implies that $y = z = 0$ (the x-axis);
❑ $z^2 + x^2 = 0$ implies that $z = x = 0$ (the y-axis).

We leave for advanced calculus the precise definition of *surface* as well as the study of conditions sufficient to imply that the graph of Eq. (1) actually is a surface.

The simplest example of a surface is a plane with linear equation $Ax + By + Cz + D = 0$. Here we discuss examples of other simple surfaces that frequently appear in multivariable calculus.

In order to sketch a surface S, it is often helpful to examine its intersections with various planes. The **trace** of the surface S in the plane $\mathcal{P}$ is the intersection of $\mathcal{P}$ and S. For example, if S is a sphere, then we can verify by the methods of elementary geometry that the trace of S in the plane $\mathcal{P}$ is a circle (Fig. 13.6.1), provided that $\mathcal{P}$ intersects the sphere but is not merely tangent to it (Problem 49). When we want to visualize a specific surface in space, it often suffices to examine its traces in the coordinate planes and possibly a few planes parallel to them.

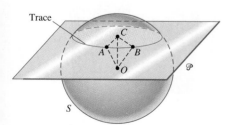

Fig. 13.6.1 The intersection of the sphere S and the plane $\mathcal{P}$ is a circle.

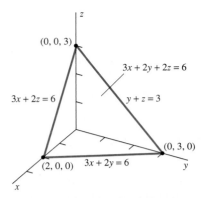

Fig. 13.6.2 Traces of the plane $3x + 2y + 2z = 6$ in the coordinate planes (Example 1)

EXAMPLE 1 Consider the plane with equation $3x + 2y + 2z = 6$. We find its trace in the xy-plane by setting $z = 0$. The equation then reduces to the equation $3x + 2y = 6$ of a straight line in the xy-plane. Similarly, when we set $y = 0$ we get the line $3x + 2z = 6$ as the trace of the given plane in the xz-plane. To find its trace in the yz-plane, we set $x = 0$, and this yields the line $y + z = 3$. Fig. 13.6.2 shows the portions of these three trace lines that lie in the first octant. Together they give us a good picture of how the plane $3x + 2y + 2z = 6$ is situated in space.

CYLINDERS

Let C be a curve in a plane and let L be a line not parallel to that plane. Then the set of all points on lines parallel to L that intersect C is called a **cylinder**. This is a generalization of the familiar right circular cylinder, for which the curve C is a circle and the line L is perpendicular to the plane of the circle. Figure 13.6.3 shows such a cylinder for which C is the circle $x^2 + y^2 = a^2$ in the xy-plane. The trace of this cylinder in any horizontal plane $z = c$ is a circle with radius a and center the point $(0, 0, c)$ on the z-axis. Thus the point (x, y, c) lies on this cylinder if and only if $x^2 + y^2 = a^2$. Hence this cylinder is the graph of the equation $x^2 + y^2 = a^2$, an equation in *three* variables—even through the variable z is technically missing.

The fact that the variable z does not appear explicity in the equation $x^2 + y^2 = a^2$ means that given any point $(x_0, y_0, 0)$ on the *circle* $x^2 + y^2 = a^2$ in the xy-plane, the point (x_0, y_0, z) lies on the cylinder for any and all values of z. The set of all such points is the vertical line through the point $(x_0, y_0, 0)$. Thus the *cylinder* $x^2 + y^2 = a^2$ in space is the union of all vertical lines through points of the *circle* $x^2 + y^2 = a^2$ in the plane (Fig. 13.6.4).

But a cylinder need not be circular—that is, the curve C can be an ellipse, a rectangle, or a quite arbitrary curve. For instance, Fig. 13.6.5 shows a vertical cylinder through a figure-eight curve in the xy-plane. (The parametric equations of this curve are $x = \sin t$, $y = \sin 2t$, $0 \leq t \leq 2\pi$.)

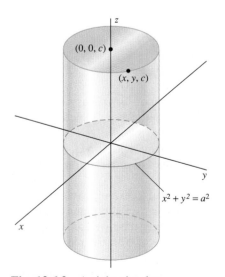

Fig. 13.6.3 A right circular cylinder

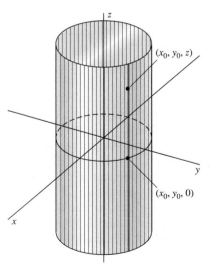

Fig. 13.6.4 The cylinder $x^2 + y^2 = a^2$; its rulings are parallel to the z-axis.

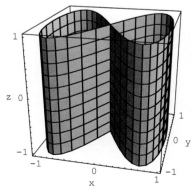

Fig. 13.6.5 The vertical cylinder through the figure-eight curve $x = \sin t$, $y = \sin 2t$

734

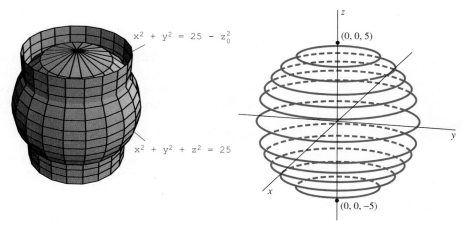

Fig. 13.6.6 The vertical cylinder of Example 2 intersects the sphere in two circles that lie in horizontal planes.

Fig. 13.6.7 A sphere as a union of circles (and two points) (Example 2)

EXAMPLE 2 Consider the sphere of radius 5 with equation $x^2 + y^2 + z^2 = 25$. If $0 < |z_0| < 5$, then the cylinder

$$x^2 + y^2 = 25 - z_0{}^2$$

intersects the sphere in two circles, each with radius

$$r_0 = \sqrt{25 - z_0{}^2}$$

(Fig. 13.6.6). These circles are the traces of the sphere in the horizontal planes $z = +z_0$ and $z = -z_0$. Figure 13.6.7 presents the sphere as the union of such circles, plus the two isolated points $(0, 0, 5)$ and $(0, 0, -5)$. For instance, $z = 3$ gives the circle with equation $x^2 + y^2 = 16$ in the horizontal plane $z = 3$.

Given a general cylinder generated by a plane curve C and a line L as in the preceding definition, the lines on the cylinder that are parallel to the line L are called **rulings** of the cylinder. Thus the rulings of the cylinder $x^2 + y^2 = a^2$ are vertical lines (parallel to the z-axis).

If the curve C in the xy-plane has equation

$$f(x, y) = 0, \tag{2}$$

then the cylinder through C with vertical rulings has the same equation in space. Thus is so because the point $P(x, y, z)$ lies on the cylinder if and only if the point $Q(x, y, 0)$ lies on the curve C. Similarly, the graph of an equation $g(x, z) = 0$ is a cylinder with rulings parallel to the y-axis, and the graph of an equation $h(y, z) = 0$ is a cylinder with rulings parallel to the x-axis. Thus the graph in space of an equation that includes only two of the three coordinate variables is always a cylinder; its rulings are parallel to the axis corresponding to the *missing* variable.

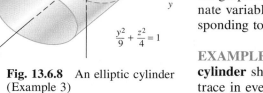

Fig. 13.6.8 An elliptic cylinder (Example 3)

EXAMPLE 3 The graph of the equation $4y^2 + 9z^2 = 36$ is the **elliptic cylinder** shown in Fig. 13.6.8. Its rulings are parallel to the x-axis, and its trace in every plane perpendicular to the x-axis is an ellipse with semiaxes of lengths 3 and 2 (just like the pictured ellipse $y^2/9 + z^2/4 = 1$ in the yz-plane).

EXAMPLE 4 The graph of the equation $z = 4 - x^2$ is the **parabolic cylinder** shown in Fig. 13.6.9. Its rulings are parallel to the y-axis, and its trace in every plane perpendicular to the y-axis is a parabola that is a parallel translate of the parabola $z = 4 - x^2$ in the xz-plane.

SURFACES OF REVOLUTION

Another way to use a plane curve C to generate a surface is to revolve the curve in space about a line L in its plane. This gives a **surface of revolution** with **axis** L. For example, Fig. 13.6.10 shows the surface generated by revolving the curve $f(x, y) = 0$ in the first quadrant of the xy-plane around the x-axis. The point $P(x, y, z)$ lies on the surface of revolution if and only if the point $Q(x, y_1, 0)$ lies on the curve, where

$$y_1 = |RQ| = |RP| = \sqrt{y^2 + z^2}.$$

Thus it is necessary that $f(x, y_1) = 0$, so the equation of the indicated surface of revolution around the x-axis is

$$f(x, \sqrt{y^2 + z^2}) = 0. \tag{3}$$

The equations of surfaces of revolution around other coordinate axes are obtained similarly. If the first-quadrant curve $f(x, y) = 0$ is revolved instead around the y-axis, then we replace x by $\sqrt{x^2 + z^2}$ to get the equation $f(\sqrt{x^2 + z^2}, y) = 0$ of the resulting surface of revolution. If the curve $g(y, z) = 0$ in the first quadrant of the yz-plane is revolved around the z-axis, we replace y by $\sqrt{x^2 + y^2}$. Thus the equation of the resulting surface about the z-axis is $g(\sqrt{x^2 + y^2}, z) = 0$. These assertions are easily verified with the aid of diagrams similar to Fig. 13.6.10.

EXAMPLE 5 Write an equation of the **ellipsoid of revolution** obtained by revolving the ellipse $4y^2 + z^2 = 4$ around the z-axis (Fig. 13.6.11).

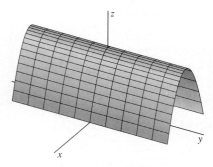

Fig. 13.6.9 The parabolic cylinder $z = 4 - x^2$ (Example 4)

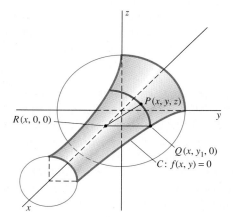

Fig. 13.6.10 The surface generated by rotating C around the x-axis. (For clarity, only a quarter of the surface is shown.)

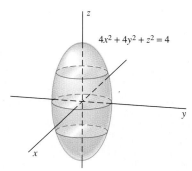

Fig. 13.6.11 The ellipsoid of revolution of Example 5

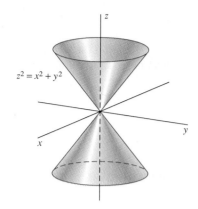

$z^2 = x^2 + y^2$

Fig. 13.6.12 The cone of Example 6

Solution We replace y by $\sqrt{x^2 + y^2}$ in the given equation. This yields $4x^2 + 4y^2 + z^2 = 4$ as an equation of the ellipsoid.

EXAMPLE 6 Determine the graph of the equation $z^2 = x^2 + y^2$.

Solution First we rewrite the given equation in the form $z = \pm\sqrt{x^2 + y^2}$. Thus the surface is symmetric about the xy-plane, and the upper half has equation $z = +\sqrt{x^2 + y^2}$. We obtained this last equation from the simple equation $z = y$ by replacing y by $\sqrt{x^2 + y^2}$. Thus we obtain the upper half of the surface by revolving the line $z = y$ (for $y \geqq 0$) around the z-axis. The graph is the **cone** shown in Fig. 13.6.12. Its upper half has equation $z = +\sqrt{x^2 + y^2}$, and its lower half has equation $z = -\sqrt{x^2 + y^2}$. The entire cone $z^2 = x^2 + y^2$ is obtained by revolving the entire line $z = y$ around the z-axis.

QUADRIC SURFACES

Cones, spheres, circular and parabolic cylinders, and ellipsoids of revolution are all surfaces that are graphs of second-degree equations in x, y, and z. The graph of a second-degree equation in three variables is called a **quadric surface**. We discuss here some important special cases of the equation

$$Ax^2 + By^2 + Cz^2 + Dx + Ey + Fz + G = 0. \qquad (4)$$

This is a special second-degree equation in that it contains no terms involving the products xy, xz, or yz.

EXAMPLE 7 The **ellipsoid**

$$\frac{x^2}{a^2} + \frac{y^2}{b^2} + \frac{z^2}{c^2} = 1 \qquad (5)$$

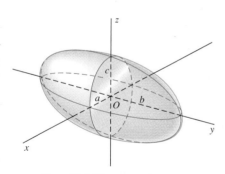

Fig. 13.6.13 The ellipsoid of Example 7

is symmetric about each of the three coordinate planes and has intercepts $(\pm a, 0, 0)$, $(0, \pm b, 0)$, and $(0, 0, \pm c)$ on the three coordinate axes. If $P(x, y, z)$ is a point of this ellipsoid, then $|x| \leqq a$, $|y| \leqq b$, and $|z| \leqq c$. Each trace in a plane parallel to one of the three coordinate planes is either a single point or an ellipse. For example, if $-c < z_0 < c$, then the trace of the ellipsoid of Eq. (5) in the plane $z = z_0$ has equation

$$\frac{x^2}{a^2} + \frac{y^2}{b^2} = 1 - \frac{z_0^2}{c^2} > 0,$$

which is the equation of an ellipse with semiaxes $(a/c)\sqrt{c^2 - z_0^2}$ and $(b/c)\sqrt{c^2 - z_0^2}$. Figure 13.6.13 shows this ellipsoid with semiaxes a, b, and c labeled. Figure 13.6.14 shows its trace ellipses in planes parallel to the three coordinate planes.

EXAMPLE 8 The **elliptic paraboloid**

$$\frac{x^2}{a^2} + \frac{y^2}{b^2} = \frac{z}{c} \qquad (6)$$

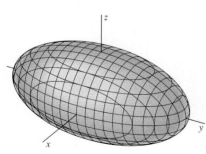

Fig. 13.6.14 The traces of the ellipsoid $\dfrac{x^2}{a^2} + \dfrac{y^2}{b^2} + \dfrac{z^2}{c^2} = 1$ (Example 7)

is shown in Fig. 13.6.15. Its trace in the horizontal plane $z = z_0 > 0$ is the ellipse $x^2/a^2 + y^2/b^2 = z_0/c$ with semiaxes $a\sqrt{z_0/c}$ and $b\sqrt{z_0/c}$. Its trace in any vertical plane is a parabola. For instance, its trace in the plane $y = y_0$

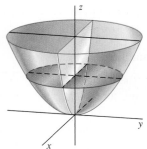

Fig. 13.6.15 An elliptic paraboloid (Example 8)

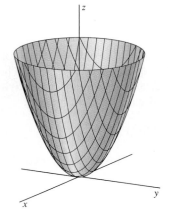

Fig. 13.6.16 Trace parabolas of a circular paraboloid (Example 8)

has equation $x^2/a^2 + y_0^2/b^2 = z/c$, which can be rewritten in the form $z - z_1 = k(x - x_1)^2$ by taking $z_1 = cy_0^2/b^2$ and $x_1 = 0$. The paraboloid opens upward if $c > 0$ and downward if $c < 0$. If $a = b$, then the paraboloid is circular. Figure 13.6.16 shows the traces of a circular paraboloid in planes parallel to the xz- and yz-planes.

EXAMPLE 9 The **elliptical cone**

$$\frac{x^2}{a^2} + \frac{y^2}{b^2} = \frac{z^2}{c^2} \tag{7}$$

is shown in Fig. 13.6.17. Its trace in the horizontal plane $z = z_0 \neq 0$ is an ellipse with semiaxes $a|z_0|/c$ and $b|z_0|/c$.

EXAMPLE 10 The **hyperboloid of one sheet** with equation

$$\frac{x^2}{a^2} + \frac{y^2}{b^2} - \frac{z^2}{c^2} = 1 \tag{8}$$

is shown in Fig. 13.6.18. Its trace in the horizontal plane $z = z_0$ is the ellipse $x^2/a^2 + y^2/b^2 = 1 + z_0^2/c^2 > 0$. Its trace in a vertical plane is a hyperbola except when the vertical plane intersects the xy-plane in a line tangent to the ellipse $x^2/a^2 + y^2/b^2 = 1$. In this special case, the trace is a degenerate hyperbola consisting of two intersecting lines. Figure 13.6.19 shows the traces (in planes parallel to the coordinate planes) of a circular ($a = b$) hyperboloid of one sheet.

The graphs of the equations

$$\frac{y^2}{b^2} + \frac{z^2}{c^2} - \frac{x^2}{a^2} = 1 \quad \text{and} \quad \frac{x^2}{a^2} + \frac{z^2}{c^2} - \frac{y^2}{b^2} = 1$$

are also hyperboloids of one sheet, opening along the x- and y-axes, respectively.

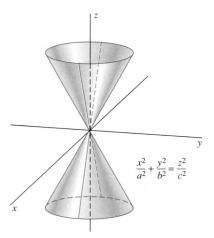

Fig. 13.6.17 An elliptic cone (Example 9)

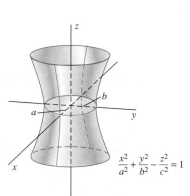

Fig. 13.6.18 A hyperboloid of one sheet (Example 10)

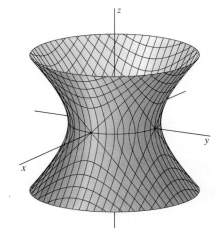

Fig. 13.6.19 A circular hyperboloid of one sheet (Example 10). Its traces in horizontal planes are circles; its traces in vertical planes are hyperbolas.

738

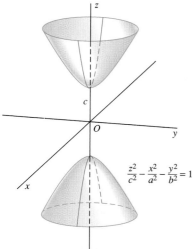

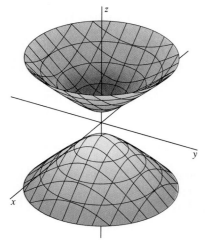

Fig. 13.6.20 A hyperboloid of two sheets (Example 11)

Fig. 13.6.21 A circular hyperboloid of two sheets (Example 11). Its (nondegenerate) traces in horizontal planes are circles; its traces in vertical planes are hyperbolas.

EXAMPLE 11 The **hyperboloid of two sheets** with equation

$$\frac{z^2}{c^2} - \frac{x^2}{a^2} - \frac{y^2}{b^2} = 1 \qquad (9)$$

consists of two separate pieces, or *sheets* (Fig. 13.6.20). The two sheets open along the positive and negative z-axis and intersect it at the points $(0, 0, \pm c)$. The trace of this hyperboloid in a horizontal plane $z = z_0$ with $|z_0| > c$ is the ellipse

$$\frac{x^2}{a^2} + \frac{y^2}{b^2} = \frac{z_0^2}{c^2} - 1 > 0.$$

Its trace in any vertical plane is a nondegenerate hyperbola. Figure 13.6.21 shows traces of a circular hyperboloid of two sheets.

The graphs of the equations

$$\frac{x^2}{a^2} - \frac{y^2}{b^2} - \frac{z^2}{c^2} = 1 \quad \text{and} \quad \frac{y^2}{b^2} - \frac{x^2}{a^2} - \frac{z^2}{c^2} = 1$$

are also hyperboloids of two sheets, opening along the x-axis and y-axis, respectively. When the equation of a hyperboloid is written in standard form with $+1$ on the right-hand side [as in Eqs. (8) and (9)], the number of sheets is equal to the number of negative terms on the left-hand side.

EXAMPLE 12 The **hyperbolic paraboloid**

$$\frac{y^2}{b^2} - \frac{x^2}{a^2} = \frac{z}{c} \qquad (c > 0) \qquad (10)$$

is saddle-shaped, as indicated in Fig. 13.6.22. Its trace in the horizontal plane $z = z_0$ is a hyperbola (or two intersecting lines if $z_0 = 0$). Its trace in a vertical plane parallel to the xz-plane is a parabola that opens downward, whereas its trace in a vertical plane parallel to the yz-plane is a parabola that opens upward. In particular, the trace of the hyperbolic paraboloid in the xz-plane is a parabola opening downward from the origin, whereas its trace in the yz-plane is a parabola opening upward from the origin. Thus the origin looks like a local maximum from one direction but like a local minimum from another. Such a point on a surface is called a **saddle point**.

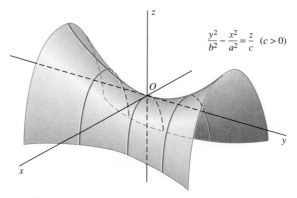

Fig. 13.6.22 The hyperbolic paraboloid is a saddle-shaped surface (Example 12).

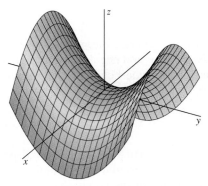

Fig. 13.6.23 The vertical traces of the hyperbolic paraboloid $z = y^2 - x^2$ (Example 12)

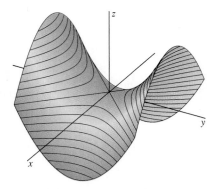

Fig. 13.6.24 The horizontal traces of the hyperbolic paraboloid $z = y^2 - x^2$ (Example 12)

Figure 13.6.23 shows the parabolic traces in vertical planes of the hyperbolic paraboloid $z = y^2 - x^2$. Figure 13.6.24 shows its hyperbolic traces in horizontal planes.

*CONIC SECTIONS AS SECTIONS OF A CONE

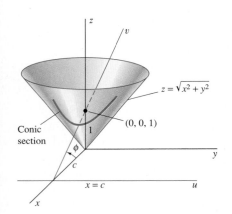

Fig. 13.6.25 Finding an equation for a conic section

The parabola, hyperbola, and ellipse that we studied in Chapter 10 were originally introduced by the ancient Greek mathematicians as plane sections (traces) of a right circular cone. Here we show that the intersection of a plane and a cone is, indeed, one of the three conic sections as defined in Chapter 10.

Figure 13.6.25 shows the cone with equation $z = \sqrt{x^2 + y^2}$ and its intersection with a plane $\mathcal{P}$ that passes through the point $(0, 0, 1)$ and the line $x = c > 0$ in the xy-plane. The equation of $\mathcal{P}$ is

$$z = 1 - \frac{x}{c}. \tag{11}$$

The angle between $\mathcal{P}$ and the xy-plane is $\phi = \tan^{-1}(1/c)$. We want to show that the conic section obtained by intersecting the cone and the plane is

$$\text{A parabola if } \phi = 45° \quad (c = 1),$$
$$\text{An ellipse if } \phi < 45° \quad (c > 1),$$
$$\text{A hyperbola if } \phi > 45° \quad (c < 1).$$

We begin by introducing uv-coordinates in the plane $\mathcal{P}$ as follows. The u-coordinate of the point (x, y, z) of $\mathcal{P}$ is $u = y$. The v-coordinate of the same point is its perpendicular distance from the line $x = c$. This explains the u- and v-axes indicated in Fig. 13.6.25. Figure 13.6.26 shows the cross section in the plane $y = 0$ exhibiting the relation between v, x, and z. We see that

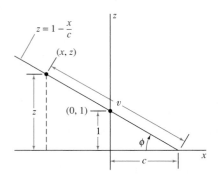

Fig. 13.6.26 Computing coordinates in the uv-plane

$$z = v \sin \phi = \frac{v}{\sqrt{1 + c^2}}. \tag{12}$$

Equations (11) and (12) give

$$x = c(1 - z) = c\left(1 - \frac{v}{\sqrt{1 + c^2}}\right) \tag{13}$$

We had $z^2 = x^2 + y^2$ for the equation of the cone. We make the following substitutions in this equation: Replace y by u, and replace z and x by the expressions on the right-hand sides of Eqs. (12) and (13), respectively. These replacements yield

$$\frac{v^2}{1 + c^2} = c^2\left(1 - \frac{v}{\sqrt{1 + c^2}}\right)^2 + u^2.$$

After we simplify, this last equation takes the form

$$u^2 + \frac{c^2 - 1}{c^2 + 1}v^2 - \frac{2c^2}{\sqrt{1 + c^2}}v + c^2 = 0. \tag{14}$$

This is the equation of the curve in the uv-plane. We examine the three cases for the angle ϕ.

Suppose first that $\phi = 45°$. Then $c = 1$, so Eq. (14) contains a term that includes u^2, another term that includes v, and a constant term. So the curve is a parabola; see Eq. (6) of Section 10.4.

Suppose next that $\phi < 45°$. Then $c > 1$, and both the coefficients of u^2 and v^2 in Eq. (14) are positive. Thus the curve is an ellipse; see Eq. (6) of Section 10.5.

Finally, if $\phi > 45°$, then $c < 1$, and the coefficients of u^2 and v^2 in Eq. (14) have opposite signs. So the curve is a hyperbola; see Eq. (8) of Section 10.6.

13.6 Problems

Describe and sketch the graphs of the equations given in Problems 1 through 30.

1. $3x + 2y + 10z = 20$
2. $3x + 2y = 30$
3. $x^2 + y^2 = 9$
4. $y^2 = x^2 - 9$
5. $xy = 4$
6. $z = 4x^2 + 4y^2$
7. $z = 4x^2 + y^2$
8. $4x^2 + 9y^2 = 36$
9. $z = 4 - x^2 - y^2$
10. $y^2 + z^2 = 1$
11. $2z = x^2 + y^2$
12. $x = 1 + y^2 + z^2$
13. $z^2 = 4(x^2 + y^2)$
14. $y^2 = 4x$
15. $x^2 = 4z + 8$
16. $x = 9 - z^2$
17. $4x^2 + y^2 = 4$
18. $x^2 + z^2 = 4$
19. $x^2 = 4y^2 + 9z^2$
20. $x^2 - 4y^2 = z$
21. $x^2 + y^2 + 4z = 0$
22. $x = \sin y$
23. $x = 2y^2 - z^2$
24. $x^2 + 4y^2 + 2z^2 = 4$
25. $x^2 + y^2 - 9z^2 = 9$
26. $x^2 - y^2 - 9z^2 = 9$
27. $y = 4x^2 + 9z^2$
28. $y^2 + 4x^2 - 9z^2 = 36$
29. $y^2 - 9x^2 - 4z^2 = 36$
30. $x^2 + 9y^2 + 4z^2 = 36$

Problems 31 through 40 give the equation of a curve in one of the coordinate planes. Write an equation for the surface generated by revolving this curve around the indicated axis. Then sketch the surface.

31. $x = 2z^2$;   the x-axis
32. $4x^2 + 9y^2 = 36$;   the y-axis
33. $y^2 - z^2 = 1$;   the z-axis
34. $z = 4 - x^2$;   the z-axis
35. $y^2 = 4x$;   the x-axis
36. $yz = 1$;   the z-axis
37. $z = \exp(-x^2)$;   the z-axis
38. $(y - z)^2 + z^2 = 1$;   the z-axis
39. The line $z = 2x$;   the z-axis
40. The line $z = 2x$;   the x-axis

In Problems 41 through 47, describe the traces of the given surfaces in planes of the indicated type.

41. $x^2 + 4y^2 = 4$;   in horizontal planes (those parallel to the xy-plane)
42. $x^2 + 4y^2 + 4z^2 = 4$;   in horizontal planes

43. $x^2 + 4y^2 + 4z^2 = 4$;   in planes parallel to the yz-plane

44. $z = 4x^2 + 9y^2$;   in horizontal planes

45. $z = 4x^2 + 9y^2$;   in planes parallel to the yz-plane

46. $z = xy$;   in horizontal planes

47. $z = xy$;   in vertical planes through the z-axis

48. Identify the surface $z = xy$ by making a suitable rotation of axes in the xy-plane (as in Section 10.7).

49. Prove that the triangles OAC and OBC in Fig. 13.6.1 are congruent, and thereby conclude that the trace of a sphere in an intersecting plane is a circle.

50. Prove that the projection into the yz-plane of the curve of intersection of the surfaces $x = 1 - y^2$ and $x = y^2 + z^2$ is an ellipse (Fig. 13.6.27).

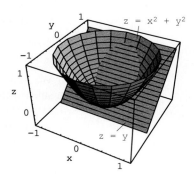

Fig. 13.6.28   The plane and paraboloid of Problem 51

52. Prove that the projection into the xz-plane of the intersection of the paraboloids $y = 2x^2 + 3z^2$ and $y = 5 - 3x^2 - 2z^2$ is a circle (Fig. 13.6.29).

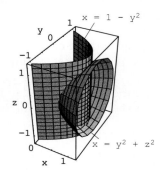

Fig. 13.6.27   The paraboloid and parabolic cylinder of Problem 50

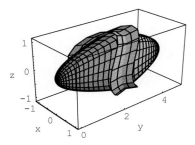

Fig. 13.6.29   The two paraboloids of Problem 52

53. Prove that the projection into the xy-plane of the intersection of the plane $x + y + z = 1$ and the ellipsoid $x^2 + 4y^2 + 4z^2 = 4$ is an ellipse.

54. Show that the curve of intersection of the plane $z = ky$ and the cylinder $x^2 + y^2 = 1$ is an ellipse. [*Suggestion:* Introduce uv-coordinates into the plane $z = ky$ as follows: Let the u-axis be the original x-axis and let the v-axis be the line $z = ky$, $x = 0$.]

51. Show that the projection into the xy-plane of the intersection of the plane $z = y$ and the paraboloid $z = x^2 + y^2$ is a circle (Fig. 13.6.28).

13.7
Cylindrical and Spherical Coordinates*

Rectangular coordinates provide only one of several useful ways of describing points, curves, and surfaces in space. Here we discuss two additional coordinate systems in three-dimensional space. Each is a generalization of polar coordinates in the coordinate plane.

Recall from Section 10.2 that the relationship between the rectangular coordinates (x, y) and the polar coordinates (r, θ) of a point in the plane is given by

$$x = r \cos \theta, \qquad y = r \sin \theta \tag{1}$$

---

*The material in this section is not required until Section 15.7 and thus may be deferred until just before that section is covered.

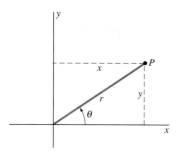

Fig. 13.7.1 The relation between rectangular and polar coordinates in the xy-plane

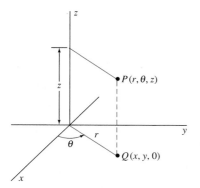

Fig. 13.7.2 Finding the cylindrical coordinates of the point P

and

$$r^2 = x^2 + y^2, \qquad \tan \theta = \frac{y}{x} \qquad \text{if } x \neq 0. \qquad (2)$$

Read these relationships directly from the triangle of Fig. 13.7.1.

CYLINDRICAL COORDINATES

The **cylindrical coordinates** (r, θ, z) of a point P in space are natural hybrids of its polar and rectangular coordinates. We use the polar coordinates (r, θ) of the point in the plane with rectangular coordinates (x, y) and use the same z-coordinate as in rectangular coordinates. (The cylindrical coordinates of a point P in space are illustrated in Fig. 13.7.2.) This means that we can obtain the relations between the rectangular coordinates (x, y, z) of the point P and its cylindrical coordinates (r, θ, z) by simply adjoining the identity $z = z$ to the equations in (1) and (2):

$$x = r \cos \theta, \qquad y = r \sin \theta, \qquad z = z \qquad (3)$$

and

$$r^2 = x^2 + y^2, \qquad \tan \theta = \frac{y}{x}, \qquad z = z. \qquad (4)$$

We can use these equations to convert from rectangular to cylindrical coordinates and vice versa. The following table lists the rectangular and corresponding cylindrical coordinates for a few points in space.

(x, y, z)	(r, θ, z)
$(1, 0, 0)$	$(1, 0, 0)$
$(-1, 0, 0)$	$(1, \pi, 0)$
$(0, 2, 3)$	$(2, \pi/2, 3)$
$(1, 1, 2)$	$(\sqrt{2}, \pi/4, 2)$
$(1, -1, 2)$	$(\sqrt{2}, 7\pi/4, 2)$
$(-1, 1, 2)$	$(\sqrt{2}, 3\pi/4, 2)$
$(-1, -1, -2)$	$(\sqrt{2}, 5\pi/4, -2)$
$(0, -3, -3)$	$(3, 3\pi/2, -3)$

The name *cylindrical coordinates* arises from the fact that the graph in space of the equation $r = c$ (c a constant) is a cylinder of radius c symmetric about the z-axis. This suggests that we should consider using cylindrical coordinates to solve problems that involve circular symmetry about the z-axis. For example, the sphere $x^2 + y^2 + z^2 = a^2$ and the cone $z^2 = x^2 + y^2$ have cylindrical equations $r^2 + z^2 = a^2$ and $z^2 = r^2$, respectively. It follows from our discussion of surfaces of revolution in Section 13.6 that if the curve $f(y, z) = 0$ in the yz-plane is revolved around the z-axis, then the cylindrical equation of the surface generated is $f(r, z) = 0$.

EXAMPLE 1 If the parabola $z = y^2$ is revolved around the z-axis, then we get the cylindrical equation of the paraboloid thus generated quite simply: We replace y with r. Hence the paraboloid has cylindrical equation

$$z = r^2.$$

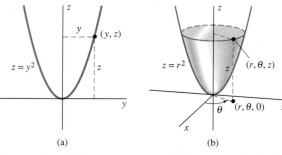

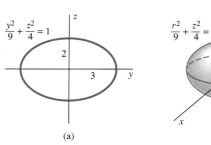

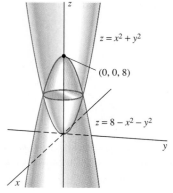

(a)

(b)

Fig. 13.7.3 (a) The parabola and (b) paraboloid of Example 1

(a)

(b)

Fig. 13.7.4 (a) The ellipse and (b) ellipsoid of Example 1

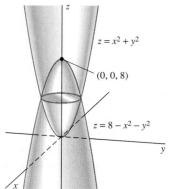

Fig. 13.7.5 The two paraboloids of Example 2

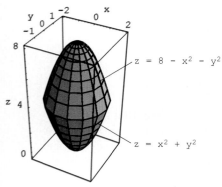

Fig. 13.7.6 The solid of Example 2

Similarly, if the ellipse $y^2/9 + z^2/4 = 1$ is revolved around the z-axis, the cylindrical equation of the resulting ellipsoid is

$$\frac{r^2}{9} + \frac{z^2}{4} = 1.$$

The parabola and paraboloid are shown in Fig. 13.7.3; the ellipse and ellipsoid are shown in Fig. 13.7.4.

EXAMPLE 2 Sketch the region that is bounded by the graphs of the cylindrical equations $z = r^2$ and $z = 8 - r^2$.

Solution We first substitute $r^2 = x^2 + y^2$ from (4) in the given equations. Thus the two surfaces of the example have rectangular equations

$$z = x^2 + y^2 \quad \text{and} \quad z = 8 - x^2 - y^2,$$

respectively. Their graphs are the two paraboloids shown in Fig. 13.7.5. The region in question, shown in Fig 13.7.6, is bounded above by the paraboloid $z = 8 - x^2 - y^2$ and below by the paraboloid $z = x^2 + y^2$.

SPHERICAL COORDINATES

Figure 13.7.7 shows the **spherical coordinates** (ρ, ϕ, θ) of the point P in space. The first spherical coordinate ρ is simply the distance $\rho = |OP|$ from the origin O to P. The second spherical coordinate ϕ is the angle between OP and the positive z-axis. Thus we may always choose ϕ in the interval $[0, \pi]$, although it is not restricted to that domain. Finally, θ is the familiar angle θ

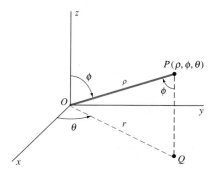

Fig. 13.7.7 Finding the spherical coordinates of the point P

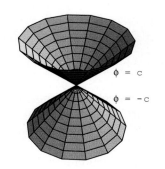

Fig. 13.7.8 The two nappes of a 45° cone; $\phi = \pi/2$ is the spherical equation of the xy-plane.

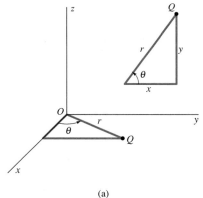

(a)

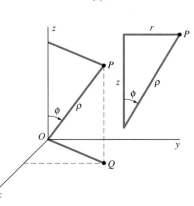

(b)

Fig. 13.7.9 Triangles used in finding spherical coordinates

of cylindrical coordinates. That is, θ is the angular coordinate of the vertical projection Q of P into the xy-plane. Thus we may always choose θ in the interval $[0, 2\pi]$, although it is not restricted to that domain. Both angles ϕ and θ are always measured in radians.

The name *spherical coordinates* is used because the graph of the equation $\rho = c$ (c is a constant) is a sphere—more precisely, a spherical surface—of radius c centered at the origin. The equation $\phi = c$ (a constant) describes (one nappe of) a cone if $0 < c < \pi/2$ or if $\pi/2 < c < \pi$. The spherical equation of the xy-plane is $\phi = \pi/2$ (Fig. 13.7.8).

From the right triangle OPQ of Fig. 13.7.7, we see that

$$r = \rho \sin \phi \quad \text{and} \quad z = \rho \cos \phi. \tag{5}$$

Indeed, these equations are most easily remembered by visualizing this triangle. Substitution of the equations in (5) into those in (3) yields

$$x = \rho \sin \phi \cos \theta, \qquad y = \rho \sin \phi \sin \theta, \qquad z = \rho \cos \phi. \tag{6}$$

These three equations give the relationship between rectangular and spherical coordinates. Also useful is the formula

$$\rho^2 = x^2 + y^2 + z^2, \tag{7}$$

a consequence of the distance formula.

It is important to note the order in which the spherical coordinates (ρ, ϕ, θ) of a point P are written—first the distance ρ of P from the origin, then the angle ϕ down from the positive z-axis, and last the counterclockwise angle θ around the positive x-axis. You may find this mnemonic device to be helpful: The consonants in the word "raft" remind us, in order, of *r*ho, *f*ee (for phi), and *t*heta.

The following table lists the rectangular coordinates and corresponding spherical coordinates of a few points in space.

(x, y, z)	(ρ, ϕ, θ)
$(1, 0, 0)$	$(1, \pi/2, 0)$
$(0, 1, 0)$	$(1, \pi/2, \pi/2)$
$(0, 0, 1)$	$(1, 0, 0)$
$(0, 0, -1)$	$(1, \pi, 0)$
$(1, 1, \sqrt{2})$	$(2, \pi/4, \pi/4)$
$(-1, -1, \sqrt{2})$	$(2, \pi/4, 5\pi/4)$
$(1, -1, -\sqrt{2})$	$(2, 3\pi/4, 7\pi/4)$
$(1, 1, \sqrt{6})$	$(2\sqrt{2}, \pi/6, \pi/4)$

Given the rectangular coordinates (x, y, z) of the point P, one systematic method for finding the spherical coordinates (ρ, ϕ, θ) of P is this. First we find the cylindrical coordinates r and θ of P with the aid of the triangle in Fig. 13.7.9(a). Then we find ρ and ϕ from the triangle in Fig. 13.7.9(b).

EXAMPLE 3 Find a spherical equation of the paraboloid with rectangular equation $z = x^2 + y^2$.

Solution We substitute $z = \rho \cos \phi$ from Eq. (5) and $x^2 + y^2 = r^2 = \rho^2 \sin^2 \phi$ from Eq. (6). This gives $\rho \cos \phi = \rho^2 \sin^2 \phi$. Cancellation of ρ gives $\cos \phi = \rho \sin^2 \phi$; that is,

$$\rho = \csc \phi \cot \phi$$

is the spherical equation of the paraboloid. We get the whole paraboloid by using ϕ in the range $0 < \phi \leq \pi/2$. Note that $\phi = \pi/2$ gives the point $\rho = 0$ that might otherwise have been lost by canceling ρ.

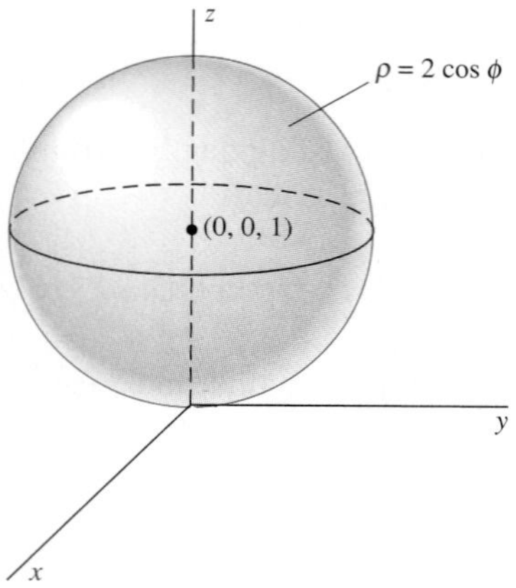

EXAMPLE 4 Determine the graph of the spherical equation $\rho = 2 \cos \phi$.

Solution Multiplication by ρ gives

$$\rho^2 = 2\rho \cos \phi;$$

then substitution of $\rho^2 = x^2 + y^2 + z^2$ and $z = \rho \cos \phi$ yields

$$x^2 + y^2 + z^2 = 2z$$

as the rectangular equation of the graph. Completion of the square in z now gives

$$x^2 + y^2 + (z - 1)^2 = 1,$$

so the graph is a sphere with center $(0, 0, 1)$ and radius 1. It is tangent to the xy-plane at the origin (Fig 13.7.10).

EXAMPLE 5 Determine the graph of the spherical equation $\rho = \sin \phi \sin \theta$.

Solution We first multiply each side by ρ and get $\rho^2 = \rho \sin \phi \sin \theta$. We then use Eqs. (6) and (7) and find that $x^2 + y^2 + z^2 = y$. This is a rectangular equation of a sphere with center $(0, \frac{1}{2}, 0)$ and radius $\frac{1}{2}$.

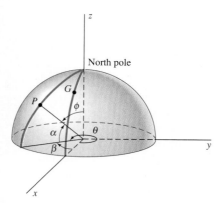

Fig. 13.7.10 The sphere of Example 4

*LATITUDE AND LONGITUDE

A **great circle** of a spherical surface is a circle formed by the intersection of the surface with a plane through the center of the sphere. Thus a great circle of a spherical surface is a circle (in the surface) that has the same radius as the sphere. Therefore, a great circle is a circle of maximum possible circumference that lies on the sphere. It's easy to see that any two points on a spherical surface lie on a great circle (uniquely determined unless the two points lie on the ends of a diameter of the sphere). In the calculus of variations, it is shown that the shortest distance between two such points—measured along the curved surface—is the shorter of the two arcs of the great circle that contains them. The surprise is that the *shortest* distance is found by using the *largest* circle.

The spherical coordinates ϕ and θ are closely related to the latitude and longitude of points on the surface of the earth. Assume that the earth is a sphere with radius $\rho = 3960$ mi. We begin with the **prime meridian** (a **meridian** is a great semicircle connecting the North and South Poles) through Greenwich, England, just outside London. This is the point marked G in Fig. 13.7.11.

Fig. 13.7.11 The relations among latitude, longitude, and spherical coordinates

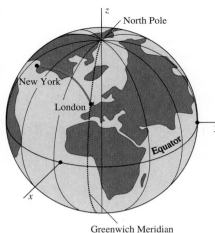

Fig. 13.7.12 Finding the great-circle distance d from New York to London (Example 6)

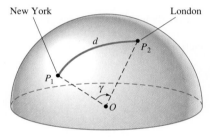

Fig. 13.7.13 The great-circle arc between New York and London (Example 6)

We take the z-axis through the North Pole and the x-axis through the point where the prime meridian intersects the equator. The **latitude** α and (west) **longitude** β of a point P in the Northern Hemisphere are given by the equations

$$\alpha = 90° - \phi° \quad \text{and} \quad \beta = 360° - \theta°, \tag{8}$$

where $\phi°$ and $\theta°$ are the angular spherical coordinates, measured in *degrees,* of P. (That is, $\phi°$ and $\theta°$ denote the degree equivalents of the angles ϕ and θ, respectively, which are measured in radians unless otherwise specified.) Thus the latitude α is measured northward from the equator and the longitude β is measured westward from the prime meridian.

EXAMPLE 6 Find the great-circle distance between New York (latitude 40.75° north, longitude 74° west) and London (latitude 51.5° north, longitude 0°). See Fig. 13.7.12.

Solution From the equations in (8) we find that $\phi° = 49.25°$, $\theta° = 286°$ for New York, whereas $\phi° = 38.5°$, $\theta° = 360°$ (or $0°$) for London. Hence the angular spherical coordinates of New York are $\phi = (49.25/180)\pi$, $\theta = (286/180)\pi$, and those of London are $\phi = (38.5/180)\pi$, $\theta = 0$. With these values of ϕ and θ and with $\rho = 3960$ (mi), the equations in (6) give the rectangular coordinates

New York: $P_1(826.90, -2883.74, 2584.93)$

and

London: $P_2(2465.16, 0.0, 3099.13)$.

The angle γ between the radius vectors $\mathbf{u} = \overrightarrow{OP_1}$ and $\mathbf{v} = \overrightarrow{OP_2}$ in Fig. 13.7.13 satisfies the equation

$$\begin{aligned}
\cos \gamma &= \frac{\mathbf{u} \cdot \mathbf{v}}{|\mathbf{u}||\mathbf{v}|} \\
&= \frac{826.90 \cdot 2465.16 - 2883.74 \cdot 0 + 2584.93 \cdot 3099.13}{(3960)^2} \approx 0.641.
\end{aligned}$$

Thus γ is approximately 0.875 (rad). Hence the great-circle distance between New York and London is close to

$$d = 3960 \cdot 0.875 \approx 3465 \quad \text{(mi)},$$

about 5576 km.

13.7 Problems

In Problems 1 through 10 find both the cylindrical coordinates and the spherical coordinates of the point P with the given rectangular coordinates.

1. $P(0, 0, 5)$ **2.** $P(0, 0, -3)$

3. $P(1, 1, 0)$ **4.** $P(2, -2, 0)$

5. $P(1, 1, 1)$ **6.** $P(-1, 1, -1)$

7. $P(2, 1, -2)$ **8.** $P(-2, -1, -2)$

9. $P(3, 4, 12)$ **10.** $P(-2, 4, -12)$

In Problems 11 through 24, describe the graph of the given equation.

11. $r = 5$ **12.** $\theta = 3\pi/4$ **13.** $\theta = \pi/4$

14. $\rho = 5$ **15.** $\phi = \pi/6$ **16.** $\phi = 5\pi/6$

17. $\phi = \pi/2$ **18.** $\phi = \pi$

19. $r = 2 \sin \theta$ **20.** $\rho = 2 \sin \phi$

21. $\cos \theta + \sin \theta = 0$ **22.** $z = 10 - 3r^2$

23. $\rho \cos \phi = 1$ **24.** $\rho = \cot \phi$

In Problems 25 through 30, convert the given equation both to cylindrical and to spherical coordinates.

25. $x^2 + y^2 + z^2 = 25$
26. $x^2 + y^2 = 2x$
27. $x + y + z = 1$
28. $x + y = 4$
29. $x^2 + y^2 + z^2 = x + y + z$
30. $z = x^2 - y^2$

31. The parabola $z = x^2$, $y = 0$ is rotated about the z-axis. Write a cylindrical equation for the surface thereby generated.

32. The hyperbola $y^2 - z^2 = 1$, $x = 0$ is rotated about the z-axis. Write a cylindrical equation for the surface thereby generated.

33. A sphere of radius 2 is centered at the origin. A hole of radius 1 is drilled through the sphere, with the axis of the hole lying on the z-axis. Describe the solid region that remains (Fig. 13.7.14) in (a) cylindrical coordinates; (b) spherical coordinates.

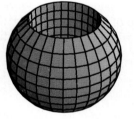

Fig. 13.7.14 The sphere-with-hole of Problem 33

34. Find the great-circle distance in miles and in kilome-

ters from Atlanta (latitude 33.75° north, longitude 84.40° west) to San Francisco (latitude 37.78° north, longitude 122.42° west).

35. Find the great-circle distance in miles and in kilometers from Fairbanks (latitude 64.80° north, longitude 147.85° west) to St. Petersburg, Russia (formerly Leningrad, U.S.S.R., latitude 59.91° north, longitude 30.43° *east* of Greenwich—alternatively, longitude 329.57° west).

36. Because Fairbanks and St. Petersburg, Russia (see Problem 35), are at almost the same latitude, a plane could fly from one to the other roughly along the 62nd parallel of latitude. Accurately estimate the length of such a trip both in kilometers and in miles.

37. In flying the great-circle route from Fairbanks to St. Petersburg, Russia (see Problem 35), how close in kilometers and in miles to the North Pole would a plane fly?

38. The vertex of a right circular cone of radius R and height H is located at the origin, and its axis lies on the nonnegative z-axis. Describe the solid cone in cylindrical coordinates.

39. Describe the cone of Problem 38 in spherical coordinates.

40. In flying the great-circle route from New York to London (Example 6), an airplane initially flies generally east-northeast. Does the plane ever fly at a latitude *higher* than that of London? [*Suggestion:* Express the z-coordinate of the plane's route as a function of x, and then maximize z.]

13.7 Project

A surface of revolution around the z-axis is readily described in cylindrical coordinates. The surface obtained by revolving the curve $r = f(z)$, $a \leq z \leq b$, around the z-axis (Fig. 13.7.15) is given by

$$x = f(z) \cos \theta, \qquad y = f(z) \sin \theta, \qquad z = z$$

for $0 \leq \theta \leq 2\pi$.

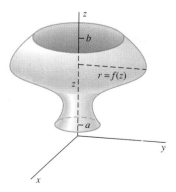

Fig. 13.7.15 A surface of revolution around the z-axis

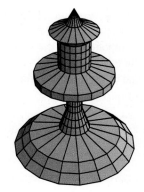

Fig. 13.7.16 The chess piece

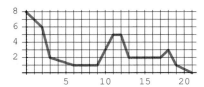

Fig. 13.7.17 The construction curve $r = f(z)$ used to generate the chess piece of Fig. 13.7.16

The chess piece shown in Fig. 13.7.16 was created by revolving the polygonal curve of Fig. 13.7.17 around the (vertical) z-axis in space. The first step in the construction was to draw the desired curve $r = f(z)$ on a piece of graph paper. The next step was to measure the coordinates $(0, 8)$, $(2, 6)$, $(3, 2)$, ..., $(18, 3)$, $(19, 1)$, and $(21, 0)$ of the vertices. Finally, the point-slope formula was used to define $r = f(z)$ on successive intervals corresponding to straight-line segments on the polygonal curve:

$$r = 8 - z \qquad \text{on } [0, 2],$$
$$r = 14 - 4z \qquad \text{on } [2, 3],$$
$$\vdots$$
$$r = 5 \qquad \text{on } [11, 12],$$
$$\vdots$$
$$r = 39 - 2z \qquad \text{on } [18, 19],$$
$$r = \tfrac{1}{2}(21 - z) \qquad \text{on } [19, 21].$$

A "construction curve" like that in Fig. 13.7.17 can be used as input to an automatic lathe, which will then follow it as a pattern for cutting the corresponding solid of revolution from a cylinder of wood or metal. Some years ago, many first-year engineering students had to produce a metal lamp stem in this fashion—from the original graph paper design to the actual operation of a manual lathe.

Your assignment in this project is simpler: Merely produce a *figure* showing such a "lathe object." Think of a lamp stem, a table leg, or perhaps a more complicated chess piece than the pawn shown in Fig. 13.7.16. Once you have defined the construction curve $f = f(z)$, generate the figure by the *Mathematica* command

```
ParametricPlot3D[ { f[z] Cos[t], f[z] Sin[t], z },
                  {z, a, b}, {t, 0, 2 Pi} ]
```

or the *Maple* command

```
plot3d( [f(z) cos(t), f(z) sin(t), z],
        z = a..b, t = 0..2Pi );
```

If you don't have access to a computer system that plots parametric surfaces, you can still draw a lathe object corresponding to a *polygonal* construction curve. Begin by drawing horizontal circles that correspond to the vertices (as ellipses in perspective).

Chapter 13 Review: DEFINITIONS, CONCEPTS, RESULTS

Use the following list as a guide to concepts that you may need to review.

1. Properties of addition of vectors in space and of multiplication of vectors by scalars

2. The dot (scalar) product of vectors—definition and geometric interpretation

3. Use of the dot product to test perpendicularity of vectors and, more generally, finding the angle between two vectors

4. The vector (cross) product of two vectors—definition and geometric interpretation

5. The scalar triple product of three vectors—definition and geometric interpretation

6. The parametric and symmetric equations of the straight line that passes through a given point and is parallel to a given vector

7. The equation of the plane through a given point normal to a given vector

8. The velocity and acceleration vectors of a particle moving along a parametric space curve

9. Arc length of a parametric space curve

10. The curvature, unit tangent vector, and principal unit normal vector of a parametric curve in the plane or in space

11. Tangential and normal components of the acceleration vector of a parametric curve

12. Equations of cylinders and of surfaces of revolution

13. The standard examples of quadric surfaces

14. Definition of the cylindrical-coordinate and spherical-coordinate systems, and the equations relating cylindrical and spherical coordinates to rectangular coordinates

Chapter 13 Miscellaneous Problems

1. Suppose that M is the midpoint of the segment PQ in space and that A is another point. Show that

$$\overrightarrow{AM} = \tfrac{1}{2}(\overrightarrow{AP} + \overrightarrow{AQ}).$$

2. Let $\mathbf{a}$ and $\mathbf{b}$ be nonzero vectors. Define

$$\mathbf{a}_\| = (\text{comp}_b\ \mathbf{a})\,\frac{\mathbf{b}}{|\mathbf{b}|} \quad \text{and} \quad \mathbf{a}_\perp = \mathbf{a} - \mathbf{a}_\|.$$

Prove that $\mathbf{a}_\perp$ is perpendicular to $\mathbf{b}$.

3. Let P and Q be different points in space. Show that the point R lies on the line through P and Q *if and only if* there exist numbers a and b such that $a + b = 1$ and $\overrightarrow{OR} = a\,\overrightarrow{OP} + b\,\overrightarrow{OQ}$. Conclude that

$$\mathbf{r}(t) = t\,\overrightarrow{OP} + (1 - t)\overrightarrow{OQ}$$

is a parametric equation of this line.

4. Conclude from the result of Problem 3 that the points P, Q, and R are collinear if and only if there exist numbers a, b, and c, not all zero, such that $a + b + c = 0$ and $a\,\overrightarrow{OP} + b\,\overrightarrow{OQ} + c\,\overrightarrow{OR} = \mathbf{0}$.

5. Let $P(x_0, y_0)$, $Q(x_1, y_1)$, and $R(x_2, y_2)$ be points in the xy-plane. Use the vector product to show that the area of the triangle PQR is

$$A = \tfrac{1}{2}\left|(x_1 - x_0)(y_2 - y_0) - (x_2 - x_0)(y_1 - y_0)\right|.$$

6. Write both symmetric and parametric equations of the line that passes through $P_1(1, -1, 0)$ and is parallel to $\mathbf{v} = \langle 2, -1, 3 \rangle$.

7. Write both symmetric and parametric equations of the line that passes through $P_1(1, -1, 2)$ and $P_2(3, 2, -1)$.

8. Write an equation of the plane through $P(3, -5, 1)$ with normal vector $\mathbf{n} = \mathbf{i} + \mathbf{j}$.

9. Show that the lines with symmetric equations

$$x - 1 = 2(y + 1) = 3(z - 2)$$

and

$$x - 3 = 2(y - 1) = 3(z + 1)$$

are parallel. Then write an equation of the plane through these two lines.

10. Let the lines L_1 and L_2 have symmetric equations

$$\frac{x - x_i}{a_i} = \frac{y - y_i}{b_i} = \frac{z - z_i}{c_i}$$

for $i = 1, 2$. Show that L_1 and L_2 are skew lines if and only if

$$\begin{vmatrix} x_1 - x_2 & y_1 - y_2 & z_1 - z_2 \\ a_1 & b_1 & c_1 \\ a_2 & b_2 & c_2 \end{vmatrix} \neq 0.$$

11. Given the four points $A(2, 3, 2)$, $B(4, 1, 0)$, $C(-1, 2, 0)$, and $D(5, 4, -2)$, find an equation of the plane that passes through A and B and is parallel to the line through C and D.

12. Given the points A, B, C, and D of Problem 11, find points P on the line AB and Q on the line CD such that the line PQ is perpendicular both to AB and to CD. What is the perpendicular distance d between the lines AB and CD?

13. Let $P_0(x_0, y_0, z_0)$ be a point of the plane with equation

$$ax + by + cz + d = 0.$$

By projecting $\overrightarrow{OP_0}$ onto the normal vector $\mathbf{n} = \langle a, b, c \rangle$, show that the distance D from the origin to this plane is

$$D = \frac{|d|}{\sqrt{a^2 + b^2 + c^2}}.$$

14. Show that the distance D from the point $P_1(x_1, y_1, z_1)$ to the plane $ax + by + cz + d = 0$ is equal to the distance from the origin to the plane with equation

$$a(x + x_1) + b(y + y_1) + c(z + z_1) + d = 0.$$

Hence conclude from the result of Problem 13 that

$$D = \frac{|ax_1 + by_1 + cz_1 + d|}{\sqrt{a^2 + b^2 + c^2}}.$$

15. Find the perpendicular distance between the parallel planes $2x - y + 2z = 4$ and $2x - y + 2z = 13$.

16. Write an equation of the plane through the point $(1, 1, 1)$ that is normal to the twisted cubic $x = t$, $y = t^2$, $z = t^3$ at this point.

17. A particle moves in space with parametric equations $x = t$, $y = t^2$, $z = \frac{4}{3}t^{3/2}$. Find the curvature of its trajectory and the tangential and normal components of its acceleration when $t = 1$.

18. The **osculating plane** to a space curve at a point P of that curve is the plane through P that is parallel to the curve's unit tangent and principal unit normal vectors at P. Write an equation of the osculating plane to the curve of Problem 17 at the point $(1, 1, \frac{4}{3})$.

19. Show that the equation of the plane that passes through the point $P_0 (x_0, y_0, z_0)$ and is parallel to the vectors $\mathbf{v}_1 = \langle a_1, b_1, c_1 \rangle$ and $\mathbf{v}_2 = \langle a_2, b_2, c_2 \rangle$ can be written in the form

$$\begin{vmatrix} x - x_0 & y - y_0 & z - z_0 \\ a_1 & b_1 & c_1 \\ a_2 & b_2 & c_2 \end{vmatrix} = 0.$$

20. Deduce from Problem 19 that the equation of the osculating plane (Problem 18) to the parametric curve $\mathbf{r}(t)$ at the point $\mathbf{r}(t_0)$ can be written in the form

$$[\mathbf{R} - \mathbf{r}(t_0)] \cdot [\mathbf{r}'(t_0) \times \mathbf{r}''(t_0)] = 0,$$

where $\mathbf{R} = \langle x, y, z \rangle$. Note first that the vectors $\mathbf{T}$ and $\mathbf{N}$ are coplanar with $\mathbf{r}'(t)$ and $\mathbf{r}''(t)$.

21. Use the result of Problem 20 to write an equation of the osculating plane to the twisted cubic $x = t$, $y = t^2$, $z = t^3$ at the point $(1, 1, 1)$.

22. Let a parametric curve in space be described by equations $r = r(t)$, $\theta = \theta(t)$, $z = z(t)$ that give the cylindrical coordinates of a moving point on the curve for $a \leqq t \leqq b$. Use the equations relating rectangular and cylindrical coordinates to show that the arc length of the curve is

$$s = \int_a^b \left[\left(\frac{dr}{dt} \right)^2 + \left(r \frac{d\theta}{dt} \right)^2 + \left(\frac{dz}{dt} \right)^2 \right]^{1/2} dt.$$

23. A point moves on the *unit* sphere $\rho = 1$ with its spherical angular coordinates at time t given by $\phi = \phi(t)$, $\theta = \theta(t)$, $a \leqq t \leqq b$. Use the equations relating rectangular and spherical coordinates to show that the arc length of its path is

$$s = \int_a^b \left[\left(\frac{d\phi}{dt} \right)^2 + (\sin^2 \phi) \left(\frac{d\theta}{dt} \right)^2 \right]^{1/2} dt.$$

24. The vector product $\mathbf{B} = \mathbf{T} \times \mathbf{N}$ of the unit tangent vector and the principal unit normal curve is the **unit binormal vector** $\mathbf{B}$ of a curve. (a) Differentiate $\mathbf{B} \cdot \mathbf{T} = 0$ to show that $\mathbf{T}$ is perpendicular to $d\mathbf{B}/ds$. (b) Differentiate $\mathbf{B} \cdot \mathbf{B} = 1$ to show that $\mathbf{B}$ is perpendicular to $d\mathbf{B}/ds$. (c) Conclude from (a) and (b) that $d\mathbf{B}/ds = -\tau \mathbf{N}$ for some number τ. Called the **torsion** of the curve, τ measures the amount that the curve twists at each point in space.

25. Show that the torsion of the helix of Example 1 in Section 13.4 is constant by showing that its value is

$$\tau = \frac{b\omega}{a^2 \omega^2 + b^2}.$$

26. Deduce from the definition of torsion (Problem 24) that $\tau \equiv 0$ for any curve such that $\mathbf{r}(t)$ lies in a fixed plane.

27. Write an equation in spherical coordinates of the sphere with radius 1 and center $x = 0 = y$, $z = 1$.

28. Let C be the circle in the yz-plane with radius 1 and center $y = 1$, $z = 0$. Write equations in both rectangular and cylindrical coordinates of the surface obtained by revolving C around the z-axis.

29. Let C be the curve in the yz-plane with equation $(y^2 + z^2)^2 = 2(z^2 - y^2)$. Write an equation in spherical coordinates of the surface obtained by revolving this curve around the z-axis. Then sketch this surface. [*Suggestion:* Remember that $r^2 = 2 \cos 2\theta$ is the polar equation of a figure-eight curve.]

30. Let A be the area of the parallelogram in space determined by the vectors $\mathbf{a} = \overrightarrow{PQ}$ and $\mathbf{b} = \overrightarrow{RS}$. Let A' be the area of the perpendicular projection of $PQRS$ into a plane that makes an acute angle γ with the plane of $PQRS$. Assuming that $A' = A \cos \gamma$ in such a situation, prove that the areas of the perpendicular projections of the parallelogram $PQRS$ into the three coordinate planes are

$$|\mathbf{i} \cdot (\mathbf{a} \times \mathbf{b})|, \qquad |\mathbf{j} \cdot (\mathbf{a} \times \mathbf{b})|, \quad \text{and} \quad |\mathbf{k} \cdot (\mathbf{a} \times \mathbf{b})|.$$

Conclude that the square of the area of a parallelogram in space is equal to the sum of the squares of the areas of its perpendicular projections into the three coordinate planes.

31. Take $\mathbf{a} = \langle a_1, a_2, a_3 \rangle$ and $\mathbf{b} = \langle b_1, b_2, b_3 \rangle$ in Problem 30. Show that

$$A^2 = \begin{vmatrix} a_2 & a_3 \\ b_2 & b_3 \end{vmatrix}^2 + \begin{vmatrix} a_3 & a_1 \\ b_3 & b_1 \end{vmatrix}^2 + \begin{vmatrix} a_1 & a_2 \\ b_1 & b_2 \end{vmatrix}^2.$$

32. Let C be a curve in a plane $\mathcal{P}$ that is not parallel to the z-axis. Suppose that the projection of C into the xy-plane is an ellipse. Introduce uv-coordinates into the plane $\mathcal{P}$ to prove that the curve C is itself an ellipse.

33. Conclude from Problem 32 that the intersection of a nonvertical plane and an elliptic cylinder with vertical sides is an ellipse.

34. Use the result of Problem 32 to prove that the intersection of the plane $z = Ax + By$ and the paraboloid $z = a^2x^2 + b^2y^2$ is either empty, a single point, or an ellipse.

35. Use the result of Problem 32 to prove that the intersection of the plane $z = Ax + By$ and the ellipsoid $x^2/a^2 + y^2/b^2 + z^2/c^2 = 1$ is either empty, a single point, or an ellipse.

36. Suppose that $y = f(x)$ is the graph of a function f for which f'' is continuous, and suppose also that the graph has an inflection point at $(a, f(a))$. Prove that the curvature of the graph at $x = a$ is zero.

37. Find the points on the curve $y = \sin x$ where the curvature is maximal and those where it is minimal.

38. The right branch of the hyperbola $x^2 - y^2 = 1$ may be parametrized by $x(t) = \cosh t$, $y(t) = \sinh t$. Find the point where its curvature is minimal.

39. Find the vectors $\mathbf{N}$ and $\mathbf{T}$ at the point of the curve $x(t) = t \cos t$, $y(t) = t \sin t$ that corresponds to $t = \pi/2$.

40. Find the points on the ellipse $x^2/a^2 + y^2/b^2 = 1$ (with $a > b$) where the curvature is maximal and those where it is minimal.

41. Suppose that the plane curve $r = f(\theta)$ is given in polar coordinates. Write r' for $f'(\theta)$ and r'' for $f''(\theta)$. Show that its curvature is given by

$$\kappa = \frac{|r^2 + 2(r')^2 - rr''|}{[r^2 + (r')^2]^{3/2}}.$$

42. Use the formula of Problem 41 to calculate the curvature $\kappa(\theta)$ at the point (r, θ) of the spiral of Archimedes with equation $r = \theta$. Then show that $\kappa(\theta) \to 0$ as $\theta \to +\infty$.

43. A railway curve must join two straight tracks, one extending due west from $(-1, -1)$ and the other extending due east from $(1, 1)$. Determine A, B, and C so that the curve $y = Ax + Bx^3 + Cx^5$ joins $(-1, -1)$ and $(1, 1)$ and so that the slope and curvature of this connecting curve are zero at both its endpoints.

44. A plane passing through the origin and not parallel to any coordinate plane has an equation of the form $Ax + By + Cz = 0$ and intersects the spherical surface $x^2 + y^2 + z^2 = R^2$ in a great circle. Find the highest point on this great circle; that is, find the coordinates of the point with the largest z-coordinate.

Partial Differentiation

Joseph Louis Lagrange (1736–1813)

❑ Joseph Louis Lagrange is remembered for his great treatises on analytical mechanics and on the theory of functions that summarized much of eighteenth-century pure and applied mathematics. These treatises—*Mécanique analytique* (1788), *Théorie des fonctions analytiques* (1797), and *Leçons sur le calcul des fonctions* (1806)—systematically developed and applied widely the differential and integral calculus of multivariable functions expressed in terms of the rectangular coordinates x, y, z in three-dimensional space. They were written and published in Paris during the last quarter-century of Lagrange's career. But he grew up and spent his first 30 years in Turin, Italy. His father pointed Lagrange toward the law, but by age 17 Lagrange had decided on a career in science and mathematics. Based on his early work in celestial mechanics (the mathematical analysis of the motions of the planets and satellites in our solar system), Lagrange in 1766 succeeded Leonhard Euler as director of the Berlin Academy in Germany.

❑ Lagrange regarded his far-reaching work on maximum–minimum problems as his best work in mathematics. This work, which continued throughout his long career, dated back to a letter to Euler that Lagrange wrote from Turin when he was only 19. This letter outlined a new approach to a certain class of optimization problems that comprise the calculus of variations. A typical example is the *isoperimetric problem*, which asks what curve of a given arc length encloses a plane region with the greatest area. (The answer: a circle.) In the *Mécanique analytique*, Lagrange applied his "method of multipliers" to investigate the motion of a particle in space that is constrained to move on a surface defined by an equation of the form

$$g(x, y, z) = 0.$$

Section 14.9 applies the Lagrange multiplier method to the problem of maximizing or minimizing a function $f(x, y, z)$ subject to a "constraint" of the form

$$g(x, y, z) = 0.$$

Today this method has applications that range from minimizing the fuel required for a spacecraft to achieve its desired trajectory to maximizing the productivity of a commercial enterprise limited by the availability of financial, natural, and personnel resources.

❑ Modern scientific visualization often employs computer graphic techniques to present different interpretations of the same data simultaneously in a single figure. This *MATLAB* 4.0 color graphic shows both a mesh graph of a surface

$$z = f(x, y)$$

and a color-coded contour map of the surface. In Section 14.5 we learn how to locate multivariable maximum–minimum points like those visible on this surface.

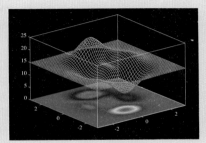

(Image created in MATLAB, courtesy of the MathWorks, Inc., Natick MA.)

14.1
Introduction

We turn our attention here and in Chapters 15 and 16 to the calculus of functions of more than one variable. Many real-world functions depend on two or more variables. For example:

❑ In physical chemistry the ideal gas law $pV = nRT$ (where n and R are constants) is used to express any one of the variables p, V, and T as a function of the other two.

❑ The altitude above sea level at a particular location on the earth's surface depends on the latitude and longitude of the location.

❑ A manufacturer's profit depends on sales, overhead costs, the cost of each raw material used, and, in some cases, additional variables.

❑ The amount of usable energy a solar panel can gather depends on its efficiency, its angle of inclination to the sun's rays, the angle of elevation of the sun above the horizon, and other factors.

A typical application may call for us to find an extreme value of a function of several variables. For example, suppose that we want to minimize the cost of making a rectangular box with a volume of 48 ft³, given that its front and back cost \$1/ft², its top and bottom cost \$2/ft², and its two ends cost \$3/ft². Figure 14.1.1 shows such a box of length x, width y, and height z. Under the conditions given, its total cost will be

$$C = 2xz + 4xy + 6yz \qquad \text{(dollars)}.$$

But x, y, and z are not independent variables, because the box has fixed volume

$$V = xyz = 48.$$

We eliminate z, for instance, from the first formula by using the second; because $z = 48/(xy)$, the cost we want to minimize is given by

$$C = 4xy + \frac{288}{x} + \frac{96}{y}.$$

Because neither of the variables x and y can be expressed in terms of the other, the single-variable maximum-minimum techniques of Chapter 3 cannot be applied here. We need new optimization techniques applicable to functions of two or more independent variables. In Section 14.5 we shall return to this problem.

The problem of optimization is merely one example. We shall see in this chapter that all the main ingredients of single-variable differential calculus—limits, derivatives and rates of change, chain rule computations, and maximum-minimum techniques—can be generalized to functions of two or more variables.

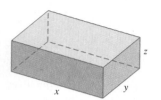

Fig. 14.1.1 A box whose total cost we want to minimize

14.2
Functions of Several Variables

Recall from Section 1.1 that a real-valued *function* is a rule or correspondence f that associates a unique real number with each element of a set D. The set D is called the *domain of definition* of the function f. The domain D has always been a subset of the real line for the functions of a single variable that we have studied up to this point. If D is a subset of the plane, then f is a function of *two* variables—for, given a point P of D, we naturally associate with P its rectangular coordinates (x, y).

> **Definition** *Functions of Two or Three Variables*
>
> A **function of two variables,** defined on the **domain** D in the plane, is a rule f that associates with each point (x, y) in D a real number, denoted by $f(x, y)$. A **function of three variables,** defined on the **domain** D in space, is a rule f that associates with each point (x, y, z) in D a real number $f(x, y, z)$.

We can typically define a function f of two (or three) variables by giving a formula that specifies $f(x, y)$ in terms of x and y (or $f(x, y, z)$ in terms of x, y, and z). In case the domain D of f is not explicitly specified, we take D to consist of all points for which the given formula is meaningful.

EXAMPLE 1 The domain of the function f with formula $f(x, y) = \sqrt{25 - x^2 - y^2}$ is the set of all (x, y) such that $25 - x^2 - y^2 \geqq 0$—that is, the circular disk $x^2 + y^2 \leqq 25$ of radius 5 centered at the origin. Similarly, the function g defined as

$$g(x, y, z) = \frac{x + y + z}{\sqrt{x^2 + y^2 + z^2}}$$

is defined at all points in space where $x^2 + y^2 + z^2 > 0$. Thus its domain consists of all points in three-dimensional space $\mathbf{R}^3$ other than the origin $(0, 0, 0)$.

EXAMPLE 2 Find the domain of definition of the function with formula

$$f(x, y) = \frac{y}{\sqrt{x - y^2}}. \tag{1}$$

Find also the points (x, y) at which $f(x, y) = \pm 1$.

Solution For $f(x, y)$ to be defined, the *radicand* $x - y^2$ must be positive—that is, $y^2 < x$. Hence the domain of f is the set of points lying strictly to the right of the parabola $x = y^2$. This domain is shaded in Fig. 14.2.1. The parabola in the figure is dotted to indicate that it is not included in the domain of f; any point for which $x = y^2$ would entail division by zero in Eq. (1). The function $f(x, y)$ has the value ± 1 whenever

$$\frac{y}{\sqrt{x - y^2}} = \pm 1;$$

that is, when $y^2 = x - y^2$, so $x = 2y^2$. Thus $f(x, y) = \pm 1$ at each point of the parabola $x = 2y^2$ [other than its vertex $(0, 0)$, which is not included in the domain of f]. This parabola is shown as a solid curve in Fig. 14.2.1.

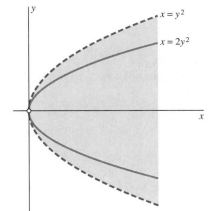

Fig. 14.2.1 The domain of $f(x, y) = \dfrac{y}{\sqrt{x - y^2}}$ (Example 2)

In a geometric, physical, or economic situation, a function typically results from expressing one descriptive variable in terms of others. As we saw in Section 14.1, the cost C of the box discussed there was given by the formula

$$C = 4xy + \frac{288}{x} + \frac{96}{y}$$

in terms of the length x and width y of the box. The value C of this function is a variable that depends on the values of x and y. Hence we call C a **dependent variable,** whereas x and y are **independent variables.** And if the temperature T at the point (x, y, z) in space is given by some formula $T = h(x, y, z)$, then the dependent variable T is a function of the three independent variables x, y, and z.

We can define a function of four or more variables by giving a formula that includes the approximate number of independent variables. For example, if an amount A of heat is released at the origin in space at time $t = 0$ in a medium with thermal diffusivity k, then—under appropriate conditions—the temperature T at the point (x, y, z) at time $t > 0$ is given by

$$T(x, y, z, t) = \frac{A}{(4\pi kt)^{3/2}} \exp\left(-\frac{x^2 + y^2 + z^2}{4kt}\right).$$

This formula gives the temperature T as a function of the four independent variables x, y, z, and t.

We shall see that the main differences between single-variable and multivariable calculus show up when only two independent variables are involved. Hence most of our results will be stated in terms of functions of two variables. Many of these results readily generalize by analogy to the case of three or more independent variables.

GRAPHS AND LEVEL CURVES

We can visualize how a function f of two variables x and y "works" in terms of its graph. The **graph** of f is the graph of the equation $z = f(x, y)$. Thus the graph of f is the set of all points in space with coordinates (x, y, z) that satisfy the equation $z = f(x, y)$ (Fig. 14.2.2).

You saw several examples of such graphs in Chapter 13. For example, the graph of the function $f(x, y) = x^2 + y^2$ is the paraboloid $z = x^2 + y^2$ shown in Fig. 14.2.3. The graph of the function

$$g(x, y) = c\sqrt{1 - \left(\frac{x}{a}\right)^2 - \left(\frac{y}{b}\right)^2}$$

is the *upper half* of the ellipsoid with equation $x^2/a^2 + y^2/b^2 + z^2/c^2 = 1$ (Fig. 14.2.4). In general, the graph of a function of two variables is a surface that lies above (or below, or both) its domain D in the xy-plane.

The intersection of the horizontal plane $z = k$ with the surface $z = f(x, y)$ is called the **contour curve** of height k on the surface (Fig.

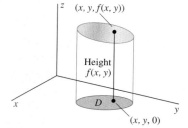

Fig. 14.2.2 The graph of a function of two variables is typically a surface "over" the domain of the function.

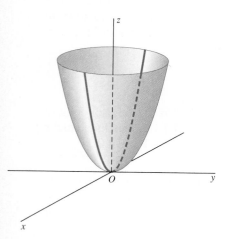

Fig. 14.2.3 The paraboloid is (part of) the graph of the function $f(x, y) = x^2 + y^2$.

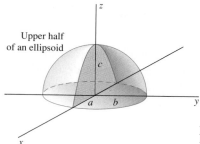

Fig. 14.2.4 The upper half of an ellipsoid is the graph of a function of two variables.

756

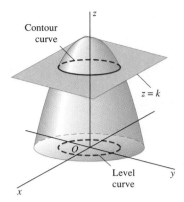

Contour
curve

$z = k$

O

Level
curve

Fig. 14.2.5 A contour curve and the corresponding level curve

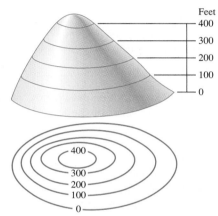

Feet
400
300
200
100
0

400
300
200
100
0

Fig. 14.2.7 Contour curves and level curves for a hill

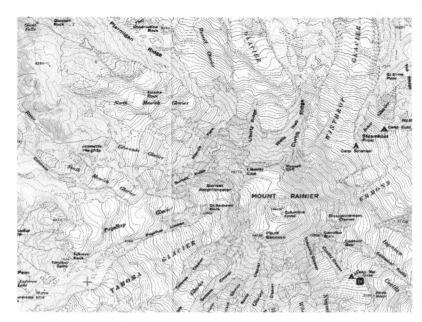

Fig. 14.2.6 The area around Mt. Rainier, Washington, showing level curves at 400-ft intervals

14.2.5). The vertical projection of this contour curve into the xy-plane is the **level curve** $f(x, y) = k$ of the function f. The level curves of f are simply the sets on which the value of f is constant. On a topographic map, such as the one in Fig. 14.2.6, the level curves are curves of constant height above sea level.

Level curves give a two-dimensional way of representing a three-dimensional surface $z = f(x, y)$, just as the two-dimensional map in Fig. 14.2.6 represents a three-dimensional mountain. We do this by drawing typical level curves of $z = f(x, y)$ in the xy-plane, labeling each with the corresponding (constant) value of z. Figure 14.2.7 illustrates this process for a simple hill.

EXAMPLE 3 Figure 14.2.8 shows some typical contour curves on the paraboloid $z = 25 - x^2 - y^2$. Figure 14.2.9 shows the corresponding level curves.

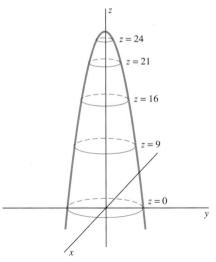

$z = 24$
$z = 21$
$z = 16$
$z = 9$
$z = 0$

Fig. 14.2.8 Contour curves on $z = 25 - x^2 - y^2$ (Example 3)

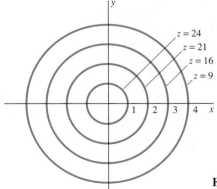

$z = 24$
$z = 21$
$z = 16$
$z = 9$

1 2 3 4 x

Fig. 14.2.9 Level curves of $f(x, y) = 25 - x^2 - y^2$ (Example 3)

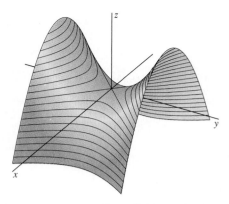

Fig. 14.2.10 Contour curves on $z = y^2 - x^2$ (Example 4)

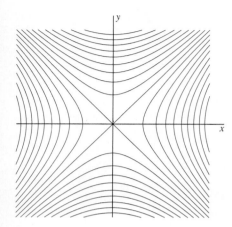

Fig. 14.2.11 Level curves of $f(x, y) = y^2 - x^2$ (Example 4)

EXAMPLE 4 Figure 14.2.10 shows contour curves on the hyperbolic paraboloid $z = y^2 - x^2$. Figure 14.2.11 shows the corresponding level curves of the function $f(x, y) = y^2 - x^2$. If $z = k > 0$, then $y^2 - x^2 = k$ is a hyperbola opening along the y-axis; if $k < 0$, it opens along the x-axis. The level curve for which $k = 0$ consists of the two straight lines $y = x$ and $y = -x$.

The graph of a function $f(x, y, z)$ of three variables cannot be drawn in three dimensions, but we can readily visualize its **level surfaces** of the form $f(x, y, z) = k$. For example, the level surfaces of the function $f(x, y, z) = x^2 + y^2 + z^2$ are spheres centered at the origin. Thus the level surfaces of f are the sets in space on which the value $f(x, y, z)$ is constant.

If the function f gives the temperature at the location (x, y) or (x, y, z), then its level curves or surfaces are called **isotherms.** A weather map typically includes level curves of the ground-level atmospheric pressure; these are called **isobars.** Even though you may be able to construct the graph of a function of two variables, that graph might be so complicated that information about the function (or the situation it describes) is obscure. Frequently the level curves themselves give more information, as in weather maps. For example, Fig. 14.2.12 shows level curves for the annual numbers of days of

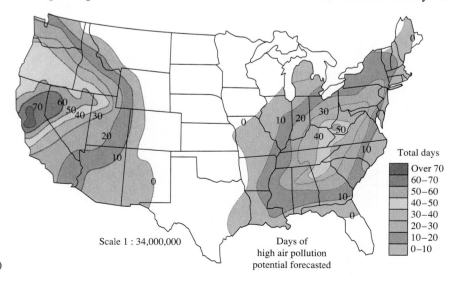

Fig. 14.2.12 Days of high air pollution forecast in the United States (from National Atlas of the United States, U.S. Department of the Interior, 1970)

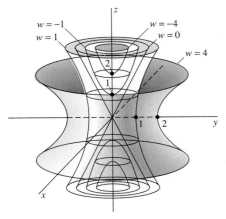

Fig. 14.2.13 Some level surfaces of the function $w = f(x, y, z) = x^2 + y^2 - z^2$ (Example 5)

high air pollution forecast at different localities in the United States. The scale of this figure does not show local variations caused by individual cities. But a glance indicates that western Colorado, southern Georgia, and central Illinois all expect the same number (10, in this case) of high-pollution days each year.

EXAMPLE 5   Figure 14.2.13 shows some level surfaces of the function

$$f(x, y, z) = x^2 + y^2 - z^2. \qquad (2)$$

If $k > 0$, then the graph of $x^2 + y^2 - z^2 = k$ is a hyperboloid of one sheet, whereas if $k < 0$, it is a hyperboloid of two sheets. The cone $x^2 + y^2 - z^2 = 0$ lies between these two types of hyperboloids.

EXAMPLE 6   The surface

$$z = \sin\sqrt{x^2 + y^2} \qquad (3)$$

is symmetrical with respect to the z-axis, because Eq. (3) reduces to the equation $z = \sin r$ (Fig. 14.2.14) in terms of the radial coordinate $r =$

Fig. 14.2.14   The curve $z = \sin r$ (Example 6)

$\sqrt{x^2 + y^2}$ that measures perpendicular distance from the z-axis. The *surface* $z = \sin r$ is generated by revolving the curve $z = \sin x$ around the z-axis. Hence its level curves are circles centered at the origin in the xy-plane. For instance, $z = 0$ if r is an integral multiple of π, whereas $z = \pm 1$ if r is any odd multiple of $\pi/2$. Figure 14.2.15 shows traces of this surface in planes parallel to the yz-plane. The "hat effect" was achieved by plotting (x, y, z) for those points (x, y) that lie within a certain ellipse in the xy-plane.

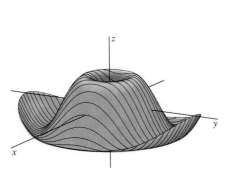

Fig. 14.2.15   The hat surface $z = \sin\sqrt{x^2 + y^2}$ (Example 6)

Given an arbitrary function $f(x, y)$, it can be quite a challenge to construct a picture of the surface $z = f(x, y)$. Example 7 illustrates some special techniques that may be useful. Additional surface-sketching techniques will appear in the remainder of this chapter.

EXAMPLE 7   Investigate the graph of the function

$$f(x, y) = \tfrac{3}{4}y^2 + \tfrac{1}{24}y^3 - \tfrac{1}{32}y^4 - x^2. \qquad (4)$$

Solution   The key feature in Eq. (4) is that the right-hand side is the *sum* of a function of x and a function of y. If we set $x = 0$, we get the curve

$$z = \tfrac{3}{4}y^2 + \tfrac{1}{24}y^3 - \tfrac{1}{32}y^4 \qquad (5)$$

in which the surface $z = f(x, y)$ intersects the yz-plane. But if we set $y = y_0$ in Eq. (4), we get

$$z = (\tfrac{3}{4}y_0^2 + \tfrac{1}{24}y_0^3 - \tfrac{1}{32}y_0^4) - x^2;$$

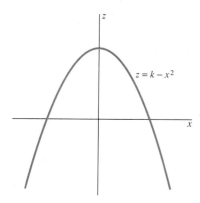

Fig. 14.2.16 The intersection of $z = f(x, y)$ and the plane $y = y_0$ (Example 7)

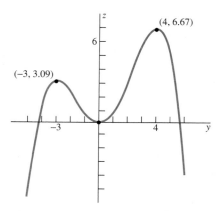

Fig. 14.2.17 The curve $z = \frac{3}{4}y^2 + \frac{1}{24}y^3 - \frac{1}{32}y^4$ (Example 7)

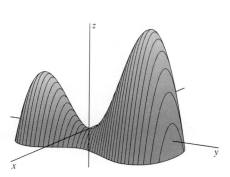

Fig. 14.2.18 Trace parabolas of $z = f(x, y)$ (Example 7)

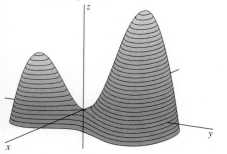

Fig. 14.2.19 Contour curves on $z = f(x, y)$ (Example 7)

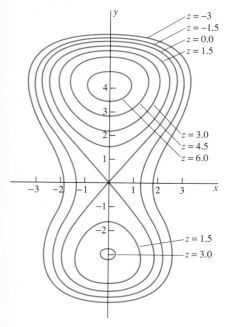

Fig. 14.2.20 Level curves of the function $f(x, y) = \frac{3}{4}y^2 + \frac{1}{24}y^3 - \frac{1}{32}y^4 - x^2$ (Example 7)

that is,

$$z = k - x^2, \tag{6}$$

which is the equation of a parabola in the xz-plane. Hence the trace of $z = f(x, y)$ in each plane $y = y_0$ is a parabola of the form in Eq. (6) (Fig. 14.2.16).

We can use the techniques of Section 4.5 to sketch the curve in Eq. (5). Calculating the derivative of z with respect to y, we get

$$\frac{dz}{dy} = \frac{3}{2}y + \frac{1}{8}y^2 - \frac{1}{8}y^3 = -\frac{1}{8}y(y^2 - y - 12) = -\frac{1}{8}y(y + 3)(y - 4).$$

Hence the critical points are $y = -3$, $y = 0$, and $y = 4$. The corresponding values of z are $f(0, -3) \approx 3.09$, $f(0, 0) = 0$, and $f(0, 4) \approx 6.67$. Because $z \to -\infty$ as $y \to \pm\infty$, it follows readily that the graph of Eq. (5) looks like that in Fig. 14.2.17.

Now we can see what the surface $z = f(x, y)$ looks like. Each vertical plane $y = y_0$ intersects the curve in Eq. (5) at a single point, and this point is the vertex of a parabola that opens downward like that in Eq. (6); this parabola is the intersection of the plane and the surface. Thus the surface $z = f(x, y)$ is generated by translating the vertex of such a parabola along the curve

$$z = \tfrac{3}{4}y^2 + \tfrac{1}{24}y^3 - \tfrac{1}{32}y^4,$$

as indicated in Fig. 14.2.18.

Figure 14.2.19 shows some typical contour curves on this surface. They indicate that the surface resembles two peaks separated by a mountain pass. To check this figure, we programmed a microcomputer to plot typical level curves of the function $f(x, y)$. The result is shown in Fig. 14.2.20. The nested level curves around the points $(0, -3)$ and $(0, 4)$ indicate the local maxima of $z = f(x, y)$. The level figure-eight curve through $(0, 0)$ marks the *saddle point* we see in Figs. 14.2.18 and 14.2.19. Local extrema and saddle points of functions of two variables are discussed in Sections 14.5 and 14.10.

760

14.2 Problems

In Problems 1 through 10, state the largest possible domain of definition of the given function f.

1. $f(x, y) = \exp(-x^2 - y^2)$ (Fig. 14.2.21)

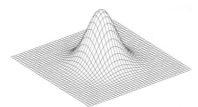

Fig. 14.2.21 The graph of the function of Problem 1

2. $f(x, y) = \ln(x^2 - y^2 - 1)$

3. $f(x, y) = \dfrac{x + y}{x - y}$ **4.** $f(x, y) = \sqrt{4 - x^2 - y^2}$

5. $f(x, y) = \dfrac{1 + \sin xy}{xy}$

6. $f(x, y) = \dfrac{1 + \sin xy}{x^2 + y^2}$ (Fig. 14.2.22)

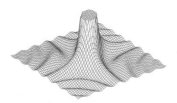

Fig. 14.2.22 The graph of the function of Problem 6

7. $f(x, y) = \dfrac{xy}{x^2 - y^2}$ **8.** $f(x, y, z) = \dfrac{1}{\sqrt{z - x^2 - y^2}}$

9. $f(x, y, z) = \exp\left(\dfrac{1}{x^2 + y^2 + z^2}\right)$

10. $f(x, y, z) = \ln(xyz)$

In Problems 11 through 20, describe the graph of the function f.

11. $f(x, y) = 10$ **12.** $f(x, y) = x$
13. $f(x, y) = x + y$ **14.** $f(x, y) = \sqrt{x^2 + y^2}$
15. $f(x, y) = x^2 + y^2$ **16.** $f(x, y) = 4 - x^2 - y^2$
17. $f(x, y) = \sqrt{4 - x^2 - y^2}$
18. $f(x, y) = 16 - y^2$
19. $f(x, y) = 10 - \sqrt{x^2 + y^2}$
20. $f(x, y) = -\sqrt{36 - 4x^2 - 9y^2}$

In Problems 21 through 30, sketch some typical level curves of the function f.

21. $f(x, y) = x - y$ **22.** $f(x, y) = x^2 - y^2$
23. $f(x, y) = x^2 + 4y^2$ **24.** $f(x, y) = y - x^2$
25. $f(x, y) = y - x^3$ **26.** $f(x, y) = y - \cos x$
27. $f(x, y) = x^2 + y^2 - 4x$
28. $f(x, y) = x^2 + y^2 - 6x + 4y + 7$
29. $f(x, y) = \exp(-x^2 - y^2)$

30. $f(x, y) = \dfrac{1}{1 + x^2 + y^2}$

In Problems 31 through 36, describe the level surfaces of the function f.

31. $f(x, y, z) = x^2 + y^2 - z$
32. $f(x, y, z) = z + \sqrt{x^2 + y^2}$
33. $f(x, y, z) = x^2 + y^2 + z^2 - 4x - 2y - 6z$
34. $f(x, y, z) = z^2 - x^2 - y^2$
35. $f(x, y, z) = x^2 + 4y^2 - 4x - 8y + 17$
36. $f(x, y, z) = x^2 + z^2 + 25$

In Problems 37 through 40, the function $f(x, y)$ is the sum of a function of x and a function of y. Hence you can use the method of Example 7 to construct a sketch of the surface $z = f(x, y)$. Match each function with its graph in Figs. 14.2.23 through 14.2.26.

37. $f(x, y) = y^3 - x^2$ **38.** $f(x, y) = y^4 + x^2$
39. $f(x, y) = y^4 - 2y^2 + x^2$
40. $f(x, y) = 2y^3 - 3y^2 - 12y + x^2$

Fig. 14.2.23

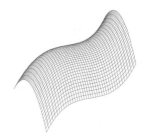

Fig. 14.2.24

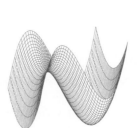

Fig. 14.2.25   **Fig. 14.2.26**

41. Figures 14.2.27 through 14.2.32 show the graphs of six functions $z = f(x, y)$. Figures 14.2.33 through 14.2.38 show level curves of the same six functions, but not in the same order. The level curves in each figure correspond to contours at equally spaced heights on the surface $z = f(x, y)$. Match each surface with its level curves.

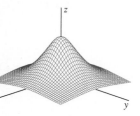

Fig. 14.2.27
$z = \dfrac{1}{1 + x^2 + y^2}$,
$|x| \leqq 2, |y| \leqq 2$

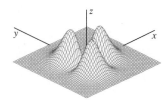

Fig. 14.2.28 $z = r^2 \exp(-r^2) \cos^2(3\theta/2)$,
$|x| \leqq 3, |y| \leqq 3$

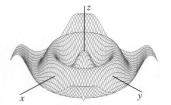

Fig. 14.2.29
$z = \cos\sqrt{x^2 + y^2}$,
$|x| \leqq 10, |y| \leqq 10$

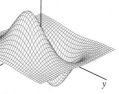

Fig. 14.2.30
$z = x \exp(-x^2 - y^2)$,
$|x| \leqq 2, |y| \leqq 2$

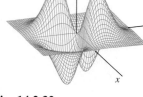

Fig. 14.2.31
$z = 3(x^2 + 3y^2) \times \exp(-x^2 - y^2)$,
$|x| \leqq 2.5, |y| \leqq 2.5$

Fig. 14.2.32
$z = xy \exp(-\frac{1}{2}(x^2 + y^2))$,
$|x| \leqq 3.5, |y| \leqq 3.5$

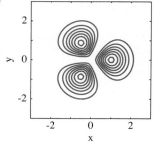

Fig. 14.2.33

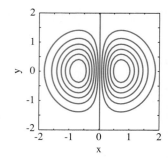

Fig. 14.2.34

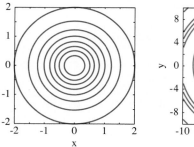

Fig. 14.2.35

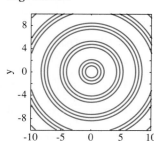

Fig. 14.2.36

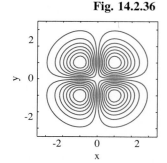

Fig. 14.2.37

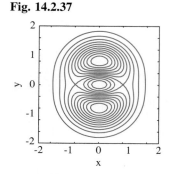

Fig. 14.2.38

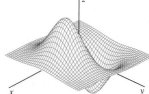

14.2 Project

Plotting surfaces with a computer graphing program can help you develop a "feel" for graphs of functions of two variables. Figure 14.2.39 lists appropriate commands in several common computer systems to plot the surface

Ch. 14 / Partial Differentiation

System	Command
Derive	`Author f(x,y) Plot Plot` `Use Length and Center to set rectangle`
Maple	`plot3d( f(x,y), x = a..b, y = c..d )`
Mathematica	`Plot3D[ f[x,y], {x,a,b}, {y,c,d} ]`
X(PLORE)	`graph3d( f(x,y), x = a to b, y = c to d )`

Fig. 14.2.39 Commands for generating three-dimensional plots in common systems

$z = f(x, y)$ over the base rectangle $a \leqq x \leqq b$, $c \leqq y \leqq d$. To begin, graph some of the following functions over rectangles of various sizes to see how the scale affects the picture.

$$f(x, y) = p \cos qx$$

$$f(x, y) = p \cos qy$$

$$f(x, y) = \sin px \sin qy$$

$$\left.\begin{array}{l} f(x, y) = p + qx^2 \\ f(x, y) = p + qy^2 \\ f(x, y) = px^2 + qy^2 \end{array}\right\} \quad \begin{array}{l} \text{(Use negative } and \text{ positive} \\ \text{values of } p \text{ and } q \text{ in} \\ \text{these three examples.)} \end{array}$$

$$f(x, y) = px^2 + qxy + ry^2$$

$$f(x, y) = \exp(-px^2 - qy^2)$$

$$f(x, y) = (px^2 + qxy + ry^2) \exp(-x^2 - y^2).$$

Similarly, vary the numerical parameters p, q, and r and note the resulting changes in the graph. Then make up some functions of your own for experimentation. If you have a computer connected to a printer, assemble a portfolio of your most interesting examples.

14.3 ▬▬▬▬

Limits and Continuity

We need limits of functions of several variables for the same reasons that we needed limits of functions of a single variable—so that we can discuss slopes and rates of change. Both the definition and the basic properties of limits of functions of several variables are essentially the same as those that we stated in Section 2.2 for functions of a single variable. For simplicity, we shall state them here only for functions of two variables x and y; for a function of three variables, the pair (x, y) should be replaced by the triple (x, y, z).

For a function f of two variables, we ask what number (if any) the values $f(x, y)$ approach as (x, y) approaches the fixed point (a, b) in the coordinate plane. For a function f of three variables, we ask what number (if any) the values $f(x, y, z)$ approach as (x, y, z) approaches the fixed point (a, b, c) in space.

x	y	$f(x, y) = xy$ (rounded)
2.2	2.5	5.50000
1.98	3.05	6.03900
2.002	2.995	5.99599
1.9998	3.0005	6.00040
2.00002	2.99995	5.99996
1.999998	3.0000005	6.00000
↓	↓	↓
2	3	6

Fig. 14.3.1 The numerical data of Example 1

EXAMPLE 1 The numerical data in the table of Fig. 14.3.1 suggest that the values of the function $f(x, y) = xy$ approach 6 as $x \to 2$ and $y \to 3$ simultaneously—that is, as (x, y) approaches the point $(2, 3)$. It therefore is natural to write

$$\lim_{(x, y) \to (2, 3)} xy = 6.$$

Our intuitive idea of the limit of a function of two variables is this. We say that the number L is the *limit* of the function $f(x, y)$ as (x, y) approaches the point (a, b) and write

$$\lim_{(x, y) \to (a, b)} f(x, y) = L, \qquad (1)$$

provided that the number $f(x, y)$ can be made as close as we please to L merely by choosing the point (x, y) sufficiently close to—but not equal to—the point (a, b).

To make this intuitive idea precise, we must specify how close to L— within the distance $\epsilon > 0$, say—we want $f(x, y)$ to be, and then how close to (a, b) the point (x, y) should be to accomplish this. We think of the point (x, y) as being close to (a, b) provided that it lies within a small square (Fig. 14.3.2) with center (a, b) and edge length 2δ, where δ is a small positive number. The point (x, y) lies within this square if and only if both

$$|x - a| < \delta \quad \text{and} \quad |y - b| < \delta. \qquad (2)$$

This observation serves as motivation for the formal definition, with two additional conditions. First, we define the limit of $f(x, y)$ as $(x, y) \to (a, b)$ *only* under the condition that the domain of definition of f contains points $(x, y) \ne (a, b)$ that lie arbitrarily close to (a, b)—that is, within *every* square of the sort shown in Fig. 14.3.2 and thus within any and every preassigned positive distance of (a, b). Hence we do not speak of the limit of f at an isolated point of its domain D. Finally, we do *not* require that f be defined at the point (a, b) itself. Hence we deliberately exclude the possibility that $(x, y) = (a, b)$.

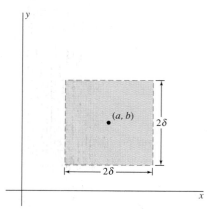

Fig. 14.3.2 The square $|x - a| < \delta,\ |y - b| < \delta$

Definition *The Limit of $f(x, y)$*

We say that the **limit of $f(x, y)$ as (x, y) approaches (a, b)** is L provided that for every number $\epsilon > 0$, there exists a number $\delta > 0$ with the following property: If (x, y) is a point of the domain of f other than (a, b) such that both

$$|x - a| < \delta \quad \text{and} \quad |y - b| < \delta, \qquad (2)$$

then it follows that

$$|f(x, y) - L| < \epsilon. \qquad (3)$$

We ordinarily shall rely on continuity rather than the formal definition of the limit to evaluate limits of functions of several variables. We say that f is **continuous at the point (a, b)** provided that $f(a, b)$ exists and $f(x, y)$ approaches $f(a, b)$ as (x, y) approaches (a, b). That is,

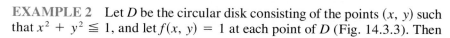

$$\lim_{(x, y) \to (a, b)} f(x, y) = f(a, b).$$

Thus f is continuous at (a, b) if it is defined there and its limit there is equal to its value there, precisely as in the case of a function of a single variable. The function f is said to be **continuous on the set D** if it is continuous at each point of D, exactly as in the single-variable case.

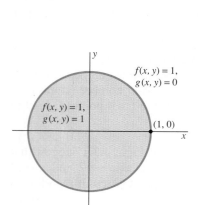

Fig. 14.3.3 The circular disk of Example 2

EXAMPLE 2 Let D be the circular disk consisting of the points (x, y) such that $x^2 + y^2 \leqq 1$, and let $f(x, y) = 1$ at each point of D (Fig. 14.3.3). Then the limit of $f(x, y)$ at each point of D is 1, so f is continuous on D. But let the new function $g(x, y)$ be defined on the entire plane R^2 as follows:

$$g(x, y) = \begin{cases} f(x, y) & \text{if } (x, y) \text{ is in } D; \\ 0 & \text{otherwise.} \end{cases}$$

Then g is *not* continuous on R^2. For instance, the limit of $g(x, y)$ as $(x, y) \to (1, 0)$ does not exist, because there exist both points within D arbitrarily close to $(1, 0)$ at which g has the value 1 and points outside of D arbitrarily close to $(1, 0)$ at which g has the value 0. Thus $g(x, y)$ cannot approach any single value as $(x, y) \to (1, 0)$. Because g has no limit at $(1, 0)$, it cannot be continuous there.

The limit laws of Section 2.2 have natural analogues for functions of several variables. If

$$\lim_{(x, y) \to (a, b)} f(x, y) = L \quad \text{and} \quad \lim_{(x, y) \to (a, b)} g(x, y) = M, \qquad (4)$$

then the sum, product, and quotient laws for limits are these:

$$\lim_{(x, y) \to (a, b)} [f(x, y) + g(x, y)] = L + M, \qquad (5)$$

$$\lim_{(x, y) \to (a, b)} [f(x, y) \cdot g(x, y)] = L \cdot M, \quad \text{and} \qquad (6)$$

$$\lim_{(x, y) \to (a, b)} \frac{f(x, y)}{g(x, y)} = \frac{L}{M} \quad \text{if } M \neq 0. \qquad (7)$$

EXAMPLE 3   Show that $\lim\limits_{(x, y) \to (a, b)} xy = ab$.

Solution   We take $f(x, y) = x$ and $g(x, y) = y$. Then it follows from the definition of limit that

$$\lim_{(x, y) \to (a, b)} f(x, y) = a \quad \text{and} \quad \lim_{(x, y) \to (a, b)} g(x, y) = b.$$

Hence the product law gives

$$\lim_{(x, y) \to (a, b)} xy = \lim_{(x, y) \to (a, b)} f(x, y)\, g(x, y)$$

$$= \left[\lim_{(x, y) \to (a, b)} f(x, y) \right]\left[\lim_{(x, y) \to (a, b)} g(x, y) \right] = ab.$$

More generally, suppose that $P(x, y)$ is a polynomial in the two variables x and y, so it can be written in the form

$$P(x, y) = \sum c_{ij} x^i y^j.$$

Then the sum and product laws imply that

$$\lim_{(x, y) \to (a, b)} P(x, y) = P(a, b).$$

An immediate but important consequence is that every polynomial in two (or more) variables is a continuous function.

Just as in the single-variable case, any composition of continuous multivariable functions is also a continuous function. For instance, suppose that the functions f and g both are continuous at (a, b) and that h is continuous at the point $(f(a, b), g(a, b))$. Then the composite function

$$H(x, y) = h(f(x, y), g(x, y))$$

is also continuous at (a, b). As a consequence, any finite combination involving sums, products, quotients, and compositions of the familiar elementary functions is continuous, except possibly at points where a denominator is zero or where the formula for the function is otherwise meaningless. This general rule suffices for the evaluation of most limits that we shall encounter.

EXAMPLE 4 By application of the limit laws, we get

$$\lim_{(x, y) \to (1, 2)} \left[e^{xy} \sin \frac{\pi y}{4} + xy \ln\sqrt{y - x} \right]$$

$$= \lim_{(x, y) \to (1, 2)} e^{xy} \sin \frac{\pi y}{4} + \lim_{(x, y) \to (1, 2)} xy \ln\sqrt{y - x}$$

$$= \left(\lim_{(x, y) \to (1, 2)} e^{xy} \right)\left(\lim_{(x, y) \to (1, 2)} \sin \frac{\pi y}{4} \right) + \left(\lim_{(x, y) \to (1, 2)} xy \right)\left(\lim_{(x, y) \to (1, 2)} \ln\sqrt{y - x} \right)$$

$$= e^2 \cdot 1 + 2 \ln 1 = e^2.$$

Examples 5 and 6 illustrate techniques that sometimes are successful in handling cases with denominators that approach zero.

EXAMPLE 5 Show that $\displaystyle\lim_{(x, y) \to (0, 0)} \frac{xy}{\sqrt{x^2 + y^2}} = 0$.

Solution Let (r, θ) be the polar coordinates of the point (x, y). Then $x = r \cos \theta$ and $y = r \sin \theta$, so

$$\frac{xy}{\sqrt{x^2 + y^2}} = \frac{(r \cos \theta)(r \sin \theta)}{\sqrt{r^2(\cos^2 \theta + \sin^2 \theta)}} = r \cos \theta \sin \theta \qquad \text{for } r > 0.$$

Because $r = \sqrt{x^2 + y^2}$, it is clear that $r \to 0$ as both x and y approach zero. It therefore follows that

$$\lim_{(x, y) \to (0, 0)} \frac{xy}{\sqrt{x^2 + y^2}} = \lim_{r \to 0} r \cos \theta \sin \theta = 0,$$

because $|\cos \theta \sin \theta| \leqq 1$ for all values of θ.

EXAMPLE 6 Show that

$$\lim_{(x, y) \to (0, 0)} \frac{xy}{x^2 + y^2}$$

does not exist.

Solution Our plan is to show that $f(x, y) = xy/(x^2 + y^2)$ approaches different values as (x, y) approaches $(0, 0)$ from different directions. Suppose that (x, y) approaches $(0, 0)$ along the straight line of slope m through the origin. On this line we have $y = mx$. So, on this line,

$$f(x, y) = \frac{x \cdot mx}{x^2 + m^2 x^2} = \frac{m}{1 + m^2}$$

if $x \neq 0$. If we take $m = 1$, we see that $f(x, y) = \frac{1}{2}$ at every point of the line $y = x$ other than $(0, 0)$. If we take $m = -1$, then $f(x, y) = -\frac{1}{2}$ at every

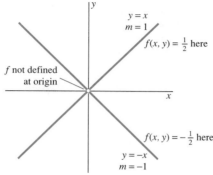

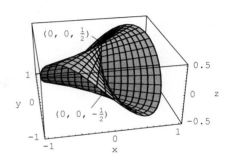

Fig. 14.3.4 The function f of Example 6 takes on both values $+\frac{1}{2}$ and $-\frac{1}{2}$ at points arbitrarily close to the origin.

Fig. 14.3.5 The graph of $f(x, y) = \dfrac{xy}{x^2 + y^2}$ (Example 6)

point of the line $y = -x$ other than $(0, 0)$. Thus $f(x, y)$ approaches two different values as (x, y) approaches $(0, 0)$ along these two lines (Fig. 14.3.4). Hence $f(x, y)$ cannot approach any *single* value as (x, y) approaches $(0, 0)$, and this implies that the limit in question cannot exist.

Figure 14.3.5 shows a computer-generated graph of the function $f(x, y) = xy/(x^2 + y^2)$. It consists of linear rays along each of which the polar angular coordinate θ is constant. For each number z between $-\frac{1}{2}$ and $\frac{1}{2}$ (inclusive), there are rays along which $f(x, y)$ has the constant value z. Hence we can make $f(x, y)$ approach any number we please in $[-\frac{1}{2}, \frac{1}{2}]$ by letting (x, y) approach $(0, 0)$ from the appropriate direction.

In order for

$$L = \lim_{(x, y) \to (a, b)} f(x, y)$$

to exist, $f(x, y)$ must approach L for *any and every* mode of approach of (x, y) to (a, b). In Problem 27 we give an example of a function f such that $f(x, y) \to 0$ as $(x, y) \to (0, 0)$ along any straight line through the origin, but $f(x, y) \to 1$ as (x, y) approaches the origin along the parabola $y = x^2$. Thus the method of Example 6 cannot be used to show that a limit exists, only that it does not. Fortunately, many important applications, including those we discuss in the remainder of this chapter, involve only functions that exhibit no such exotic behavior as the function of Problem 27.

14.3 Problems

Use the limit laws and consequences of continuity to evaluate the limits in Problems 1 through 15.

1. $\lim_{(x, y) \to (0, 0)} (7 - x^2 + 5xy)$

2. $\lim_{(x, y) \to (1, -2)} (3x^2 - 4xy + 5y^2)$

3. $\lim_{(x, y) \to (1, -1)} e^{-xy}$

4. $\lim_{(x, y) \to (0, 0)} \dfrac{x + y}{1 + xy}$

5. $\lim_{(x, y) \to (0, 0)} \dfrac{5 - x^2}{3 + x + y}$

6. $\lim_{(x, y) \to (2, 3)} \dfrac{9 - x^2}{1 + xy}$

7. $\lim_{(x, y) \to (0, 0)} \ln\sqrt{1 - x^2 - y^2}$

8. $\lim_{(x, y) \to (2, -1)} \ln\dfrac{1 + x + 2y}{3y^2 - x}$

9. $\lim_{(x, y) \to (0, 0)} \dfrac{e^{xy} \sin xy}{xy}$

10. $\lim_{(x, y) \to (0, 0)} \exp\left(-\dfrac{1}{x^2 + y^2}\right)$

11. $\lim_{(x, y, z) \to (1, 1, 1)} \dfrac{x^2 + y^2 + z^2}{1 - x - y - z}$

12. $\displaystyle\lim_{(x,y,z)\to(1,1,1)} (x + y + z) \ln xyz$

13. $\displaystyle\lim_{(x,y,z)\to(1,1,0)} \frac{xy - z}{\cos xyz}$ **14.** $\displaystyle\lim_{(x,y,z)\to(2,-1,3)} \frac{x + y + z}{x^2 + y^2 + z^2}$

15. $\displaystyle\lim_{(x,y,z)\to(2,8,1)} \sqrt{xy} \tan\frac{3\pi z}{4}$

In Problems 16 through 20, evaluate the limits

$$\lim_{h\to 0} \frac{f(x + h, y) - f(x, y)}{h} \quad \text{and}$$

$$\lim_{k\to 0} \frac{f(x, y + k) - f(x, y)}{k}.$$

16. $f(x, y) = x + y$ **17.** $f(x, y) = xy$

18. $f(x, y) = x^2 + y^2$ **19.** $f(x, y) = xy^2 - 2$

20. $f(x, y) = x^2 y^3 - 10$

In Problems 21 through 23, use the method of Example 5 to verify the given limit.

21. $\displaystyle\lim_{(x,y)\to(0,0)} \frac{x^2 - y^2}{\sqrt{x^2 + y^2}} = 0$ **22.** $\displaystyle\lim_{(x,y)\to(0,0)} \frac{x^3 - y^3}{x^2 + y^2} = 0$

23. $\displaystyle\lim_{(x,y)\to(0,0)} \frac{x^4 + y^4}{(x^2 + y^2)^{3/2}} = 0$

24. Use the method of Example 6 to show that

$$\lim_{(x,y)\to(0,0)} \frac{x^2 - y^2}{x^2 + y^2}$$

does not exist. The graph of $f(x, y) = \dfrac{x^2 - y^2}{x^2 + y^2}$ is shown in Fig. 14.3.6.

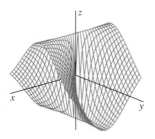

Fig. 14.3.6 The graph of $f(x, y) = \dfrac{x^2 - y^2}{x^2 + y^2}$ of Problem 24

25. Substitute spherical coordinates $x = \rho \sin \phi \cos \theta$, $y = \rho \sin \phi \sin \theta$, $z = \rho \cos \phi$ to show that

$$\lim_{(x,y,z)\to(0,0,0)} \frac{xyz}{x^2 + y^2 + z^2} = 0.$$

26. Determine whether or not

$$\lim_{(x,y,z)\to(0,0,0)} \frac{xy + xz + yz}{x^2 + y^2 + z^2}$$

exists.

27. Let

$$f(x, y) = \frac{2x^2 y}{x^4 + y^2}.$$

(a) Show that $f(x, y) \to 0$ as $(x, y) \to (0, 0)$ along *any* and every straight line through the origin. (b) Show that $f(x, y) \to 1$ as $(x, y) \to (0, 0)$ along the parabola $y = x^2$. Conclude that the limit of $f(x, y)$ as $(x, y) \to (0, 0)$ does not exist. The graph of f is shown in Fig. 14.3.7.

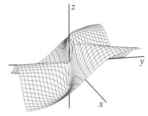

Fig. 14.3.7 The graph of the function of Problem 27

28. Suppose that $f(x, y) = (x - y)/(x^3 - y)$ except at points of the curve $y = x^3$, where we *define* $f(x, y)$ to be 1. Show that f is not continuous at the point $(1, 1)$. Evaluate the limits of $f(x, y)$ as $(x, y) \to (1, 1)$ along the vertical line $x = 1$ and along the horizontal line $y = 1$. [*Suggestion:* Recall that $a^3 - b^3 = (a - b)(a^2 + ab + b^2)$.]

29. Locate and identify the extrema (local or global, maximum or minimum) of the function $f(x, y) = x^2 - x + y^2 + 2y + 1$. [*Note:* The point of this problem is that you do *not* need calculus to solve it.]

30. Sketch enough level curves of the function $h(x, y) = y - x^2$ to show that the function h has *no* extreme values—no local or global maxima or minima.

14.4
Partial Derivatives

Suppose that $y = f(x)$ is a function of *one* real variable. Its first derivative

$$\frac{dy}{dx} = D_x f(x) = \lim_{h\to 0} \frac{f(x + h) - f(x)}{h} \qquad (1)$$

can be interpreted as the instantaneous rate of change of y with respect to x. For a function $z = f(x, y)$ of two variables, we need a similar understanding of the rate at which z changes as x and y vary (either singly or simultaneously).

To reach this more complicated concept, we adopt a divide-and-conquer strategy.

First we hold y fixed and let x vary. The rate of change of z with respect to x is then denoted by $\partial z/\partial x$ and has the value

$$\frac{\partial z}{\partial x} = \lim_{h \to 0} \frac{f(x + h, y) - f(x, y)}{h}. \tag{2}$$

The value of this limit—if it exists—is called the **partial derivative of f with respect to x.** In like manner, we may hold x fixed and let y vary. The rate of change of z with respect to y is then the **partial derivative of f with respect to y,** defined to be

$$\frac{\partial z}{\partial y} = \lim_{k \to 0} \frac{f(x, y + k) - f(x, y)}{k} \tag{3}$$

for all (x, y) for which this limit exists. Note the symbol ∂ that is used instead of d to indicate the partial derivatives of a function of two variables. A function of three or more variables has a partial derivative (defined similarly) with respect to each of its independent variables. Some other common notations for partial derivatives are

$$\frac{\partial z}{\partial x} = \frac{\partial f}{\partial x} = f_x(x, y) = D_x f(x, y) = D_1 f(x, y), \tag{4}$$

$$\frac{\partial z}{\partial y} = \frac{\partial f}{\partial y} = f_y(x, y) = D_y f(x, y) = D_2 f(x, y). \tag{5}$$

Note that if we delete the symbol y in Eq. (2), the result is the limit in Eq. (1). This means that we can calculate $\partial z/\partial x$ as an "ordinary" derivative with respect to x simply by regarding y as a constant during the process of differentiation. Similarly, we can compute $\partial z/\partial y$ as an ordinary derivative by thinking of y as the *only* variable and treating x as a constant during the computation.

EXAMPLE 1   Compute the partial derivatives $\partial f/\partial x$ and $\partial f/\partial y$ of the function $f(x, y) = x^2 + 2xy^2 - y^3$.

Solution   To compute the partial of f with respect to x, we regard y as a constant. Then we differentiate normally and find that

$$\frac{\partial f}{\partial x} = 2x + 2y^2.$$

When we regard x as a constant and differentiate with respect to y, we find that

$$\frac{\partial f}{\partial y} = 4xy - 3y^2.$$

To get an intuitive feel for the meaning of partial derivatives, we can think of $f(x, y)$ as the temperature at the point (x, y) of the plane. Then $f_x(x, y)$ is the instantaneous rate of change of temperature at (x, y) per unit increase in x (with y held constant). Similarly, $f_y(x, y)$ is the instantaneous rate of change of temperature per unit increase in y (with x held constant). For example, with the temperature function $f(x, y) = x^2 + 2xy^2 - y^3$ of Example 1, the rate of change of temperature at the point $(1, -1)$ is $+4°$ per unit distance in the positive x-direction and $-7°$ per unit distance in the positive y-direction.

EXAMPLE 2 Find $\partial z/\partial x$ and $\partial z/\partial y$ if $z = (x^2 + y^2)e^{-xy}$.

Solution Because $\partial z/\partial x$ is calculated as if it were an ordinary derivative with respect to x, with y held constant, we use the product rule. This gives

$$\frac{\partial z}{\partial x} = (2x)(e^{-xy}) + (x^2 + y^2)(-ye^{-xy}) = (2x - x^2y - y^3)e^{-xy}.$$

Because x and y appear symmetrically in the expression for z, we get $\partial z/\partial y$ when we interchange x and y in the expression for $\partial z/\partial x$:

$$\frac{\partial z}{\partial y} = (2y - xy^2 - x^3)e^{-xy}.$$

Check this result by differentiating z with respect to y.

EXAMPLE 3 The volume V (in cubic centimeters) of 1 mole (mol) of an ideal gas is given by

$$V = \frac{(82.06)\,T}{p},$$

where p is the pressure (in atmospheres) and T is the absolute temperature [in kelvins (K), where K = °C + 273]. Find the rates of change of the volume of 1 mol of an ideal gas with respect to pressure and with respect to temperature when $T = 300K$ and $p = 5$ atm.

Solution The partial derivatives of V with respect to its two variables are

$$\frac{\partial V}{\partial p} = -\frac{(82.06)T}{p^2} \quad \text{and} \quad \frac{\partial V}{\partial T} = \frac{82.06}{p}.$$

With $T = 300$ and $p = 5$, we have the two values $\partial V/\partial p = -984.72$ (cm³/atm) and $\partial V/\partial T = 16.41$ (cm³/K). These partial derivatives allow us to estimate the effect of a change in temperature or in pressure on the volume V of the gas, as follows. We are given $T = 300$ and $p = 5$, so the volume of gas with which we are dealing is

$$V = \frac{(82.06)(300)}{5} = 4923.60 \quad (\text{cm}^3).$$

We would expect an increase in pressure of 1 atm (with T held constant) to decrease the volume of gas by approximately 1 L (1000 cm³), because $-984.72 \approx -1000$. An increase in temperature of 1 K (or 1°C) would, with p held constant, increase the volume by about 16 cm³, because $16.41 \approx 16$.

GEOMETRIC INTERPRETATION OF PARTIAL DERIVATIVES

The partial derivatives f_x and f_y are the slopes of lines tangent to certain curves on the surface $z = f(x, y)$. Figure 14.4.1 illustrates the intersection of this surface with a vertical plane $y = b$ that is parallel to the xz-coordinate plane.

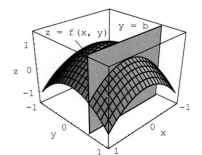

Fig. 14.4.1 A vertical plane parallel to the xz-plane intersects the surface $z = f(x, y)$ in an x-curve.

Ch. 14 / Partial Differentiation

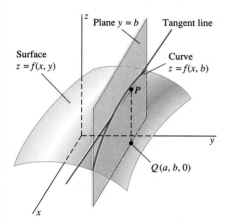

Fig. 14.4.2 An x-curve and its tangent line

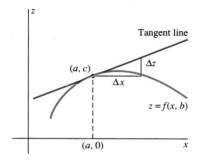

Fig. 14.4.3 Projection into the xz-plane of the x-curve through $P(a, b, c)$ and its tangent line

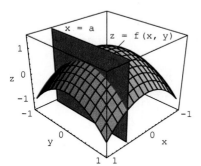

Fig. 14.4.4 A vertical plane parallel to the yz-plane intersects the surface $z = f(x, y)$ in a y-curve.

Along this intersection curve, the x-coordinate varies but the y-coordinate is constant: $y = b$ at each point, because the curve lies in the vertical plane $y = b$. A curve of intersection of $z = f(x, y)$ with a vertical plane parallel to the xz-plane is therefore called an **x-curve** on the surface.

Figure 14.4.2 shows a point $P(a, b, c)$ in the surface $z = f(x, y)$, the x-curve through P, and the line tangent to this x-curve at P. Figure 14.4.3 shows the parallel projection of the vertical plane $y = b$ into the xz-plane itself. We can now "ignore" the presence of $y = b$ and regard $z = f(x, b)$ as a function of the *single* variable x. The slope of the line tangent to the original x-curve through P (see Fig. 14.4.2) is equal to the slope

$$\left(\frac{\text{Rise}}{\text{Run}} = \frac{\Delta z}{\Delta x} \right)$$

of the tangent line in Fig. 14.4.3. But by familiar single-variable calculus, this latter slope is given by

$$\lim_{h \to 0} \frac{f(a + h, b) - f(a, b)}{h} = f_x(a, b).$$

Thus we see that the geometric meaning of f_x is this:

Geometric Interpretation of $\dfrac{\partial z}{\partial x}$

The value $f_x(a, b)$ is the slope of the tangent line at $P(a, b, c)$ to the x-curve through P on the surface $z = f(x, y)$.

We proceed in much the same way to investigate the geometric meaning of partial differentiation with respect to y. Figure 14.4.4 illustrates the intersection with the surface $z = f(x, y)$ of a vertical plane $x = a$ that is parallel to the yz-coordinate plane. Now the curve of intersection is a **y-curve** along which y varies but $x = a$ is constant. Figure 14.4.5 shows this y-curve $z = f(a, y)$ and its tangent line at P. The projection of the tangent line into the yz-plane (in Fig. 14.4.6) has slope $\partial z / \partial y = f_y(a, b)$. Thus we see that the geometric meaning of f_y is this:

Geometric Interpretation of $\dfrac{\partial z}{\partial y}$

The value $f_y(a, b)$ is the slope of the tangent line at $P(a, b, c)$ to the y-curve through P in the surface $z = f(x, y)$.

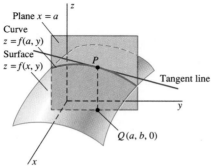

Fig. 14.4.5 A y-curve and its tangent line

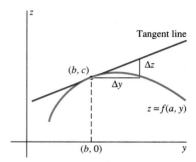

Fig. 14.4.6 Projection into the yz-plane of the y-curve through $P(a, b, c)$ and its tangent line

TANGENT PLANES TO SURFACES

The two tangent lines we have just found determine a unique plane through the point $P(a, b, f(a, b))$. We will see in Section 14.7 that if the partial derivatives f_x and f_y are continuous functions of x and y, then this plane contains the tangent line at P to *every* smooth curve on the surface $z = f(x, y)$ that passes through P. This plane is therefore (by definition) the plane tangent to the surface at P.

> **Definition** *Plane Tangent to $z = f(x, y)$*
> Suppose that the function $f(x, y)$ has continuous partial derivatives on a rectangle in the xy-plane containing (a, b) in its interior. Then the **plane tangent** to the surface $z = f(x, y)$ at the point $P(a, b, f(a, b))$ is the plane through P that contains the lines tangent to the two curves
>
> $$z = f(x, b), \qquad y = b \qquad (x\text{-curve}) \tag{6}$$
>
> and
>
> $$z = f(a, y), \qquad x = a \qquad (y\text{-curve}). \tag{7}$$

To write an equation of this tangent plane, all we need is a vector $\mathbf{n}$ normal to the plane. One way to get such a vector is to find the vector product of the tangent vectors to the curves in Eqs. (6) and (7). Figures 14.4.7 and 14.4.8 show these two curves. As we saw before, the y-curve in Eq. (7) has slope $f_y(a, b)$ at P, so we can take

$$\mathbf{u} = \mathbf{j} + \mathbf{k} f_y(a, b) \tag{8}$$

as its tangent vector at P. The x-curve in Eq. (6) has slope $f_x(a, b)$ at P, so we can take

$$\mathbf{v} = \mathbf{i} + \mathbf{k} f_x(a, b) \tag{9}$$

as its tangent vector. Using these two tangent vectors, we obtain the normal vector

$$\mathbf{n} = \mathbf{u} \times \mathbf{v} = \begin{vmatrix} \mathbf{i} & \mathbf{j} & \mathbf{k} \\ 0 & 1 & f_y(a, b) \\ 1 & 0 & f_x(a, b) \end{vmatrix},$$

so

$$\mathbf{n} = \mathbf{i} f_x(a, b) + \mathbf{j} f_y(a, b) - \mathbf{k}. \tag{10}$$

Note that

$$\mathbf{n} = \left\langle \frac{\partial z}{\partial x}, \frac{\partial z}{\partial y}, -1 \right\rangle \tag{11}$$

is a downward-pointing vector; its negative $-\mathbf{n}$ is the normal vector shown in Fig. 14.4.9.

Finally, we use the normal vector $\mathbf{n}$ of Eq. (10) to find an equation of the plane tangent to the surface $z = f(x, y)$ at the point $P(a, b, f(a, b))$. This equation is

$$f_x(a, b)(x - a) + f_y(a, b)(y - b) - [z - f(a, b)] = 0. \tag{12}$$

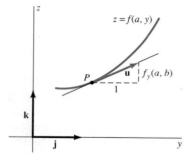

Fig. 14.4.7 The curve $z = f(a, y)$ in the plane $x = a$

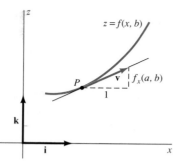

Fig. 14.4.8 The curve $z = f(x, b)$ in the plane $y = b$

Ch. 14 / Partial Differentiation

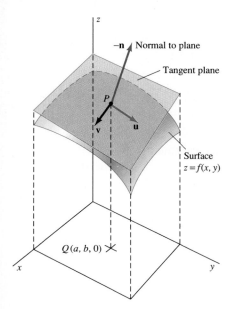

Fig. 14.4.9 A vector normal to the tangent plane can be found by forming the vector product of two tangent vectors.

An equivalent form of this equation is

$$z - c = \frac{\partial z}{\partial x}(x - a) + \frac{\partial z}{\partial y}(y - b), \tag{13}$$

where $c = f(a, b)$ and we remember that the partial derivatives $\partial z/\partial x$ and $\partial z/\partial y$ are evaluated at the point (a, b).

EXAMPLE 4   Write an equation of the plane tangent to the paraboloid $z = x^2 + y^2$ at the point $(2, -1, 5)$.

Solution   We begin by computing $\partial z/\partial x = 2x$ and $\partial z/\partial y = 2y$. The values of these partial derivatives at $(x, y) = (2, -1)$ are 4 and -2, respectively. So Eq. (12) gives

$$4(x - 2) - 2(y + 1) - (z - 5) = 0$$

(when simplified, $4x - 2y - z = 5$) as an equation of the indicated tangent plane.

HIGHER-ORDER PARTIAL DERIVATIVES

The **first-order partial derivatives** f_x and f_y are themselves functions of x and y, so they may be differentiated with respect to x or to y. The partial derivatives of $f_x(x, y)$ and $f_y(x, y)$ are called the **second-order partial derivatives** of f. There are four of them, because there are four possibilities in the order of differentiation:

$$(f_x)_x = f_{xx} = \frac{\partial f_x}{\partial x} = \frac{\partial}{\partial x}\left(\frac{\partial f}{\partial x}\right) = \frac{\partial^2 f}{\partial x^2},$$

$$(f_x)_y = f_{xy} = \frac{\partial f_x}{\partial y} = \frac{\partial}{\partial y}\left(\frac{\partial f}{\partial x}\right) = \frac{\partial^2 f}{\partial y\, \partial x},$$

$$(f_y)_x = f_{yx} = \frac{\partial f_y}{\partial x} = \frac{\partial}{\partial x}\left(\frac{\partial f}{\partial y}\right) = \frac{\partial^2 f}{\partial x\, \partial y},$$

$$(f_y)_y = f_{yy} = \frac{\partial f_y}{\partial y} = \frac{\partial}{\partial y}\left(\frac{\partial f}{\partial y}\right) = \frac{\partial^2 f}{\partial y^2}.$$

If we write $z = f(x, y)$, then we can replace each occurrence of the symbol f here by z.

NOTE   The function f_{xy} is the second-order partial derivative of f with respect to x first and then to y; f_{yx} is the result of differentiating with respect to y first and then to x. Although f_{xy} and f_{yx} are not necessarily equal, it is proved in advanced calculus that these two "mixed" second-order partial derivatives are equal if they are both continuous. More precisely, if f_{xy} and f_{yx} are continuous on a circular disk centered at the point (a, b), then

$$f_{xy}(a, b) = f_{yx}(a, b). \tag{14}$$

[If both f_{xy} and f_{yx} are continuous merely at (a, b), they may well be unequal there.] Because most functions of interest to us have second-order partial derivatives that are continuous everywhere they are defined, we will ordinarily need to deal with only three distinct second-order partial derivatives rather than with four. Similarly, if $f(x, y, z)$ is a function of three variables with continuous second-order partial derivatives, then

$$\frac{\partial^2 f}{\partial x\, \partial y} = \frac{\partial^2 f}{\partial y\, \partial x}, \qquad \frac{\partial^2 f}{\partial x\, \partial z} = \frac{\partial^2 f}{\partial z\, \partial x}, \quad \text{and} \quad \frac{\partial^2 f}{\partial y\, \partial z} = \frac{\partial^2 f}{\partial z\, \partial y}.$$

Third-order and higher-order partial derivatives are defined similarly, and the order in which the differentiations are performed is unimportant as long as all derivatives involved are continuous. For example, the distinct third-order partial derivatives of a function $z = f(x, y)$ are

$$f_{xxx} = \frac{\partial}{\partial x}\left(\frac{\partial^2 f}{\partial x^2}\right) = \frac{\partial^3 f}{\partial x^3},$$

$$f_{xxy} = \frac{\partial}{\partial y}\left(\frac{\partial^2 f}{\partial x^2}\right) = \frac{\partial^3 f}{\partial y\, \partial x^2},$$

$$f_{xyy} = \frac{\partial}{\partial y}\left(\frac{\partial^2 f}{\partial y\, \partial x}\right) = \frac{\partial^3 f}{\partial y^2\, \partial x}, \quad \text{and}$$

$$f_{yyy} = \frac{\partial}{\partial y}\left(\frac{\partial^2 f}{\partial y^2}\right) = \frac{\partial^3 f}{\partial y^3}.$$

EXAMPLE 5   Show that the partial derivatives of third and higher orders of the function $f(x, y) = x^2 + 2xy^2 - y^3$ are constant.

Solution   We find that

$$f_x(x, y) = 2x + 2y^2 \quad \text{and} \quad f_y(x, y) = 4xy - 3y^2.$$

So

$$f_{xx}(x, y) = 2, \qquad f_{xy}(x, y) = 4y, \quad \text{and} \quad f_{yy}(x, y) = 4x - 6y.$$

Finally,

$$f_{xxx}(x, y) = 0, \quad f_{xxy}(x, y) = 0, \quad f_{xyy}(x, y) = 4, \quad \text{and} \quad f_{yyy}(x, y) = -6.$$

The function f is a polynomial, so all its partial derivatives are polynomials and are, therefore, continuous everywhere. Hence we need not compute any other third-order partial derivatives; each is equal to one of these four. Moreover, because the third-order partial derivatives are all constant, all higher-order partial derivatives of f are zero.

14.4 Problems

In Problems 1 through 20, compute the first-order partial derivatives of each function.

1. $f(x, y) = x^4 - x^3y + x^2y^2 - xy^3 + y^4$

2. $f(x, y) = x \sin y$

3. $f(x, y) = e^x(\cos y - \sin y)$

4. $f(x, y) = x^2 e^{xy}$

5. $f(x, y) = \dfrac{x + y}{x - y}$

6. $f(x, y) = \dfrac{xy}{x^2 + y^2}$

7. $f(x, y) = \ln(x^2 + y^2)$

8. $f(x, y) = (x - y)^{14}$

9. $f(x, y) = x^y$

10. $f(x, y) = \tan^{-1} xy$

11. $f(x, y, z) = x^2 y^3 z^4$

12. $f(x, y, z) = x^2 + y^3 + z^4$

13. $f(x, y, z) = e^{xyz}$

14. $f(x, y, z) = x^4 - 16yz$

15. $f(x, y, z) = x^2 e^y \ln z$

16. $f(u, v) = (2u^2 + 3v^2) \exp(-u^2 - v^2)$

17. $f(r, s) = \dfrac{r^2 - s^2}{r^2 + s^2}$

18. $f(u, v) = e^{uv}(\cos uv + \sin uv)$

19. $f(u, v, w) = ue^v + ve^w + we^u$

20. $f(r, s, t) = (1 - r^2 - s^2 - t^2)e^{-rst}$

In Problems 21 through 30, verify that $z_{xy} = z_{yx}$.

21. $z = x^2 - 4xy + 3y^2$

22. $z = 2x^3 + 5x^2y - 6y^2 + xy^4$

23. $z = x^2 \exp(-y^2)$ **24.** $z = xye^{-xy}$

25. $z = \ln(x + y)$ **26.** $z = (x^3 + y^3)^{10}$

27. $z = e^{-3x} \cos y$ **28.** $z = (x + y)\sec xy$

29. $z = x^2 \cosh(1/y^2)$ **30.** $z = \sin xy + \tan^{-1} xy$

In Problems 31 through 40, find an equation of the plane tangent to the given surface $z = f(x, y)$ at the indicated point P.

31. $z = x^2 + y^2$; $P = (3, 4, 25)$

32. $z = \sqrt{25 - x^2 - y^2}$; $P = (4, -3, 0)$

33. $z = \sin \dfrac{\pi xy}{2}$; $P = (3, 5, -1)$

34. $z = \dfrac{4}{\pi} \tan^{-1} xy$; $P = (1, 1, 1)$

35. $z = x^3 - y^3$; $P = (3, 2, 19)$

36. $z = 3x + 4y$; $P = (1, 1, 7)$

37. $z = xy$; $P = (1, -1, -1)$

38. $z = \exp(-x^2 - y^2)$; $P = (0, 0, 1)$

39. $z = x^2 - 4y^2$; $P = (5, 2, 9)$

40. $z = \sqrt{x^2 + y^2}$; $P = (3, -4, 5)$

41. Verify that the mixed second-order partial derivatives f_{xy} and f_{yx} are equal if $f(x, y) = x^m y^n$, where m and n are positive integers.

42. Suppose that $z = e^{x+y}$. Show that e^{x+y} is the result of differentiating z first m times with respect to x, then n times with respect to y.

43. Let $f(x, y, z) = e^{xyz}$. Calculate the distinct second-order partial derivatives of f and the third-order partial derivative f_{xyz}.

44. Suppose that $g(x, y) = \sin xy$. Verify that $g_{xy} = g_{yx}$ and that $g_{xxy} = g_{xyx} = g_{yxx}$.

45. It is shown in physics that the temperature $u(x, t)$ at time t at the point x of a long, insulated rod that lies along the x-axis satisfies the *one-dimensional heat equation*

$$\frac{\partial u}{\partial t} = k \frac{\partial^2 u}{\partial x^2} \qquad (k \text{ is a constant}).$$

Show that the function

$$u = u(x, t) = \exp(-n^2 kt) \sin nx$$

satisfies the one-dimensional heat equation for any choice of the constant n.

46. The *two-dimensional heat equation* for an insulated plane is

$$\frac{\partial u}{\partial t} = k\left(\frac{\partial^2 u}{\partial x^2} + \frac{\partial^2 u}{\partial y^2}\right).$$

Show that the function

$$u = u(x, y, t) = \exp(-[m^2 + n^2]kt) \sin mx \cos ny$$

satisfies this equation for any choice of the constants m and n.

47. A string is stretched along the x-axis, fixed at each end, and then set into vibration. It is shown in physics that the displacement $y = y(x, t)$ of the point of the string at location x at time t satisfies the *one-dimensional wave equation*

$$\frac{\partial^2 y}{\partial t^2} = a^2 \frac{\partial^2 y}{\partial x^2},$$

where the constant a depends on the density and tension of the string. Show that the following functions satisfy the one-dimensional wave equation: (a) $y = \sin(x + at)$; (b) $y = \cosh(3[x - at])$; (c) $y = \sin kx \cos kat$ (k is a constant).

48. A steady-state temperature function of $u = u(x, y)$ for a thin, flat plate satisfies *Laplace's equation*

$$\frac{\partial^2 u}{\partial x^2} + \frac{\partial^2 u}{\partial y^2} = 0.$$

Determine which of the following functions satisfy Laplace's equation: (a) $u = \ln(\sqrt{x^2 + y^2})$; (b) $u = \sqrt{x^2 + y^2}$; (c) $u = \arctan(y/x)$; (d) $u = e^{-x} \sin y$.

49. The **ideal gas law** $pV = nRT$ (n is the number of moles of the gas, R is a constant) determines each of the three variables p, V, and T (pressure, volume, and temperature) as functions of the other two. Show that

$$\frac{\partial p}{\partial V} \cdot \frac{\partial V}{\partial T} \cdot \frac{\partial T}{\partial p} = -1.$$

50. It is geometrically evident that every plane tangent to the cone $z^2 = x^2 + y^2$ passes through the origin. Show this by methods of calculus.

51. There is only one point at which the plane tangent to the surface $z = x^2 + 2xy + 2y^2 - 6x + 8y$ is horizontal. Find it.

52. Show that the plane tangent to the paraboloid with equation $z = x^2 + y^2$ at the point (a, b, c) intersects the xy-plane in the line with equation $2ax + 2by = a^2 + b^2$.

Then show that this line is tangent to the circle with equation $4x^2 + 4y^2 = a^2 + b^2$.

53. According to van der Waals' equation, 1 mol of a gas satisfies the equation

$$\left(p + \frac{a}{V^2}\right)(V - b) = (82.06)T$$

where p, V, and T are as in Example 2. For carbon dioxide, $a = 3.59 \times 10^6$ and $b = 42.7$, and V is 25,600 cm³ when p is 1 atm and $T = 313$ K. (a) Compute $\partial V/\partial p$ by differentiating van der Waals' equation with T held constant. Then estimate the change in volume that would result from an increase of 0.1 atm of pressure with T at 313 K. (b) Compute $\partial V/\partial T$ by differentiating van der Waals' equation with p held constant. Then estimate the change in volume that would result from an increase of 1 K in temperature with p held at 1 atm.

54. A *minimal surface* has the least surface area of all surfaces with the same boundary. Figure 14.4.10 shows *Scherk's minimal surface*. It has the equation

$$z = \ln(\cos x) - \ln(\cos y).$$

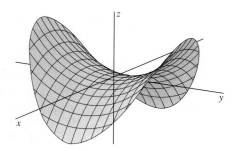

Fig. 14.4.10 Scherk's minimal surface (Problem 54)

A minimal surface $z = f(x, y)$ is known to satisfy the partial differential equation

$$(1 + z_y^2)z_{xx} - 2z_x z_y z_{xy} + (1 + z_x^2)z_{yy} = 0.$$

Verify this in the case of Scherk's minimal surface.

55. We say that the function $z = f(x, y)$ is **harmonic** if it satisfies Laplace's equation $z_{xx} + z_{yy} = 0$ (see Problem 48). Show that each of the functions in (a) through (d) is harmonic:

(a) $f_1(x, y) = \sin x \sinh(\pi - y)$;

(b) $f_2(x, y) = \sinh 2x \sin 2y$;

(c) $f_3(x, y) = \sin 3x \sinh 3y$;

(d) $f_4(x, y) = \sinh 4(\pi - x) \sin 4y$.

56. Figure 14.4.11 shows the graph of the function defined by

$$z = \sum_{i=1}^{4} f_i(x, y)$$

for $0 \leq x \leq \pi$, $0 \leq y \leq \pi$. Explain why z is harmonic (see Problem 55).

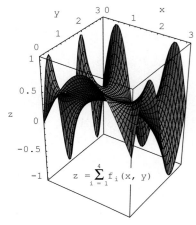

Fig. 14.4.11 The surface $z = f(x, y)$ of Problem 56

14.5

Maxima and Minima of Functions of Several Variables

The single-variable maximum-minimum techniques of Section 3.5 generalize readily to functions of several variables. We consider first a function f of two variables. Suppose that we are interested in the extreme values attained by $f(x, y)$ on a plane region R that consists of the points on and within a simple closed curve C (Fig. 14.5.1). We say that the function f attains its **absolute,** or **global, maximum value** M on R at the point (a, b) of R provided that

$$f(x, y) \leq M = f(a, b)$$

for all points (x, y) of R. Similarly, f attains its **absolute,** or **global, minimum value** m on R at the point (c, d) of R provided that $f(x, y) \geq m = f(c, d)$ for all points (x, y) of R. Theorem 1, proved in advanced calculus courses, guarantees the existence of absolute maximum and minimum values in many situations of practical interest.

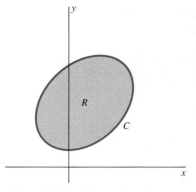

Fig. 14.5.1 A bounded plane region R whose boundary is the simple closed curve C

Theorem 1 *Existence of Extreme Values*

Suppose that the function f is continuous on the region R that consists of the points on and within a simple closed curve C in the plane. Then f attains an absolute maximum value at some point (a, b) of R and attains an absolute minimum value at some point (c, d) of R.

Theorem 1 is the two-dimensional analogue of the theorem we used so much in Chapters 3 and 4: A continuous function (of one variable) defined on the closed and bounded interval $[a, b]$ attains its absolute maximum value at some point of $[a, b]$ as well as its absolute minimum at some point of $[a, b]$.

But, as in single-variable calculus, the next natural question is this: How do we *find* these extrema? We are interested mainly in the case in which the function f attains its absolute maximum (or minimum) value at an interior point of R. The point (a, b) of R is called an **interior point** of R provided that some circular disk centered at (a, b) lies wholly within R. The interior points of a region R of the sort described in Theorem 1 are precisely those that do *not* lie on the boundary curve C.

An absolute extreme value attained by the function f at an *interior* point of R is necessarily a local extreme value. We say that $f(a, b)$ is a **local maximum value** of f if there is a circular disk D centered at (a, b) such that f is defined on D and $f(x, y) \leqq f(a, b)$ for all points (x, y) of D. If the inequality is reversed, then $f(a, b)$ is a **local minimum value** of f.

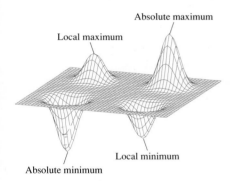

Fig. 14.5.2 Local extrema contrasted with global extrema (Example 1)

EXAMPLE 1 Figure 14.5.2 shows the graph of a certain function $f(x, y)$ defined on a region R in the xy-plane bounded by a simple closed curve C. We can think of the surface $z = f(x, y)$ as formed by starting with a stretched elastic membrane with fixed boundary C and then poking two fingers upward and two downward. The two fingers in both directions have different lengths. Looking at the four extreme values of $f(x, y)$, we see

❑ A local maximum that is not an absolute maximum,

❑ A local maximum that is also an absolute maximum,

❑ A local minimum that is not an absolute minimum, and

❑ A local minimum that is also an absolute minimum.

We saw in Theorem 2 of Section 3.5 that if the function f has a local maximum or local minimum value at a point $x = c$ where f is differentiable, then $f'(c) = 0$. We shall now show that an analogous event occurs in the two-dimensional situation: If $f(a, b)$ is either a local maximum value or a local minimum value of the function $f(x, y)$, then both $f_x(a, b)$ and $f_y(a, b)$ are zero, provided that these two partial derivatives exist at the point (a, b).

Suppose, for example, that $f(a, b)$ is a local maximum value of $f(x, y)$ and that both the partial derivatives $f_x(a, b)$ and $f_y(a, b)$ exist. We look at cross sections of the graphs of $z = f(x, y)$ in the same way as when we defined these partial derivatives in Section 14.4. Let

$$G(x) = f(x, b) \quad \text{and} \quad H(y) = f(a, y).$$

Because f is defined on a circular disk centered at (a, b), it follows that $G(x)$ is defined on some open interval containing the point $x = a$ and that $H(y)$ is defined on some open interval containing the point $y = b$. But we are assuming that f has a local maximum at (a, b), so it follows that $G(x)$ has a local maximum at $x = a$ and that $H(y)$ has a local maximum at $y = b$. The single-variable maximum-minimum result just cited therefore implies that $G'(a) = 0$ and that $H'(b) = 0$. But

$$G'(a) = \lim_{h \to 0} \frac{G(a + h) - G(a)}{h} = \lim_{h \to 0} \frac{f(a + h, b) - f(a, b)}{h} = f_x(a, b)$$

and

$$H'(b) = \lim_{k \to 0} \frac{H(b + k) - H(b)}{k} = \lim_{k \to 0} \frac{f(a, b + k) - f(a, b)}{k} = f_y(a, b).$$

Hence we conclude that $f_x(a, b) = 0 = f_y(a, b)$. A similar argument gives the same conclusion if $f(a, b)$ is a local minimum value of f. This discussion establishes Theorem 2.

Theorem 2   *Necessary Conditions for Local Extrema*
Suppose that $f(x, y)$ attains a local maximum value or a local minimum value at the point (a, b) and that both the partial derivatives $f_x(a, b)$ and $f_y(a, b)$ exist. Then

$$f_x(a, b) = 0 = f_y(a, b). \tag{1}$$

Equations (1) imply that the plane tangent to the surface $z = f(x, y)$ must be horizontal at any local maximum or local minimum point $(a, b, f(a, b))$, in perfect analogy to the single-variable case (in which the tangent line is horizontal at any local maximum or minimum point on the graph of a differentiable function).

EXAMPLE 2   Consider the three familiar surfaces

$$z = f(x, y) = x^2 + y^2,$$

$$z = g(x, y) = -x^2 - y^2, \quad \text{and}$$

$$z = h(x, y) = y^2 - x^2$$

shown in Fig. 14.5.3. In each case $\partial z / \partial x = \pm 2x$ and $\partial z / \partial y = \pm 2y$. Thus both partial derivatives are zero at the origin $(0, 0)$ (and only there). It is clear from the figure that $f(x, y) = x^2 + y^2$ has a local minimum at $(0, 0)$. In fact, because a square cannot be negative, $z = x^2 + y^2$ has the global minimum value 0 at $(0, 0)$. Similarly, $g(x, y)$ has a local (indeed, global) maximum value at $(0, 0)$, whereas $h(x, y)$ has neither a local minimum nor a local maximum value there—the origin is a *saddle point* of h. This example shows that a point (a, b) where

$$\frac{\partial z}{\partial x} = 0 = \frac{\partial z}{\partial y}$$

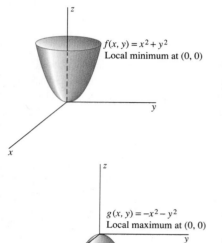

$f(x, y) = x^2 + y^2$
Local minimum at $(0, 0)$

$g(x, y) = -x^2 - y^2$
Local maximum at $(0, 0)$

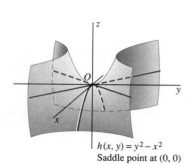

$h(x, y) = y^2 - x^2$
Saddle point at $(0, 0)$

Fig. 14.5.3 When both partial derivatives are zero, there may be (a) a minimum, (b) a maximum, or (c) neither (Example 2).

may correspond to either a local minimum, a local maximum, or neither. Thus the necessary condition in Eq. (1) is *not* a sufficient condition for a local extremum.

EXAMPLE 3 Find all points on the surface

$$z = \tfrac{3}{4}y^2 + \tfrac{1}{24}y^3 - \tfrac{1}{32}y^4 - x^2$$

at which the tangent plane is horizontal.

Solution We first calculate the partial derivatives $\partial z/\partial x$ and $\partial z/\partial y$:

$$\frac{\partial z}{\partial x} = -2x,$$

$$\frac{\partial z}{\partial y} = \frac{3}{2}y + \frac{1}{8}y^2 - \frac{1}{8}y^3 = -\frac{1}{8}y(y^2 - y - 12) = -\frac{1}{8}y(y + 3)(y - 4).$$

We next equate both $\partial z/\partial x$ and $\partial z/\partial y$ to zero. This yields

$$-2x = 0 \quad \text{and} \quad -\tfrac{1}{8}y(y + 3)(y - 4) = 0.$$

Simultaneous solution of these equations yields exactly three points where both partial derivatives are zero: $(0, -3)$, $(0, 0)$, and $(0, 4)$. The three corresponding points on the surface where the tangent plane is horizontal are $(0, -3, \tfrac{99}{32})$, $(0, 0, 0)$, and $(0, 4, \tfrac{20}{3})$. These three points are indicated on the graph in Fig. 14.5.4 of this surface. (Recall that we constructed this surface in Example 7 of Section 14.2.)

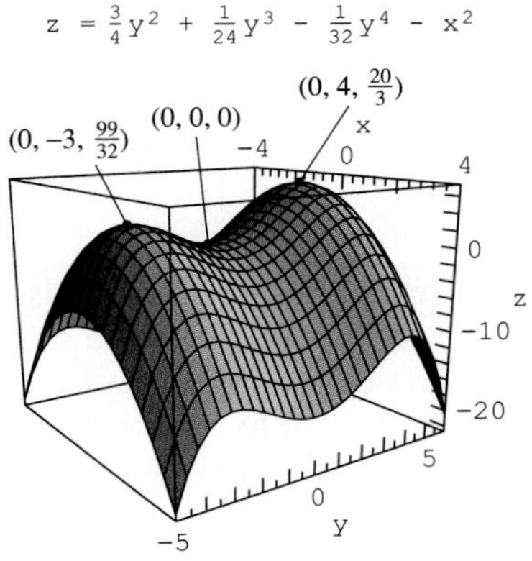

Fig. 14.5.4 The surface of Example 3

Theorem 2 is a very useful tool for finding the absolute maximum and absolute minimum values attained by a continuous function f on a region R of the type described in Theorem 1. If $f(a, b)$ is the absolute maximum value, for example, then (a, b) is either an interior point of R or a point of the boundary curve C. If (a, b) is an interior point and both the partial derivatives $f_x(a, b)$ and $f_y(a, b)$ exist, then Theorem 2 implies that both these partial derivatives must be zero. Thus we have the following result.

Theorem 3 *Types of Absolute Extrema*

Suppose that f is continuous on the plane region R consisting of the points on and within a simple closed curve C. If $f(a, b)$ is either the absolute maximum or the absolute minimum value of $f(x, y)$ on R, then (a, b) is either

1. An interior point of R at which

$$\frac{\partial f}{\partial x} = \frac{\partial f}{\partial y} = 0,$$

2. An interior point of R where not both partial derivatives exist, or

3. A point of the boundary curve C of R.

A point (a, b) where either condition (1) or (2) holds is called a **critical point** of the function f. Thus Theorem 3 says that *any extreme value of the* *continuous function f on the plane region R must occur at an interior critical point or at a boundary point.* Note the analogy with Theorem 3 of Section 3.5, which implies that an extreme value of a single-variable function $f(x)$ on a closed and bounded interval I must occur either at an interior critical point of I or at an endpoint (boundary point) of I.

As a consequence of Theorem 3, we can find the absolute maximum and minimum values of $f(x, y)$ on R as follows:

1. First, locate the interior critical points.
2. Next, find the possible extreme values of f on the boundary curve C.
3. Finally, compare the values of f at the points found in steps 1 and 2.

The technique to be used in the second step will depend on the nature of the boundary curve C, as illustrated in Example 5.

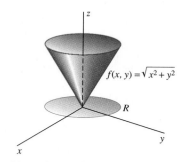

Fig. 14.5.5 The graph of the function of Example 4

EXAMPLE 4 Let $f(x, y) = \sqrt{x^2 + y^2}$ on the region R consisting of the points on and within the circle $x^2 + y^2 = 1$ in the xy-plane. The graph of f is shown in Fig. 14.5.5. We see that the minimum value 0 of f occurs at the origin $(0, 0)$, where both the partial derivatives f_x and f_y fail to exist (why?), whereas the maximum value 1 of f on R occurs at *each and every* point of the boundary circle.

EXAMPLE 5 Find the maximum and minimum values attained by the function

$$f(x, y) = xy - x - y + 3$$

at points of the triangular region R in the xy-plane with vertices at $(0, 0)$, $(2, 0)$, and $(0, 4)$.

Solution The region R is shown in Fig. 14.5.6. Its boundary "curve" C consists of the segment $0 \leq x \leq 2$ on the x-axis, the segment $0 \leq y \leq 4$ on the y-axis, and the part of the line $2x + y = 4$ that lies in the first quadrant. Any interior extremum must occur at a point where both

$$\frac{\partial f}{\partial x} = y - 1 \quad \text{and} \quad \frac{\partial f}{\partial y} = x - 1$$

are zero. Hence the only interior critical point is $(1, 1)$.

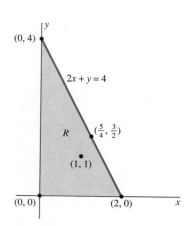

Fig. 14.5.6 The triangular region of Example 5

Along the edge where $y = 0$: The function $f(x, y)$ takes the form

$$\alpha(x) = f(x, 0) = 3 - x, \qquad 0 \leq x \leq 2.$$

Because $\alpha(x)$ is a decreasing function, its extrema for $0 \leq x \leq 2$ occur at the endpoints $x = 0$ and $x = 2$. This gives the two possibilities $(0, 0)$ and $(2, 0)$ for locations of extrema of $f(x, y)$.

Along the edge where $x = 0$: The function $f(x, y)$ takes the form

$$\beta(y) = f(0, y) = 3 - y, \qquad 0 \leq y \leq 4.$$

The endpoints of this interval yield the possibilities $(0, 0)$ and $(0, 4)$ as possibilities for locations of extrema of $f(x, y)$.

Ch. 14 / Partial Differentiation

On the edge of R where $y = 4 - 2x$: We may substitute $4 - 2x$ for y into the formula for $f(x, y)$ and thus express f as a function of a single variable:

$$\gamma(x) = x(4 - 2x) - x - (4 - 2x) + 3$$

$$= -2x^2 + 5x - 1, \qquad 0 \leq x \leq 2.$$

To find the extreme values of $\gamma(x)$, we first calculate

$$\gamma'(x) = -4x + 5;$$

$\gamma'(x) = 0$ where $x = \frac{5}{4}$. Thus each extreme value of $\gamma(x)$ on $[0, 2]$ must occur either at the interior point $x = \frac{5}{4}$ of the interval $[0, 2]$ or at one of the endpoints $x = 0$ and $x = 2$. This gives the possibilities $(0, 4)$, $(\frac{5}{4}, \frac{3}{2})$, and $(2, 0)$ for locations of extrema of $f(x, y)$.

We conclude by evaluating f at each of the points we have found:

$$f(0, 0) = 3, \qquad \longleftarrow \quad \text{maximum}$$

$$f(\tfrac{5}{4}, \tfrac{3}{2}) = 2.125,$$

$$f(1, 1) = 2,$$

$$f(2, 0) = 1,$$

$$f(0, 4) = -1. \qquad \longleftarrow \quad \text{minimum}$$

Thus the maximum value of $f(x, y)$ on the region R is $f(0, 0) = 3$, and the minimum value is $f(0, 4) = -1$.

Note the terminology used throughout this section. In Example 5, the maximum *value* of f is 3, the maximum value *occurs* at the point $(0, 0)$ in the domain of f, and the *highest point* on the graph of f is $(0, 0, 3)$.

HIGHEST AND LOWEST POINTS OF SURFACES

In applied problems we frequently know in advance that the absolute maximum (or minimum) value of $f(x, y)$ on R occurs at an *interior* point of R where both partial derivatives of f exist. In this important case, Theorem 3 tells us that we can locate every possible point at which the maximum might occur by simultaneously solving the two equations

$$f_x(x, y) = 0 \quad \text{and} \quad f_y(x, y) = 0. \tag{2}$$

If we are lucky, these equations have only one simultaneous solution (x, y) interior to R. If so, then *that* solution must be the location of the desired maximum. If we find that the equations in (2) have several simultaneous solutions interior to R, then we simply evaluate f at each solution to determine which yields the largest value of $f(x, y)$ and is therefore the desired maximum point.

We can use this method to find the lowest point on a surface $z = f(x, y)$ that opens upward, as in Fig. 14.5.7. If R is a sufficiently large rectangle, $f(x, y)$ attains large positive values everywhere on the boundary of R but smaller values at interior points. It follows that the minimum value of $f(x, y)$ must be attained at an interior point of R.

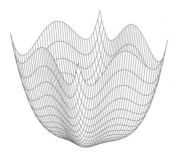

Fig. 14.5.7 The surface $z = x^4 + y^4 - x^2 y^2$ opens upward.

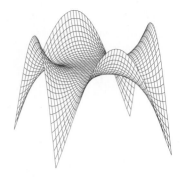

Fig. 14.5.8 The surface $z = x^4 + y^4 - 4x^2y^2$ opens both upward and downward.

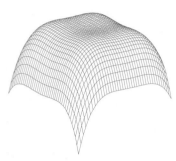

Fig. 14.5.9 The surface $z = \frac{8}{3}x^3 + 4y^3 - x^4 - y^4$ opens downward (Example 6).

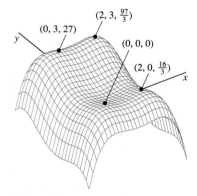

Fig. 14.5.10 The critical points of Example 6

The question of a highest or lowest point is not pertinent for a surface that opens both upward and downward, as in Fig. 14.5.8.

EXAMPLE 6 Find the highest point on the surface

$$z = \tfrac{8}{3}x^3 + 4y^3 - x^4 - y^4. \tag{3}$$

Solution Because of the negative fourth-degree terms in Eq. (3) that predominate when $|x|$ and/or $|y|$ is large, this surface opens downward (Fig. 14.5.9). We can verify this fact by writing

$$z = (x^4 + y^4)\left(-1 + \frac{\frac{8}{3}x^3 + 4y^3}{x^4 + y^4}\right)$$

and then substituting $x = r\cos\theta$, $y = r\sin\theta$:

$$z = (x^4 + y^4)\left(-1 + \frac{\frac{8}{3}\cos^3\theta + 4\sin^3\theta}{r(\cos^4\theta + \sin^4\theta)}\right).$$

It is now clear that the fraction approaches zero as $r \to \infty$ and hence that $z < 0$ if either $|x|$ or $|y|$ is large.

But $z = z(x, y)$ does attain positive values, such as $z(1, 1) = \frac{14}{3}$. So let us find the maximum value of z.

Because the partial derivatives of z with respect to x and y exist everywhere, Theorem 3 implies that we need only solve the equations $\partial z/\partial x = 0$ and $\partial z/\partial y = 0$ in Eq. (2)—that is,

$$\frac{\partial z}{\partial x} = 8x^2 - 4x^3 = 4x^2(2 - x) = 0,$$

$$\frac{\partial z}{\partial y} = 12y^2 - 4y^3 = 4y^2(3 - y) = 0.$$

If these two equations are satisfied, then

Either $x = 0$ or $x = 2$ and either $y = 0$ or $y = 3$.

It follows that either

| $x = 0$ and $y = 0$ | or | $x = 0$ and $y = 3$ | or | $x = 2$ and $y = 0$ | or | $x = 2$ and $y = 3$. |

Consequently, we need only inspect the values

$$z(0, 0) = 0,$$

$$z(2, 0) = \tfrac{16}{3} = 5.333\,333\,333\ldots,$$

$$z(0, 3) = 27,$$

$$z(2, 3) = \tfrac{97}{3} = 32.333\,333\,333\ldots \longleftarrow \text{maximum}$$

Thus the highest point on the surface is the point $(2, 3, \frac{97}{3})$. The four critical points on the surface are indicated in Fig. 14.5.10.

782

APPLIED MAXIMUM-MINIMUM PROBLEMS

The analysis of a multivariable applied maximum-minimum problem involves the same general steps that we listed at the beginning of Section 3.6. Here, however, we will express the dependent variable—the quantity to be maximized or minimized—as a function $f(x, y)$ of *two* independent variables. Once we have identified the appropriate region in the xy-plane as the domain of f, the methods of this section are applicable. We often find that a preliminary step is required: If the meaningful domain of definition of f is an unbounded region, then we first restrict f to a *bounded* plane region R on which we know the desired extreme value occurs. This procedure is similar to the one we used with open-interval maximum-minimum problems in Section 4.4.

EXAMPLE 7 Find the minimum cost of a rectangular box with volume 48 ft³ if the front and back cost \$1/ft², top and bottom cost \$2/ft², and two ends cost \$3/ft². (We first discussed such a box in Section 14.1.) This box is shown in Fig. 14.5.11.

Solution We found in Section 14.1 that the cost C (in dollars) of this box is given by

$$C(x, y) = 4xy + \frac{288}{x} + \frac{96}{y}$$

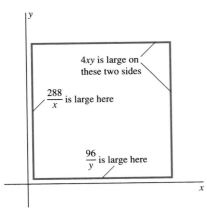

Fig. 14.5.11 A box whose total cost we want to minimize (Example 7)

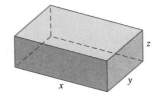

4xy is large on these two sides

$\frac{288}{x}$ is large here

$\frac{96}{y}$ is large here

Fig. 14.5.12 The cost function C of Example 7 takes on large positive values on the boundary of the square.

in terms of its length x and width y. Let R be a square such as the one shown in Fig. 14.5.12. Two sides of R are so close to the coordinate axes that $288/x > 1000$ on the side nearest the y-axis and $96/y > 1000$ on the side nearest the x-axis. Also, the square is so large that $4xy > 1000$ on both of the other two sides. This means that $C(x, y) > 1000$ at every point (x, y) of the first quadrant that lies on or outside the boundary of the square R. Because $C(x, y)$ attains reasonably small values within R (for instance, $C(1, 1) = 388$), it is clear that the absolute minimum of C must occur at an interior point of R. Thus, although the natural domain of the cost function $C(x, y)$ is the entire first quadrant, we have succeeded in restricting its domain to a region R of the sort to which Theorem 3 applies.

We therefore solve the equations

$$\frac{\partial C}{\partial x} = 4y - \frac{288}{x^2} = 0,$$

$$\frac{\partial C}{\partial y} = 4x - \frac{96}{y^2} = 0.$$

We multiply the first equation by x and the second by y. (*Ad hoc* methods are frequently required in the solution of simultaneous nonlinear equations.) This procedure gives

$$\frac{288}{x} = 4xy = \frac{96}{y},$$

so that $x = 288y/96 = 3y$. We substitute $x = 3y$ into the equation $\partial C/\partial y = 0$ and find that

$$12y - \frac{96}{y^2} = 0, \quad \text{so} \quad 12y^3 = 96.$$

Hence $y = \sqrt[3]{8} = 2$, so $x = 6$. Therefore, the minimum cost of this box is $C(6, 2) = 144$ (dollars). Because the volume of the box is $V = xyz = 48$, its height is $z = 48/(6 \cdot 2) = 4$ when $x = 6$ and $y = 2$. Thus the optimal box is 6 ft wide, 2 ft deep, and 4 ft high.

REMARK   As a check, note that the cheapest surfaces (front and back) are the largest, whereas the most expensive surfaces (the sides) are the smallest.

We have seen that if $f_x(a, b) = 0 = f_y(a, b)$, then $f(a, b)$ may be either a maximum value, a minimum value, or neither. In Section 14.10 we will discuss sufficient conditions for $f(a, b)$ to be either a local maximum or a local minimum. These conditions involve the second-order partial derivatives of f at (a, b).

The methods of this section generalize readily to functions of three or more variables. For example, if the function $f(x, y, z)$ has a local extremum at the point (a, b, c) where its three first-order partial derivatives exist, then all three must be zero there. That is,

$$f_x(a, b, c) = f_y(a, b, c) = f_z(a, b, c) = 0. \qquad (4)$$

Example 8 illustrates a "line-through-the-point" method that we can sometimes use to show that a point (a, b, c) where the conditions in (4) hold is neither a local maximum nor a local minimum point. (The method is also applicable to functions of two or of more than three variables.)

EXAMPLE 8   Determine whether the function $f(x, y, z) = xy + yz - xz$ has any local extrema.

Solution   The necessary conditions in Eq. (4) give the equations

$$f_x(x, y, z) = y - z = 0,$$
$$f_y(x, y, z) = x + z = 0,$$
$$f_z(x, y, z) = y - x = 0.$$

We easily find that the simultaneous solution of these equations is $x = y = z = 0$. On the line $x = y = z$ through $(0, 0, 0)$, the function $f(x, y, z)$ reduces to x^2, which is minimal at $x = 0$. But on the line $x = -y = z$, it reduces to $-3x^2$, which is maximal when $x = 0$. Hence f can have neither a local maximum nor a local minimum at $(0, 0, 0)$. Therefore, it has no extrema, local or global.

14.5 Problems

In Problems 1 through 12, find every point on the given surface $z = f(x, y)$ at which the tangent plane is horizontal.

1. $z = x - 3y + 5$
2. $z = 4 - x^2 - y^2$
3. $z = xy + 5$
4. $z = x^2 + y^2 + 2x$
5. $z = x^2 + y^2 - 6x + 2y + 5$
6. $z = 10 + 8x - 6y - x^2 - y^2$
7. $z = x^2 + 4x + y^3$
8. $z = x^4 + y^3 - 3y$

9. $z = 3x^2 + 12x + 4y^3 - 6y^2 + 5$ (Fig. 14.5.13)

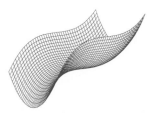

Fig. 14.5.13   The surface of Problem 9

784

10. $z = \dfrac{1}{1 - 2x + 2y + x^2 + y^2}$

11. $z = (2x^2 + 3y^2) \exp(-x^2 - y^2)$ (Fig. 14.5.14)

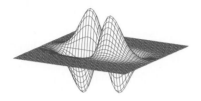

Fig. 14.5.14   The surface of Problem 11

12. $z = 2xy \exp(-\frac{1}{8}(4x^2 + y^2))$ (Fig. 14.5.15)

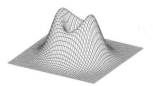

Fig. 14.5.15   The surface of Problem 12

In Problems 13 through 20, find either the highest point or the lowest point (whichever is applicable) on the surface with the given equation.

13. $z = f(x, y) = x^2 - 2x + y^2 - 2y + 3$

14. $z = f(x, y) = 6x - 8y - x^2 - y^2$

15. $z = f(x, y) = 2x - x^2 + 2y^2 - y^4$

16. $z = f(x, y) = 3x^4 + 4x^3 + 6y^4 - 16y^3 + 12y^2 + 7$

17. $z = f(x, y) = 2x^2 + 8xy + y^4$

18. $z = f(x, y) = \dfrac{1}{10 - 2x - 4y + x^2 + y^4}$

19. $z = f(x, y) = \exp(2x - 4y - x^2 - y^2)$

20. $z = f(x, y) = (1 + x^2) \exp(-x^2 - y^2)$

In Problems 21 through 26, find the maximum and minimum values attained by the given function $f(x, y)$ on the given region R.

21. $f(x, y) = x + 2y$; R is the square with vertices at $(\pm 1, \pm 1)$.

22. $f(x, y) = x^2 + y^2 - x$; R is the square of Problem 21.

23. $f(x, y) = x^2 + y^2 - 2x$; R is the triangular region with vertices at $(0, 0)$, $(2, 0)$, and $(0, 2)$.

24. $f(x, y) = x^2 + y^2 - x - y$; R is the region of Problem 23.

25. $f(x, y) = 2xy$; R is the circular disk $x^2 + y^2 \leqq 1$.

26. $f(x, y) = xy^2$; R is the circular disk $x^2 + y^2 \leqq 3$.

27. Find the dimensions x, y, and z of a rectangular box with fixed volume $V = 1000$ and minimum total surface area A.

28. Find the points on the surface $xyz = 1$ that are closest to the origin.

29. Find the dimensions of the rectangular box with maximum volume that has total surface area 600 cm^2.

30. A rectangular box without a lid is to have fixed volume 4000 cm^3. What dimensions would minimize its total surface area?

31. A rectangular box is placed in the first octant with one of its corners at the origin and three of its sides lying in the three coordinate planes. The vertex opposite the origin lies on the plane with equation $x + 2y + 3z = 6$. What is the maximum possible volume of such a box? What are the dimensions of that box?

32. The sum of three positive numbers is 120. What is the maximum possible value of their product?

33. A rectangular building is to have a volume of 8000 ft^3. Annual heating and cooling costs will amount to \$2/ft^2 for its top, front, and back, and \$4/ft^2 for the two end walls. What dimensions of the building would minimize these annual costs?

34. You must make a rectangular box with no top from materials that cost \$3/ft^2 for the bottom and \$2/ft^2 for the four sides. The box is to have volume 48 ft^3. What dimensions would minimize its cost?

35. A rectangular shipping crate is to have volume 12 m^3. Its bottom costs *twice* as much (per square meter) as its top and four sides. What dimensions would minimize the total cost of the crate?

36. Use the maximum-minimum methods of this section to find the point of the plane $2x - 3y + z = 1$ that is closest to the point $(3, -2, 1)$. [*Suggestion:* A positive quantity is minimized when its square is minimized.]

37. Find the maximum volume of a rectangular box that a post office will send if the sum of its *length* and *girth* cannot exceed 108 in.

38. Repeat Problem 37 for the case of a cylindrical box—one shaped like a hatbox or a fat mailing tube.

39. A rectangular box with its base in the xy-plane is inscribed under the graph of the paraboloid $z = 1 - x^2 - y^2$, $z \geqq 0$. Find the maximum possible volume of the box. [*Suggestion:* You may assume that the sides of the box are parallel to the vertical coordinate planes, and it follows that the box is symmetrically placed around these planes.]

40. What is the maximum possible volume of a rectangular box inscribed in a *hemisphere* of radius R? You may assume that one face of the box lies in the planar base of the hemisphere.

41. A wire 120 cm long is cut into three *or fewer* pieces, and each piece is bent into the shape of a square. How should this be done in order to minimize the total area of these squares? To maximize it?

42. You must divide a lump of putty of fixed volume V into three or fewer pieces and form the pieces into cubes. How should you do this to maximize the total surface area of the cubes? To minimize it?

43. Consider the function $f(x, y) = (y - x^2)(y - 3x^2)$.

(a) Show that $f_x(0,0) = 0 = f_y(0,0)$. (b) Show that for every straight line $y = mx$ through $(0,0)$, the function $f(x, mx)$ has a local minimum at $x = 0$. (c) Examine the values of f at points of the parabola $y = 2x^2$ to show that f does *not* have a local minimum at $(0,0)$. This tells us we cannot use the line-through-the-point method of Example 8 to show that a point *is* a local extremum.

44. A very long rectangle of sheet metal has width L and is to be folded to make a rain gutter (Fig. 14.5.16). Maximize its volume by maximizing the cross-sectional area shown in the figure. [*Suggestion:* Use the two independent variables x and θ indicated in the figure. Although the equations you obtain may appear at first to be intractable, you can, in fact, rather easily solve them for the exact values of x and θ that maximize the area.]

Fig. 14.5.16 Cross section of the rain gutter of Problem 44

45. Locate and identify the extrema of $f(x, y) = x^2 - 2xy + y^3 - y$.

46. A rectangular box with no top is to have a bottom made from material that costs $3/ft^2$ and sides made from material that costs $1/ft^2$. If the box is to have total volume 12 ft^3, what is its minimum possible cost?

47. Locate and identity the extrema of $f(x, y) = x^2y$ on the square in the plane with vertices at $(\pm 1, \pm 1)$.

48. Locate and identify the extrema of $g(x, y) = x^4 + 4xy + y^4$.

49. What is the maximum possible volume of a rectangular box if the sum of the lengths of its 12 edges is 12 m?

50. A rectangular box is inscribed in the first octant with three of its sides in the coordinate planes, their common vertex at the origin, and the opposite vertex on the plane with equation $x + 3y + 7z = 11$. What is the maximum possible volume of such a box?

51. Three sides of a rectangular box lie in the coordinate planes, their common vertex at the origin; the opposite vertex is on the plane with equation

$$\frac{x}{a} + \frac{y}{b} + \frac{z}{c} = 1$$

(a, b, and c are positive constants). In terms of a, b, and c, what is the maximum possible volume of such a box?

52. A buoy is to have the shape of a right circular cylinder capped at each end by identical right circular cones with the same radius as the cylinder. Find the minimum possible surface area of the buoy, given that it has fixed volume V.

53. You want to build a rectangular aquarium with a bottom made of slate costing 28¢/in.^{2} Its sides will be glass, which costs 5¢/in.^{2}, and its top will be stainless steel, which costs 2¢/in.^{2} The volume of this aquarium is to be 24,000 in.^{3} What are the dimensions of the least expensive such aquarium?

54. A pentagonal window is to have the shape of a rectangle surmounted by an isosceles triangle (with horizontal base), and its perimeter is to be 24 ft. What are the dimensions of such a window that will admit the most light (because its area is the greatest)?

55. Find the point (x, y) in the plane for which the sum of squares of its distances from $(0, 1)$, $(0, 0)$, and $(2, 0)$ is a minimum.

56. Find the point (x, y) in the plane for which the sum of the squares of its distances from (a_1, b_1), (a_2, b_2), and (a_3, b_3) is a minimum.

57. An A-frame house is to have fixed volume V. Its front and rear walls are in the shape of equal, parallel isosceles triangles with horizontal bases. The roof consists of two rectangles that connect pairs of upper sides of the triangles. To minimize heating and cooling costs, the total area of the A-frame (excluding the floor) is to be minimized. Describe the shape of the A-frame of minimal area.

58. What is the maximum possible volume of a rectangular box whose longest diagonal has fixed length L?

14.5 Project

The problems in this project require the use of a graphics calculator or a computer with graphing utility.

Let R denote the unit circular disk $x^2 + y^2 \leq 1$ bounded by the unit

786

circle C in the xy-plane. Suppose that we want to find the minimum and maximum values of the function

$$z = f(x, y) = 3x^2 + 4xy - 5y^2 \qquad (5)$$

at points of R.

We look first for interior possibilities. The equations

$$\frac{\partial f}{\partial x} = 6x + 4y = 0, \qquad \frac{\partial f}{\partial y} = 4x - 10y = 0$$

have only the trivial solution $x = y = 0$, so the only interior possibility is $f(0, 0) = 0$.

To investigate the boundary possibilities we use the polar-coordinates parametrization

$$x = \cos t, \qquad y = \sin t \qquad (0 \leqq t \leqq 2\pi) \qquad (6)$$

of the boundary circle C. Substitution of (6) into Eq. (5) then yields the single-variable function

$$z = g(t) = 3\cos^2 t + 4\cos t \sin t - 5\sin^2 t, \qquad (7)$$

whose extreme values we now seek.

The graph $z = g(t)$ plotted in Fig. 14.5.17 reveals a positive maximum value and a negative minimum value, both occurring at two different points in $[0, 2\pi]$. The zooms indicated in Figs. 14.5.18 and 14.5.19 yield (accurate to two decimal places) the approximate maximum value $g(0.23) \approx 3.47$ and the approximate minimum value $g(1.80) \approx -5.47$ attained by $f(x, y)$ at points of the disk R.

In Problems 1 through 3, find similarly the approximate maximum and minimum values attained by the indicated function $f(x, y)$ at points of the unit disk R: $x^2 + y^2 \leqq 1$. Let p, q, and r denote three selected integers, such as the last three nonzero digits of your student I.D. number.

1. $f(x, y) = px + qy + r$
2. $f(x, y) = px^2 + qxy + ry^2$
3. $f(x, y) = px^4 + qy^4 - rx^2y^2$

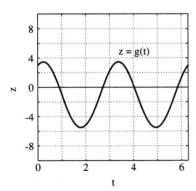

Fig. 14.5.17 $z = g(t)$ on the interval $0 \leqq t \leqq 2\pi$

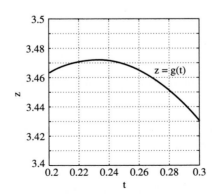

Fig. 14.5.18 A high point of Fig. 14.5.17

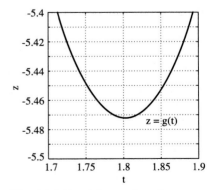

Fig. 14.5.19 A low point of Fig. 14.5.17

14.6

Increments and Differentials

In Section 4.2 we used the *differential*

$$df = f'(x)\,\Delta x \tag{1}$$

to approximate the *increment,* or actual change,

$$\Delta f = f(x + \Delta x) - f(x) \tag{2}$$

in the value of a single-variable function that results from the change Δx in the independent variable. Thus

$$\Delta f = f(x + \Delta x) - f(x) \approx f'(x)\,\Delta x = df. \tag{3}$$

We now describe the use of the partial derivatives $\partial f/\partial x$ and $\partial f/\partial y$ to approximate the **increment**

$$\Delta f = f(x + \Delta x, y + \Delta y) - f(x, y) \tag{4}$$

in the value of a two-variable function that results when its independent variables are changed simultaneously. If only x were changed and y were held constant, we could temporarily regard $f(x, y)$ as a function of x alone. Then, with $f_x(x, y)$ playing the role of $f'(x)$, the linear approximation in Eq. (3) would give

$$f(x + \Delta x, y) - f(x, y) \approx f_x(x, y)\,\Delta x \tag{5}$$

for the change in f corresponding to the change Δx in x. Similarly, if only y were changed and x were held constant, then—temporarily regarding $f(x, y)$ as a function of y alone—we would get

$$f(x, y + \Delta y) - f(x, y) \approx f_y(x, y)\,\Delta y \tag{6}$$

for the change in f corresponding to the change Δy in y.

If both x and y are changed simultaneously, we expect the *sum* of the approximations in (5) and (6) to be a good estimate of the resulting increment in the value of f. On this basis we define the **differential**

$$df = f_x(x, y)\,\Delta x + f_y(x, y)\,\Delta y \tag{7}$$

of a function $f(x, y)$ of two independent variables.

EXAMPLE 1   The differential of

$$f(x, y) = x^2 + 3xy - 2y^2$$

is

$$df = \frac{\partial f}{\partial x}\Delta x + \frac{\partial f}{\partial y}\Delta y = (2x + 3y)\,\Delta x + (3x - 4y)\,\Delta y.$$

At the point $P(3, 5)$ this differential is

$$df = 21\,\Delta x - 11\,\Delta y.$$

With $\Delta x = 0.2$ and $\Delta y = -0.1$, corresponding to a change from $P(3, 5)$ to the nearby point $Q(3.2, 4.9)$, we get

$$df = 21 \cdot 0.2 - 11 \cdot (-0.1) = 5.3.$$

The actual change in the value of f from P to Q is the increment

$$\Delta f = f(3.2, 4.9) - f(3, 5) = 9.26 - 4 = 5.26,$$

so in this example the differential seems to be a good approximation to the increment.

At the fixed point $P(a, b)$ the differential

$$df = f_x(a, b)\,\Delta x + f_y(a, b)\,\Delta y \qquad (8)$$

is a *linear* function of Δx and Δy; the coefficients $f_x(a, b)$ and $f_y(a, b)$ in this linear function depend on a and b. Thus the differential df is a **linear approximation** to the actual increment Δf. At the end of this section, we shall show that df is a very good approximation to Δf when Δx and Δy are both small, in the sense that

$$\Delta f - df = \epsilon_1\,\Delta x + \epsilon_2\,\Delta y, \qquad (9)$$

where ϵ_1 and ϵ_2 are functions of Δx and Δy that approach zero as $\Delta x \to 0$ and $\Delta y \to 0$. Hence we write

$$\Delta f - df \approx 0, \quad \text{or} \quad \Delta f \approx df$$

when Δx and Δy are small. The approximation

$$f(a + \Delta x, \ b + \Delta y) = f(a, b) + \Delta f \approx f(a, b) + df;$$

$$f(a + \Delta x, \ b + \Delta y) \approx f(a, b) + f_x(a, b)\,\Delta x + f_y(a, b)\,\Delta y \qquad (10)$$

may then be used to estimate the value of $f(a + \Delta x, b + \Delta y)$ when Δx and Δy are small and the values $f(a, b)$, $f_x(a, b)$, and $f_y(a, b)$ are all known.

EXAMPLE 2 Use Eq. (10) to estimate $\sqrt{2 \cdot (2.02)^3 + (2.97)^2}$. Note that $\sqrt{2 \cdot 2^3 + 3^2} = \sqrt{25} = 5$.

Solution We take $f(x, y) = \sqrt{2x^3 + y^2}$. Then

$$\frac{\partial f}{\partial x} = \frac{3x^2}{\sqrt{2x^3 + y^2}} \quad \text{and} \quad \frac{\partial f}{\partial y} = \frac{y}{\sqrt{2x^3 + y^2}}.$$

Now let $a = 2$, $b = 3$, $\Delta x = 0.02$, and $\Delta y = -0.03$. Then $f(2, 3) = 5$,

$$f_x(2, 3) = \tfrac{12}{5}, \quad \text{and} \quad f_y(2, 3) = \tfrac{3}{5}.$$

Hence Eq. (10) gives

$$\sqrt{2 \cdot (2.02)^3 + (2.97)^2} = f(2.02, 2.97)$$

$$\approx f(2, 3) + f_x(2, 3) \cdot (0.02) + f_y(2, 3) \cdot (-0.03)$$

$$= 5 + \tfrac{12}{5}(0.02) + \tfrac{3}{5}(-0.03) = 5.03.$$

The actual value to four decimal places is 5.0305.

If $z = f(x, y)$, we often write dz in place of df. So the differential of the dependent variable z at the point (a, b) is $dz = f_x(a, b)\,\Delta x + f_y(a, b)\,\Delta y$. At the arbitrary point (x, y), the differential of z takes the form

$$dz = f_x(x, y)\,\Delta x + f_y(x, y)\,\Delta y.$$

More simply, we can write

$$dz = \frac{\partial z}{\partial x}\,\Delta x + \frac{\partial z}{\partial y}\,\Delta y. \qquad (11)$$

It is customary to write dx for Δx and dy for Δy in this formula. When this is done, Eq. (11) takes the form

$$dz = \frac{\partial z}{\partial x}\,dx + \frac{\partial z}{\partial y}\,dy. \tag{12}$$

When we use this notation, we must realize that dx and dy have *no* connotation of being "infinitesimal" or even small. The differential dz is still simply a linear function of the ordinary real variables dx and dy, a function that gives a linear approximation to the change in z when x and y are changed by the amounts dx and dy, respectively.

EXAMPLE 3 In Example 3 of Section 14.4, we considered 1 mol of an ideal gas—its volume V in cubic centimeters given in terms of its pressure p in atmospheres and temperature T in kelvins by the formula $V = 82.06 \cdot T/p$. Approximate the change in V when p is increased from 5 atm to 5.2 atm and T is increased from 300 K to 310 K.

Solution The differential of $V = V(p, T)$ is

$$dV = \frac{\partial V}{\partial p}\,dp + \frac{\partial V}{\partial T}\,dT = -\frac{82.06 \cdot T}{p^2}\,dp + \frac{82.06}{p}\,dT.$$

With $p = 5$, $T = 300$, $dp = 0.2$, and $dT = 10$, we compute

$$dV = -\frac{82.06 \cdot 300}{5^2} \cdot 0.2 + \frac{82.06}{5} \cdot 10 = -32.8 \quad (\text{cm}^3).$$

This indicates that the gas will decrease in volume by about 33 cm³. The actual change is

$$\Delta V = \frac{82.06 \cdot 310}{5.2} - \frac{82.06 \cdot 300}{5} = 4892.0 - 4923.6 = -31.6\ (\text{cm}^3).$$

EXAMPLE 4 The point $(1, 2)$ lies on the curve with equation

$$f(x, y) = 2x^3 + y^3 - 5xy = 0.$$

Approximate the y-coordinate of the nearby point $(1.1, y)$.

Solution First we compute the differential

$$df = \frac{\partial f}{\partial x}\,dx + \frac{\partial f}{\partial y}\,dy = (6x^2 - 5y)\,dx + (3y^2 - 5x)\,dy = 0.$$

When we substitute $x = 1$, $y = 2$, and $dx = 0.1$, we find that $dy \approx 0.06$. This yields $(1.1, 2.06)$ for the approximate coordinates of the nearby point. As a check on the accuracy of this approximation, we can substitute $x = 1.1$ into the original equation to get

$$2 \cdot (1.1)^3 + y^3 - 5 \cdot 1.1 \cdot y = 0.$$

Then we solve for y, using Newton's method. This technique gives $y \approx 2.05$.

Increments and differentials of functions of more than two variables are defined similarly. A function $w = f(x, y, z)$ has *increment*

$$\Delta w = \Delta f = f(x + \Delta x, y + \Delta y, z + \Delta z) - f(x, y, z)$$

and *differential*

$$dw = df = \frac{\partial f}{\partial x} \Delta x + \frac{\partial f}{\partial y} \Delta y + \frac{\partial f}{\partial z} \Delta z;$$

that is,

$$dw = \frac{\partial w}{\partial x} dx + \frac{\partial w}{\partial y} dy + \frac{\partial w}{\partial z} dz,$$

if [as in Eq. (11)] we write dx for Δx, dy for Δy, and dz for Δz.

EXAMPLE 5   You have constructed a metal cube that is supposed to have edge length 100 mm, but each of its three measured dimensions x, y, and z may be in error by as much as a millimeter. Use differentials to estimate the maximum resulting error in its calculated volume $V = xyz$.

Solution   We need to approximate the increment

$$\Delta V = V(100 + dx, 100 + dy, 100 + dz) - V(100, 100, 100)$$

when the errors dx, dy, and dz in x, y, and z are maximal. The differential of $V = xyz$ is

$$dV = yz \, dx + xz \, dy + xy \, dz.$$

When we substitute $x = y = z = 100$ and $dx = \pm 1$, $dy = \pm 1$, and $dz = \pm 1$, we get

$$dV = 100 \cdot 100 \cdot (\pm 1) + 100 \cdot 100 \cdot (\pm 1) + 100 \cdot 100 \cdot (\pm 1) = \pm 30{,}000.$$

It may surprise you to find that an error of only a millimeter in each dimension of a cube can result in an error of 30,000 mm³ in its volume. (For a cube made of precious metal, an error of 30 cm³ in its volume could correspond to a difference of hundreds or thousands of dollars in its cost.)

THE LINEAR APPROXIMATION THEOREM

The differential $df = f_x \, dx + f_y \, dy$ is defined provided that both partial derivatives f_x and f_y exist. Theorem 1 gives sufficient conditions for df to be a good approximation to the increment Δf when Δx and Δy are small.

Theorem 1   *Linear Approximation*
Suppose that $f(x, y)$ has continuous first-order partial derivatives in a rectangular region that has horizontal and vertical sides and that contains the points $P(a, b)$ and $Q(a + \Delta x, b + \Delta y)$ in its interior. Let

$$\Delta f = f(a + \Delta x, b + \Delta y) - f(a, b)$$

be the corresponding increment in the value of f. Then

$$\Delta f = f_x(a, b) \, \Delta x + f_y(a, b) \, \Delta y + \epsilon_1 \, \Delta x + \epsilon_2 \, \Delta y, \qquad (13)$$

where ϵ_1 and ϵ_2 are functions of Δx and Δy that approach zero as $\Delta x \to 0$ and $\Delta y \to 0$.

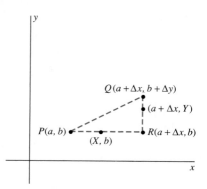

Fig. 14.6.1 Illustration of the proof of the linear approximation theorem

Proof   If R is the point $(a + \Delta x, b)$ indicated in Fig. 14.6.1, then

$$\Delta f = f(Q) - f(P) = [f(R) - f(P)] + [f(Q) - f(R)]$$
$$= [f(a + \Delta x, b) - f(a, b)]$$
$$+ [f(a + \Delta x, b + \Delta y) - f(a + \Delta x, b)]. \qquad (14)$$

We consider separately the two terms on the right in Eq. (14). For the first term, we define the single-variable function

$$g(x) = f(x, b) \qquad \text{for } x \text{ in } [a, a + \Delta x].$$

Then the mean value theorem gives

$$f(a + \Delta x, b) - f(a, b) = g(a + \Delta x) - g(a) = g'(X) \, \Delta x = f_x(X, b) \, \Delta x$$

for some number X in the open interval $(a, a + \Delta x)$.

For the second term on the right in Eq. (14), we define the single-variable function

$$h(y) = f(a + \Delta x, y) \qquad \text{for } y \text{ in } [b, b + \Delta y].$$

The mean value theorem now yields

$$f(a + \Delta x, b + \Delta y) - f(a + \Delta x, b) = h(b + \Delta y) - h(b)$$
$$= h'(Y) \, \Delta y = f_y(a + \Delta x, Y) \, \Delta y$$

for some number Y in the open interval $(b, b + \Delta y)$.

When we substitute these two results into Eq. (14), we find that

$$\Delta f = f_x(X, b) \, \Delta x + f_y(a + \Delta x, Y) \, \Delta y$$
$$= [f_x(a, b) + f_x(X, b) - f_x(a, b)] \, \Delta x$$
$$+ [f_y(a, b) + f_y(a + \Delta x, Y) - f_y(a, b)] \, \Delta y.$$

So

$$\Delta f = f_x(a, b) \, \Delta x + f_y(a, b) \, \Delta y + \epsilon_1 \, \Delta x + \epsilon_2 \, \Delta y,$$

where

$$\epsilon_1 = f_x(X, b) - f_x(a, b) \quad \text{and} \quad \epsilon_2 = f_y(a + \Delta x, Y) - f_y(a, b).$$

Finally, because both the points (X, b) and $(a + \Delta x, Y)$ approach (a, b) as $\Delta x \to 0$ and $\Delta y \to 0$, it follows from the continuity of f_x and f_y that both ϵ_1 and ϵ_2 approach zero as Δx and Δy approach zero. This completes the proof.   ❑

The function defined in Problem 39 illustrates the fact that a function $f(x, y)$ of two variables can have partial derivatives at a point without even being continuous there. Thus the mere existence of partial derivatives means much less for a function of two (or more) variables than it does for a single-variable function. But it follows from the linear approximation theorem here that a function with partial derivatives that are *continuous* at every point within a circle is itself continuous within that circle (see Problem 40).

A function f of two variables is said to be **differentiable** at the point (a, b) if both $f_x(a, b)$ and $f_y(a, b)$ exist *and* there exist functions ϵ_1 and ϵ_2 of Δx and Δy that approach zero as Δx and Δy do and for which Eq. (13) holds.

Ch. 14 / Partial Differentiation

We shall have little need for this concept of differentiability, because it will always suffice for us to assume that our functions have continuous partial derivatives.

We can generalize Theorem 1 to functions of three or more variables. For example, if $w = f(x, y, z)$, then the analogue of Eq. (13) is

$$\Delta f = f_x(a, b, c) \, \Delta x + f_y(a, b, c) \, \Delta y + f_z(a, b, c) \, \Delta z$$
$$+ \, \epsilon_1 \, \Delta x + \epsilon_2 \, \Delta y + \epsilon_3 \, \Delta z,$$

where ϵ_1, ϵ_2, and ϵ_3 all approach zero as Δx, Δy, and Δz approach zero. The proof for the three-variable case is like that given here for two variables.

14.6 Problems

Find the differential dw in Problems 1 through 16.

1. $w = 3x^2 + 4xy - 2y^3$ **2.** $w = \exp(-x^2 - y^2)$

3. $w = \sqrt{1 + x^2 + y^2}$ **4.** $w = xye^{x+y}$

5. $w = \arctan\left(\dfrac{y}{x}\right)$

6. $w = xz^2 - yx^2 + zy^2$

7. $w = \ln(x^2 + y^2 + z^2)$ **8.** $w = \sin xyz$

9. $w = x \tan yz$ **10.** $w = xye^{uv}$

11. $w = e^{-xyz}$ **12.** $w = \ln(1 + rs)$

13. $w = u^2 \exp(-v^2)$ **14.** $w = \dfrac{s + t}{s - t}$

15. $w = \sqrt{x^2 + y^2 + z^2}$

16. $w = pqr \exp(-p^2 - q^2 - r^2)$

In Problems 17 through 23, use differentials to approximate $\Delta f = f(Q) - f(P)$.

17. $f(x, y) = \sqrt{x^2 + y^2}$; $P(3, 4)$, $Q(2.97, 4.04)$

18. $f(x, y) = \sqrt{x^2 - y^2}$; $P(13, 5)$, $Q(13.2, 4.9)$

19. $f(x, y) = \dfrac{1}{1 + x + y}$; $P(3, 6)$, $Q(3.02, 6.05)$

20. $f(x, y, z) = \sqrt{xyz}$; $P(1, 3, 3)$, $Q(0.9, 2.9, 3.1)$

21. $f(x, y, z) = \sqrt{x^2 + y^2 + z^2}$; $P(3, 4, 12)$, $Q(3.03, 3.96, 12.05)$

22. $f(x, y, z) = \dfrac{xyz}{x + y + z}$; $P(2, 3, 5)$, $Q(1.98, 3.03, 4.97)$

23. $f(x, y, z) = e^{-xyz}$; $P(1, 0, -2)$, $Q(1.02, 0.03, -2.02)$

In Problems 24 through 29, use differentials to approximate the indicated number.

24. $(\sqrt{26})(\sqrt[3]{28})(\sqrt[4]{17})$ **25.** $(\sqrt{15} + \sqrt{99})^2$

26. $\dfrac{\sqrt[3]{25}}{\sqrt[5]{30}}$ **27.** $e^{0.4} = \exp(1.1^2 - 0.9^2)$

28. The y-coordinate of the point P near $(1, 2)$ on the curve $2x^3 + 2y^3 = 9xy$, if the x-coordinate of P is 1.1

29. The x-coordinate of the point P near $(2, 4)$ on the curve $4x^4 + 4y^4 = 17x^2y^2$, if the y-coordinate of P is 3.9

30. The base and height of a rectangle are measured as 10 cm and 15 cm, respectively, with a possible error of as much as 0.1 cm in each measurement. Use differentials to estimate the maximum resulting error in computing the area of the rectangle.

31. The base radius r and height h of a right circular cone are measured as 5 in. and 10 in., respectively. There is a possible error of as much as $\frac{1}{16}$ in. in each measurement. Use differentials to estimate the maximum resulting error that might occur in computing the volume of the cone.

32. The dimensions of a closed rectangular box are found by measurement to be 10 cm by 15 cm by 20 cm, but there is a possible error of 0.1 cm in each. Use differentials to estimate the maximum resulting error in computing the total surface area of the box.

33. A surveyor wants to find the area in acres of a certain field (1 acre is 43,560 ft²). She measures two adjacent sides, finding them to be $a = 500$ ft and $b = 700$ ft, with a possible error of as much as 1 ft in each measurement. She finds the angle between these two sides to be $\theta = 30°$, with a possible error of as much as 0.25°. The field is triangular, so its area is given by $A = \frac{1}{2}ab \sin \theta$. Use differentials to estimate the maximum resulting error, in acres, in computing the area of the field by this formula.

34. Use differentials to estimate the change in the volume of the gas of Example 3 if its pressure is decreased from 5 atm to 4.9 atm and its temperature is decreased from 300 K to 280 K.

35. The period of oscillation of a pendulum of length L is given (approximately) by the formula $T = 2\pi \sqrt{L/g}$. Estimate the change in the period of a pendulum if its length is increased from 2 ft to 2 ft 1 in. and it is simultaneously moved from a location where g is exactly 32 ft/s² to one where $g = 32.2$ ft/s².

36. Given the pendulum of Problem 35, show that the relative error in the determination of T is half the difference of the relative errors in measuring L and g—that is, that

$$\frac{dT}{T} = \frac{1}{2}\left(\frac{dL}{L} - \frac{dg}{g}\right).$$

37. The range of a projectile fired (in a vacuum) with initial velocity v_0 and inclination angle α from the horizontal is $R = \frac{1}{32}v_0^2 \sin 2\alpha$. Use differentials to approximate the change in range if v_0 is increased from 400 ft/s to 410 ft/s and α is increased from $30°$ to $31°$.

38. A horizontal beam is supported at both ends and supports a uniform load. The deflection, or sag, at its midpoint is given by

$$S = \frac{k}{wh^3}, \qquad (15)$$

where w and h are the width and height, respectively, of the beam and k is a constant that depends on the length and composition of the beam and the amount of the load. Show that

$$dS = -S\left(\frac{1}{w}\,dw + \frac{3}{h}\,dh\right).$$

If $S = 1$ in. when $w = 2$ in. and $h = 4$ in., approximate the sag when $w = 2.1$ in. and $h = 4.1$ in. Compare your approximation with the actual value you compute from Eq. (15).

39. Let the function f be defined on the whole xy-plane by $f(x, y) = 1$ if $x = y \neq 0$, whereas $f(x, y) = 0$ otherwise. (a) Show that f is not continuous at $(0, 0)$. (b) Show that both partial derivatives f_x and f_y exist at $(0, 0)$.

40. Deduce from Eq. (13) that under the hypotheses of the linear approximation theorem, $\Delta f \to 0$ as $\Delta x \to 0$ and $\Delta y \to 0$. What does this imply about the continuity of f at the point (a, b)?

14.7
The Chain Rule

The single-variable chain rule expresses the derivative of a composite function $f(g(t))$ in terms of the derivatives of f and g:

$$D_t f(g(t)) = f'(g(t)) \cdot g'(t). \qquad (1)$$

With $w = f(x)$ and $x = g(t)$, the chain rule says that

$$\frac{dw}{dt} = \frac{dw}{dx}\frac{dx}{dt}. \qquad (2)$$

The simplest multivariable chain rule situation involves a function $w = f(x, y)$, where both x and y are functions of the same single variable t: $x = g(t)$ and $y = h(t)$. The composite function $f(g(t), h(t))$ is then a single-variable function of t, and Theorem 1 expresses its derivative in terms of the partial derivatives of f and the ordinary derivatives of g and h. We assume that the stated hypotheses hold on suitable domains such that the composite function is defined.

Theorem 1   The Chain Rule

Suppose that $w = f(x, y)$ has continuous first-order partial derivatives and that $x = g(t)$ and $y = h(t)$ are differentiable functions. Then w is a differentiable function of t, and

$$\frac{dw}{dt} = \frac{\partial w}{\partial x}\cdot\frac{dx}{dt} + \frac{\partial w}{\partial y}\cdot\frac{dy}{dt}. \qquad (3)$$

The variable notation of Eq. (3) ordinarily will be more useful than function notation. Remember, in any case, that the partial derivatives in Eq. (3) are to be evaluated at the point $(g(t), h(t))$, so in function notation Eq. (3) is

$$D_t [f(g(t), h(t))] = f_x(g(t), h(t)) \cdot g'(t) + f_y(g(t), h(t)) \cdot h'(t). \qquad (4)$$

EXAMPLE 1 Suppose that $w = e^{xy}$, $x = t^2$, and $y = t^3$. Then

$$\frac{\partial w}{\partial x} = ye^{xy}, \qquad \frac{\partial w}{\partial y} = xe^{xy}, \qquad \frac{dx}{dt} = 2t, \quad \text{and} \quad \frac{dy}{dt} = 3t^2.$$

So Eq. (3) yields

$$\frac{dw}{dt} = \frac{\partial w}{\partial x} \cdot \frac{dx}{dt} + \frac{\partial w}{\partial y} \cdot \frac{dy}{dt} = (ye^{xy})(2t) + (xe^{xy})(3t^2)$$

$$= (t^3 e^{t^5})(2t) + (t^2 e^{t^5})(3t^2) = 5t^4 e^{t^5}.$$

Had our purposes not been to illustrate the multivariable chain rule, we could have obtained the same result more simply by writing

$$w = e^{xy} = e^{(t^2)(t^3)} = e^{t^5}$$

and then differentiating w as a single-variable function of t.

A proof of the chain rule is included at the end of this section. In outline, it consists of beginning with the linear approximation

$$\Delta w \approx \frac{\partial w}{\partial x} \Delta x + \frac{\partial w}{\partial y} \Delta y$$

of Section 14.6 and dividing by Δt:

$$\frac{\Delta w}{\Delta t} \approx \frac{\partial w}{\partial x} \frac{\Delta x}{\Delta t} + \frac{\partial w}{\partial y} \frac{\Delta y}{\Delta t}.$$

Then we take the limit as $\Delta t \to 0$ to obtain

$$\frac{dw}{dt} = \frac{\partial w}{\partial x} \frac{dx}{dt} + \frac{\partial w}{\partial y} \frac{dy}{dt}.$$

In the situation of Theorem 1, we may refer to w as the **dependent variable,** x and y as **intermediate variables,** and t as the **independent variable.** Then note that the right-hand side of Eq. (3) has two terms, one for each intermediate variable, both like the right-hand side of the single-variable chain rule of Eq. (2). If there are more than two intermediate variables, then there is still one term on the right-hand side for each intermediate variable. For example, if $w = f(x, y, z)$ with x, y, and z each a function of t, then the chain rule takes the form

$$\frac{dw}{dt} = \frac{\partial w}{\partial x} \cdot \frac{dx}{dt} + \frac{\partial w}{\partial y} \cdot \frac{dy}{dt} + \frac{\partial w}{\partial z} \cdot \frac{dz}{dt}. \tag{5}$$

The proof of Eq. (5) is essentially the same as the proof of Eq. (3); it requires the linear approximation theorem for three variables rather than for two variables.

You may find it useful to envision the three types of variables—dependent, intermediate, and independent—as though they were lying at three different levels, as in Fig. 14.7.1, with the dependent variable at the top and the independent variable at the bottom. Each variable then depends (either directly or indirectly) on those that lie below it.

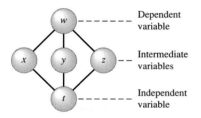

Fig. 14.7.1 Levels of chain rule variables

EXAMPLE 2 Find dw/dt if $w = x^2 + ze^y + \sin xz$ and $x = t$, $y = t^2$, $z = t^3$.

Solution Equation (5) gives

$$\frac{dw}{dt} = \frac{\partial w}{\partial x} \cdot \frac{dx}{dt} + \frac{\partial w}{\partial y} \cdot \frac{dy}{dt} + \frac{\partial w}{\partial z} \cdot \frac{dz}{dt}$$

$$= (2x + z \cos xz)(1) + (ze^y)(2t) + (e^y + x \cos xz)(3t^2)$$

$$= 2t + (3t^2 + 2t^4)e^{t^2} + 4t^3 \cos t^4.$$

In Example 2 we could check the result given by the chain rule by first writing w as an explicit function of t and then computing the ordinary single-variable derivative of w with respect to t.

SEVERAL INDEPENDENT VARIABLES

There may be several independent variables as well as several intermediate variables. For example, if $w = f(x, y, z)$, where $x = g(u, v)$, $y = h(u, v)$, and $z = k(u, v)$, so that

$$w = f(x, y, z) = f(g(u, v), h(u, v), k(u, v)),$$

then we have the three intermediate variables x, y, and z and the two independent variables u and v. In this case we would need to compute the *partial* derivatives $\partial w/\partial u$ and $\partial w/\partial v$ of the composite function. The general chain rule in Theorem 2 says that each partial derivative of the dependent variable w is given by a chain rule formula such as Eq. (3) or (5). The only difference is that the derivatives with respect to the independent variables are partial derivatives. For instance,

$$\frac{\partial w}{\partial u} = \frac{\partial w}{\partial x} \cdot \frac{\partial x}{\partial u} + \frac{\partial w}{\partial y} \cdot \frac{\partial y}{\partial u} + \frac{\partial w}{\partial z} \cdot \frac{\partial z}{\partial u}.$$

The "molecular model" in Fig. 14.7.2 illustrates this formula. The "atom" at the top represents the dependent variable w. The atoms at the next level represent the intermediate variables x, y, and z. The atoms at the bottom represent the independent variables u and v. Each "bond" in the model represents a partial derivative involving the two variables (the atoms joined by that bond). Finally, note that the formula displayed before this paragraph expresses $\partial w/\partial u$ as the sum of the products of the partial derivatives taken along all paths from w to u. Similarly, the sum of the products of the partial derivatives along all paths from w to v yields the correct formula

$$\frac{\partial w}{\partial v} = \frac{\partial w}{\partial x} \cdot \frac{\partial x}{\partial v} + \frac{\partial w}{\partial y} \cdot \frac{\partial y}{\partial v} + \frac{\partial w}{\partial z} \cdot \frac{\partial z}{\partial v}.$$

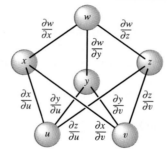

Fig. 14.7.2 Diagram for $w = w(x, y, z)$, where $x = x(u, v)$, $y = y(u, v)$, and $z = z(u, v)$

Theorem 2 describes the most general such situation.

Thus there is a formula in Eq. (6) for *each* of the independent variables $t_1, t_2, \ldots, t_n$, and the right-hand side of each such formula contains one typical chain rule term for each of the intermediate variables $x_1, x_2, \ldots, x_m$.

EXAMPLE 3   Suppose that

$$z = f(u, v), \qquad u = 2x + y, \qquad v = 3x - 2y.$$

Given the values $\partial z/\partial u = 3$ and $\partial z/\partial v = -2$ at the point $(u, v) = (3, 1)$, find the values $\partial z/\partial x$ and $\partial z/\partial y$ at the corresponding point $(x, y) = (1, 1)$.

Solution   The relationships among the variables are shown in Fig. 14.7.3. The chain rule gives

$$\frac{\partial z}{\partial x} = \frac{\partial z}{\partial u} \cdot \frac{\partial u}{\partial x} + \frac{\partial z}{\partial v} \cdot \frac{\partial v}{\partial x} = 3 \cdot 2 + (-2) \cdot 3 = 0$$

and

$$\frac{\partial z}{\partial y} = \frac{\partial z}{\partial u} \cdot \frac{\partial u}{\partial y} + \frac{\partial z}{\partial v} \cdot \frac{\partial v}{\partial y} = 3 \cdot 1 + (-2) \cdot (-2) = 7$$

at the indicated point $(x, y) = (1, 1)$.

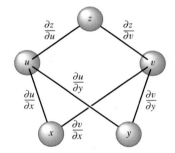

Fig. 14.7.3   Diagram for $z = z(u, v)$, where $u = u(x, y)$ and $v = v(x, y)$ (Example 3)

EXAMPLE 4   Let $w = f(x, y)$ where x and y are given in polar coordinates by the equations $x = r \cos \theta$ and $y = r \sin \theta$. Calculate

$$\frac{\partial w}{\partial r}, \qquad \frac{\partial w}{\partial \theta}, \quad \text{and} \quad \frac{\partial^2 w}{\partial r^2}$$

in terms of r and θ and the partial derivatives of w with respect to x and y (Fig. 14.7.4).

Solution   Here x and y are intermediate variables; the independent variables are r and θ. First note that

$$\frac{\partial x}{\partial r} = \cos \theta, \qquad \frac{\partial y}{\partial r} = \sin \theta, \qquad \frac{\partial x}{\partial \theta} = -r \sin \theta, \quad \text{and} \quad \frac{\partial y}{\partial \theta} = r \cos \theta.$$

Then

$$\frac{\partial w}{\partial r} = \frac{\partial w}{\partial x} \cdot \frac{\partial x}{\partial r} + \frac{\partial w}{\partial y} \cdot \frac{\partial y}{\partial r} = \frac{\partial w}{\partial x} \cos \theta + \frac{\partial w}{\partial y} \sin \theta \tag{7a}$$

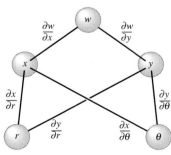

Fig. 14.7.4   Diagram for $w = w(x, y)$, where $x = x(r, \theta)$ and $y = y(r, \theta)$ (Example 4)

and

$$\frac{\partial w}{\partial \theta} = \frac{\partial w}{\partial x} \cdot \frac{\partial x}{\partial \theta} + \frac{\partial w}{\partial y} \cdot \frac{\partial y}{\partial \theta} = -r \frac{\partial w}{\partial x} \sin \theta + r \frac{\partial w}{\partial y} \cos \theta. \qquad (7b)$$

Next,

$$\frac{\partial^2 w}{\partial r^2} = \frac{\partial}{\partial r} \left(\frac{\partial w}{\partial r} \right) = \frac{\partial}{\partial r} \left(\frac{\partial w}{\partial x} \cos \theta + \frac{\partial w}{\partial y} \sin \theta \right)$$

$$= \frac{\partial w_x}{\partial r} \cos \theta + \frac{\partial w_y}{\partial r} \sin \theta,$$

where $w_x = \partial w / \partial x$ and $w_y = \partial w / \partial y$. We apply Eq. (7a) to calculate $\partial w_x / \partial r$ and $\partial w_y / \partial r$, and we obtain

$$\frac{\partial^2 w}{\partial r^2} = \left(\frac{\partial w_x}{\partial x} \cdot \frac{\partial x}{\partial r} + \frac{\partial w_x}{\partial y} \cdot \frac{\partial y}{\partial r} \right) \cos \theta + \left(\frac{\partial w_y}{\partial x} \cdot \frac{\partial x}{\partial r} + \frac{\partial w_y}{\partial y} \cdot \frac{\partial y}{\partial r} \right) \sin \theta$$

$$= \left(\frac{\partial^2 w}{\partial x^2} \cos \theta + \frac{\partial^2 w}{\partial y \, \partial x} \sin \theta \right) \cos \theta + \left(\frac{\partial^2 w}{\partial x \, \partial y} \cos \theta + \frac{\partial^2 w}{\partial y^2} \sin \theta \right) \sin \theta.$$

Finally, because $w_{yx} = w_{xy}$, we get

$$\frac{\partial^2 w}{\partial r^2} = \frac{\partial^2 w}{\partial x^2} \cos^2 \theta + 2 \frac{\partial^2 w}{\partial x \, \partial y} \cos \theta \sin \theta + \frac{\partial^2 w}{\partial y^2} \sin^2 \theta. \qquad (8)$$

EXAMPLE 5 Suppose that $w = f(u, v, x, y)$, where u and v are functions of x and y. Here x and y play dual roles as intermediate and independent variables. The chain rule yields

$$\frac{\partial w}{\partial x} = \frac{\partial f}{\partial u} \cdot \frac{\partial u}{\partial x} + \frac{\partial f}{\partial v} \cdot \frac{\partial v}{\partial x} + \frac{\partial f}{\partial x} \cdot \frac{\partial x}{\partial x} + \frac{\partial f}{\partial y} \cdot \frac{\partial y}{\partial x}$$

$$= \frac{\partial f}{\partial u} \cdot \frac{\partial u}{\partial x} + \frac{\partial f}{\partial v} \cdot \frac{\partial v}{\partial x} + \frac{\partial f}{\partial x},$$

because $\partial x / \partial x = 1$ and $\partial y / \partial x = 0$. Similarly,

$$\frac{\partial w}{\partial y} = \frac{\partial f}{\partial u} \cdot \frac{\partial u}{\partial y} + \frac{\partial f}{\partial v} \cdot \frac{\partial v}{\partial y} + \frac{\partial f}{\partial y}.$$

These results are consistent with the paths from w to x and from w to y in the molecular model shown in Fig. 14.7.5.

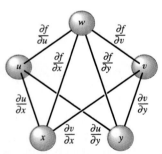

Fig. 14.7.5 Diagram for $w = f(u, v, x, y)$, where $u = u(x, y)$ and $v = v(x, y)$ (Example 5)

EXAMPLE 6 Consider a parametric curve $x = x(t)$, $y = y(t)$, $z = z(t)$ that lies on the surface $z = f(x, y)$ in space. Recall that if

$$\mathbf{T} = \left\langle \frac{dx}{dt}, \frac{dy}{dt}, \frac{dz}{dt} \right\rangle \quad \text{and} \quad \mathbf{N} = \left\langle \frac{\partial z}{\partial x}, \frac{\partial z}{\partial y}, -1 \right\rangle,$$

then $\mathbf{T}$ is tangent to the curve and $\mathbf{N}$ is normal to the surface. Show that $\mathbf{T}$ and $\mathbf{N}$ are everywhere perpendicular.

Solution The chain rule in Eq. (3) tells us that

$$\frac{dz}{dt} = \frac{\partial z}{\partial x} \cdot \frac{dx}{dt} + \frac{\partial z}{\partial y} \cdot \frac{dy}{dt}.$$

But this equation is equivalent to the vector equation

$$\left\langle \frac{\partial z}{\partial x}, \frac{\partial z}{\partial y}, -1 \right\rangle \cdot \left\langle \frac{dx}{dt}, \frac{dy}{dt}, \frac{dz}{dt} \right\rangle = 0.$$

Thus $\mathbf{N} \cdot \mathbf{T} = 0$, so $\mathbf{N}$ and $\mathbf{T}$ are indeed perpendicular.

Theorem 3 *Implicit Partial Differentiation*

Suppose that the function $F(x, y, z)$ has continuous first partial derivatives and that the equation $F(x, y, z) = 0$ implicitly defines a function $z = f(x, y)$ that has continuous first-order partial derivatives. Then

$$\frac{\partial z}{\partial x} = -\frac{F_x}{F_z} \quad \text{and} \quad \frac{\partial z}{\partial y} = -\frac{F_y}{F_z} \tag{9}$$

wherever $F_z = \partial F / \partial z \neq 0$.

Proof Because $w = F(x, y, f(x, y))$ is identically zero, differentiation with respect to x yields

$$0 = \frac{\partial w}{\partial x} = \frac{\partial F}{\partial x} \cdot \frac{\partial x}{\partial x} + \frac{\partial F}{\partial y} \cdot \frac{\partial y}{\partial x} + \frac{\partial F}{\partial z} \cdot \frac{\partial z}{\partial x}$$

$$= 1 \cdot F_x + 0 \cdot F_y + \frac{\partial z}{\partial x} \cdot F_z,$$

so

$$F_x + \frac{\partial z}{\partial x} \cdot F_z = 0.$$

This gives the first formula in (9). The second is obtained similarly by differentiating w with respect to y. ❑

REMARK It is usually simpler in a specific example to differentiate the equation $F(x, y, f(x, y)) = 0$ implicitly than to apply the formulas in (9).

EXAMPLE 7 Find the tangent plane at the point $(1, 3, 2)$ to the surface with equation

$$z^3 + xz - y^2 = 1.$$

Solution Implicit partial differentiation of the given equation with respect to x and with respect to y yields the equations

$$3z^2 \frac{\partial z}{\partial x} + z + x \frac{\partial z}{\partial x} = 0 \quad \text{and} \quad 3z^2 \frac{\partial z}{\partial y} + x \frac{\partial z}{\partial y} - 2y = 0.$$

When we substitute $x = 1$, $y = 3$, and $z = 2$, we find that $\partial z / \partial x = -\frac{2}{13}$ and $\partial z / \partial y = \frac{6}{13}$. Hence an equation of the tangent plane in question is

$$z - 2 = -\tfrac{2}{13}(x - 1) + \tfrac{6}{13}(y - 3);$$

that is,

$$2x - 6y + 13z = 10.$$

Proof of the Chain Rule Given that $w = f(x, y)$ satisfies the hypotheses of Theorem 1, we choose a point t_0 at which we wish to compute dw/dt and write

$$a = g(t_0), \qquad b = h(t_0).$$

Let

$$\Delta x = g(t_0 + \Delta t) - g(t_0), \qquad \Delta y = h(t_0 + \Delta t) - h(t_0).$$

Then

$$g(t_0 + \Delta t) = a + \Delta x \quad \text{and} \quad h(t_0 + \Delta t) = b + \Delta y.$$

If

$$\Delta w = f(g(t_0 + \Delta t), h(t_0 + \Delta t)) - f(g(t_0), h(t_0))$$
$$= f(a + \Delta x, b + \Delta y) - f(a, b),$$

then what we need to compute is

$$\frac{dw}{dt} = \lim_{\Delta t \to 0} \frac{\Delta w}{\Delta t}.$$

The linear approximation theorem of Section 14.6 gives

$$\Delta w = f_x(a, b) \, \Delta x + f_y(a, b) \, \Delta y + \epsilon_1 \, \Delta x + \epsilon_2 \, \Delta y,$$

where ϵ_1 and ϵ_2 approach zero as $\Delta x \to 0$ and $\Delta y \to 0$. We note that both Δx and Δy approach zero as $\Delta t \to 0$, because both the derivatives

$$\frac{dx}{dt} = \lim_{\Delta t \to 0} \frac{\Delta x}{\Delta t} \quad \text{and} \quad \frac{dy}{dt} = \lim_{\Delta t \to 0} \frac{\Delta y}{\Delta t}$$

exist. Therefore,

$$\frac{dw}{dt} = \lim_{\Delta t \to 0} \frac{\Delta w}{\Delta t} = \lim_{\Delta t \to 0} \left[f_x(a, b) \frac{\Delta x}{\Delta t} + f_y(a, b) \frac{\Delta y}{\Delta t} + \epsilon_1 \frac{\Delta x}{\Delta t} + \epsilon_2 \frac{\Delta y}{\Delta t} \right]$$

$$= f_x(a, b) \frac{dx}{dt} + f_y(a, b) \frac{dy}{dt} + 0 \frac{dx}{dt} + 0 \frac{dy}{dt}.$$

Hence

$$\frac{dw}{dt} = \frac{\partial w}{\partial x} \cdot \frac{dx}{dt} + \frac{\partial w}{\partial y} \cdot \frac{dy}{dt}.$$

Thus we have established Eq. (3), writing $\partial w / \partial x$ and $\partial w / \partial y$ for the partial derivatives $f_x(a, b)$ and $f_y(a, b)$ in the final step. ❏

14.7 Problems

In Problems 1 through 4, find dw/dt both by using the chain rule and by expressing w explicitly as a function of t before differentiating.

1. $w = \exp(-x^2 - y^2)$, $x = t$, $y = \sqrt{t}$

2. $w = \dfrac{1}{u^2 + v^2}$, $u = \cos 2t$, $v = \sin 2t$

3. $w = \sin xyz$, $x = t$, $y = t^2$, $z = t^3$

4. $w = \ln(u + v + z)$, $u = \cos^2 t$, $v = \sin^2 t$, $z = t^2$

In Problems 5 through 8, find $\partial w / \partial s$ and $\partial w / \partial t$.

5. $w = \ln(x^2 + y^2 + z^2)$, $x = s - t$, $y = s + t$, $z = 2\sqrt{st}$

6. $w = pq \sin r$, $p = 2s + t$, $q = s - t$, $r = st$

7. $w = \sqrt{u^2 + v^2 + z^2}$, $u = 3e^t \sin s$, $v = 3e^t \cos s$, $z = 4e^t$

8. $w = yz + zx + xy$, $x = s^2 - t^2$, $y = s^2 + t^2$, $z = s^2 t^2$

In Problems 9 and 10, find $\partial r/\partial x$, $\partial r/\partial y$, and $\partial r/\partial z$.

9. $r = e^{u+v+w}$, $u = yz$, $v = xz$, $w = xy$

10. $r = uvw - u^2 - v^2 - w^2$, $u = y + z$, $v = x + z$, $w = x + y$

In Problems 11 through 15, find $\partial z/\partial x$ and $\partial z/\partial y$ as functions of x, y, and z, assuming that $z = f(x, y)$ satisfies the given equation.

11. $x^{2/3} + y^{2/3} + z^{2/3} = 1$ **12.** $x^3 + y^3 + z^3 = xyz$

13. $xe^{xy} + ye^{zx} + ze^{xy} = 3$

14. $x^5 + xy^2 + yz = 5$ **15.** $\dfrac{x^2}{a^2} + \dfrac{y^2}{b^2} + \dfrac{z^2}{c^2} = 1$

In Problems 16 through 19, use the method of Example 5 to find $\partial w/\partial x$ and $\partial w/\partial y$ as functions of x and y.

16. $w = u^2 + v^2 + x^2 + y^2$, $u = x - y$, $v = x + y$

17. $w = \sqrt{uvxy}$, $u = \sqrt{x - y}$, $v = \sqrt{x + y}$

18. $w = xy \ln(u + v)$, $u = (x^2 + y^2)^{1/3}$, $v = (x^3 + y^3)^{1/2}$

19. $w = uv - xy$, $u = \dfrac{x}{x^2 + y^2}$, $v = \dfrac{y}{x^2 + y^2}$

In Problems 20 through 23, write an equation for the tangent plane at the point P to the surface with the given equation.

20. $x^2 + y^2 + z^2 = 9$; $P(1, 2, 2)$

21. $x^2 + 2y^2 + 2z^2 = 14$; $P(2, 1, -2)$

22. $x^3 + y^3 + z^3 = 5xyz$; $P(2, 1, 1)$

23. $z^3 + (x + y)z^2 + x^2 + y^2 = 13$; $P(2, 2, 1)$

24. Suppose that $y = g(x, z)$ satisfies the equation $F(x, y, z) = 0$ and that $F_y \neq 0$. Show that

$$\frac{\partial y}{\partial x} = -\frac{\partial F/\partial x}{\partial F/\partial y}.$$

25. Suppose that $x = h(y, z)$ satisfies the equation $F(x, y, z) = 0$ and that $F_x \neq 0$. Show that

$$\frac{\partial x}{\partial y} = -\frac{\partial F/\partial y}{\partial F/\partial x}.$$

26. Suppose that $w = f(x, y)$, $x = r \cos \theta$,   and $y = r \sin \theta$. Show that

$$\left(\frac{\partial w}{\partial x}\right)^2 + \left(\frac{\partial w}{\partial y}\right)^2 = \left(\frac{\partial w}{\partial r}\right)^2 + \frac{1}{r^2}\left(\frac{\partial w}{\partial \theta}\right)^2.$$

27. Suppose that $w = f(u)$ and that $u = x + y$. Show that $\partial w/\partial x = \partial w/\partial y$.

28. Suppose that $w = f(u)$ and that $u = x - y$. Show that $\partial w/\partial x = -\partial w/\partial y$ and that

$$\frac{\partial^2 w}{\partial x^2} = \frac{\partial^2 w}{\partial y^2} = -\frac{\partial^2 w}{\partial x\,\partial y}.$$

29. Suppose that $w = f(x, y)$, where $x = u + v$ and $y = u - v$. Show that

$$\frac{\partial^2 w}{\partial x^2} - \frac{\partial^2 w}{\partial y^2} = \frac{\partial^2 w}{\partial u\,\partial v}.$$

30. Assume that $w = f(x, y)$, where $x = 2u + v$ and $y = u - v$. Show that

$$5\frac{\partial^2 w}{\partial x^2} + 2\frac{\partial^2 w}{\partial x\,\partial y} + 2\frac{\partial^2 w}{\partial y^2} = \frac{\partial^2 w}{\partial u^2} + \frac{\partial^2 w}{\partial v^2}.$$

31. Suppose   that $w = f(x, y)$, $x = r \cos \theta$,   and $y = r \sin \theta$. Show that

$$\frac{\partial^2 w}{\partial x^2} + \frac{\partial^2 w}{\partial y^2} = \frac{\partial^2 w}{\partial r^2} + \frac{1}{r}\frac{\partial w}{\partial r} + \frac{1}{r^2}\frac{\partial^2 w}{\partial \theta^2}.$$

[*Suggestion:* First find $\partial^2 w/\partial\theta^2$ by the method of Example 4. Then combine the result with Eqs. (7) and (8).]

32. Suppose that

$$w = \frac{1}{r}f\left(t - \frac{r}{a}\right)$$

and that $r = \sqrt{x^2 + y^2 + z^2}$. Show that

$$\frac{\partial^2 w}{\partial x^2} + \frac{\partial^2 w}{\partial y^2} + \frac{\partial^2 w}{\partial z^2} = \frac{1}{a^2}\frac{\partial^2 w}{\partial t^2}.$$

33. Suppose   that $w = f(r)$ and   that $r = \sqrt{x^2 + y^2 + z^2}$. Show that

$$\frac{\partial^2 w}{\partial x^2} + \frac{\partial^2 w}{\partial y^2} + \frac{\partial^2 w}{\partial z^2} = \frac{d^2 w}{dr^2} + \frac{2}{r}\frac{dw}{dr}.$$

34. Suppose that $w = f(u) + g(v)$, that $u = x - at$, and that $v = x + at$. Show that

$$\frac{\partial^2 w}{\partial t^2} = a^2 \frac{\partial^2 w}{\partial x^2}.$$

35. Assume that $w = f(u, v)$, where $u = x + y$ and $v = x - y$. Show that

$$\frac{\partial w}{\partial x}\frac{\partial w}{\partial y} = \left(\frac{\partial w}{\partial u}\right)^2 - \left(\frac{\partial w}{\partial v}\right)^2.$$

36. Given: $w = f(x, y)$, $x = e^u \cos v$, and $y = e^u \sin v$. Show that

$$\left(\frac{\partial w}{\partial x}\right)^2 + \left(\frac{\partial w}{\partial y}\right)^2 = e^{-2u}\left[\left(\frac{\partial w}{\partial u}\right)^2 + \left(\frac{\partial w}{\partial v}\right)^2\right].$$

37. Assume that $w = f(x, y)$ and that there is a constant α such that $x = u \cos \alpha - v \sin \alpha$ and $y = u \sin \alpha + v \cos \alpha$. Show that

$$\left(\frac{\partial w}{\partial u}\right)^2 + \left(\frac{\partial w}{\partial v}\right)^2 = \left(\frac{\partial w}{\partial x}\right)^2 + \left(\frac{\partial w}{\partial y}\right)^2.$$

38. Suppose that $w = f(u)$, where

$$u = \frac{x^2 - y^2}{x^2 + y^2}.$$

Show that $xw_x + yw_y = 0$.

Suppose that the equation $F(x, y, z) = 0$ defines implicitly the three functions $z = f(x, y)$, $y = g(x, z)$, and $x = h(y, z)$. To keep track of the various partial derivatives, we use the notation

$$\left(\frac{\partial z}{\partial x}\right)_y = \frac{\partial f}{\partial x}, \qquad \left(\frac{\partial z}{\partial y}\right)_x = \frac{\partial f}{\partial y}, \qquad (10\text{a})$$

$$\left(\frac{\partial y}{\partial x}\right)_z = \frac{\partial g}{\partial x}, \qquad \left(\frac{\partial y}{\partial z}\right)_x = \frac{\partial g}{\partial z}, \qquad (10\text{b})$$

$$\left(\frac{\partial x}{\partial y}\right)_z = \frac{\partial h}{\partial y}, \qquad \left(\frac{\partial x}{\partial z}\right)_y = \frac{\partial h}{\partial z}. \qquad (10\text{c})$$

In short, the general symbol $(\partial w/\partial u)_v$ denotes the derivative of w with respect to u, where w is regarded as a function of the independent variables u and v.

39. Using the notation in Eqs. (10), show that

$$\left(\frac{\partial x}{\partial y}\right)_z \left(\frac{\partial y}{\partial z}\right)_x \left(\frac{\partial z}{\partial x}\right)_y = -1.$$

[*Suggestion:* Find the three partial derivatives on the right-hand side in Eqs. (10) in terms of F_x, F_y, and F_z.]

40. Verify the result of Problem 39 for the equation

$$F(x, y, z) = x^2 + y^2 + z^2 - 1 - 0.$$

41. Verify the result of Problem 39 (with p, V, and T in place of x, y, and z) for the equation

$$F(p, V, T) = pV - nRT = 0$$

(n and R are constants), which expresses the ideal gas law.

14.8
Directional Derivatives and the Gradient Vector

The change in the value of the function $w = f(x, y, z)$ from the point $P(x, y, z)$ to the nearby point $Q(x + \Delta x, y + \Delta y, z + \Delta z)$ is given by the increment

$$\Delta w = f(Q) - f(P). \qquad (1)$$

The linear approximation theorem of Section 14.6 yields

$$\Delta w \approx \frac{\partial f}{\partial x} \Delta x + \frac{\partial f}{\partial y} \Delta y + \frac{\partial f}{\partial z} \Delta z. \qquad (2)$$

We can express this approximation concisely in terms of the **gradient vector** ∇f (read as "del f") of the function f, which is defined to be

$$\nabla f(x, y, z) = \mathbf{i} f_x(x, y, z) + \mathbf{j} f_y(x, y, z) + \mathbf{k} f_z(x, y, z). \qquad (3)$$

We also write

$$\nabla f = \left\langle \frac{\partial f}{\partial x}, \frac{\partial f}{\partial y}, \frac{\partial f}{\partial z} \right\rangle = \frac{\partial f}{\partial x} \mathbf{i} + \frac{\partial f}{\partial y} \mathbf{j} + \frac{\partial f}{\partial z} \mathbf{k}.$$

Then Eq. (2) implies that the increment $\Delta w = f(Q) - f(P)$ is given approximately by

$$\Delta w \approx \nabla f(P) \cdot \mathbf{v}, \qquad (4)$$

where $\mathbf{v} = \overrightarrow{PQ} = \langle \Delta x, \Delta y, \Delta z \rangle$ is the *displacement vector* from P to Q.

EXAMPLE 1 If $f(x, y, z) = x^2 + yz - 2xy - z^2$, then the definition of the gradient vector in Eq. (3) yields

$$\nabla f(x, y, z) = \frac{\partial f}{\partial x} \mathbf{i} + \frac{\partial f}{\partial y} \mathbf{j} + \frac{\partial f}{\partial z} \mathbf{k}$$

$$= (2x - 2y)\mathbf{i} + (z - 2x)\mathbf{j} + (y - 2z)\mathbf{k}.$$

For instance, the value of ∇f at the point $P(2, 1, 3)$ is

$$\nabla f(P) = \nabla f(2, 1, 3) = 2\mathbf{i} - \mathbf{j} - 5\mathbf{k}.$$

To apply Eq. (4), we first calculate

$$f(P) = f(2, 1, 3) = 2^2 + 1 \cdot 3 - 2 \cdot 2 \cdot 1 - 3^2 = -6.$$

If Q is the nearby point $(1.9, 1.2, 3.1)$, then $\vec{PQ} = \mathbf{v} = \langle -0.1, 0.2, 0.1 \rangle$, so the approximation in (4) gives

$$f(Q) - f(P) \approx \nabla f(P) \cdot \mathbf{v} = \langle 2, -1, -5 \rangle \cdot \langle -0.1, 0.2, 0.1 \rangle = -0.9.$$

Hence $f(Q) \approx -6 + (-0.9) = -6.9$. In this case we can readily calculate, for comparison, the exact value of $f(Q) = -6.84$.

DIRECTIONAL DERIVATIVES

We know that the partial derivatives $f_x(x, y, z), f_y(x, y, z)$, and $f_z(x, y, z)$ give the rates of change of $w = f(x, y, z)$ at the point $P(x, y, z)$ in the x-, y-, and z-directions, respectively. We can now use the gradient vector ∇f to calculate the rate of change of w at P in an *arbitrary* direction. Recall that a "direction" is prescribed by a *unit* vector $\mathbf{u}$.

Let Q be a point on the ray in the direction of $\mathbf{u}$ from the point P (Fig. 14.8.1). The **average rate of change of w with respect to distance between P and Q** is

$$\frac{f(Q) - f(P)}{|\vec{PQ}|} = \frac{\Delta w}{\Delta s},$$

where $\Delta s = |\vec{PQ}| = |\mathbf{v}|$ is the distance from P to Q. Then the approximation in (4) yields

$$\frac{\Delta w}{\Delta s} \approx \frac{\nabla f(P) \cdot \mathbf{v}}{|\mathbf{v}|} = \nabla f(P) \cdot \mathbf{u}, \tag{5}$$

where $\mathbf{u} = \mathbf{v}/|\mathbf{v}|$ is the *unit* vector in the direction from P to Q. When we take the limit of the average rate of change $\Delta w/\Delta s$ as $\Delta s \to 0$, we get the *instantaneous* rate of change

$$\frac{dw}{ds} = \lim_{\Delta s \to 0} \frac{\Delta w}{\Delta s} = \nabla f(P) \cdot \mathbf{u}. \tag{6}$$

This computation motivates the *definition*

$$D_{\mathbf{u}} f(P) = \nabla f(P) \cdot \mathbf{u} \tag{7}$$

of the **directional derivative of f at $P(x, y, z)$ in the direction $\mathbf{u}$.** Physics and engineering texts may use the notation

$$\left. \frac{df}{ds} \right|_P = D_{\mathbf{u}} f(P),$$

or simply dw/ds as in Eq. (6), for the rate of change of the function $w = f(x, y, z)$ *with respect to distance s in the direction of the unit vector $\mathbf{u}$.*

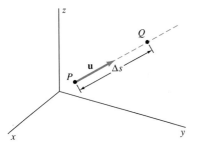

Fig. 14.8.1 First step in computing the rate of change of $f(x, y, z)$ in the direction of the unit vector $\mathbf{u}$

REMARK Remember that the vector **u** in Eq. (7) is a *unit* vector: $|\mathbf{u}| = 1$. If $\mathbf{u} = \langle a, b, c \rangle$, then Eq. (7) implies simply that

$$D_{\mathbf{u}} f = a \frac{\partial f}{\partial x} + b \frac{\partial f}{\partial y} + c \frac{\partial f}{\partial z}. \tag{8}$$

EXAMPLE 2 Suppose that the temperature at the point (x, y, z), with distance measured in kilometers, is given by

$$w = f(x, y, z) = 10 + xy + xz + yz$$

(in degrees Celsius). Find the rate of change (in degrees per kilometer) of temperature at the point $P(1, 2, 3)$ in the direction of the vector

$$\mathbf{v} = \mathbf{i} + 2\mathbf{j} - 2\mathbf{k}.$$

Solution Because **v** is not a unit vector, we must replace it by a unit vector with the same direction before we can use the formulas of this section. So we take

$$\mathbf{u} = \frac{\mathbf{v}}{|\mathbf{v}|} = \left\langle \frac{1}{3}, \frac{2}{3}, -\frac{2}{3} \right\rangle.$$

The gradient vector of f is

$$\nabla f = (y + z)\mathbf{i} + (x + z)\mathbf{j} + (x + y)\mathbf{k},$$

so $\nabla f(1, 2, 3) = 5\mathbf{i} + 4\mathbf{j} + 3\mathbf{k}$. Hence Eq. (7) gives

$$D_{\mathbf{u}} f(P) = \langle 5, 4, 3 \rangle \cdot \langle \tfrac{1}{3}, \tfrac{2}{3}, -\tfrac{2}{3} \rangle = \tfrac{7}{3}$$

(degrees per kilometer) for the desired rate of change of temperature with respect to distance.

THE VECTOR CHAIN RULE

The directional derivative $D_{\mathbf{u}} f$ is closely related to a version of the multivariable chain rule. Suppose that the first-order partial derivatives of f are continuous and that

$$\mathbf{r}(t) = x(t)\mathbf{i} + y(t)\mathbf{j} + z(t)\mathbf{k}$$

is a differentiable vector-valued function. Then

$$f(\mathbf{r}(t)) = f(x(t)), y(t), z(t))$$

is a differentiable function of t, and its (ordinary) derivative with respect to t is

$$D_t f(\mathbf{r}(t)) = D_t[f(x(t), y(t), z(t))]$$

$$= \frac{\partial f}{\partial x} \cdot \frac{dx}{dt} + \frac{\partial f}{\partial y} \cdot \frac{dy}{dt} + \frac{\partial f}{\partial z} \cdot \frac{dz}{dt}.$$

Hence

$$D_t f(\mathbf{r}(t)) = \nabla f(\mathbf{r}(t)) \cdot \mathbf{r}'(t), \tag{9}$$

where $\mathbf{r}'(t) = \langle x'(t), y'(t), z'(t) \rangle$ is the velocity vector of the parametric curve $\mathbf{r}(t)$. Equation (9) is the **vector chain rule**. The operation on the right-hand side of Eq. (9) is the *dot* product, because both the gradient of f and the derivative of $\mathbf{r}$ are *vector*-valued functions.

If the velocity vector $\mathbf{v}(t) = \mathbf{r}'(t) \neq \mathbf{0}$, then $\mathbf{v} = v\mathbf{u}$, where $v = |\mathbf{v}|$ is the speed and $\mathbf{u} = \mathbf{v}/v$ is the unit vector tangent to the curve. Then Eq. (9) implies that

$$D_t\, f(\mathbf{r}(t)) = vD_\mathbf{u}\, f(\mathbf{r}(t)). \tag{10}$$

With $w = f(\mathbf{r}(t))$, $D_\mathbf{u}\, f = dw/ds$, and $v = ds/dt$, Eq. (10) takes the simple chain rule form

$$\frac{dw}{dt} = \frac{dw}{ds} \cdot \frac{ds}{dt}. \tag{11}$$

EXAMPLE 3   If the function

$$w = f(x, y, z) = 10 + xy + xz + yz$$

of Example 2 gives the temperature, what time rate of change (degrees per minute) will a hawk observe as it flies through $P(1, 2, 3)$ at a speed of 2 km/min, heading directly toward the point $Q(3, 4, 4)$?

Solution   In Example 2 we calculated $\nabla f(P) = \langle 5, 4, 3 \rangle$, and the unit vector in the direction from P to Q is

$$\mathbf{u} = \frac{\overrightarrow{PQ}}{|\overrightarrow{PQ}|} = \left\langle \frac{2}{3}, \frac{2}{3}, \frac{1}{3} \right\rangle.$$

Then

$$D_\mathbf{u}\, f(P) = \nabla f(P) \cdot \mathbf{u} = \langle 5, 4, 3 \rangle \cdot \left(\tfrac{2}{3}, \tfrac{2}{3}, \tfrac{1}{3} \right) = 7$$

(degrees per kilometer). Hence Eq. (11) yields

$$\frac{dw}{dt} = \frac{dw}{ds} \cdot \frac{ds}{dt} = \left(7\, \frac{\deg}{\mathrm{km}} \right)\left(2\, \frac{\mathrm{km}}{\min} \right) = 14\, \frac{\deg}{\min}$$

as the hawk's time rate of change of temperature.

INTERPRETATION OF THE GRADIENT VECTOR

As yet we have discussed directional derivatives only for functions of three variables. The formulas for a function of two (or more than three) variables are analogous:

$$\nabla f(x, y) = \left\langle \frac{\partial f}{\partial x}, \frac{\partial f}{\partial y} \right\rangle = \frac{\partial f}{\partial x}\, \mathbf{i} + \frac{\partial f}{\partial y}\, \mathbf{j} \tag{12}$$

and

$$D_\mathbf{u}\, f(x, y) = \nabla f(x, y) \cdot \mathbf{u} = a\, \frac{\partial f}{\partial x} + b\, \frac{\partial f}{\partial y} \tag{13}$$

if $\mathbf{u} = \langle a, b \rangle$ is a unit vector. If α is the angle of inclination of $\mathbf{u}$ (measured counterclockwise from the positive x-axis, as in Fig. 14.8.2), then $a = \cos \alpha$ and $b = \sin \alpha$, so Eq. (13) takes the form

$$D_\mathbf{u}\, f(x, y) = \frac{\partial f}{\partial x} \cos \alpha + \frac{\partial f}{\partial y} \sin \alpha. \tag{14}$$

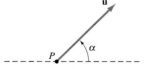

Fig. 14.8.2   The unit vector of $\mathbf{u}$ of Eq. (13)

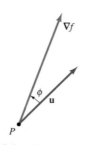

Fig. 14.8.3 The angle ϕ between ∇f and the unit vector $\mathbf{u}$

The gradient vector ∇f has an important interpretation that involves the *maximal* directional derivative of f. If ϕ is the angle between ∇f at the point P and the unit vector $\mathbf{u}$ (Fig. 14.8.3), then the formula in Eq. (7) gives

$$D_{\mathbf{u}} f(P) = \nabla f(P) \cdot \mathbf{u} = |\nabla f(P)| \cos \phi,$$

because $|\mathbf{u}| = 1$. The maximum value of $\cos \phi$ is 1, and this occurs when $\phi = 0$. This is so when $\mathbf{u}$ is the particular unit vector $\nabla f(P)/|\nabla f(P)|$ that points in the direction of the gradient vector itself. In this case the previous formula yields

$$D_{\mathbf{u}} f(P) = |\nabla f(P)|,$$

so the value of the directional derivative is the length of the gradient vector. We have therefore proved Theorem 1.

Theorem 1 *Significance of the Gradient Vector*
The maximum value of the directional derivative $D_{\mathbf{u}} f(P)$ is obtained when $\mathbf{u}$ is the vector in the direction of the gradient vector $\nabla f(P)$—that is, when $\mathbf{u} = \nabla f(P)/|\nabla f(P)|$. The value of the maximum directional derivative is $|\nabla f(P)|$, the length of the gradient vector.

Thus *the gradient vector ∇f points in the direction in which the function f increases the most rapidly, and its length is the rate of increase of f (with respect to distance) in that direction.* For instance, if the function f gives the temperature in space, then the gradient vector $\nabla f(P)$ points in the direction in which a bumblebee at P should initially fly to get warmer the fastest.

EXAMPLE 4   Suppose that the temperature w (in degrees Celsius) at the point (x, y) is given by

$$w = f(x, y) = 10 + (0.003)x^2 - (0.004)y^2.$$

In what direction $\mathbf{u}$ should a bumblebee at the point $(40, 30)$ initially fly in order to get warmer fastest? Find the directional derivative $D_{\mathbf{u}} f(40, 30)$ in this optimal direction $\mathbf{u}$.

Solution   The gradient vector is

$$\nabla f = \frac{\partial f}{\partial x} \mathbf{i} + \frac{\partial f}{\partial y} \mathbf{j} = (0.006x)\mathbf{i} - (0.008y)\mathbf{j},$$

so

$$\nabla f(40, 30) = (0.24)\mathbf{i} - (0.24)\mathbf{j} = (0.24\sqrt{2})\mathbf{u}.$$

The unit vector

$$\mathbf{u} = \frac{\nabla f(40, 30)}{|\nabla f(40, 30)|} = \frac{\mathbf{i} - \mathbf{j}}{\sqrt{2}}$$

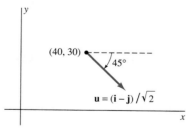

Fig. 14.8.4 The unit vector $\mathbf{u} = \dfrac{\nabla f}{|\nabla f|}$ of Example 4

points southeast (Fig. 14.8.4); this is the direction in which the bumblebee should initially fly. And, according to Theorem 1, the derivative of f in this optimal direction is

$$D_{\mathbf{u}} f(40, 30) = |\nabla f(40, 30)| = (0.24)\sqrt{2} \approx 0.34$$

degrees per unit of distance.

THE GRADIENT VECTOR AS A NORMAL VECTOR

Consider the graph of the equation

$$F(x, y, z) = 0, \tag{15}$$

where F is a function with continuous first-order partial derivatives. According to the **implicit function theorem** of advanced calculus, near every point where $\nabla F \neq \mathbf{0}$—that is, at least one of the partial derivatives of F is nonzero—the graph of Eq. (15) coincides with the graph of an equation of one of the forms

$$z = f(x, y), \qquad y = g(x, z), \qquad x = h(y, z).$$

Because of this, we are justified in general in referring to the graph of Eq. (15) as a "surface." The gradient vector ∇F is normal to this surface, in the sense of Theorem 2.

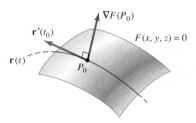

Fig. 14.8.5 The gradient vector ∇F is normal to every curve in the surface $F(x, y, z) = 0$.

> **Theorem 2** *Gradient Vector as Normal Vector*
>
> Suppose that $F(x, y, z)$ has continuous first-order partial derivatives, and let $P_0(x_0, y_0, z_0)$ be a point of the graph of the equation $F(x, y, z) = 0$ at which $\nabla F(P_0) \neq \mathbf{0}$. If $\mathbf{r}(t)$ is a differentiable curve on this surface with $\mathbf{r}(t_0) = \langle x_0, y_0, z_0 \rangle$, then
>
> $$\nabla F(P_0) \cdot \mathbf{r}'(t_0) = 0. \tag{16}$$
>
> Thus $\nabla F(P_0)$ is perpendicular to the tangent vector $\mathbf{r}'(t_0)$, as indicated in Fig. 14.8.5.

Proof The statement that $\mathbf{r}(t)$ lies on the surface $F(x, y, z) = 0$ means that $F(\mathbf{r}(t)) = 0$ for all t. Hence

$$0 = D_t F(\mathbf{r}(t_0)) = \nabla F(\mathbf{r}(t_0)) \cdot \mathbf{r}'(t_0) = \nabla F(P_0) \cdot \mathbf{r}'(t_0)$$

by the chain rule in the form of Eq. (9). Therefore, the vectors $\nabla F(P_0)$ and $\mathbf{r}'(t_0)$ are perpendicular. ❑

Because $\nabla F(P_0)$ is perpendicular to every curve on the surface $F(x, y, z) = 0$ through the point P_0, it is a *normal vector* to the surface at P_0,

$$\mathbf{n} = \frac{\partial F}{\partial x} \mathbf{i} + \frac{\partial F}{\partial y} \mathbf{j} + \frac{\partial F}{\partial z} \mathbf{k}. \tag{17}$$

If we rewrite the equation $z = f(x, y)$ in the form $F(x, y, z) = f(x, y) - z = 0$, then

$$\left\langle \frac{\partial F}{\partial x}, \frac{\partial F}{\partial y}, \frac{\partial F}{\partial z} \right\rangle = \left\langle \frac{\partial f}{\partial x}, \frac{\partial f}{\partial y}, -1 \right\rangle.$$

Thus Eq. (17) agrees with the definition of normal vector that we gave in Section 14.4 [Eq. (1) there].

The **tangent plane** to the surface $F(x, y, z) = 0$ at the point $P_0(x_0, y_0, z_0)$ is the plane through P_0 that is perpendicular to the normal vector $\mathbf{n}$ of Eq. (13). Its equation is

$$F_x(x_0, y_0, z_0)(x - x_0) + F_y(x_0, y_0, z_0)(y - y_0)$$
$$+ F_z(x_0, y_0, z_0)(z - z_0) = 0. \tag{18}$$

EXAMPLE 5 Write an equation of the tangent plane to the ellipsoid $2x^2 + 4y^2 + z^2 = 45$ at the point $(2, -3, -1)$.

Solution　If we write

$$F(x, y, z) = 2x^2 + 4y^2 + z^2 - 45,$$

then $F(x, y, z) = 0$ is the equation of the ellipsoid. Hence a normal vector is $\nabla F(x, y, z) = \langle 4x, 8y, 2z \rangle$, so

$$\nabla F(2, -3, -1) = 8\mathbf{i} - 24\mathbf{j} - 2\mathbf{k}$$

is normal to the ellipsoid at $(2, -3, -1)$. Equation (18) then gives the answer in the form

$$8(x - 2) - 24(y + 3) - 2(z + 1) = 0;$$

that is,

$$4x - 12y - z = 45.$$

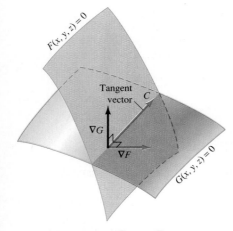

Fig. 14.8.6 $\nabla F \times \nabla G$ is tangent to the curve C of intersection.

The intersection of the two surfaces $F(x, y, z) = 0$ and $G(x, y, z) = 0$ will generally be some sort of curve in space. By the implicit function theorem, we can represent this curve in parametric fashion near every point where the gradient vectors ∇F and ∇G are *not* parallel. This curve C is perpendicular to both normal vectors ∇F and ∇G. That is, if P is a point of C, then the tangent vector to C at P is perpendicular to both vectors $\nabla F(P)$ and $\nabla G(P)$ (Fig. 14.8.6). It follows that the vector

$$\mathbf{T} = \nabla F \times \nabla G \qquad (19)$$

is tangent to the curve of intersection of the surfaces $F(x, y, z) = 0$ and $G(x, y, z) = 0$.

EXAMPLE 6　The point $P(1, -1, 2)$ lies on both the paraboloid

$$F(x, y, z) = x^2 + y^2 - z = 0$$

and the ellipsoid

$$G(x, y, z) = 2x^2 + 3y^2 + z^2 - 9 = 0.$$

Write an equation of the plane through P that is normal to the curve of intersection of these two surfaces (Fig. 14.8.7).

Solution　First we compute

$$\nabla F = \langle 2x, 2y, -1 \rangle \quad \text{and} \quad \nabla G = \langle 4x, 6y, 2z \rangle.$$

(a) Paraboloid

(b) Ellipsoid

(c) Intersection of paraboloid and ellipsoid

(d) Cutaway view

Fig. 14.8.7 Example 6

At $P(1, -1, 2)$ these two vectors are

$$\nabla F(1, -1, 2) = \langle 2, -2, -1 \rangle \quad \text{and} \quad \nabla G(1, -1, 2) = \langle 4, -6, 4 \rangle.$$

Hence a tangent vector to the curve of intersection of the paraboloid and the ellipsoid is

$$\mathbf{T} = \nabla F \times \nabla G = \begin{vmatrix} \mathbf{i} & \mathbf{j} & \mathbf{k} \\ 2 & -2 & -1 \\ 4 & -6 & 4 \end{vmatrix} = \langle -14, -12, -4 \rangle.$$

A slightly simpler vector parallel to $\mathbf{T}$ is $\mathbf{n} = \langle 7, 6, 2 \rangle$, and $\mathbf{n}$ is also normal to the desired plane through $(1, -1, 2)$. Therefore, an equation of the plane is

$$7(x - 1) + 6(y + 1) + 2(z - 2) = 0;$$

that is, $7x + 6y + 2z = 5$.

A result analogous to Theorem 2 holds in two dimensions. The graph of the equation $F(x, y) = 0$ looks like a *curve* near each point at which $\nabla F \neq \mathbf{0}$, and ∇F is normal to the curve in such cases.

EXAMPLE 7 Write an equation of the line tangent at the point $(1, 2)$ to the folium of Descartes with equation

$$F(x, y) = 2x^3 + 2y^3 - 9xy = 0$$

(Fig. 14.8.8).

Solution The gradient of F is

$$\nabla F(x, y) = (6x^2 - 9y)\mathbf{i} + (6y^2 - 9x)\mathbf{j}.$$

So a normal vector to the folium at $(1, 2)$ is $\nabla F(1, 2) = -12\mathbf{i} + 15\mathbf{j}$. Hence the tangent line has equation $-12(x - 1) + 15(y - 2) = 0$. Simplified, this is $4x - 5y + 6 = 0$.

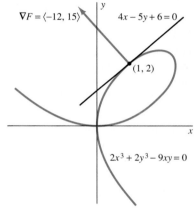

Fig. 14.8.8 The folium and its tangent (Example 7)

14.8 Problems

In Problems 1 through 10, find the gradient vector ∇f at the indicated point P.

1. $f(x, y) = 3x - 7y$; $P(17, 39)$
2. $f(x, y) = 3x^2 - 5y^2$; $P(2, -3)$
3. $f(x, y) = \exp(-x^2 - y^2)$; $P(0, 0)$
4. $f(x, y) = \sin \frac{1}{4}\pi xy$; $P(3, -1)$
5. $f(x, y, z) = y^2 - z^2$; $P(17, 3, 2)$
6. $f(x, y, z) = \sqrt{x^2 + y^2 + z^2}$; $P(12, 3, 4)$
7. $f(x, y, z) = e^x \sin y + e^y \sin z + e^z \sin x$; $P(0, 0, 0)$
8. $f(x, y, z) = x^2 - 3yz + z^3$; $P(2, 1, 0)$
9. $f(x, y, z) = 2\sqrt{xyz}$; $P(3, -4, -3)$
10. $f(x, y, z) = (2x - 3y + 5z)^5$; $P(-5, 1, 3)$

In Problems 11 through 20, find the directional derivative of f at P in the direction of $\mathbf{v}$; that is, find

$$D_\mathbf{u} f(P), \quad \text{where} \quad \mathbf{u} = \frac{\mathbf{v}}{|\mathbf{v}|}.$$

11. $f(x, y) = x^2 + 2xy + 3y^2$; $P(2, 1)$, $\mathbf{v} = \langle 1, 1 \rangle$
12. $f(x, y) = e^x \sin y$; $P(0, \pi/4)$, $\mathbf{v} = \langle 1, -1 \rangle$
13. $f(x, y) = x^3 - x^2 y + xy^2 + y^3$; $P(1, -1)$, $\mathbf{v} = 2\mathbf{i} + 3\mathbf{j}$
14. $f(x, y) = \tan^{-1}\left(\frac{y}{x}\right)$; $P(-3, 3)$, $\mathbf{v} = 3\mathbf{i} + 4\mathbf{j}$
15. $f(x, y) = \sin x \cos y$; $P\left(\frac{\pi}{3}, \frac{-2\pi}{3}\right)$, $\mathbf{v} = \langle 4, -3 \rangle$
16. $f(x, y, z) = xy + yz + zx$; $P(1, -1, 2)$, $\mathbf{v} = \langle 1, 1, 1 \rangle$

17. $f(x, y, z) = \sqrt{xyz}$; $P(2, -1, -2)$,
$\mathbf{v} = \mathbf{i} + 2\mathbf{j} - 2\mathbf{k}$

18. $f(x, y, z) = \ln(1 + x^2 + y^2 - z^2)$; $P(1, -1, 1)$,
$\mathbf{v} = 2\mathbf{i} - 2\mathbf{j} + 3\mathbf{k}$

19. $f(x, y, z) = e^{xyz}$; $P(4, 0, -3)$, $\mathbf{v} = \mathbf{j} - \mathbf{k}$

20. $f(x, y, z) = \sqrt{10 - x^2 - y^2 - z^2}$; $P(1, 1, -2)$,
$\mathbf{v} = \langle 3, 4, -12 \rangle$

In Problems 21 through 25, find the maximum directional derivative of f at P and the direction in which it occurs.

21. $f(x, y) = 2x^2 + 3xy + 4y^2$; $P(1, 1)$

22. $f(x, y) = \arctan\left(\dfrac{y}{x}\right)$; $P(1, -2)$

23. $f(x, y, z) = 3x^2 + y^2 + 4z^2$; $P(1, 5, -2)$

24. $f(x, y, z) = \exp(x - y - z)$; $P(5, 2, 3)$

25. $f(x, y, z) = \sqrt{xy^2 z^3}$; $P(2, 2, 2)$

In Problems 26 through 30, write an equation of the line (or plane) tangent to the given curve (or surface) at the given point P.

26. $2x^2 + 3y^2 = 35$; $P(2, 3)$

27. $x^4 + xy + y^2 = 19$; $P(2, -3)$

28. $3x^2 + 4y^2 + 5z^2 = 73$; $P(2, 2, 3)$

29. $x^{1/3} + y^{1/3} + z^{1/3} = 1$; $P(1, -1, 1)$

30. $xyz + x^2 - 2y^2 + z^3 = 14$; $P(5, -2, 3)$

31. Show that the gradient operator ∇ has the following formal properties that exhibit its close analogy with the single-variable derivative operator D.

(a) If a and b are constants, then

$$\nabla(au + bv) = a\nabla u + b\nabla v.$$

(b) $\nabla(uv) = u\nabla v + v\nabla u.$

(c) $\nabla\left(\dfrac{u}{v}\right) = \dfrac{v\nabla u - u\nabla v}{v^2}$ if $v \neq 0.$

(d) If n is a positive integer, then

$$\nabla u^n = nu^{n-1}\nabla u.$$

32. Suppose that f is a function of three independent variables x, y, and z. Show that $D_{\mathbf{i}}f = f_x$, $D_{\mathbf{j}}f = f_y$, and $D_{\mathbf{k}}f = f_z$.

33. Show that the equation of the line tangent to the conic section $Ax^2 + Bxy + Cy^2 = D$ at the point (x_0, y_0) is

$$(Ax_0)x + \tfrac{1}{2}B(y_0 x + x_0 y) + (Cy_0)y = D.$$

34. Show that the equation of the plane tangent to the quadric surface $Ax^2 + By^2 + Cz^2 = D$ at the point (x_0, y_0, z_0) is

$$(Ax_0)x + (By_0)y + (Cz_0)z = D.$$

35. Suppose that the temperature W (in degrees Celsius) at the point (x, y, z) in space is given by $W = 50 + xyz$. (a) Find the rate of change of temperature with respect to distance at the point $P(3, 4, 1)$ in the direction of the vector $\mathbf{v} = \langle 1, 2, 2 \rangle$. (The units of distance in space are feet.) (b) Find the maximal directional derivative $D_{\mathbf{u}}W$ at the point $P(3, 4, 1)$ and the direction $\mathbf{u}$ in which that maximum occurs.

36. Suppose that the temperature at the point (x, y, z) in space (in degrees Celsius) is given by the formula $W = 100 - x^2 - y^2 - z^2$. The units in space are meters. (a) Find the rate of change of temperature at the point $P(3, -4, 5)$ in the direction of the vector $\mathbf{v} = 3\mathbf{i} - 4\mathbf{j} + 12\mathbf{k}$. (b) In what direction does W increase most rapidly at P? What is the value of the maximal directional derivative at P?

37. Suppose that the altitude z (in miles above sea level) of a certain hill is described by the equation $z = f(x, y)$, where

$$f(x, y) = (0.1)(x^2 - xy + 2y^2).$$

(a) Write an equation (in the form $z = ax + by + c$) of the plane tangent to the hillside at the point $P(2, 1, 0.4)$. (b) Use $\nabla f(P)$ to approximate the altitude of the hill above the point $(2.2, 0.9)$ in the xy-plane. Compare your result with the actual altitude at this point.

38. Find an equation for the plane tangent to the paraboloid $z = 2x^2 + 3y^2$ and, simultaneously, parallel to the plane $4x - 3y - z = 10$.

39. The cone with equation $z^2 = x^2 + y^2$ and the plane with equation $2x + 3y + 4z + 2 = 0$ intersect in an ellipse. Write an equation of the plane normal to this ellipse at the point $P(3, 4, -5)$ (Fig. 14.8.9).

Fig. 14.8.9 The cone and plane of Problems 39 and 40

40. It is apparent from geometry that the highest and lowest points of the ellipse of Problem 39 are those points where its tangent line is horizontal. Find those points.

41. Show that the sphere $x^2 + y^2 + z^2 = r^2$ and the cone $z^2 = a^2 x^2 + b^2 y^2$ are orthogonal (that is, have perpendicular tangent planes) at every point of their intersection (Fig. 14.8.10).

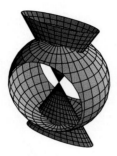

Fig. 14.8.10 A cutaway view of the cone and sphere of Problem 41

In Problems 42 through 46, the function $z = f(x, y)$ describes the shape of a hill; $f(P)$ is the altitude of the hill above the point $P(x, y)$ in the xy-plane. If you start at the point $(P, f(P))$ on this hill, then $D_{\mathbf{u}} f(P)$ is your rate of climb (rise per unit of horizontal distance) as you proceed in the horizontal *direction* $\mathbf{u} = a\mathbf{i} + b\mathbf{j}$. *And the angle at which you climb while you walk in this direction is* $\gamma = \tan^{-1}(D_{\mathbf{u}} f(P))$, *as shown in Fig. 14.8.11.*

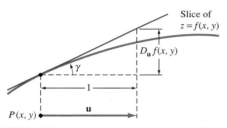

Fig. 14.8.11 The cross section of the part of the graph above $\mathbf{u}$ (Problems 42 through 46)

42. You are standing at the point $(-100, -100, 430)$ on a hill that has the shape of the graph of

$$z = 500 - (0.003)x^2 - (0.004)y^2,$$

with x, y, and z given in feet. (a) What will be your rate of climb (*rise* over *run*) if you head northwest? At what angle from the horizontal will you be climbing? (b) Repeat part (a), except now you head northeast.

43. You are standing at the point $(-100, -100, 430)$ on the hill of Problem 42. In what direction (that is, with what compass heading) should you proceed in order to climb the most steeply? At what angle from the horizontal will you initially be climbing?

44. Repeat Problem 42, but now you are standing at the point $P(100, 100, 500)$ on the hill described by

$$z = \frac{1000}{1 + (0.00003)x^2 + (0.00007)y^2}.$$

45. Repeat Problem 43, except begin at the point $P(100, 100, 500)$ on the hill of Problem 44.

46. You are standing at the point $(30, 20, 5)$ on a hill with the shape of the surface

$$z = 100 \exp\left(-\frac{x^2 + 3y^2}{701}\right).$$

(a) In what direction (with what compass heading) should you proceed in order to climb the most steeply? At what angle from the horizontal will you initially be climbing? (b) If, instead of climbing as in part (a), you head directly west (the negative x-direction), then at what angle will you be climbing initially?

14.9
Lagrange Multipliers and Constrained Maximum-Minimum Problems

In Section 14.5 we discussed the problem of finding the maximum and minimum values attained by a function $f(x, y)$ at points of the plane region R, in the simple case in which R consists of the points on and within the simple closed curve C. We saw that any local maximum or minimum in the *interior* of R occurs at a point where $f_x = 0 = f_y$ or at a point where f is not differentiable (the latter usually signaled by the failure of f_x or f_y to exist). Here we discuss the very different matter of finding the maximum and minimum values attained by f at points of the *boundary* curve C.

If the curve C is the graph of the equation $g(x, y) = 0$, then our task is to maximize or minimize the function $f(x, y)$ subject to the **constraint**, or **side condition**,

$$g(x, y) = 0. \tag{1}$$

We could in principle try to solve this constraint equation for $y = \phi(x)$ and then maximize or minimize the single-variable function $f(x, \phi(x))$ by the standard method of finding its critical points. But what if it is impractical or impossible to solve Eq. (1) explicitly for y in terms of x? An alternative approach that does not require that we first solve this equation is the **method** of **Lagrange multipliers**. It is named for its discoverer, the Italian-born French mathematician Joseph Louis Lagrange (1736–1813). The method is based on Theorem 1.

Theorem 1 *Lagrange Multipliers (one constraint)*

Let $f(x, y)$ and $g(x, y)$ be functions with continuous first-order partial derivatives. If the maximum (or minimum) value of f subject to the condition

$$g(x, y) = 0 \qquad (1)$$

occurs at a point P where $\nabla g(P) \neq \mathbf{0}$, then

$$\nabla f(P) = \lambda \nabla g(P) \qquad (2)$$

for some constant λ.

Proof By the implicit function theorem mentioned in Section 14.8, the fact that $\nabla g(P) \neq \mathbf{0}$ allows us to represent the curve $g(x, y) = 0$ near P by a parametric curve $\mathbf{r}(t)$, and in such fashion that $\mathbf{r}$ has a nonzero tangent vector near P. Thus $\mathbf{r}'(t) \neq \mathbf{0}$ (Fig. 14.9.1). Let t_0 be the value of t such that $\mathbf{r}(t_0) = \overrightarrow{OP}$. If $f(x, y)$ attains its maximum value at P, then the composite function $f(\mathbf{r}(t))$ attains its maximum value at $t = t_0$, so

$$D_t f(\mathbf{r}(t)) \Big|_{t=t_0} = \nabla f(\mathbf{r}(t_0)) \cdot \mathbf{r}'(t_0) = \nabla f(P) \cdot \mathbf{r}'(t_0) = 0. \qquad (3)$$

Here we have used the vector chain rule, Eq. (9) of Section 14.8.

Because $\mathbf{r}(t)$ lies on the curve $g(x, y) = 0$, the composite function $g(\mathbf{r}(t))$ is a constant function. Therefore,

$$D_t g(\mathbf{r}(t)) \Big|_{t=t_0} = \nabla g(\mathbf{r}(t_0)) \cdot \mathbf{r}'(t_0) = \nabla g(P) \cdot \mathbf{r}'(t_0) = 0. \qquad (4)$$

Equations (3) and (4) together imply that both vectors $\nabla f(P)$ and $\nabla g(P)$ are perpendicular to the (nonzero) tangent vector $\mathbf{r}'(t_0)$. Hence $\nabla f(P)$ must be a scalar multiple of $\nabla g(P)$, and this is exactly the meaning of Eq. (2). This concludes the proof of the theorem. $\square$

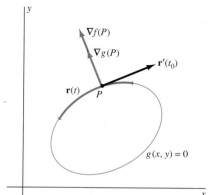

Fig. 14.9.1 The conclusion of Theorem 1 illustrated

GEOMETRIC INTERPRETATION

Let $f(P) = m$ be a maximum or minimum value of $f(x, y)$ subject to the constraint $g(x, y) = 0$. Then by Theorem 2 of Section 14.8, the gradient vectors $\nabla f(P)$ and $\nabla g(P)$ are normal to the curves $f(x, y) = m$ and $g(x, y) = 0$ at the point P. Hence the condition in Eq. (2) implies that

If $f(P) = m$ is a maximum or minimum value subject to $g(x, y) = 0$, then the curves $f(x, y) = m$ and $g(x, y) = 0$ are tangent at P.

Figure 14.9.2 illustrates this interpretation with

$$f(x, y) = x^2 + y^2 \qquad \text{(square of distance),}$$

$$g(x, y) = xy - 1 = 0 \qquad \text{(rectangular hyperbola).}$$

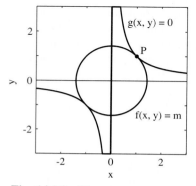

Fig. 14.9.2 The curves $f(x, y) = m$ and $g(x, y) = 0$ are tangent at a constrained maximum or minimum point P of $f(x, y)$.

In this case $f(x, y)$ is minimal at the points of the hyperbola closest to the origin. By geometry we see that these are the points $(1, 1)$ and $(-1, -1)$ where the circle $x^2 + y^2 = 2$ and the hyperbola $xy = 1$ are indeed tangent.

Let's see what steps we should follow to solve a problem by using Theorem 1—the method of Lagrange multipliers. First we need to identify a quantity $z = f(x, y)$ to be maximized or minimized, subject to the constraint $g(x, y) = 0$. Then Eq. (1) and the two scalar components of Eq. (2) yield three equations:

$$g(x, y) = 0, \tag{1}$$

$$f_x(x, y) = \lambda g_x(x, y), \quad \text{and} \tag{2a}$$

$$f_y(x, y) = \lambda g_y(x, y). \tag{2b}$$

Thus we have three equations that we can attempt to solve for the three unknowns x, y, and λ. The points (x, y) that we find (assuming our efforts are successful) are the only possible locations for the extrema of f subject to the constraint $g(x, y) = 0$. The associated values of λ, called **Lagrange multipliers**, may be revealed as well but often are not of much interest. Finally, we calculate the value $f(x, y)$ at each of the solution points (x, y) in order to spot its maximum and minimum values.

We must bear in mind the additional possibility that the maximum or minimum (or both) of f might occur at a point where $g_x = 0 = g_y$. The Lagrange multiplier method may fail to locate these exceptional points, but they can usually be recognized as points where the graph $g(x, y) = 0$ fails to be a smooth curve.

EXAMPLE 1 Let us return to the sawmill problem of Example 5 of Section 3.6: to maximize the cross-sectional area of a rectangular beam cut from a circular log. With $r = \sqrt{2}$ as the given radius of the log, we want to show by using Lagrange multipliers that the optimal beam has a square cross section.

Solution With the coordinate system indicated in Fig. 14.9.3, we want to maximize the area

$$A = f(x, y) = 4xy \tag{5}$$

of the beam's rectangular cross section subject to the constraint

$$g(x, y) = x^2 + y^2 - 2 = 0 \tag{6}$$

that describes the circular log. Because

$$\frac{\partial f}{\partial x} = 4y, \quad \frac{\partial g}{\partial x} = 2x, \quad \frac{\partial f}{\partial y} = 4x, \quad \text{and} \quad \frac{\partial g}{\partial y} = 2y,$$

Eqs. (2a) and (2b) become

$$4y = 2\lambda x \quad \text{and} \quad 4x = 2\lambda y.$$

It is clear that neither $x = 0$ nor $y = 0$ gives the maximum area, so $\lambda \neq 0$ as well. Hence we can divide the first equation by the second to obtain

$$\frac{y}{x} = \frac{x}{y},$$

and it follows that $x^2 = y^2$.

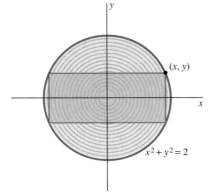

Fig. 14.9.3 Cutting a rectangular beam from a circular log (Example 1)

We substitute this consequence of the Lagrange multiplier equations into the constraint equation $x^2 + y^2 = 2$ and get $2x^2 = 2$; therefore $x^2 = y^2 = 1$. We seek (as in Fig. 14.9.3) a first-quadrant solution point, so we conclude that $x = y = 1$ gives the maximum. Thus the optimal beam does, indeed, have a square cross section with edges $2x = 2y = 2$ ft. Its cross-sectional area of 4 ft^2 is about 64% of the total cross-sectional area 2π of the original circular log.

REMARK 1   Note that $f(x, y) = 4xy$ attains its maximum value of 4 at both $(1, 1)$ and $(-1, -1)$ and its minimum value of -4 at both $(1, -1)$ and $(-1, 1)$. The Lagrange multiplier method actually locates all four of these points for us.

REMARK 2   In the applied maximum-minimum problems of Section 3.6, we typically started with a *formula* such as Eq. (5) of this section, expressing the quantity to be optimized in terms of *two* variables x and y, for example. We then used some available *relation* such as Eq. (6) between the variables x and y to eliminate one of them, such as y. Thus we finally obtained a single-variable *function* by substituting for y in terms of x in the original formula. As in Example 1, the Lagrange multiplier method frees us from the algebraic process of substitution and elimination.

EXAMPLE 2   After the square beam of Example 1 has been cut from the circular log of radius $\sqrt{2}$ ft, let us cut four planks from the remaining pieces, each of dimensions $2x$ by $y - 1$, as shown in Fig. 14.9.4. How should we do this in order to maximize the combined cross-sectional area of all four planks, thereby using as efficiently as possible what might otherwise be scrap lumber?

Solution   Because the point (x, y) lies on the circle $x^2 + y^2 = 2$, we want to maximize the total (four planks) area function

$$f(x, y) = 4 \cdot 2x \cdot (y - 1) = 8xy - 8x$$

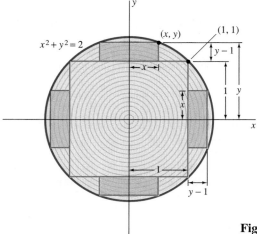

Fig. 14.9.4 Cutting four more planks from the log (Example 2)

Ch. 14 / Partial Differentiation

subject to the constraint condition

$$g(x, y) = x^2 + y^2 - 2 = 0.$$

The Lagrange multiplier equations $f_x = \lambda g_x$ and $f_y = \lambda g_y$ then take the form

$$8y - 8 = 2\lambda x, \qquad 8x = 2\lambda y.$$

Because $x \neq 0$ and $y \neq 0$ (is this obvious?), we can solve these equations for

$$\lambda = \frac{4y - 4}{x} = \frac{4x}{y}.$$

We forget λ and multiply by the common denominator xy to obtain $x^2 = y^2 - y$.

We substitute this consequence of the Lagrange multiplier equations into the constraint equation $x^2 + y^2 - 2 = 0$. This finally yields the quadratic equation

$$2y^2 - y - 2 = 0,$$

whose only positive solution is $y \approx 1.2808$, so $x = \sqrt{y^2 - y} \approx 0.5997$. Hence our planks should be $2x \approx 1.1994$ ft ≈ 14.4 in. wide and $y - 1 \approx 0.2808$ ft ≈ 3.4 in. thick. Their combined cross-sectional area will then be

$$f(0.5997, 1.2808) \approx 4 \cdot (1.1994) \cdot (0.2808) \approx 1.3472 \quad (\text{ft}^2).$$

This is approximately 21% of the original circular cross-sectional area, so the beam and four planks account for about 85% of the lumber in the original log. (Would anyone like to try for a dozen more rectangular planks made from the 12 pieces that yet remain?)

REMARK   The solution of Example 2 follows a typical pattern. The value of the Lagrange parameter λ itself frequently is of no interest to us. By eliminating λ from the Lagrange multiplier equations, we get a relation between x and y. If we substitute this relation into the constraint equation $g(x, y) = 0$, we are finally able to solve explicitly for x and y.

LAGRANGE MULTIPLIERS IN THREE DIMENSIONS

Now suppose that $f(x, y, z)$ and $g(x, y, z)$ have continuous first-order partial derivatives and that we want to find the points of the *surface*

$$g(x, y, z) = 0 \tag{7}$$

at which the function $f(x, y, z)$ attains its maximum and minimum values. With functions of three rather than two variables, Theorem 1 holds precisely as we stated it, with the z-direction taken into account. We leave the details to Problem 31, but an argument similar to the proof of Theorem 1 shows that at a maximum or minimum point P of $f(x, y, z)$ on the surface in Eq. (7), both gradient vectors $\nabla f(P)$ and $\nabla g(P)$ are normal to the surface (Fig. 14.9.5). It follows that

$$\nabla f(P) = \lambda \nabla g(P) \tag{8}$$

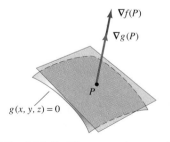

Fig. 14.9.5 The natural generalization of Theorem 1 holds for functions of three variables.

for some scalar λ. This vector equation corresponds to three scalar equations. To find the possible locations of the extrema of f subject to the constraint g, we can attempt to solve simultaneously the four equations

$$g(x, y, z) = 0, \tag{7}$$

$$f_x(x, y, z) = \lambda g_x(x, y, z), \tag{8a}$$

$$f_y(x, y, z) = \lambda g_y(x, y, z), \tag{8b}$$

$$f_z(x, y, z) = \lambda g_z(x, y, z) \tag{8c}$$

for the four unknowns x, y, z, and λ. If successful, we then evaluate $f(x, y, z)$ at each of the solution points (x, y, z) to see at which it attains its maximum and minimum values. In analogy to the two-dimensional case, we also check points at which the surface $g(x, y, z) = 0$ fails to be smooth. Thus the Lagrange multiplier method with one constraint is essentially the same in dimension three as in dimension two.

EXAMPLE 3 Find the maximum volume of a rectangular box inscribed in the ellipsoid $x^2/a^2 + y^2/b^2 + z^2/c^2 = 1$ with its faces parallel to the coordinate planes.

Solution Let (x, y, z) be the vertex of the box that lies in the first octant (where x, y, and z are all positive). We want to maximize the volume $V(x, y, z) = 8xyz$ subject to the constraint

$$g(x, y, z) = \frac{x^2}{a^2} + \frac{y^2}{b^2} + \frac{z^2}{c^2} - 1 = 0.$$

Equations (8a), (8b), and (8c) give

$$8yz = \frac{2\lambda x}{a^2}, \qquad 8xz = \frac{2\lambda y}{b^2}, \qquad 8xy = \frac{2\lambda z}{c^2}.$$

Part of the art of mathematics lies in pausing for a moment to find an elegant way to solve a problem rather than rushing in headlong with brute force methods. Here, if we multiply the first equation by x, the second by y, and the third by z, we find that

$$2\lambda \frac{x^2}{a^2} = 2\lambda \frac{y^2}{b^2} = 2\lambda \frac{z^2}{c^2} = 8xyz.$$

Now $\lambda \neq 0$ because (at maximum volume) x, y, and z are nonzero. We conclude that

$$\frac{x^2}{a^2} = \frac{y^2}{b^2} = \frac{z^2}{c^2}.$$

The sum of the last three expressions is 1, because that is precisely the constraint condition in this problem. Thus each of these three expressions is equal to $\frac{1}{3}$. All three of x, y, and z are positive, and therefore

$$x = \frac{a}{\sqrt{3}}, \qquad y = \frac{b}{\sqrt{3}}, \quad \text{and} \quad z = \frac{c}{\sqrt{3}}.$$

Therefore, the box of maximum volume has volume

$$V = V_{\text{max}} = \frac{8}{3\sqrt{3}} abc.$$

The answer is dimensionally correct (the product of the three lengths a, b, and c yields a volume), and it is plausible—the maximum box occupies about 37% of the volume $\frac{4}{3}\pi abc$ of the circumscribed ellipsoid.

PROBLEMS THAT HAVE TWO CONSTRAINTS

Suppose that we want to find the maximum and minimum values of the function $f(x, y, z)$ at points of the curve of intersection of the two surfaces

$$g(x, y, z) = 0 \quad \text{and} \quad h(x, y, z) = 0. \tag{9}$$

This is a maximum-minimum problem with *two* constraints. The Lagrange multiplier method for such situations is based on Theorem 2.

Theorem 2 *Lagrange Multipliers (two constraints)*

Let $f(x, y, z)$, $g(x, y, z)$, and $h(x, y, z)$ be functions with continuous first-order partial derivatives. If the maximum (or minimum) value of f subject to the two conditions

$$g(x, y, z) = 0 \quad \text{and} \quad h(x, y, z) = 0 \tag{9}$$

occurs at a point P where the vectors $\nabla g(P)$ and $\nabla h(P)$ are nonzero and nonparallel, then

$$\nabla f(P) = \lambda_1 \nabla g(P) + \lambda_2 \nabla h(P) \tag{10}$$

for some two constants λ_1 and λ_2.

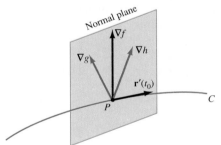

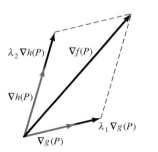

Fig. 14.9.6 The relation between the gradient vectors in the proof of Theorem 2

Fig. 14.9.7 Geometry of the equation $\nabla f(P) = \lambda_1 \nabla g(P) + \lambda_2 \nabla h(P)$

Outline of Proof By an appropriate version of the implicit function theorem, the curve C of intersection of the two surfaces (Fig. 14.9.6) may be represented near P by a parametric curve $\mathbf{r}(t)$ with nonzero tangent vector $\mathbf{r}'(t)$. Let t_0 be the value of t such that $\mathbf{r}(t_0) = \overrightarrow{OP}$. We compute the derivatives at t_0 of the composite functions $f(\mathbf{r}(t))$, $g(\mathbf{r}(t))$, and $h(\mathbf{r}(t))$. We find—exactly as in the proof of Theorem 1—that

$$\nabla f(P) \cdot \mathbf{r}'(t_0) = 0, \qquad \nabla g(P) \cdot \mathbf{r}'(t_0) = 0, \quad \text{and} \quad \nabla h(P) \cdot \mathbf{r}'(t_0) = 0.$$

These three equations imply that all three gradient vectors are perpendicular to the curve C at P and thus that they all lie in a single plane, the plane normal to the curve C at the point P.

Now $\nabla g(P)$ and $\nabla h(P)$ are nonzero and nonparallel, so $\nabla f(P)$ is the sum of its projections onto $\nabla g(P)$ and $\nabla h(P)$ (see Problem 47 of Section 13.1). As illustrated in Fig. 14.9.7, this fact implies Eq. (8). ☐

In examples we prefer to avoid subscripts by writing λ and μ for the Lagrange multipliers λ_1 and λ_2 in the statement of Theorem 2. The equations in (9) and the three scalar components of the vector equation in (10) then give rise to the five simultaneous equations

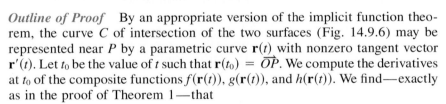

$$g(x, y, z) = 0, \tag{9a}$$

$$h(x, y, z) = 0, \tag{9b}$$

$$f_x(x, y, z) = \lambda g_x(x, y, z) + \mu h_x(x, y, z), \tag{10a}$$

$$f_y(x, y, z) = \lambda g_y(x, y, z) + \mu h_y(x, y, z), \tag{10b}$$

$$f_z(x, y, z) = \lambda g_z(x, y, z) + \mu h_z(x, y, z) \tag{10c}$$

in the five unknowns x, y, z, λ, and μ.

EXAMPLE 4 The plane $x + y + z = 12$ intersects the paraboloid $z = x^2 + y^2$ in an ellipse (Fig. 14.9.8). Find the highest and lowest points on this ellipse.

Solution The height of the point (x, y, z) is z, so we want to find the maximum and minimum values of

$$f(x, y, z) = z \tag{11}$$

subject to the two conditions

$$g(x, y, z) = x + y + z - 12 = 0 \tag{12}$$

and

$$h(x, y, z) = x^2 + y^2 - z = 0. \tag{13}$$

The conditions in (10a) through (10c) yield

$$0 = \lambda + 2\mu x, \tag{14a}$$

$$0 = \lambda + 2\mu y, \quad \text{and} \tag{14b}$$

$$1 = \lambda - \mu. \tag{14c}$$

If μ were zero, then Eq. (14a) would imply that $\lambda = 0$, which contradicts Eq. (14c). Hence $\mu \neq 0$, and therefore the equations

$$2\mu x = -\lambda = 2\mu y$$

imply that $x = y$. Substitution of $x = y$ into Eq. (13) gives $z = 2x^2$, and then Eq. (12) yields

$$2x^2 + 2x - 12 = 0;$$

$$x^2 + x - 6 = 0;$$

$$(x + 3)(x - 2) = 0.$$

Thus we obtain the two solutions $x = -3$ and $x = 2$. Because $y = x$ and $z = 2x^2$, the corresponding points of the ellipse are $P_1(2, 2, 8)$ and $P_2(-3, -3, 18)$. It's clear which is the lowest and which is the highest.

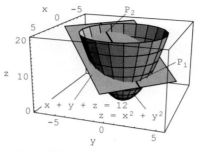

Fig. 14.9.8 The plane and paraboloid intersecting in the ellipse of Example 4

14.9 Problems

In Problems 1 through 10, find the maximum and minimum values—if any—of the given function f subject to the given constraint or constraints.

1. $f(x, y) = x^2 - y^2$; $\quad x^2 + y^2 = 4$

2. $f(x, y) = x^2 + y^2$; $\quad 2x + 3y = 6$

3. $f(x, y) = xy$; $\quad 4x^2 + 9y^2 = 36$

4. $f(x, y) = 4x^2 + 9y^2$; $\quad x^2 + y^2 = 1$

5. $f(x, y, z) = x^2 + y^2 + z^2$; $\quad 3x + 2y + z = 6$

6. $f(x, y, z) = 3x + 2y + z$; $\quad x^2 + y^2 + z^2 = 1$

7. $f(x, y, z) = x + y + z$; $\quad x^2 + 4y^2 + 9z^2 = 36$

8. $f(x, y, z) = xyz$; $\quad x^2 + y^2 + z^2 = 1$

9. $f(x, y, z) = x^2 + y^2 + z^2$; $x + y + z = 1$ and $x + 2y + 3z = 6$

10. $f(x, y, z) = z$; $x^2 + y^2 = 1$ and $2x + 2y + z = 5$

In Problems 11 through 20, use Lagrange multipliers to solve the indicated problem from Section 14.5.

11. Problem 27   **12.** Problem 28

13. Problem 29   **14.** Problem 30

15. Problem 31   **16.** Problem 32

17. Problem 34   **18.** Problem 35

19. Problem 40   **20.** Problem 41

21. Find the point or points of the surface $z = xy + 5$ closest to the origin. [*Suggestion:* Minimize the *square* of the distance.]

22. A triangle with sides x, y, and z has fixed perimeter $2s = x + y + z$. Its area A is given by *Heron's formula:*

$$A = \sqrt{s(s - x)(s - y)(s - z)}.$$

Use the method of Lagrange multipliers to show that, among all triangles with the given perimeter, the one of largest area is equilateral. [*Suggestion:* Consider maximizing A^2 rather than A.]

23. Use the method of Lagrange multipliers to show that, of all triangles inscribed in the unit circle, the one of greatest area is equilateral. [*Suggestion:* Use Fig. 14.9.9 and the fact that the area of a triangle with sides a and b and included angle θ is given by the formula $A = \frac{1}{2}ab \sin \theta$.]

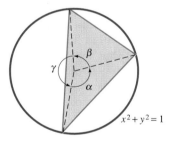

Fig. 14.9.9   A triangle inscribed in a circle (Problem 23)

24. Find the points on the rotated ellipse $x^2 + xy + y^2 = 3$ that are closest to and farthest from the origin. [*Suggestion:* Write the Lagrange multiplier equations in the form

$$ax + by = 0,$$

$$cx + dy = 0.$$

These equations have a nontrivial solution *only if* $ad - bc = 0$. Use this fact to solve first for λ.]

25. Use the method of Problem 24 to find the points of the rotated hyperbola $x^2 + 12xy + 6y^2 = 130$ that are closest to the origin.

26. Find the points of the ellipse $4x^2 + 9y^2 = 36$ that are closest to the point $(1, 1)$ as well as the point or points farthest from it.

27. Find the highest and lowest points on the ellipse of intersection of the cylinder $x^2 + y^2 = 1$ and the plane $2x + y - z = 4$.

28. Apply the method of Example 4 to find the highest and lowest points on the ellipse of intersection of the cone $z^2 = x^2 + y^2$ and the plane $x + 2y + 3z = 3$.

29. Find the points on the ellipse of Problem 28 that are nearest the origin and those that are farthest from it.

30. The ice tray shown in Fig. 14.9.10 is to be made from material that costs 1¢/in.² Minimize the cost function $f(x, y, z) = xy + 3xz + 7yz$ subject to the constraints that each of the 12 compartments is to have a square horizontal cross section and that the total volume (ignoring the partitions) is to be 12 in.³.

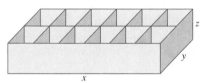

Fig. 14.9.10   The ice tray of Problem 30

31. Prove Theorem 1 for functions of three variables by showing that both of the vectors $\nabla f(P)$ and $\nabla g(P)$ are perpendicular at P to every curve on the surface $g(x, y, z) = 0$.

32. Find the lengths of the semiaxes of the ellipse of Example 4.

33. Figure 14.9.11 shows a right triangle with sides x, y, and z and fixed perimeter P. Maximize its area $A = \frac{1}{2}xy$ subject to the constraints $x + y + z = P$ and $x^2 + y^2 = z^2$. In particular, show that the optimal such triangle is isosceles (by showing that $x = y$).

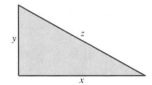

Fig. 14.9.11   A right triangle with fixed perimeter P (Problem 33)

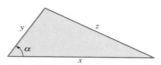

Fig. 14.9.12 A general triangle with fixed perimeter P (Problem 34)

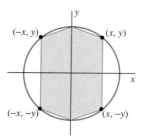

Fig. 14.9.13 The inscribed hexagon of Problem 35

34. Figure 14.9.12 shows a general triangle with sides x, y, and z and fixed perimeter P. Maximize its area

$$A = f(x, y, z, \alpha) = \tfrac{1}{2} xy \sin \alpha$$

subject to the constraints $x + y + z = P$ and to the law of cosines

$$z^2 = x^2 + y^2 - 2xy \cos \alpha.$$

In particular, show that the optimal such triangle is equilateral (by showing that $x = y = z$). [*Note:* The Lagrange multiplier equations for optimizing $f(x, y, z, w)$ subject to the constraint $g(x, y, z, w)$ take the form

$$f_x = \lambda g_x, \qquad f_y = \lambda g_y, \qquad f_z = \lambda g_z, \qquad f_w = \lambda g_w;$$

that is, $\nabla f = \lambda \nabla g$ in terms of the gradient vectors with four components.]

35. Figure 14.9.13 shows a hexagon with vertices $(0, \pm 1)$ and $(\pm x, \pm y)$ inscribed in the unit circle $x^2 + y^2 = 1$.

Show that its area is maximal when it is a *regular* hexagon with equal sides and angles.

36. When the hexagon of Fig. 14.9.13 is rotated around the y-axis, it generates a solid of revolution consisting of a cylinder and two cones (Fig. 14.9.14). What radius and cylinder height maximize the volume of this solid?

Fig. 14.9.14 The solid of Problem 36

14.9 Projects

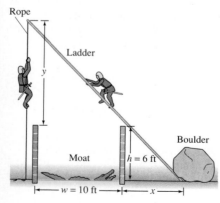

Fig. 14.9.15 The alligator-filled moat of Project A

PROJECT A You will need a graphics calculator or computer with graphing utility for this project. Figure 14.9.15 shows an alligator-filled moat of width $w = 10$ ft bounded on each side by a wall of height $h = 6$ ft. Soldiers plan to bridge this moat by scaling a ladder that is placed across the wall as indicated and anchored at the ground by a handy boulder, with the upper end directly above the wall on the opposite side of the moat. What is the minimal length L of a ladder that will suffice for this purpose? We outline two approaches.

With a Single Constraint Apply the Pythagorean theorem and the proportionality theorem for similar triangles to show that you need to minimize the (ladder-length-squared) function

$$f(x, y) = (x + 10)^2 + (y + 6)^2$$

subject to the constraint

$$g(x, y) = xy - 60 = 0.$$

Then apply the Lagrange multiplier method to derive the fourth-degree equation

$$x^4 + 10x^3 - 360x - 3600 = 0. \tag{15}$$

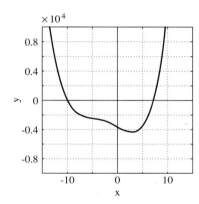

Fig. 14.9.16 $y = x^4 + 10x^3 - 360x - 3600$ (Project A)

You can approximate the pertinent solution of this equation graphically (Fig. 14.9.16). You may even be able to solve this equation manually—if you can first spot an integer solution (which must be an integral factor of the constant term 3600).

With Two Constraints   You can avoid the manual algebra involved in deriving and solving the quartic equation in Eq. (15) if a computer algebra system is available to you. With $z = L$ for the length of the ladder, observe directly from Fig. 14.9.15 that you need to minimize the function

$$f(x, y, z) = z$$

subject to the two constraints

$$g(x, y, z) = xy - 60 = 0,$$
$$h(x, y, z) = (x + 10)^2 + (y + 6)^2 - z^2 = 0.$$

This leads to a system of five equations in five unknowns (x, y, z, and the two Lagrange multipliers). You might investigate the *Maple* command

```
solve({equations}, {unknowns})
```

or the *Mathematica* command

```
Solve[ {equations}, {unknowns} ]
```

to solve a list of equations for its list of unknowns. The *Derive* command

```
NEWTONS([equations], [unknowns], [initial guesses], n)
```

produces n successive numerical approximations to the solution.

For your own personal moat problem, you might choose w and $h < w$ as the two largest distinct digits in your student I.D. number.

PROJECT B   Figure 14.9.17 shows a 14-sided polygon that is almost inscribed in the unit circle. It has vertices $(0, \pm 1)$, $(\pm x, \pm y)$, $(\pm u, \pm v)$, and $(\pm u, \pm y)$. When this polygon is revolved around the y-axis, it generates the "spindle" solid illustrated in Fig. 14.9.18, which consists of a solid cylinder of radius x, two solid cylinders of radius u, and two solid cones. The problem is to determine x, y, u, and v in order to *maximize* the volume of this spindle.

First express the volume V of the spindle as a function

$$V = f(x, y, u, v)$$

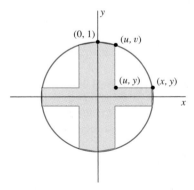

Fig. 14.9.17   The polygon of Project B

of four variables. The problem then is to maximize $f(x, y, u, v)$ subject to the two constraints

$$g(x, y, u, v) = x^2 + y^2 - 1 = 0,$$
$$h(x, y, u, v) = u^2 + v^2 - 1 = 0.$$

The corresponding Lagrange multiplier condition takes the form

$$\nabla f = \lambda \nabla g + \mu \nabla h,$$

where $\nabla f = \langle f_x, f_y, f_u, f_v \rangle$ and ∇g and ∇h are similar 4-vectors of partial derivatives.

All this results in a system of six equations in the six unknowns x, y, u, v, λ, and μ. Set up this system, but you probably should attempt to solve it only if a computer algebra system is available to you.

Fig. 14.9.18   The spindle of Project B

14.10

The Second Derivative Test for Functions of Two Variables

We saw in Section 14.5 that in order for the differentiable function $f(x, y)$ to have either a local minimum or a local maximum at an interior point $P(a, b)$ of its domain, it is a *necessary* condition that P be a *critical point* of f—that is, that

$$f_x(a, b) = 0 = f_y(a, b).$$

Here we give conditions *sufficient* to ensure that f has a local extremum at a critical point. The criterion stated in Theorem 1 involves the second-order partial derivatives of f at (a, b) and plays the role of the single-variable second derivative test (Section 4.6) for functions of two variables. To simplify the statement of this result, we shall use the following abbreviations:

$$A = f_{xx}(a, b), \qquad B = f_{xy}(a, b), \qquad C = f_{yy}(a, b), \tag{1}$$

and

$$\Delta = AC - B^2 = f_{xx}(a, b)\, f_{yy}(a, b) - [f_{xy}(a, b)]^2. \tag{2}$$

We outline a proof of Theorem 1 at the end of this section.

Theorem 1 *Sufficient Conditions for Local Extrema*

Let (a, b) be a critical point of the function $f(x, y)$, and suppose that f has continuous first-order and second-order partial derivatives in some circular disk centered at (a, b).

1. If $\Delta > 0$ and $A > 0$, then f has a local minimum at (a, b).
2. If $\Delta > 0$ and $A < 0$, then f has a local maximum at (a, b).
3. If $\Delta < 0$, then f has neither a local minimum nor a local maximum at (a, b). Instead it has a saddle point there.

Thus f has *either* a local maximum *or* a local minimum at the critical point (a, b) provided that the **discriminant** $\Delta = AC - B^2$ is *positive*. In this case, $A = f_{xx}(a, b)$ plays the role of the second derivative of a single-variable function: There is a local minimum at (a, b) if $A > 0$ and a local maximum if $A < 0$.

If $\Delta < 0$, then f has *neither* a local maximum *nor* a local minimum at (a, b). In this case we call (a, b) a **saddle point** for f, thinking of the appearance of the hyperbolic paraboloid $f(x, y) = x^2 - y^2$ (Fig. 14.10.1), a typical example of this case.

Theorem 1 does not answer the question of what happens when $\Delta = 0$. In this case, the two-variable second derivative test fails—it gives no information. Moreover, at such point (a, b), *anything* can happen, ranging from the local (indeed global) minimum of $f(x, y) = x^4 + y^4$ at $(0, 0)$ to the "monkey saddle" of Example 2.

In the case of a function $f(x, y)$ with several critical points, we must compute the quantities $A, B, C,$ and Δ separately at each critical point in order to apply the test.

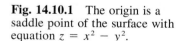

Fig. 14.10.1 The origin is a saddle point of the surface with equation $z = x^2 - y^2$.

EXAMPLE 1 Locate and classify the critical points of

$$f(x, y) = 3x - x^3 - 3xy^2.$$

Fig. 14.10.2 Critical-point analysis for the function of Example 1

Critical point	A	B	C	Δ	Type of extremum
$(1, 0)$	-6	0	-6	36	Local maximum
$(-1, 0)$	6	0	6	36	Local minimum
$(0, 1)$	0	-6	0	-36	Not an extremum
$(0, -1)$	0	6	0	-36	Not an extremum

Solution This function is a polynomial, so all its partial derivatives exist and are continuous everywhere. When we equate its first partial derivatives to zero (to locate the critical points of f), we get

$$f_x(x, y) = 3 - 3x^2 - 3y^2 = 0, \qquad f_y(x, y) = -6xy = 0.$$

The second of these equations implies that x or y must be zero; then the first implies that the other must be ± 1. Thus there are four critical points: $(1, 0)$, $(-1, 0)$, $(0, 1)$, and $(0, -1)$.

The second-order partial derivatives of f are

$$A = f_{xx} = -6x, \qquad B = f_{xy} = -6y, \qquad C = f_{yy} = -6x.$$

Hence $\Delta = 36(x^2 - y^2)$ at each of the critical points. The table in Fig. 14.10.2 summarizes the situation at each of the four critical points. The graph of f is shown in Figs. 14.10.3 and 14.10.4.

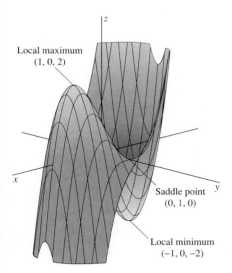

Local maximum $(1, 0, 2)$

Saddle point $(0, 1, 0)$

Local minimum $(-1, 0, -2)$

Fig. 14.10.3 Graph of the function of Example 1

EXAMPLE 2 Find and classify the critical points of the function

$$f(x, y) = 6xy^2 - 2x^3 - 3y^4.$$

Solution When we equate the first-order partial derivatives to zero, we get the equations

$$f_x(x, y) = 6y^2 - 6x^2 = 0 \quad \text{and} \quad f_y(x, y) = 12xy - 12y^3 = 0.$$

It follows that

$$x^2 = y^2 \quad \text{and} \quad y(x - y^2) = 0.$$

The first of these equations gives $x = \pm y$. If $x = y$, the second equation implies that $y = 0$ or $y = 1$. If $x = -y$, the second equation implies that $y = 0$ or $y = -1$. Hence there are three critical points: $(0, 0)$, $(1, 1)$, and $(1, -1)$.

The second-order partial derivatives of f are

$$A = f_{xx} = -12x, \qquad B = f_{xy} = 12y, \qquad C = f_{yy} = 12x - 36y^2.$$

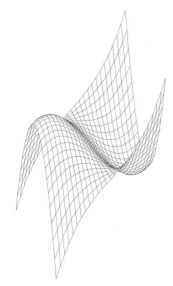

Fig. 14.10.4 Another view of the surface of Example 1

These expressions give the data shown in the table in Fig. 14.10.5. The critical point test fails at $(0, 0)$, so we must find another way to test this point.

Fig. 14.10.5 Critical-point analysis for the function of Example 2

Critical point	A	B	C	Δ	Type of extremum
$(0, 0)$	0	0	0	0	Test fails
$(1, 1)$	-12	12	-24	144	Local maximum
$(1, -1)$	-12	-12	-24	144	Local maximum

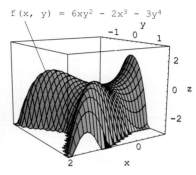

$f(x, y) = 6xy^2 - 2x^3 - 3y^4$

Fig. 14.10.6 The monkey saddle of Example 2

Fig. 14.10.7 The monkey in its saddle (Example 2)

We observe that $f(x, 0) = -2x^3$ and that $f(0, y) = -3y^4$. Hence, as we move away from the origin in the

Positive x-direction:	f decreases;
Negative x-direction:	f increases;
Positive y-direction:	f decreases;
Negative y-direction:	f decreases.

Consequently, f has neither a local maximum nor a local minimum at the origin. The graph of f is shown in Fig. 14.10.6. If a monkey were to sit with its rump at the origin and face the negative x-direction, then the directions in which $f(x, y)$ decreases would provide places for both its tail and its two legs to hang. That's why this particular surface is called a *monkey saddle* (Fig. 14.10.7).

EXAMPLE 3 Find and classify the critical points of the function

$$f(x, y) = \tfrac{1}{3}x^4 + \tfrac{1}{2}y^4 - 4xy^2 + 2x^2 + 2y^2 + 3.$$

Solution When we equate to zero the first-order partial derivatives of f, we obtain the equations

$$f_x(x, y) = \tfrac{4}{3}x^3 - 4y^2 + 4x = 0, \tag{3}$$

$$f_y(x, y) = 2y^3 - 8xy + 4y = 0, \tag{4}$$

which are not as easy to solve as the corresponding equations in Examples 1 and 2. But if we write Eq. (4) in the form

$$2y(y^2 - 4x + 2) = 0,$$

we see that either $y = 0$ or

$$y^2 = 4x - 2. \tag{5}$$

If $y = 0$, then Eq. (3) reduces to the equation

$$\tfrac{4}{3}x^3 + 4x = \tfrac{4}{3}x(x^2 + 3) = 0,$$

whose only solution is $x = 0$. Thus one critical point of f is $(0, 0)$.

If $y \neq 0$, we substitute $y^2 = 4x - 2$ into Eq. (3) to obtain

$$\tfrac{4}{3}x^3 - 4(4x - 2) + 4x = 0;$$

that is,

$$\tfrac{4}{3}x^3 - 12x + 8 = 0.$$

Thus we need to solve the cubic equation

$$\phi(x) = x^3 - 9x + 6 = 0. \tag{6}$$

The graph of $\phi(x)$ in Fig. 14.10.8 shows that this equation has three real solutions with approximate values $x \approx -3$, $x \approx 1$, and $x \approx +3$. Using either graphical techniques or Newton's method (Section 3.9), you can obtain the values

$$x \approx -3.2899, \qquad x \approx 0.7057, \qquad x \approx 2.5842, \tag{7}$$

accurate to four decimal places. The corresponding values of y are given from Eq. (5) by

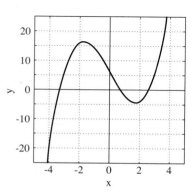

Fig. 14.10.8 The graph of $\phi(x) = x^3 - 9x + 6$ (Example 3)

Ch. 14 / Partial Differentiation

$$y = \pm\sqrt{4x - 2}, \tag{8}$$

but the first value of x in Eq. (7) yields *no* real value at all for y. Thus the two positive values of x in Eq. (7) add *four* critical points of $f(x, y)$ to the one critical point $(0, 0)$ already found.

Critical point	1	2	3	4	5
x	0.0000	0.7057	0.7057	2.5842	2.5842
y	0.0000	0.9071	−0.9071	2.8874	2.8874
z	3.0000	3.7402	3.7402	−3.5293	−3.5293
A	4.00	5.99	5.99	30.71	30.71
B	0.00	−7.26	7.26	−23.10	23.10
C	4.00	3.29	3.29	33.35	33.35
Δ	16.00	−32.94	−32.94	490.64	490.64
Type	Local minimum	Saddle point	Saddle point	Local minimum	Local minimum

Fig. 14.10.9 Classification of the critical points in Example 3

These five critical points are listed in the table in Fig. 14.10.9, together with the corresponding values of

$$A = f_{xx} = 4x^2 + 4, \qquad B = f_{xy} = f_{yx} = -8y.$$
$$C = f_{yy} = 6y^2 - 8x + 4, \qquad \Delta = AC - B^2$$

(rounded to two decimal places) at each of these critical points. We see that $\Delta > 0$ and $A > 0$ at $(0, 0)$ and at $(2.5482, \pm 2.8874)$, so these points are local minimum points. But $\Delta < 0$ at $(0.7057, \pm 0.9071)$, so these two are saddle points. The level curve diagram in Fig. 14.10.10 shows how these five critical points fit together.

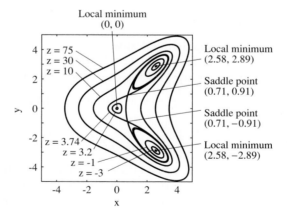

Fig. 14.10.10 Level curves for the function of Example 3

Finally, we observe that the behavior of $f(x, y)$ is approximately that of $\frac{1}{3}x^4 + \frac{1}{2}y^4$ when $|x|$ or $|y|$ is large, so the surface $z = f(x, y)$ must open upward and, therefore, have a global low point (but no global high point). Examining the values

$$f(0, 0) = 3 \quad \text{and} \quad f(2.5842, \pm 2.8874) \approx -3.5293,$$

we see that the global minimum value of $f(x, y)$ is approximately -3.5293.

DISCUSSION OF THEOREM 1

A complete proof of Theorem 1 (providing sufficient conditions for local extrema of functions of two variables) is best left for advanced calculus. Here, however, we provide an outline of the main ideas. Given a function $f(x, y)$ with critical point (a, b) that we wish to investigate, the function $f(x - a, y - b)$ would have a critical point of the same type at $(0, 0)$, so let's assume that $a = b = 0$.

To analyze the behavior of $f(x, y)$ near $(0, 0)$, we fix x and y and introduce the single-variable function

$$g(t) = f(xt, yt), \tag{9}$$

whose values agree with those of f on the straight line through $(0, 0)$ and (x, y) in the xy-plane. Its second-degree Taylor formula at $t = 0$ is

$$g(t) = g(0) + g'(0) \cdot t + \tfrac{1}{2} g''(0) \cdot t^2 + R, \tag{10}$$

where the remainder term is of the form $R = g^{(3)}(\tau) \cdot t^3/3!$ for some τ between 0 and t. With $t = 1$ we get

$$g(1) = g(0) + g'(0) + \tfrac{1}{2} g''(0) + R. \tag{11}$$

But

$$g(0) = f(0, 0) \quad \text{and} \quad g(1) = f(x, y) \tag{12}$$

by Eq. (9), and the chain rule gives

$$g'(0) = \frac{\partial f}{\partial x} \frac{dx}{dt} + \frac{\partial f}{\partial y} \frac{dy}{dt} = x f_x + y f_y \tag{13}$$

and

$$g''(0) = \frac{\partial}{\partial x}(x f_x + y f_y) \frac{dx}{dt} + \frac{\partial}{\partial y}(x f_x + y f_y) \frac{dy}{dt}$$

$$= x^2 f_{xx} + 2xy f_{xy} + y^2 f_{yy}, \tag{14}$$

where the partial derivatives of f are to be evaluated at the point $(0, 0)$.

Because $f_x(0, 0) = f_y(0, 0) = 0$, substitution of Eqs. (12), (13), and (14) into Eq. (11) yields the two-variable Taylor expansion

$$f(x, y) = f(0, 0) + \tfrac{1}{2}(Ax^2 + 2Bxy + Cy^2) + R, \tag{15}$$

where

$$A = f_{xx}(0, 0), \qquad B = f_{xy}(0, 0), \qquad C = f_{yy}(0, 0). \tag{16}$$

If $|x|$ and $|y|$ are sufficiently small, then the remainder term R is negligible, so the behavior of $f(x, y)$ near the critical point $(0, 0)$ is determined by the behavior near $(0, 0)$ of the *quadratic form*

$$q(x, y) = Ax^2 + 2Bxy + Cy^2. \tag{17}$$

But the shape of the *surface*

$$z = q(x, y) \tag{18}$$

is determined by the nature of the (rotated) *conic section*

$$q(x, y) = Ax^2 + 2Bxy + Cy^2 = 1 \tag{19}$$

that we discussed in Section 10.7. (There we wrote B in place of $2B$ and hence $4AC - B^2$ instead of $\Delta = AC - B^2$, as we have here.) We consider the three cases separately.

If $\Delta > 0$ and $A > 0$, then Eq. (19) is the equation of a rotated *ellipse*. It follows that $z = q(x, y)$ is a paraboloid that opens *upward*. Assuming that $z = f(x, y)$ has much the same shape near the origin, it follows that the critical point $(0, 0)$ is a *local minimum* point for f.

If $\Delta > 0$ and $A < 0$, the situation is the same except with the signs of the coefficients in Eq. (19) reversed. In this case $z = q(x, y)$ is a paraboloid that opens *downward,* so the critical point $(0, 0)$ is a *local maximum* point for f.

If $\Delta < 0$, then Eq. (19) is the equation of a rotated *hyperbola,* so it follows that $z = q(x, y)$ is a hyperboloid with a *saddle point* at the origin. Thus $(0, 0)$ is in this case a *saddle point* for f.

Finally, we observe that these possibilities correspond directly to the conclusions of Theorem 1.

14.10 Problems

Find and classify the critical points of the functions in Problems 1 through 22.

1. $f(x, y) = 2x^2 + y^2 + 4x - 4y + 5$

2. $f(x, y) = 10 + 12x - 12y - 3x^2 - 2y^2$

3. $f(x, y) = 2x^2 - 3y^2 + 2x - 3y + 7$
(Fig. 14.10.11)

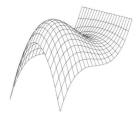

Fig. 14.10.11   Graph for Problem 3

4. $f(x, y) = xy + 3x - 2y + 4$

5. $f(x, y) = 2x^2 + 2xy + y^2 + 4x - 2y + 1$

6. $f(x, y) = x^2 + 4xy + 2y^2 + 4x - 8y + 3$

7. $f(x, y) = x^3 + y^3 + 3xy + 3$ (Fig. 14.10.12)

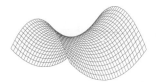

Fig. 14.10.12   Graph for Problem 7

8. $f(x, y) = x^2 - 2xy + y^3 - y$

9. $f(x, y) = 6x - x^3 - y^3$

10. $f(x, y) = 3xy - x^3 - y^3$

11. $f(x, y) = x^4 + y^4 - 4xy$

12. $f(x, y) = x^3 + 6xy + 3y^2$

13. $f(x, y) = x^3 + 6xy + 3y^2 - 9x$

14. $f(x, y) = x^3 + 6xy + 3y^2 + 6x$

15. $f(x, y) = 3x^2 + 6xy + 2y^3 + 12x - 24y$
(Fig. 14.10.13)

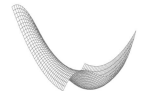

Fig. 14.10.13   Graph for Problem 15

16. $f(x, y) = 3x^2 + 12xy + 2y^3 - 6x + 6y$

17. $f(x, y) = 4xy - 2x^4 - y^2$

18. $f(x, y) = 8xy - 2x^2 - y^4$

19. $f(x, y) = 2x^3 - 3x^2 + y^2 - 12x + 10$
(Fig. 14.10.14)

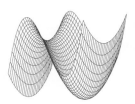

Fig. 14.10.14   Graph for Problem 19

20. $f(x, y) = 2x^3 + y^3 - 3x^2 - 12x - 3y$
(Fig. 14.10.15)

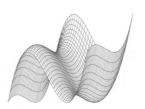 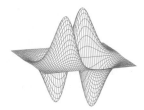

Fig. 14.10.15 Graph for Problem 20

Fig. 14.10.16 Graph for Problem 21

21. $f(x, y) = xy \exp(-x^2 - y^2)$ (Fig. 14.10.16)
22. $f(x, y) = (x^2 + y^2) \exp(x^2 - y^2)$

In Problems 23 through 25, first show that $\Delta = f_{xx}f_{yy} - (f_{xy})^2$ is zero at the origin. Then classify this critical point by imagining what the surface $z = f(x, y)$ looks like.

23. $f(x, y) = x^4 + y^4$
24. $f(x, y) = x^3 + y^3$
25. $f(x, y) = \exp(-x^4 - y^4)$

26. Let $f(s, t)$ denote the *square* of the distance between a typical point of the line $x = t$, $y = t + 1$, $z = 2t$ and a typical point of the line $x = 2s$, $y = s - 1$, $z = s + 1$. Show that the single critical point of f is a local minimum. Hence find the closest points on these two skew lines.

27. Let $f(x, y)$ denote the square of the distance from $(0, 0, 2)$ to a typical point of the surface $z = xy$. Find and classify the critical points of f.

28. Show that the surface

$$z = (x^2 + 2y^2) \exp(1 - x^2 - y^2)$$

looks like two mountain peaks joined by two ridges with a pit between them.

29. A wire 120 cm long is cut into three pieces of lengths x, y, and $120 - x - y$, and each piece is bent into the shape of a square. Let $f(x, y)$ denote the sum of the areas of these squares. Show that the single critical point of f is a local minimum. But surely it is possible to *maximize* the sum of the areas. Explain.

30. Show that the graph of the function

$$f(x, y) = xy \exp(\tfrac{1}{8}[x^2 + 4y^2])$$

had a saddle point, two local minima, and two global minima.

31. Find and classify the critical points of the function

$$f(x, y) = \sin\frac{\pi x}{2} \sin\frac{\pi y}{2}.$$

32. Let $f(x, y) = x^3 - 3xy^2$. (a) Show that its only critical point is $(0, 0)$ and that $\Delta = 0$ there. (b) By examining the behavior of $x^3 - 3xy^2$ on straight lines through the origin, show that the surface $z = x^3 - 3xy^2$ qualifies as a monkey saddle (Fig. 14.10.17).

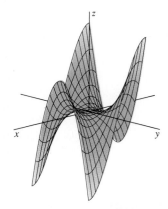

Fig. 14.10.17 The monkey saddle of Problem 32

33. Repeat Problem 32 with $f(x, y) = 4xy(x^2 - y^2)$. Show that near the critical point $(0, 0)$ the surface $z = f(x, y)$ qualifies as a "dog saddle" (Fig. 14.10.18).

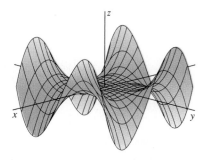

Fig. 14.10.18 The dog saddle of Problem 33

34. Let

$$f(x, y) = \frac{xy(x^2 - y^2)}{x^2 + y^2}.$$

Classify the behavior of f near the critical point $(0, 0)$.

In Problems 35 through 39, use graphical or numerical methods to find the critical points of f to four-place accuracy. Then classify them.

35. $f(x, y) = 2x^4 - 12x^2 + y^2 + 8x$
36. $f(x, y) = x^4 + 4x^2 - y^2 - 16x$
37. $f(x, y) = x^4 + 12xy + 6y^2 + 4x + 10$
38. $f(x, y) = x^4 + 8xy - 4y^2 - 16x + 10$
39. $f(x, y) = x^4 + 2y^4 - 12xy^2 - 20y^2$

14.10 Projects

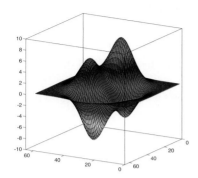

Fig. 14.10.19 Graph for Project A (*MATLAB* 4.0 color graphic courtesy of The MathWorks, Inc.)

PROJECT A This project is computationally intensive, and it will help to have a calculator or computer that can calculate derivatives symbolically. Consider the function f of two variables defined by

$$f(x, y) = 10(x^3 + y^5 + \tfrac{1}{5}x) \exp(-x^2 - y^2)$$
$$+ \tfrac{1}{3}\exp(-[x - 1]^2 - y^2), \qquad (20)$$

whose graph is shown in Fig. 14.10.19. Because $f(x, y) \to 0$ as $x, y \to \pm\infty$, the surface $z = f(x, y)$ must have a highest point and a lowest point. Your task is to find them.

Six critical points are visible: two local maxima, two saddle points, and two local minima. The problem is to calculate the locations of these critical points. Show first that when you calculate the two partial derivatives f_x and f_y, equate both to zero, and then remove the common factor $\exp(-x^2 - y^2)$ from each resulting equation, you get the two equations

$$-\tfrac{2}{3}(x - 1)e^{2x-1} - 20x^4 + 26x^2 - 20xy^5 + 2 = 0, \qquad (21)$$
$$-\tfrac{2}{3}ye^{2x-1} - 20x^3y - 4xy - 20y^6 + 50y^4 = 0, \qquad (22)$$

to be solved for the coordinates x and y of the critical points of function f.

Although these equations look intimidating, they have two redeeming features that permit you to solve them:

❑ Note that $y = 0$ satisfies Eq. (22). Thus you can find the critical points that lie on the x-axis by substituting $y = 0$ into Eq. (21) and then solving the remaining equation in x by graph-and-zoom techniques.

❑ You can solve Eq. (21) for y in terms of x. When you substitute the result into Eq. (22), you get another equation in x that can be solved by graph-and-zoom techniques. This gives the remaining critical points.

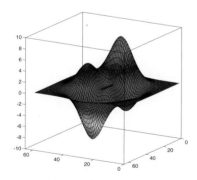

Fig. 14.10.20 Graph for Project B (*MATLAB* 4.0 color graphic courtesy of The MathWorks, Inc.)

PROJECT B Figure 14.10.20 shows the graph of the new function $g(x, y)$ defined by the formula

$$g(x, y) = 10(x^3 + y^5 - \tfrac{1}{5}x) \exp(-x^2 - y^2)$$
$$+ \tfrac{1}{3} \exp(-[x - 1]^2 - y^2), \qquad (23)$$

obtained by changing a single sign (that of $\tfrac{1}{5}x$) in Eq. (20). Now it is apparent that there is some additional "action" near the origin. With persistence, you can carry out the procedure outlined above to locate and analyze all the critical points of this altered function. You should find that g has ten critical points—three local maxima, three local minima, and four saddle points. Six of these critical points are located roughly as in Project A; the four new ones are located within a unit square centered at the origin.

REMARK We first saw the function $g(x, y)$ in a brochure describing the *MATLAB* software system for interactive numeric computation. The discovery of the function $f(x, y)$, whose critical point structure is somewhat simpler, was the serendipitous result of a typographical error made the first time we attempted to investigate the function $g(x, y)$.

Chapter 14 Review: DEFINITIONS, CONCEPTS, RESULTS

Use the following list as a guide to concepts that you may need to review.

1. Graphs and level curves of functions of two variables
2. Limits and continuity of functions of two or three variables
3. Partial derivatives—definition and computation
4. Geometric interpretation of partial derivatives and the plane tangent to the surface $z = f(x, y)$
5. Absolute and local maxima and minima
6. Necessary conditions for a local extremum
7. Increments and differentials of functions of two or three variables

8. The linear approximation theorem
9. The chain rule for functions of several variables
10. Directional derivatives—definition and computation
11. The gradient vector and the vector chain rule
12. Significance of the length and direction of the gradient vector
13. The gradient vector as a normal vector; tangent plane to a surface $F(x, y, z) = 0$
14. Constrained maximum-minimum problems and the Lagrange multiplier method
15. Sufficient conditions for a local extremum of a function of two variables

Chapter 14 Miscellaneous Problems

1. Use the method of Example 5 of Section 14.3 to show that

$$\lim_{(x, y) \to (0, 0)} \frac{x^2 y^2}{x^2 + y^2} = 0.$$

2. Use spherical coordinates to show that

$$\lim_{(x, y, z) \to (0, 0, 0)} \frac{x^3 + y^3 - z^3}{x^2 + y^2 + z^2} = 0.$$

3. Suppose that

$$g(x, y) = \frac{xy}{x^2 + y^2}$$

if $(x, y) \neq (0, 0)$; we *define* $g(0, 0)$ to be zero. Show that g is not continuous at $(0, 0)$.

4. Compute $g_x(0, 0)$ and $g_y(0, 0)$ for the function g of Problem 3.

5. Find a function $f(x, y)$ such that

$$f_x(x, y) = 2xy^3 + e^x \sin y$$

and

$$f_y(x, y) = 3x^2 y^2 + e^x \cos y + 1.$$

6. Prove that there is *no* function f with continuous second-order partial derivatives such that $f_x(x, y) = 6xy^2$ and $f_y(x, y) = 8x^2 y$.

7. Find the points on the paraboloid $z = x^2 + y^2$ at which the normal line passes through the point $(0, 0, 1)$.

8. Write an equation of the plane tangent to the surface $\sin xy + \sin yz + \sin xz = 1$ at the point $(1, \pi/2, 0)$.

9. Prove that every normal line to the cone with equation $z = \sqrt{x^2 + y^2}$ intersects the z-axis.

10. Show that the function

$$u(x, t) = \frac{1}{\sqrt{4\pi kt}} \exp\left(-\frac{x^2}{4kt}\right)$$

satisfies the one-dimensional heat equation

$$\frac{\partial u}{\partial t} = k \frac{\partial^2 u}{\partial x^2}.$$

11. Show that the function

$$u(x, y, t) = \frac{1}{4\pi kt} \exp\left(-\frac{x^2 + y^2}{4kt}\right)$$

satisfies the two-dimensional heat equation

$$\frac{\partial u}{\partial t} = k \left(\frac{\partial^2 u}{\partial x^2} + \frac{\partial^2 u}{\partial y^2} \right).$$

12. Let

$$f(x, y) = \frac{xy(x^2 - y^2)}{x^2 + y^2}$$

unless $(x, y) = (0, 0)$; we *define* $f(0, 0)$ to be zero. Show that the second-order partial derivatives $f_{xx}, f_{xy}, f_{yx},$ and f_{yy} all exist at $(0, 0)$ but that $f_{xy}(0, 0) \neq f_{yx}(0, 0)$.

13. Define the partial derivatives $\mathbf{r}_x$ and $\mathbf{r}_y$ of the vector-valued function $\mathbf{r}(x, y) = \mathbf{i}x + \mathbf{j}y + \mathbf{k}f(x, y)$ by componentwise partial differentiation. Then show that the vector $\mathbf{r}_x \times \mathbf{r}_y$ is normal to the surface $z = f(x, y)$.

14. An open-topped rectangular box is to have total surface area 300 cm². Find the dimensions that maximize its volume.

15. You must build a rectangular shipping crate with volume 60 ft³. Its sides cost $1/ft², its top costs $2/ft², and its bottom costs $3/ft². What dimensions would minimize the cost of the box?

16. A pyramid is bounded by the three coordinate planes and by the plane tangent to the surface $xyz = 1$ at a point in the first octant. Find the volume of this pyramid (it is independent of the point of tangency).

17. Two resistors have resistances R_1 and R_2, respectively. When they are connected in parallel, the total resistance R of the resulting circuit satisfies the equation

$$\frac{1}{R} = \frac{1}{R_1} + \frac{1}{R_2}.$$

Suppose that R_1 and R_2 are measured to be 300 and 600 Ω (ohms), respectively, with a maximum error of 1% in each measurement. Use differentials to estimate the maximum error (in ohms) in the calculated value of R.

18. Consider a gas that satisfies van der Waals' equation (see Problem 53 of Section 14.4). Use differentials to approximate the change in its volume if p is increased from 1 atm to 1.1 atm and T is decreased from 313 K to 303 K.

19. Each of the semiaxes a, b, and c of an ellipsoid with volume $V = \frac{4}{3}\pi abc$ is measured with a maximum percentage error of 1%. Use differentials to estimate the maximum percentage error in the calculated value of V.

20. Two spheres have radii a and b, and the distance between their centers is $c < a + b$. Thus the spheres meet in a common circle. Let P be a point on this circle, and let $\mathcal{P}_1$ and $\mathcal{P}_2$ be the tangent planes at P to the two spheres. Find the angle between $\mathcal{P}_1$ and $\mathcal{P}_2$ in terms of a, b, and c. [*Suggestion:* Recall that the angle between two planes is, by definition, the angle between their normal vectors.]

21. Find every point on the surface of the ellipsoid $x^2 + 4y^2 + 9z^2 = 16$ at which the normal line at the point passes through the center $(0, 0, 0)$ of the ellipsoid.

22. Suppose that $F(x) = \int_{g(x)}^{h(x)} f(t)\, dt$. Show that

$$F'(x) = f(h(x))h'(x) - f(g(x))g'(x).$$

[*Suggestion:* Write $w = \int_u^v f(t)\, dt$, where $u = g(x)$ and $v = h(x)$.]

23. Suppose that **a**, **b**, and **c** are mutually perpendicular unit vectors in space and that f is a function of the three independent variables x, y, and z. Show that

$$\nabla f = \mathbf{a}(D_{\mathbf{a}} f) + \mathbf{b}(D_{\mathbf{b}} f) + \mathbf{c}(D_{\mathbf{c}} f).$$

24. Let $\mathbf{R} = \langle \cos \theta, \sin \theta, 0 \rangle$ and $\mathbf{\Theta} = \langle -\sin \theta, \cos \theta, 0 \rangle$ be the polar-coordinates unit vectors. Given $f(x, y, z) = w(r, \theta, z)$, show that

$$D_{\mathbf{R}} f = \frac{\partial w}{\partial r} \quad \text{and} \quad D_{\mathbf{\Theta}} f = \frac{1}{r}\frac{\partial w}{\partial \theta}.$$

Then conclude from Problem 23 that the gradient vector is given in cylindrical coordinates by

$$\nabla f = \frac{\partial w}{\partial r}\mathbf{R} + \frac{1}{r}\frac{\partial w}{\partial \theta}\mathbf{\Theta} + \frac{\partial w}{\partial z}\mathbf{k}.$$

25. Suppose that you are standing at the point with coordinates $(-100, -100, 430)$ on a hill that has the shape of the graph of

$$z = 500 - (0.003)x^2 - (0.004)y^2$$

(in units of meters). In what (horizontal) direction should you move in order to maintain a constant altitude—that is, to neither climb nor descend the hill?

26. Suppose that the blood concentration in the ocean at the point (x, y) is given by

$$f(x, y) = A \exp(-k[x^2 + 2y^2]),$$

where A and k are positive constants. A shark always swims in the direction of ∇f. Show that its path is a parabola $y = cx^2$. [*Suggestion:* Show that the condition that $\langle dx/dt, dy/dt \rangle$ is a multiple of ∇f implies that

$$\frac{1}{x}\frac{dx}{dt} = \frac{1}{2y}\frac{dy}{dt}.$$

Then antidifferentiate this equation.]

27. Consider a plane tangent to the surface with equation $x^{2/3} + y^{2/3} + z^{2/3} = 1$. Find the sum of the squares of the x-, y-, and z-intercepts of this plane.

28. Find the points on the ellipse $x^2/a^2 + y^2/b^2 = 1$ (with $a \ne b$) where the normal line passes through the origin.

29. (a) Show that the origin is a critical point of the function f of Problem 12. (b) Show that f does not have a local extremum at $(0, 0)$.

30. Find the point of the surface $z = xy + 1$ that is closest to the origin.

31. Use the method of Problem 24 in Section 14.9 to find the semiaxes of the rotated ellipse

$$73x^2 + 72xy + 52y^2 = 100.$$

32. Use the Lagrange multiplier method to show that the longest chord of the sphere $x^2 + y^2 + z^2 = 1$ has length 2. [*Suggestion:* There is no loss of generality in assuming that $(1, 0, 0)$ is one endpoint of the chord.]

33. Use the method of Lagrange multipliers, the law of cosines, and Fig. 14.9.9 to find the triangle of minimum perimeter inscribed in the unit circle.

34. When a current I enters two resistors, with resistances R_1 and R_2, that are connected in parallel, it splits into two currents I_1 and I_2 (with $I = I_1 + I_2$) so as to minimize the total power $R_1 I_1^2 + R_2 I_2^2$. Express I_1 and I_2 in terms of R_1, R_2, and I. Then derive the formula of Problem 17.

35. Use the method of Lagrange multipliers to find the points of the ellipse $x^2 + 2y^2 = 1$ that are closest to and farthest from the line $x + y = 2$. [*Suggestion:* Let $f(x, y, u, v)$ denote the square of the distance between the point (x, y) of the ellipse and the point (u, v) of the line.]

36. (a) Show that the maximum of

$$f(x, y, z) = x + y + z$$

at points of the sphere $x^2 + y^2 + z^2 = a^2$ is $a\sqrt{3}$.

(b) Conclude from the result of part (a) that

$$(x + y + z)^2 \leq 3(x^2 + y^2 + z^2)$$

for any three numbers x, y, and z.

37. Generalize the method of Problem 36 to show that

$$\left(\sum_{i=1}^{n} x_i\right)^2 \leq n \sum_{i=1}^{n} x_i^2$$

for any n real numbers $x_1, x_2, \ldots, x_n$.

38. Find the maximum and minimum values of $f(x, y) = xy - x - y$ at points on and within the triangle with vertices $(0, 0)$, $(0, 1)$, and $(3, 0)$.

39. Find the maximum and minimum values of $f(x, y, z) = x^2 - yz$ at points of the sphere $x^2 + y^2 + z^2 = 1$.

40. Find the maximum and minimum values of $f(x, y) = x^2 y^2$ at points of the ellipse $x^2 + 4y^2 = 24$.

Locate and classify the critical points (local maxima, local minima, saddle points, and other points at which the tangent plane is horizontal) of the functions in Problems 41 through 50.

41. $f(x, y) = x^3 y - 3xy + y^2$

42. $f(x, y) = x^2 + xy + y^2 - 6x + 2$

43. $f(x, y) = x^3 - 6xy + y^3$

44. $f(x, y) = x^2 y + xy^2 + x + y$

45. $f(x, y) = x^3 y^2 (1 - x - y)$

46. $f(x, y) = x^4 - 2x^2 + y^2 + 4y + 3$

47. $f(x, y) = e^{xy} - 2xy$

48. $f(x, y) = x^3 - y^3 + x^2 + y^2$

49. $f(x, y) = (x - y)(xy - 1)$

50. $f(x, y) = (2x^2 + y^2) \exp(-x^2 - y^2)$

51. Given the data points (x_i, y_i) for $i = 1, 2, \ldots, n$, the **least-squares straight line** $y = mx + b$ is the line that best fits these data in the following sense. Let $d_i = y_i - (mx_i + b)$ be the *deviation* of the predicted value $mx_i + b$ from the true value y_i. Let

$$f(m, b) = d_1^2 + d_2^2 + \cdots + d_n^2 = \sum_{i=1}^{n} [y_i - (mx_i + b)]^2$$

be the sum of the squares of the deviations. The least-squares straight line is the one that minimizes this sum (Fig. 14.MP.1). Show how to choose m and b by minimizing f. [*Note:* The only variables in this computation are m and b.]

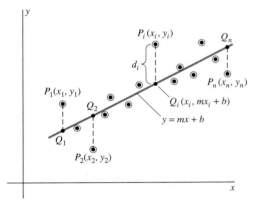

Fig. 14.MP.1 Fitting the best straight line to the data points (x_i, y_i), $1 \leq i \leq n$ (Problem 51)

Multiple Integrals

Henri Lebesgue (1875–1941)

❑ Geometric problems of *measure*—dealing with concepts of length, area, and volume—can be traced back 4000 years to the rise of civilizations in the fertile river valleys of Africa and Asia, when such issues as areas of fields and volumes of granaries became important. These problems led ultimately to the *integral,* which is used to calculate (among other things) areas and volumes of curvilinear figures. But only in the early twentieth century were certain long-standing difficulties with measure and integration finally resolved, largely as a consequence of the work of the French mathematician Henri Lebesgue.

❑ In his 1902 thesis presented at the Sorbonne in Paris, Lebesgue presented a new definition of the integral, gener-

alizing Riemann's definition. In essence, to define the integral of the function f from $x = a$ to $x = b$, Lebesgue replaced Riemann's subdivision of the interval $[a, b]$ into nonoverlapping subintervals with a partition of $[a, b]$ into disjoint measurable sets $\{E_i\}$. The Riemann sum $\sum f(x_i^*) \Delta x$ was thereby replaced with a sum of the form $\sum f(x_i^*) m_i$, where m_i is the measure of the ith set E_i containing x_i^*. To see the advantage of the "Lebesgue integral," consider the fact that there exist differentiable functions whose derivatives are not integrable in the sense of Riemann. For such a function, the fundamental theorem of calculus in the form

$$\int_a^b f'(x)\,dx = f(b) - f(a)$$

fails to hold. But with his new definition of the integral, Lebesgue showed that a derivative function f' is integrable and

that the fundamental theorem holds. Similarly, the equality of double and iterated integrals (Section 15.1) holds only under rather drastic restrictions if the Riemann definition of multiple integrals is used, but the Lebesgue integral resolves the difficulty.

❑ For such reasons, the Lebesgue theory of measure and integration predominates in modern mathematical research, both pure and applied. For instance, the Lebesgue integral is basic to such diverse realms as applied probability and mathematical biology, the quantum theory of atoms and nuclei, and the information theory and electric signals processing of modern computer technology.

❑ The Section 15.5 Project illustrates the application of multiple integrals to such concrete problems as the optimal design of race-car wheels.

We could use multiple integrals to determine the best design for the wheels of these soapbox derby cars.

15.1
Double Integrals

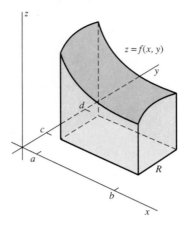

Fig. 15.1.1 We will use a double integral to compute the volume V.

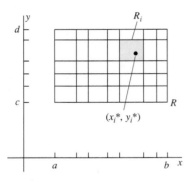

Fig. 15.1.2 A partition $\mathcal{P}$ of the rectangle R

This chapter is devoted to integrals of functions of two or three variables. Such integrals are called **multiple integrals**. The applications of multiple integrals include computation of area, volume, mass, and surface area in a wider variety of situations than can be handled with the single integral of Chapters 5 and 6.

The simplest sort of multiple integral is the *double integral*

$$\iint_R f(x, y) \, dA$$

of a continuous function $f(x, y)$ over the *rectangle*

$$R = [a, b] \times [c, d]$$
$$= \{(x, y) \mid a \leq x \leq b, \quad c \leq y \leq d\}$$

in the xy-plane. Just as the definition of the single integral is motivated by the problem of computing areas, the definition of the double integral is motivated by the problem of computing the volume V of the solid of Fig. 15.1.1—a solid bounded above by the graph $z = f(x, y)$ of the nonnegative function f over the rectangle R in the xy-plane.

To define the *value*

$$V = \iint_R f(x, y) \, dA$$

of such a double integral, we begin with an approximation to V. To obtain this approximation, the first step is to construct a **partition** $\mathcal{P}$ of R into subrectangles $R_1, R_2, \ldots, R_k$ determined by the partitions

$$a = x_0 < x_1 < x_2 < \cdots < x_m = b$$

of $[a, b]$ and

$$c = y_0 < y_1 < y_2 < \cdots < y_n = d$$

of $[c, d]$. Such a partition of R into $k = mn$ rectangles is shown in Fig. 15.1.2. The order in which these rectangles are labeled makes no difference.

Next we choose an arbitrary point (x_i^*, y_i^*) of the ith subrectangle R_i for each i $(1 \leq i \leq k)$. The collection of points $S = \{(x_i^*, y_i^*) \mid 1 \leq i \leq k\}$ is called a **selection** for the partition $\mathcal{P} = \{R_i \mid 1 \leq i \leq k\}$. As a measure of the size of the subrectangles of the partition $\mathcal{P}$, we define its **mesh** $|\mathcal{P}|$ to be the maximum of the lengths of the diagonals of the rectangles $\{R_i\}$.

Now consider a rectangular column that rises straight up from the xy-plane. Its base is the subrectangle R_i and its height is the value $f(x_i^*, y_i^*)$ of f at the selected point (x_i^*, y_i^*) of R_i. One such column is shown in Fig. 15.1.3. If ΔA_i denotes the area of R_i, then the volume of the ith column is $f(x_i^*, y_i^*) \, \Delta A_i$. The sum of the volumes of all such columns (Fig. 15.1.4) is the **Riemann sum**

$$\sum_{i=1}^{k} f(x_i^*, y_i^*) \, \Delta A_i, \tag{1}$$

an approximation to the volume V of the solid region that lies above the rectangle R and under the graph $z = f(x, y)$.

We would expect to determine the exact volume V by taking the limit of the Riemann sum in Eq. (1) as the mesh $|\mathcal{P}|$ of the partition $\mathcal{P}$ approaches

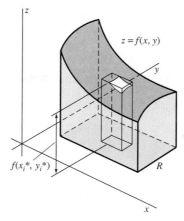

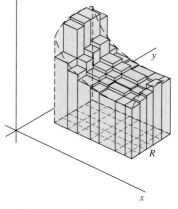

Fig. 15.1.3 Approximating the volume under the surface by summing volumes of towers with rectangular bases

Fig. 15.1.4 Columns corresponding to a partition of the rectangle R

zero. We therefore define the **(double) integral** of the function f over the rectangle R to be

$$\iint_R f(x, y)\, dA = \lim_{|\mathcal{P}| \to 0} \sum_{i=1}^{k} f(x_i^*, y_i^*)\, \Delta A_i, \tag{2}$$

provided that this limit exists (we will make the concept of the existence of such a limit more precise in Section 15.2). It is proved in advanced calculus that the limit in Eq. (2) *does* exist if f is continuous on R. To motivate the introduction of the Riemann sum in Eq. (1), we assumed that f was nonnegative on R, but Eq. (2) serves to define the double integral over a rectangle whether or not f is nonnegative.

ITERATED INTEGRALS

The direct evaluation of the limit in Eq. (2) is generally even less practical than the direct evaluation of the limit we used in Section 5.4 to define the single-variable integral. In practice, we shall calculate double integrals over rectangles by means of the **iterated integrals** that appear in Theorem 1.

> **Theorem 1** *Double Integrals as Iterated Single Integrals*
> Suppose that $f(x, y)$ is continuous on the rectangle $R = [a, b] \times [c, d]$. Then
> $$\iint_R f(x, y)\, dA = \int_a^b \left(\int_c^d f(x, y)\, dy \right) dx = \int_c^d \left(\int_a^b f(x, y)\, dx \right) dy. \tag{3}$$

Theorem 1 tells us how to compute a double integral by means of two successive (or *iterated*) single-variable integrations, both of which we can compute by using the fundamental theorem of calculus (if the function f is sufficiently well behaved on R).

Let us explain what we mean by the parentheses in the iterated integral

$$\int_a^b \int_c^d f(x, y) \, dy \, dx = \int_a^b \left(\int_c^d f(x, y) \, dy \right) dx. \tag{4}$$

First we hold x constant and integrate with respect to y, from $y = c$ to $y = d$. The result of this first integration is the **partial integral of f with respect to y**, denoted by

$$\int_c^d f(x, y) \, dy,$$

and it is a function of x alone. Then we integrate this latter function with respect to x, from $x = a$ to $x = b$.

Similarly, we calculate the iterated integral

$$\int_c^d \int_a^b f(x, y) \, dx \, dy = \int_c^d \left(\int_a^b f(x, y) \, dx \right) dy \tag{5}$$

by first integrating from a to b with respect to x (while holding y fixed) and then integrating the result from c to d with respect to y. The order of integration (either first with respect to x and then with respect to y, or the reverse) is determined by the order in which the differentials dx and dy appear in the iterated integrals in Eqs. (4) and (5). We always work "from the inside out." Theorem 1 guarantees that the value obtained is independent of the order of integration provided that f is continuous on R.

EXAMPLE 1   Compute the iterated integrals in Eqs. (4) and (5) for the function $f(x, y) = 4x^3 + 6xy^2$ on the rectangle $R = [1, 3] \times [-2, 1]$.

Solution   The rectangle R is shown in Fig. 15.1.5, where the vertical segment (on which x is constant) corresponds to the inner integral of Eq. (4). Its endpoints lie at heights $y = -2$ and $y = 1$, which are, therefore, the limits on the inner integral. So Eq. (4) yields

$$\int_1^3 \left(\int_{-2}^1 (4x^3 + 6xy^2) \, dy \right) dx = \int_1^3 \left[4x^3 y + 2xy^3 \right]_{y=-2}^1 dx$$

$$= \int_1^3 \left[(4x^3 + 2x) - (-8x^3 - 16x) \right] dx$$

$$= \int_1^3 (12x^3 + 18x) \, dx$$

$$= \left[3x^4 + 9x^2 \right]_1^3 = 312.$$

The horizontal segment (on which y is constant) in Fig. 15.1.6 corresponds to the inner integral of Eq. (5). Its endpoints lie at $x = 1$ and $x = 3$ (the limits on x), so Eq. (5) gives

$$\int_{-2}^1 \left(\int_1^3 (4x^3 + 6xy^2) \, dx \right) dy = \int_{-2}^1 \left[x^4 + 3x^2 y^2 \right]_{x=1}^3 dy$$

$$= \int_{-2}^1 \left[(81 + 27y^2) - (1 + 3y^2) \right] dy$$

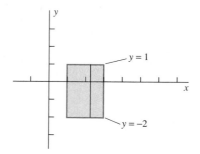

Fig. 15.1.5 The inner limits of the first iterated integral (Example 1)

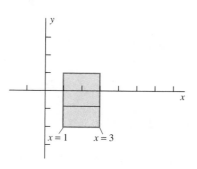

Fig. 15.1.6 The inner limits of the second iterated integral (Example 1)

$$= \int_{-2}^{1} (80 + 24y^2) \, dy$$

$$= \left[80y + 8y^3 \right]_{-2}^{1} = 312.$$

When we note that iterated double integrals are always evaluated from the inside out, it becomes clear that the parentheses appearing on the right-hand sides in Eqs. (4) and (5) are unnecessary. They are, therefore, generally omitted, as in Examples 2 and 3. When $dy \, dx$ appears in the integrand, we integrate first with respect to y, whereas the appearance of $dx \, dy$ tells us to integrate first with respect to x.

EXAMPLE 2 See Fig. 15.1.7.

$$\int_{0}^{\pi} \int_{0}^{\pi/2} \cos x \cos y \, dy \, dx = \int_{0}^{\pi} \left[\cos x \sin y \right]_{y=0}^{\pi/2} dx$$

$$= \int_{0}^{\pi} \cos x \, dx = \left[\sin x \right]_{0}^{\pi} = 0.$$

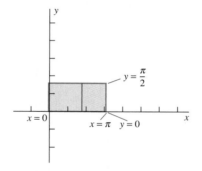

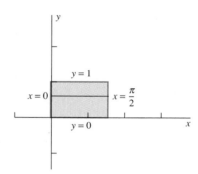

Fig. 15.1.7 Example 2   **Fig. 15.1.8** Example 3

EXAMPLE 3 See Fig. 15.1.8.

$$\int_{0}^{1} \int_{0}^{\pi/2} (e^y + \sin x) \, dx \, dy = \int_{0}^{1} \left[xe^y - \cos x \right]_{x=0}^{\pi/2} dy$$

$$= \int_{0}^{1} \left(\frac{1}{2} \pi e^y + 1 \right) dy$$

$$= \left[\frac{1}{2} \pi e^y + y \right]_{0}^{1} = \frac{\pi(e - 1)}{2} + 1.$$

ITERATED INTEGRALS AND CROSS SECTIONS

An outline of the proof of Theorem 1 illuminates the relationship between iterated integrals and the method of cross sections (for computing volumes) discussed in Section 6.2. First we divide $[a, b]$ into n equal subintervals, each of length $\Delta x = (b - a)/n$, and we also divide $[c, d]$ into n equal subintervals, each of length $\Delta y = (d - c)/n$. This gives n^2 subrectangles, each of which has area $\Delta A = \Delta x\,\Delta y$. Choose a point x_i^* in $[x_{i-1}, x_i]$ for each i, $1 \leq i \leq n$. Then the average value theorem for single integrals (Section 5.6) gives a point y_{ij}^* in $[y_{j-1}, y_j]$ such that

$$\int_{y_{j-1}}^{y_j} f(x_i^*, y)\,dy = f(x_i^*, y_{ij}^*)\,\Delta y.$$

This gives us the selected point (x_i^*, y_{ij}^*) in the subrectangle $[x_{i-1}, x_i] \times [y_{i-1}, y_i]$. Then

$$\iint_R f(x, y)\,dA \approx \sum_{i,j=1}^{n} f(x_i^*, y_{ij}^*)\,\Delta A = \sum_{i=1}^{n}\sum_{j=1}^{n} f(x_i^*, y_{ij}^*)\,\Delta y\,\Delta x$$

$$= \sum_{i=1}^{n}\left(\sum_{j=1}^{n}\int_{y_{j-1}}^{y_j} f(x_i^*, y)\,dy\right)\Delta x$$

$$= \sum_{i=1}^{n}\left(\int_c^d f(x_i^*, y)\,dy\right)\Delta x$$

$$= \sum_{i=1}^{n} A(x_i^*)\,\Delta x,$$

where

$$A(x) = \int_c^d f(x, y)\,dy.$$

This last sum is a Riemann sum for the integral

$$\int_a^b A(x)\,dx,$$

so the result of our computation is

$$\iint_R f(x, y)\,dA \approx \sum_{i=1}^{n} A(x_i^*)\,\Delta x$$

$$\approx \int_a^b A(x)\,dx = \int_a^b\left(\int_c^d f(x, y)\,dy\right)dx.$$

We can convert this outline into a complete proof of Theorem 1 by showing that the preceding approximations become equalities when we take limits as $n \to \infty$.

In case the function f is nonnegative on R, the function $A(x)$ introduced here gives the area of the vertical cross section perpendicular to the x-axis (Fig. 15.1.9). Thus the iterated integral in Eq. (4) expresses the volume V as the integral from $x = a$ to $x = b$ of the cross-sectional area function $A(x)$. Similarly, the iterated integral in Eq. (5) expresses V as the integral from $y = c$ to $y = d$ of the function

$$A(y) = \int_a^b f(x, y)\,dx,$$

which gives the area of a vertical cross section in a plane perpendicular to the y-axis. [Although it seems appropriate to use the notation $A(y)$ here, note that $A(x)$ and $A(y)$ are by no means the same function!]

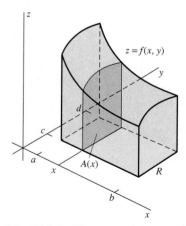

Fig. 15.1.9 The area of the cross section at x is

$$A(x) = \int_c^d f(x, y)\,dy.$$

15.1 Problems

Evaluate the iterated integrals in Problems 1 through 20.

1. $\int_0^2 \int_0^4 (3x + 4y) \, dx \, dy$

2. $\int_0^3 \int_0^2 x^2 y \, dx \, dy$

3. $\int_{-1}^2 \int_1^3 (2x - 7y) \, dy \, dx$

4. $\int_{-2}^1 \int_2^4 x^2 y^3 \, dy \, dx$

5. $\int_0^3 \int_0^3 (xy + 7x + y) \, dx \, dy$

6. $\int_0^2 \int_2^4 (x^2 y^2 - 17) \, dx \, dy$

7. $\int_{-1}^2 \int_{-1}^2 (2xy^2 - 3x^2 y) \, dy \, dx$

8. $\int_1^3 \int_{-3}^{-1} (x^3 y - xy^3) \, dy \, dx$

9. $\int_0^{\pi/2} \int_0^{\pi/2} (\sin x \cos y) \, dx \, dy$

10. $\int_0^{\pi/2} \int_0^{\pi/2} (\cos x \sin y) \, dy \, dx$

11. $\int_0^1 \int_0^1 xe^y \, dy \, dx$

12. $\int_0^1 \int_{-2}^2 x^2 e^y \, dx \, dy$

13. $\int_0^1 \int_0^\pi e^x \sin y \, dy \, dx$

14. $\int_0^1 \int_0^1 e^{x+y} \, dx \, dy$

15. $\int_0^\pi \int_0^\pi (xy + \sin x) \, dx \, dy$

16. $\int_0^{\pi/2} \int_0^{\pi/2} (y - 1) \cos x \, dx \, dy$

17. $\int_0^{\pi/2} \int_1^e \frac{\sin y}{x} \, dx \, dy$

18. $\int_1^e \int_1^e \frac{1}{xy} \, dy \, dx$

19. $\int_0^1 \int_0^1 \left(\frac{1}{x+1} + \frac{1}{y+1} \right) dx \, dy$

20. $\int_1^2 \int_1^3 \left(\frac{x}{y} + \frac{y}{x} \right) dy \, dx$

In Problems 21 through 24, verify that the values of

$$\iint_R f(x, y) \, dA$$

given by the iterated integrals in Eqs. (4) and (5) are indeed equal.

21. $f(x, y) = 2xy - 3y^2;\quad R = [-1, 1] \times [-2, 2]$

22. $f(x, y) = \sin x \cos y;\quad R = [0, \pi] \times [-\pi/2, \pi/2]$

23. $f(x, y) = \sqrt{x + y};\quad R = [0, 1] \times [1, 2]$

24. $f(x, y) = e^{x+y};\quad R = [0, \ln 2] \times [0, \ln 3]$

25. Prove that

$$\lim_{n \to \infty} \int_0^1 \int_0^1 x^n y^n \, dx \, dy = 0.$$

15.1 Project

This project explores the *midpoint approximation* to the double integral

$$I = \iint_R f(x, y) \, dA \tag{6}$$

of the function $f(x, y)$ over the plane rectangle $R = [a, b] \times [c, d]$. To define the midpoint approximation, let $[a, b]$ be divided into m subintervals, all with the same length $h = \Delta x = (b - a)/m$, and let $[c, d]$ be divided into n subintervals, all with the same length $k = \Delta y = (d - c)/n$. For each i and j ($1 \leq i \leq m$ and $1 \leq j \leq n$), let u_i and v_j denote the *midpoints* of the ith subinterval $[x_{i-1}, x_i]$ and the jth subinterval $[y_{j-1}, y_j]$, respectively. Then the corresponding **midpoint approximation** to the double integral I is the sum

$$S_{mn} = \sum_{i=1}^m \sum_{j=1}^n f(u_i, v_j) \, hk. \tag{7}$$

Figure 15.1.10 illustrates the case $m = 3$, $n = 2$, in which $h = (b - a)/3$, $k = (d - c)/2$, and

$$S_{32} = hk[f(u_1, v_1) \\ + f(u_2, v_1) + f(u_3, v_1) + f(u_1, v_2) + f(u_2, v_2) + f(u_3, v_2)].$$

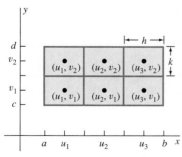

Fig. 15.1.10 The points used in the midpoint approximation

You can calculate numerically the double sum in Eq. (7) by using BASIC or graphics calculator programs. With a computer algebra system or a calculator such as the HP-48 or TI-85 that has a SUM function, you can use commands like those listed in Fig. 5.4.11. For example, the *Mathematica* command

```
Sum[f[x,y], {x, a + h/2, b - h/2, h},
            {y, c + k/2, d - k/2, k} ] h k
```

is appropriate.

For each of the double integrals in Problems 1 through 6, first calculate the midpoint approximation S_{mn} with the indicated values of m and n. Then try larger values. Compare each numerical approximation with the exact value of the integral.

1. $\displaystyle\int_0^1 \int_0^1 (x + y)\, dy\, dx, \quad m = n = 2$

2. $\displaystyle\int_0^3 \int_0^2 (2x + 3y)\, dy\, dx, \quad m = 3, n = 2$

3. $\displaystyle\int_0^2 \int_0^2 xy\, dy\, dx, \quad m = n = 2$

4. $\displaystyle\int_0^1 \int_0^1 x^2 y\, dy\, dx, \quad m = n = 3$

5. $\displaystyle\int_0^{\pi/2} \int_0^{\pi/2} \sin x \sin y\, dy\, dx, \quad m = n = 2$

6. $\displaystyle\int_0^{\pi/2} \int_0^1 \frac{\cos x}{1 + y^2}\, dy\, dx, \quad m = n = 2$

15.2

Double Integrals over More General Regions

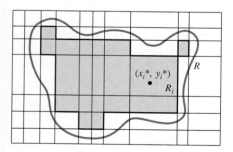

Fig. 15.2.1 The rectangular partition of S produces an associated inner partition (shown shaded) of the region R.

Now we want to define and compute double integrals over regions more general than rectangles. Let the function f be defined on the plane region R, and suppose that R is **bounded**—that is, that R lies within some rectangle S. To define the (double) integral of f over R, we begin with a partition $\mathcal{Q}$ of the rectangle S into subrectangles. Some of the rectangles of $\mathcal{Q}$ will lie wholly within R, some will be outside R, and some will lie partly within and partly outside R. We consider the collection $\mathcal{P} = \{R_1, R_2, \ldots, R_k\}$ of all those subrectangles of $\mathcal{Q}$ that lie *completely within* the region R. This collection $\mathcal{P}$ is called the **inner partition** of the region R determined by the partition $\mathcal{Q}$ of the rectangle S (Fig. 15.2.1). By the **mesh** $|\mathcal{P}|$ of the inner partition $\mathcal{P}$ we mean the mesh of the partition $\mathcal{Q}$ that determines $\mathcal{P}$. Note that $|\mathcal{P}|$ depends not only on $\mathcal{P}$ but on $\mathcal{Q}$ as well.

Using the inner partition $\mathcal{P}$ of the region R, we can proceed in much the same way as in Section 15.1. By choosing an arbitrary point (x_i^*, y_i^*) in the ith subrectangle R_i of $\mathcal{P}$ for $i = 1, 2, 3, \ldots, k$, we obtain a **selection** for the inner partition $\mathcal{P}$. Let us denote by ΔA_i the area of R_i. Then this selection gives the **Riemann sum**

$$\sum_{i=1}^k f(x_i^*, y_i^*)\, \Delta A_i$$

associated with the inner partition $\mathcal{P}$. In case f is nonnegative on R, this Riemann sum approximates the volume of the three-dimensional region that lies under the surface $z = f(x, y)$ and above the region R in the xy-plane. We therefore define the double integral of f over the region R by taking the limit of this Riemann sum as the mesh $|\mathcal{P}|$ approaches zero. Thus

$$\iint\limits_{R} f(x, y) \, dA = \lim_{|\mathcal{P}| \to 0} \sum_{i=1}^{k} f(x_i^*, y_i^*) \, \Delta A_i \tag{1}$$

provided that this limit exists in the sense of the following definition.

Definition *The Double Integral*
The **double integral** of the bounded function f over the plane region R is the number

$$I = \iint\limits_{R} f(x, y) \, dA$$

provided that, for every $\epsilon > 0$, there exists a number $\delta > 0$ such that

$$\left| \sum_{i=1}^{k} f(x_i^*, y_i^*) \, \Delta A_i - I \right| < \epsilon$$

for every inner partition $\mathcal{P} = \{R_1, R_2, \ldots, R_k\}$ of R that has mesh $|\mathcal{P}| < \delta$ and every selection of points (x_i^*, y_i^*) in R_i $(i = 1, 2, \ldots, k)$.

Thus the meaning of the limit in Eq. (1) is that the Riemann sum can be made arbitrary close to the number

$$I = \iint\limits_{R} f(x, y) \, dA$$

merely by choosing the mesh of the inner partition $\mathcal{P}$ sufficiently small.

NOTE   If R is a rectangle and we choose $S = R$ (so that an inner partition of R is simply a partition of R), then the preceding definition reduces to our earlier definition of a double integral over a rectangle. In advanced calculus the double integral of the function f over the bounded plane region R is shown to exist provided that f is continuous on R and the *boundary* of R is reasonably well behaved. In particular, it suffices for the boundary of R to consist of a finite number of piecewise smooth, simple closed curves (that is, each boundary curve consists of a finite number of smooth arcs).

EVALUATION OF DOUBLE INTEGRALS

For certain common types of regions, we can evaluate double integrals by using iterated integrals in much the same way as when the region is a rectangle. The plane region R is called **vertically simple** if it is described by means of the inequalities

$$a \leqq x \leqq b, \qquad y_1(x) \leqq y \leqq y_2(x), \tag{2}$$

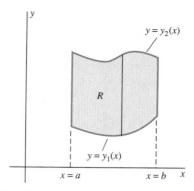

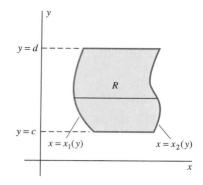

Fig. 15.2.2 A vertically simple region R

Fig. 15.2.3 A horizontally simple region R

where $y_1(x)$ and $y_2(x)$ are continuous functions of x on $[a, b]$. Such a region appears in Fig. 15.2.2. The region R is called **horizontally simple** if it is described by the inequalities

$$c \leqq y \leqq d, \qquad x_1(y) \leqq x \leqq x_2(y), \tag{3}$$

where $x_1(y)$ and $x_2(y)$ are continuous functions of y on $[c, d]$. The region in Fig. 15.2.3 is horizontally simple.

Theorem 1 tells us how to compute by iterated integration a double integral over a region R that is either vertically simple or horizontally simple.

Theorem 1   Evaluation of Double Integrals

Suppose that $f(x, y)$ is continuous on the region R. If R is the vertically simple region given in (2), then

$$\iint_R f(x, y)\, dA = \int_a^b \int_{y_1(x)}^{y_2(x)} f(x, y)\, dy\, dx. \tag{4}$$

If R is the horizontally simple region given in (3), then

$$\iint_R f(x, y)\, dA = \int_c^d \int_{x_1(y)}^{x_2(y)} f(x, y)\, dx\, dy. \tag{5}$$

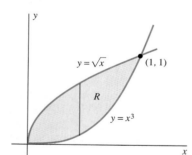

Fig. 15.2.4 The vertically simple region of Example 1

Theorem 1 includes Theorem 1 of Section 15.1 as a special case (when R is a rectangle), and it can be proved by a generalization of the argument we outlined there.

EXAMPLE 1 Compute in two different ways the integral

$$\iint_R xy^2\, dA,$$

where R is the first-quadrant region bounded by the two curves $y = \sqrt{x}$ and $y = x^3$.

Solution Always sketch the region R of integration before attempting to evaluate a double integral. As indicated in Figs. 15.2.4 and 15.2.5, the given region R is both vertically and horizontally simple. The vertical segment in

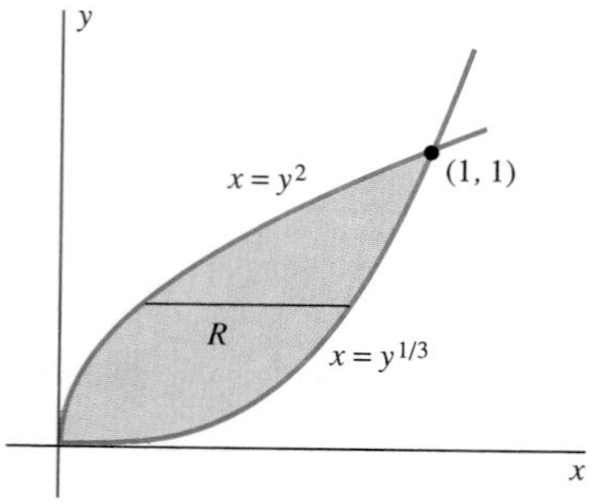

Fig. 15.2.5 The horizontally simple region of Example 1

Fig. 15.2.4 with endpoints on the curves $y = x^3$ and $y = \sqrt{x}$ corresponds to integrating first with respect to y:

$$\iint_R xy^2 \, dA = \int_0^1 \int_{x^3}^{\sqrt{x}} xy^2 \, dy \, dx = \int_0^1 \left[\tfrac{1}{3} xy^3 \right]_{y=x^3}^{\sqrt{x}} dx$$

$$= \int_0^1 (\tfrac{1}{3} x^{5/2} - \tfrac{1}{3} x^{10}) \, dx = \tfrac{2}{21} - \tfrac{1}{33} = \tfrac{5}{77}.$$

We get $x = y^2$ and $x = y^{1/3}$ when we solve the equations $y = \sqrt{x}$ and $y = x^3$ for x in terms of y. The horizontal segment in Fig. 15.2.5 corresponds to integrating first with respect to x:

$$\iint_R xy^2 \, dA = \int_0^1 \int_{y^2}^{y^{1/3}} xy^2 \, dx \, dy = \int_0^1 \left[\tfrac{1}{2} x^2 y^2 \right]_{x=y^2}^{y^{1/3}} dy$$

$$= \int_0^1 (\tfrac{1}{2} y^{8/3} - \tfrac{1}{2} y^6) \, dy = \tfrac{3}{22} - \tfrac{1}{14} = \tfrac{5}{77}.$$

EXAMPLE 2 Evaluate

$$\iint_R (6x + 2y^2) \, dA,$$

where R is the region bounded by the parabola $x = y^2$ and by the straight line $x + y = 2$.

Solution The region R appears in Fig. 15.2.6. It is both horizontally and vertically simple. If we wished to integrate first with respect to y and then with respect to x, we would need to evaluate two integrals:

$$\iint_R f(x, y) \, dA = \int_0^1 \int_{-\sqrt{x}}^{\sqrt{x}} (6x + 2y^2) \, dy \, dx + \int_1^4 \int_{-\sqrt{x}}^{2-x} (6x + 2y^2) \, dy \, dx.$$

The reason is that the formula of the function $y = y_2(x)$ describing the "top boundary curve" of R changes at the point $(1, 1)$, from $y = \sqrt{x}$ on the left to $y = 2 - x$ on the right. But as we see in Fig. 15.2.7, every *horizontal* segment in R extends from $x = y^2$ on the left to $x = 2 - y$ on the right. Therefore, integration first with respect to x requires us to evaluate only *one* double integral:

$$\iint_R f(x, y) \, dA = \int_{-2}^1 \int_{y^2}^{2-y} (6x + 2y^2) \, dx \, dy$$

$$= \int_{-2}^1 \left[3x^2 + 2xy^2 \right]_{x=y^2}^{2-y} dy$$

$$= \int_{-2}^1 [3(2 - y)^2 + 2(2 - y)y^2 - 3(y^2)^2 - 2y^4] \, dy$$

$$= \int_{-2}^1 (12 - 12y + 7y^2 - 2y^3 - 5y^4) \, dy$$

$$= \left[12y - 6y^2 + \tfrac{7}{3} y^3 - \tfrac{1}{2} y^4 - y^5 \right]_{-2}^1 = \tfrac{99}{2}.$$

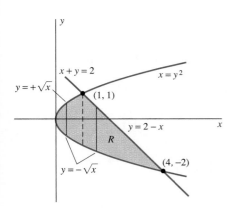

Fig. 15.2.6 The vertically simple region of Example 2

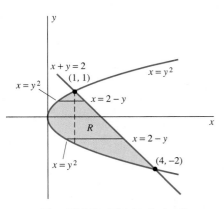

Fig. 15.2.7 The horizontally simple region of Example 2

Example 2 indicates that even when the region R is both vertically and horizontally simple, it may be easier to integrate in one order rather than the other because of the shape of R. We naturally prefer the easier route. The choice of the preferable order of integration may be influenced also by the nature of the function $f(x, y)$. It may be difficult—or even impossible—to compute a given iterated integral but easy to do so *after we reverse the order of integration*. Example 3 shows that the key to reversing the order of integration is this: Find (and sketch) the region R over which the integration is to be performed.

EXAMPLE 3 Evaluate

$$\int_0^2 \int_{y/2}^1 y e^{x^3} \, dx \, dy.$$

Solution We cannot integrate first with respect to x, as indicated, because $\exp(x^3)$ is known to have no elementary antiderivative. So we try to evaluate the integral by first reversing the order of integration. To do so, we sketch the region of integration specified by the limits in the given iterated integral.

The region R is determined by the inequalities

$$\tfrac{1}{2} y \leqq x \leqq 1 \quad \text{and} \quad 0 \leqq y \leqq 2.$$

Thus all points (x, y) of R lie between the horizontal lines $y = 0$ and $y = 2$ and between the two lines $x = y/2$ and $x = 1$. We draw the four lines $y = 0$, $y = 2$, $x = y/2$, and $x = 1$ and find that the region of integration is the shaded triangle that appears in Fig. 15.2.8.

Integrating first with respect to y, from $y_1(x) = 0$ to $y_2(x) = 2x$, we obtain

$$\int_0^2 \int_{y/2}^1 y e^{x^3} \, dx \, dy = \int_0^1 \int_0^{2x} y e^{x^3} \, dy \, dx = \int_0^1 \left[\tfrac{1}{2} y^2 \right]_{y=0}^{2x} e^{x^3} \, dx$$

$$= \int_0^1 2x^2 e^{x^3} \, dx = \left[\tfrac{2}{3} e^{x^3} \right]_{x=0}^1 = \tfrac{2}{3}(e - 1).$$

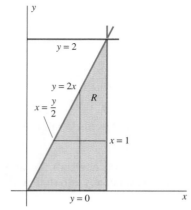

y = 2
y = 2x
x = y/2
R
x = 1
y = 0

Fig. 15.2.8 The region of Example 3

We conclude this section by listing some useful formal properties of double integrals. Let c be a constant and f and g be continuous functions on a region R on which $f(x, y)$ attains a minimum value m and a maximum value M. Let $a(R)$ denote the area of the region R. If all the indicated integrals exist, then:

$$\iint_R cf(x, y) \, dA = c \iint_R f(x, y) \, dA, \tag{6}$$

$$\iint_R [f(x, y) + g(x, y)] \, dA = \iint_R f(x, y) \, dA + \iint_R g(x, y) \, dA, \tag{7}$$

$$m \cdot a(R) \leqq \iint_R f(x, y) \, dA \leqq M \cdot a(R), \tag{8}$$

$$\iint_R f(x, y) \, dA = \iint_{R_1} f(x, y) \, dA + \iint_{R_2} f(x, y) \, dA. \tag{9}$$

Ch. 15 / Multiple Integrals

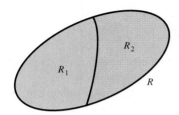

In Eq. (9), R_1 and R_2 are simply two nonoverlapping regions (regions with disjoint interiors) with union R (Fig. 15.2.9). We indicate in Problems 25 through 28 proofs of the properties in (6) through (9) for the special case in which R is a rectangle.

The property in Eq. (9) enables us to evaluate double integrals over a region R that is neither vertically nor horizontally simple. All that is necessary is to divide R into a finite number of simple regions $R_1, R_2, \ldots, R_n$. Then we integrate over each (converting each double integral into an iterated integral, as in the examples of this section) and add the results.

Fig. 15.2.9 The regions of Eq. (9)

15.2 Problems

Evaluate the iterated integrals in Problems 1 through 14.

1. $\displaystyle\int_0^1 \int_0^x (1 + x)\, dy\, dx$

2. $\displaystyle\int_0^2 \int_0^{2x} (1 + y)\, dy\, dx$

3. $\displaystyle\int_0^1 \int_y^1 (x + y)\, dx\, dy$ (Fig. 15.2.10)

9. $\displaystyle\int_0^1 \int_{x^4}^x (y - x)\, dy\, dx$

10. $\displaystyle\int_{-1}^2 \int_{-y}^{y+2} (x + 2y^2)\, dx\, dy$ (Figure 15.2.14)

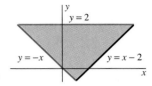

Fig. 15.2.14   Problem 10

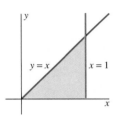

Fig. 15.2.10   Problem 3

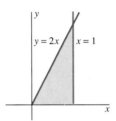

Fig. 15.2.11   Problem 4

11. $\displaystyle\int_0^1 \int_0^{x^3} e^{y/x}\, dy\, dx$

12. $\displaystyle\int_0^\pi \int_0^{\sin x} y\, dy\, dx$ (Fig. 15.2.15)

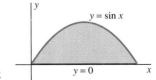

Fig. 15.2.15   Problem 12

4. $\displaystyle\int_0^2 \int_{y/2}^1 (x + y)\, dx\, dy$ (Fig. 15.2.11)

5. $\displaystyle\int_0^1 \int_0^{x^2} xy\, dy\, dx$

6. $\displaystyle\int_0^1 \int_y^{\sqrt{y}} (x + y)\, dx\, dy$

7. $\displaystyle\int_0^1 \int_x^{\sqrt{x}} (2x - y)\, dy\, dx$ (Fig. 15.2.12)

13. $\displaystyle\int_0^3 \int_0^y \sqrt{y^2 + 16}\, dx\, dy$ **14.** $\displaystyle\int_1^{e^2} \int_0^{1/y} e^{xy}\, dx\, dy$

In Problems 15 through 24, first sketch the region of integration, reverse the order of integration as in Examples 2 and 3 and finally, evaluate the resulting integral.

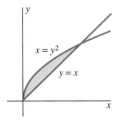

Fig. 15.2.12   Problem 7

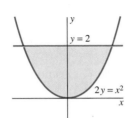

Fig. 15.2.13   Problem 8

15. $\displaystyle\int_{-2}^2 \int_{x^2}^4 x^2 y\, dy\, dx$

16. $\displaystyle\int_0^1 \int_{x^4}^x (x - 1)\, dy\, dx$

17. $\displaystyle\int_{-1}^3 \int_{x^2}^{2x+3} x\, dy\, dx$

18. $\displaystyle\int_{-2}^2 \int_{y^2-4}^{4-y^2} y\, dx\, dy$

19. $\displaystyle\int_0^2 \int_{2x}^{4x-x^2} 1\, dy\, dx$

20. $\displaystyle\int_0^1 \int_y^1 e^{-x^2}\, dx\, dy$

8. $\displaystyle\int_0^2 \int_{-\sqrt{2y}}^{\sqrt{2y}} (3x + 2y)\, dx\, dy$ (Fig. 15.2.13)

21. $\displaystyle\int_0^\pi \int_x^\pi \frac{\sin y}{y}\, dy\, dx$ **22.** $\displaystyle\int_0^{\sqrt\pi} \int_y^{\sqrt\pi} \sin x^2\, dx\, dy$

23. $\displaystyle\int_0^1 \int_y^1 \frac{1}{1+x^4}\, dx\, dy$ **24.** $\displaystyle\int_0^1 \int_{\tan^{-1} y}^{\pi/4} \sec x\, dx\, dy$

25. Use Riemann sums to prove Eq. (6) for the case in which R is a rectangle with sides parallel to the coordinate axes.

26. Use iterated integrals and familiar properties of single integrals to prove Eq. (7) for the case in which R is a rectangle with sides parallel to the coordinate axes.

27. Use Riemann sums to prove the inequalities in (8) for the case in which R is a rectangle with sides parallel to the coordinate axes.

28. Use iterated integrals and familiar properties of single integrals to prove Eq. (9) if R_1 and R_2 are rectangles with sides parallel to the coordinate axes and the right-hand edge of R_1 is the left-hand edge of R_2.

29. Use Riemann sums to prove that

$$\iint_R f(x, y)\, dA \leqq \iint_R g(x, y)\, dA$$

if $f(x, y) \leqq g(x, y)$ at each point of the region R, a rectangle with sides parallel to the coordinate axes.

30. Suppose that the continuous function f is integrable on the plane region R and that f attains a minimum value m and a maximum value M on R. Assume that R is *connected* in the following sense: For any two points (x_0, y_0) and (x_1, y_1) of R, there is a continuous parametric curve $\mathbf{r}(t)$ in R for which $\mathbf{r}(0) = \langle x_0, y_0\rangle$ and $\mathbf{r}(1) = \langle x_1, y_1\rangle$. Then deduce from (8) the *average value property* of double integrals:

$$\iint_R f(x, y)\, dA = f(\hat{x}, \hat{y}) \cdot a(R)$$

for some point $(\hat{x}, \hat{y})$ of R. [*Suggestion:* If $m = f(x_0, y_0)$ and $M = f(x_1, y_1)$, then you may apply the intermediate value property of the continuous function $f(\mathbf{r}(t))$.]

15.3

Area and Volume by Double Integration

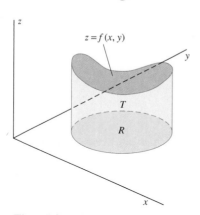

Fig. 15.3.1 A solid region T with vertical sides and base R in the xy-plane

Our definition of $\iint_R f(x, y)\, dA$ was *motivated* in Section 15.2 by the problem of computing the volume of the solid

$$T = \{(x, y, z)\,|\,(x, y) \in R \quad \text{and} \quad 0 \leqq z \leqq f(x, y)\}$$

that lies below the surface $z = f(x, y)$ and above the region R in the xy-plane. Such a solid appears in Fig. 15.3.1. Despite this geometric motivation, the actual definition of the double integral as a limit of Riemann sums does not depend on the concept of volume. We may, therefore, turn matters around and use the double integral to *define* volume.

> **Definition** *Volume below* $z = f(x, y)$
> Suppose that the function f is continuous and nonnegative on the bounded plane region R. Then the **volume** V of the solid that lies below the surface $z = f(x, y)$ and above the region R is defined to be
>
> $$V = \iint_R f(x, y)\, dA, \tag{1}$$
>
> provided that this integral exists.

It is of interest to note the connection between this definition and the cross-sections approach to volume that we discussed in Section 6.2. If, for example, the region R is vertically simple, then the volume integral of Eq. (1) takes the form

$$V = \iint_R z\, dA = \int_a^b \int_{y_1(x)}^{y_2(x)} f(x, y)\, dy\, dx$$

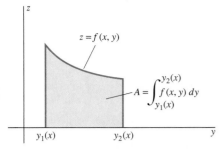

Fig. 15.3.2 The inner integral in Eq. (1) as the area of a region in the yz-plane

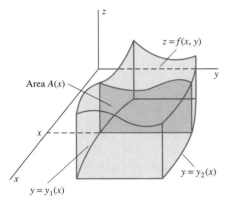

Fig. 15.3.3 The cross-sectional area is $A = \int_{y_1(x)}^{y_2(x)} f(x, y)\, dy$.

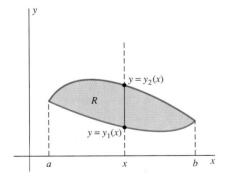

Fig. 15.3.4 A vertically simple region

in terms of iterated integrals. The inner integral

$$A(x) = \int_{y_1(x)}^{y_2(x)} f(x, y)\, dy$$

is equal to the area of the region in the yz-plane that lies below the curve

$$z = f(x, y) \qquad (x \text{ fixed})$$

and above the interval $y_1(x) \leqq y \leqq y_2(x)$ (Fig. 15.3.2). But this is the projection of the cross section shown in Fig. 15.3.3. Hence the value of the inner integral is simply the area of the cross section of the solid region T in a plane perpendicular to the x-axis. Thus

$$V = \int_a^b A(x)\, dx,$$

and so in this case Eq. (1) reduces to "volume is the integral of cross-sectional area."

EXAMPLE 1 The rectangle R in the xy-plane consists of those points (x, y) for which $0 \leqq x \leqq 2$ and $0 \leqq y \leqq 1$. Find the volume V of the solid that lies below the surface $z = 1 + xy$ and above R.

Solution Here $f(x, y) = 1 + xy$, so Eq. (1) yields

$$V = \iint_R z\, dA = \int_0^2 \int_0^1 (1 + xy)\, dy\, dx$$

$$= \int_0^2 \left[y + \tfrac{1}{2}xy^2 \right]_{y=0}^1 dx = \int_0^2 (1 + \tfrac{1}{2}x)\, dx = \left[x + \tfrac{1}{4}x^2 \right]_0^2 = 3.$$

VOLUME BY ITERATED INTEGRALS

A three-dimensional region T is typically described in terms of the surfaces that bound it. The first step in applying Eq. (1) to compute the volume V of such a region is to determine the region R in the xy-plane over which T lies. The second step is to determine the appropriate order of integration. This may be done in the following way:

If each vertical line in the xy-plane meets R in a *single* line segment, then R is vertically simple, and you may integrate first with respect to y. The limits on y will be the y-coordinates $y_1(x)$ and $y_2(x)$ of the endpoints of this line segment (Fig. 15.3.4). The limits on x will be the endpoints a and b of the interval on the x-axis onto which R projects. Theorem 1 of Section 15.2 then gives

$$V = \iint_R f(x, y)\, dA = \int_a^b \int_{y_1(x)}^{y_2(x)} f(x, y)\, dy\, dx. \tag{2}$$

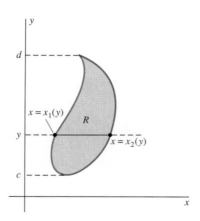

Fig. 15.3.5 A horizontally simple region

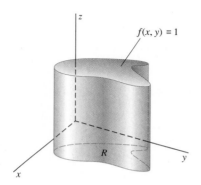

Fig. 15.3.6 The mesa

Alternatively,

If each horizontal line in the xy-plane meets R in a *single* line segment, then R is horizontally simple, and you may integrate with respect to x first. In this case,

$$V = \iint_R f(x, y) \, dA = \int_c^d \int_{x_1(y)}^{x_2(y)} f(x, y) \, dx \, dy. \qquad (3)$$

As indicated in Fig. 15.3.5, $x_1(y)$ and $x_2(y)$ are the x-coordinates of the endpoints of this horizontal line segment, and c and d are the endpoints of the corresponding interval on the y-axis.

If the region R is both vertically simple and horizontally simple, then you have the pleasant option of choosing the order of integration that will lead to the simpler subsequent computations. If R is neither vertically simple nor horizontally simple, then you must first subdivide R into simple regions before you proceed with iterated integration.

The special case $f(x, y) \equiv 1$ in Eq. (1) gives the area

$$A = a(R) = \iint_R 1 \, dA = \iint_R dA \qquad (4)$$

of the plane region R. In this case the solid region T resembles a desert mesa (Fig. 15.3.6)—a solid cylinder with base R of area A and height 1. The volume of any such cylinder—not necessarily circular—is the product of its height and the area of its base. In this case, the iterated integrals in Eqs. (2) and (3) reduce to

$$A = \int_a^b \int_{y_{\text{bot}}}^{y_{\text{top}}} 1 \, dy \, dx \quad \text{and} \quad A = \int_c^d \int_{x_{\text{left}}}^{x_{\text{right}}} 1 \, dx \, dy,$$

respectively.

EXAMPLE 2 Compute by double integration the area A of the region R in the xy-plane that is bounded by the line $y = x$ and by the parabola $y = x^2 - 2x$.

Solution As indicated in Fig. 15.3.7, the line $y_{\text{top}} = x$ and the parabola $y_{\text{bot}} = x^2 - 2x$ intersect in the points $(0, 0)$ and $(3, 3)$. (These coordinates are easy to find by solving the equation $y_{\text{top}} = y_{\text{bot}}$.) Therefore,

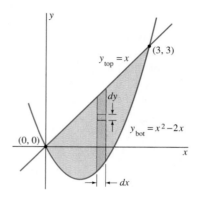

Fig. 15.3.7 The region R of Example 2

848

$$A = \int_a^b \int_{y_{bot}}^{y_{top}} 1 \; dy \; dx = \int_0^3 \int_{x^2-2x}^x 1 \; dy \; dx$$

$$= \int_0^3 \left[y \right]_{y=x^2-2x}^x dx = \int_0^3 (3x - x^2) \; dx = \left[\tfrac{3}{2}x^2 - \tfrac{1}{3}x^3 \right]_0^3 = \tfrac{9}{2}.$$

EXAMPLE 3 Find the volume of the wedge-shaped solid T that lies above the xy-plane, below the plane $z = x$, and within the cylinder $x^2 + y^2 = 4$. This wedge is diagrammed in Fig. 15.3.8.

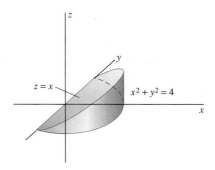

Fig. 15.3.8 The wedge of Example 3

Fig. 15.3.9 *Half* of the base R of the wedge (Example 3)

Solution The base region R is a semicircle of radius 2, but by symmetry we may integrate over the first-quadrant quarter-circle S alone and then double the result. A sketch of the quarter circle (Fig. 15.3.9) helps establish the limits of integration. We could integrate in either order, but integration with respect to x first gives a slightly simpler computation of the volume V:

$$V = \iint_S z \; dA = 2 \int_0^2 \int_0^{\sqrt{4-y^2}} x \; dx \; dy = 2 \int_0^2 \left[\tfrac{1}{2}x^2 \right]_{x=0}^{\sqrt{4-y^2}} dy$$

$$= \int_0^2 (4 - y^2) \; dy = \left[4y - \tfrac{1}{3}y^3 \right]_0^2 = \tfrac{16}{3}.$$

As an exercise, you should integrate in the other order and compare the results.

VOLUME BETWEEN TWO SURFACES

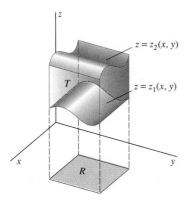

Fig. 15.3.10 The solid T has vertical sides and is bounded above and below by surfaces.

Suppose now that the solid region T lies above the plane region R, as before, but *between* the surfaces $z = z_1(x, y)$ and $z = z_2(x, y)$, where $z_1(x, y) \leq z_2(x, y)$ for all (x, y) in R (Fig. 15.3.10). Then we get the volume V of T by subtracting the volume below $z = z_1(x, y)$ from the volume below $z = z_2(x, y)$, so

$$V = \iint_R [z_2(x, y) - z_1(x, y)] \; dA. \tag{5}$$

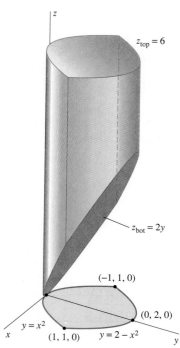

Fig. 15.3.11 The solid T of Example 4

More briefly,

$$V = \iint_R (z_{\text{top}} - z_{\text{bot}}) \, dA$$

where $z_{\text{top}} = z_2(x, y)$ describes the top surface and $z_{\text{bot}} = z_1(x, y)$ the bottom surface of T. This is a natural generalization of the formula for the area of the plane region between the curves $y = f_1(x)$ and $y = f_2(x)$ over the interval $[a, b]$. Moreover, like that formula, Eq. (5) is valid even if $f_1(x, y)$, or both $f_1(x, y)$ and $f_2(x, y)$, are negative over part or all of the region R.

EXAMPLE 4 Find the volume V of the solid T bounded by the planes $z = 6$ and $z = 2y$ and by the parabolic cylinders $y = x^2$ and $y = 2 - x^2$. This solid is sketched in Fig. 15.3.11.

Solution Because the parabolic cylinders given are perpendicular to the xy-plane, the solid T has vertical sides. Thus we may think of T as lying between the planes $z_{\text{top}} = 6$ and $z_{\text{bot}} = 2y$ and above the xy-plane region R that is bounded by the parabolas $y = x^2$ and $y = 2 - x^2$. As indicated in Fig. 15.3.12, these parabolas intersect at the points $(-1, 1)$ and $(1, 1)$.

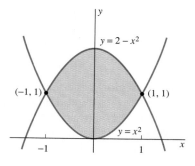

Fig. 15.3.12 The region R of Example 4

Integrating first with respect to y (for otherwise we would need two integrals), we get

$$V = \iint_R (z_{\text{top}} - z_{\text{bot}}) \, dA = \int_{-1}^{1} \int_{x^2}^{2-x^2} (6 - 2y) \, dy \, dx$$

$$= 2 \int_0^1 \left[6y - y^2 \right]_{y=x^2}^{2-x^2} dx \quad \text{(by symmetry)}$$

$$= 2 \int_0^1 ([6(2 - x^2) - (2 - x^2)^2] - [6x^2 - x^4]) \, dx$$

$$= 2 \int_0^1 (8 - 8x^2) \, dx = 16 \left[x - \tfrac{1}{3}x^3 \right]_0^1 = \tfrac{32}{3}.$$

15.3 Problems

In Problems 1 through 10, use double integration to find the area of the region in the xy-plane bounded by the given curves.

1. $y = x$, $y^2 = x$

2. $y = x$, $y = x^4$

3. $y = x^2$, $y = 2x + 3$ (Fig. 15.3.13)

850

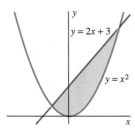

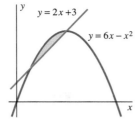

Fig. 15.3.13   Problem 3   **Fig. 15.3.14**   Problem 4

4. $y = 2x + 3$, $y = 6x - x^2$ (Fig. 15.3.14)
5. $y = x^2$, $x + y = 2$, $y = 0$
6. $y = (x - 1)^2$, $y = (x + 1)^2$, $y = 0$
7. $y = x^2 + 1$, $y = 2x^2 - 3$ (Fig. 15.3.15)

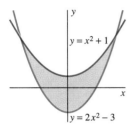

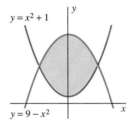

Fig. 15.3.15   Problem 7   **Fig. 15.3.16**   Problem 8

8. $y = x^2 + 1$, $y = 9 - x^2$ (Fig. 15.3.16)
9. $y = x$, $y = 2x$, $xy = 2$
10. $y = x^2$, $y = \dfrac{2}{1 + x^2}$

In Problems 11 through 26, find the volume of the solid that lies below the surface $z = f(x, y)$ and above the region in the xy-plane bounded by the given curves.

11. $z = 1 + x + y$; $x = 0$, $x = 1$, $y = 0$, $y = 1$
12. $z = 2x + 3y$; $x = 0$, $x = 3$, $y = 0$, $y = 2$
13. $z = y + e^x$; $x = 0$, $x = 1$, $y = 0$, $y = 2$
14. $z = 3 + \cos x + \cos y$; $x = 0$, $x = \pi$, $y = 0$, $y = \pi$ (Fig. 15.3.17)

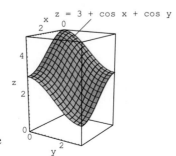

Fig. 15.3.17   The surface of Problem 14

15. $z = x + y$; $x = 0$, $y = 0$, $x + y = 1$
16. $z = 3x + 2y$; $x = 0$, $y = 0$, $x + 2y = 4$
17. $z = 1 + x + y$; $x = 1$, $y = 0$, $y = x^2$
18. $z = 2x + y$; $x = 0$, $y = 1$, $x = \sqrt{y}$
19. $z = x^2$; $y = x^2$, $y = 1$
20. $z = y^2$; $x = y^2$, $x = 4$
21. $z = x^2 + y^2$; $x = 0$, $y = 0$, $x = 1$, $y = 2$
22. $z = 1 + x^2 + y^2$; $y = x$, $y = 2 - x^2$
23. $z = 9 - x - y$; $y = 0$, $x = 3$, $y = 2x/3$
24. $z = 10 + y - x^2$; $y = x^2$, $x = y^2$
25. $z = 4x^2 + y^2$; $x = 0$, $y = 0$, $2x + y = 2$
26. $z = 2x + 3y$; $y = x^2$, $y = x^3$

27. Use double integration to find the volume of the tetrahedron in the first octant that is bounded by the coordinate planes and by the plane with equation $x/a + y/b + z/c = 1$ (Fig. 15.3.18). The numbers a, b, and c are positive constants.

Fig. 15.3.18   The tetrahedron of Problem 27

28. Suppose that $h > a > 0$. Show that the volume of the solid bounded by the cylinder $x^2 + y^2 = a^2$, by the plane $z = 0$, and by the plane $z = x + h$ is $\pi a^2 h$.

29. Find the volume of the first octant part of the solid bounded by the cylinders $x^2 + y^2 = 1$ and $y^2 + z^2 = 1$ (Fig. 15.3.19). [*Suggestion:* One order of integration is considerably easier than the other.]

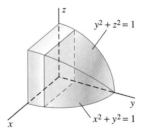

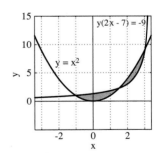

Fig. 15.3.19   The solid of Problem 29

Fig. 15.3.20   The two regions of Problem 30

30. Find the areas of the two regions bounded by the parabola $y = x^2$ and by the curve $y(2x - 7) = -9$, a translated rectangular hyperbola (Fig. 15.3.20). [*Suggestion:* $x = -1$ is one root of the cubic equation you will need to solve.]

In Problems 31 through 38, you may consult Chapter 9 or the integral table inside the covers of this book to find antiderivatives of such expressions as $(a^2 - u^2)^{3/2}$.

31. Find the volume of a sphere of radius a by double integration.

32. Use double integration to find the formula $V = V(a, b, c)$ for the volume of an ellipsoid with semiaxes a, b, and c.

33. Find the volume of the solid bounded by the xy-plane and the paraboloid $z = 25 - x^2 - y^2$ by evaluating a double integral (Fig. 15.3.21).

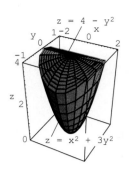

Fig. 15.3.23 The solid of Problem 36

36. Find the volume of the solid bounded by the two surfaces $z = x^2 + 3y^2$ and $z = 4 - y^2$ (Fig. 15.3.23).

37. Find the volume V of the solid T bounded by the parabolic cylinders $z = x^2$, $z = 2x^2$, $y = x^2$, and $y = 8 - x^2$.

38. Suppose that a square hole with sides of length 2 is cut symmetrically through the center of a sphere of radius 2. Show that the volume removed is given by

$$V = \int_0^1 F(x)\, dx,$$

where

$$F(x) = 4\sqrt{3 - x^2} + 4(4 - x^2) \arcsin \frac{1}{\sqrt{4 - x^2}}.$$

Use Simpson's rule (or the integration key or subroutine on a calculator) to approximate this integral. Is your numerical result consistent with the exact value

$$V = \tfrac{4}{3}(19\pi + 2\sqrt{2} - 54 \arctan \sqrt{2})?$$

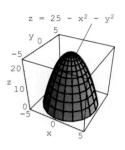

Fig. 15.3.21 The solid paraboloid of Problem 33

Fig. 15.3.22 The solid of Problem 34

34. Find the volume of the solid bounded by the paraboloids $z = x^2 + 2y^2$ and $z = 12 - 2x^2 - y^2$ (Fig. 15.3.22).

35. Find the volume removed when a vertical square hole of edge length R is cut directly through the center of a long horizontal cylinder of radius R.

15.4

Double Integrals in Polar Coordinates

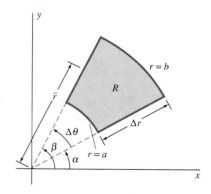

Fig. 15.4.1 A polar rectangle

A double integral may be easier to evaluate after it has been transformed from rectangular xy-coordinates into polar $r\theta$-coordinates. This is likely to be the case when the region R of integration is a *polar rectangle*. A **polar rectangle** is a region described in polar coordinates by the inequalities

$$a \leqq r \leqq b, \qquad \alpha \leqq \theta \leqq \beta. \tag{1}$$

This polar rectangle is shown in Fig. 15.4.1. If $a = 0$, it is a sector of a circular disk of radius b. If $0 < a < b$, $\alpha = 0$, and $\beta = 2\pi$, it is an annular ring of inner radius a and outer radius b. Because the area of a circular sector with radius r and central angle θ is $\frac{1}{2} r^2 \theta$, the area of the polar rectangle in (1) is

$$A = \tfrac{1}{2} b^2(\beta - \alpha) - \tfrac{1}{2} a^2(\beta - \alpha)$$
$$= \tfrac{1}{2}(a + b)(b - a)(\beta - \alpha) = \bar{r}\, \Delta r\, \Delta \theta, \tag{2}$$

where $\Delta r = b - a$, $\Delta \theta = \beta - \alpha$, and $\bar{r} = \frac{1}{2}(a + b)$ is the *average radius* of the polar rectangle.

Suppose that we want to compute the value of the double integral

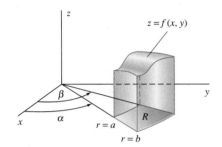

Fig. 15.4.2 Solid region whose base is the polar rectangle R

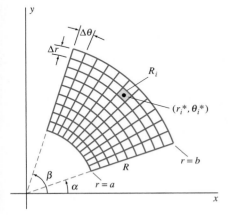

Fig. 15.4.3 A polar partition of the polar rectangle R

$$\iint_R f(x, y) \, dA,$$

where R is the polar rectangle in (1). Thus we want the volume of the solid with base R that lies below the surface $z = f(x, y)$ (Fig. 15.4.2). We defined in Section 15.1 the double integral as a limit of Riemann sums associated with partitions consisting of ordinary rectangles. We can define the double integral in terms of *polar partitions* as well, made up of polar rectangles. We begin with a partition

$$a = r_0 < r_1 < r_2 < \cdots < r_m = b$$

of $[a, b]$ into m equal subintervals each of length $\Delta r = (b - a)/m$ and a partition

$$\alpha = \theta_0 < \theta_1 < \theta_2 < \cdots < \theta_n = \beta$$

of $[\alpha, \beta]$ into n equal subintervals each of length $\Delta\theta = (\beta - \alpha)/n$. This gives the **polar partition** $\mathcal{P}$ of R into the $k = mn$ polar rectangles $R_1, R_2, \ldots, R_k$ indicated in Fig. 15.4.3. The mesh $|\mathcal{P}|$ of this polar partition is the maximum of the lengths of the diagonals of its polar subrectangles.

Let the center point of R_i have polar coordinates (r_i^*, θ_i^*), where r_i^* is the average radius of R_i. Then the rectangular coordinates of this point are $x_i^* = r_i^* \cos \theta_i^*$ and $y_i^* = r_i^* \sin \theta_i^*$. Therefore the Riemann sum for the function $f(x, y)$ associated with the polar partition $\mathcal{P}$ is

$$\sum_{i=1}^{k} f(x_i^*, y_i^*) \, \Delta A_i,$$

where $\Delta A_i = r_i^* \Delta r \Delta\theta$ is the area of the polar rectangle R_i [in part a consequence of Eq. (2)]. When we express this Riemann sum in polar coordinates, we obtain

$$\sum_{i=1}^{k} f(x_i^*, y_i^*) \, \Delta A_i = \sum_{i=1}^{k} f(r_i^* \cos \theta_i^*, r_i^* \sin \theta_i^*) \, r_i^* \, \Delta r \, \Delta\theta$$

$$= \sum_{i=1}^{k} g(r_i^*, \theta_i^*) \, \Delta r \, \Delta\theta,$$

where $g(r, \theta) = r f(r \cos \theta, r \sin \theta)$. This last sum is simply a Riemann sum for the double integral

$$\int_{\alpha}^{\beta} \int_{a}^{b} g(r, \theta) \, dr \, d\theta = \int_{\alpha}^{\beta} \int_{a}^{b} f(r \cos \theta, r \sin \theta) \, r \, dr \, d\theta,$$

so it finally follows that

$$\iint_R f(x, y) \, dA = \lim_{|\mathcal{P}| \to 0} \sum_{i=1}^{k} f(x_i^*, y_i^*) \, \Delta A_i$$

$$= \lim_{\Delta r, \Delta\theta \to 0} \sum_{i=1}^{k} g(r_i^*, \theta_i^*) \, \Delta r \, \Delta\theta = \int_{\alpha}^{\beta} \int_{a}^{b} g(r, \theta) \, dr \, d\theta.$$

That is,

$$\iint_R f(x, y) \, dA = \int_{\alpha}^{\beta} \int_{a}^{b} f(r \cos \theta, r \sin \theta) \, r \, dr \, d\theta. \qquad (3)$$

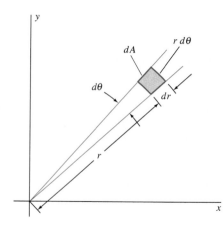

Fig. 15.4.4 The dimensions of the small polar rectangle suggest that $dA = r\,dr\,d\theta$.

Thus we formally transform into polar coordinates a double integral over a polar rectangle of the form in (1) by substituting

$$x = r \cos \theta, \qquad y = r \sin \theta, \qquad dA = r\,dr\,d\theta \qquad (4)$$

and inserting the appropriate limits of integration on r and θ. In particular, *note the "extra" r on the right-hand side of Eq. (3).* You may remember it by visualizing the "infinitesimal polar rectangle" of Fig. 15.4.4, with "area" $dA = r\,dr\,d\theta$ (formally).

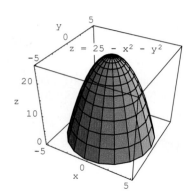

Fig. 15.4.5 The paraboloid of Example 1

EXAMPLE 1 Find the volume V of the solid shown in Fig. 15.4.5. This is the figure bounded below by the xy-plane and above by the paraboloid $z = 25 - x^2 - y^2$.

Solution The paraboloid intersects the xy-plane in the circle $x^2 + y^2 = 25$. We can compute the volume of the solid by integrating over the quarter of that circle that lies in the first quadrant (Fig. 15.4.6) and then multiplying the result by 4. Thus

$$V = 4 \int_0^5 \int_0^{\sqrt{25-x^2}} (25 - x^2 - y^2)\,dy\,dx.$$

There is no difficulty in performing the integration with respect to y, but then we are confronted with the integrals

$$\int \sqrt{25 - x^2}\,dx, \qquad \int x^2\sqrt{25 - x^2}\,dx, \quad \text{and} \quad \int (25 - x^2)^{3/2}\,dx.$$

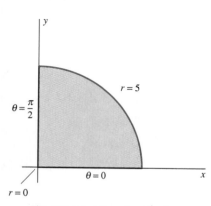

Fig. 15.4.6 One-fourth the domain of the integral of Example 1

Let us instead transform the original integral into polar coordinates. Because $25 - x^2 - y^2 = 25 - r^2$ and because the quarter of the circular disk in the first quadrant is described by

$$0 \le r \le 5, \qquad 0 \le \theta \le \pi/2,$$

Eq. (3) yields the volume

$$V = 4 \int_0^{\pi/2} \int_0^5 (25 - r^2)\,r\,dr\,d\theta$$

$$= 4 \int_0^{\pi/2} \left[\frac{25}{2} r^2 - \frac{1}{4} r^4 \right]_{r=0}^5 d\theta = 4 \cdot \frac{625}{4} \cdot \frac{\pi}{2} = \frac{625\pi}{2}.$$

854

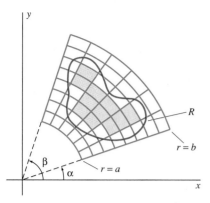

Fig. 15.4.7 A polar inner partition of the region R

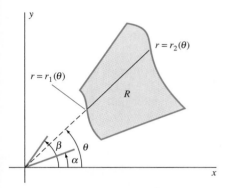

Fig. 15.4.8 A radially simple region R

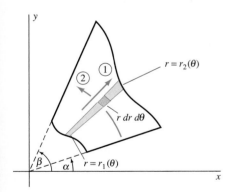

Fig. 15.4.9 Integrating first with respect to r and then with respect to θ

MORE GENERAL POLAR-COORDINATE REGIONS

If R is a more general region, then we can transform into polar coordinates the double integral

$$\iint_R f(x, y)\, dA$$

by expressing it as a limit of Riemann sums associated with "polar inner partitions" of the sort indicated in Fig. 15.4.7. Instead of giving the detailed derivation—a generalization of the preceding derivation of Eq. (3)—we shall simply give the results in one special case of practical importance.

Figure 15.4.8 shows a *radially simple* region R consisting of those points with polar coordinates that satisfy the inequalities

$$\alpha \leqq \theta \leqq \beta, \qquad r_1(\theta) \leqq r \leqq r_2(\theta).$$

In this case, the formula

$$\iint_R f(x, y)\, dA = \int_\alpha^\beta \int_{r_1(\theta)}^{r_2(\theta)} f(r \cos\theta, r \sin\theta)\, r\, dr\, d\theta \qquad (5)$$

gives the evaluation in polar coordinates of a double integral over R (under the usual assumption that the indicated integrals exist). Note that we integrate first with respect to r, with the limits $r_1(\theta)$ and $r_2(\theta)$ being the r-coordinates of the endpoints of a typical radial segment in R (Fig. 15.4.8).

Figure 15.4.9 shows how we can set up the iterated integral on the right-hand side of Eq. (5) in a formal way. First, a typical area element $dA = r\, dr\, d\theta$ is swept radially from $r = r_1(\theta)$ to $r = r_2(\theta)$. Second, the resulting strip is rotated from $\theta = \alpha$ to $\theta = \beta$ to sweep out the region R. Equation (5) yields the volume formula

$$V = \int_\alpha^\beta \int_{r_{\text{inner}}}^{r_{\text{outer}}} zr\, dr\, d\theta \qquad (6)$$

for the volume V of the solid that lies above the region R of Fig. 15.4.8 and below the surface $z = f(x, y) = f(r \cos\theta, r \sin\theta)$.

Observe that Eqs. (3) and (5) for the evaluation of a double integral in polar coordinates take the form

$$\iint_R f(x, y)\, dA = \iint_S f(r \cos\theta, r \sin\theta)\, r\, dr\, d\theta. \qquad (7)$$

The symbol S on the right-hand side represents the appropriate limits on r and θ such that the region R is swept out in the manner indicated in Fig. 15.4.9. With $f(x, y) \equiv 1$, Eq. (7) reduces to the formula

$$A = a(R) = \iint_S r\, dr\, d\theta \qquad (8)$$

for computing the area of R by double integration in polar coordinates. Note again that the symbol S refers not to a new region in the xy-plane, but to a new description—in terms of polar coordinates—of the original region R.

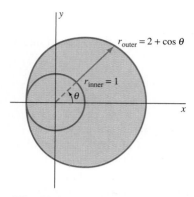

Fig. 15.4.10 The region R of Example 2

EXAMPLE 2 Figure 15.4.10 shows the region R bounded on the inside by the circle $r = 1$ and on the outside by the limaçon $r = 2 + \cos \theta$. By following a typical radial line outward from the origin, we see that $r_{inner} = 1$ and $r_{outer} = 2 + \cos \theta$. Hence the area of R is

$$A = \int_\alpha^\beta \int_{r_{inner}}^{r_{outer}} r \, dr \, d\theta$$

$$= 2 \int_0^\pi \int_1^{2+\cos\theta} r \, dr \, d\theta \qquad \text{(symmetry)}$$

$$= 2 \int_0^\pi \tfrac{1}{2}[(2 + \cos \theta)^2 - (1)^2] \, d\theta = \int_0^\pi (3 + 4 \cos \theta + \cos^2 \theta) \, d\theta$$

$$= \int_0^\pi (3 + 4 \cos \theta + \tfrac{1}{2} + \tfrac{1}{2}\cos 2\theta) \, d\theta = \int_0^\pi (3 + \tfrac{1}{2}) \, d\theta = \tfrac{7}{2}\pi.$$

The cosine terms in the next-to-last integral contribute nothing, because upon integration they yield sine terms that are zero at both limits.

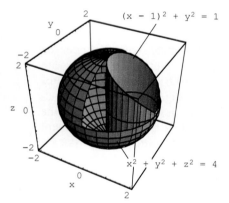

Fig. 15.4.11 The sphere with off-center hole (Example 3)

EXAMPLE 3 Find the volume of the solid region that is interior to both the sphere $x^2 + y^2 + z^2 = 4$ of radius 2 and the cylinder $(x - 1)^2 + y^2 = 1$. This is the volume of material removed when an off-center hole of radius 1 is bored just tangent to a diameter all the way through a sphere of radius 2 (Fig. 15.4.11).

Solution We need to integrate the function $f(x, y) = \sqrt{4 - x^2 - y^2}$ over the disk R that is bounded by the circle with center $(1, 0)$ and radius 1 (Fig. 15.4.12). The desired volume V is twice that of the part above the xy-plane, so

$$V = 2 \iint_R \sqrt{4 - x^2 - y^2} \, dA.$$

But this integral would be troublesome to evaluate in rectangular coordinates, so we change to polar coordinates.

The circle of radius 1 in Fig. 15.4.12 is familiar from Chapter 10; its polar equation is $r = 2 \cos \theta$. Therefore, the region R is described by the inequalities

$$0 \leqq r \leqq 2 \cos \theta, \qquad -\pi/2 \leqq \theta \leqq \pi/2.$$

We shall integrate only over the upper half of R, taking advantage of the symmetry of the sphere-with-hole. This involves doubling, for a second time, the integral we write. So—using Eq. (5)—we find that

$$V = 4 \int_0^{\pi/2} \int_0^{2\cos\theta} \sqrt{4 - r^2} \, r \, dr \, d\theta$$

$$= 4 \int_0^{\pi/2} \left[-\tfrac{1}{3}(4 - r^2)^{3/2} \right]_{r=0}^{2\cos\theta} d\theta = \tfrac{32}{3} \int_0^{\pi/2} (1 - \sin^3 \theta) \, d\theta.$$

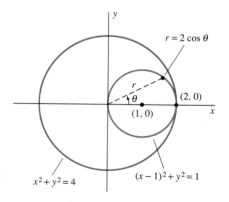

Fig. 15.4.12 The small circle is the domain of the integral of Example 3.

Now we see from Formula (113) inside the back cover that

$$\int_0^{\pi/2} \sin^3 \theta \, d\theta = \tfrac{2}{3},$$

and therefore

$$V = \tfrac{16}{3}\pi - \tfrac{64}{9} \approx 9.64405.$$

In Example 4 we use a polar-coordinates version of the familiar volume formula

$$V = \iint_R (z_{\text{top}} - z_{\text{bot}}) \, dA.$$

EXAMPLE 4 Find the volume of the solid that is bounded above by the paraboloid $z = 8 - r^2$ and below by the paraboloid $z = r^2$ (Fig. 15.4.13).

Solution The curve of intersection of the two paraboloids is found by simultaneously solving the equations of the two surfaces. We eliminate z to obtain

$$r^2 = 8 - r^2; \quad \text{that is,} \quad r^2 = 4.$$

Hence the solid lies above the plane circular disk D with polar description $r \leqq 2$, and so its volume is

$$V = \iint_D (z_{\text{top}} - z_{\text{bot}}) \, dA = \int_0^{2\pi} \int_0^2 [(8 - r^2) - r^2] \, r \, dr \, d\theta$$

$$= \int_0^{2\pi} \int_0^2 (8r - 2r^3) \, dr \, d\theta = 2\pi \left[4r^2 - \tfrac{1}{2}r^4 \right]_0^2 = 16\pi.$$

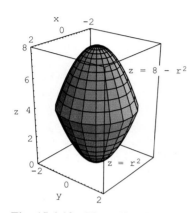

Fig. 15.4.13 The solid of Example 4

EXAMPLE 5 Here we apply a standard polar-coordinates technique to show that

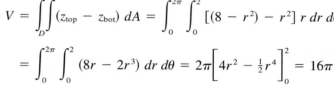

$$I = \int_0^\infty e^{-x^2} \, dx = \frac{\sqrt{\pi}}{2}. \tag{9}$$

This important improper integral converges because

$$\int_1^b e^{-x^2} \, dx \leqq \int_1^b e^{-x} \, dx \leqq \int_1^\infty e^{-x} \, dx = \frac{1}{e}.$$

(The first inequality is valid because $e^{-x^2} \leqq e^{-x}$ for $x \geqq 1$.) It follows that

$$\int_1^b e^{-x^2} \, dx$$

is a bounded and increasing function of b.

Solution Let V_b denote the volume of the region that lies below the surface $z = e^{-x^2-y^2}$ and above the square with vertices $(\pm b, \pm b)$ in the xy-plane (Fig. 15.4.14). Then

$$V_b = \int_{-b}^b \int_{-b}^b e^{-x^2-y^2} \, dx \, dy = \int_{-b}^b e^{-y^2} \left(\int_{-b}^b e^{-x^2} \, dx \right) dy$$

$$= \left(\int_{-b}^b e^{-x^2} \, dx \right) \left(\int_{-b}^b e^{-y^2} \, dy \right) = \left(\int_{-b}^b e^{-x^2} \, dx \right)^2 = 4 \left(\int_0^b e^{-x^2} \, dx \right)^2.$$

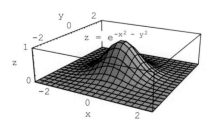

Fig. 15.4.14 The surface $z = e^{-x^2-y^2}$ (Example 5)

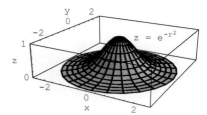

Fig. 15.4.15 The surface $z = e^{-r^2}$ (Example 5)

It follows that the volume below $z = e^{-x^2-y^2}$ and above the entire xy-plane is

$$V = \lim_{b \to \infty} V_b = \lim_{b \to \infty} 4\left(\int_0^b e^{-x^2}\, dx\right)^2 = 4\left(\int_0^\infty e^{-x^2}\, dx\right)^2 = 4I^2.$$

Now we compute V by another method: by using polar coordinates. We take the limit, as $b \to \infty$, of the volume below $z = e^{-x^2-y^2} = e^{-r^2}$ and above the circular disk with center $(0, 0)$ and radius b (Fig. 15.4.15). This disk is described by $0 \leqq r \leqq b$, $0 \leqq \theta \leqq 2\pi$, so we obtain

$$V = \lim_{b \to \infty} \int_0^{2\pi} \int_0^b re^{-r^2}\, dr\, d\theta = \lim_{b \to \infty} \int_0^{2\pi} \left[-\tfrac{1}{2}e^{-r^2}\right]_{r=0}^b d\theta$$

$$= \lim_{b \to \infty} \int_0^{2\pi} \tfrac{1}{2}\left[1 - e^{-b^2}\right] d\theta = \lim_{b \to \infty} \pi\left[1 - e^{-b^2}\right] = \pi.$$

We equate these two values of V, and it follows that $4I^2 = \pi$. Therefore, $I = \tfrac{1}{2}\sqrt{\pi}$, as desired.

15.4 Problems

In Problems 1 through 7, find the indicated area by double integration in polar coordinates.

1. The area bounded by the circle $r = 1$

2. The area bounded by the circle $r = 3 \sin \theta$

3. The area bounded by the cardioid $r = 1 + \cos \theta$ (Fig. 15.4.16)

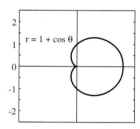

Fig. 15.4.16 The cardioid of Problem 3

4. The area bounded by one loop of $r = 2 \cos 2\theta$ (Fig. 15.4.17)

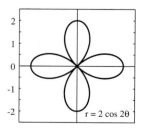

Fig. 15.4.17 The rose of Problem 4

5. The area inside both the circles $r = 1$ and $r = 2 \sin \theta$

6. The area inside $r = 2 + \cos \theta$ and outside the circle $r = 2$.

7. The area inside the smaller loop of $r = 1 - 2 \sin \theta$ (Fig. 15.4.18)

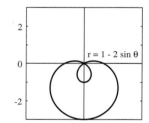

Fig. 15.4.18 The limaçon of Problem 7

In Problems 8 through 12, use double integration in polar coordinates to find the volume of the solid that lies below the given surface and above the plane region R bounded by the given curve.

8. $z = x^2 + y^2$; $\quad r = 3$

9. $z = \sqrt{x^2 + y^2}$; $\quad r = 2$

10. $z = x^2 + y^2$; $\quad r = 2 \cos \theta$

11. $z = 10 + 2x + 3y$; $\quad r = \sin \theta$

12. $z = a^2 - x^2 - y^2$; $\quad r = a$

In Problems 13 through 18, evaluate the given integral by first converting to polar coordinates.

13. $\displaystyle\int_0^1 \int_0^{\sqrt{1-y^2}} \frac{1}{1 + x^2 + y^2}\, dx\, dy$ (Fig. 15.4.19)

14. $\displaystyle\int_0^1 \int_0^{\sqrt{1-x^2}} \frac{1}{\sqrt{4 - x^2 - y^2}}\, dy\, dx$ (Fig. 15.4.19)

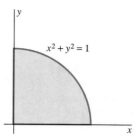

Fig. 15.4.19 The quarter-circle of Problems 13 and 14

15. $\int_0^2 \int_0^{\sqrt{4-x^2}} (x^2 + y^2)^{3/2} \, dy \, dx$

16. $\int_0^1 \int_x^1 x^2 \, dy \, dx$

17. $\int_0^1 \int_0^{\sqrt{1-y^2}} \sin(x^2 + y^2) \, dx \, dy$

18. $\int_1^2 \int_0^{\sqrt{2x-x^2}} \dfrac{1}{\sqrt{x^2 + y^2}} \, dy \, dx$ (Fig. 15.4.20)

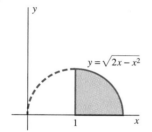

Fig. 15.4.20 The quarter-circle of Problem 18

In Problems 19 through 22, find the volume of the solid that is bounded above and below by the given surfaces $z = z_1(x, y)$ and $z = z_2(x, y)$ and lies above the plane region R that is bounded by the given curve $r = g(\theta)$.

19. $z = 1, z = 3 + x + y;$ $\quad r = 1$

20. $z = 2 + x, z = 4 + 2x;$ $\quad r = 2$

21. $z = 0, z = 3 + x + y;$ $\quad r = 2 \sin \theta$

22. $z = 0, z = 1 + x;$ $\quad r = 1 + \cos \theta$

Solve Problems 23 through 32 by double integration in polar coordinates.

23. Problem 31, Section 15.3

24. Problem 34, Section 15.3

25. Problem 28, Section 15.3

26. Find the volume of the wedge-shaped solid described in Example 3 of Section 15.3 (Fig. 15.4.21).

27. Find the volume bounded by the paraboloids $z = x^2 + y^2$ and $z = 4 - 3x^2 - 3y^2$.

28. Find the volume bounded by the paraboloids $z = x^2 + y^2$ and $z = 2x^2 + 2y^2 - 1$.

29. Find the volume of the "ice-cream cone" bounded by the sphere $x^2 + y^2 + z^2 = a^2$ and by the cone $z =$

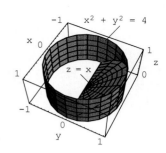

Fig. 15.4.21 The wedge of Problem 26

$\sqrt{x^2 + y^2}$ (Fig. 15.4.22, where $a = 1$).

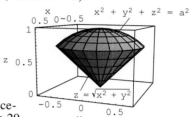

Fig. 15.4.22 The fat ice-cream cone of Problem 29

30. Find the volume bounded by the paraboloid $z = r^2$, by the cylinder $r = 2a \sin \theta$, and by the plane $z = 0$.

31. Find the volume that lies below the paraboloid $z = r^2$ and above one loop of the lemniscate with equation $r^2 = 2 \sin \theta$.

32. Find the volume that lies inside both the cylinder $x^2 + y^2 = 4$ and the ellipsoid $2x^2 + 2y^2 + z^2 = 18$.

33. If $0 < h < a$, then the plane $z = a - h$ cuts off a spherical segment of height h and radius b from the sphere $x^2 + y^2 + z^2 = a^2$ (Fig. 15.4.23). (a) Show that $b^2 = 2ah - h^2$. (b) Show that the volume of the spherical segment is $V = \frac{1}{6} \pi h (3b^2 + h^2)$.

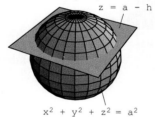

Fig. 15.4.23 The spherical segment of Problem 33

Fig. 15.4.24 The torus of Problem 35 (with $a = 1$ and $b = 2$)

34. Show by the method of Example 5 that

$$\int_0^\infty \int_0^\infty \frac{dx \, dy}{(1 + x^2 + y^2)^2} = \frac{\pi}{4}.$$

35. Find the volume of the solid torus obtained by revolving the disk $r \leqq a$ around the line $x = b > a$ (Fig. 15.4.24). [*Suggestion:* If the area element $dA = r \, dr \, d\theta$ is revolved around the line, the volume generated is $dV = 2\pi(b - x) \, dA$. Express everything in polar coordinates.]

15.5

Applications of Double Integrals

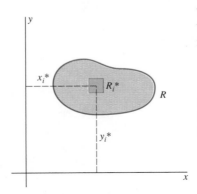

Fig. 15.5.1 The area element $\Delta A_i = a(R_i)$

Centroids are related to the stability of certain mechanical systems.

Fig. 15.5.2 A lamina balanced on its centroid

We can use the double integral to find the mass m and the *centroid* $(\bar{x}, \bar{y})$ of a plane *lamina,* or thin plate, that occupies a bounded region R in the xy-plane. We suppose that the density of the lamina (in units of mass per unit *area*) at the point (x, y) is given by the continuous function $\rho(x, y)$.

Let $\mathcal{P} = \{R_1, R_2, \ldots, R_n\}$ be an inner partition of R, and choose a point (x_i^*, y_i^*) in each subrectangle R_i (Fig. 15.5.1). Then the mass of the piece of the lamina occupying R_i is approximately $\rho(x_i^*, y_i^*) \, \Delta A_i$, where ΔA_i denotes the area $a(R_i)$ of R_i. Hence the mass of the entire lamina is given approximately by

$$m \approx \sum_{i=1}^{n} \rho(x_i^*, y_i^*) \, \Delta A_i.$$

As the mesh $|\mathcal{P}|$ of the inner partition $\mathcal{P}$ approaches zero, this Riemann sum approaches the corresponding double integral over R. We therefore *define* the **mass** m of the lamina by means of the formula

$$m = \iint_R \rho(x, y) \, dA. \tag{1}$$

In brief,

$$m = \iint_R \rho \, dA = \iint_R dm$$

in terms of the density ρ and the mass element

$$dm = \rho \, dA.$$

The coordinates $(\bar{x}, \bar{y})$ of the **centroid,** or *center of mass,* of the lamina are defined to be

$$\bar{x} = \frac{1}{m} \iint_R x\rho(x, y) \, dA, \tag{2}$$

$$\bar{y} = \frac{1}{m} \iint_R y\rho(x, y) \, dA. \tag{3}$$

You may remember these formulas in the form

$$\bar{x} = \frac{1}{m} \iint_R x \, dm, \qquad \bar{y} = \frac{1}{m} \iint_R y \, dm.$$

Thus $\bar{x}$ and $\bar{y}$ are the *average values* of x and y *with respect to mass* in the region R. The centroid $(\bar{x}, \bar{y})$ is the point of the lamina where it would balance horizontally if placed on the point of an ice pick (Fig. 15.5.2).

If the density function ρ has the *constant* value $k > 0$, then the coordinates of $\bar{x}$ and $\bar{y}$ are independent of the specific value of k. (Why?) In such a case we will generally take $\rho = 1$ in our computations. Moreover, in this case m will have the same numerical value as the area A of R, and $(\bar{x}, \bar{y})$ is then called the **centroid of the plane region R.**

Generally, we must calculate all three integrals in Eqs. (1) through (3) in order to find the centroid of a lamina. But sometimes we can take advantage of the following *symmetry principle:* If the plane region R (considered to be a lamina of constant density) is symmetric with respect to the line L —that

860

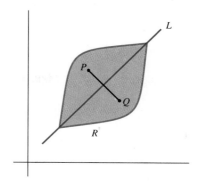

Fig. 15.5.3 A line of symmetry

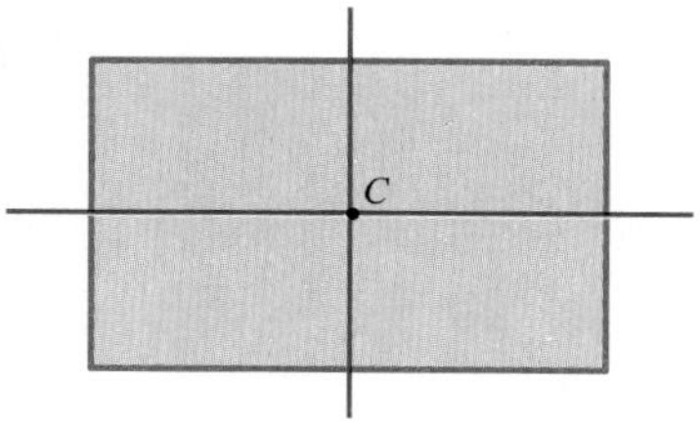

Fig. 15.5.4 The centroid of a rectangle

is, if R is carried onto itself when the plane is rotated through an angle of 180° about the line L —then the centroid of R lies on L (Fig. 15.5.3). For example, the centroid of a rectangle (Fig. 15.5.4) is the point where the perpendicular bisectors of its sides meet, because these bisectors are also lines of symmetry.

In the case of a nonconstant density function ρ, we require (for symmetry) that ρ —as well as the region itself—be symmetric about the geometric line L of symmetry. That is, $\rho(P) = \rho(Q)$ if, as in Fig. 15.5.3, the points P and Q are symmetrically located with respect to L. Then the centroid of the lamina R will lie on the line L of symmetry.

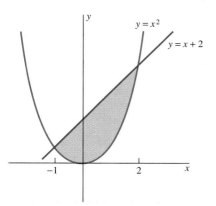

Fig. 15.5.5 The centroid of a semicircular disk (Example 1)

EXAMPLE 1 Consider the semicircular disk of radius a shown in Fig. 15.5.5. If it has constant density $\rho \equiv 1$, then its mass is $m = \frac{1}{2}\pi a^2$ (numerically equal to its area), and by symmetry its centroid $C(0, \bar{y})$ lies on the y-axis. Hence we need only compute

$$\bar{y} = \frac{1}{m} \iint_R y \, dm$$

$$= \frac{2}{\pi a^2} \int_0^\pi \int_0^a (r \sin \theta) \, r \, dr \, d\theta \qquad \text{(polar coordinates)}$$

$$= \frac{2}{\pi a^2} \left[-\cos \theta \right]_0^\pi \left[\frac{1}{3}r^3 \right]_0^a = \frac{2}{\pi a^2} \cdot 2 \cdot \frac{a^3}{3} = \frac{4a}{3\pi}.$$

Thus the centroid of the semicircular lamina is located at the point $(0, 4a/3\pi)$. Note that the computed value for $\bar{y}$ has the dimensions of length (because a is a length), as it should. Any answer that has other dimensions would be suspect.

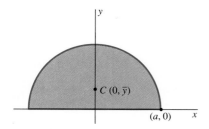

Fig. 15.5.6 The lamina of Example 2

EXAMPLE 2 A lamina occupies the region bounded by the line $y = x + 2$ and by the parabola $y = x^2$ (Fig. 15.5.6). The density of the lamina at the point $P(x, y)$ is proportional to the square of the distance of P from the y-axis—thus $\rho(x, y) = kx^2$ (where k is a positive constant). Find the mass and centroid of the lamina.

Solution The line and the parabola intersect in the two points $(-1, 1)$ and $(2, 4)$, so Eq. (1) gives mass

$$m = \int_{-1}^{2} \int_{x^2}^{x+2} kx^2 \, dy \, dx = k \int_{-1}^{2} \left[x^2 y \right]_{y=x^2}^{x+2} dx$$

$$= k \int_{-1}^{2} (x^3 + 2x^2 - x^4) \, dx = \frac{63k}{20}.$$

Then Eqs. (2) and (3) give

$$\bar{x} = \frac{20}{63k} \int_{-1}^{2} \int_{x^2}^{x+2} kx^3 \, dy \, dx = \frac{20}{63} \int_{-1}^{2} \left[x^3 y \right]_{y=x^2}^{x+2} dx$$

$$= \frac{20}{63} \int_{-1}^{2} (x^4 + 2x^3 - x^5) \, dx = \frac{20}{63} \cdot \frac{18}{5} = \frac{8}{7};$$

$$\bar{y} = \frac{20}{63k} \int_{-1}^{2} \int_{x^2}^{x+2} kx^2 y \, dy \, dx = \frac{20}{63} \int_{-1}^{2} \left[\frac{1}{2} x^2 y^2 \right]_{y=x^2}^{x+2} dx$$

$$= \frac{10}{63} \int_{-1}^{2} (x^4 + 4x^3 + 4x^2 - x^6) \, dx = \frac{10}{63} \cdot \frac{531}{35} = \frac{118}{49}.$$

Thus the lamina of this example has mass $63k/20$, and its centroid is located at the point $(8/7, 118/49)$.

EXAMPLE 3   A lamina is shaped like the first-quadrant quarter-circle of radius a shown in Fig. 15.5.7. Its density is proportional to distance from the origin—that is, its density at (x, y) is $\rho(x, y) = k\sqrt{x^2 + y^2} = kr$ (where k is a positive constant). Find its mass and centroid.

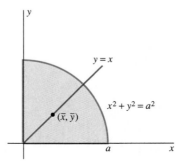

Fig. 15.5.7 Finding mass and centroid (Example 3)

Solution   First we change to polar coordinates, because both the shape of the boundary of the lamina and the formula for its density suggest that this will make the computations much simpler. Equation (1) then yields the mass to be

$$m = \iint_{R} \rho \, dA = \int_{0}^{\pi/2} \int_{0}^{a} kr^2 \, dr \, d\theta$$

$$= k \int_{0}^{\pi/2} \left[\frac{1}{3} r^3 \right]_{r=0}^{a} d\theta = k \int_{0}^{\pi/2} \frac{1}{3} a^3 \, d\theta = \frac{k\pi a^3}{6}.$$

By symmetry of the lamina and its density function, the centroid lies on the line $y = x$. So Eq. (3) gives

Ch. 15 / Multiple Integrals

$$\overline{x} = \overline{y} = \frac{1}{m} \iint_R y\rho \, dA = \frac{6}{k\pi a^3} \int_0^{\pi/2} \int_0^a kr^3 \sin\theta \, dr \, d\theta$$

$$= \frac{6}{\pi a^3} \int_0^{\pi/2} \left[\tfrac{1}{4} r^4 \sin\theta \right]_{r=0}^a d\theta = \frac{6}{\pi a^3} \cdot \frac{a^4}{4} \int_0^{\pi/2} \sin\theta \, d\theta = \frac{3a}{2\pi}.$$

Thus the given lamina has mass $\frac{1}{6} k\pi a^3$, and its centroid is located at the point $(3a/2\pi, 3a/2\pi)$.

VOLUME AND THE FIRST THEOREM OF PAPPUS

An important theorem relating centroids and volumes of revolution is named for the Greek mathematician who stated it during the third century A.D.

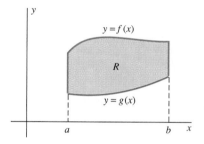

Axis of revolution

Area A

Centroid

Fig. 15.5.8 A solid of volume $V = A \cdot d$ is generated by the area A as its centroid travels the distance $d = 2\pi r$ around a circle of radius r.

First Theorem of Pappus: *Volume of Revolution*

Suppose that a plane region R is revolved around an axis in its plane (Fig. 15.5.8), generating a solid of revolution with volume V. Assume that the axis does not intersect the interior of R. Then the volume

$$V = A \cdot d$$

is the product of the area A of R and the distance d traveled by the centroid of R.

Proof for the Special Case of a Region Like that in Fig. 15.5.9 This is the region between the two graphs $y = f(x)$ and $y = g(x)$ for $a \leqq x \leqq b$, with the axis of revolution being the y-axis. Then, in a revolution about the y-axis, the distance traveled by the centroid of R is $d = 2\pi\overline{x}$. By the method of cylindrical shells [see Eq. (4) of Section 6.3 and Fig. 15.5.10], the volume of the solid generated is

$$V = \int_a^b 2\pi x [f(x) - g(x)] \, dx = \int_a^b \int_{g(x)}^{f(x)} 2\pi x \, dy \, dx$$

$$= 2\pi \iint_R x \, dA = 2\pi\overline{x} \cdot A$$

[by Eq. (2), with $\rho \equiv 1$]. Thus $V = d \cdot A$, as desired. ☐

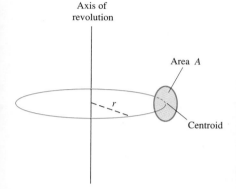

Fig. 15.5.9 A region R between the graphs of two functions

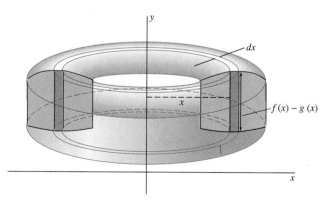

Fig. 15.5.10 A solid of revolution consisting of cylindrical shells

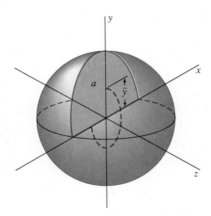

Fig. 15.5.11 A sphere of radius a generated by revolving a semicircle of area $A = \frac{1}{2}\pi a^2$ around its diameter on the x-axis (Example 4). The centroid of the semicircle travels along a circle of circumference $d = 2\pi\bar{y}$.

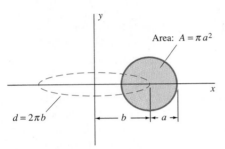

Fig. 15.5.12 Rotating the circular disk around the y-axis to generate a torus (Example 5)

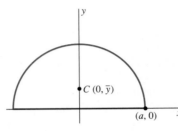

Fig. 15.5.13 The semicircular arc of Example 6

EXAMPLE 4 Find the volume V of the sphere of radius a generated by revolving around the x-axis the semicircle D of Example 1. See Fig. 15.5.11.

Solution The area of D is $A = \frac{1}{2}\pi a^2$, and we found in Example 1 that $\bar{y} = 4a/3\pi$. Hence Pappus's theorem gives

$$V = 2\pi\bar{y}A = 2\pi \cdot \frac{4a}{3\pi} \cdot \frac{\pi a^2}{2} = \frac{4}{3}\pi a^3.$$

EXAMPLE 5 Consider the circular disk of Fig. 15.5.12, with radius a and center at the point $(b, 0)$ with $0 < a < b$. Find the volume V of the solid torus generated by revolving this disk around the y-axis. Such a torus is shown in Fig. 15.4.24.

Solution The centroid of the circle is at its center $(b, 0)$, so $\bar{x} = b$. Hence the centroid is revolved through the distance $d = 2\pi b$. Consequently,

$$V = d \cdot A = 2\pi b \cdot \pi a^2 = 2\pi^2 a^2 b.$$

Note that this result is dimensionally correct.

SURFACE AREA AND THE SECOND THEOREM OF PAPPUS

Centroids of plane *curves* are defined in analogy with the method for plane regions, so we shall present this topic in less detail. It will suffice for us to treat only the case of constant density $\rho \equiv 1$ (like a wire with unit mass per unit length). Then the centroid $(\bar{x}, \bar{y})$ of the plane curve C is defined by the formulas

$$\bar{x} = \frac{1}{s}\int_C x\,ds, \qquad \bar{y} = \frac{1}{s}\int_C y\,ds \qquad (4)$$

where s is the arc length of C.

The meaning of the integrals in Eq. (4) is that of the notation of Section 6.4. That is, ds is a symbol to be replaced (before the integral is evaluated) by either

$$ds = \sqrt{1 + \left(\frac{dy}{dx}\right)^2}\,dx \quad \text{or} \quad ds = \sqrt{1 + \left(\frac{dx}{dy}\right)^2}\,dy,$$

depending on whether C is a smooth arc of the form $y = f(x)$ or one of the form $x = g(y)$. Alternatively, we may have

$$ds = \sqrt{(dx)^2 + (dy)^2} = \sqrt{\left(\frac{dx}{dt}\right)^2 + \left(\frac{dy}{dt}\right)^2}\,dt$$

if C is presented in parametric form, as in Section 12.2.

EXAMPLE 6 Let J denote the upper half of the *circle* (not the disk) of radius a and center $(0, 0)$, represented parametrically by

$$x = a\cos t, \qquad y = a\sin t, \qquad 0 \leqq t \leqq \pi.$$

The arc J is shown in Fig. 15.5.13. Find its centroid.

Solution Note first that $\bar{x} = 0$ by symmetry. The arc length of J is $s = \pi a$; the arc-length element is

$$ds = \sqrt{(-a \sin t\, dt)^2 + (a \cos t\, dt)^2} = a\, dt.$$

Hence the second formula in (4) yields

$$\bar{y} = \frac{1}{\pi a} \int_0^\pi (a \sin t)(a\, dt) = \frac{a}{\pi}\Big[-\cos t \Big]_0^\pi = \frac{2a}{\pi}.$$

Thus the centroid of the semicircular arc is located at the point $(0, 2a/\pi)$ on the y-axis. Note that the answer is both plausible and dimensionally correct.

The first theorem of Pappus has an analogue for surface area of revolution.

Second Theorem of Pappus: *Surface Area of Revolution*
Let the plane curve C be revolved about an axis in its plane that does not intersect the curve. Then the area

$$A = s \cdot d$$

of the surface of revolution generated is equal to the product of the length s of C and the distance d traveled by the centroid of C.

Proof for the Special Case in which C Is a Smooth Arc Described by $y = f(x)$, $a \leqq x \leqq b$, *and the Axis of Revolution Is the* y-*axis* The distance traveled by the centroid of C is $d = 2\pi \bar{x}$. By Eq. (11) of Section 6.4, the area of the surface of revolution is

$$A = \int_*^{**} 2\pi x\, ds = \int_a^b 2\pi x \sqrt{1 + [f'(x)]^2}\, dx = 2\pi s \cdot \frac{1}{s} \int_C x\, ds = 2\pi s \bar{x}$$

by Eq. (4). Therefore, $A = d \cdot s$, as desired. ❑

EXAMPLE 7 Find the surface area A of the sphere of radius a generated by revolving around the x-axis the semicircular arc of Example 6.

Solution Because we found that $\bar{y} = 2a/\pi$ and we know that $s = \pi a$, the second theorem of Pappus gives

$$A = 2\pi \bar{y} s = 2\pi \cdot \frac{2a}{\pi} \cdot \pi a = 4\pi a^2.$$

EXAMPLE 8 Find the surface area A of the torus of Example 5.

Solution Now we think of revolving around the y-axis the circle (*not* the disk) of radius a centered at the point $(b, 0)$. Of course, the centroid of the circle is located at its center $(b, 0)$; this follows from the symmetry principle or can be verified by using computations such as those in Example 6. Hence the distance traveled by the centroid is $d = 2\pi b$. Because the circumference of the circle is $s = 2\pi a$, the second theorem of Pappus gives

$$A = 2\pi b \cdot 2\pi a = 4\pi^2 ab.$$

MOMENTS OF INERTIA

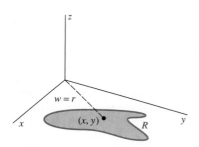

Fig. 15.5.14 A lamina in the xy-plane in space

Let R be a plane lamina and L a straight line that may or may not lie in the xy-plane. The the **moment of inertia** I of R about the axis L is defined to be

$$I = \iint_R w^2 \, dm, \tag{5}$$

where $w = w(x, y)$ denotes the (perpendicular) distance to L from a typical point (x, y) of R.

The most important case is that in which the axis is the z-axis, so $w = r = \sqrt{x^2 + y^2}$ (Fig. 15.5.14). In this case we call $I = I_0$ the **polar moment of inertia** of the lamina R. Thus the polar moment of inertia of R is defined to be

$$I_0 = \iint_R r^2 \rho(x, y) \, dA = \iint_R (x^2 + y^2) \, dm. \tag{6}$$

It follows that

$$I_0 = I_x + I_y,$$

where

$$I_x = \iint_R y^2 \, dm = \iint_R y^2 \rho \, dA \tag{7}$$

and

$$I_y = \iint_R x^2 \, dm = \iint_R x^2 \rho \, dA. \tag{8}$$

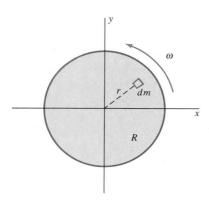

Fig. 15.5.15 The rotating disk

Here I_x is the moment of inertia of the lamina about the x-axis, whereas I_y is the moment of inertia about the y-axis.

An important application of moments of inertia involves *kinetic energy of rotation*. Consider a circular disk that is revolving around its center (the origin) with angular speed ω radians per second. A mass element dm at distance r from the origin is moving with (linear) velocity $v = r\omega$ (Fig. 15.5.15). Thus the kinetic energy of this mass element is

$$\tfrac{1}{2}(dm)v^2 = \tfrac{1}{2}\omega^2 r^2 \, dm.$$

Summing by integration over the whole disk, we find that its kinetic energy due to rotation at angular speed ω is

$$\text{KE}_{\text{rot}} = \iint_R \tfrac{1}{2}\omega^2 r^2 \, dm = \tfrac{1}{2}\omega^2 \iint_R r^2 \, dm;$$

that is,

$$\text{KE}_{\text{rot}} = \tfrac{1}{2} I_0 \omega^2. \tag{9}$$

Because linear kinetic energy has the formula $\text{KE} = \tfrac{1}{2} m v^2$, Eq. (9) suggests that moment of inertia is the rotational analogue of mass.

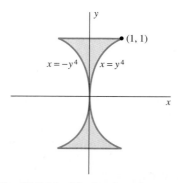

Fig. 15.5.16 The lamina of Example 9

EXAMPLE 9 Compute I_x for a lamina of constant density $\rho \equiv 1$ that occupies the region bounded by the curves $x = \pm y^4$, $-1 \leq y \leq 1$ (Fig. 15.5.16).

Solution Equation (7) gives

$$I_x = \int_{-1}^{1} \int_{-y^4}^{y^4} y^2 \, dx \, dy = \int_{-1}^{1} \left[xy^2 \right]_{x=-y^4}^{y^4} dy = \int_{-1}^{1} 2y^6 \, dy = \tfrac{4}{7}.$$

The region of Example 9 resembles the cross section of an I beam. It is known that the stiffness, or resistance to bending, of a horizontal beam is proportional to the moment of inertia of its cross section with respect to a horizontal axis through the centroid of the cross section. Let us compare our I beam with a rectangular beam of equal height 2 and equal area

$$A = \int_{-1}^{1} \int_{-y^4}^{y^4} 1 \, dx \, dy = \tfrac{4}{5}.$$

The cross section of such a rectangular beam is shown in Fig. 15.5.17. Its width is $\tfrac{2}{5}$, and the moment of inertia of its cross section is

$$I_x = \int_{-1}^{1} \int_{-1/5}^{1/5} y^2 \, dx \, dy = \tfrac{4}{15}.$$

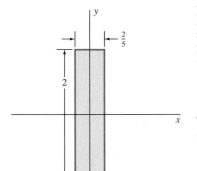

Fig. 15.5.17 A rectangular beam for comparison with the I beam of Example 9

Because the ratio of $\tfrac{4}{7}$ to $\tfrac{4}{15}$ is $\tfrac{15}{7}$, we see that the I-beam is more than twice as strong as a rectangular beam of the same cross-sectional area. This strength is why I beams are commonly used in construction.

EXAMPLE 10 Find the polar moment of inertia of a circular lamina R of radius a and constant density ρ centered at the origin.

Solution In Cartesian coordinates, the lamina R occupies the plane region $x^2 + y^2 \leq a^2$; in polar coordinates, this region has the much simpler description $0 \leq r \leq a$, $0 \leq \theta \leq 2\pi$. Equation (6) then gives

$$I_0 = \iint_R r^2 \rho \, dA = \int_0^{2\pi} \int_0^a \rho r^3 \, dr \, d\theta = \frac{\rho \pi a^4}{2} = \frac{1}{2} ma^2,$$

where $m = \rho \pi a^2$ is the mass of the circular lamina.

Finally, the **radius of gyration** r of a lamina of mass m about an axis is defined to be

$$\hat{r} = \sqrt{\frac{I}{m}}, \tag{10}$$

where I is the moment of inertia of the lamina about that axis. For example, the radii of gyration $\hat{x}$ and $\hat{y}$ about the y-axis and x-axis, respectively, are given by

$$\hat{x} = \sqrt{\frac{I_y}{m}} \quad \text{and} \quad \hat{y} = \sqrt{\frac{I_x}{m}}. \tag{11}$$

Now suppose that this lamina lies in the right half-plane $x > 0$ and is symmetric about the x-axis. If it represents the face of a tennis racquet whose handle (considered of negligible weight) extends along the x-axis from the origin to the face, then the point $(\hat{x}, 0)$ is a plausible candidate for the racquet's "sweet spot," that delivers the maximum impact and control (see Problem 56).

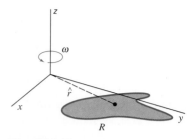

Fig. 15.5.18 A plane lamina rotating around the *z*-axis

The definition in Eq. (10) is motivated by consideration of a plane lamina R rotating with angular speed ω about the z-axis (Fig. 15.5.18). Then Eq. (10) yields

$$I_0 = m\hat{r}^2,$$

so it follows from Eq. (9) that the kinetic energy of the lamina is

$$\text{KE} = \tfrac{1}{2}m(\hat{r}\omega)^2.$$

Thus the kinetic energy of the rotating lamina equals that of a single particle of mass m revolving at the distance $\hat{r}$ from the axis of revolution.

15.5 Problems

In Problems 1 through 10, find the centroid of the plane region bounded by the given curves. Assume that the density is $\rho \equiv 1$ for each region.

1. $x = 0, x = 4, y = 0, y = 6$

2. $x = 1, x = 3, y = 2, y = 4$

3. $x = -1, x = 3, y = -2, y = 4$

4. $x = 0, y = 0, x + y = 3$

5. $x = 0, y = 0, x + 2y = 4$

6. $y = 0, y = x, x + y = 2$

7. $y = 0, y = x^2, x = 2$

8. $y = x^2, y = 9$

9. $y = 0, y = x^2 - 4$

10. $x = -2, x = 2, y = 0, y = x^2 + 1$

In Problems 11 through 30, find the mass and centroid of a plane lamina with the indicated shape and density.

11. The triangular region bounded by $x = 0, y = 0$, and $x + y = 1$, with $\rho(x, y) = xy$

12. The triangular region of Problem 11, with $\rho(x, y) = x^2$

13. The region bounded by $y = 0$, and $y = 4 - x^2$, with $\rho(x, y) = y$

14. The region bounded by $x = 0$ and $x = 9 - y^2$, with $\rho(x, y) = x^2$

15. The region bounded by the parabolas $y = x^2$ and $x = y^2$, with $\rho(x, y) = xy$

16. The region of Problem 15, with $\rho(x, y) = x^2 + y^2$

17. The region bounded by the parabolas $y = x^2$ and $y = 2 - x^2$, with $\rho(x, y) = y$

18. The region bounded by $x = 0, x = e, y = 0$, and $y = \ln x$ for $1 \leq x \leq e$, with $\rho(x, y) \equiv 1$

19. The region bounded by $y = 0$ and $y = \sin x$ for $0 \leq x \leq \pi$, with $\rho(x, y) \equiv 1$

20. The region bounded by $y = 0, x = -1, x = 1$, and $y = \exp(-x^2)$, with $\rho(x, y) = |xy|$

21. The square with vertices $(0, 0), (0, a), (a, a)$, and $(a, 0)$, with $\rho(x, y) = x + y$

22. The triangular region bounded by the coordinate axes and the line $x + y = a$; $\rho(x, y) = x^2 + y^2$

23. The region bounded by $y = x^2$ and $y = 4$; $\rho(x, y) = y$

24. The region bounded by $y = x^2$ and $y = 2x + 3$; $\rho(x, y) = x^2$

25. The region of Problem 19; $\rho(x, y) = x$

26. The semicircular region $x^2 + y^2 \leq a^2$, $y \geq 0$; $\rho(x, y) = y$

27. The region of Problem 26; $\rho(x, y) = r$ (the radial polar coordinate)

28. The region bounded by the cardioid with polar equation $r = 1 + \cos\theta$; $\rho = r$ (Fig. 15.5.19)

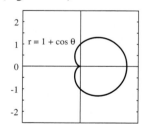

Fig. 15.5.19 The cardioid of Problem 28

29. The region inside the circle $r = 2\sin\theta$ and outside the circle $r = 1$; $\rho(x, y) = y$

30. The region inside the limaçon $r = 1 + 2\cos\theta$ and outside the circle $r = 2$; $\rho(x, y) = r$ (Fig. 15.5.20)

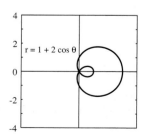

Fig. 15.5.20 The limaçon of Problem 30

In Problems 31 through 35, find the polar moment of inertia I_0 of the indicated lamina.

31. The region bounded by the circle $r = a$; $\rho(x, y) = r^n$, where n is a fixed positive integer

32. The lamina of Problem 26

33. The disk bounded by $r = 2 \cos \theta$; $\rho(x, y) = k$ (a positive constant)

34. The lamina of Problem 29

35. The region bounded by the right-hand loop of the lemniscate $r^2 = \cos 2\theta$; $\rho(x, y) = r^2$ (Fig. 15.5.21)

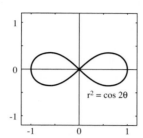

Fig. 15.5.21 The lemniscate of Problem 35

In Problems 36 through 40, find the radii of gyration $\hat{x}$ and $\hat{y}$ of the indicated lamina about the coordinate axes.

36. The lamina of Problem 21

37. The lamina of Problem 23

38. The lamina of Problem 24

39. The lamina of Problem 27

40. The lamina of Problem 33

41. Find the centroid of the first quadrant of the circular disk $x^2 + y^2 \leq r^2$ by direct computation, as in Example 1.

42. Apply the first theorem of Pappus to find the centroid of the first quadrant of the circular disk $x^2 + y^2 \leq r^2$. Use the facts that $\bar{x} = \bar{y}$ (by symmetry) and that revolution of this quarter-disk about either coordinate axis gives a solid hemisphere with volume $V = \frac{2}{3} \pi r^3$.

43. Find the centroid of the arc that consists of the first-quadrant portion of the circle $x^2 + y^2 = r^2$ by direct computation, as in Example 6.

44. Apply the second theorem of Pappus to find the centroid of the quarter-circle arc of Problem 43. Note that $\bar{x} = \bar{y}$ (by symmetry) and that revolution of this arc about either coordinate axis gives a hemisphere with surface area $A = 2\pi r^2$.

45. Show by direct computation that the centroid of the triangle with vertices $(0, 0)$, $(r, 0)$, and $(0, h)$ is the point $(r/3, h/3)$. Verify that this point lies on the line from the vertex $(0, 0)$ to the midpoint of the opposite side of the triangle and two-thirds of the way from the vertex to the midpoint.

46. Apply the first theorem of Pappus and the result of Problem 45 to verify the formula $V = \frac{1}{3} \pi r^2 h$ for the volume of the cone obtained by revolving the triangle around the y-axis.

47. Apply the second theorem of Pappus to show that the lateral surface area of the cone of Problem 46 is $A = \pi r L$, where $L = \sqrt{r^2 + h^2}$ is the slant height of the cone.

48. (a) Find the centroid of the trapezoid shown in Fig. 15.5.22. (b) Apply the first theorem of Pappus and the result of part (a) to show that the volume of the conical frustum generated by revolving the trapezoid around the y-axis is

$$V = \frac{\pi h}{3}(r_1{}^2 + r_1 r_2 + r_2{}^2).$$

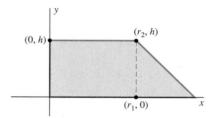

Fig. 15.5.22 The trapezoid of Problem 48

49. Apply the second theorem of Pappus to show that the lateral surface area of the conical frustum of Problem 48 is $A = \pi(r_1 + r_2)L$, where

$$L = \sqrt{(r_1 - r_2)^2 + h^2}$$

is its slant height.

50. (a) Apply the second theorem of Pappus to verify that the curved surface area of a right circular cylinder of height h and base radius r is $A = 2\pi r h$. (b) Explain how this follows also from the result of Problem 49.

51. (a) Find the centroid of the plane region shown in Fig. 15.5.23, which consists of a semicircular region of radius a sitting atop a rectangular region of width $2a$ and height b whose base is on the x-axis. (b) Then apply the first theorem of Pappus to find the volume generated by rotating this region around the x-axis.

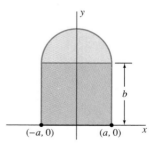

Fig. 15.5.23 The plane region of Problem 51(a)

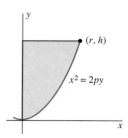

Fig. 15.5.24 The region of Problem 52

52. (a) Consider the plane region of Fig. 15.5.24, bounded by $x^2 = 2py$, $x = 0$, and $y = h = r^2/2p$ $(p > 0)$. Show that its area is $A = \frac{2}{3}rh$ and that the x-coordinate of its centroid is $\bar{x} = 3r/8$. (b) Use Pappus's theorem and the result of part (a) to show that the volume of a paraboloid of revolution with radius r and height h is $V = \frac{1}{2}\pi r^2 h$.

53. Find the centroid of the unbounded region between the graph of $y = e^{-x}$ and the x-axis for $x \geq 0$.

54. The centroid of a uniform plane region is at $(0, 0)$, and the region has total mass m. Show that its moment of inertia about an axis perpendicular to the xy-plane at the point (x_0, y_0) is

$$I = I_0 + m(x_0{}^2 + y_0{}^2).$$

55. Suppose that a plane lamina consists of two nonoverlapping laminae. Show that its polar moment of inertia is the sum of theirs. Use this fact together with the results of Problems 53 and 54 to find the polar moment of inertia of the T-shaped lamina of constant density $\rho = k > 0$ shown in Fig. 15.5.25.

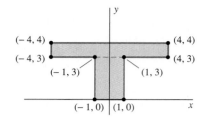

Fig. 15.5.25 One lamina made of two simpler ones (Problem 55)

56. A racquet consists of a uniform lamina that occupies the region inside the right-hand loop of $r^2 = \cos 2\theta$ on the end of a handle (assumed to be of negligible mass) corresponding to the interval $-1 \leq x \leq 0$ (Fig. 15.5.26). Find the radius of gyration of the racquet about the line $x = -1$. Where is its sweet spot?

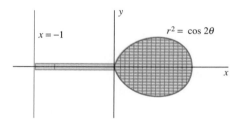

Fig. 15.5.26 The racquet of Problem 56

15.5 Project

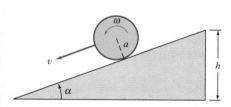

Fig. 15.5.27 A circular object rolling down an incline

To see moments of inertia in action, suppose that your club is designing an unpowered race car for the annual downhill derby. You have a choice of solid wheels, bicycle wheels with thin spokes, or even solid spherical wheels (like giant ball bearings). Which wheels will make the race car go fastest?

Imagine an experiment in which you roll various types of wheels down an incline to see which reaches the bottom the fastest (Fig. 15.5.27). Suppose that a wheel of radius a and mass M starts from rest at the top with potential energy PE $= Mgh$ and reaches the bottom with angular speed ω and (linear) velocity $v = a\omega$. Then (by conservation of energy) the wheel's initial potential energy has been transformed to a sum $\mathrm{KE_{tr}} + \mathrm{KE_{rot}}$ of translational kinetic energy $\mathrm{KE_{tr}} = \frac{1}{2}Mv^2$ and rotational kinetic energy

$$\mathrm{KE_{rot}} = \frac{1}{2}I_0\omega^2 = \frac{I_0 v^2}{2a^2}, \tag{12}$$

a consequence of Eq. (9) of this section. Thus

$$Mgh = \frac{1}{2}Mv^2 + \frac{I_0 v^2}{2a^2}. \tag{13}$$

Problems 1 through 8 explore the implications of this formula.

1. Suppose that the wheel's (polar) moment of inertia is given by

$$I_0 = kMa^2 \tag{14}$$

for some constant k. (For instance, Example 10 gives $k = \frac{1}{2}$ for a wheel in the shape of a uniform solid disk.) Then deduce from Eq. (13) that

$$v = \sqrt{\frac{2gh}{1 + k}}. \tag{15}$$

Thus the smaller k is (and hence the smaller the wheel's moment of inertia), the faster the wheel will roll down the incline.

In Problems 2 through 8, take $g = 32$ ft/s² and assume that the vertical height of the incline is $h = 100$ ft.

2. Why does it follow from Eq. (4) that, whatever the wheel's design, the maximum velocity a circular wheel can attain on this incline is 80 ft/s (just under 55 mi/h)?

3. If the wheel is a uniform solid disk (like an old-fashioned wooden wagon wheel) with $I_0 = \frac{1}{2}Ma^2$, what is its speed v at the bottom of the incline?

4. Answer Problem 3 if the wheel is shaped like a narrow bicycle tire, with its entire mass, in effect, concentrated at the distance a from its center. In this case, $I_0 = Ma^2$. (Why?)

5. Answer Problem 3 if the wheel is shaped like an annular ring (or washer) with outer radius a and inner radius b.

Do not attempt Problems 6 through 8 until you have studied Example 3 of Section 15.7. In Problems 6 through 8, what is the velocity of the wheel when it reaches the bottom of the incline?

6. The wheel is a uniform solid sphere of radius a.

7. The wheel is a very thin, spherical shell whose entire mass is, in effect, concentrated at the distance a from its center.

8. The wheel is a spherical shell with outer radius a and inner radius $b = \frac{1}{2}a$.

Finally, what is your conclusion? What is the shape of the wheels that will yield the fastest downhill race car?

15.6
Triple Integrals

The definition of the triple integral is the three-dimensional version of the definition of the double integral of Section 15.2. Let $f(x, y, z)$ be continuous on the bounded space region T, and suppose that T lies inside the rectangular block R determined by the inequalities $a \leq x \leq b$, $c \leq y \leq d$, and $p \leq z \leq q$. We divide $[a, b]$ into subintervals of equal length Δx, $[c, d]$ into subintervals of equal length Δy, and $[p, q]$ into subintervals of equal length Δz. This generates a partition of R into smaller rectangular blocks (as in Fig. 15.6.1), each of volume $\Delta V = \Delta x\, \Delta y\, \Delta z$. Let $\mathcal{P} = \{T_1, T_2, \ldots, T_n\}$ be the collection of these smaller blocks that lie wholly within T. Then $\mathcal{P}$ is called

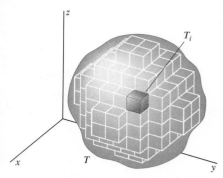

Fig. 15.6.1 One small block in an inner partition of the bounded space region T

an **inner partition** of the region T. The mesh $|\mathcal{P}|$ of $\mathcal{P}$ is the length of a longest diagonal of any of the blocks T_i. If (x_i^*, y_i^*, z_i^*) is an arbitrarily selected point of T_i (for each $i = 1, 2, \ldots, n$), then the **Riemann sum**

$$\sum_{i=1}^{n} f(x_i^*, y_i^*, z_i^*) \, \Delta V$$

is an approximation to the triple integral of f over the region T.

For example, if T is a solid body with density function f, then such a Riemann sum approximates its total mass. We define the **triple integral of f over T** by means of the equation

$$\iiint_T f(x, y, z) \, dV = \lim_{|\mathcal{P}| \to 0} \sum_{i=1}^{n} f(x_i^*, y_i^*, z_i^*) \, \Delta V. \tag{1}$$

It is proved in advanced calculus that this limit of Riemann sums exists as the mesh $|\mathcal{P}|$ approaches zero provided that f is continuous on T and that the boundary of the region T is reasonably well behaved. For instance, it suffices for the boundary of T to consist of a finite number of smooth surfaces.

Just as with double integrals, we ordinarily compute triple integrals by means of iterated integrals. If the region of integration is a rectangular block, as in Example 1, then we can integrate in any order we wish.

EXAMPLE 1   If $f(x, y, z) = xy + yz$ and T consists of those points (x, y, z) in space that satisfy the inequalities $-1 \leq x \leq 1$, $2 \leq y \leq 3$, and $0 \leq z \leq 1$, then

$$\begin{aligned}
\iiint_T f(x, y, z) \, dV &= \int_{-1}^{1} \int_{2}^{3} \int_{0}^{1} (xy + yz) \, dz \, dy \, dx \\
&= \int_{-1}^{1} \int_{2}^{3} \left[xyz + \tfrac{1}{2} yz^2 \right]_{z=0}^{1} dy \, dx \\
&= \int_{-1}^{1} \int_{2}^{3} (xy + \tfrac{1}{2} y) \, dy \, dx \\
&= \int_{-1}^{1} \left[\tfrac{1}{2} xy^2 + \tfrac{1}{4} y^2 \right]_{y=2}^{3} dx \\
&= \int_{-1}^{1} (\tfrac{5}{2} x + \tfrac{5}{4}) \, dx = \left[\tfrac{5}{4} x^2 + \tfrac{5}{4} x \right]_{-1}^{1} = \tfrac{5}{2}.
\end{aligned}$$

The applications of double integrals that we saw in earlier sections generalize immediately to triple integrals. If T is a solid body with the density function $\rho(x, y, z)$, then its **mass m** is given by

$$m = \iiint_T \rho \, dV. \tag{2}$$

The case $\rho \equiv 1$ gives the **volume**

$$V = \iiint_T dV \tag{3}$$

of T. The coordinates of its **centroid** are

$$\bar{x} = \frac{1}{m} \iiint_T x\rho \ dV, \tag{4a}$$

$$\bar{y} = \frac{1}{m} \iiint_T y\rho \ dV, \quad \text{and} \tag{4b}$$

$$\bar{z} = \frac{1}{m} \iiint_T z\rho \ dV. \tag{4c}$$

The **moments of inertia** of T about the three coordinate axes are

$$I_x = \iiint_T (y^2 + z^2)\rho \ dV, \tag{5a}$$

$$I_y = \iiint_T (x^2 + z^2)\rho \ dV, \quad \text{and} \tag{5b}$$

$$I_z = \iiint_T (x^2 + y^2)\rho \ dV. \tag{5c}$$

As indicated previously, we almost always evaluate triple integrals by iterated single integration. Suppose that the nice region T is **z-simple**: Each line parallel to the z-axis intersects T (if at all) in a single line segment. In effect, this means that T can be described by the inequalities

$$z_1(x, y) \leqq z \leqq z_2(x, y), \qquad (x, y) \text{ in } R,$$

where R is the vertical projection of T into the xy-plane. Then

$$\iiint_T f(x, y, z) \ dV = \iint_R \left(\int_{z_1(x,y)}^{z_2(x,y)} f(x, y, z) \ dz \right) dA. \tag{6}$$

In Eq. (6), we take $dA = dx \ dy$ or $dA = dy \ dx$, depending on the preferred order of integration over the set R. The limits $z_1(x, y)$ and $z_2(x, y)$ for z are the z-coordinates of the endpoints of the line segment in which the vertical line at (x, y) meets T (Fig. 15.6.2).

If the region R has the description

$$y_1(x) \leqq y \leqq y_2(x), \qquad a \leqq x \leqq b,$$

then (integrating last with respect to x),

$$\iiint_T f(x, y, z) \ dV = \int_a^b \int_{y_1(x)}^{y_2(x)} \int_{z_1(x,y)}^{z_2(x,y)} f(x, y, z) \ dz \ dy \ dx.$$

Thus the triple integral reduces in this case to three iterated single integrals. These can (in principle) be evaluated by using the fundamental theorem of calculus.

If the solid T is bounded by the *two* surfaces $z = z_1(x, y)$ and $z = z_2(x, y)$, as in Fig. 15.6.3, we can find the region R of Eq. (6) as follows. The equation $z_1(x, y) = z_2(x, y)$ determines a vertical cylinder that passes through the curve of intersection of the two surfaces. So the cylinder intersects the xy-plane in the boundary curve of R. For example, we shall see that the solid of Example 3 is bounded by the paraboloid $z = x^2 + y^2$ and by the

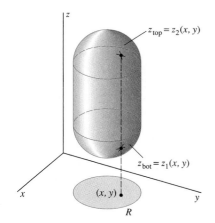

Fig. 15.6.2 Obtaining the limits of integration for z

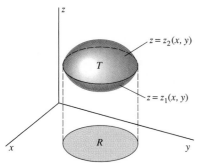

Fig. 15.6.3 To find the boundary of R, solve the equation $z_1(x, y) = z_2(x, y)$.

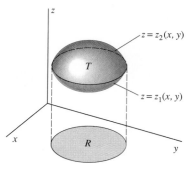

(a) T is z-simple

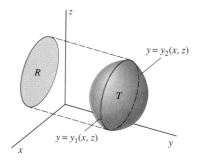

(b) T is y-simple

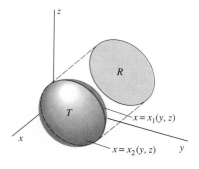

(c) T is x-simple

Fig. 15.6.4 Solids that are (a) z-simple, (b) y-simple, and (c) x-simple

plane $z = y + 2$. And the graph of the equation $x^2 + y^2 = y + 2$ is a circle that bounds the region R above which the solid T lies.

We proceed similarly if T is either **x-simple** or **y-simple**. Such situations, as well as a z-simple solid, appear in Fig. 15.6.4. For example, suppose that T is y-simple, so that it has a description of the form

$$y_1(x, z) \leqq y \leqq y_2(x, z), \qquad (x, z) \text{ in } R,$$

where R is the projection of T into the xz-plane. Then

$$\iiint_T f(x, y, z) \, dV = \iint_R \left(\int_{y_1(x, z)}^{y_2(x, z)} f(x, y, z) \, dy \right) dA, \qquad (7)$$

where $dA = dx \, dz$ or $dA = dz \, dx$ and the limits $y_1(x, z)$ and $y_2(x, z)$ are the y-coordinates of the endpoints of the line segment in which a typical line parallel to the y-axis intersects T. If T is x-simple, we have

$$\iiint_T f(x, y, z) \, dV = \iint_R \left(\int_{x_1(y, z)}^{x_2(y, z)} f(x, y, z) \, dx \right) dA, \qquad (8)$$

where $dA = dy \, dz$ or $dA = dz \, dy$ and R is the projection of T into the yz-plane.

EXAMPLE 2 Compute by triple integration the volume of the region T that is bounded by the parabolic cylinder $x = y^2$ and the planes $z = 0$ and $x + z = 1$. Also find the centroid of T given that it has constant density $\rho \equiv 1$.

COMMENT The three segments in Fig. 15.6.5 parallel to the coordinate axes indicate that the region T is simultaneously x-simple, y-simple, and z-simple. We may therefore integrate in any order we choose, so there are six ways to evaluate the integral. Here are three computations of the volume V of T.

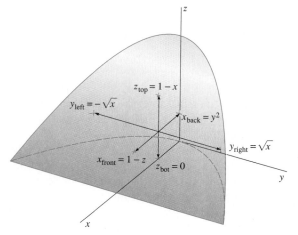

Fig. 15.6.5 The region T of Example 2 is x-simple, y-simple, and z-simple.

874

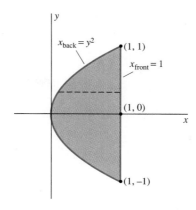

Fig. 15.6.6 The vertical projection of the solid region T into the xy-plane (Example 2, Solution 1)

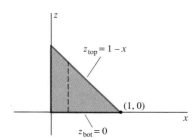

Fig. 15.6.7 The vertical projection of the solid region T into the xz-plane (Example 2, Solution 2)

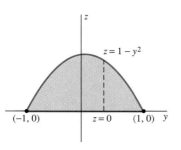

Fig. 15.6.8 The vertical projection of the solid region T into the yz-plane (Example 2, Solution 3)

Solution 1 The projection of T into the xy-plane is the region shown in Fig. 15.6.6, bounded by $x = y^2$ and $x = 1$. So Eq. (6) gives

$$V = \int_{-1}^{1} \int_{y^2}^{1} \int_{0}^{1-x} dz\, dx\, dy = 2\int_{0}^{1} \int_{y^2}^{1} (1 - x)\, dx\, dy$$

$$= 2\int_{0}^{1} \left[x - \tfrac{1}{2}x^2 \right]_{x=y^2}^{1} dy = 2\int_{0}^{1} \left(\tfrac{1}{2} - y^2 + \tfrac{1}{2}y^4\right) dy = \tfrac{8}{15}.$$

Solution 2 The projection of T into the xz-plane is the triangle bounded by the coordinate axes and by the line $x + z = 1$ (Fig. 15.6.7), so Eq. (7) gives

$$V = \int_{0}^{1} \int_{0}^{1-x} \int_{-\sqrt{x}}^{\sqrt{x}} dy\, dz\, dx = 2\int_{0}^{1} \int_{0}^{1-x} \sqrt{x}\, dz\, dx$$

$$= 2\int_{0}^{1} (x^{1/2} - x^{3/2})\, dx = \tfrac{8}{15}.$$

Solution 3 The projection of T into the yz-plane is bounded by the y-axis and by the parabola $z = 1 - y^2$ (Fig. 15.6.8), so Eq. (8) yields

$$V = \int_{-1}^{1} \int_{0}^{1-y^2} \int_{y^2}^{1-z} dx\, dz\, dy,$$

and evaluation of this integral again gives $V = \tfrac{8}{15}$.

Now for the centroid of T. Because the region T is symmetric about the xz-plane, its centroid lies in this plane, and so $\overline{y} = 0$. We compute $\overline{x}$ and $\overline{z}$ by integrating first with respect to y:

$$\overline{x} = \frac{1}{V} \iiint_T x\, dV = \frac{15}{8} \int_{0}^{1} \int_{0}^{1-x} \int_{-\sqrt{x}}^{\sqrt{x}} x\, dy\, dz\, dx$$

$$= \frac{15}{4} \int_{0}^{1} \int_{0}^{1-x} x^{3/2}\, dz\, dx = \frac{15}{4} \int_{0}^{1} (x^{3/2} - x^{5/2})\, dx = \frac{3}{7},$$

and similarly,

$$\overline{z} = \frac{1}{V} \iiint_T z\, dV = \frac{15}{8} \int_{0}^{1} \int_{0}^{1-x} \int_{-\sqrt{x}}^{\sqrt{x}} z\, dy\, dz\, dx = \frac{2}{7}.$$

Thus the centroid of T is located at the point $(\tfrac{3}{7}, 0, \tfrac{2}{7})$.

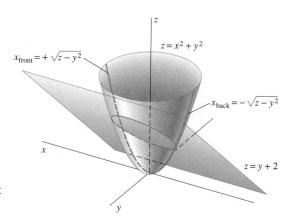

Fig. 15.6.9 An oblique segment of a paraboloid (Example 3)

EXAMPLE 3 Find the volume of the *oblique segment of a paraboloid* bounded by the paraboloid $z = x^2 + y^2$ and the plane $z = y + 2$ (Fig. 15.6.9).

Solution The given region T is z-simple, but its projection into the xy-plane is bounded by the graph of the equation $x^2 + y^2 = y + 2$, which is a translated circle. It would be possible to integrate first with respect to z, but perhaps another choice will yield a simpler integral.

The region T also is x-simple, so we may integrate first with respect to x. The projection of T into the yz-plane is bounded by the line $z = y + 2$ and by the parabola $z = y^2$, which intersect at the points $(-1, 1)$ and $(2, 4)$ (Fig. 15.6.10). The endpoints of a line segment in T parallel to the x-axis have x-coordinates $x = \pm\sqrt{z - y^2}$. Because T is symmetric about the yz-plane, we can integrate from $x = 0$ to $x = \sqrt{z - y^2}$ and double the result. Hence T has volume

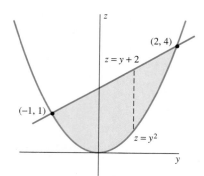

Fig. 15.6.10 Projection of the segment of the paraboloid into the yz-plane (Example 3)

$$
\begin{aligned}
V &= 2 \int_{-1}^{2} \int_{y^2}^{y+2} \int_{0}^{\sqrt{z - y^2}} dx\, dz\, dy = 2 \int_{-1}^{2} \int_{y^2}^{y+2} \sqrt{z - y^2}\, dz\, dy \\
&= 2 \int_{-1}^{2} \left[\frac{2}{3}\left(z - y^2\right)^{3/2} \right]_{z=y^2}^{y+2} dy = \frac{4}{3} \int_{-1}^{2} (2 + y - y^2)^{3/2}\, dy \\
&= \frac{4}{3} \int_{-3/2}^{3/2} \left(\frac{9}{4} - u^2 \right)^{3/2} du \qquad \text{(completing the square; } u = y - \tfrac{1}{2}\text{)} \\
&= \frac{27}{4} \int_{-\pi/2}^{\pi/2} \cos^4 \theta\, d\theta \qquad (u = \tfrac{3}{2} \sin \theta) \\
&= \frac{27}{4} \cdot 2 \cdot \frac{1}{2} \cdot \frac{3}{4} \cdot \frac{\pi}{2} = \frac{81\pi}{32}.
\end{aligned}
$$

In the final evaluation, we used Formula (113) (inside the back cover).

15.6 Problems

In Problems 1 through 10, compute the value of the triple integral

$$\iiint_T f(x, y, z)\, dV.$$

1. $f(x, y, z) = x + y + z$; T is the rectangular box $0 \le x \le 2, 0 \le y \le 3, 0 \le z \le 1$.

2. $f(x, y, z) = xy \sin z$; T is the cube $0 \le x \le \pi$, $0 \le y \le \pi, 0 \le z \le \pi$.

876

3. $f(x, y, z) = xyz$; T is the rectangular block $-1 \leqq x \leqq 3, 0 \leqq y \leqq 2, -2 \leqq z \leqq 6$.

4. $f(x, y, z) = x + y + z$; T is the rectangular block of Problem 3.

5. $f(x, y, z) = x^2$; T is the tetrahedron bounded by the coordinate planes and the first octant part of the plane with equation $x + y + z = 1$.

6. $f(x, y, z) = 2x + 3y$; T is a first-octant tetrahedron as in Problem 5, except that the plane has equation $2x + 3y + z = 6$.

7. $f(x, y, z) = xyz$; T lies below the surface $z = 1 - x^2$ and above the rectangle $-1 \leqq x \leqq 1, 0 \leqq y \leqq 2$ in the xy-plane.

8. $f(x, y, z) = 2y + z$; T lies below the surface with equation $z = 4 - y^2$ and above the rectangle $-1 \leqq x \leqq 1, -2 \leqq y \leqq 2$ in the xy-plane.

9. $f(x, y, z) = x + y$; T is the region between the surfaces $z = 2 - x^2$ and $z = x^2$ for $0 \leqq y \leqq 3$ (Fig. 15.6.11).

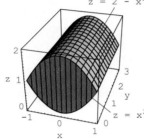

Fig. 15.6.11 The solid of Problem 9

10. $f(x, y, z) = z$; T is the region between the surfaces $z = y^2$ and $z = 8 - y^2$ for $-1 \leqq x \leqq 1$.

In Problems 11 through 20, sketch the solid bounded by the graphs of the given equations. Then find its volume by triple integration.

11. $2x + 3y + z = 6, x = 0, y = 0, z = 0$

12. $z = y, y = x^2, y = 4, z = 0$ (Fig. 15.6.12)

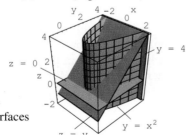

Fig. 15.6.12 The surfaces of Problem 12

13. $y + z = 4, y = 4 - x^2, y = 0, z = 0$

14. $z = x^2 + y^2, z = 0, x = 0, y = 0, x + y = 1$

15. $z = 10 - x^2 - y^2, y = x^2, x = y^2, z = 0$

16. $x = z^2, x = 8 - z^2, y = -1, y = -3$

17. $z = x^2, y + z = 4, y = 0, z = 0$

18. $z = 1 - y^2, z = y^2 - 1, x + z = 1, x = 0$ (Fig. 15.6.13)

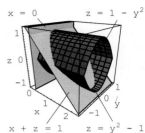

Fig. 15.6.13 The surfaces of Problem 18

19. $y = z^2, z = y^2, x + y + z = 2, x = 0$

20. $y = 4 - x^2 - z^2, x = 0, y = 0, z = 0, x + z = 2$

In Problems 21 through 32, assume that the indicated solid has constant density $\rho \equiv 1$.

21. Find the centroid of the solid of Problem 12.

22. Find the centroid of the hemisphere $x^2 + y^2 + z^2 \leqq R^2, z \geqq 0$.

23. Find the centroid of the solid of Problem 17.

24. Find the centroid of the solid bounded by $z = 1 - x^2$, $z = 0, y = -1$, and $y = 1$.

25. Find the centroid of the solid bounded by $z = \cos x$, $x = -\pi/2, x = \pi/2, y = 0, z = 0$, and $y + z = 1$.

26. Find the moment of inertia about the z-axis of the solid of Problem 12.

27. Find the moment of inertia about the y-axis of the solid of Problem 24.

28. Find the moment of inertia about the z-axis of the solid cylinder $x^2 + y^2 \leqq R^2, 0 \leqq z \leqq H$.

29. Find the moment of inertia about the z-axis of the solid bounded by $x + y + z = 1, x = 0, y = 0$, and $z = 0$.

30. Find the moment of inertia about the z-axis of the cube with the vertices $(\pm 0.5, 3, \pm 0.5)$ and $(\pm 0.5, 4, \pm 0.5)$.

31. Consider the solid paraboloid bounded by $z = x^2 + y^2$ and by the plane $z = h > 0$. Show that its centroid lies on its axis of symmetry, two-thirds of the way from its "vertex" $(0, 0, 0)$ to its base.

32. Show that the centroid of a right circular cone lies on the axis of the cone and three-fourths of the way from the vertex to the base.

In Problems 33 through 40, the indicated solid has uniform density $\rho \equiv 1$ unless otherwise indicated.

33. For a cube with edge length a, find the moment of inertia about one of its edges.

34. The density at $P(x, y, z)$ of the first-octant cube with edge length a, faces parallel to the coordinate planes, and opposite vertices $(0, 0, 0)$ and (a, a, a) is proportional to the square of the distance from P to the origin. Find the coordinates of its centroid.

35. Find the moment of inertia about the z-axis of the cube of Problem 34.

36. The cube bounded by the coordinate planes and by the planes $x = 1$, $y = 1$, and $z = 1$ has density $\rho = kz$ at the point $P(x, y, z)$ (k is a positive constant). Find its centroid.

37. Find the moment of inertia about the z-axis of the cube of Problem 36.

38. Find the moment of inertia about a diameter of a solid sphere of radius a.

39. Find the centroid of the first-octant region that is interior to the two cylinders $x^2 + z^2 = 1$ and $y^2 + z^2 = 1$ (Figs. 15.6.14 and 15.6.15).

40. Find the moment of inertia about the z-axis of the solid of Problem 39.

41. Find the volume bounded by the elliptic paraboloids $z = 2x^2 + y^2$ and $z = 12 - x^2 - 2y^2$. Note that this solid projects onto a circular disk in the xy-plane.

42. Find the volume bounded by the elliptic paraboloid $y = x^2 + 4z^2$ and by the plane $y = 2x + 3$.

43. Find the volume of the elliptical cone bounded by $z = \sqrt{x^2 + 4y^2}$ and by the plane $z = 1$. [*Suggestion:* Integrate first with respect to x.]

44. Find the volume of the region bounded by the paraboloid $x = y^2 + 2z^2$ and by the parabolic cylinder $x = 2 - y^2$ (Fig. 15.6.16).

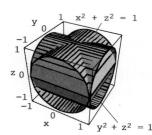

Fig. 15.6.14 The intersecting cylinders of Problem 39

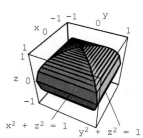

Fig. 15.6.15 The solid of intersection of Problem 39

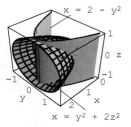

Fig. 15.6.16 The surfaces of Problem 44

15.7

Integration in Cylindrical and Spherical Coordinates

Suppose that $f(x, y)$ is a continuous function defined on the z-simple region T, which—because it is z-simple—can be described by

$$z_1(x, y) \leq z \leq z_2(x, y) \qquad \text{for } (x, y) \text{ in } R$$

(where R is the projection of T into the xy-plane, as usual). We saw in Section 15.6 that

$$\iiint_T f(x, y, z) \, dV = \iint_R \left(\int_{z_1(x, y)}^{z_2(x, y)} f(x, y, z) \, dz \right) dA. \qquad (1)$$

If we can describe the region R more naturally in polar coordinates than in rectangular coordinates, then it is likely that the integration over the plane region R will be simpler if it is carried out in polar coordinates.

We first express the inner partial integral of Eq. (1) in terms of r and θ by writing

$$\int_{z_1(x, y)}^{z_2(x, y)} f(x, y, z) \, dz = \int_{Z_1(r, \theta)}^{Z_2(r, \theta)} F(r, \theta, z) \, dz, \qquad (2)$$

where

$$F(r, \theta, z) = f(r \cos \theta, r \sin \theta, z) \qquad (3a)$$

and

$$Z_i(r, \theta) = z_i(r \cos \theta, r \sin \theta) \qquad (3b)$$

878

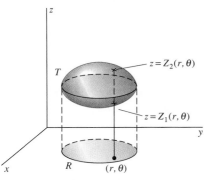

Fig. 15.7.1 The limits on z in a triple integral in cylindrical coordinates are determined by the lower and upper surfaces.

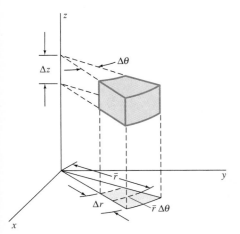

Fig. 15.7.2 The volume of the cylindrical block is $\Delta V = \bar{r}\,\Delta z\,\Delta r\,\Delta\theta$.

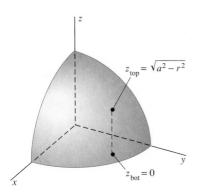

Fig. 15.7.3 The first octant of the sphere (Example 1)

for $i = 1, 2$. Substitution of Eq. (2) into Eq. (1) with **(important)** $dA = r\,dr\,d\theta$ gives

$$\iiint\limits_{T} f(x, y, z)\,dV = \iint\limits_{S}\left(\int_{Z_1(r,\theta)}^{Z_2(r,\theta)} F(r, \theta, z)\,dz\right) r\,dr\,d\theta, \qquad (4)$$

where F, Z_1, and Z_2 are the functions given in (3) and S represents the appropriate limits on r and θ needed to describe the plane region R in polar coordinates (as discussed in Section 15.4). The limits on z are simply the z-coordinates (in terms of r and θ) of a typical line segment joining the lower and upper boundary surfaces of T, as indicated in Fig. 15.7.1.

Thus the general formula for **triple integration in cylindrical coordinates** is

$$\iiint\limits_{T} f(x, y, z)\,dV = \iiint\limits_{U} f(r\cos\theta,\, r\sin\theta,\, z)\, r\,dz\,dr\,d\theta, \qquad (5)$$

where U is not a region in space, but—as in Section 15.4—a representation of limits on z, r, and θ appropriate to describe the space region T in cylindrical coordinates. Before we integrate, we must replace the variables x and y by $r\cos\theta$ and $r\sin\theta$, respectively, but z is left unchanged. The cylindrical-coordinates volume element

$$dV = r\,dz\,dr\,d\theta$$

may be regarded formally as the product of dz and the polar-coordinates area element $dA = r\,dr\,d\theta$. It is a consequence of the formula $\Delta V = \bar{r}\,\Delta z\,\Delta r\,\Delta\theta$ for the volume of the *cylindrical block* shown in Fig. 15.7.2.

Integration in cylindrical coordinates is particularly useful for computations associated with solids of revolution. So that the limits of integration will be the simplest, the solid should usually be placed so that the axis of revolution is the z-axis.

EXAMPLE 1 Find the centroid of the first-octant portion T of the solid ball bounded by the sphere $r^2 + z^2 = a^2$. The solid T appears in Fig. 15.7.3.

Solution The volume of the first octant of the solid ball is $V = \frac{1}{8}\cdot\frac{4}{3}\pi a^3 = \frac{1}{6}\pi a^3$. Because $\bar{x} = \bar{y} = \bar{z}$ by symmetry, we need calculate only

$$\bar{z} = \frac{1}{V}\iiint\limits_{T} z\,dV = \frac{6}{\pi a^3}\int_0^{\pi/2}\int_0^a\int_0^{\sqrt{a^2-r^2}} zr\,dz\,dr\,d\theta$$

$$= \frac{6}{\pi a^3}\int_0^{\pi/2}\int_0^a \frac{1}{2}r(a^2 - r^2)\,dr\,d\theta$$

$$= \frac{3}{\pi a^3}\int_0^{\pi/2}\left[\frac{1}{2}a^2r^2 - \frac{1}{4}r^4\right]_{r=0}^a d\theta = \frac{3}{\pi a^3}\cdot\frac{\pi}{2}\cdot\frac{a^4}{4} = \frac{3a}{8}.$$

Thus the centroid is located at the point $(3a/8,\ 3a/8,\ 3a/8)$. Observe that the answer is both plausible and dimensionally correct.

EXAMPLE 2 Find the volume and centroid of the solid T that is bounded by the paraboloid $z = b(x^2 + y^2)$ $(b > 0)$ and by the plane $z = h$ $(h > 0)$.

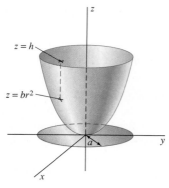

Fig. 15.7.4 The paraboloid of Example 2

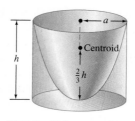

Fig. 15.7.5 Volume and centroid of a right circular paraboloid in terms of the circumscribed cylinder

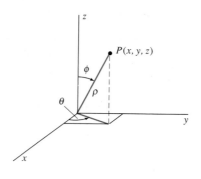

Fig. 15.7.6 The spherical coordinates (ρ, ϕ, θ) of the point P

Solution Figure 15.7.4 makes it clear that we get the radius of the circular top of T by equating $z = b(x^2 + y^2) = br^2$ and $z = h$. This gives $a = \sqrt{h/b}$ for the radius of the circle over which the solid lies. Hence Eq. (4), with $f(x, y, z) \equiv 1$, gives the volume:

$$V = \iiint_T dV = \int_0^{2\pi} \int_0^a \int_{br^2}^h r \, dz \, dr \, d\theta = \int_0^{2\pi} \int_0^a (hr - br^3) \, dr \, d\theta$$

$$= 2\pi \left(\frac{1}{2} ha^2 - \frac{1}{4} ba^4 \right) = \frac{\pi h^2}{2b} = \frac{1}{2} \pi a^2 h$$

(because $a^2 = h/b$).

By symmetry, the centroid of T lies on the z-axis, so all that remains is to compute $\bar{z}$:

$$\bar{z} = \frac{1}{V} \iiint_T z \, dV = \frac{2}{\pi a^2 h} \int_0^{2\pi} \int_0^a \int_{br^2}^h rz \, dz \, dr \, d\theta$$

$$= \frac{2}{\pi a^2 h} \int_0^{2\pi} \int_0^a \left(\frac{1}{2} h^2 r - \frac{1}{2} b^2 r^5 \right) dr \, d\theta$$

$$= \frac{4}{a^2 h} \left(\frac{1}{4} h^2 a^2 - \frac{1}{12} b^2 a^6 \right) = \frac{2}{3} h,$$

again using the fact that $a^2 = h/b$. Therefore the centroid of T is located at the point $(0, 0, 2h/3)$. Again, this answer is both plausible and dimensionally correct.

We can summarize the results of Example 2 as follows: The volume of a right circular paraboloid is *half* that of the circumscribed cylinder (Fig. 15.7.5), and its centroid lies on its axis of symmetry *two-thirds* of the way from the "vertex" at $(0, 0, 0)$ to the top.

SPHERICAL COORDINATE INTEGRALS

When the boundary surfaces of the region T of integration are spheres, cones, or other surfaces with simple descriptions in spherical coordinates, it is generally advantageous to transform a triple integral over T into spherical coordinates. Recall from Section 13.7 that the relationship between spherical coordinates (ρ, ϕ, θ) (shown in Fig. 15.7.6) and rectangular coordinates (x, y, z) is given by

$$x = \rho \sin \phi \cos \theta, \quad y = \rho \sin \phi \sin \theta, \quad z = \rho \cos \phi. \quad (6)$$

Suppose, for example, that T is the **spherical block** determined by the simple inequalities

$$\rho_1 \leqq \rho \leqq \rho_2 = \rho_1 + \Delta\rho,$$

$$\phi_1 \leqq \phi \leqq \phi_2 = \phi_1 + \Delta\phi, \quad (7)$$

$$\theta_1 \leqq \theta \leqq \theta_2 = \theta_1 + \Delta\theta.$$

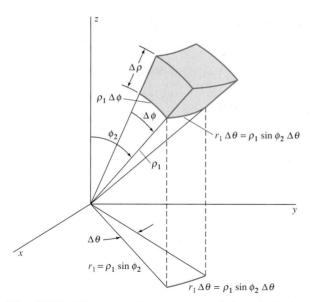

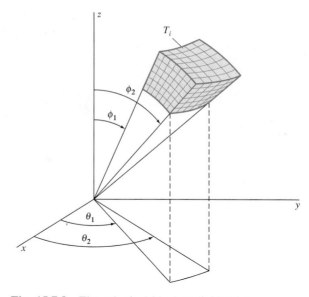

Fig. 15.7.7 The approximate volume of the spherical block is $\rho_1{}^2 \sin \phi_2 \, \Delta\rho \, \Delta\phi \, \Delta\theta$.

Fig. 15.7.8 The spherical block T divided into n smaller spherical blocks

As indicated by the dimensions labeled in Fig. 15.7.7, this spherical block is (if $\Delta\rho$, $\Delta\phi$, and $\Delta\theta$ are small) *approximately* a rectangular block with dimensions $\Delta\rho$, $\rho_1 \, \Delta\phi$, and $\rho_1 \sin \phi_2 \, \Delta\theta$. Thus its volume is approximately $\rho_1{}^2 \sin \phi_2 \, \Delta\rho \, \Delta\phi \, \Delta\theta$. It can be shown (see Problem 19 of Section 15.8) that the *exact* volume of the spherical block described in (7) is

$$\Delta V = \hat{\rho}^2 \sin \hat{\phi} \, \Delta\rho \, \Delta\phi \, \Delta\theta \tag{8}$$

for certain numbers $\hat{\rho}$ and $\hat{\phi}$ such that $\rho_1 < \hat{\rho} < \rho_2$ and $\phi_1 < \hat{\phi} < \phi_2$.

Now suppose that we partition each of the intervals $[\rho_1, \rho_2]$, $[\phi_1, \phi_2]$, and $[\theta_1, \theta_2]$ into n equal subintervals of lengths

$$\Delta\rho = \frac{\rho_2 - \rho_1}{n}, \qquad \Delta\phi = \frac{\phi_2 - \phi_1}{n}, \quad \text{and} \quad \Delta\theta = \frac{\theta_2 - \theta_1}{n},$$

respectively. This produces a **spherical partition** $\mathcal{P}$ of the spherical block T into $k = n^3$ smaller spherical blocks $T_1, T_2, \ldots, T_k$; see Fig. 15.7.8. By Eq. (8), there exists a point $(\hat{\rho}_i, \hat{\phi}_i, \hat{\theta}_i)$ of the spherical block T_i such that its volume $\Delta V_i = \hat{\rho}_i{}^2 \sin \hat{\phi}_i \, \Delta\rho \, \Delta\phi \, \Delta\theta$. The mesh $|\mathcal{P}|$ of $\mathcal{P}$ is the length of the longest diagonal of any of the small spherical blocks $T_1, T_2, \ldots, T_k$.

If (x_i^*, y_i^*, z_i^*) are the rectangular coordinates of the point with spherical coordinates $(\hat{\rho}_i, \hat{\phi}_i, \hat{\theta}_i)$, then the definition of the triple integral as a limit of Riemann sums as the mesh $|\mathcal{P}|$ approaches zero gives

$$\iiint_T f(x, y, z) \, dV = \lim_{|\mathcal{P}| \to 0} \sum_{i=1}^{k} f(x_i^*, y_i^*, z_i^*) \, \Delta V_i$$

$$= \lim_{|\mathcal{P}| \to 0} \sum_{i=1}^{k} F(\hat{\rho}_i, \hat{\phi}_i, \hat{\theta}_i) \, \hat{\rho}_i{}^2 \sin \hat{\phi}_i \, \Delta\rho \, \Delta\phi \, \Delta\theta, \tag{9}$$

where

$$F(\rho, \phi, \theta) = f(\rho \sin \phi \cos \theta, \rho \sin \phi \sin \theta, \rho \cos \phi) \tag{10}$$

is the result of substituting Eq. (6) into $f(x, y, z)$. But the right-hand sum in Eq. (9) is simply a Riemann sum for the triple integral

$$\int_{\theta_1}^{\theta_2} \int_{\phi_1}^{\phi_2} \int_{\rho_1}^{\rho_2} F(\rho, \phi, \theta) \, \rho^2 \sin \phi \, d\rho \, d\phi \, d\theta.$$

It therefore follows that

$$\iiint_T f(x, y, z) \, dV = \int_{\theta_1}^{\theta_2} \int_{\phi_1}^{\phi_2} \int_{\rho_1}^{\rho_2} F(\rho, \phi, \theta) \, \rho^2 \sin \phi \, d\rho \, d\phi \, d\theta. \quad (11)$$

Thus we transform the integral

$$\iiint_T f(x, y, z) \, dV$$

into spherical coordinates by replacing the rectangular-coordinate variables x, y, and z by their expressions in Eq. (6) in terms of the spherical-coordinate variables ρ, ϕ and θ. In addition, we write

$$dV = \rho^2 \sin \phi \, d\rho \, d\phi \, d\theta$$

for the volume element in spherical coordinates.

More generally, we can transform the triple integral

$$\iiint_T f(x, y, z) \, dV$$

into spherical coordinates whenever the region T is **centrally simple**—that is, whenever it has a spherical-coordinates description of the form

$$\rho_1(\phi, \theta) \leq \rho \leq \rho_2(\phi, \theta), \qquad \phi_1 \leq \phi \leq \phi_2, \qquad \theta_1 \leq \theta \leq \theta_2. \quad (12)$$

If so, then

$$\iiint_T f(x, y, z) \, dV = \int_{\theta_1}^{\theta_2} \int_{\phi_1}^{\phi_2} \int_{\rho_1(\phi, \theta)}^{\rho_2(\phi, \theta)} F(\rho, \phi, \theta) \, \rho^2 \sin \phi \, d\rho \, d\phi \, d\theta. \quad (13)$$

The limits on ρ are simply the ρ-coordinates (in terms of ϕ and θ) of the endpoints of a typical radial segment that joins the "inner" and "outer" parts of the boundary of T (Fig. 15.7.9). Thus the general formula for **triple integration in spherical coordinates** is

$$\iiint_T f(x, y, z) \, dV$$

$$= \iiint_U f(\rho \sin \phi \cos \theta, \rho \sin \phi \sin \theta, \rho \cos \phi) \, \rho^2 \sin \phi \, d\rho \, d\phi \, d\theta, \quad (14)$$

where, as before, U does not denote a region in space but rather indicates limits on ρ, ϕ, and θ appropriate to describe the region T in spherical coordinates.

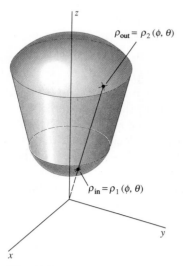

Fig. 15.7.9 A centrally simple region

EXAMPLE 3 A solid ball T with constant density δ is bounded by the spherical surface with equation $\rho = a$. Use spherical coordinates to compute its volume V and its moment of inertia I_z about the z-axis.

Solution   The points of the ball T are described by the inequalities

$$0 \leqq \rho \leqq a, \qquad 0 \leqq \phi \leqq \pi, \qquad 0 \leqq \theta \leqq 2\pi.$$

We take $f = F \equiv 1$ in Eq. (11) and thereby obtain

$$V = \iiint\limits_{T} dV = \int_0^{2\pi} \int_0^{\pi} \int_0^{a} \rho^2 \sin\phi \, d\rho \, d\phi \, d\theta$$

$$= \tfrac{1}{3} a^3 \int_0^{2\pi} \int_0^{\pi} \sin\phi \, d\phi \, d\theta$$

$$= \tfrac{1}{3} a^3 \int_0^{2\pi} \Big[-\cos\phi \Big]_{\phi=0}^{\pi} d\theta = \tfrac{2}{3} a^3 \int_0^{2\pi} d\theta = \tfrac{4}{3} \pi a^3.$$

The distance from the typical point (ρ, ϕ, θ) of the sphere to the z-axis is $r = \rho \sin\phi$, so the moment of inertia of the sphere about that axis is

$$I_z = \iiint\limits_{T} r^2 \delta \, dV = \int_0^{2\pi} \int_0^{\pi} \int_0^{a} \delta \rho^4 \sin^3\phi \, d\rho \, d\phi \, d\theta$$

$$= \tfrac{1}{5} \delta a^5 \int_0^{2\pi} \int_0^{\pi} \sin^3\phi \, d\phi \, d\theta$$

$$= \tfrac{2}{5} \pi \delta a^5 \int_0^{\pi} \sin^3\phi \, d\phi = \tfrac{2}{5} \pi \delta a^5 \cdot 2 \cdot \tfrac{2}{3} = \tfrac{2}{5} m a^2,$$

where $m = \tfrac{4}{3} \pi a^3 \delta$ is the mass of the ball. The answer is dimensionally correct because it is the product of mass and the square of a distance. The answer is plausible because it implies that, for purposes of rotational inertia, the sphere acts as if its mass were concentrated about 63% of the way from the axis to the equator.

EXAMPLE 4   Find the volume and centroid of the "ice-cream cone" C that is bounded by the cone $\phi = \pi/6$ and by the sphere $\rho = 2a \cos\phi$ of radius a. The sphere and the part of the cone within it are shown in Fig. 15.7.10.

Solution   The ice-cream cone is described by the inequalities

$$0 \leqq \theta \leqq 2\pi, \qquad 0 \leqq \phi \leqq \frac{\pi}{6}, \qquad 0 \leqq \rho \leqq 2a \cos\phi.$$

Using Eq. (13) to compute its volume, we get

$$V = \int_0^{2\pi} \int_0^{\pi/6} \int_0^{2a\cos\phi} \rho^2 \sin\phi \, d\rho \, d\phi \, d\theta$$

$$= \tfrac{8}{3} a^3 \int_0^{2\pi} \int_0^{\pi/6} \cos^3\phi \sin\phi \, d\phi \, d\theta$$

$$= \tfrac{16}{3} \pi a^3 \Big[-\tfrac{1}{4} \cos^4\phi \Big]_0^{\pi/6} = \tfrac{7}{12} \pi a^3.$$

Now for the centroid. It is clear by symmetry that $\bar{x} = \bar{y} = 0$. Because $z = \rho \cos\phi$, the z-coordinate of the centroid of C is

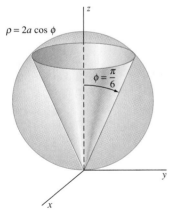

$\rho = 2a \cos\phi$

$\phi = \dfrac{\pi}{6}$

Fig. 15.7.10   The ice-cream cone of Example 4 is the part of the cone that lies within the sphere.

$$\bar{z} = \frac{1}{V} \iiint\limits_{C} z \, dV = \frac{12}{7\pi a^3} \int_0^{2\pi} \int_0^{\pi/6} \int_0^{2a\cos\phi} \rho^3 \cos\phi \sin\phi \, d\rho \, d\phi \, d\theta$$

$$= \frac{48a}{7\pi} \int_0^{2\pi} \int_0^{\pi/6} \cos^5\phi \sin\phi \, d\phi \, d\theta = \frac{96a}{7} \left[-\frac{1}{6}\cos^6\phi \right]_0^{\pi/6} = \frac{37a}{28}.$$

Hence the centroid of the ice-cream cone is located at the point $(0, 0, 37a/28)$.

15.7 Problems

Solve Problems 1 through 20 by triple integration in cylindrical coordinates. Assume throughout that each solid has unit density unless another density function is specified.

1. Find the volume of the solid bounded above by the plane $z = 4$ and below by the paraboloid $z = r^2$.

2. Find the centroid of the solid of Problem 1.

3. Derive the formula for the volume of a sphere of radius a.

4. Find the moment of inertia about the z-axis of the solid sphere of Problem 3 given that the z-axis passes through its center.

5. Find the volume of the region that lies inside both the sphere $x^2 + y^2 + z^2 = 4$ and the cylinder $x^2 + y^2 = 1$.

6. Find the centroid of the half of the region of Problem 5 that lies on or above the xy-plane.

7. Find the mass of the cylinder $0 \le r \le a, 0 \le z \le h$ if its density at (x, y, z) is z.

8. Find the centroid of the cylinder of Problem 7.

9. Find the moment of inertia about the z-axis of the cylinder of Problem 7.

10. Find the volume of the region that lies inside both the sphere $x^2 + y^2 + z^2 = 4$ and the cylinder $x^2 + y^2 - 2x = 0$ (Fig. 15.7.11).

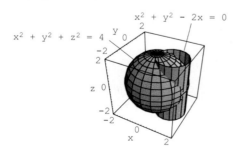

Fig. 15.7.11 The sphere and cylinder of Problem 10

11. Find the volume and centroid of the region bounded by the plane $z = 0$ and by the paraboloid $z = 9 - x^2 - y^2$.

12. Find the volume and centroid of the region bounded by the paraboloids $z = x^2 + y^2$ and $z = 12 - 2x^2 - 2y^2$.

13. Find the volume of the region bounded by the paraboloids $z = 2x^2 + y^2$ and $z = 12 - x^2 - 2y^2$.

14. Find the volume of the region bounded below by the paraboloid $z = x^2 + y^2$ and above by the plane $z = 2x$ (Fig. 15.7.12).

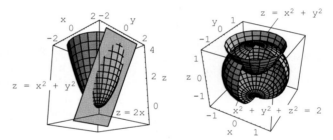

Fig. 15.7.12 The plane and paraboloid of Problem 14

Fig. 15.7.13 The sphere and paraboloid of Problem 15

15. Find the volume of the region bounded above by the spherical surface $x^2 + y^2 + z^2 = 2$ and below by the paraboloid $z = x^2 + y^2$ (Fig. 15.7.13).

16. A homogeneous solid cylinder has mass m and radius a. Show that its moment of inertia about its axis of symmetry is $\frac{1}{2}ma^2$.

17. Find the moment of inertia I of a homogeneous solid cylinder about a diameter of its base. Express I in terms of the radius a, the height h, and the (constant) density δ of the cylinder.

18. Find the centroid of a homogeneous solid cylinder of radius a.

19. Find the volume of the region bounded by the plane $z = 1$ and by the cone $z = r$.

20. Show that the centroid of a homogeneous solid right circular cone lies on its axis three-quarters of the way from the vertex to its base.

Solve Problems 21 through 30 by triple integration in spherical coordinates.

21. Find the centroid of a homogeneous solid hemisphere of radius a.

22. Find the mass and centroid of a solid hemisphere of radius a if its density δ is proportional to distance z from its base—so $\delta = kz$ (where k is a positive constant).

23. Solve Problem 19 by triple integration in spherical coordinates.

24. Solve Problem 20 by triple integration in spherical coordinates.

25. Find the volume and centroid of the solid that lies inside the sphere $\rho = a$ and above the cone $r = z$.

26. Find the moment of inertia I_z of the solid of Problem 25 under the assumption that it has constant density δ.

27. Find the moment of inertia about a tangent line of a solid homogeneous sphere of radius a and total mass m.

28. A spherical shell of mass m is bounded by the spheres $\rho = a$ and $\rho = 2a$, and its density function is $\delta = \rho^2$. Find its moment of inertia about a diameter.

29. Describe the surface $\rho = 2a \sin \phi$, and compute the volume of the region it bounds.

30. Describe the surface $\rho = 1 + \cos \phi$, and compute the volume of the region it bounds. Figure 15.7.14 may be useful.

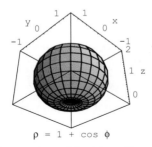

$\rho = 1 + \cos \phi$

Fig. 15.7.14 The surface of Problem 30

31. Find the moment of inertia about the x-axis of the region that lies inside both the cylinder $r = a$ and the sphere $\rho = 2a$.

32. Find the moment of inertia about the z-axis of the ice-cream cone of Example 4.

33. Find the mass and centroid of the ice-cream cone of Example 4 if its density is given by $\delta = z$.

34. Consider a homogeneous spherical ball of radius a centered at the origin, with density δ and mass $M = \frac{4}{3}\pi a^3 \delta$. Show that the gravitational force $\mathbf{F}$ exerted by this ball upon a point mass m located at the point $(0, 0, c)$, where $c > a$ (Fig. 15.7.15), is the same as though all the mass of the ball were concentrated at its center $(0, 0, 0)$. That is, show that $|\mathbf{F}| = GMm/c^2$. [*Suggestion:* By symmetry you may assume that the force is vertical, so that $\mathbf{F} = F_z\mathbf{k}$. Set up the integral

$$F_z = -\int_0^{2\pi} \int_0^a \int_0^\pi \frac{Gm\delta \cos \alpha}{w^2} \rho^2 \sin \phi \, d\phi \, d\rho \, d\theta.$$

Change the first variable of integration from ϕ to w by using the law of cosines:

$$w^2 = \rho^2 + c^2 - 2\rho c \cos \phi.$$

Then $2w \, dw = 2\rho c \sin \phi \, d\phi$ and $w \cos \alpha + \rho \cos \phi = c$. (Why?)]

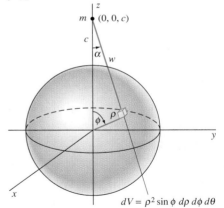

$dV = \rho^2 \sin \phi \, d\rho \, d\phi \, d\theta$

Fig. 15.7.15 The system of Problem 34

35. Consider now the spherical shell $a \leqq r \leqq b$ with uniform density δ. Show that this shell exerts *no* net force on a point mass m located at the point $(0, 0, c)$ *inside* it—that is, $|c| < a$. The computation will be the same as in Problem 34 except for the limits of integration on ρ and w.

15.7 Project

If the earth were a perfect sphere with radius $R = 6370$ km, *uniform* density δ, and mass $M = \frac{4}{3}\delta \pi R^3$, then (according to Example 3) its moment of inertia about its polar axis would be $I = \frac{2}{5}MR^2$. In actuality, however, it turns out that

$$I = kMR^2, \tag{15}$$

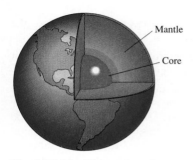

Fig. 15.7.16 The core and mantle of the earth

where $k < 0.4 = \frac{2}{5}$. The reason is that, instead of having a uniform interior, the earth has a dense core covered with a lighter mantle a few thousand kilometers thick (Fig. 15.7.16). The density of the core is

$$\delta_1 \approx 11 \times 10^3 \quad (\text{kg/m}^3)$$

and that of the mantle is

$$\delta_2 \approx 5 \times 10^3 \quad (\text{kg/m}^3).$$

The numerical value of k in Eq. (15) can be determined from certain earth satellite observations. If the earth's polar moment of inertia I and mass M (for the core-mantle model) are expressed in terms of the unknown radius x of the spherical core, then substitution of these expressions into Eq. (15) yields an equation that can be solved for x.

Show that this equation can be written in the form

$$2(\delta_1 - \delta_2)x^5 - 5k(\delta_1 - \delta_2)R^2x^3 + (2 - 5k)\delta_2 R^5 = 0. \quad (16)$$

Given the measured numerical value $k = 0.371$, solve this equation (graphically or numerically) to find x and from this solution determine the thickness of the earth's mantle.

15.8
Surface Area

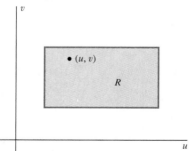

Fig. 15.8.1 The uv-region R on which the transformation $\mathbf{r}$ is defined

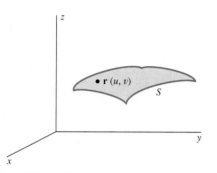

Fig. 15.8.2 The parametric surface S in xyz-space

Until now our concept of a surface has been the graph $z = f(x, y)$ of a function of two variables. Occasionally we have seen such a surface defined implicitly by an equation of the form $F(x, y, z) = 0$. Now we want to introduce the more precise concept of a *parametric surface*—the two-dimensional analogue of a parametric curve.

A **parametric surface** S is the *image* of a function or transformation $\mathbf{r}$ that is defined on a region R in the uv-plane (Fig. 15.8.1) and has values in xyz-space (Fig. 15.8.2). The **image** under $\mathbf{r}$ of each point (u, v) in R is the point in xyz-space with position vector

$$\mathbf{r}(u, v) = \langle x(u, v), y(u, v), z(u, v) \rangle. \quad (1)$$

We shall assume throughout this section that the component functions of $\mathbf{r}$ have continuous partial derivatives with respect to u and v and also that the vectors

$$\mathbf{r}_u = \frac{\partial \mathbf{r}}{\partial u} = \langle x_u, y_u, z_u \rangle = \frac{\partial x}{\partial u}\mathbf{i} + \frac{\partial y}{\partial u}\mathbf{j} + \frac{\partial z}{\partial u}\mathbf{k} \quad (2)$$

and

$$\mathbf{r}_v = \frac{\partial \mathbf{r}}{\partial v} = \langle x_v, y_v, z_v \rangle = \frac{\partial x}{\partial v}\mathbf{i} + \frac{\partial y}{\partial v}\mathbf{j} + \frac{\partial z}{\partial v}\mathbf{k} \quad (3)$$

are nonzero and nonparallel at each interior point of R. (Compare this with the definition of *smooth* parametric curve $\mathbf{r}(t)$ in Section 12.1.) We call the variables u and v the *parameters* for the surface S (in analogy with the single parameter t for a parametric curve).

EXAMPLE 1 (a) We may regard the graph $z = f(x, y)$ of a function as a parametric surface with parameters x and y. In this case the transformation $\mathbf{r}$ from the xy-plane to xyz-space has the component functions

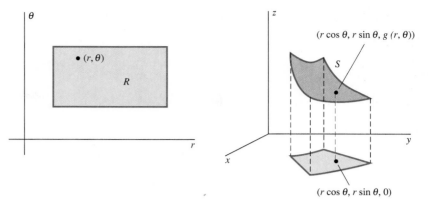

Fig. 15.8.3 A rectangle in the $r\theta$-plane; the domain of the function $z = g(r, \theta)$ (Example 1)

Fig. 15.8.4 A cylindrical-coordinates surface in xyz-space (Example 1)

$$x = x, \qquad y = y, \qquad z = f(x, y). \tag{4}$$

(b) Similarly, we may regard a surface given in cylindrical coordinates by the graph of $z = g(r, \theta)$ as a parametric surface with parameters r and θ. The transformation $\mathbf{r}$ from the $r\theta$-plane (Fig. 15.8.3) to xyz-space (Fig. 15.8.4) is then given by

$$x = r \cos \theta, \qquad y = r \sin \theta, \qquad z = g(r, \theta). \tag{5}$$

(c) We may regard a surface given in spherical coordinates by $\rho = h(\phi, \theta)$ as a parametric surface with parameters ϕ and θ, and the corresponding transformation from the $\phi\theta$-plane to xyz-space is then given by

$$x = h(\phi, \theta) \sin \phi \cos \theta, \quad y = h(\phi, \theta) \sin \phi \sin \theta, \quad z = h(\phi, \theta) \cos \phi. \tag{6}$$

The concept of a parametric surface lets us treat all these special cases, and many others, with the same techniques.

Now we want to define the *surface area* of the general parametric surface given in Eq. (1). We begin with an inner partition of the region R—the domain of $\mathbf{r}$ in the uv-plane—into rectangles $R_1, R_2, \ldots, R_n$, each with dimensions Δu and Δv. Let (u_i, v_i) be the lower left-hand corner of R_i (as in Fig. 15.8.5). The image S_i of R_i under $\mathbf{r}$ will not generally be a rectangle in xyz-space; it will look more like a *curvilinear figure* on the image surface

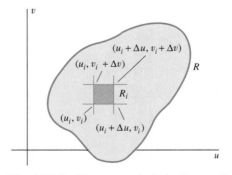

Fig. 15.8.5 The rectangle R_i in the uv-plane

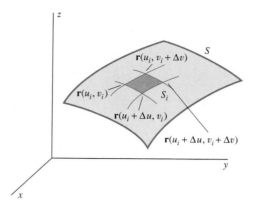

Fig. 15.8.6 The image of R_i is a curvilinear figure.

S, with $\mathbf{r}(u_i, v_i)$ as one "vertex" (Fig. 15.8.6). Let ΔS_i denote the area of this curvilinear figure S_i.

The parametric curves $\mathbf{r}(u, v_i)$ and $\mathbf{r}(u_i, v)$—with parameters u and v, respectively—lie on the surface S and meet at the point $\mathbf{r}(u_i, v_i)$. At this point of intersection, these two curves have the tangent vectors $\mathbf{r}_u(u_i, v_i)$ and $\mathbf{r}_v(u_i, v_i)$ shown in Fig. 15.8.7. Hence their vector product

$$\mathbf{N}(u_i, v_i) = \mathbf{r}_u(u_i, v_i) \times \mathbf{r}_v(u_i, v_i) \tag{7}$$

is a normal vector to S at the point $\mathbf{r}(u_i, v_i)$.

Now suppose that both Δu and Δv are small. Then the area ΔS_i of the curvilinear figure S_i will be approximately equal to the area ΔP_i of the parallelogram with adjacent sides $\mathbf{r}_u(u_i, v_i)\,\Delta u$ and $\mathbf{r}_v(u_i, v_i)\,\Delta v$ (Fig. 15.8.8). But the area of this parallelogram is

$$\Delta P_i = |\,\mathbf{r}_u(u_i, v_i)\,\Delta u \times \mathbf{r}_v(u_i, v_i)\,\Delta v\,| = |\,\mathbf{N}(u_i, v_i)\,|\,\Delta u\,\Delta v.$$

This means that the area $a(S)$ of the surface S is given approximately by

$$a(S) = \sum_{i=1}^{n} \Delta S_i \approx \sum_{i=1}^{n} \Delta P_i,$$

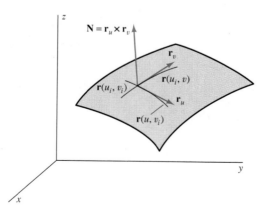

Fig. 15.8.7 The normal vector $\mathbf{N}$ to the surface at $\mathbf{r}(u_i, v_i)$

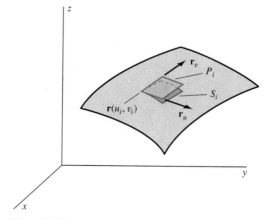

Fig. 15.8.8 The area of the parallelogram P_i is an approximation to the area of the curvilinear figure S_i.

so

$$a(S) \approx \sum_{i=1}^{n} |\mathbf{N}(u_i, v_i)| \, \Delta u \, \Delta v.$$

But this last sum is a Riemann sum for the double integral

$$\iint_R |\mathbf{N}(u, v)| \, du \, dv.$$

We are therefore motivated to *define* the **surface area** A of the parametric surface S by

$$A = a(S) = \iint_R |\mathbf{N}(u, v)| \, du \, dv = \iint_R \left| \frac{\partial \mathbf{r}}{\partial u} \times \frac{\partial \mathbf{r}}{\partial v} \right| \, du \, dv. \qquad (8)$$

SURFACE AREA IN RECTANGULAR COORDINATES

In the case of the surface $z = f(x, y)$, for (x, y) in the region R in the xy-plane, the component functions of $\mathbf{r}$ are given by the equations in (4) with parameters x and y (in place of u and v). Then

$$\mathbf{N} = \frac{\partial \mathbf{r}}{\partial x} \times \frac{\partial \mathbf{r}}{\partial y} = \begin{vmatrix} \mathbf{i} & \mathbf{j} & \mathbf{k} \\ 1 & 0 & \dfrac{\partial f}{\partial x} \\ 0 & 1 & \dfrac{\partial f}{\partial y} \end{vmatrix} = -\frac{\partial f}{\partial x}\mathbf{i} - \frac{\partial f}{\partial y}\mathbf{j} + \mathbf{k},$$

so Eq. (8) takes the special form

$$A = a(S) = \iint_R \sqrt{1 + \left(\frac{\partial f}{\partial x}\right)^2 + \left(\frac{\partial f}{\partial x}\right)^2} \, dx \, dy$$

$$= \iint_R \sqrt{1 + z_x^2 + z_y^2} \, dx \, dy. \qquad (9)$$

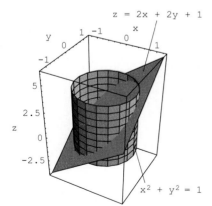

Fig. 15.8.9 The cylinder and plane of Example 2

EXAMPLE 2 Find the area of the ellipse cut from the plane $z = 2x + 2y + 1$ by the cylinder $x^2 + y^2 = 1$ (Fig. 15.8.9).

Solution Here, R is the unit circle in the xy-plane with area

$$\iint_R 1 \, dx \, dy = \pi,$$

so Eq. (9) gives the area of the ellipse to be

$$A = \iint_R \sqrt{1 + z_x^2 + z_y^2} \, dx \, dy$$

$$= \iint_R \sqrt{1 + 2^2 + 2^2} \, dx \, dy = \iint_R 3 \, dx \, dy = 3\pi.$$

REMARK   Computer-generated figures such as Fig. 15.8.9 could not be constucted without using parametric surfaces, For example, the vertical cylinder in Fig. 15.8.9 was generated by instructing the computer to plot the parametric surface defined on the $z\theta$-rectangle

$$-5 \leqq z \leqq 5, \qquad 0 \leqq \theta \leqq 2\pi$$

by

$$\mathbf{r}(z, \theta) = \langle \cos \theta, \sin \theta, z \rangle.$$

Is it clear to you that the image of this transformation is the cylinder $x^2 + y^2 = 1$, $-5 \leqq z \leqq 5$?

SURFACE AREA IN CYLINDRICAL COORDINATES

Now consider a cylindrical-coordinates surface $z = g(r, \theta)$ parametrized by the equations in (5) for (r, θ) in a region R of the $r\theta$-plane. Then the normal vector is

$$\mathbf{N} = \frac{\partial \mathbf{r}}{\partial r} \times \frac{\partial \mathbf{r}}{\partial \theta} = \begin{vmatrix} \mathbf{i} & \mathbf{j} & \mathbf{k} \\ \cos \theta & \sin \theta & \dfrac{\partial z}{\partial r} \\ -r \sin \theta & r \cos \theta & \dfrac{\partial z}{\partial \theta} \end{vmatrix}$$

$$= \mathbf{i}\left(\frac{\partial z}{\partial \theta} \sin \theta - r \frac{\partial z}{\partial r} \cos \theta\right) - \mathbf{j}\left(\frac{\partial z}{\partial \theta} \cos \theta + r \frac{\partial z}{\partial r} \sin \theta\right) + r\mathbf{k}.$$

After some simplifications, we find that

$$|\mathbf{N}| = \sqrt{r^2 + r^2\left(\frac{\partial z}{\partial r}\right)^2 + \left(\frac{\partial z}{\partial \theta}\right)^2}.$$

Then Eq. (8) yields the formula

$$A = \iint_R \sqrt{r^2 + (rz_r)^2 + (z_\theta)^2} \; dr \, d\theta \qquad (10)$$

for surface area in cylindrical coordinates.

EXAMPLE 3   Find the surface area cut from the paraboloid $z = r^2$ by the cylinder $r = 1$ (Fig. 15.8.10).

Solution   Equation (10) gives area

$$A = \int_0^{2\pi} \int_0^1 \sqrt{r^2 + r^2(2r)^2} \; dr \, d\theta = 2\pi \int_0^1 r\sqrt{1 + 4r^2} \; dr$$

$$= 2\pi\left[\frac{2}{3} \cdot \frac{1}{8}(1 + 4r^2)^{3/2}\right]_0^1 = \frac{\pi}{6}(5\sqrt{5} - 1) \approx 5.3304.$$

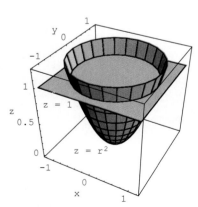

Fig. 15.8.10   The part of the paraboloid $z = r^2$ inside the cylinder $r = 1$ (Example 3) is the same as the part beneath the plane $z = 1$. (Why?)

In Example 3, you would get the same result if you first wrote $z = x^2 + y^2$, used Eq. (9), which gives

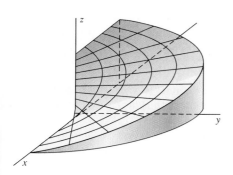

Fig. 15.8.11 The spiral ramp of Example 4

Fig. 15.8.12 The torus of Example 5

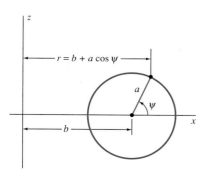

Fig. 15.8.13 The circle that generates the torus (Example 5)

$$A = \iint_R \sqrt{1 + 4x^2 + 4y^2}\, dx\, dy,$$

and then changed to polar coordinates. In Example 4 it would be less convenient to begin with rectangular coordinates.

EXAMPLE 4 Find the area of the *spiral ramp* $z = \theta$, $0 \leq r \leq 1$, $0 \leq \theta \leq \pi$. This is the upper surface of the solid shown in Fig. 15.8.11.

Solution Equation (10) gives area

$$A = \int_0^\pi \int_0^1 \sqrt{r^2 + 1}\, dr\, d\theta = \frac{\pi}{2}[\sqrt{2} + \ln(1 + \sqrt{2})] \approx 3.6059.$$

We avoided a trigonometric substitution by using the table of integrals inside the back cover.

EXAMPLE 5 Find the surface area of the torus generated by revolving the circle $(x - b)^2 + z^2 = a^2$ $(0 < a < b)$ in the xz-plane around the z-axis (Fig. 15.8.12).

Solution With the ordinary polar coordinate θ and the angle ψ of Fig. 15.8.13, the torus is described for $0 \leq \theta \leq 2\pi$ and $0 \leq \psi \leq 2\pi$ by the parametric equations

$$x = r \cos \theta = (b + a \cos \psi) \cos \theta,$$
$$y = r \sin \theta = (b + a \cos \psi) \sin \theta,$$
$$z = a \sin \psi.$$

When we compute $\mathbf{N} = \mathbf{r}_\theta \times \mathbf{r}_\psi$ and simplify, we find that

$$|\mathbf{N}| = a(b + a \cos \psi).$$

Hence the general surface-area formula, Eq. (8), gives area

$$A = \int_0^{2\pi} \int_0^{2\pi} a(b + a \cos \psi)\, d\theta\, d\psi = 2\pi a \left[b\psi + a \sin \psi \right]_0^{2\pi} = 4\pi^2 ab.$$

We obtained the same result in Section 15.5 with the aid of Pappus's theorem.

15.8 Problems

1. Find the area of the portion of the plane $z = x + 3y$ that lies inside the elliptical cylinder with equation $x^2/4 + y^2/9 = 1$.

2. Find the area of the region in the plane $z = 1 + 2x + 2y$ that lies directly above the region in the xy-plane bounded by the parabolas $y = x^2$ and $x = y^2$.

3. Find the area of the part of the paraboloid $z = 9 - x^2 - y^2$ that lies above the plane $z = 5$.

4. Find the area of the part of the surface $2z = x^2$ that lies directly above the triangle in the xy-plane with vertices at $(0, 0)$, $(1, 0)$, and $(1, 1)$.

5. Find the area of the surface that is the graph of $z = x + y^2$ for $0 \leq x \leq 1$, $0 \leq y \leq 2$.

6. Find the area of that part of the surface of Problem 5 that lies above the triangle in the xy-plane with vertices at $(0, 0)$, $(0, 1)$, and $(1, 1)$.

7. Find by integration the area of the part of the plane $2x + 3y + z = 6$ that lies in the first octant.

8. Find the area of the ellipse that is cut from the plane $2x + 3y + z = 6$ by the cylinder $x^2 + y^2 = 2$.

9. Find the area that is cut from the saddle-shaped surface $z = xy$ by the cylinder $x^2 + y^2 = 1$.

10. Find the area that is cut from the surface $z = x^2 - y^2$ by the cylinder $x^2 + y^2 = 4$.

11. Find the surface area of the part of the paraboloid $z = 16 - x^2 - y^2$ that lies above the xy-plane.

12. Show by integration that the surface area of the conical surface $z = br$ between the planes $z = 0$ and $z = h = ab$ is given by $A = \pi aL$, where L is the slant height $\sqrt{a^2 + h^2}$ and a is the radius of the base of the cone.

13. Let the part of the cylinder $x^2 + y^2 = a^2$ between the planes $z = 0$ and $z = h$ be parametrized by $x = a \cos \theta$, $y = a \sin \theta$, $z = z$. Apply Eq. (8) to show that the area of this zone is $A = 2\pi ah$.

14. Consider the meridional zone of height $h = c - b$ that lies on the sphere $r^2 + z^2 = a^2$ between the planes $z = b$ and $z = c$, where $0 \leq b < c \leq a$. Apply Eq. (10) to show that the area of this zone is $A = 2\pi ah$.

15. Find the area of the part of the cylinder $x^2 + z^2 = a^2$ that lies within the cylinder $r^2 = x^2 + y^2 = a^2$.

16. Find the area of the part of the sphere $r^2 + z^2 = a^2$ that lies within the cylinder $r = a \sin \theta$.

17. (a) Apply Eq. (8) to show that the surface area of the surface $y = f(x, z)$, for (x, z) in the region R of the xz-plane, is given by

$$A = \iint_R \sqrt{1 + \left(\frac{\partial f}{\partial x}\right)^2 + \left(\frac{\partial f}{\partial z}\right)^2}\, dx\, dz.$$

(b) State and derive a similar formula for the area of the surface $x = f(y, z)$, (y, z) in R.

18. Suppose that R is a region in the $\phi\theta$-plane. Consider the part of the sphere $\rho = a$ that corresponds to (ϕ, θ) in R, parametrized by the equations in (6) with $h(\phi, \theta) = a$. Apply Eq. (8) to show that the surface area of this part of the sphere is

$$A = \iint_R a^2 \sin \phi\, d\phi\, d\theta.$$

19. (a) Consider the "spherical rectangle" defined by $\rho = a$, $\phi_1 \leq \phi \leq \phi_2 = \phi_1 + \Delta\phi$, $\theta_1 \leq \theta \leq \theta_2 = \theta_1 + \Delta\theta$. Apply the formula of Problem 18 and the average value property (see Problem 30 in Section 15.2) to show that the area of this spherical rectangle is $A = a^2 \sin \hat{\phi}\, \Delta\phi\, \Delta\theta$ for some $\hat{\phi}$ in (ϕ_1, ϕ_2). (b) Conclude

from the result of part (a) that the volume of the spherical block defined by $\rho_1 \leq \rho \leq \rho_2 = \rho_1 + \Delta\rho$ and $\phi_1 \leq \phi \leq \phi_2$, $\theta_1 \leq \theta \leq \theta_2$ is

$$\Delta V = \tfrac{1}{3}(\rho_2{}^3 - \rho_1{}^3) \sin \hat{\phi}\, \Delta\phi\, \Delta\theta.$$

Finally, derive Eq. (8) of Section 15.7 by applying the mean value theorem to the function $f(\rho) = \rho^3$ on the interval $[\rho_1, \rho_2]$.

20. Describe the surface $\rho = 2a \sin \phi$. Why is it called a *pinched torus*? It is parametrized as in Eq. (6) with $h(\phi, \theta) = 2a \sin \phi$. Show that its surface area is $A = 4\pi^2 a^2$. Figure 15.8.14 may be helpful.

Fig. 15.8.14   Cutaway view of the pinched torus of Problem 20

21. The surface of revolution obtained when we revolve the curve $x = f(z)$, $a \leq z \leq b$, around the z-axis is parametrized in terms of θ $(0 \leq \theta \leq 2\pi)$ and z $(a \leq z \leq b)$ by $x = f(z) \cos \theta$, $y = f(z) \sin \theta$, $z = z$. From Eq. (8) derive the surface-area formula

$$A = \int_0^{2\pi} \int_a^b f(z)\sqrt{1 + [f'(z)]^2}\, dz\, d\theta.$$

This formula agrees with the area of a surface of revolution as defined in Section 6.4.

22. Apply the formula of Problem 18 in both parts of this problem. (a) Verify the formula $A = 4\pi a^2$ for the surface area of a sphere of radius a. (b) Find the area of that part of a sphere of radius a and center $(0, 0, 0)$ that lies inside the cone $\phi = \pi/6$.

23. Apply the result of Problem 21 to verify the formula $A = 2\pi rh$ for the lateral surface area of a right circular cylinder of radius r and height h.

24. Apply Eq. (9) to verify the formula $A = 2\pi rh$ for the lateral surface area of the cylinder $x^2 + z^2 = r^2$, $0 \leq y \leq h$ of radius r and height h.

15.8 Projects

Parametric surfaces are used in most serious computer graphics work. Common computer systems provide parametric plotting instructions similar to the *Maple* command

```
plot3d( [f(u,v), g(u,v), h(u,v)], u = a..b, v = c..d )
```

or the *Mathematica* command

```
ParametricPlot3D[ {f[u,v], g[u,v], h[u,v]},
                    {u, a, b}, {v, c, d} ]
```

to plot the parametric surface defined by

$$x = f(u, v), \qquad y = g(u, v), \qquad z = h(u, v)$$

for $a \leq u \leq b$, $c \leq v \leq d$. The following projects list some possibilities for you to try if such a system is available to you.

PROJECT A   Use the cylindrical-coordinate parametrization

$$x = r \cos \theta, \qquad y = r \sin \theta, \qquad z = f(r, \theta)$$

of the surface $z = f(r, \theta)$ to plot some cones $(z = kr)$ and paraboloids $(z = kr^2)$. Then try something more exotic, perhaps $z = (\sin r)/r$.

PROJECT B   Show that the elliptic cylinder $(x/a)^2 + (y/b)^2 = 1$ centered along the z-axis is parametrized by

$$x = a \cos \theta, \qquad y = b \sin \theta, \qquad z = z$$

for $0 \leq \theta \leq 2\pi$. Plot cylinders with varying semiaxes a and b.

PROJECT C   Show that the parametrization

$$x = ar \cos \theta, \qquad y = br \sin \theta, \qquad z = r^2$$

produces the elliptic paraboloid $z = (x/a)^2 + (y/b)^2$. Plot a few such paraboloids with different values of a and b.

PROJECT D   Show that the parametrization

$$x = a \sin \phi \cos \theta, \qquad y = b \sin \phi \sin \theta, \qquad z = c \cos \phi$$

produces an ellipsoid. Plot some with a few different semiaxes a, b, and c.

PROJECT E   Show that

$$x = x, \qquad y = f(x) \cos \theta, \qquad z = f(x) \sin \theta$$

for $a \leq x \leq b$, $0 \leq \theta \leq 2\pi$, parametrizes the surface obtained by revolving the curve $y = f(x)$, $a \leq x \leq b$ around the x-axis in xyz-space. Plot a few such surfaces generated by a variety of curves $y = f(x)$.

PROJECT F   Explain how to alter the parametrization of Example 5 in this section to graph a torus with an elliptical (rather than a circular) cross section.

---

*15.9

Change of Variables in Multiple Integrals

We have seen in preceding sections that we can evaluate certain multiple integrals by transforming them from rectangular coordinates into polar or spherical coordinates. The technique of changing coordinate systems to evaluate a multiple integral is the multivariable analogue of substitution in a single integral. Recall from Section 5.7 that if $x = g(u)$, then

$$\int_a^b f(x) \, dx = \int_c^d f(g(u))g'(u) \, du, \tag{1}$$

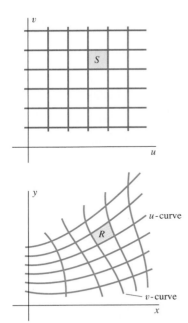

Fig. 15.9.1 The transformation T turns the rectangle S into the curvilinear figure R.

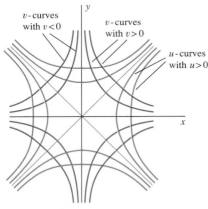

Fig. 15.9.2 The u-curves and v-curves of Example 1

where $a = g(c)$ and $b = g(d)$. The method of substitution involves a "change of variables" that is tailored to the evaluation of a given integral.

Suppose that we want to evaluate the double integral

$$\iint_R F(x, y) \, dx \, dy.$$

A change of variables for this integral is determined by a **transformation** T from the uv-plane to the xy-plane—that is, a function T that associates with the point (u, v) a point $(x, y) = T(u, v)$ given by equations of the form

$$x = f(u, v), \qquad y = g(u, v). \tag{2}$$

The point (x, y) is called the **image** of the point (u, v) under the transformation T. If no two different points in the uv-plane have the same image point in the xy-plane, then the transformation T is said to be **one-to-one**. In this case it may be possible to solve the equations in (2) for u and v in terms of x and y and thus obtain the equations

$$u = h(x, y), \qquad v = k(x, y) \tag{3}$$

of the **inverse transformation** T^{-1} from the xy-plane to the uv-plane.

It is often convenient to visualize the transformation T geometrically in terms of its u-curves and v-curves. The **u-curves** of T are the images of horizontal lines in the uv-plane and the **v-curves** of T are the images of vertical lines in the uv-plane. Note that the image under T of a rectangle bounded by horizontal and vertical lines in the uv-plane is a *curvilinear figure* bounded by u-curves and v-curves in the xy-plane (Fig. 15.9.1). If we know the equations in (3) of the inverse transformation, then we can find the u-curves and the v-curves quite simply by writing the equations

$$k(x, y) = C_1 \quad \text{and} \quad h(x, y) = C_2,$$

respectively, where C_1 and C_2 are constants.

EXAMPLE 1 Determine the u-curves and the v-curves of the transformation T whose inverse T^{-1} is specified by the equations $u = xy, v = x^2 - y^2$.

Solution The u-curves are the rectangular hyperbolas

$$xy = u = C_1 \qquad \text{(constant)},$$

whereas the v-curves are the hyperbolas

$$x^2 - y^2 = v = C_2 \qquad \text{(constant)}.$$

These two familiar families of hyperbolas are shown in Fig. 15.9.2.

Now we shall describe the change of variables in a double integral that corresponds to the transformation T specified by the equations in (2). Let the region R in the xy-plane be the image under T of the region S in the uv-plane. Let $F(x, y)$ be continuous on R and let $\{S_1, S_2, \ldots, S_n\}$ be an inner partition of S into rectangles each with dimensions Δu by Δv. Each rectangle S_i is transformed by T into a curvilinear figure R_i in the xy-plane (Fig. 15.9.3). The images $\{R_1, R_2, \ldots, R_n\}$ under T of the rectangles S_i then constitute an inner partition of the region R (though into curvilinear figures rather than rectangles).

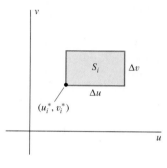

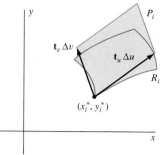

Fig. 15.9.3 The effect of the transformation T; we estimate the area of $R_i = T(S_i)$ by computing the area of P_i.

Let (u_i^*, v_i^*) be the lower left-hand corner point of S_i, and write

$$(x_i^*, y_i^*) = (f(u_i^*, v_i^*), g(u_i^*, v_i^*))$$

for its image under T. The u-curve through (x_i^*, y_i^*) has velocity vector

$$\mathbf{t}_u = \mathbf{i} f_u(u_i^*, v_i^*) + \mathbf{j} g_u(u_i^*, v_i^*) = \frac{\partial x}{\partial u}\mathbf{i} + \frac{\partial y}{\partial u}\mathbf{j},$$

whereas the v-curve through (x_i^*, y_i^*) has velocity vector

$$\mathbf{t}_v = \mathbf{i} f_v(u_i^*, v_i^*) + \mathbf{j} g_v(u_i^*, v_i^*) = \frac{\partial x}{\partial v}\mathbf{i} + \frac{\partial y}{\partial v}\mathbf{j}.$$

Thus we can approximate the curvilinear figure R_i by a parallelogram P_i with edges that are "copies" of the vectors $\mathbf{t}_u\,\Delta u$ and $\mathbf{t}_v\,\Delta v$. These edges and the approximating parallelogram appear in Fig. 15.9.3.

Now the area ΔA_i of R_i is also approximated by the area of the parallelogram P_i, and we can compute the latter. Indeed,

$$\Delta A_i \approx a(P_i) = |(\mathbf{t}_u\,\Delta u) \times (\mathbf{t}_v\,\Delta v)| = |\mathbf{t}_u \times \mathbf{t}_v|\,\Delta u\,\Delta v.$$

But

$$\mathbf{t}_u \times \mathbf{t}_v = \begin{vmatrix} \mathbf{i} & \mathbf{j} & \mathbf{k} \\ \dfrac{\partial x}{\partial u} & \dfrac{\partial y}{\partial u} & 0 \\ \dfrac{\partial x}{\partial v} & \dfrac{\partial y}{\partial v} & 0 \end{vmatrix} = \begin{vmatrix} \dfrac{\partial x}{\partial u} & \dfrac{\partial x}{\partial v} \\ \dfrac{\partial y}{\partial u} & \dfrac{\partial y}{\partial v} \end{vmatrix}\mathbf{k}.$$

The two-by-two determinant on the right is called the **Jacobian** of the transformation T, after the German mathematician Carl Jacobi (1804–1851), who first investigated general changes of variables in multiple integrals. The Jacobian of the transformation T is a function of u and v, and we denote it by $J_T = J_T(u, v)$. Thus

$$J_T(u, v) = \begin{vmatrix} f_u(u, v) & f_v(u, v) \\ g_u(u, v) & g_v(u, v) \end{vmatrix}. \tag{4}$$

A common and particularly suggestive notation for the Jacobian is

$$J_T = \frac{\partial(x, y)}{\partial(u, v)}.$$

The computation preceding Eq. (4) shows that the area ΔA_i of R_i is given approximately by

$$\Delta A_i \approx |J_T(u_i^*, v_i^*)|\,\Delta u\,\Delta v.$$

Therefore, when we set up Riemann sums for approximating double integrals, we find that

$$\iint_R F(x, y)\,dx\,dy \approx \sum_{i=1}^n F(x_i^*, y_i^*)\,\Delta A_i$$

$$\approx \sum_{i=1}^n F(f(u_i^*, v_i^*), g(u_i^*, v_i^*))\,|J_T(u_i^*, v_i^*)|\,\Delta u\,\Delta v$$

$$\approx \iint_R F(f(u, v), g(u, v))\,|J_T(u, v)|\,du\,dv.$$

This discussion is, in fact, an outline of a proof of the following general **change-of-variables** theorem. We assume that T transforms the bounded region S in the uv-plane into the bounded region R in the xy-plane and that T is one-to-one from the interior of S to the interior of R. Suppose that the function $F(x, y)$ and the first-order partial derivatives of the component functions of T are continuous functions. Finally, to ensure the existence of the indicated double integrals, we assume that the boundaries of the regions S and R consist of a finite number of piecewise smooth simple closed curves.

Theorem 1 *Change of Variables*

If the transformation T with component functions $x = f(u, v)$, $y = g(u, v)$ satisfies the conditions in the previous paragraph, then

$$\iint_R F(x, y) \, dx \, dy = \iint_S F(f(u, v), g(u, v)) \left| J_T(u, v) \right| \, du \, dv. \qquad (5)$$

If we write $G(u, v) = F(f(u, v), g(u, v))$, then the change-of-variables formula, Eq. (5), becomes

$$\iint_R F(x, y) \, dx \, dy = \iint_S G(u, v) \left| \frac{\partial(x, y)}{\partial(u, v)} \right| \, du \, dv. \qquad (5a)$$

Thus we formally transform $\iint_R F(x, y) \, dA$ by replacing the variables x and y with $f(u, v)$ and $g(u, v)$, respectively, and writing

$$dA = \left| \frac{\partial(x, y)}{\partial(u, v)} \right| \, du \, dv$$

for the area element in terms of u and v. Note the analogy between Eq. (5a) and the single-variable formula in Eq. (1). In fact, if $g'(x) \neq 0$ on $[c, d]$ and we denote by α the smaller and by β the larger of the two limits c and d in Eq. (1), then Eq. (1) takes the form

$$\int_a^b f(x) \, dx = \int_\alpha^\beta f(g(u)) \left| g'(u) \right| \, du. \qquad (1a)$$

Thus the Jacobian in Eq. (5a) plays the role of the derivative $g'(u)$ in Eq. (1).

EXAMPLE 2   Suppose that the transformation T from the $r\theta$-plane to the xy-plane is determined by the polar equations

$$x = f(r, \theta) = r \cos \theta, \qquad y = g(r, \theta) = r \sin \theta.$$

The Jacobian of T is

$$\frac{\partial(x, y)}{\partial(r, \theta)} = \begin{vmatrix} \cos \theta & -r \sin \theta \\ \sin \theta & r \cos \theta \end{vmatrix} = r > 0,$$

so Eq. (5) or (5a) reduces to the familiar formula

$$\iint_R F(x, y) \, dx \, dy = \iint_S F(r \cos \theta, r \sin \theta) \, r \, dr \, d\theta.$$

Given a particular double integral $\iint_R f(x, y)\, dx\, dy$, how do we find a *productive* change of variables? One standard approach is to choose a transformation T such that the boundary of R consists of u-curves and v-curves. In case it is more convenient to express u and v in terms of x and y, we can first compute $\partial(u, v)/\partial(x, y)$ explicitly and then find the needed Jacobian $\partial(x, y)/\partial(u, v)$ from the formula

$$\frac{\partial(x, y)}{\partial(u, v)} \cdot \frac{\partial(u, v)}{\partial(x, y)} = 1. \tag{6}$$

Equation (6) is a consequence of the chain rule (see Problem 18).

EXAMPLE 3 Suppose that R is the plane region of unit density that is bounded by the hyperbolas

$$xy = 1, \quad xy = 3 \quad \text{and} \quad x^2 - y^2 = 1, \quad x^2 - y^2 = 4.$$

Find the polar moment of inertia

$$I_0 = \iint_R (x^2 + y^2)\, dx\, dy$$

of this region.

Solution The hyperbolas bounding R are u-curves and v-curves if $u = xy$ and $v = x^2 - y^2$, as in Example 1. We can most easily write the integrand $x^2 + y^2$ in terms of u and v by first noting that

$$4u^2 + v^2 = 4x^2y^2 + (x^2 - y^2)^2 = (x^2 + y^2)^2,$$

so $x^2 + y^2 = \sqrt{4u^2 + v^2}$. Now

$$\frac{\partial(u, v)}{\partial(x, y)} = \begin{vmatrix} y & x \\ 2x & -2y \end{vmatrix} = -2(x^2 + y^2).$$

Hence Eq. (6) gives

$$\frac{\partial(x, y)}{\partial(u, v)} = -\frac{1}{2(x^2 + y^2)} = -\frac{1}{2\sqrt{4u^2 + v^2}}.$$

We are now ready to apply the change-of-variables theorem, with the regions S and R as shown in Fig. 15.9.4. With $F(x, y) = x^2 + y^2$, Eq. (5a) gives

$$I_0 = \iint_R (x^2 + y^2)\, dx\, dy = \int_1^4 \int_1^3 \sqrt{4u^2 + v^2}\, \frac{1}{2\sqrt{4u^2 + v^2}}\, du\, dv$$

$$= \int_1^4 \int_1^3 \frac{1}{2}\, du\, dv = 3.$$

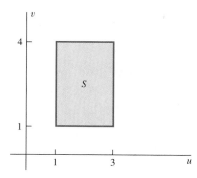

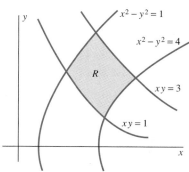

Fig. 15.9.4 The transformation T and the new region S constructed in Example 3

Example 4 is motivated by an important application. Consider an engine with an operating cycle that consists of alternate expansion and compression of gas in a piston. During one cycle the point (p, V), which gives the pressure and volume of this gas, traces a closed curve in the pV-plane. The work done by the engine—ignoring friction and related losses—is then equal (in appropriate units) to the area *enclosed by this curve,* called the *indicator diagram* of the engine. The indicator diagram for an ideal *Carnot engine* consists of two *isotherms* $xy = a$, $xy = b$ and two *adiabatics* $xy^\gamma = c$,

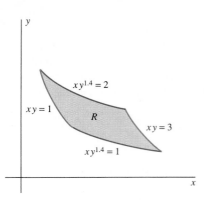

Fig. 15.9.5 Finding the area of the region R (Example 4)

$xy^\gamma = d$, where γ is the heat capacity ratio of the working gas in the piston. A typical value is $\gamma = 1.4$.

EXAMPLE 4 Find the area of the region R bounded by the curves $xy = 1$, $xy = 3$, and $xy^{1.4} = 1$, $xy^{1.4} = 2$ (Fig. 15.9.5).

Solution To force the given curves to be u-curves and v-curves, we define our change of variables transformation by $u = xy$ and $v = xy^{1.4}$. Then

$$\frac{\partial(u, v)}{\partial(x, y)} = \begin{vmatrix} y & x \\ y^{1.4} & (1.4)xy^{0.4} \end{vmatrix} = (0.4)xy^{1.4} = (0.4)v.$$

So

$$\frac{\partial(x, y)}{\partial(u, v)} = \frac{1}{\partial(u, v)/\partial(x, y)} = \frac{2.5}{v}.$$

Consequently, the change-of-variables theorem gives the formula

$$A = \iint_R 1 \, dx \, dy = \int_1^2 \int_1^3 \frac{2.5}{v} \, du \, dv = 5 \ln 2.$$

CHANGE OF VARIABLES IN TRIPLE INTEGRALS

The change-of-variables formula for triple integrals is similar to Eq. (5). Let S and R be regions that correspond under the one-to-one transformation T from uvw-space to xyz-space, where the coordinate functions that comprise T are

$$x = f(u, v, w), \qquad y = g(u, v, w), \qquad z = h(u, v, w). \tag{7}$$

The Jacobian of T is

$$J_T(u, v, w) = \frac{\partial(x, y, z)}{\partial(u, v, w)} = \begin{vmatrix} \dfrac{\partial x}{\partial u} & \dfrac{\partial x}{\partial v} & \dfrac{\partial x}{\partial w} \\ \dfrac{\partial y}{\partial u} & \dfrac{\partial y}{\partial v} & \dfrac{\partial y}{\partial w} \\ \dfrac{\partial z}{\partial u} & \dfrac{\partial z}{\partial v} & \dfrac{\partial z}{\partial w} \end{vmatrix}. \tag{8}$$

Then the change-of-variables formula for triple integrals is

$$\iiint_R F(x, y, z) \, dx \, dy \, dz = \iiint_S G(u, v, w) \left| \frac{\partial(x, y, z)}{\partial(u, v, w)} \right| du \, dv \, dw, \tag{9}$$

where $G(u, v, w) = F(f(u, v, w), g(u, v, w), h(u, v, w))$ is the function obtained from $F(x, y, z)$ by expressing the variables x, y, and z in terms of u, v, and w.

EXAMPLE 5 If T is the spherical-coordinates transformation given by

$$x = \rho \sin \phi \cos \theta, \qquad y = \rho \sin \phi \sin \theta, \qquad z = \rho \cos \phi,$$

then the Jacobian of T is

$$\frac{\partial(x, y, z)}{\partial(\rho, \phi, \theta)} = \begin{vmatrix} \sin\phi\cos\theta & \rho\cos\phi\cos\theta & -\rho\sin\phi\sin\theta \\ \sin\phi\sin\theta & \rho\cos\phi\sin\theta & \rho\sin\phi\cos\theta \\ \cos\phi & -\rho\sin\phi & 0 \end{vmatrix} = \rho^2\sin\phi.$$

Thus Eq. (9) reduces to the familiar formula

$$\iiint\limits_R F(x, y, z)\, dx\, dy\, dz = \iiint\limits_S G(\rho, \phi, \theta)\, \rho^2 \sin\phi\, d\rho\, d\phi\, d\theta.$$

The sign is correct because $\rho^2 \sin\phi \geqq 0$ for ϕ in $[0, \pi]$.

EXAMPLE 6 Find the volume of the solid torus R obtained by revolving around the z-axis the circular disk

$$(x - b)^2 + z^2 \leqq a^2, \qquad 0 < a < b \tag{10}$$

in the xz-plane.

Solution This is the torus of Example 5 of Section 15.8. Let us write u for the ordinary polar coordinate angle θ, v for the angle ψ of Fig. 15.8.12, and w for the distance from the center of the circular disk described by the inequality in (10). We then define the transformation T by means of the equations

$$x = (b + w\cos v)\cos u, \qquad y = (b + w\cos v)\sin u, \qquad z = w\sin v.$$

Then the solid torus R is the image under T of the region in uvw-space described by the inequalities $0 \leqq u \leqq 2\pi$, $0 \leqq v \leqq 2\pi$, $0 \leqq w \leqq a$. By a routine computation, we find that the Jacobian of T is

$$\frac{\partial(x, y, z)}{\partial(u, v, w)} = w(b + w\cos v).$$

Hence Eq. (9) with $F(x, y, z) \equiv 1$ yields volume

$$V = \iiint\limits_T dx\, dy\, dz = \int_0^{2\pi}\int_0^{2\pi}\int_0^a (bw + w^2\cos v)\, dw\, du\, dv$$

$$= 2\pi \int_0^{2\pi} (\tfrac{1}{2}a^2 b + \tfrac{1}{3}a^3\cos v)\, dv = 2\pi^2 a^2 b,$$

which agrees with the value $V = 2\pi b \cdot \pi a^2$ given by Pappus's first theorem (Section 15.5).

15.9 Problems

In Problems 1 through 6, solve for x and y in terms of u and v. Then compute the Jacobian $\partial(x, y)/\partial(u, v)$.

1. $u = x + y$, $v = x - y$

2. $u = x - 2y$, $v = 3x + y$

3. $u = xy$, $v = y/x$

4. $u = 2(x^2 + y^2)$, $v = 2(x^2 - y^2)$

5. $u = x + 2y^2$, $v = x - 2y^2$

6. $u = \dfrac{2x}{x^2 + y^2}$, $v = \dfrac{-2y}{x^2 + y^2}$

7. Let R be the parallelogram bounded by the lines $x + y = 1$, $x + y = 2$ and $2x - 3y = 2$, $2x - 3y = 5$. Substitute $u = x + y$, $v = 2x - 3y$ to find its area

$$A = \iint\limits_R dx\, dy.$$

8. Substitute $u = xy$, $v = y/x$ to find the area of the first-quadrant region bounded by the lines $y = x$, $y = 2x$ and by the hyperbolas $xy = 1$, $xy = 2$ (Fig. 15.9.6).

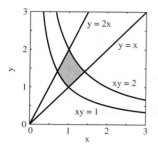

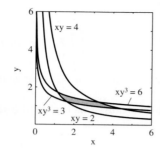

Fig. 15.9.6 The region of Problem 8

Fig. 15.9.7 The region of Problem 9

9. Substitute $u = xy$, $v = xy^3$ to find the area of the first-quadrant region bounded by the curves $xy = 2$, $xy = 4$ and $xy^3 = 3$, $xy^3 = 6$ (Fig. 15.9.7).

10. Find the area of the first-quadrant region bounded by the curves $y = x^2$, $y = 2x^2$ and $x = y^2$, $x = 4y^2$ (Fig. 15.9.8). [*Suggestion:* Let $y = ux^2$ and $x = vy^2$.]

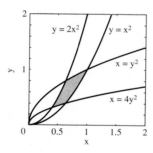

Fig. 15.9.8 The region of Problem 10

Use the method of Problem 10 to find the area of the first-quadrant region bounded by the curves $y = x^3$, $y = 2x^3$ and $x = y^3$, $x = 4y^3$.

12. Let R be the first-quadrant region bounded by the circles $x^2 + y^2 = 2x$, $x^2 + y^2 = 6x$ and by the circles $x^2 + y^2 = 2y$, $x^2 + y^2 = 8y$. Use the transformation $u = 2x/(x^2 + y^2)$, $v = 2y/(x^2 + y^2)$ to evaluate the integral

$$\iint_R \frac{1}{(x^2 + y^2)^2} \, dx \, dy.$$

13. Use elliptical coordinates $x = 3r \cos \theta$, $y = 2r \sin \theta$ to find the volume of the region bounded by the xy-plane, by the paraboloid $z = x^2 + y^2$, and by the elliptic cylinder $x^2/9 + y^2/4 = 1$.

14. Let R be the solid ellipsoid with outer boundary surface $x^2/a^2 + y^2/b^2 + z^2/c^2 = 1$. Use the transformation $x = au$, $y = bv$, $z = cw$ to show that the volume of this ellipsoid is

$$V = \iiint_T 1 \, dx \, dy \, dz = \tfrac{4}{3}\pi abc.$$

15. Find the volume of the region in the first octant that is bounded by the hyperbolic cylinders $xy = 1$, $xy = 4$; $xz = 1$, $xz = 9$; and $yz = 4$, $yz = 9$. [*Suggestion:* Let $u = xy$, $v = xz$, $w = yz$, and note that $uvw = x^2y^2z^2$.]

16. Use the transformation

$$x = \frac{r}{t} \cos \theta, \qquad y = \frac{r}{t} \sin \theta, \qquad z = r^2$$

to find the volume of the region R that lies between the paraboloids $z = x^2 + y^2$, $z = 4(x^2 + y^2)$ and between the planes $z = 1$, $z = 4$.

17. Let R be the rotated elliptical region bounded by the graph of $x^2 + xy + y^2 = 3$. Let $x = u + v$ and $y = u - v$. Show that

$$\iint_R \exp(-x^2 - xy - y^2) \, dx \, dy$$

$$= 2 \iint_S \exp(-3u^2 - v^2) \, du \, dv.$$

Then substitute $u = r \cos \theta$, $v = \sqrt{3}(r \sin \theta)$ to evaluate the latter integral.

18. From the chain rule and from the following property of determinants, derive the relation in Eq. (6) between the Jacobians of a transformation and its inverse.

$$\begin{vmatrix} a_1 & b_1 \\ c_1 & d_1 \end{vmatrix} \cdot \begin{vmatrix} a_2 & b_2 \\ c_2 & d_2 \end{vmatrix} = \begin{vmatrix} a_1 a_2 + b_1 c_2 & a_1 b_2 + b_1 d_2 \\ a_2 c_1 + c_2 d_1 & b_2 c_1 + d_1 d_2 \end{vmatrix}.$$

19. Change to spherical coordinates to show that, for $k > 0$,

$$\int_{-\infty}^{+\infty} \int_{-\infty}^{+\infty} \int_{-\infty}^{+\infty} \sqrt{x^2 + y^2 + z^2} \, e^{-k(x^2+y^2+z^2)} \, dx \, dy \, dz = \frac{2\pi}{k^2}.$$

20. Let R be the solid ellipsoid with constant density δ and boundary surface $x^2/a^2 + y^2/b^2 + z^2/c^2 = 1$. Use ellipsoidal coordinates $x = a\rho \sin \phi \cos \theta$, $y = b\rho \sin \phi \sin \theta$, $z = c\rho \cos \phi$ to show that its mass is $M = \tfrac{4}{3}\pi\delta abc$.

21. Show that the moment of inertia of the ellipsoid of Problem 20 about the z-axis is $I_z = \tfrac{1}{5}M(a^2 + b^2)$.

Chapter 15 Review: DEFINITIONS, CONCEPTS, RESULTS

Use the following list as a guide to concepts that you may need to review.

1. Definition of the double integral as a limit of Riemann sums

2. Evaluation of double integrals by iterated single integrals

3. Use of the double integral to find the volume between two surfaces above a given plane region

4. Transformation of the double integral $\iint_R f(x, y)\, dA$ into polar coordinates

5. Application of double integrals to find mass, centroids, and moments of inertia of plane laminae

6. The two theorems of Pappus

7. Definition of the triple integral as a limit of Riemann sums

8. Evaluation of triple integrals by iterated single integrals

9. Application of triple integrals to find volume, mass, centroids, and moments of inertia

10. Transformation of the triple integral $\iiint_T f(x, y, z)\, dV$ into cylindrical and spherical coordinates

11. The surface area of a parametric surface

12. The area of a surface $z = f(x, y)$ for (x, y) in the plane region R

13. The Jacobian of a transformation of coordinates

14. The transformation of a double or triple integral corresponding to a given change of variables

Chapter 15 Miscellaneous Problems

In Problems 1 through 5, evaluate the given integral by first reversing the order of integration.

1. $\displaystyle\int_0^1 \int_{y^{1/3}}^1 \frac{1}{\sqrt{1 + x^2}}\, dx\, dy$

2. $\displaystyle\int_0^1 \int_y^1 \frac{\sin x}{x}\, dx\, dy$

3. $\displaystyle\int_0^1 \int_x^1 \exp(-y^2)\, dy\, dx$

4. $\displaystyle\int_0^8 \int_{x^{2/3}}^4 x \cos y^4\, dy\, dx$

5. $\displaystyle\int_0^4 \int_{\sqrt{y}}^2 \frac{y \exp(x^2)}{x^3}\, dx\, dy$

6. The double integral

$$\int_0^\infty \int_x^\infty \frac{e^{-y}}{y}\, dy\, dx$$

is an improper integral over the unbounded region in the first quadrant that is bounded by the lines $y = x$ and $x = 0$. Assuming that it is valid to reverse the order of integration, evaluate this integral by integrating first with respect to x.

7. Find the volume of the solid T that lies below the paraboloid $z = x^2 + y^2$ and above the triangle R in the xy-plane that has vertices at $(0, 0, 0)$, $(1, 1, 0)$, and $(2, 0, 0)$.

8. Find by integration in cylindrical coordinates the volume bounded by the paraboloids $z = 2x^2 + 2y^2$ and $z = 48 - x^2 - y^2$.

9. Use integration in spherical coordinates to find the volume and centroid of the solid region that is inside the sphere $\rho = 3$, below the cone $\phi = \pi/3$, and above the xy-plane $\phi = \pi/2$.

10. Find the volume of the solid bounded by the elliptic paraboloids $z = x^2 + 3y^2$ and $z = 8 - x^2 - 5y^2$.

11. Find the volume bounded by the paraboloid $y = x^2 + 3z^2$ and by the parabolic cylinder $y = 4 - z^2$.

12. Find the volume of the region bounded by the parabolic cylinders $z = x^2$, $z = 2 - x^2$ and by the planes $y = 0$, $y + z = 4$.

13. Find the volume of the region bounded by the elliptical cylinder $y^2 + 4z^2 = 4$ and by the planes $x = 0$, $x = y + 2$.

14. Show that the volume of the solid bounded by the elliptical cylinder $x^2/a^2 + y^2/b^2 = 1$ and by the planes $z = 0$, $z = h + x$ (where $h > a > 0$) is $V = \pi abh$.

15. Let R be the first-quadrant region bounded by the line $y = x$ and by the curve $x^4 + x^2y^2 = y^2$. Use polar coordinates to evaluate

$$\iint_R \frac{1}{(1 + x^2 + y^2)^2}\, dA.$$

In Problems 16 through 20, find the mass and centroid of a plane lamina with the given shape and density ρ.

16. The region bounded by $y = x^2$ and $x = y^2$; $\rho = x^2 + y^2$

17. The region bounded by $x = 2y^2$ and $y^2 = x - 4$; $\rho = y^2$

18. The region between $y = \ln x$ and the x-axis over the interval $1 \leqq x \leqq 2$; $\rho = 1/x$

19. The circle bounded by $r = 2 \cos \theta$; $\rho = k$ (a constant)

20. The region of Problem 19; $\rho = r$

21. Use the first theorem of Pappus to find the y-coordinate of the centroid of the upper half of the ellipse $(x/a)^2 + (y/b)^2 = 1$. Employ the facts that the area of this semiellipse is $A = \pi a b/2$, whereas the volume of the ellipsoid it generates when rotated around the x-axis is $V = \frac{4}{3} \pi a b^2$.

22. (a) Use the first theorem of Pappus to find the centroid of the first-quadrant portion of the annular ring with boundary circles $x^2 + y^2 = a^2$ and $x^2 + y^2 = b^2$ (where $0 < a < b$). (b) Show that the limiting position of this centroid as $b \to a$ is the centroid of a quarter-circular arc, as we found in Problem 44 of Section 15.5.

23. Find the centroid of the region in the xy-plane bounded by the x-axis and by the parabola $y = 4 - x^2$.

24. Find the volume of the solid that lies below the parabolic cylinder $z = x^2$ and above the triangle in the xy-plane bounded by the x-axis, the y-axis, and the line $x + y = 1$.

25. Use cylindrical coordinates to find the volume of the ice-cream cone bounded above by the sphere $x^2 + y^2 + z^2 = 5$ and below by the cone $z = 2\sqrt{x^2 + y^2}$.

26. Find the volume and centroid of the ice-cream cone bounded above by the sphere $\rho = a$ and below by the cone $\phi = \pi/3$.

27. A homogeneous solid circular cone has mass M and base radius a. Find its moment of inertia about its axis of symmetry.

28. Find the mass of the first octant of the ball $\rho \leq a$ if its density at (x, y, z) is $\delta = xyz$.

29. Find the moment of inertia about the x-axis of the homogeneous solid ellipsoid with unit density and boundary surface $(x/a)^2 + (y/b)^2 + (z/c)^2 = 1$.

30. Find the volume of the region in the first octant that is bounded by the sphere $\rho = a$, the cylinder $r = a$, the plane $z = a$, the xz-plane, and the yz-plane.

31. Find the moment of inertia about the z-axis of the homogeneous region of unit density that lies inside both the sphere $\rho = 2$ and the cylinder $r = 2 \cos \theta$.

In Problems 32 through 34, a volume is generated by revolving a plane region R around an axis. To find the volume, set up a double integral over R by revolving an area element dA around the indicated axis to generate a volume element dV.

32. Find the volume of the solid obtained by revolving around the y-axis the region inside the circle $r = 2a \cos \theta$.

33. Find the volume of the solid obtained by revolving around the x-axis the region enclosed by the cardioid $r = 1 + \cos \theta$.

34. Find the volume of the solid torus obtained by revolving the disk $0 \leq r \leq a$ around the line $x = -b, |b| \geq a$.

35. Refer to the oblique segment of a paraboloid discussed in Example 3 of Section 15.6 and shown in Fig. 15.6.9. (a) Show first that its centroid is at the point $C(0, \frac{1}{2}, \frac{7}{4})$. (b) Show that the center of the elliptical upper "base" of the solid paraboloid is at the point $Q(0, \frac{1}{2}, \frac{5}{2})$. (c) Verify that the point $V(0, \frac{1}{2}, \frac{1}{4})$ is the point where the plane tangent to the paraboloid is parallel to the upper base. The point V is called the **vertex** of the oblique segment, and the line segment VQ is its **principal axis**. (d) Show that C lies on the principal axis two-thirds of the way from the vertex to the upper base. Archimedes showed that this is true for any segment cut off from a paraboloid by a plane. This was a key step in his determination of the equilibrium position of a floating right circular paraboloid, in terms of its density and dimensions. The possible positions are shown in Fig. 15.MP.1. The principles he introduced for the solution of this problem are still important in naval architecture.

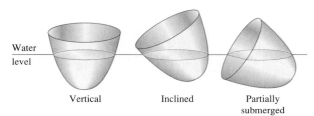

Fig. 15.MP.1 How a uniform solid paraboloid might float (Problem 35)

Problems 36 through 42 deal with average distance. The ***average distance*** $\overline{d}$ *of the point* (x_0, y_0) *from the points of the plane region R with area A is defined to be*

$$\overline{d} = \frac{1}{A} \iint_R \sqrt{(x - x_0)^2 + (y - y_0)^2} \, dA.$$

The average distance of a point (x_0, y_0, z_0) *from the points of a space region is defined analogously.*

36. Show that the average distance of the points of a disk of radius a from its center is $2a/3$.

37. Show that the average distance of the points of a disk of radius a from a fixed point on its boundary is $32a/9\pi$.

38. A circle of radius 1 is interior to and tangent to a circle of radius 2. Find the average distance of the point of tangency from the points that lie between the two circles.

39. Show that the average distance of the points of a spherical ball of radius a from its center is $3a/4$.

40. Show that the average distance of the points of a

spherical ball of radius a from a fixed point on its surface is $6a/5$.

41. A sphere of radius 1 is interior to and tangent to a sphere of radius 2. Find the average distance of the point of tangency from the set of all points between the two spheres.

42. A right circular cone has radius R and height H. Find the average distance of points of the cone from its vetex.

43. Find the surface area of the part of the paraboloid $z = 10 - r^2$ that lies between the two planes $z = 1$ and $z = 6$.

44. Find the surface area of the part of the surface $z = y^2 - x^2$ that is inside the cylinder $x^2 + y^2 = 4$.

45. Use the formula of Problem 18 of Section 15.8 that the surface area of the zone on the sphere $\rho = a$ between the planes $z = z_1$ and $z = z_2$ (where $-a \leqq z_1 < z_2 \leqq a$) is $A = 2\pi ah$, where $h = z_2 - z_1$.

46. Find the surface area of the part of the sphere $\rho = 2$ that is inside the cylinder $x^2 + y^2 = 2x$.

47. A square hole with side length 2 is cut through a cone of height 2 and base radius 2; the centerline of the hole is the axis of symmetry of the cone. Find the area of the surface removed from the cone.

48. Numerically approximate the surface area of the part of the parabolic cylinder $2z = x^2$ that lies inside the cylinder $x^2 + y^2 = 1$.

49. A "fence" of variable height $h(t)$ stands above the plane curve $(x(t), y(t))$. Thus the fence has the parametrization $x = x(t)$, $y = y(t)$, $z = z$ for $a \leqq t \leqq b$, $0 \leqq z \leqq h(t)$. Apply Eq. (8) of Section 15.8 to show that the area of the fence is

$$A = \int_a^b \int_0^{h(t)} \left[\left(\frac{dx}{dt} \right)^2 + \left(\frac{dy}{dt} \right)^2 \right]^{1/2} dz \, dt.$$

50. Apply the formula of Problem 49 to compute the area of the part of the cylinder $r = a \sin \theta$ that lies inside the sphere $r^2 + z^2 = a^2$.

51. Find the polar moment of inertia of the first-quadrant region of constant density δ that is bounded by the hyperbolas $xy = 1$, $xy = 3$ and $x^2 - y^2 = 1$, $x^2 - y^2 = 4$.

52. Substitute $u = x - y$ and $v = x + y$ to evaluate

$$\iint_R \exp\left(\frac{x - y}{x + y} \right) dx \, dy,$$

where R is bounded by the coordinate axes and by the line $x + y = 1$.

53. Use ellipsoidal coordinates $x = a\rho \sin \phi \cos \theta$, $y = b\rho \sin \phi \sin \theta$, $z = c\rho \cos \phi$ to find the mass of the solid ellipsoid $(x/a)^2 + (y/b)^2 + (z/c)^2 \leqq 1$ if its density at the point (x, y, z) is given by $\delta = 1 - (x/a)^2 - (y/b)^2 - (z/c)^2$.

54. Let R be the first-quadrant region bounded by the lemniscates $r^2 = 3 \cos 2\theta$, $r^2 = 4 \cos 2\theta$ and $r^2 = 3 \sin 2\theta$, $r^2 = 4 \sin 2\theta$ (Fig. 15.MP.2). Show that its area is $A = (10 - 7\sqrt{2})/4$. [*Suggestion:* Define the transformation T from the uv-plane to the $r\theta$-plane by $r^2 = u^{1/2} \cos 2\theta$, $r^2 = v^{1/2} \sin 2\theta$. Show first that

$$r^4 = \frac{uv}{u + v}, \qquad \theta = \frac{1}{2} \arctan \frac{u^{1/2}}{v^{1/2}}.$$

Then show that

$$\frac{\partial(r, \theta)}{\partial(u, v)} = -\frac{1}{16r(u + v)^{3/2}}.$$

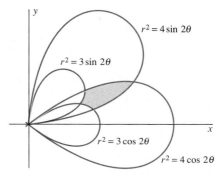

Fig. 15.MP.2   The region R of Problem 54

55. A 2-by-2 square hole is cut symmetrically through a sphere by radius $\sqrt{3}$ (Fig. 15.MP.3). (a) Show that the total surface area of the two pieces cut from the sphere is

$$A = \int_0^1 8\sqrt{3} \arcsin\left(\frac{1}{\sqrt{3 - x^2}} \right) dx.$$

Then use Simpson's rule to approximate this integral.

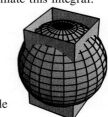

Fig. 15.MP.3   Cutting a square hole through the sphere of Problem 55

(b) (Difficult!) Show that the exact value of the integral in part (a) is $A = 4\pi(\sqrt{3} - 1)$. [*Suggestion:* First integrate by parts, and then substitute $x = \sqrt{2} \sin \theta$.]

56. Show that the volume enclosed by the surface

$$x^{2/3} + y^{2/3} + z^{2/3} = a^{2/3}$$

is $V = 4\pi a^3/35$. [*Suggestion:* Substitute $y = b \sin^3 \theta$.]

57. Show that the volume enclosed by the surface

$$x^{1/3} + y^{1/3} + z^{1/3} = a^{1/3}$$

is $V = a^3/210$. [*Suggestion:* Substitute $y = b \sin^6 \theta$.]

16 Vector Analysis

C. F. Gauss (1777–1855)

❑ It is customary to list Archimedes, Newton, and Carl Friedrich Gauss as history's three pre-eminent mathematicians. Gauss was a precocious infant in a poor and uneducated family. He learned to calculate before he could talk and taught himself to read before beginning school in his native Brunswick, Germany. From the age of 14 he received a stipend from the Duke of Brunswick that enabled Gauss to continue his education. By then he was already familiar with elementary geometry, algebra, and analysis. By age 18, when he entered the University of Göttingen, he had discovered empirically the "prime number theorem," which implies that the number of primes $p \leqq n$ is about $n/(\ln n)$. This theorem was not proved rigorously until a century later.

❑ During his first year at university, Gauss discovered conditions for the ruler-and-compass construction of regular polygons and demonstrated the constructability of the regular 17-gon (the first advance in this area since the similar construction of the regular pentagon in Euclid's *Elements* 2000 years earlier). In 1801 Gauss published his great treatise *Disquisitiones arithmeticae*, which summarized number theory to that time and set the pattern for nineteenth-century research in that area. This book established Gauss as a mathematician of uncommon stature, but another event thrust him into the public eye. On January 1, 1801, the new asteroid Ceres was observed, but it disappeared behind the sun a month later. In the following months, astronomers searched the skies in vain for Ceres' reappearance. It was Gauss who developed the new method of least-squares approximations to predict the asteroid's future orbit on the basis of a handful of observations. When Gauss's three-month long computation was finished, Ceres was soon spotted in the precise location he had predicted. All this made Gauss famous as a mathematician and astronomer at the age of 25.

❑ In 1807 Gauss became director of the astronomical observatory in Göttingen, where he remained until his death. His published work thereafter dealt mainly with physical science, although his unpublished papers show that he continued to work on theoretical mathematics ranging from infinite series and special functions to non-Euclidean geometry. His work on the shape of the earth's surface established the new subject of differential geometry, and his studies of the earth's magnetic and gravitational fields involved results such as the divergence theorem (Section 16.6), which is sometimes called Gauss's theorem.

❑ The concept of curved space–time in Albert Einstein's general relativity traces back to the discovery of non-Euclidean geometry and Gauss's early investigations of differential geometry. A current application of relativity theory is the study of black holes. Space is itself thought to be severely warped in the region of a black hole, with its immense gravitational attraction, and the mathematics required to analyze such a situation begins with the vector analysis of Chapter 16.

Artist's conception of a black hole (lower left); gas from the atmosphere of the "normal star" (upper left) spirals into the black hole due to gravitational attraction.

16.1
Vector Fields

This chapter is devoted to topics in the calculus of vector fields of importance in science and engineering. A **vector field** defined on a region T in space is a vector-valued function $\mathbf{F}$ that associates with each point (x, y, z) of T a vector

$$\mathbf{F}(x, y, z) = \mathbf{i}P(x, y, z) + \mathbf{j}Q(x, y, z) + \mathbf{k}R(x, y, z). \qquad (1)$$

We may more briefly describe the vector field $\mathbf{F}$ in terms of its *component functions* P, Q, and R by writing $\mathbf{F} = \langle P, Q, R \rangle$. Note that P, Q, and R are scalar (real-valued) functions.

A **vector field** in the plane is similar except that neither z-components nor z-coordinates are involved. Thus a vector field on the plane region R is a vector-valued function $\mathbf{F}$ that associates with each point (x, y) of R a vector

$$\mathbf{F}(x, y) = \mathbf{i}P(x, y) + \mathbf{j}Q(x, y). \qquad (2)$$

It is useful to be able to visualize a given vector field $\mathbf{F}$. One common way is to sketch a collection of typical vectors $\mathbf{F}(x, y)$, each represented by an arrow of length $|\mathbf{F}(x, y)|$ and placed with (x, y) as its initial point. This procedure is illustrated in Example 1.

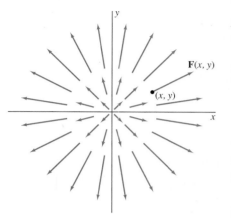

Fig. 16.1.1 The vector field $\mathbf{F}(x, y) = x\mathbf{i} + y\mathbf{j}$

EXAMPLE 1 Describe the vector field $\mathbf{F}(x, y) = x\mathbf{i} + y\mathbf{j}$.

Solution For each point (x, y) in the coordinate plane, $\mathbf{F}(x, y)$ is simply its position vector. It points directly away from the origin and has length

$$|\mathbf{F}(x, y)| = |x\mathbf{i} + y\mathbf{j}| = \sqrt{x^2 + y^2} = r,$$

equal to the distance from the origin to (x, y). Figure 16.1.1 shows some typical vectors representing this vector field.

Velocity vector field for airflow around an automobile

Among the most important vector fields in applications are velocity vector fields. Imagine the steady flow of a fluid, such as the water in a river or the solar wind. By a *steady flow* we mean that the velocity vector $\mathbf{v}(x, y, z)$ of the fluid flowing through each point (x, y, z) is independent of time (although not necessarily independent of x, y, and z), so the pattern of the flow remains constant. Then $\mathbf{v}(x, y, z)$ is the **velocity vector field** of the fluid flow.

EXAMPLE 2 Suppose that the horizontal xy-plane is covered with a thin sheet of water that is revolving (rather like a whirlpool) around the origin with constant angular speed ω radians per second in the counterclockwise direction. Describe the associated velocity vector field.

Solution In this case we have a two-dimensional vector field $\mathbf{v}(x, y)$. At each point (x, y) the water is moving with speed $v = r\omega$ and tangential to the circle of radius $r = \sqrt{x^2 + y^2}$. The vector field

$$\mathbf{v}(x, y) = \omega(-y\mathbf{i} + x\mathbf{j}) \qquad (3)$$

has length $r\omega$ and points in a generally counterclockwise direction, and

$$\mathbf{v} \cdot \mathbf{r} = \omega(-y\mathbf{i} + x\mathbf{j}) \cdot (x\mathbf{i} + y\mathbf{j}) = 0,$$

so $\mathbf{v}$ is tangent to the circle just mentioned. The velocity field determined by Eq. (3) is illustrated in Fig. 16.1.2.

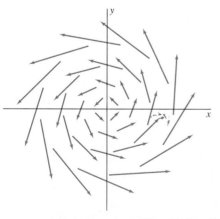

Fig. 16.1.2 The velocity vector field $\mathbf{v}(x, y) = \omega(-y\mathbf{i} + x\mathbf{j})$, drawn for $\omega = 1$ (Example 2)

Equally important in physical applications are *force fields*. Suppose that some circumstance (perhaps gravitational or electrical in character) causes a force $\mathbf{F}(x, y, z)$ to act on a particle when it is placed at the point (x, y, z). Then we have a force field $\mathbf{F}$. Example 3 deals with what is perhaps the most common force field perceived by human beings.

EXAMPLE 3 Suppose that a mass M is fixed at the origin in space. When a particle of unit mass is placed at the point (x, y, z) other than the origin, it is subjected to a force $\mathbf{F}(x, y, z)$ of gravitational attraction directed toward the mass M at the origin. By Newton's inverse-square law of gravitation, the magnitude of $\mathbf{F}$ is $F = GM/r^2$, where $r = \sqrt{x^2 + y^2 + z^2}$ is the length of the position vector $\mathbf{r} = x\mathbf{i} + y\mathbf{j} + z\mathbf{k}$. It follows immediately that

$$\mathbf{F}(x, y, z) = -\frac{k\,\mathbf{r}}{r^3}, \tag{4}$$

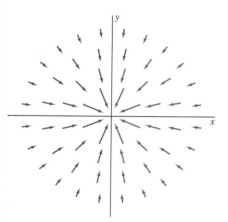

Fig. 16.1.3 An inverse-square force field (Example 3)

where $k = GM$, because this vector has both the correct magnitude and the correct direction (toward the origin, for $\mathbf{F}$ is a multiple of $-\mathbf{r}$). A force field of the form in Eq. (4) is called an *inverse-square* force field. Note that $\mathbf{F}(x, y, z)$ is not defined at the origin and that $|\mathbf{F}| \to \infty$ as $r \to 0^+$. Figure 16.1.3 illustrates an inverse-square force field.

THE GRADIENT VECTOR FIELD

In Section 14.8 we introduced the gradient vector of the real-valued function $f(x, y, z)$. It is the vector ∇f defined as follows:

$$\nabla f = \mathbf{i}\frac{\partial f}{\partial x} + \mathbf{j}\frac{\partial f}{\partial y} + \mathbf{k}\frac{\partial f}{\partial z}. \tag{5}$$

The partial derivatives on the right-hand side of Eq. (5) are evaluated at the point (x, y, z). Thus $\nabla f(x, y, z)$ is a vector field: It is the **gradient vector field** of the function f. According to Theorem 1 of Section 14.8, the vector $\nabla f(x, y, z)$ points in the direction in which the maximal directional derivative of f at (x, y, z) is obtained. For example if $f(x, y, z)$ is the temperature at the point (x, y, z) in space, then you should move in the direction $\nabla f(x, y, z)$ in order to warm up the most quickly.

The notation in Eq. (5) suggests the formal expression

$$\nabla = \mathbf{i}\frac{\partial}{\partial x} + \mathbf{j}\frac{\partial}{\partial y} + \mathbf{k}\frac{\partial}{\partial z}. \tag{6}$$

It is fruitful to think of ∇ as a *vector differential operator*. That is, ∇ is the operation that, when applied to the scalar function f, yields its gradient vector field ∇f. This operation behaves in several familiar and important ways like the operation of single-variable differentiation. For a familiar example of this phenomenon, recall that in Chapter 14 we found the critical points of a function f of several variables to be those points at which $\nabla f(x, y, z) = \mathbf{0}$ and those at which $\nabla f(x, y, z)$ does not exist. As a computationally useful instance, suppose that f and g are functions and that a and b are constants. It then follows readily from (5) and from the linearity of partial differentiation that

$$\nabla(af + bg) = a\nabla f + b\nabla g. \tag{7}$$

Thus the gradient operation is *linear*. It also satisfies the product rule, as demonstrated in Example 4.

EXAMPLE 4 Given the differentiable functions $f(x, y, z)$ and $g(x, y, z)$, show that

$$\nabla(fg) = f\nabla g + g\nabla f. \tag{8}$$

Solution We apply the definition in Eq. (5) and the product rule for partial differentiation. Thus

$$\nabla(fg) = \mathbf{i}\frac{\partial(fg)}{\partial x} + \mathbf{j}\frac{\partial(fg)}{\partial y} + \mathbf{k}\frac{\partial(fg)}{\partial z}$$

$$= \mathbf{i}(fg_x + gf_x) + \mathbf{j}(fg_y + gf_y) + \mathbf{k}(fg_z + gf_z)$$

$$= f \cdot (\mathbf{i}g_x + \mathbf{j}g_y + \mathbf{k}g_z) + g \cdot (\mathbf{i}f_x + \mathbf{j}f_y + \mathbf{k}f_z) = f\nabla g + g\nabla f,$$

as desired.

THE DIVERGENCE OF A VECTOR FIELD

Suppose that we are given the vector-valued function

$$\mathbf{F}(x, y, z) = \mathbf{i}P(x, y, z) + \mathbf{j}Q(x, y, z) + \mathbf{k}R(x, y, z)$$

with differentiable component functions P, Q, and R. Then the **divergence** of $\mathbf{F}$ is the scalar function div $\mathbf{F}$ defined as follows:

$$\text{div } \mathbf{F} = \nabla \cdot \mathbf{F} = \frac{\partial P}{\partial x} + \frac{\partial Q}{\partial y} + \frac{\partial R}{\partial z}. \tag{9}$$

Here *div* is an abbreviation for "divergence," and the alternative notation $\nabla \cdot \mathbf{F}$ is consistent with the formal expression for ∇ in Eq. (6). That is,

$$\nabla \cdot \mathbf{F} = \left\langle \frac{\partial}{\partial x}, \frac{\partial}{\partial y}, \frac{\partial}{\partial z} \right\rangle \cdot \langle P, Q, R \rangle = \frac{\partial P}{\partial x} + \frac{\partial Q}{\partial y} + \frac{\partial R}{\partial z}.$$

We will see in Section 16.7 that if $\mathbf{v}$ is the velocity vector field of a steady fluid flow, then the value of div $\mathbf{v}$ at a point (x, y, z) is essentially the net rate per unit volume at which fluid mass is flowing away (or "diverging") from the point (x, y, z).

EXAMPLE 5 If the vector field $\mathbf{F}$ is given by

$$\mathbf{F}(x, y, z) = (xe^y)\mathbf{i} + (z \sin y)\mathbf{j} + (xy \ln z)\mathbf{k},$$

then $P(x, y, z) = xe^y$, $Q(x, y, z) = z \sin y$, and $R(x, y, z) = xy \ln z$. Hence Eq. (9) yields

$$\text{div } \mathbf{F} = \frac{\partial}{\partial x}(xe^y) + \frac{\partial}{\partial y}(z \sin y) + \frac{\partial}{\partial z}(xy \ln z) = e^y + z \cos y + \frac{xy}{z}.$$

For instance, the value of div **F** at the point $(-3, 0, 2)$ is

$$\nabla \cdot \mathbf{F}(-3, 0, 2) = e^0 + 2 \cos 0 + 0 = 3.$$

The analogues of Eqs. (7) and (8) for divergence are the formulas

$$\nabla \cdot (a\mathbf{F} + b\mathbf{G}) = a\nabla \cdot \mathbf{F} + b\nabla \cdot \mathbf{G} \qquad (10)$$

and

$$\nabla \cdot (f\mathbf{G}) = (f)(\nabla \cdot \mathbf{G}) + (\nabla f) \cdot \mathbf{G}. \qquad (11)$$

We ask you to verify these formulas in the problems. Note that Eq. (11)—in which f is a scalar function and **G** is a vector field—is consistent in that f and $\nabla \cdot G$ are scalar functions, whereas ∇f and **G** are vector fields, so the sum on the right-hand makes sense (and is a scalar function).

THE CURL OF A VECTOR FIELD

The **curl** of the vector field $\mathbf{F} = P\mathbf{i} + Q\mathbf{j} + R\mathbf{k}$ is the following vector field, abbreviated as curl **F**:

$$\text{curl } \mathbf{F} = \nabla \times \mathbf{F} = \begin{vmatrix} \mathbf{i} & \mathbf{j} & \mathbf{k} \\ \dfrac{\partial}{\partial x} & \dfrac{\partial}{\partial y} & \dfrac{\partial}{\partial z} \\ P & Q & R \end{vmatrix}. \qquad (12)$$

When we evaluate the formal determinant expression in Eq. (12), we obtain

$$\text{curl } \mathbf{F} = \mathbf{i}\left(\frac{\partial R}{\partial y} - \frac{\partial Q}{\partial z}\right) + \mathbf{j}\left(\frac{\partial P}{\partial z} - \frac{\partial R}{\partial x}\right) + \mathbf{k}\left(\frac{\partial Q}{\partial x} - \frac{\partial P}{\partial y}\right). \qquad (13)$$

Although you may wish to memorize this formula, we recommend—because you will generally find it simpler—that in practice you set up and evaluate directly the formal determinant in Eq. (12). Example 6 shows how easy this is.

EXAMPLE 6   For the vector field **F** of Example 5, Eq. (12) yields

$$\text{curl } \mathbf{F} = \begin{vmatrix} \mathbf{i} & \mathbf{j} & \mathbf{k} \\ \dfrac{\partial}{\partial x} & \dfrac{\partial}{\partial y} & \dfrac{\partial}{\partial z} \\ xe^y & z \sin y & xy \ln z \end{vmatrix}$$

$$= \mathbf{i}(x \ln z - \sin y) + \mathbf{j}(-y \ln z) + \mathbf{k}(-xe^y).$$

In particular, the value of curl **F** at the point $(3, \pi/2, e)$ is

$$\nabla \times \mathbf{F}(3, \pi/2, e) = 2\mathbf{i} - \tfrac{1}{2}\pi\mathbf{j} - 3e^{\pi/2}\mathbf{k}.$$

We will see in Section 16.7 that if **v** is the velocity vector of a fluid flow, then the value of the vector curl **v** at the point (x, y, z) (where that vector is nonzero) determines the axis through (x, y, z) about which the fluid is rotating (or whirling or "curling") as well as the angular velocity of the rotation.

The analogues of Eqs. (10) and (11) for curl are the formulas

$$\nabla \times (a\mathbf{F} + b\mathbf{G}) = a(\nabla \times \mathbf{F}) + b(\nabla \times \mathbf{G}) \tag{14}$$

and

$$\nabla \times (f\mathbf{G}) = (f)(\nabla \times \mathbf{G}) + (\nabla f) \times \mathbf{G} \tag{15}$$

that we ask you to verify in the problems.

EXAMPLE 7 If the function $f(x, y, z)$ has continuous second-order partial derivatives, show that

$$\text{curl (grad } f) = \mathbf{0}.$$

Solution Direct computation yields

$$\nabla \times \nabla f = \begin{vmatrix} \mathbf{i} & \mathbf{j} & \mathbf{k} \\ \dfrac{\partial}{\partial x} & \dfrac{\partial}{\partial y} & \dfrac{\partial}{\partial z} \\ \dfrac{\partial f}{\partial x} & \dfrac{\partial f}{\partial y} & \dfrac{\partial f}{\partial z} \end{vmatrix}$$

$$= \mathbf{i}\left(\frac{\partial^2 f}{\partial y\,\partial z} - \frac{\partial^2 f}{\partial z\,\partial y}\right) + \mathbf{j}\left(\frac{\partial^2 f}{\partial z\,\partial x} - \frac{\partial^2 f}{\partial x\,\partial z}\right) + \mathbf{k}\left(\frac{\partial^2 f}{\partial x\,\partial y} - \frac{\partial^2 f}{\partial y\,\partial x}\right).$$

Therefore,

$$\nabla \times \nabla f = \mathbf{0}$$

because of the equality of continuous mixed second-order partial derivatives.

In Section 16.2 we shall define line integrals, which are used (for example) to compute the work done by a force field in moving a particle along a curved path. In Section 16.5 we shall discuss surface integrals, which are used (for example) to compute the rate at which a fluid with a known velocity vector field is moving across a given surface. The three basic integral theorems of vector calculus—Green's theorem (Section 16.4), the divergence theorem (Section 16.6), and Stokes' theorem (Section 16.7)—play much the same role for line and surface integrals that the fundamental theorem of calculus plays for ordinary single-variable integrals.

16.1 Problems

In Problems 1 through 10, illustrate the given vector field $\mathbf{F}$ *by sketching several typical vectors in the field.*

1. $\mathbf{F}(x, y) = \mathbf{i} + \mathbf{j}$
2. $\mathbf{F}(x, y) = 3\mathbf{i} - 2\mathbf{j}$
3. $\mathbf{F}(x, y) = x\mathbf{i} - y\mathbf{j}$
4. $\mathbf{F}(x, y) = 2\mathbf{i} + x\mathbf{j}$
5. $\mathbf{F}(x, y) = (x^2 + y^2)^{1/2}(x\mathbf{i} + y\mathbf{j})$
6. $\mathbf{F}(x, y) = (x^2 + y^2)^{-1/2}(x\mathbf{i} + y\mathbf{j})$
7. $\mathbf{F}(x, y, z) = \mathbf{j} + \mathbf{k}$

8. $\mathbf{F}(x, y, z) = \mathbf{i} + \mathbf{j} - \mathbf{k}$
9. $\mathbf{F}(x, y, z) = -x\mathbf{i} - y\mathbf{j}$
10. $\mathbf{F}(x, y, z) = x\mathbf{i} + y\mathbf{j} + z\mathbf{k}$

In Problems 11 through 20, calculate the divergence and curl of the given vector field $\mathbf{F}$.

11. $\mathbf{F}(x, y, z) = x\mathbf{i} + y\mathbf{j} + z\mathbf{k}$
12. $\mathbf{F}(x, y, z) = 3x\mathbf{i} - 2y\mathbf{j} - 4z\mathbf{k}$
13. $\mathbf{F}(x, y, z) = yz\mathbf{i} + xz\mathbf{j} + xy\mathbf{k}$

14. $\mathbf{F}(x, y, z) = x^2\mathbf{i} + y^2\mathbf{j} + z^2\mathbf{k}$

15. $\mathbf{F}(x, y, z) = xy^2\mathbf{i} + yz^2\mathbf{j} + zx^2\mathbf{k}$

16. $\mathbf{F}(x, y, z) = (2x - y)\mathbf{i} + (3y - 2z)\mathbf{j} + (7z - 3x)\mathbf{k}$

17. $\mathbf{F}(x, y, z) = (y^2 + z^2)\mathbf{i} + (x^2 + z^2)\mathbf{j} + (x^2 + y^2)\mathbf{k}$

18. $\mathbf{F}(x, y, z) = (e^{xz} \sin y)\mathbf{j} + (e^{xy} \cos z)\mathbf{k}$

19. $\mathbf{F}(x, y, z) = (x + \sin yz)\mathbf{i} + (y + \sin xz)\mathbf{j} + (z + \sin xy)\mathbf{k}$

20. $\mathbf{F}(x, y, z) = (x^2e^{-z})\mathbf{i} + (y^3 \ln x)\mathbf{j} + (z \cosh y)\mathbf{k}$

*Apply the definitions of gradient, divergence, and curl to establish the identities in Problems 21 through 27, where a and b denote constants, f and g denote differentiable scalar functions, and **F** and **G** denote differentiable vector fields.*

21. $\nabla(af + bg) = a\,\nabla f + b\,\nabla g$

22. $\nabla \cdot (a\mathbf{F} + b\mathbf{G}) = a\nabla \cdot \mathbf{F} + b\nabla \cdot \mathbf{G}$

23. $\nabla \times (a\mathbf{F} + b\mathbf{G}) = a(\nabla \times \mathbf{F}) + b(\nabla \times \mathbf{G})$

24. $\nabla \cdot (f\mathbf{G}) = (f)(\nabla \cdot \mathbf{G}) + (\nabla f) \cdot \mathbf{G}$

25. $\nabla \times (f\mathbf{G}) = (f)(\nabla \times \mathbf{G}) + (\nabla f) \times \mathbf{G}$

26. $\nabla\left(\dfrac{f}{g}\right) = \dfrac{g\,\nabla f - f\,\nabla g}{g^2}$

27. $\nabla \cdot (\mathbf{F} \times \mathbf{G}) = \mathbf{G} \cdot (\nabla \times \mathbf{F}) - \mathbf{F} \cdot (\nabla \times \mathbf{G})$

*Establish the identities in Problems 28 through 30 under the assumption that the scalar functions f and g and the vector field **F** are twice differentiable.*

28. $\operatorname{div}(\operatorname{curl} \mathbf{F}) = 0$

29. $\operatorname{div}(\nabla fg) = f\operatorname{div}(\nabla g) + g\operatorname{div}(\nabla f) + 2(\nabla f) \cdot (\nabla g)$

30. $\operatorname{div}(\nabla f \times \nabla g) = 0$

*Verify the identities in Problems 31 through 40, in which **a** is a constant vector, $\mathbf{r} = x\mathbf{i} + y\mathbf{j} + z\mathbf{k}$, and $r = |\mathbf{r}|$. Problems 33 and 34 imply that both the divergence and curl of an inverse-square vector field vanish identically.*

31. $\nabla \cdot \mathbf{r} = 3$ and $\nabla \times \mathbf{r} = \mathbf{0}$

32. $\nabla \cdot (\mathbf{a} \times \mathbf{r}) = 0$ and $\nabla \times (\mathbf{a} \times \mathbf{r}) = 2\mathbf{a}$

33. $\nabla \cdot \dfrac{\mathbf{r}}{r^3} = 0$

34. $\nabla \times \dfrac{\mathbf{r}}{r^3} = \mathbf{0}$

35. $\nabla r = \dfrac{\mathbf{r}}{r}$

36. $\nabla\left(\dfrac{1}{r}\right) = -\dfrac{\mathbf{r}}{r^3}$

37. $\nabla \cdot (r\mathbf{r}) = 4r$

38. $\nabla \cdot (\nabla r) = 0$

39. $\nabla(\ln r) = \dfrac{\mathbf{r}}{r^2}$

40. $\nabla(r^{10}) = 10r^8\mathbf{r}$

16.2
Line Integrals

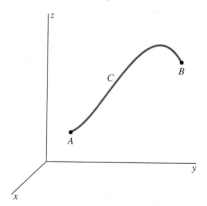

To motivate the definition of the line integral, we imagine a thin wire shaped like the smooth curve C with endpoints A and B (Fig. 16.2.1). Suppose that the wire has variable density given at the point (x, y, z) by the known continuous function $f(x, y, z)$, in units such as grams per (linear) centimeter. Let

$$x = x(t), \qquad y = y(t), \qquad z = z(t), \qquad t \text{ in } [a, b] \qquad (1)$$

be a smooth parametrization of the curve C, with $t = a$ corresponding to the initial point A of the curve and $t = b$ to its terminal point B.

To *approximate* the total mass m of the curved wire, we begin with a partition

$$a = t_0 < t_1 < t_2 \cdots t_{n-1} < t_n = b$$

of $[a, b]$ into n subintervals, all with the same length $\Delta t = (b - a)/n$. These subdivision points of $[a, b]$ produce, via our parametrization, a physical division of the wire into short curve segments (Fig. 16.2.2). We let P_i denote the point $(x(t_i), y(t_i), z(t_i))$ for $i = 0, 1, 2, \ldots, n$. Then the points $P_0, P_1, \ldots, P_n$ are the subdivision points of C.

From our study of arc length in Sections 12.2 and 13.4, we know that the arc length Δs_i of the segment of C from P_{i-1} to P_i is

$$\Delta s_i = \int_{t_{i-1}}^{t_i} \sqrt{[x'(t)]^2 + [y'(t)]^2 + [z'(t)]^2} \, dt$$

$$= \sqrt{[x'(t_i^*)]^2 + [y'(t_i^*)]^2 + [z'(t_i^*)]^2} \, \Delta t \qquad (2)$$

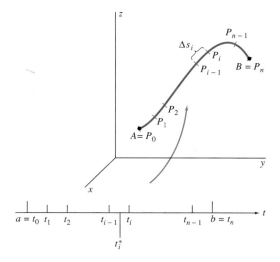

Fig. 16.2.2 The partition of the interval $[a, b]$ determines a related partition of the curve C into short arcs.

for some number t_i^* in the interval $[t_{i-1}, t_i]$. This is a consequence of the average value theorem for integrals of Section 5.5.

If we multiply the density at the point (x_i^*, y_i^*, z_i^*) by the length Δs_i of the segment of C containing that point, we obtain an estimate of the mass of that segment of C. So, after we sum over all the segments, we have an estimate of the total mass m of the wire:

$$m \approx \sum_{i=1}^{n} f(x(t_i^*), y(t_i^*), z(t_i^*)) \, \Delta s_i.$$

The limit of this sum as $\Delta t \to 0$ should be the actual mass m. This is our motivation for the definition of the line integral of the function f along the curve C, denoted by

$$\int_C f(x, y, z) \, ds.$$

Definition   *Line Integral with Respect to Arc Length*
Suppose that the function $f(x, y, z)$ is continuous at each point of the smooth parametric curve C from A to B, as given in (1). Then the **line integral of f along C from A to B with respect to arc length** is defined to be

$$\int_C f(x, y, z) \, ds = \lim_{\Delta t \to 0} \sum_{i=1}^{n} f(x(t_i^*), y(t_i^*), z(t_i^*)) \, \Delta s_i. \tag{3}$$

When we substitute Eq. (2) into Eq. (3), we recognize the result as the limit of a Riemann sum. Therefore,

$$\int_C f(x, y, z) \, ds = \int_a^b f(x(t), y(t), z(t)) \sqrt{[x'(t)]^2 + [y'(t)]^2 + [z'(t)]^2} \, dt. \tag{4}$$

Thus we may evaluate the line integral $\int_C f(x, y, z) \, ds$ by expressing everything in terms of the parameter t, including the symbolic arc-length element

$$ds = \sqrt{[x'(t)]^2 + [y'(t)]^2 + [z'(t)]^2}\; dt.$$

The result—the right-hand side of Eq. (4)—is an **ordinary integral with respect to the single real variable t.**

A curve C that lies in the xy-plane may be regarded as a space curve for which z [and $z'(t)$] are zero. In this case we simply suppress the variable z in Eq. (4) and write

$$\int_C f(x, y)\; ds = \int_a^b f(x(t), y(t))\sqrt{[x'(t)]^2 + [y'(t)]^2}\; dt. \qquad (5)$$

EXAMPLE 1 Evaluate the line integral

$$\int_C xy\; ds,$$

where C is the first-quadrant quarter-circle parametrized by $x = \cos t$, $y = \sin t$, $0 \le t \le \pi/2$.

Solution Here

$$ds = \sqrt{(-\sin t)^2 + (\cos t)^2}\; dt = dt,$$

so Eq. (5) yields

$$\int_C xy\; ds = \int_0^{\pi/2} \cos t \sin t\; dt = \left[\tfrac{1}{2}\sin^2 t\right]_0^{\pi/2} = \tfrac{1}{2}.$$

Let us now return to the physical wire and denote its density function by $\rho(x, y, z)$. The mass of a small piece of length Δs is $\Delta m = \rho\, \Delta s$, so we write

$$dm = \rho(x, y, z)\; ds$$

for its (symbolic) element of mass. Then the **mass** m of the wire and its **centroid** $(\bar{x}, \bar{y}, \bar{z})$ are defined as follows:

$$m = \int_C dm = \int_C \rho\; ds, \qquad \bar{y} = \frac{1}{m}\int_C y\; dm,$$

$$\bar{x} = \frac{1}{m}\int_C x\; dm, \qquad \bar{z} = \frac{1}{m}\int_C z\; dm. \qquad (6)$$

Note the analogy with Eqs. (2) and (4) of Section 15.6. The **moment of inertia** of the wire about a given axis is

$$I = \int_C w^2\; dm, \qquad (7)$$

where $w = w(x, y, z)$ denotes the perpendicular distance from the point (x, y, z) of the wire to the axis in question.

EXAMPLE 2 Find the centroid of a wire that has density $\rho = kz$ and the shape of the helix C with parametrization

$$x = 3\cos t, \qquad y = 3\sin t, \qquad z = 4t, \qquad 0 \le t \le \pi.$$

Solution The mass element of the wire is

$$dm = \rho \, ds = kz \, ds = 4kt \sqrt{(-3 \sin t)^2 + (3 \cos t)^2 + 4^2} \, dt = 20kt \, dt.$$

Hence the formulas in (6) yield

$$m = \int_C \rho \, ds = \int_0^\pi 20kt \, dt = 10k\pi^2;$$

$$\bar{x} = \frac{1}{m} \int_C \rho x \, ds = \frac{1}{10k\pi^2} \int_a^b 60kt \cos t \, dt$$

$$= \frac{6}{\pi^2} \left[\cos t + t \sin t \right]_0^\pi = -\frac{12}{\pi^2} \approx -1.22;$$

$$\bar{y} = \frac{1}{m} \int_C \rho y \, ds = \frac{1}{10k\pi^2} \int_a^b 60kt \sin t \, dt$$

$$= \frac{6}{\pi^2} \left[\sin t - t \cos t \right]_0^\pi = \frac{6}{\pi} \approx 1.91;$$

$$\bar{z} = \frac{1}{m} \int_C \rho z \, ds = \frac{1}{10k\pi^2} \int_a^b 80kt^2 \, dt = \frac{8}{\pi^2} \left[\frac{1}{3} t^3 \right]_0^\pi = \frac{8\pi}{3} \approx 8.38.$$

So the centroid of the wire is located at the point with approximate coordinates $(-1.22, 1.91, 8.38)$.

LINE INTEGRALS WITH RESPECT TO COORDINATE VARIABLES

We obtain a different type of line integral by replacing Δs_i in Eq. (3) by

$$\Delta x_i = x(t_i) - x(t_{i-1}) = x'(t_i^*) \, \Delta t.$$

The **line integral of f along C with respect to x** is defined to be

$$\int_C f(x, y, z) \, dx = \lim_{\Delta t \to 0} \sum_{i=1}^n f(x(t_i^*), y(t_i^*), z(t_i^*)) \, \Delta x_i.$$

Thus

$$\int_C f(x, y, z) \, dx = \int_a^b f(x(t), y(t), z(t)) \, x'(t) \, dt. \tag{8a}$$

Similarly, the **line integrals of f along C with respect to y** and **with respect to z** are given by

$$\int_C f(x, y, z) \, dy = \int_a^b f(x(t), y(t), z(t)) \, y'(t) \, dt \tag{8b}$$

and

$$\int_C f(x, y, z) \, dz = \int_a^b f(x(t), y(t), z(t)) \, z'(t) \, dt. \tag{8c}$$

The three integrals in (8) typically occur together. If P, Q, and R are continuous functions of the variables x, y, and z, then we write (indeed, *define*)

$$\int_C P\,dx + Q\,dy + R\,dz = \int_C P\,dx + \int_C Q\,dy + \int_C R\,dz. \qquad (9)$$

The line integrals in Eqs. (8) and (9) are evaluated by expressing x, y, z, dx, dy, and dz in terms of t as determined by a suitable parametrization of the curve C. The result is an ordinary single-variable integral. For instance, if C is a parametric plane curve parametrized over the interval $[a, b]$ by $\mathbf{r}(t) = \langle x(t), y(t) \rangle$, then

$$\int_C P\,dx + Q\,dy = \int_a^b [P(x(t), y(t))\, x'(t) + Q(x(t), y(t))\, y'(t)]\, dt.$$

EXAMPLE 3 Evaluate the line integral

$$\int_C y\,dx + z\,dy + x\,dz,$$

where C is the parametric curve $x = t$, $y = t^2$, $z = t^3$, $0 \leq t \leq 1$.

Solution Because $dx = dt$, $dy = 2t\,dt$, and $dz = 3t^2\,dt$, substitution in terms of t yields

$$\int_C y\,dx + z\,dy + x\,dz = \int_0^1 t^2\,dt + t^3(2t\,dt) + t(3t^2\,dt)$$

$$= \int_0^1 (t^2 + 3t^3 + 2t^4)\,dt$$

$$= \left[\tfrac{1}{3}t^3 + \tfrac{3}{4}t^4 + \tfrac{2}{5}t^5 \right]_0^1 = \tfrac{89}{60}.$$

There is an important difference between the line integral in Eq. (4) with respect to arc length s and the line integrals in Eq. (9) with respect to the coordinate variables x, y, and z. Suppose that the *orientation* of the curve C (the direction in which it is traced as t increases) is reversed. Then, because of the terms $x'(t)$, $y'(t)$, and $z'(t)$ in Eqs. (8), the *sign* of the line integral in Eq. (9) is changed. But this reversal of orientation does *not* change the value of the line integral in Eq. (4). We may express this by writing

$$\int_{-C} f\,ds = \int_C f\,ds, \qquad (10)$$

in contrast with the formula

$$\int_{-C} P\,dx + Q\,dy + R\,dz = -\int_C P\,dx + Q\,dy + R\,dz. \qquad (11)$$

Here the symbol $-C$ denotes the curve C with its orientation reversed (from B to A rather than from A to B). It is proved in advanced calculus that for either type of line integral, two one-to-one parametrizations of the smooth curve C *that agree in orientation* give the same value.

If the curve C consists of a finite number of smooth curves joined at consecutive corner points, then we say that C is **piecewise smooth**. Then the value of a line integral along C is defined to be the sum of its values along the smooth segments of C.

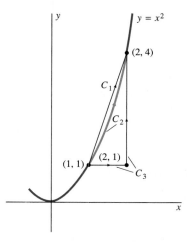

Fig. 16.2.3 The three arcs of Example 4

EXAMPLE 4 Evaluate the line integral

$$\int_C y \, dx + 2x \, dy$$

for each of these three curves C (Fig. 16.2.3):

C_1: The straight line segment in the plane from $A(1, 1)$ to $B(2, 4)$;

C_2: The plane path from $A(1, 1)$ to $B(2, 4)$ along the graph of the parabola $y = x^2$; and

C_3: The straight line in the plane from $A(1, 1)$ to $Q(2, 1)$ followed by the straight line from $Q(2, 1)$ to $B(2, 4)$.

Solution   The straight line segment C_1 from A to B can be parametrized by $x = 1 + t$, $y = 1 + 3t$, $0 \leq t \leq 1$. Hence

$$\int_{C_1} y \, dx + 2x \, dy = \int_0^1 (1 + 3t) \, dt + 2(1 + t)(3 \, dt)$$

$$= \int_0^1 (7 + 9t) \, dt = \tfrac{23}{2}.$$

Next, the arc C_2 of the parabola $y = x^2$ from A to B is "self-parametrizing": It has the parametrization $x = x$, $y = x^2$, $1 \leq x \leq 2$. So

$$\int_{C_2} y \, dx + 2x \, dy = \int_1^2 (x^2)(dx) + 2(x)(2x \, dx) = \int_1^2 5x^2 \, dx = \tfrac{35}{3}.$$

Finally, along the straight line segment from $(1, 1)$ to $(2, 1)$ we have $y \equiv 1$ and (because y is a constant) $dy = 0$. Along the vertical segment from $(2, 1)$ to $(2, 4)$ we have $x \equiv 2$ and $dx = 0$. Therefore,

$$\int_{C_3} y \, dx + 2x \, dy = \int_1^2 [(1)(dx) + (2x)(0)] + \int_1^4 [(y)(0) + (4)(dy)]$$

$$= \int_1^2 1 \, dx + \int_1^4 4 \, dy = 13.$$

Example 4 shows that we may well obtain different values for the line integral from A to B if we evaluate it along different curves from A to B. Thus this line integral is *path-dependent*. We shall give in Section 16.3 a sufficient condition for the line integral

$$\int_C P \, dx + Q \, dy + R \, dz$$

to have the same value for *all* smooth curves C from A to B, and thus for the integral to be *independent of path*.

LINE INTEGRALS AND WORK

Suppose now that $\mathbf{F} = P\mathbf{i} + Q\mathbf{j} + R\mathbf{k}$ is a force field defined on a region that contains the curve C from the point A to the point B. Suppose also that C has a parametrization

$$\mathbf{r}(t) = \mathbf{i}x(t) + \mathbf{j}y(t) + \mathbf{k}z(t), \qquad t \text{ in } [a, b],$$

with a *nonzero* velocity vector

$$\mathbf{v} = \mathbf{i}\frac{dx}{dt} + \mathbf{j}\frac{dy}{dt} + \mathbf{k}\frac{dz}{dt}.$$

The speed associated with this velocity vector is

$$v = |\mathbf{v}| = \sqrt{\left(\frac{dx}{dt}\right)^2 + \left(\frac{dy}{dt}\right)^2 + \left(\frac{dz}{dt}\right)^2}.$$

Recall from Section 13.5 that the *unit tangent vector* to the curve C is

$$\mathbf{T} = \frac{\mathbf{v}}{v} = \frac{1}{v}\left(\frac{dx}{dt}\mathbf{i} + \frac{dy}{dt}\mathbf{j} + \frac{dz}{dt}\mathbf{k}\right).$$

We want to approximate the work W done by the force field $\mathbf{F}$ in moving a particle along the curve C from A to B. Subdivide C as indicated in Fig. 16.2.4. Think of $\mathbf{F}$ moving the particle from P_{i-1} to P_i, two consecutive division points of C. The work ΔW_i done is approximately the product of the distance Δs_i from P_{i-1} to P_i (measured along C) and the tangential component $\mathbf{F} \cdot \mathbf{T}$ of the force $\mathbf{F}$ at a typical point $(x(t_i^*), y(t_i^*), z(t_i^*))$ between P_{i-1} and P_i. Thus

$$\Delta W_i \approx \mathbf{F}(x(t_i^*), y(t_i^*), z(t_i^*)) \cdot \mathbf{T}(t_i^*)\, \Delta s_i,$$

so the total work W is given approximately by

$$W \approx \sum_{i=1}^{n} \mathbf{F}(x(t_i^*), y(t_i^*), z(t_i^*)) \cdot \mathbf{T}(t_i^*)\, \Delta s_i.$$

This approximation suggests that we *define* the **work** W as

$$W = \int_C \mathbf{F} \cdot \mathbf{T}\, ds. \tag{12}$$

Thus *work is the integral with respect to arc length of the tangential component of the force.* Intuitively, we may regard $dW = \mathbf{F} \cdot \mathbf{T}\, ds$ as the infinitesimal element of work done by the tangential component $\mathbf{F} \cdot \mathbf{T}$ of the force in moving the particle along the arc-length element ds. The line integral in Eq. (12) is then the "sum" of all these infinitesimal elements of work.

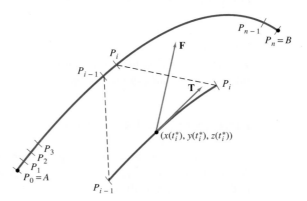

Fig. 16.2.4 The component of $\mathbf{F}$ along C from P_{i-1} to P_i is $\mathbf{F} \cdot \mathbf{T}$.

It is customary to write formally

$$\mathbf{r} = x\mathbf{i} + y\mathbf{j} + z\mathbf{k}, \qquad d\mathbf{r} = \mathbf{i}\,dx + \mathbf{j}\,dy + \mathbf{k}\,dz,$$

and

$$\mathbf{T}\,ds = \left(\frac{dx}{ds}\mathbf{i} + \frac{dy}{ds}\mathbf{j} + \frac{dz}{ds}\mathbf{k}\right)ds = d\mathbf{r}.$$

With this notation, Eq. (12) takes the form

$$W = \int_C \mathbf{F} \cdot d\mathbf{r} \tag{13}$$

that is common in engineering and physics texts.

To evaluate the line integral in Eqs. (12) or (13), we express its integrand in terms of the parameter t, as usual. Thus

$$W = \int_C \mathbf{F} \cdot \mathbf{T}\,ds$$
$$= \int_a^b (P\mathbf{i} + Q\mathbf{j} + R\mathbf{k}) \cdot \frac{1}{v}\left(\frac{dx}{dt}\mathbf{i} + \frac{dy}{dt}\mathbf{j} + \frac{dz}{dt}\mathbf{k}\right)v\,dt$$
$$= \int_a^b \left(P\frac{dx}{dt} + Q\frac{dy}{dt} + R\frac{dz}{dt}\right)dt.$$

Therefore,

$$W = \int_C P\,dx + Q\,dy + R\,dz. \tag{14}$$

This computation reveals an important relation between the two types of line integrals we have defined here.

Theorem 1 *Equivalent Line Integrals*

Suppose that the vector field $\mathbf{F} = P\mathbf{i} + Q\mathbf{j} + R\mathbf{k}$ has continuous component functions and that $\mathbf{T}$ is the unit tangent vector to the smooth curve C. Then

$$\int_C \mathbf{F} \cdot \mathbf{T}\,ds = \int_C P\,dx + Q\,dy + R\,dz. \tag{15}$$

REMARK   If the orientation of the curve C is reversed, then the sign of the right-hand integral of Eq. (15) is changed according to Eq. (11), whereas the sign of the left-hand integral is changed because $\mathbf{T}$ is replaced by $-\mathbf{T}$.

EXAMPLE 5   The work done by the force field $\mathbf{F} = y\mathbf{i} + z\mathbf{j} + x\mathbf{k}$ in moving a particle from $(0, 0, 0)$ to $(1, 1, 1)$ along the twisted cubic $x = t$, $y = t^2$, $z = t^3$ is given by the line integral

$$W = \int_C \mathbf{F} \cdot \mathbf{T}\,ds = \int_C y\,dx + z\,dy + x\,dz,$$

and we computed the value of this integral in Example 3. Hence $W = \frac{89}{60}$.

EXAMPLE 6 Find the work done by the inverse-square force field

$$\mathbf{F}(x, y, z) = \frac{k\mathbf{r}}{r^3} = \frac{k(x\mathbf{i} + y\mathbf{j} + z\mathbf{k})}{(x^2 + y^2 + z^2)^{3/2}}$$

in moving a particle along the straight line segment C from $(0, 4, 0)$ to $(0, 4, 3)$.

Solution Along C we have $x = 0$, $y = 4$, and z varying from 0 to 3. Thus we choose z as the parameter:

$$x \equiv 0, \qquad y \equiv 4, \quad \text{and} \quad z = z, \qquad 0 \le z \le 3.$$

Because $dx = 0 = dy$, Eq. (14) gives

$$W = \int_C \frac{k(x\,dx + y\,dy + z\,dz)}{(x^2 + y^2 + z^2)^{3/2}} = \int_0^3 \frac{kz}{(16 + z^2)^{3/2}}\,dz$$

$$= \left[\frac{-k}{\sqrt{16 + z^2}}\right]_0^3 = \frac{k}{20}.$$

16.2 Problems

In Problems 1 through 5, evaluate the line integrals

$$\int_C f(x, y)\;ds, \qquad \int_C f(x, y)\,dx, \quad and \quad \int_C f(x, y)\,dy$$

along the indicated parametric curve.

1. $f(x, y) = x^2 + y^2$; $x = 4t - 1, y = 3t + 1$, $-1 \le t \le 1$

2. $f(x, y) = x$; $x = t, y = t^2, 0 \le t \le 1$

3. $f(x, y) = x + y$; $x = e^t + 1, y = e^t - 1$, $0 \le t \le \ln 2$

4. $f(x, y) = 2x - y$; $x = \sin t, y = \cos t$, $0 \le t \le \pi/2$

5. $f(x, y) = xy$; $x = 3t, y = t^4, 0 \le t \le 1$

6. Evaluate $\int_C xy\,dx + (x + y)\,dy$, where C is the part of the graph of $y = x^2$ from $(-1, 1)$ to $(2, 4)$.

7. Evaluate $\int_C y^2\,dx + x\,dy$, where C is the part of the graph of $x = y^3$ from $(-1, -1)$ to $(1, 1)$.

8. Evaluate $\int_C y\sqrt{x}\,dx + x\sqrt{x}\,dy$, where C is the part of the graph of $y^2 = x^3$ from $(1, 1)$ to $(4, 8)$.

9. Evaluate the line integral $\int_C x^2y\,dx + xy^3\,dy$, where C consists of the line segments from $(-1, 1)$ to $(2, 1)$ and from $(2, 1)$ to $(2, 5)$.

10. Evaluate $\int_C (x + 2y)\,dx + (2x - y)\,dy$, where C consists of the line segments from $(3, 2)$ to $(3, -1)$ and from $(3, -1)$ to $(-2, -1)$.

In Problems 11 through 15, evaluate the line integral $\int_C \mathbf{F} \cdot \mathbf{T}\,ds$ along the indicated path C.

11. $\mathbf{F} = z\mathbf{i} + x\mathbf{j} - y\mathbf{k}$; C is parametrized by $x = t$, $y = t^2, z = t^3, 0 \le t \le 1$.

12. $\mathbf{F} = yz\mathbf{i} + xz\mathbf{j} + xy\mathbf{k}$; C is the straight line segment from $(2, -1, 3)$ to $(4, 2, -1)$.

13. $\mathbf{F} = y\mathbf{i} - x\mathbf{j} + z\mathbf{k}$; $x = \sin t, y = \cos t, z = 2t$, $0 \le t \le \pi$.

14. $\mathbf{F} = (2x + 3y)\mathbf{i} + (3x + 2y)\mathbf{j} + 3z^2\mathbf{k}$; C is the path from $(0, 0, 0)$ to $(4, 2, 3)$ that consists of three line segments parallel to the x-axis, the y-axis, and the z-axis, in that order.

15. $\mathbf{F} = yz^2\mathbf{i} + xz^2\mathbf{j} + 2xyz\mathbf{k}$; C is the path from $(-1, 2, -2)$ to $(1, 5, 2)$ consisting of three line segments parallel to the z-axis, the x-axis, and the y-axis, in that order.

In Problems 16 through 18, evaluate $\int_C f(x, y, z)\,ds$ for the given function $f(x, y, z)$ and the given path C.

16. $f(x, y, z) = xyz$; C is the straight line segment from $(1, -1, 2)$ to $(3, 2, 5)$.

17. $f(x, y, z) = 2x + 9xy$; C is the curve $x = t$, $y = t^2, z = t^3, 0 \le t \le 1$.

18. $f(x, y, z) = xy$; C is the elliptical helix $x = 4\cos t$, $y = 9\sin t, z = 7t, 0 \le t \le 5\pi/2$.

19. Find the centroid of a uniform thin wire shaped like the semicircle $x^2 + y^2 = a^2, y \ge 0$.

20. Find the moments of inertia about the x- and y-axes of the wire of Problem 19.

21. Find the mass and centroid of a wire that has constant density $\rho = k$ and is shaped like the helix $x = 3\cos t$, $y = 3\sin t, z = 4t, 0 \le t \le 2\pi$.

22. Find the moment of inertia about the z-axis of the wire of Example 1 of this section.

23. A wire shaped like the first-quadrant portion of the circle $x^2 + y^2 = a^2$ has density $\rho = kxy$ at the point (x, y). Find its mass, centroid, and moment of inertia about each coordinate axis.

24. Find the work done by the inverse-square force field of Example 6 in moving a particle from $(1, 0, 0)$ to $(0, 3, 4)$. Integrate first along the line segment from $(1, 0, 0)$ to $(5, 0, 0)$ and then along a path on the sphere with equation $x^2 + y^2 + z^2 = 25$. The second integral is automatically zero. (Why?)

25. Imagine an infinitely long and uniformly charged wire that coincides with the z-axis. The electric force that it exerts on a unit charge at the point $(x, y) \neq (0, 0)$ in the xy-plane is

$$\mathbf{F}(x, y) = \frac{k(x\mathbf{i} + y\mathbf{j})}{x^2 + y^2}.$$

Find the work done by $\mathbf{F}$ in moving a unit charge along the straight line segment from (a) $(1, 0)$ to $(1, 1)$; (b) $(1, 1)$ to $(0, 1)$.

26. Show that if $\mathbf{F}$ is a *constant* force field, then it does zero work on a particle that moves once uniformly counterclockwise around the unit circle in the xy-plane.

27. Show that if $\mathbf{F} = k\mathbf{r} = k(x\mathbf{i} + y\mathbf{j})$, then $\mathbf{F}$ does zero work on a particle that moves once uniformly counterclockwise around the unit circle in the xy-plane.

28. Find the work done by the force field $\mathbf{F} = -y\mathbf{i} + x\mathbf{j}$ in moving a particle counterclockwise once around the unit circle in the xy-plane.

29. Let C be a curve on the unit sphere $x^2 + y^2 + z^2 = 1$. Explain why the inverse-square force field of Example 6 does zero work in moving a particle along C.

In Problems 30 through 32, the given curve C joins the points P and Q in the xy-plane. The point P represents the top of a ten-story building, and Q is a point on the ground 100 ft from the base of the building. A 150-lb person slides down a frictionless slide shaped like the curve from P to Q under the influence of the gravitational force $\mathbf{F} = -150\mathbf{j}$. In each problem show that $\mathbf{F}$ does the same amount of work on the person, $W = 15{,}000$ ft·lb, as if he or she had dropped straight down to the ground.

30. C is the straight line segment $y = x$ from $P(100, 100)$ to $Q(0, 0)$.

31. C is the circular arc $x = 100 \sin t$, $y = 100 \cos t$ from $P(0, 100)$ to $Q(100, 0)$.

32. C is the parabolic arc $y = x^2/100$ from $P(100, 100)$ to $Q(0, 0)$.

16.3
Independence of Path

Let $\mathbf{F} = P\mathbf{i} + Q\mathbf{j} + R\mathbf{k}$ be a vector field with continuous component functions. We know by Theorem 1 in Section 16.2 that

$$\int_C \mathbf{F} \cdot \mathbf{T} \, ds = \int_C P \, dx + Q \, dy + R \, dz \tag{1}$$

for any piecewise smooth curve C. Thus the two sides of Eq. (1) are two different ways of writing the same line integral. Here we discuss the question of whether this line integral has the *same value* for *any* two curves with the same endpoints (the same initial point and the same terminal point).

Definition *Independence of Path*

The line integral in Eq. (1) is said to be **independent of the path in the region D** provided that, given any two points A and B of D, the integral has the same value along every piecewise smooth curve, or **path**, in D from A to B. In this case we may write

$$\int_C \mathbf{F} \cdot \mathbf{T} \, ds = \int_A^B \mathbf{F} \cdot \mathbf{T} \, ds \tag{2}$$

because the value of the integral depends only on the points A and B and not on the particular choice of the path C joining them.

For a tangible interpretation of independence of path, let us think of walking along the curve C from point A to point B in the plane where a wind with velocity vector $\mathbf{w}(x, y)$ is blowing. Suppose that when we are at (x, y), the wind exerts a force $\mathbf{F} = k\mathbf{w}(x, y)$ on us, k being a constant that depends on our size and shape (and perhaps other factors as well). Then, by Eq. (12) of Section 16.2, the amount of work the wind does on us as we walk along C is given by

$$W = \int_C \mathbf{F} \cdot \mathbf{T} \, ds = k \int_C \mathbf{w} \cdot \mathbf{T} \, ds. \tag{3}$$

This is the wind's contribution to our trip from A to B. In this context, the question of independence of path is whether or not the wind's work W depends on *which* path we choose from point A to point B.

EXAMPLE 1 Suppose that a steady wind blows toward the northeast with velocity vector $\mathbf{w} = 10\mathbf{i} + 10\mathbf{j}$ in fps units; its speed is $|\mathbf{w}| = 10\sqrt{2} \approx 14$ ft/s—about 10 mi/h. Assume that $k = 0.5$, so the wind exerts 0.5 lb of force for each foot per second of its velocity. Then $\mathbf{F} = 5\mathbf{i} + 5\mathbf{j}$, so Eq. (3) yields

$$W = \int_C \langle 5, 5 \rangle \cdot \mathbf{T} \, ds = \int_C 5 \, dx + 5 \, dy \tag{4}$$

for the work done on us by the wind as we walk along C.

For instance, if C is the straight path $x = 10t$, $y = 10t$ $(0 \le t \le 1)$ from $(0, 0)$ to $(10, 10)$, then Eq. (4) gives

$$W = \int_0^1 5 \cdot 10 \, dt + 5 \cdot 10 \, dt = 100 \int_0^1 1 \, dt = 100$$

ft·lb of work. Or, if C is the parabolic path $y = \frac{1}{10}x^2$, $0 \le x \le 10$ from the same initial point $(0, 0)$ to be the same terminal point $(10, 10)$, then Eq. (4) yields

$$W = \int_0^{10} 5 \cdot dx + 5 \cdot \tfrac{1}{5} x \, dx = \int_0^{10} (5 + x) \, dx$$

$$= \left[5x + \tfrac{1}{2}x^2 \right]_0^{10} = 100$$

ft·lb of work, the same as before. We shall see that it follows from Theorem 1 of this section that the line integral in Eq. (4) is independent of path, so the wind does 100 ft·lb of work along any path from $(0, 0)$ to $(10, 10)$.

EXAMPLE 2 Suppose that $\mathbf{w} = -2y\mathbf{i} + 2x\mathbf{j}$. This wind is blowing counterclockwise around the origin, as in a hurricane with its eye at $(0, 0)$. With $k = 0.5$ as before, $\mathbf{F} = -y\mathbf{i} + x\mathbf{j}$, so the work integral is

$$W = \int_C \mathbf{F} \cdot \mathbf{T} \, ds = \int_C -y \, dx + x \, dy. \tag{5}$$

If we walk from $(10, 0)$ to $(-10, 0)$ along the straight path C_1 (through the eye of the hurricane!), then the wind is always perpendicular to our unit tangent vector $\mathbf{T}$ (Fig. 16.3.1). Hence $\mathbf{F} \cdot \mathbf{T} = 0$, and therefore

$$W = \int_{C_1} \mathbf{F} \cdot \mathbf{T} \, ds = \int_{C_1} -y \, dx + x \, dy = 0.$$

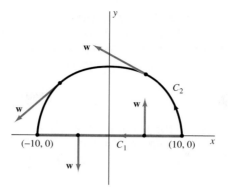

Fig. 16.3.1 Around and through the eye of the hurricane (Example 2)

But if we walk along the semicircular path C_2 shown in Fig. 16.3.1, then **w** remains tangent to our path, so $\mathbf{F} \cdot \mathbf{T} = |\mathbf{F}| = 10$ at each point. In this case,

$$W = \int_{C_2} -y\,dx + x\,dy = \int_{C_2} \mathbf{F} \cdot \mathbf{T}\,ds = 10 \cdot 10\pi = 100\pi.$$

The fact that we get different values along different paths from $(10, 0)$ to $(-10, 0)$ shows that the line integral in Eq. (5) is *not* independent of path.

Theorem 1 tells when a given line integral is independent of path and when it is not.

> **Theorem 1   *Independence of Path***
> The line integral $\int_C \mathbf{F} \cdot \mathbf{T}\,ds$ is independent of path in the plane region D if and only if $\mathbf{F} = \nabla f$ for some function f defined on D.

Proof   Suppose that $\mathbf{F} = \nabla f = \langle \partial f/\partial x, \partial f/\partial y, \partial f/\partial z \rangle$ and that C is a path from A to B in D parametrized as usual with parameter t in $[a, b]$. Then, by Eq. (1),

$$\int_C \mathbf{F} \cdot \mathbf{T}\,ds = \int_C \frac{\partial f}{\partial x}\,dx + \frac{\partial f}{\partial y}\,dy + \frac{\partial f}{\partial z}\,dz$$

$$= \int_a^b \left(\frac{\partial f}{\partial x} \cdot \frac{dx}{dt} + \frac{\partial f}{\partial y} \cdot \frac{dy}{dt} + \frac{\partial f}{\partial z} \cdot \frac{dz}{dt} \right) dt$$

$$= \int_a^b D_t[f(x(t), y(t), z(t))]\,dt$$

$$= f(x(b), y(b), z(b)) - f(x(a), y(a), z(a))$$

by the fundamental theorem of calculus. Therefore,

$$\int_C \mathbf{F} \cdot \mathbf{T}\,ds = f(B) - f(A). \tag{6}$$

Equation (6) shows that the value of the line integral depends only on the points A and B and is therefore independent of the choice of the particular path C. This proves the *if* part of Theorem 1.

To prove the *only if* part of Theorem 1, we suppose that the line integral is independent of path in D. Choose a *fixed* point $A_0 = A_0(x_0, y_0, z_0)$ in D, and let $B = B(x, y, z)$ be an arbitrary point in D. Given any path C from A_0 to B in D, we *define* the function f by means of the equation

$$f(x, y, z) = \int_C \mathbf{F} \cdot \mathbf{T} \, ds = \int_{(x_0, y_0, z_0)}^{(x, y, z)} \mathbf{F} \cdot \mathbf{T} \, ds. \tag{7}$$

Because of the hypothesis of independence of path, the resulting value of $f(x, y, z)$ depends only on (x, y, z) and not on the particular path C used. We shall omit the verification that $\nabla f = \mathbf{F}$ (see Problem 21 of Section 16.7). ❏

As an application of Theorem 1, we see that the vector field $\mathbf{F} = -y\mathbf{i} + x\mathbf{j}$ of Example 2 is not the gradient of any scalar function f because $\int \mathbf{F} \cdot \mathbf{T} \, ds$ is not independent of path. More precisely, $\int \mathbf{F} \cdot \mathbf{T} \, ds$ is not independent of path in any region that either includes or encloses the origin.

Definition   *Conservative Fields and Potential Functions*
The vector field $\mathbf{F}$ defined on a region D is **conservative** provided that there exists a scalar function f defined on D such that

$$\mathbf{F} = \nabla f \tag{8}$$

at each point of D. In this case f is called a **potential function** for the vector field $\mathbf{F}$.[†]

Note that Eq. (6) has the form of a "fundamental theorem of calculus for line integrals," with the potential function f playing the role of an antiderivative. If the line integral $\int \mathbf{F} \cdot \mathbf{T} \, ds$ is known to be independent of path, then Theorem 1 guarantees that the vector field $\mathbf{F}$ is conservative and that Eq. (7) yields a potential function for $\mathbf{F}$.

EXAMPLE 3   Find a potential function for the conservative vector field

$$\mathbf{F}(x, y) = (6xy - y^3)\mathbf{i} + (4y + 3x^2 - 3xy^2)\mathbf{j}. \tag{9}$$

Solution   Because we are given the information that $\mathbf{F}$ is a conservative field, the line integral $\int \mathbf{F} \cdot \mathbf{T} \, ds$ is independent of path by Theorem 1. Therefore, we may apply Eq. (7) to find a scalar potential function $\mathbf{F}$. Let C be the straight-line path from $A(0, 0)$ to $B(x_1, y_1)$ parametrized by $x = x_1 t, y = y_1 t$, $0 \leqq t \leqq 1$. Then Eq. (7) yields

$$f(x_1, y_1) = \int_A^B \mathbf{F} \cdot \mathbf{T} \, ds$$

$$= \int_A^B (6xy - y^3) \, dx + (4y + 3x^2 - 3xy^2) \, dy$$

$$= \int_0^1 (6x_1 y_1 t^2 - y_1^3 t^3)(x_1 \, dt) + (4y_1 t + 3x_1^2 t^2 - 3x_1 y_1^2 t^3)(y_1 \, dt)$$

[†] In some physical applications the scalar function f is called a *potential function* for the vector field $\mathbf{F}$ provided that $\mathbf{F} = -\nabla f$.

$$= \int_0^1 (4y_1^2 t + 9x_1^2 y_1 t^2 - 4x_1 y_1^3 t^3) \, dt$$

$$= \left[2y_1^2 t^2 + 3x_1^2 y_1 t^3 - x_1 y_1^3 t^4 \right]_0^1 = 2y_1^2 + 3x_1^2 y_1 - x_1 y_1^3.$$

At this point we delete the subscripts, because (x_1, y_1) is an arbitrary point of the plane. Thus we obtain the potential function

$$f(x, y) = 2y^2 + 3x^2 y - xy^3$$

for the vector field $\mathbf{F}$ in Eq. (9). As a check, we can differentiate f to obtain

$$\frac{\partial f}{\partial x} = 6xy - y^3, \qquad \frac{\partial f}{\partial y} = 4y + 3x^2 - 3xy^2.$$

But how did we know in advance that the vector field $\mathbf{F}$ was conservative? The answer is provided by Theorem 2.[†]

Theorem 2   Conservative Fields and Potential Functions

Suppose that the functions $P(x, y)$ and $Q(x, y)$ are continuous and have continuous first-order partial derivatives in the open rectangle $R = \{(x, y) \mid a < x < b, c < y < d\}$. Then the vector field $\mathbf{F} = P\mathbf{i} + Q\mathbf{j}$ is conservative in R—and hence has a potential function $f(x, y)$ defined on R—if and only if

$$\frac{\partial P}{\partial y} = \frac{\partial Q}{\partial x} \tag{10}$$

at each point of R.

Observe that the vector field $\mathbf{F}$ in Eq. (9), where $P(x, y) = 6xy - y^3$ and $Q(x, y) = 4y + 3x^2 - 3xy^2$, satisfies the criterion in Eq. (10) because

$$\frac{\partial P}{\partial y} = 6x - 3y^2 = \frac{\partial Q}{\partial x}.$$

When this sufficient condition for the existence of a potential function is satisfied, the method illustrated in Example 4 is usually an easier way to find a potential function than the evaluation of the line integral in Eq. (7)—the method used in Example 3.

EXAMPLE 4   Given

$$P(x, y) = 6xy - y^3 \quad \text{and} \quad Q(x, y) = 4y + 3x^2 - 3xy^2,$$

note that P and Q satisfy the condition $\partial P / \partial y = \partial Q / \partial x$. Find a potential function $f(x, y)$ such that

$$\frac{\partial f}{\partial x} = 6xy - y^3 \quad \text{and} \quad \frac{\partial f}{\partial y} = 4y + 3x^2 - 3xy^2. \tag{11}$$

[†] For a proof, see Section 1.7 of C. H. Edwards, Jr., and David E. Penney, *Elementary Differential Equations with Boundary Value Problems* (Englewood Cliffs, N.J.: Prentice Hall, 1993).

Solution   Upon integrating the first of these two equations with respect to x, we get

$$f(x, y) = 3x^2y - xy^3 + \xi(y), \tag{12}$$

where $\xi(y)$ is an "arbitrary function" of y alone; it acts as a "constant of integration" with respect to x, because its derivative with respect to x is zero. We next determine $\xi(y)$ by imposing the second condition in (11):

$$\frac{\partial f}{\partial y} = 3x^2 - 3xy^2 + \xi'(y) = 4y + 3x^2 - 3xy^2.$$

It follows that $\xi'(y) = 4y$, so $\xi(y) = 2y^2 + C$. When we set $C = 0$ and substitute the result into Eq. (12), we get the same potential function

$$f(x, y) = 3x^2y - xy^3 + 2y^2$$

that we found by entirely different methods in Example 3.

CONSERVATIVE FORCE FIELDS

Given a conservative force field $\mathbf{F}$, it is customary in physics to introduce a minus sign and write $\mathbf{F} = -\nabla V$. Then $V(x, y, z)$ is called the **potential energy** at the point (x, y, z). With $f = -V$ in Eq. (6), we have

$$W = \int_A^B \mathbf{F} \cdot \mathbf{T}\, ds = V(A) - V(B), \tag{13}$$

and this means that the work W done by $\mathbf{F}$ in moving a particle from A to B is equal to the *decrease* in potential energy.

For example, a brief computation shows that

$$\nabla\left(-\frac{k}{\sqrt{x^2 + y^2 + z^2}}\right) = \frac{k(x\mathbf{i} + y\mathbf{j} + z\mathbf{k})}{(x^2 + y^2 + z^2)^{3/2}} = \mathbf{F}$$

for the inverse-square force field of Example 6 in Section 16.2. With $V = k/(x^2 + y^2 + z^2)^{1/2}$, Eq. (16) then gives

$$\int_{(0,4,0)}^{(0,4,3)} \mathbf{F} \cdot \mathbf{T}\, ds = \frac{k}{\sqrt{0^2 + 4^2 + 0^2}} - \frac{k}{\sqrt{0^2 + 4^2 + 3^2}} = \frac{k}{20},$$

as we found by direct integration.

Here is the reason why the expression *conservative field* is used. Suppose that a particle of mass m moves from A to B under the influence of the conservative force $\mathbf{F}$, with position vector $\mathbf{r}(t)$, $a \leqq t \leqq b$. Then Newton's law $\mathbf{F}(\mathbf{r}(t)) = m\mathbf{r}''(t) = m\mathbf{v}'(t)$ gives

$$\int_A^B \mathbf{F} \cdot \mathbf{T}\, ds = \int_a^b m\mathbf{v}'(t) \cdot \frac{\mathbf{v}(t)}{v}\, v\, dt$$

$$= \int_a^b mD_t[\tfrac{1}{2}\mathbf{v}(t) \cdot \mathbf{v}(t)]\, dt = \left[\tfrac{1}{2}m[v(t)]^2\right]_a^b.$$

Thus with the abbreviations v_A for $v(a)$ and v_B for $v(b)$, we see that

$$\int_A^B \mathbf{F} \cdot \mathbf{T} \, ds = \tfrac{1}{2} m(v_B)^2 - \tfrac{1}{2} m(v_A)^2. \tag{14}$$

By equating the right-hand sides of Eqs. (13) and (14), we get the formula

$$\tfrac{1}{2} m(v_A)^2 + V(A) = \tfrac{1}{2} m(v_B)^2 + V(B). \tag{15}$$

This is the law of **conservation of mechanical energy** for a particle moving under the influence of a *conservative* force field: The sum of its kinetic energy and its potential energy remains constant.

16.3 Problems

Apply the method of Example 4 to find a potential function for the vector fields in Problems 1 through 12.

1. $\mathbf{F}(x, y) = (2x + 3y)\mathbf{i} + (3x + 2y)\mathbf{j}$

2. $\mathbf{F}(x, y) = (4x - y)\mathbf{i} + (6y - x)\mathbf{j}$

3. $\mathbf{F}(x, y) = (3x^2 + 2y^2)\mathbf{i} + (4xy + 6y^2)\mathbf{j}$

4. $\mathbf{F}(x, y) = (2xy^2 + 3x^2)\mathbf{i} + (2x^2y + 4y^3)\mathbf{j}$

5. $\mathbf{F}(x, y) = \left(x^3 + \dfrac{y}{x}\right)\mathbf{i} + (y^2 + \ln x)\mathbf{j}$

6. $\mathbf{F}(x, y) = (1 + ye^{xy})\mathbf{i} + (2y + xe^{xy})\mathbf{j}$

7. $\mathbf{F}(x, y) = (\cos x + \ln y)\mathbf{i} + \left(\dfrac{x}{y} + e^y\right)\mathbf{j}$

8. $\mathbf{F}(x, y) = (x + \arctan y)\mathbf{i} + \dfrac{x + y}{1 + y^2}\mathbf{j}$

9. $\mathbf{F}(x, y) = (3x^2y^3 + y^4)\mathbf{i} + (3x^3y^2 + y^4 + 4xy^3)\mathbf{j}$

10. $\mathbf{F}(x, y) = (e^x \sin y + \tan y)\mathbf{i} + (e^x \cos y + x \sec^2 y)\mathbf{j}$

11. $\mathbf{F}(x, y) = \left(\dfrac{2x}{y} - \dfrac{3y^2}{x^4}\right)\mathbf{i} + \left(\dfrac{2y}{x^3} - \dfrac{x^2}{y^2} + \dfrac{1}{\sqrt{y}}\right)\mathbf{j}$

12. $\mathbf{F}(x, y) = \dfrac{2x^{5/2} - 3y^{5/3}}{2x^{5/2}y^{2/3}}\mathbf{i} + \dfrac{3y^{5/3} - 2x^{5/2}}{3x^{3/2}y^{5/3}}\mathbf{j}$

In Problems 13 through 16, apply the method of Example 3 to find a potential function for the indicated vector field.

13. The vector field of Problem 3

14. The vector field of Problem 4

15. The vector field of Problem 9

16. The vector field of Problem 6

In Problems 17 through 22, show that the given line integral is independent of path in the entire xy-plane, and then calculate the value of the line integral.

17. $\displaystyle\int_{(0,0)}^{(1,2)} (y^2 + 2xy) \, dx + (x^2 + 2xy) \, dy$

18. $\displaystyle\int_{(0,0)}^{(1,1)} (2x - 3y) \, dx + (2y - 3x) \, dy$

19. $\displaystyle\int_{(0,0)}^{(1,-1)} 2xe^y \, dx + x^2 e^y \, dy$

20. $\displaystyle\int_{(0,0)}^{(2,\pi)} \cos y \, dx - x \sin y \, dy$

21. $\displaystyle\int_{(\pi/2,\,\pi/2)}^{(\pi,\pi)} (\sin y + y \cos x) \, dx + (\sin x + x \cos y) \, dy$

22. $\displaystyle\int_{(0,0)}^{(1,-1)} (e^y + ye^x) \, dx + (e^x + xe^y) \, dy$

Find a potential function for each of the conservative vector fields in Problems 23 through 25.

23. $\mathbf{F}(x, y, z) = yz\mathbf{i} + xz\mathbf{j} + xy\mathbf{k}$

24. $\mathbf{F}(x, y, z) = (2x - y - z)\mathbf{i} + (2y - x)\mathbf{j} + (2z - x)\mathbf{k}$

25. $\mathbf{F}(x, y, z) = (y \cos z - yze^x)\mathbf{i} + (x \cos z - ze^x)\mathbf{j} - (xy \sin z + ye^x)\mathbf{k}$

26. Let $\mathbf{F}(x, y) = (-y\mathbf{i} + x\mathbf{j})/(x^2 + y^2)$ for x and y not both zero. Calculate the values of

$$\int_C \mathbf{F} \cdot \mathbf{T} \, ds$$

along both the upper and lower halves of the circle $x^2 + y^2 = 1$ from $(1, 0)$ to $(-1, 0)$. Is there a function $f = f(x, y)$ defined for x and y not both zero such that $\nabla f = \mathbf{F}$? Why?

27. Show that if the force field $\mathbf{F} = P\mathbf{i} + Q\mathbf{j}$ is conservative, then $\partial P/\partial y = \partial Q/\partial x$. Show that the force field of Problem 26 satisfies the condition $\partial P/\partial y = \partial Q/\partial x$ but nevertheless is *not* conservative.

28. Suppose that the force field $\mathbf{F} = P\mathbf{i} + Q\mathbf{j} + R\mathbf{k}$ is conservative. Show that $\partial P/\partial y = \partial Q/\partial x$, $\partial P/\partial z = \partial R/\partial x$, and $\partial Q/\partial z = \partial R/\partial y$.

29. Apply Theorem 1 and the result of Problem 28 to show that

$$\int_C 2xy\, dx + x^2\, dy + y^2\, dz$$

is not independent of path.

30. Let $\mathbf{F}(x, y, z) = yz\mathbf{i} + (xz + y)\mathbf{j} + (xy + 1)\mathbf{k}$.

Define the function f by

$$f(x, y, z) = \int_C \mathbf{F} \cdot \mathbf{T}\, ds,$$

where C is the line segment from $(0, 0, 0)$ to (x, y, z). Determine f by evaluating this line integral, and then show that $\nabla f = \mathbf{F}$.

16.4
Green's Theorem

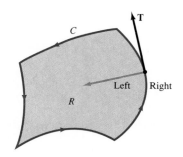

Fig. 16.4.1 Positive orientation of the curve C: The region R within C is to the *left* of the unit tangent vector $\mathbf{T}$.

Green's theorem relates a line integral around a simple closed plane curve C to an ordinary double integral over the plane region R bounded by C. Suppose that the curve C is piecewise smooth—it consists of a finite number of parametric arcs with continuous nonzero velocity vectors. Then C has a unit tangent vector $\mathbf{T}$ everywhere except possibly at a finite number of *corner points*. The **positive**, or **counterclockwise**, direction along C is the direction determined by a parametrization $\mathbf{r}(t)$ of C such that the region R remains on the *left* as the point $\mathbf{r}(t)$ traces the boundary curve C. That is, the vector obtained from the unit tangent vector $\mathbf{T}$ by a counterclockwise rotation through $90°$ always points *into* the region R (Fig. 16.4.1). The symbol

$$\oint_C P\, dx + Q\, dy$$

denotes a line integral along or around C in this positive direction. A reversed arrow on the circle through the integral sign, $\oint$, indicates a line integral around C in the opposite direction, which we call the **negative**, or the **clockwise**, direction.

The following result first appeared (in an equivalent form) in a booklet on the applications of mathematics to electricity and magnetism, published privately in 1828 by the self-taught English mathematical physicist George Green (1793–1841).

Green's Theorem

Let C be a piecewise smooth simple closed curve that bounds the region R in the plane. Suppose that the functions $P(x, y)$ and $Q(x, y)$ are continuous and have continuous first-order partial derivatives on R. Then

$$\oint_C P\, dx + Q\, dy = \iint_R \left(\frac{\partial Q}{\partial x} - \frac{\partial P}{\partial y} \right) dA. \tag{1}$$

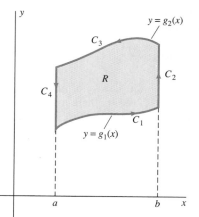

Fig. 16.4.2 The boundary curve C is the union of the four arcs $C_1, C_2, C_3,$ and C_4.

Proof First we give a proof for the case in which the region R is both horizontally simple and vertically simple. Then we indicate how to extend the result to more general regions.

Recall from Section 15.2 that if R is vertically simple, then it has a description of the form $g_1(x) \leq y \leq g_2(x), a \leq x \leq b$. The boundary curve C is then the union of the four arcs $C_1, C_2, C_3,$ and C_4 of Fig. 16.4.2, positively oriented as indicated there. Hence

Ch. 16 / Vector Analysis

$$\oint_C P \, dx = \int_{C_1} P \, dx + \int_{C_2} P \, dx + \int_{C_3} P \, dx + \int_{C_4} P \, dx.$$

The integrals along both C_2 and C_4 are zero, because on those two curves $x(t)$ is constant, so $dx = x'(t) \, dt = 0$. Hence we must compute only the integrals along C_1 and C_3.

The point $(x, g_1(x))$ traces C_1 as x increases from a to b, whereas the point $(x, g_2(x))$ traces C_3 as x *decreases* from b to a. Hence

$$\oint_C P \, dx = \int_a^b P(x, g_1(x)) \, dx + \int_b^a P(x, g_2(x)) \, dx$$

$$= -\int_a^b [P(x, g_2(x)) - P(x, g_1(x))] \, dx = -\int_a^b \int_{g_1(x)}^{g_2(x)} \frac{\partial P}{\partial y} \, dy \, dx$$

by the fundamental theorem of calculus. Thus

$$\oint_C P \, dx = -\iint_R \frac{\partial P}{\partial y} \, dA. \tag{2}$$

In Problem 28 we ask you to show in a similar way that

$$\oint_C Q \, dy = +\iint_R \frac{\partial Q}{\partial x} \, dA \tag{3}$$

if the region R is horizontally simple. We then obtain Eq. (1), the conclusion of Green's theorem, simply by adding Eqs. (2) and (3). ❑

The complete proof of Green's theorem for more general regions is beyond the scope of an elementary text. But the typical region R that appears in practice can be divided into smaller regions $R_1, R_2, \ldots, R_k$ that are both vertically and horizontally simple. Green's theorem for the region R then follows from the fact that it holds for each of the regions $R_1, R_2, \ldots, R_k$ (see Problem 29).

For example, we can divide the horseshoe-shaped region R of Fig. 16.4.3 into the two regions R_1 and R_2, both of which are horizontally simple and vertically simple. We also subdivide the boundary C of R and write $C_1 \cup D_1$ for the boundary of R_1 and $C_2 \cup D_2$ for the boundary of R_2 (Fig. 16.4.3). Applying Green's theorem separately to the regions R_1 and R_2, we get

$$\oint_{C_1 \cup D_1} P \, dx + Q \, dy = \iint_{R_1} \left(\frac{\partial Q}{\partial x} - \frac{\partial P}{\partial y} \right) dA$$

and

$$\oint_{C_2 \cup D_2} P \, dx + Q \, dy = \iint_{R_2} \left(\frac{\partial Q}{\partial x} - \frac{\partial P}{\partial y} \right) dA.$$

When we add these two equations, the result is Eq. (1), Green's theorem for the region R, because the two line integrals along D_1 and D_2 cancel. This occurs because D_1 and D_2 represent the same curve with opposite orientations, so

$$\int_{D_2} P \, dx + Q \, dy = -\int_{D_1} P \, dx + Q \, dy$$

by Eq. (11) in Section 16.2. It therefore follows that

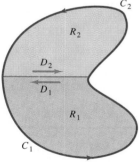

Fig. 16.4.3 Decomposing the region R into two horizontally and vertically simple regions by using a crosscut

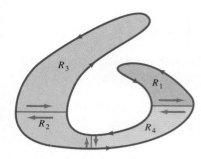

Fig. 16.4.4 Many important regions can be decomposed into simple regions by using one or more crosscuts.

$$\int_{C_1 \cup D_1 \cup C_2 \cup D_2} P\,dx + Q\,dy = \oint_{C_1 \cup C_2} P\,dx + Q\,dy = \oint_C P\,dx + Q\,dy.$$

Similarly, we could establish Green's theorem for the region shown in Fig. 16.4.4 by dividing it into the four simple regions indicated.

EXAMPLE 1 Use Green's theorem to evaluate the line integral

$$\oint_C (2y + \sqrt{9 + x^3})\,dx + (5x + e^{\arctan y})\,dy,$$

where C is the circle $x^2 + y^2 = 4$.

Solution With $P(x, y) = 2y + \sqrt{9 + x^3}$ and $Q(x, y) = 5x + e^{\arctan y}$, we see that

$$\frac{\partial Q}{\partial x} - \frac{\partial P}{\partial y} = 5 - 2 = 3.$$

Because C bounds R, a circular disk with area 4π, Green's theorem therefore implies that the given line integral is equal to

$$\iint_R 3\,dA = 3 \cdot 4\pi = 12\pi.$$

EXAMPLE 2 Evaluate the line integral

$$\oint_C 3xy\,dx + 2x^2\,dy,$$

where C is the boundary of the region R shown in Fig. 16.4.5. It is bounded above by the line $y = x$ and below by the parabola $y = x^2 - 2x$.

Solution To evaluate the line integral directly, we would need to parametrize separately the line and the parabola. Instead, we apply Green's theorem with $P = 3xy$ and $Q = 2x^2$, so

$$\frac{\partial Q}{\partial x} - \frac{\partial P}{\partial y} = 4x - 3x = x.$$

Then

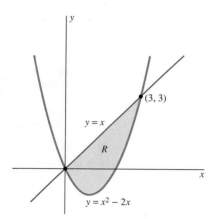

Fig. 16.4.5 The region of Example 2

$$\oint_C 3xy\,dx + 2x^2\,dy = \iint_R x\,dA$$

$$= \int_0^3 \int_{x^2-2x}^x x\,dy\,dx = \int_0^3 \Big[xy\Big]_{y=x^2-2x}^x dx$$

$$= \int_0^3 (3x^2 - x^3)\,dx = \Big[x^3 - \tfrac{1}{4}x^4\Big]_0^3 = \tfrac{27}{4}.$$

In Examples 1 and 2 we found the double integral easier to evaluate directly than the line integral. Sometimes the situation is the reverse. The following consequence of Green's theorem illustrates the technique of evaluating a double integral $\iint_R f(x, y)\,dA$ by converting it into a line integral

928

$$\oint_C P\,dx + Q\,dy.$$

To do this we must be able to find functions $P(x, y)$ and $Q(x, y)$ such that $\partial Q/\partial x - \partial P/\partial y = f(x, y)$. As in the proof of the following result, this is sometimes easy.

Corollary to Green's Theorem
The area A of the region R bounded by the piecewise smooth simple closed curve C is given by

$$A = \tfrac{1}{2}\oint_C -y\,dx + x\,dy = -\oint_C y\,dx = \oint_C x\,dy. \qquad (4)$$

Proof   With $P(x, y) = -y$ and $Q(x, y) \equiv 0$, Green's theorem gives

$$-\oint_C y\,dx = \iint_R 1\,dA = A.$$

Similarly, with $P(x, y) \equiv 0$ and $Q(x, y) = x$, we obtain

$$\oint_C x\,dy = \iint_R 1\,dA = A.$$

The third result may be obtained by averaging the left- and right-hand sides in the last two equations. Alternatively, with $P(x, y) = -y/2$ and $Q(x, y) = x/2$, Green's theorem gives

$$\tfrac{1}{2}\oint_C -y\,dx + x\,dy = \iint_R (\tfrac{1}{2} + \tfrac{1}{2})\,dA = A. \qquad \square$$

EXAMPLE 3   Apply the corollary to Green's theorem to find the area A bounded by the ellipse $x^2/a^2 + y^2/b^2 = 1$.

Solution   With the parametrization $x = a\cos t$, $y = b\sin t$, $0 \leqq t \leqq 2\pi$, Eq. (4) gives

$$A = \oint_{\text{ellipse}} x\,dy = \int_0^{2\pi} (a\cos t)(b\cos t\,dt)$$

$$= \tfrac{1}{2}ab\int_0^{2\pi} (1 + \cos 2t)\,dt = \pi ab.$$

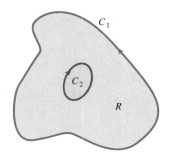

Fig. 16.4.6   An annular region— the boundary consists of two simple closed curves, one within the other.

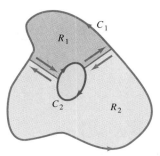

Fig. 16.4.7   Two crosscuts convert the annular region into the union of two ordinary regions.

By using the technique of subdividing a region into simpler ones, we can extend Green's theorem to regions with boundaries that consist of two or more simple closed curves. For example, consider the annular region R of Fig. 16.4.6, with boundary C consisting of the two simple closed curves C_1 and C_2. The positive direction along C—the direction for which the region R always lies on the left—is counterclockwise on the outer curve C_1 but clockwise on the inner curve C_2.

We divide R into two regions R_1 and R_2 by using two crosscuts, as shown in Fig. 16.4.7. Applying Green's theorem to each of these subregions, we get

$$\iint_R (Q_x - P_y)\, dA = \iint_{R_1} (Q_x - P_y)\, dA + \iint_{R_2} (Q_x - P_y)\, dA$$

$$= \oint_{C_1} (P\, dx + Q\, dy) + \oint_{C_2} (P\, dx + Q\, dy)$$

$$= \oint_C P\, dx + Q\, dy.$$

Thus we obtain Green's theorem for the given region R. What makes this proof work is that the opposite line integrals along the two crosscuts cancel each other. You may, of course, use any finite number of crosscuts.

EXAMPLE 4 Suppose that C is a smooth simple closed curve that encloses the origin $(0, 0)$. Show that

$$\oint_C \frac{-y\, dx + x\, dy}{x^2 + y^2} = 2\pi,$$

but that this integral is zero if C does *not* enclose the origin.

Solution With $P(x, y) = -y/(x^2 + y^2)$ and $Q(x, y) = x/(x^2 + y^2)$, a brief computation gives $\partial Q/\partial x - \partial P/\partial y \equiv 0$ when x and y are not both zero. If the region R bounded by C does not contain the origin, then P and Q and their derivatives are continuous on R. Hence Green's theorem implies that the integral in question is zero.

If C does enclose the origin, then we enclose the origin in a small circle C_a of radius a so small that C_a lies wholly within C (Fig. 16.4.8). We parametrize this circle by $x = a \cos t$, $y = a \sin t$, $0 \le t \le 2\pi$. Then Green's theorem, applied to the region R between C and C_a, gives

$$\oint_C \frac{-y\, dx + x\, dy}{x^2 + y^2} + \oint_{C_a} \frac{-y\, dx + x\, dy}{x^2 + y^2} = \iint_R 0\, dA = 0.$$

IMPORTANT Note the *reversed* arrow in the second line integral, required because the parametrization we chose is the clockwise (*negative*) orientation for C_a. Therefore,

$$\oint_C \frac{-y\, dx + x\, dy}{x^2 + y^2} = \oint_{C_a} \frac{-y\, dx + x\, dy}{x^2 + y^2}$$

$$= \int_0^{2\pi} \frac{(-a \sin t)(-a \sin t\, dt) + (a \cos t)(a \cos t\, dt)}{(a \cos t)^2 + (a \sin t)^2}$$

$$= \int_0^{2\pi} 1\, dt = 2\pi.$$

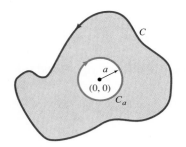

Fig. 16.4.8 Use the small circle C_a if C encloses the origin (Example 4).

The result of Example 4 can be interpreted in terms of the polar-coordinate angle $\theta = \arctan(y/x)$. Because

$$d\theta = \frac{-y\, dx + x\, dy}{x^2 + y^2},$$

the line integral of Example 4 measures the net change in θ as we go around the curve C once in a counterclockwise direction. This net change is 2π if C encloses the origin and is zero otherwise.

THE DIVERGENCE AND FLUX OF A VECTOR FIELD

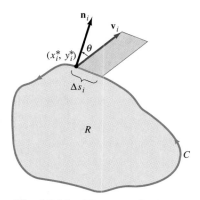

Fig. 16.4.9 The area of the parallelogram approximates the fluid flow across Δs_i in unit time.

Now let us consider the steady flow of a thin layer of fluid in the plane (perhaps like a sheet of water spreading across a floor). Let $\mathbf{v}(x, y)$ be its velocity vector field and $\rho(x, y)$ be the density of the fluid at the point (x, y). The term *steady flow* means that $\mathbf{v}$ and ρ depend only on x and y and *not* on time t. We want to compute the rate at which the fluid flows out of the region R bounded by a simple closed curve C (Fig. 16.4.9). We seek the net rate of outflow—the actual outflow minus the inflow.

Let Δs_i be a short segment of the curve C, and let (x_i^*, y_i^*) be an endpoint of Δs_i. Then the area of the portion of the fluid that flows out of R across Δs_i per unit time is approximately the area of the parallelogram in Fig. 16.4.9. This is the parallelogram spanned by the segment Δs_i and the vector $\mathbf{v}_i = \mathbf{v}(x_i^*, y_i^*)$. Suppose that $\mathbf{n}_i$ is the unit normal vector to C at the point (x_i^*, y_i^*), the normal pointing *out* of R. Then the area of this parallelogram is

$$(|\mathbf{v}_i|\cos\theta)\,\Delta s_i = \mathbf{v}_i \cdot \mathbf{n}_i\,\Delta s_i,$$

where θ is the angle between $\mathbf{n}_i$ and $\mathbf{v}_i$.

We multiply this area by the density $\rho_i = \rho(x_i^*, y_i^*)$ and then add these terms over those values of i that correspond to a subdivision of the entire curve C. This gives the (net) total mass of fluid leaving R per unit of time; it is approximately

$$\sum_{i=1}^{n} \rho_i \mathbf{v}_i \cdot \mathbf{n}_i\,\Delta s_i = \sum_{i=1}^{n} \mathbf{F}_i \cdot \mathbf{n}_i\,\Delta s_i,$$

where $\mathbf{F} = \rho\mathbf{v}$. The line integral around C that this sum approximates is called the **flux of the vector field $\mathbf{F}$ across the curve C**. Thus the flux ϕ of $\mathbf{F}$ across C is given by

$$\phi = \oint_C \mathbf{F} \cdot \mathbf{n}\,ds, \tag{5}$$

where $\mathbf{n}$ is the *outer* unit normal vector to C.

In the present case of fluid flow with velocity vector $\mathbf{v}$, the flux ϕ of $\mathbf{F} = \rho\mathbf{v}$ is the rate at which the fluid is flowing out of R across the boundary curve C, in units of mass per unit of time. But the same terminology is used for an arbitrary vector field $\mathbf{F} = M\mathbf{i} + N\mathbf{j}$. For example, we may speak of the flux of an electric or gravitational field across a curve C.

From Fig. 16.4.10 we see that the outer unit normal vector $\mathbf{n}$ is equal to $\mathbf{T} \times \mathbf{k}$. The unit tangent vector $\mathbf{T}$ to the curve C is

$$\mathbf{T} = \frac{1}{v}\left(\mathbf{i}\frac{dx}{dt} + \mathbf{j}\frac{dy}{dt}\right) = \mathbf{i}\frac{dx}{ds} + \mathbf{j}\frac{dy}{ds}$$

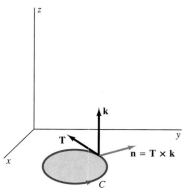

Fig. 16.4.10 Computing the outer unit normal vector $\mathbf{n}$ from the unit tangent vector $\mathbf{T}$

because $v = ds/dt$. Hence

$$\mathbf{n} = \mathbf{T} \times \mathbf{k} = \left(\mathbf{i} \frac{dx}{ds} + \mathbf{j} \frac{dy}{ds} \right) \times \mathbf{k}.$$

But $\mathbf{i} \times \mathbf{k} = -\mathbf{j}$ and $\mathbf{j} \times \mathbf{k} = \mathbf{i}$. Thus we find that

$$\mathbf{n} = \mathbf{i} \frac{dy}{ds} - \mathbf{j} \frac{dx}{ds}. \tag{6}$$

Substitution of the expression in Eq. (6) into the flux integral of Eq. (5) gives

$$\oint_C \mathbf{F} \cdot \mathbf{n} \, ds = \oint_C (M\mathbf{i} + N\mathbf{j}) \cdot \left(\mathbf{i} \frac{dy}{ds} - \mathbf{j} \frac{dx}{ds} \right) ds = \oint_C -N \, dx + M \, dy.$$

Applying Green's theorem to the last line integral with $P = -N$ and $Q = M$, we get

$$\oint_C \mathbf{F} \cdot \mathbf{n} \, ds = \iint_R \left(\frac{\partial M}{\partial x} + \frac{\partial N}{\partial y} \right) dA \tag{7}$$

for the flux of $\mathbf{F} = M\mathbf{i} + N\mathbf{j}$ across C.

The scalar function $\partial M / \partial x + \partial N / \partial y$ that appears in Eq. (7) is the **divergence** of the two-dimensional vector field $\mathbf{F} = M\mathbf{i} + N\mathbf{j}$ as defined in Section 16.1 and denoted by

$$\text{div } \mathbf{F} = \nabla \cdot \mathbf{F} = \frac{\partial M}{\partial x} + \frac{\partial N}{\partial y}. \tag{8}$$

When we substitute Eq. (8) into Eq. (7), we obtain a **vector form of Green's theorem**:

$$\oint_C \mathbf{F} \cdot \mathbf{n} \, ds = \iint_R \nabla \cdot \mathbf{F} \, dA, \tag{9}$$

with the understanding that $\mathbf{n}$ is the *outer* unit normal vector to C. Thus the flux of a vector field across a simple closed curve C is equal to the double integral of its divergence over the region R bounded by C.

If the region R is bounded by a circle C_r of radius r centered at the point (x_0, y_0), then

$$\oint_{C_r} \mathbf{F} \cdot \mathbf{n} \, ds = \iint_R \nabla \cdot \mathbf{F} \, dA = (\pi r^2) \, \nabla \cdot \mathbf{F}(\bar{x}, \bar{y})$$

for some point $(\bar{x}, \bar{y})$ in R; this is a consequence of the average value property of double integrals (see Problem 30 of Section 15.2). We divide by πr^2 and then let r approach zero. Thus we find that

$$\nabla \cdot \mathbf{F}(x_0, y_0) = \lim_{r \to 0} \frac{1}{\pi r^2} \oint_{C_r} \mathbf{F} \cdot \mathbf{n} \, ds \tag{10}$$

because $(\bar{x}, \bar{y}) \to (x_0, y_0)$ as $r \to 0$.

In the case of our original fluid flow, with $\mathbf{F} = \rho \mathbf{v}$, Eq. (10) implies that the value of $\nabla \cdot \mathbf{F}$ at (x_0, y_0) is a measure of the rate at which the fluid is "diverging away" from the point (x_0, y_0).

EXAMPLE 5 The vector field $\mathbf{F} = -y\mathbf{i} + x\mathbf{j}$ is the velocity field of a steady-state counterclockwise rotation about the origin. Show that the flux of $\mathbf{F}$ across any simple closed curve C is zero.

Solution This follows immediately from Eq. (9) because

$$\nabla \cdot \mathbf{F} = \frac{\partial}{\partial x}(-y) + \frac{\partial}{\partial y}(x) = 0.$$

16.4 Problems

In Problems 1 through 12, apply Green's theorem to evaluate the integral

$$\oint_C P\,dx + Q\,dy$$

around the specified closed curve C.

1. $P = x + y^2$, $Q = y + x^2$; C is the square with vertices $(\pm 1, \pm 1)$.

2. $P = x^2 + y^2$, $Q = -2xy$; C is the boundary of the triangle bounded by the lines $x = 0$, $y = 0$, and $x + y = 1$.

3. $P = y + e^x$, $Q = 2x^2 + \cos y$; C is the boundary of the triangle with vertices $(0, 0)$, $(1, 1)$, and $(2, 0)$.

4. $P = x^2 - y^2$, $Q = xy$; C is the boundary of the region bounded by the line $y = x$ and by the parabola $y = x^2$.

5. $P = -y^2 + \exp(e^x)$, $Q = \arctan y$; C is the boundary of the region between the parabolas $y = x^2$ and $x = y^2$.

6. $P = y^2$, $Q = 2x - 3y$; C is the circle $x^2 + y^2 = 9$.

7. $P = x - y$, $Q = y$; C is the boundary of the region between the x-axis and the graph of $y = \sin x$ for $0 \leq x \leq \pi$.

8. $P = e^x \sin y$, $Q = e^x \cos y$; C is the right-hand loop of the graph of the polar equation $r^2 = 4 \cos \theta$.

9. $P = y^2$, $Q = xy$; C is the ellipse with equation $x^2/9 + y^2/4 = 1$.

10. $P = y/(1 + x^2)$, $Q = \arctan x$; C is the oval with equation $x^4 + y^4 = 1$.

11. $P = xy$, $Q = x^2$; C is the first-quadrant loop of the graph of the polar equation $r = \sin 2\theta$.

12. $P = x^2$, $Q = -y^2$; C is the cardioid with polar equation $r = 1 + \cos \theta$.

In Problems 13 through 16, use the corollary to Green's theorem to find the area of the indicated region.

13. The circle bounded by $x = a \cos t$, $y = a \sin t$, $0 \leq t \leq 2\pi$

14. The region under one arch of the cycloid with parametric equations $x = a(t - \sin t)$, $y = a(1 - \cos t)$

15. The region bounded by the astroid with parametric equations $x = \cos^3 t$, $y = \sin^3 t$, $0 \leq t \leq 2\pi$

16. The region bounded by $y = x^2$ and by $y = x^3$

17. Suppose that f is a twice-differentiable scalar function of x and y. Show that

$$\nabla^2 f = \operatorname{div}(\nabla f) = \frac{\partial^2 f}{\partial x^2} + \frac{\partial^2 f}{\partial y^2}.$$

18. Show that $f(x, y) = \ln(x^2 + y^2)$ satisfies **Laplace's equation** $\nabla^2 f = 0$, except at the point $(0, 0)$.

19. Suppose that f and g are twice-differentiable functions. Show that $\nabla^2(fg) = f\nabla^2 g + g\nabla^2 f + 2\nabla f \cdot \nabla g$.

20. Suppose that the function $f(x, y)$ is twice continuously differentiable in the region R bounded by the piecewise smooth curve C. Prove that

$$\oint_C \frac{\partial f}{\partial x}\,dy - \frac{\partial f}{\partial y}\,dx = \iint_R \nabla^2 f\,dx\,dy.$$

21. Let R be the plane region with area A enclosed by the piecewise smooth simple closed curve C. Use Green's theorem to show that the coordinates of the centroid of R are

$$\bar{x} = \frac{1}{2A}\oint_C x^2\,dy, \qquad \bar{y} = -\frac{1}{2A}\oint_C y^2\,dx.$$

22. Use the result of Problem 21 to find the centroid of (a) a semicircular region of radius a; (b) a quarter-circular region of radius a.

23. Suppose that a lamina shaped like the region of Problem 21 has constant density ρ. Show that its moments of inertia about the coordinate axes are

$$I_x = -\frac{\rho}{3}\oint_C y^3\,dx \qquad I_y = \frac{\rho}{3}\oint_C x^3\,dy.$$

24. Use the result of Problem 23 to show that the polar moment of inertia $I_0 = I_x + I_y$ of a circular lamina of radius a, centered at the origin and of constant density ρ, is $\frac{1}{2}Ma^2$, where M is the mass of the lamina.

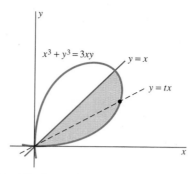

$$\oint_C f\,\nabla g\cdot\mathbf{n}\,ds = \iint_R [(f)(\nabla\cdot\nabla g) + \nabla f\cdot\nabla g]\,dA.$$

It was this formula rather than Green's theorem itself that appeared in Green's book of 1828.

28. Complete the proof of the simple case of Green's theorem by showing directly that

$$\oint_C Q\,dy = \iint_R \frac{\partial Q}{\partial x}\,dA$$

if the region R is horizontally simple.

29. Suppose that the bounded plane region R is divided into the nonoverlapping subregions $R_1, R_2, \ldots, R_k$. If Green's theorem, Eq. (1), holds for each of these subregions, explain why it follows that Green's theorem holds for R. State carefully any assumptions that you need to make.

Fig. 16.4.11 The loop of Problem 25

25. The loop of the folium of Descartes (with equation $x^3 + y^3 = 3xy$) appears in Fig. 16.4.11. Apply the corollary to Green's theorem to find the area of this loop. [*Suggestion:* Set $y = tx$ to discover a parametrization of the loop. To obtain the area of the loop, use values of t that lie in the interval $[0, 1]$. This gives the half of the loop that lies below the line $y = x$.]

26. Find the area bounded by one loop of the curve $x = \sin 2t, y = \sin t$.

27. Let f and g be functions with continuous second-order partial derivatives in the region R bounded by the piecewise smooth simple closed curve C. Apply Green's theorem in vector form to show that

In Problems 30 through 32, find the flux of the given vector field $\mathbf{F}$ *outward across the ellipse* $x = 4\cos t$, $y = 3\sin t$, $0 \le t \le 2\pi$.

30. $\mathbf{F} = x\mathbf{i} + 2y\mathbf{j}$

31. $\mathbf{F} = \dfrac{x\mathbf{i} + y\mathbf{j}}{x^2 + y^2}$

32. $\mathbf{F} = -y\mathbf{i} + x\mathbf{j}$

16.5
Surface Integrals

A surface integral is to surfaces in space what a line (or "curve") integral is to curves in the plane. Consider a curved, thin metal sheet shaped like the surface S. Suppose that this sheet has variable density, given at the point (x, y, z) by the known continuous function $f(x, y, z)$, in units such as grams per square centimeter of surface. We want to define the surface integral

$$\iint_S f(x, y, z)\,dS$$

in such a way that—upon evaluation—it gives the total mass of the thin metal sheet. In case $f(x, y, z) \equiv 1$, the numerical value of the integral should also equal the surface area of S.

As in Section 15.8, we assume that S is a parametric surface described by the function or transformation

$$\mathbf{r}(u, v) = \langle x(u, v), y(u, v), z(u, v)\rangle$$

for (u, v) in a region D in the uv-plane. We suppose throughout that the component functions of $\mathbf{r}$ have continuous partial derivatives and also that the vectors $\mathbf{r}_u = \partial\mathbf{r}/\partial u$ and $\mathbf{r}_v = \partial\mathbf{r}/\partial v$ are nonzero and nonparallel at each interior point of D.

Recall how we computed the surface area A of S in Section 15.8. We began with an inner partition of D consisting of n rectangles $R_1, R_2, \ldots, R_n$, each Δu by Δv in size. The images under $\mathbf{r}$ of these rectangles are curvilinear

934

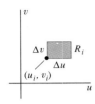

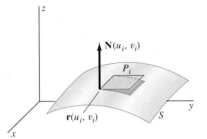

Fig. 16.5.1 Approximating surface area with parallelograms

figures filling up most of the surface S, and these pieces of S are themselves approximated by parallelograms P_i of the sort shown in Fig. 16.5.1. This gave us the approximation

$$A \approx \sum_{i=1}^{n} \Delta P_i = \sum_{i=1}^{n} | \mathbf{N}(u_i, v_i)| \, \Delta u \, \Delta v. \tag{1}$$

Here, $\Delta P_i = | \mathbf{N}(u_i, v_i)| \, \Delta u \, \Delta v$ is the area of the parallelogram P_i that is tangent to the surface S at the point $\mathbf{r}(u_i, v_i)$. The vector

$$\mathbf{N} = \frac{\partial \mathbf{r}}{\partial u} \times \frac{\partial \mathbf{r}}{\partial v} \tag{2}$$

is normal to S at $\mathbf{r}(u, v)$. If the surface S now has a density function $f(x, y, z)$, then we can approximate the total mass m of the surface by first multiplying each parallelogram area ΔP_i in Eq. (1) by the density $f(\mathbf{r}(u_i, v_i))$ at $\mathbf{r}(u_i, v_i)$ and then summing these estimates over all such parallelograms. Thus we obtain the approximation

$$m \approx \sum_{i=1}^{n} f(\mathbf{r}(u_i, v_i)) \, \Delta P_i = \sum_{i=1}^{n} f(\mathbf{r}(u_i, v_i))| \mathbf{N}(u_i, v_i)| \, \Delta u \, \Delta v.$$

This approximation is a Riemann sum for the **surface integral of the function f over the surface S**, denoted by

$$\iint_S f(x, y, z) \, dS = \iint_D f(\mathbf{r}(u, v))| \mathbf{N}(u, v)| \, du \, dv$$

$$= \iint_D f(\mathbf{r}(u, v)) \left| \frac{\partial \mathbf{r}}{\partial u} \times \frac{\partial \mathbf{r}}{\partial v} \right| du \, dv. \tag{3}$$

To evaluate the surface integral $\iint_S f(x, y, z) \, dS$, we simply use the parametrization $\mathbf{r}$ to express the variables x, y, and z in terms of u and v and formally replace the **surface area element** dS by

$$dS = | \mathbf{N}(u, v)| \, du \, dv = \left| \frac{\partial \mathbf{r}}{\partial u} \times \frac{\partial \mathbf{r}}{\partial v} \right| du \, dv. \tag{4}$$

This converts the surface integral into an *ordinary double integral* over the region D in the uv-plane.

In the important special case of a surface S described by $z = h(x, y)$, (x, y) in D in the xy-plane, we may use x and y as the parameters (rather than u and v). The surface area element then takes the form

$$dS = \sqrt{1 + \left(\frac{\partial h}{\partial x}\right)^2 + \left(\frac{\partial h}{\partial y}\right)^2} \, dx \, dy. \tag{5}$$

[We saw this expression for dS first in Eq. (9) of Section 15.8.] The surface integral of f over S is then given by

$$\iint_S f(x, y, z) \, dS = \iint_D f(x, y, h(x, y)) \sqrt{1 + \left(\frac{\partial h}{\partial x}\right)^2 + \left(\frac{\partial h}{\partial y}\right)^2} \, dx \, dy. \tag{6}$$

Centroids and moments of inertia for surfaces are computed in much the same way as for curves (see Section 16.2, using surface integrals in place of line integrals). For example, suppose that the surface S has density

$\rho(x, y, z)$ at the point (x, y, z) and total mass m. Then the z-component $\bar{z}$ of its centroid and its moment of inertia I_z about the z-axis are given by

$$\bar{z} = \frac{1}{m} \iint_S z\rho(x, y, z)\, dS \quad \text{and} \quad I_z = \iint_S (x^2 + y^2)\rho(x, y, z)\, dS.$$

EXAMPLE 1 Find the centroid of the unit-density hemispherical surface $z = \sqrt{a^2 - x^2 - y^2}$, $x^2 + y^2 \le a^2$.

Solution By symmetry, $\bar{x} = 0 = \bar{y}$. A simple computation gives $\partial z/\partial x = -x/z$ and $\partial z/\partial y = -y/z$, so Eq. (5) takes the form

$$dS = \sqrt{1 + \left(\frac{\partial z}{\partial x}\right)^2 + \left(\frac{\partial z}{\partial y}\right)^2}\, dx\, dy = \sqrt{1 + \left(\frac{x}{z}\right)^2 + \left(\frac{y}{z}\right)^2}\, dx\, dy$$

$$= \frac{1}{z}\sqrt{x^2 + y^2 + z^2}\, dx\, dy = \frac{a}{z}\, dx\, dy.$$

Hence

$$\bar{z} = \frac{1}{2\pi a^2} \iint_D z \cdot \frac{a}{z}\, dx\, dy = \frac{1}{2\pi a} \iint_D 1\, dx\, dy = \frac{a}{2}.$$

Note in the final step that D is a circle of radius a in the xy-plane. This simplifies the computation of the last integral.

EXAMPLE 2 Find the moment of inertia about the z-axis of the spherical surface $x^2 + y^2 + z^2 = a^2$, assuming that it has constant density $\rho = k$.

Solution The spherical surface of radius a is most easily parametrized in spherical coordinates:

$$x = a \sin \phi \cos \theta, \qquad y = a \sin \phi \sin \theta, \qquad z = a \cos \phi$$

for $0 \le \phi \le \pi$ and $0 \le \theta \le 2\pi$. By Problem 18 of Section 15.8, the surface element is then $dS = a^2 \sin \phi\, d\phi\, d\theta$. We first note that

$$x^2 + y^2 = a^2 \sin^2 \phi \cos^2 \theta + a^2 \sin^2 \phi \sin^2 \theta = a^2 \sin^2 \phi,$$

and it then follows that

$$I_z = \iint_S (x^2 + y^2)\rho\, dS = \int_0^{2\pi} \int_0^{\pi} k(a^2 \sin^2 \phi)\, a^2 \sin \phi\, d\phi\, d\theta$$

$$= 2\pi k a^4 \int_0^{\pi} \sin^3 \phi\, d\phi = \tfrac{8}{3}\pi k a^4 = \tfrac{2}{3}ma^2.$$

In the last step we used the fact that the mass m of the spherical surface with density k is $4\pi k a^2$. Finally, the answer is both dimensionally correct and numerically plausible.

A SECOND TYPE OF SURFACE INTEGRAL

The surface integral $\iint_S f(x, y, z)\, dS$ is analogous to the line integral $\int_C f(x, y)\, ds$. There is a second type of surface integral that is analogous to the line integral of the form $\int_C P\, dx + Q\, dy$. To define the surface integral

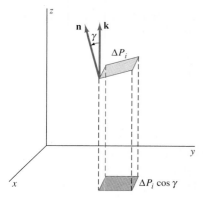

Fig. 16.5.2 Finding the area of the projected parallelogram

$$\iint_S f(x, y, z)\ dx\ dy,$$

with $dx\ dy$ in place of dS, we replace the parallelogram area ΔP_i in Eq. (1) by the area of its projection into the xy-plane (Fig. 16.5.2). To see how this works out, consider the *unit* normal vector to S,

$$\mathbf{n} = \frac{\mathbf{N}}{|\mathbf{N}|} = \mathbf{i} \cos \alpha + \mathbf{j} \cos \beta + \mathbf{k} \cos \gamma. \tag{7}$$

Because

$$\mathbf{N} = \begin{vmatrix} \mathbf{i} & \mathbf{j} & \mathbf{k} \\ \dfrac{\partial x}{\partial u} & \dfrac{\partial y}{\partial u} & \dfrac{\partial z}{\partial u} \\ \dfrac{\partial x}{\partial v} & \dfrac{\partial y}{\partial v} & \dfrac{\partial z}{\partial v} \end{vmatrix}$$

or—in the Jacobian notation of Section 15.9—

$$\mathbf{N} = \mathbf{i}\frac{\partial(y, z)}{\partial(u, v)} + \mathbf{j}\frac{\partial(z, x)}{\partial(u, v)} + \mathbf{k}\frac{\partial(x, y)}{\partial(u, v)},$$

the components of the unit normal vector $\mathbf{n}$ are

$$\cos \alpha = \frac{1}{|\mathbf{N}|}\frac{\partial(y, z)}{\partial(u, v)}, \qquad \cos \beta = \frac{1}{|\mathbf{N}|}\frac{\partial(z, x)}{\partial(u, v)}, \qquad \text{and}$$

$$\cos \gamma = \frac{1}{|\mathbf{N}|}\frac{\partial(x, y)}{\partial(u, v)}. \tag{8}$$

We see from Fig. 16.5.2 that the (signed) projection of the area ΔP_i into the xy-plane is $\Delta P_i \cos \gamma$. The corresponding Riemann sum motivates the *definition*

$$\iint_S f(x, y, z)\ dx\ dy = \iint_S f(x, y, z) \cos \gamma\ dS$$

$$= \iint_S f(\mathbf{r}(u, v))\frac{\partial(x, y)}{\partial(u, v)}\ du\ dv, \tag{9}$$

where the last integral is obtained by substituting for $\cos \gamma$ from Eq. (8) and for dS from Eq. (4). Similarly, we *define*

$$\iint_S f(x, y, z)\ dy\ dz = \iint_S f(x, y, z) \cos \alpha\ dS$$

$$= \iint_S f(\mathbf{r}(u, v))\frac{\partial(y, z)}{\partial(u, v)}\ du\ dv \tag{10}$$

and

$$\iint_S f(x, y, z)\ dz\ dx = \iint_S f(x, y, z) \cos \beta\ dS$$

$$= \iint_S f(\mathbf{r}(u, v))\frac{\partial(z, x)}{\partial(u, v)}\ du\ dv. \tag{11}$$

The symbols z and x in Eq. (11) appear in the reverse of alphabetical order. It is important to write them in the correct order because

$$\frac{\partial(x, z)}{\partial(u, v)} = -\frac{\partial(z, x)}{\partial(u, v)}.$$

This implies that

$$\iint_S f(x, y, z)\, dx\, dz = -\iint_S f(x, y, z)\, dz\, dx.$$

In an ordinary *double integral,* the order in which the differentials are written simply indicates the order of integration. But in a *surface integral,* it instead indicates the order of appearance of the corresponding variables in the Jacobians in Eqs. (9) through (11).

The general surface integral of the second type is the sum

$$\iint_S P\, dy\, dz + Q\, dz\, dx + R\, dx\, dy$$

$$= \iint_S (P \cos \alpha + Q \cos \beta + R \cos \gamma)\, dS \tag{12}$$

$$= \iint_S \left(P\frac{\partial(y, z)}{\partial(u, v)} + Q\frac{\partial(z, x)}{\partial(u, v)} + R\frac{\partial(x, y)}{\partial(u, v)} \right) du\, dv. \tag{13}$$

Here, P, Q, and R are continuous functions of x, y, and z.

Suppose that $\mathbf{F} = P\mathbf{i} + Q\mathbf{j} + R\mathbf{k}$. Then the integrand in Eq. (12) is simply $\mathbf{F} \cdot \mathbf{n}$, so we obtain the basic relation

$$\iint_S \mathbf{F} \cdot \mathbf{n}\, dS = \iint_S P\, dy\, dz + Q\, dz\, dx + R\, dx\, dy \tag{14}$$

between these two types of surface integrals. This formula is analogous to the earlier formula

$$\int_C \mathbf{F} \cdot \mathbf{T}\, ds = \int_C P\, dx + Q\, dy + R\, dz$$

for line integrals. However, Eq. (13) gives the evaluation procedure for the surface integral in Eq. (12): Substitute for x, y, z and their derivatives in terms of u and v, then integrate over the appropriate region D in the uv-plane.

Only the *sign* of the right-hand surface integral in Eq. (14) depends on the parametrization of S. The unit normal vector on the left-hand side is the vector provided by the parametrization of S via the equations in (8). In the case of a surface given by $z = h(x, y)$, with x and y used as the parameters u and v, this will be the *upper* normal, as you will see in Example 3.

EXAMPLE 3   Suppose that S is the surface $z = h(x, y)$, (x, y) in D. Then show that

$$\iint_S P\, dy\, dz + Q\, dz\, dx + R\, dx\, dy$$

$$= \iint_D \left(-P\frac{\partial z}{\partial x} - Q\frac{\partial z}{\partial y} + R \right) dx\, dy, \tag{15}$$

where P, Q, and R in the second integral are evaluated at $(x, y, h(x, y))$.

Solution   This is simply a matter of computing the three Jacobians in Eq. (13) with the parameters x and y. We note first that $\partial x/\partial x = 1 = \partial y/\partial y$ and that $\partial x/\partial y = 0 = \partial y/\partial x$. Hence

$$\frac{\partial(y, z)}{\partial(x, y)} = \begin{vmatrix} \dfrac{\partial y}{\partial x} & \dfrac{\partial y}{\partial y} \\[2mm] \dfrac{\partial z}{\partial x} & \dfrac{\partial z}{\partial y} \end{vmatrix} = -\frac{\partial z}{\partial x},$$

$$\frac{\partial(z, x)}{\partial(x, y)} = \begin{vmatrix} \dfrac{\partial z}{\partial x} & \dfrac{\partial z}{\partial y} \\[2mm] \dfrac{\partial x}{\partial x} & \dfrac{\partial x}{\partial y} \end{vmatrix} = -\frac{\partial z}{\partial y},$$

and

$$\frac{\partial(x, y)}{\partial(x, y)} = \begin{vmatrix} \dfrac{\partial x}{\partial x} & \dfrac{\partial x}{\partial y} \\[2mm] \dfrac{\partial y}{\partial x} & \dfrac{\partial y}{\partial y} \end{vmatrix} = 1.$$

Equation (15) is an immediate consequence.

THE FLUX OF A VECTOR FIELD

One of the most important applications of surface integrals involves the computation of the flux of a vector field. To define the flux of the vector field **F** across the surface S, we assume that S has a unit normal vector field **n** that varies *continuously* from point to point of S. This condition excludes from our consideration one-sided (*nonorientable*) surfaces, such as the Möbius strip of Fig. 16.5.3. If S is a two-sided (*orientable*) surface, then there are two possible choices for **n**. For example, if S is a closed surface (such as a torus or sphere) that separates space, then we may choose for **n** either the outer normal vector (at each point of S) or the inner normal vector (Fig. 16.5.4). The unit normal vector defined in Eq. (7) may be either the outer normal or the inner normal; which of the two it is depends on how S has been parametrized.

Now we turn our attention to the flux of a vector field. Suppose that we are given the vector field **F,** the orientable surface S, and a continuous unit normal vector field **n** on S. We define the **flux ϕ of F across S in the direction of n** in analogy with Eq. (5) of Section 16.4:

$$\phi = \iint_S \mathbf{F} \cdot \mathbf{n} \, dS. \qquad (16)$$

For example, if $\mathbf{F} = \rho\mathbf{v}$, where **v** is the velocity vector field corresponding to the steady flow in space of a fluid of density ρ and **n** is the *outer* unit normal vector for a closed surface S that bounds the space region T, then the flux determined by Eq. (16) is the net rate of flow of the fluid *out of T* across its boundary surface S in units such as grams per second.

A similar application is to the flow of heat, which is mathematically quite similar to the flow of a fluid. Suppose that a body has temperature

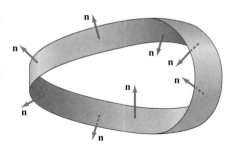

Fig. 16.5.3 The Möbius strip is an example of a one-sided surface.

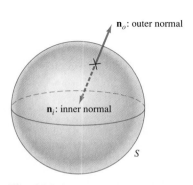

Fig. 16.5.4 Inner and outer normal vectors to a two-sided closed surface

Thermogram indicating heat flow across the surface of a house

$u = u(x, y, z)$ at the point (x, y, z). Experiments indicate that the flow of heat in the body is described by the heat-flow vector

$$\mathbf{q} = -K \nabla u. \tag{17}$$

The number K—normally, but not always, a constant—is the *heat conductivity* of the body. The vector $\mathbf{q}$ points in the direction of heat flow, and its length is the rate of flow of heat across a unit area normal to $\mathbf{q}$. This flow rate is measured in units such as calories per second per square centimeter. If S is a closed surface within the body bounding the solid region T and $\mathbf{n}$ denotes the outer unit normal vector to S, then

$$\iint_S \mathbf{q} \cdot \mathbf{n} \, dS = -\iint_S K \nabla u \cdot \mathbf{n} \, dS \tag{18}$$

is the net rate of heat flow (in calories per second, for example) out of the region T across its boundary surface S.

EXAMPLE 4 Calculate the flux $\iint_S \mathbf{F} \cdot \mathbf{n} \, dS$, where $\mathbf{F} = v_0 \mathbf{k}$ and S is the hemispherical surface of radius a and equation $z = \sqrt{a^2 - x^2 - y^2}$ shown in Fig. 16.5.5, with outer unit normal vector $\mathbf{n}$.

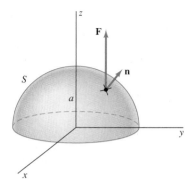

Fig. 16.5.5 The hemisphere S of Example 4

Solution If we think of $\mathbf{F} = v_0 \mathbf{k}$ as the velocity vector field of a fluid that is flowing upward with constant speed v_0, then we can interpret the flux in question as the rate of flow (in cubic centimeters per second, for example) of the fluid across S. To calculate this flux, we note that

$$\mathbf{n} = \frac{x\mathbf{i} + y\mathbf{j} + z\mathbf{k}}{\sqrt{x^2 + y^2 + z^2}} = \frac{1}{a}(x\mathbf{i} + y\mathbf{j} + z\mathbf{k}).$$

Hence

$$\mathbf{F} \cdot \mathbf{n} = v_0 \mathbf{k} \cdot \frac{1}{a}(x\mathbf{i} + y\mathbf{j} + z\mathbf{k}) = \frac{v_0}{a} z,$$

so

$$\iint_S \mathbf{F} \cdot \mathbf{n} \, dS = \iint_S \frac{v_0}{a} z \, dS.$$

If we introduce spherical coordinates $z = a \cos \phi$, $dS = a^2 \sin \phi \, d\phi \, d\theta$ on the sphere, we get

$$\iint_S \mathbf{F} \cdot \mathbf{n} \, dS = \frac{v_0}{a} \int_0^{2\pi} \int_0^{\pi/2} (a \cos \phi)(a^2 \sin \phi) \, d\phi \, d\theta$$

$$= 2\pi a^2 v_0 \int_0^{\pi/2} \cos \phi \sin \phi \, d\phi = 2\pi a^2 v_0 \left[\tfrac{1}{2} \sin^2 \phi \right]_0^{\pi/2};$$

thus

$$\iint_S \mathbf{F} \cdot \mathbf{n} \, dS = \pi a^2 v_0.$$

This last quantity is equal to the flux of $\mathbf{F} = v_0 \mathbf{k}$ across the disk $x^2 + y^2 \leqq a^2$ of area πa^2. If we think of the hemispherical region T bounded by the hemisphere S and by the circular disk D that forms its base, it should be no surprise that the rate of inflow of an incompressible fluid across the disk D is equal to its rate of outflow across the hemisphere S.

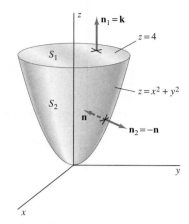

Fig. 16.5.6 The surface of Example 5

EXAMPLE 5 Find the flux of the vector field $\mathbf{F} = x\mathbf{i} + y\mathbf{j} + 3\mathbf{k}$ out of the region T bounded by the paraboloid $z = x^2 + y^2$ and by the plane $z = 4$ (Fig. 16.5.6).

Solution Let S_1 denote the circular top, which has outer unit normal vector $\mathbf{n}_1 = \mathbf{k}$. Let S_2 be the parabolic part of this surface, with outer unit normal vector $\mathbf{n}_2$. The flux across S_1 is

$$\iint_{S_1} \mathbf{F} \cdot \mathbf{n}_1 \; dS = \iint_{S_1} 3 \; dS = 12\pi$$

because S_1 is a circle of radius 2.

Next, the computation of Example 3 gives

$$\mathbf{N} = \left\langle -\frac{\partial z}{\partial x}, -\frac{\partial z}{\partial y}, 1 \right\rangle = \langle -2x, -2y, 1 \rangle$$

for a normal vector to the paraboloid $z = x^2 + y^2$. Then $\mathbf{n} = \mathbf{N}/|\mathbf{N}|$ is an upper—and thus an *inner*—unit normal vector to the surface S_2. The unit *outer* normal vector is, therefore, $\mathbf{n}_2 = -\mathbf{n}$, opposite to the direction of $\mathbf{N} = \langle -2x, -2y, 1 \rangle$. With parameters (x, y) in the circular disk $x^2 + y^2 \le 4$ in the xy-plane, the surface-area element is $dS = |\mathbf{N}| \; dx \; dy$. Therefore, the outward flux across S_2 is

$$\iint_{S_2} \mathbf{F} \cdot \mathbf{n}_2 \; dS = -\iint_{S_2} \mathbf{F} \cdot \mathbf{n} \; dS = -\iint_{D} \mathbf{F} \cdot \frac{\mathbf{N}}{|\mathbf{N}|} |\mathbf{N}| \; dx \; dy$$

$$= -\iint_{D} [(x)(-2x) + (y)(-2y) + (3)(1)] \; dx \; dy.$$

We change to polar coordinates and find that

$$\iint_{S_2} \mathbf{F} \cdot \mathbf{n}_2 \; dS = \int_0^{2\pi} \int_0^2 (2r^2 - 3) \, r \, dr \, d\theta = 2\pi \left[\tfrac{1}{2} r^4 - \tfrac{3}{2} r^2 \right]_0^2 = 4\pi.$$

Hence the total flux of $\mathbf{F}$ out of T is $16\pi \approx 50.27$.

16.5 Problems

In Problems 1 through 5, evaluate the surface integral

$$\iint_S f(x, y, z) \; dS.$$

1. $f(x, y, z) = xyz$; S is the first-octant part of the plane $x + y + z = 1$.

2. $f(x, y, z) = x + y$; S is the part of the plane $z = 2x + 3y$ that lies within the cylinder $x^2 + y^2 = 9$.

3. $f(x, y, z) = z$; S is the part of the paraboloid $z = r^2$ that lies within the cylinder $r = 2$.

4. $f(x, y, z) = (x^2 + y^2)z$; S is the hemispherical surface $\rho = 1$, $z \ge 0$. [*Suggestion:* Use spherical coordinates.]

5. $f(x, y, z) = z^2$; S is the cylindrical surface parametrized by $x = \cos \theta$, $y = \sin \theta$, $z = z$, $0 \le \theta \le 2\pi$, $0 \le z \le 2$.

In Problems 6 through 10, use Eqs. (13) and (14) to evaluate the surface integral

$$\iint_S \mathbf{F} \cdot \mathbf{n} \; dS,$$

where $\mathbf{n}$ is the upward-pointing unit normal vector to the given surface S.

6. $\mathbf{F} = x\mathbf{i} + y\mathbf{j} + z\mathbf{k}$; S is the first-octant part of the plane $2x + 2y + z = 3$.

7. $\mathbf{F} = 2y\mathbf{i} + 3z\mathbf{k}$; S is the part of the plane $z = 3x + 2$ that lies within the cylinder $x^2 + y^2 = 4$.

8. $\mathbf{F} = z\mathbf{k}$; S is the upper half of the spherical surface $\rho = 2$. [*Suggestion:* Use spherical coordinates.]

9. $\mathbf{F} = y\mathbf{i} - x\mathbf{j}$; S is the part of the cone $z = r$ that lies within the cylinder $r = 3$.

10. $\mathbf{F} = 2x\mathbf{i} + 2y\mathbf{j} + 3\mathbf{k}$; S is the part of the paraboloid $z = 4 - x^2 - y^2$ that lies above the xy-plane.

11. The first-octant part of the spherical surface $\rho = a$ has unit density. Find its centroid.

12. The conical surface $z = r$, $r \leq a$, has constant density $\delta = k$. Find its centroid and its moment of inertia about the z-axis.

13. The paraboloid $z = r^2$, $r \leq a$, has constant density δ. Find its centroid and moment of inertia about the z-axis.

14. Find the centroid of the part of the spherical surface $\rho = a$ that lies within the cone $r = z$.

15. Find the centroid of the part of the spherical surface $x^2 + y^2 + z^2 = 4$ that lies both inside the cylinder $x^2 + y^2 = 2x$ and above the xy-plane.

16. Suppose that the toroidal surface ("spiral ramp") of Example 4 of Section 15.8 has uniform density and total mass M. Show that its moment of inertia about the z-axis is $\frac{1}{2}M(3a^2 + 2b^2)$.

17. Let S denote the boundary of the region bounded by the coordinate planes and by the plane $2x + 3y + z = 6$.

Find the flux of $\mathbf{F} = x\mathbf{i} - y\mathbf{j}$ across S in the direction of its outer normal vector.

18. Let S denote the boundary of the solid bounded by the paraboloid $z = 4 - x^2 - y^2$ and by the xy-plane. Find the flux of the vector field $\mathbf{F} = 2x\mathbf{i} + 2y\mathbf{j} + 3\mathbf{k}$ across S in the direction of the outer normal vector.

19. Let S denote the boundary of the solid bounded by the paraboloids $z = x^2 + y^2$ and $z = 18 - x^2 - y^2$. Find the flux of $\mathbf{F} = z^2\mathbf{k}$ across S in the direction of the outer normal vector.

20. Let S denote the surface $z = h(x, y)$ for (x, y) in the region D in the xy-plane, and let γ be the angle between $\mathbf{k}$ and the upper normal vector $\mathbf{N}$ to S. Prove that

$$\iint_S f(x, y, z) \, dS = \iint_D f(x, y, h(x, y)) \sec \gamma \, dx \, dy.$$

21. Consider a homogeneous thin spherical shell S of radius a centered at the origin, with density δ and total mass $M = 4\pi a^2 \delta$. A particle of mass m is located at the point $(0, 0, c)$ with $c > a$. Use the method and notation of Problem 34 of Section 15.7 to show that the gravitational force of attraction between the particle and the spherical shell is

$$F = \iint_S \frac{Gm\delta}{w^2} \, dS = \frac{GMm}{c^2}.$$

16.5 Projects

Suppose that we revolve a line segment around the z-axis in space while we simultaneously rotate the segment in its own vertical plane (Fig. 16.5.7). If the line segment rotates vertically through the angle $p\theta$ as it revolves through the angle θ around the z-axis, then the surface generated is a ribbon with p *twists*. Suppose also that the line segment has length 2 and that its midpoint

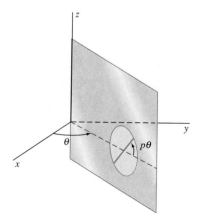

Fig. 16.5.7 The line segment rotates in a vertical plane through the angle $p\theta$ as it revolves through the angle θ around the z-axis.

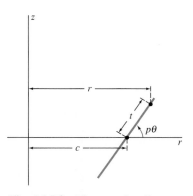

Fig. 16.5.8 The rotating line segment in its own vertical plane

describes a circle of radius c as it revolves around the z-axis. Use Fig. 16.5.8 to show that this ribbon is parametrized by the equations

$$x = (c + t \cos p\theta) \cos \theta,$$
$$y = (c + t \cos p\theta) \sin \theta, \quad \text{and} \tag{19}$$
$$z = t \sin p\theta$$

for $-1 \leqq t \leqq 1$, $0 \leqq \theta \leqq 2\pi$.

You can construct such a ribbon if you have access to a computer system with parametric plotting commands (see Projects 15.8). With $c = 4$ and $p = \frac{1}{2}$ twists, you should get the one-sided Möbius strip of Fig. 16.5.9. Figure 16.5.10 shows the ribbon constructed with a full twist ($p = 1$). Can you see that it is two-sided? What would happen if $p = \frac{3}{2}$?

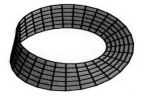

Fig. 16.5.9 The Möbius strip with a half twist ($p = \frac{1}{2}$)

Fig. 16.5.10 The ribbon with a full twist ($p = 1$)

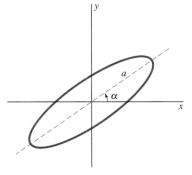

Fig. 16.5.11 The rotated ellipse

PROJECT A (Twisted Tubes) When a closed curve is revolved and twisted, we get a *twisted tube* instead of a ribbon. Figure 16.5.11 shows the result of rotating the ellipse with parametric equations

$$x = a \cos t, \qquad y = b \sin t \tag{20}$$

through the angle α. Apply the formulas in Eq. (7) of Section 10.7 (rotation of axes) to show that the rotated ellipse is parametrized by the equations

$$x = a \cos t \cos \alpha - b \sin t \sin \alpha,$$
$$y = a \cos t \cos \alpha + b \sin t \cos \alpha, \tag{21}$$

Now suppose that an ellipse with semiaxes a and b is revolved with twist p around a circle of radius c. If

$$r(t, \theta) = c + a \cos t \cos p\theta - b \sin t \sin p\theta, \tag{22}$$

use Eqs. (21) to show that the resulting twisted tube is parametrized by the equations

$$x = r(t, \theta) \cos \theta,$$
$$y = r(t, \theta) \sin \theta, \qquad \text{and} \tag{23}$$
$$z = a \cos t \sin p\theta + b \sin t \cos p\theta$$

for $0 \leqq t \leqq 2\pi$, $0 \leqq \theta \leqq 2\pi$. When you parametrically plot these equations with $a = b = p = 1$ and $c > 1$, you should get an ordinary torus. The *cruller* shown in Fig. 16.5.12 is obtained with $a = 3$, $b = 1$, $c = 7$, and $p = \frac{3}{2}$.

Fig. 16.5.12 The cruller

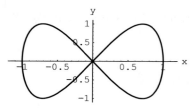

Fig. 16.5.13 The figure-eight curve

Fig. 16.5.14 The Klein bottle

PROJECT B (The Klein Bottle) The parametric equations

$$x = \sin t, \qquad y = \sin 2t \tag{24}$$

for $0 \leqq t \leqq 2\pi$ describe a figure-eight curve (Fig. 16.5.13). If you use Eqs. (24) rather than Eqs. (20) to generate a twisted tube with twist $p = \frac{1}{2}$, you should obtain the *Klein bottle* of Fig. 16.5.14. It is a representation in space of a closed surface that is one-sided. A true Klein bottle does not intersect itself; it is a surface much like a torus, but it is one-sided. It cannot exist in ordinary three-dimensional space. This is indicated by the fact that its representation in Fig. 16.5.14 intersects itself in a simple closed curve.

If twisted tubes interest you, you can find additional suggestions in C. Henry Edwards, "Twisted Tubes," *The Mathematica Journal* **3** (Winter 1993), 10–13.

16.6
The Divergence Theorem

The *divergence theorem* is to surface integrals what Green's theorem is to line integrals. It lets us convert a surface integral over a closed surface into a triple integral over the enclosed region, or vice versa. The divergence theorem is known also as *Gauss's theorem,* and in some eastern European countries it is called *Ostrogradski's theorem.* The German "prince of mathematics" Carl Friedrich Gauss (1777–1855) used it to study inverse-square force fields; the Russian Michel Ostrogradski (1801–1861) used it to study heat flow. Both did their work in the 1830s.

The surface S is called **piecewise smooth** if it consists of a finite number of smooth parametric surfaces. It is called **closed** if it is the boundary of a bounded region in space. For example, the boundary of a cube is a closed piecewise smooth surface, as are the boundary of a pyramid and the boundary of a solid cylinder.

> **The Divergence Theorem**
>
> Suppose that S is a closed piecewise smooth surface that bounds the space region T. Let $\mathbf{F} = P\mathbf{i} + Q\mathbf{j} + R\mathbf{k}$ be a vector field with component functions that have continuous first-order partial derivatives on T. Let $\mathbf{n}$ be the *outer* unit normal vector to S. Then
>
> $$\iint_S \mathbf{F} \cdot \mathbf{n} \, dS = \iiint_T \mathbf{\nabla} \cdot \mathbf{F} \, dV. \tag{1}$$

Equation (1) is a three-dimensional analogue of the vector form of Green's theorem that we saw in Eq. (9) of Section 16.4:

$$\oint_C \mathbf{F} \cdot \mathbf{n} \, ds = \iint_R \mathbf{\nabla} \cdot \mathbf{F} \, dA,$$

where $\mathbf{F}$ is a vector field in the plane, C is a piecewise smooth curve that bounds the plane region R, and $\mathbf{n}$ is the outer unit normal vector to C. The left-hand side of Eq. (1) is the flux of $\mathbf{F}$ across S in the direction of the outer unit normal vector $\mathbf{n}$.

Recall from Section 16.1 that the *divergence* $\mathbf{\nabla} \cdot \mathbf{F}$ of the vector field $\mathbf{F}$ is given in the three-dimensional case by

$$\text{div } \mathbf{F} = \mathbf{\nabla} \cdot \mathbf{F} = \frac{\partial P}{\partial x} + \frac{\partial Q}{\partial y} + \frac{\partial R}{\partial z}. \tag{2}$$

If $\mathbf{n}$ is given in terms of its direction cosines, as $\mathbf{n} = \langle \cos \alpha, \cos \beta, \cos \gamma \rangle$, then we can write the divergence theorem in scalar form:

$$\iint_S (P \cos \alpha + Q \cos \beta + R \cos \gamma)\, dS = \iiint_T \left(\frac{\partial P}{\partial x} + \frac{\partial Q}{\partial y} + \frac{\partial R}{\partial z} \right) dV. \tag{3}$$

It is best to parametrize S so that the normal vector given by the parametrization is the outer normal. Then we can write Eq. (3) entirely in Cartesian form:

$$\iint_S P\, dy\, dz + Q\, dz\, dx + R\, dx\, dy = \iiint_T \left(\frac{\partial P}{\partial x} + \frac{\partial Q}{\partial y} + \frac{\partial R}{\partial z} \right) dV. \tag{4}$$

Proof of the Divergence Theorem   We shall prove the divergence theorem only for the case in which the region T is simultaneously x-simple, y-simple, and z-simple. This guarantees that every straight line parallel to a coordinate axis intersects T, if at all, in a single point or a line segment. It suffices for us to derive separately the equations

$$\begin{aligned} \iint_S P\, dy\, dz &= \iiint_T \frac{\partial P}{\partial x}\, dV, \\ \iint_S Q\, dz\, dx &= \iiint_T \frac{\partial Q}{\partial y}\, dV, \quad \text{and} \\ \iint_S R\, dx\, dy &= \iiint_T \frac{\partial R}{\partial z}\, dV. \end{aligned} \tag{5}$$

Then the sum of Eqs. (5) is Eq. (4).

Because T is z-simple, it has the description

$$z_1(x, y) \leqq z \leqq z_2(x, y)$$

for (x, y) in D, the projection of T into the xy-plane. As in Fig. 16.6.1, we denote the lower surface $z = z_1(x, y)$ of T by S_1, the upper surface

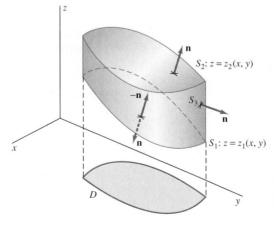

Fig. 16.6.1 A z-simple space region bounded by the surfaces S_1, S_2, and S_3

$z = z_2(x, y)$ by S_2, and the lateral surface between S_1 and S_2 by S_3. In the case of some simple surfaces, such as a spherical surface, there may be no S_3 to consider. But even if there is,

$$\iint_{S_3} R\, dx\, dy = \iint_{S_3} R \cos \gamma\, dS = 0, \qquad (6)$$

because $\gamma = 90°$ at each point of the vertical cylinder S_3.

On the upper surface S_2, the unit upper normal vector corresponding to the parametrization $z = z_2(x, y)$ is the given outer unit normal vector $\mathbf{n}$, so Eq. (15) of Section 16.5 yields

$$\iint_{S_2} R\, dx\, dy = \iint_{D} R(x, y, z_2(x, y))\, dx\, dy. \qquad (7)$$

But on the lower surface S_1, the unit upper normal vector corresponding to the parametrization $z = z_1(x, y)$ is the inner normal vector $-\mathbf{n}$, so we must reverse the sign. Thus

$$\iint_{S_1} R\, dx\, dy = -\iint_{D} R(x, y, z_1(x, y))\, dx\, dy. \qquad (8)$$

We add Eqs. (6), (7), and (8). The result is that

$$\iint_{S} R\, dx\, dy = \iint_{D} [R(x, y, z_2(x, y)) - R(x, y, z_1(x, y))]\, dx\, dy$$

$$= \iint_{D} \left(\int_{z_1(x, y)}^{z_2(x, y)} \frac{\partial R}{\partial z}\, dz \right) dx\, dy.$$

Therefore,

$$\iint_{S} R\, dx\, dy = \iiint_{T} \frac{\partial R}{\partial z}\, dV.$$

This is the third equation in (5), and we can derive the other two in much the same way. ❑

We can establish the divergence theorem for more general regions by the device of subdividing T into simpler regions, regions for which the preceding proof holds. For example, suppose that T is the shell between the concentric spherical surfaces S_a and S_b of radii a and b, with $0 < a < b$. The coordinate planes separate T into eight regions $T_1, T_2, \ldots, T_8$, each shaped as in Fig. 16.6.2. Let Σ_i denote the boundary of T_i and let $\mathbf{n}_i$ be the outer unit normal vector to Σ_i. We apply the divergence theorem to each of these eight regions and obtain

$$\iiint_{T} \nabla \cdot \mathbf{F}\, dV = \sum_{i=1}^{8} \iiint_{T_i} \nabla \cdot \mathbf{F}\, dV$$

$$= \sum_{i=1}^{8} \iint_{\Sigma_i} \mathbf{F} \cdot \mathbf{n}_i\, dS \qquad \text{(divergence theorem)}$$

$$= \iint_{S_a} \mathbf{F} \cdot \mathbf{n}_a\, dS + \iint_{S_b} \mathbf{F} \cdot \mathbf{n}_b\, dS.$$

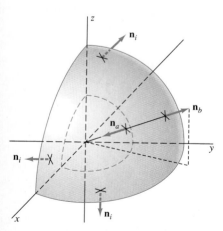

Fig. 16.6.2 One octant of the shell between S_a and S_b

Here we write $\mathbf{n}_a$ for the inner normal vector on S_a and $\mathbf{n}_b$ for the outer normal vector on S_b. The last equality holds because the surface integrals over the internal boundary surfaces (the surfaces in the coordinate planes) cancel in pairs—the normals are oppositely oriented there. As the boundary S of T is the union of the spherical surfaces S_a and S_b, it now follows that

$$\iiint_T \mathbf{\nabla} \cdot \mathbf{F}\, dV = \iint_S \mathbf{F} \cdot \mathbf{n}\, dS.$$

This is the divergence theorem for the spherical shell T.

EXAMPLE 1 Let S be the surface (with outer unit normal vector $\mathbf{n}$) of the region T bounded by the planes $z = 0$, $y = 0$, $y = 2$, and by the paraboloid $z = 1 - x^2$ (Fig. 16.6.3). Apply the divergence theorem to compute

$$\iint_S \mathbf{F} \cdot \mathbf{n}\, dS$$

given $\mathbf{F} = (x + \cos y)\mathbf{i} + (y + \sin z)\mathbf{j} + (z + e^x)\mathbf{k}$.

Solution To evaluate the surface integral directly would be a lengthy project. But div $\mathbf{F} = 1 + 1 + 1 = 3$, so we can apply the divergence theorem easily:

$$\iint_S \mathbf{F} \cdot \mathbf{n}\, dS = \iiint_T \text{div } \mathbf{F}\, dV = \iiint_T 3\, dV.$$

We examine Fig. 16.6.3 to find the limits for the volume integral and thus obtain

$$\iint_S \mathbf{F} \cdot \mathbf{n}\, dS = \int_{-1}^1 \int_0^2 \int_0^{1-x^2} 3\, dz\, dy\, dx = 12 \int_0^1 (1 - x^2)\, dx = 8.$$

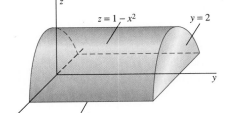

Fig. 16.6.3 The region of Example 1

(Figure labels: z, $z = 1 - x^2$, $y = 2$, y, $x = 1$, x)

EXAMPLE 2 Let S be the surface of the solid cylinder T bounded by the planes $z = 0$ and $z = 3$ and by the cylinder $x^2 + y^2 = 4$. Calculate the outward flux

$$\iint_S \mathbf{F} \cdot \mathbf{n}\, dS$$

given $\mathbf{F} = (x^2 + y^2 + z^2)(x\mathbf{i} + y\mathbf{j} + z\mathbf{k})$.

Solution If we denote by P, Q, and R the component functions of the vector field $\mathbf{F}$, we find that

$$\frac{\partial P}{\partial x} = 2x \cdot x + (x^2 + y^2 + z^2) \cdot 1 = 3x^2 + y^2 + z^2.$$

Similarly,

$$\frac{\partial Q}{\partial y} = 3y^2 + z^2 + x^2 \quad \text{and} \quad \frac{\partial R}{\partial z} = 3z^2 + x^2 + y^2,$$

so

$$\text{div } \mathbf{F} = 5(x^2 + y^2 + z^2).$$

Therefore, the divergence theorem yields

$$\iint_S \mathbf{F} \cdot \mathbf{n} \, dS = \iiint_T 5(x^2 + y^2 + z^2) \, dV.$$

Using cylindrical coordinates to evaluate the triple integral, we get

$$\iint_S \mathbf{F} \cdot \mathbf{n} \, dS = \int_0^{2\pi} \int_0^2 \int_0^3 5(r^2 + z^2) r \, dz \, dr \, d\theta$$

$$= 10\pi \int_0^2 \left[r^3 z + \tfrac{1}{3} r z^3 \right]_{z=0}^3 dr$$

$$= 10\pi \int_0^2 (3r^3 + 9r) \, dr = 10\pi \left[\tfrac{3}{4} r^4 + \tfrac{9}{2} r^2 \right]_0^2 = 300\pi.$$

EXAMPLE 3 Suppose that the region T is bounded by the closed surface S with a parametrization that gives the outer unit normal vector. Show that the volume V of T is given by

$$V = \tfrac{1}{3} \iint_S x \, dy \, dz + y \, dz \, dx + z \, dx \, dy. \tag{9}$$

Solution Equation (9) follows immediately from Eq. (4) if we take $P(x, y, z) = x$, $Q(x, y, z) = y$, and $R(x, y, z) = z$. For example, if S is the spherical surface $x^2 + y^2 + z^2 = a^2$ with volume V, surface area A, and outer unit normal vector

$$\mathbf{n} = \langle \cos \alpha, \cos \beta, \cos \gamma \rangle = \left\langle \frac{x}{a}, \frac{y}{a}, \frac{z}{a} \right\rangle,$$

then Eq. (9) yields

$$V = \frac{1}{3} \iint_S x \, dy \, dz + y \, dz \, dx + z \, dx \, dy$$

$$= \frac{1}{3} \iint_S (x \cos \alpha + y \cos \beta + z \cos \gamma) \, dS$$

$$= \frac{1}{3} \iint_S \frac{x^2 + y^2 + z^2}{a} \, dS = \frac{1}{3} a \iint_S 1 \, dS = \frac{1}{3} aA.$$

You should confirm that this result is consistent with the familiar formulas $V = \tfrac{4}{3} \pi a^3$ and $A = 4\pi a^2$.

EXAMPLE 4 Show that the divergence of the vector field $\mathbf{F}$ at the point P is given by

$$\{\operatorname{div} \mathbf{F}\}(P) = \lim_{r \to 0} \frac{1}{V_r} \iint_{S_r} \mathbf{F} \cdot \mathbf{n} \, dS, \tag{10}$$

where S_r is the sphere of radius r centered at P and $V_r = \tfrac{4}{3} \pi r^3$ is the volume of the ball B_r that the sphere bounds.

948

Solution   The divergence theorem gives

$$\iint_{S_r} \mathbf{F} \cdot \mathbf{n} \, dS = \iiint_{B_r} \operatorname{div} \mathbf{F} \, dV.$$

Then we apply the average value property of triple integrals, a result analogous to the double integral result of Problem 30 in Section 15.2. This yields

$$\iiint_{B_r} \operatorname{div} \mathbf{F} \, dV = V_r \cdot \{\operatorname{div} \mathbf{F}\}(P^*)$$

for some point P^* of B_r; here, we write $\{\operatorname{div} \mathbf{F}\}(P^*)$ for the value of div $\mathbf{F}$ at the point P^*. We assume that the component functions of $\mathbf{F}$ have continuous first-order partial derivatives at P, so it follows that

$$\{\operatorname{div} \mathbf{F}\}(P^*) \to \{\operatorname{div} \mathbf{F}\}(P) \quad \text{as} \quad P^* \to P.$$

Equation (10) follows after we divide both sides by V_r and then take the limit as $r \to 0$.

For instance, suppose that $\mathbf{F} = \rho \mathbf{v}$ is a fluid flow vector field. We can interpret Eq. (10) as saying that $\{\operatorname{div} \mathbf{F}\}(P)$ is the net rate per unit volume that fluid mass is flowing away (or "diverging") from the point P. For this reason the point P is called a **source** if $\{\operatorname{div} \mathbf{F}\}(P) > 0$ but a **sink** if $\{\operatorname{div} \mathbf{F}\}(P) < 0$.

Heat in a conducting body can be treated mathematically as though it were a fluid flowing through the body. Miscellaneous Problems 25 through 27 at the end of this chapter ask you to apply the divergence theorem to show that if $u = u(x, y, z, t)$ is the temperature at the point (x, y, z) at the time t in a body through which heat is flowing, then the function u must satisfy the equation

$$\frac{\partial^2 u}{\partial x^2} + \frac{\partial^2 u}{\partial y^2} + \frac{\partial^2 u}{\partial z^2} = \frac{1}{k} \cdot \frac{\partial u}{\partial t}, \tag{11}$$

where k is a constant (the *thermal diffusivity* of the body). This is a *partial differential equation* called the **heat equation**. If both the initial temperature $u(x, y, z, 0)$ and the temperature on the boundary of the body are given, then its interior temperatures at future times are determined by the heat equation. A large part of advanced applied mathematics consists of techniques for solving such partial differential equations.

Another impressive consequence of the divergence theorem is Archimedes' law of buoyancy; see Problem 21 here and Problem 22 of Section 16.7.

16.6 Problems

In Problems 1 through 5, verify the divergence theorem by direct computation of both the surface integral and the triple integral of Eq. (1).

1. $\mathbf{F} = x\mathbf{i} + y\mathbf{j} + z\mathbf{k}$; S is the spherical surface with equation $x^2 + y^2 + z^2 = 1$.

2. $\mathbf{F} = |\mathbf{r}|\mathbf{r}$, where $\mathbf{r} = x\mathbf{i} + y\mathbf{j} + z\mathbf{k}$; S is the spherical surface with equation $x^2 + y^2 + z^2 = 9$.

3. $\mathbf{F} = x\mathbf{i} + y\mathbf{j} + z\mathbf{k}$; S is the surface of the cube bounded by the three coordinate planes and by the three planes $x = 2$, $y = 2$, and $z = 2$.

4. $\mathbf{F} = xy\mathbf{i} + yz\mathbf{j} + xz\mathbf{k}$; S is the surface of Problem 3.

5. $\mathbf{F} = (x + y)\mathbf{i} + (y + z)\mathbf{j} + (z + x)\mathbf{k}$; S is the surface of the tetrahedron bounded by the three coordinate planes and by the plane $x + y + z = 1$.

In Problems 6 through 14, use the divergence theorem to evaluate $\iint_S \mathbf{F} \cdot \mathbf{n} \, dS$, where $\mathbf{n}$ is the outer unit normal vector to the surface S.

6. $\mathbf{F} = x^2\mathbf{i} + y^2\mathbf{j} + z^2\mathbf{k}$; S is the surface of Problem 3.

7. $\mathbf{F} = x^3\mathbf{i} + y^3\mathbf{j} + z^3\mathbf{k}$; S is the surface of the cylinder bounded by $x^2 + y^2 = 9$, $z = -1$, and $z = 4$.

8. $\mathbf{F} = (x^2 + y^2)(x\mathbf{i} + y\mathbf{j})$; S is the surface of the region bounded by the plane $z = 0$ and by the paraboloid $z = 25 - x^2 - y^2$.

9. $\mathbf{F} = (x^2 + e^{-yz})\mathbf{i} + (y + \sin xz)\mathbf{j} + (\cos xy)\mathbf{k}$; S is the surface of Problem 5.

10. $\mathbf{F} = (xy^2 + e^{-y} \sin z)\mathbf{i} + (x^2y + e^{-x} \cos z)\mathbf{j} + (\tan^{-1}xy)\mathbf{k}$; S is the surface of the region bounded by the paraboloid $z = x^2 + y^2$ and by the plane $z = 9$.

11. $\mathbf{F} = (x^2 + y^2 + z^2)(x\mathbf{i} + y\mathbf{j} + z\mathbf{k})$; S is the surface of Problem 8.

12. $\mathbf{F} = \mathbf{r}/|\mathbf{r}|$, where $\mathbf{r} = x\mathbf{i} + y\mathbf{j} + z\mathbf{k}$; S is the sphere $\rho = 2$ of radius 2 centered at the origin.

13. $\mathbf{F} = x\mathbf{i} + y\mathbf{j} + 3\mathbf{k}$; S is the boundary of the region bounded by the paraboloid $z = x^2 + y^2$ and by the plane $z = 4$.

14. $\mathbf{F} = (x^3 + e^z)\mathbf{i} + x^2y\mathbf{j} + (\sin xy)\mathbf{k}$; S is the boundary of the region bounded by the paraboloid $z = 4 - x^2$ and by the planes $y = 0$, $z = 0$, and $y + z = 5$.

15. The **Laplacian** of the twice-differentiable scalar function f is defined to be $\nabla^2 f = \operatorname{div}(\operatorname{grad} f) = \nabla \cdot \nabla f$. Show that

$$\nabla^2 f = \frac{\partial^2 f}{\partial x^2} + \frac{\partial^2 f}{\partial y^2} + \frac{\partial^2 f}{\partial z^2}.$$

16. Let $\partial f/\partial n = \nabla f \cdot \mathbf{n}$ denote the directional derivative of the scalar function f in the direction of the outer unit normal vector $\mathbf{n}$ to the surface S that bounds the region T. Show that

$$\iint_S \frac{\partial f}{\partial n} \, dS = \iiint_T \nabla^2 f \, dV.$$

17. Suppose that $\nabla^2 f \equiv 0$ in the region T with boundary surface S. Show that

$$\iint_S f \frac{\partial f}{\partial n} \, dS = \iiint_T |\nabla f|^2 \, dV.$$

(See Problems 15 and 16 for the notation.)

18. Apply the divergence theorem to $\mathbf{F} = f\nabla g$ to establish **Green's first identity**,

$$\iint_S f \frac{\partial g}{\partial n} \, dS = \iiint_T (f \nabla^2 g + \nabla f \cdot \nabla g) \, dV.$$

19. Interchange f and g in Green's first identity (Problem 18) to establish **Green's second identity**,

$$\iint_S \left(f \frac{\partial g}{\partial n} - g \frac{\partial f}{\partial n} \right) dS = \iiint_T (f \nabla^2 g - g \nabla^2 f) \, dV.$$

20. Suppose that f is a differentiable scalar function defined on the region T of space and that S is the boundary of T. Prove that

$$\iint_S f\mathbf{n} \, dS = \iiint_T \nabla f \, dV.$$

[*Suggestion:* Apply the divergence theorem to $\mathbf{F} = f\mathbf{a}$, where $\mathbf{a}$ is an arbitrary constant vector. *Note:* Integrals of vector-valued functions are defined by componentwise integration.]

21. *Archimedes' Law of Buoyancy*   Let S be the surface of a body T submerged in a fluid of constant density ρ. Set up coordinates so that $z = 0$ corresponds to the surface of the fluid and so that positive values of z are measured *downward* from the surface. Then the pressure at depth z is $p = \rho g z$. The buoyant force exerted on the body by the fluid is

$$\mathbf{B} = -\iint_S p\mathbf{n} \, dS.$$

(Why?) Apply the result of Problem 20 to show that $\mathbf{B} = -W\mathbf{k}$, where W is the weight of the fluid displaced by the body. Because z is measured downward, the vector $\mathbf{B}$ is directed upward.

22. Let $\mathbf{r} = \langle x, y, z \rangle$, let $\mathbf{r}_0 = \langle x_0, y_0, z_0 \rangle$ be a fixed point, and suppose that

$$\mathbf{F}(x, y, z) = \frac{\mathbf{r} - \mathbf{r}_0}{|\mathbf{r} - \mathbf{r}_0|}.$$

Show that $\operatorname{div} \mathbf{F} = 0$ except at the point $\mathbf{r}_0$.

23. Apply the divergence theorem to compute the outward flux

$$\iint_S \mathbf{F} \cdot \mathbf{n} \, dS,$$

where $\mathbf{F} = |\mathbf{r}|\mathbf{r}, \mathbf{r} = x\mathbf{i} + y\mathbf{j} + z\mathbf{k}$, and S is the surface of Problem 8. [*Suggestion:* Integrate in cylindrical coordinates, first with respect to r and then with respect to z. For the latter integration, make a trigonometric substitution and then consult Eq. (8) of Section 9.4 for the antiderivative of $\sec^5 \theta$.]

16.7
Stokes' Theorem

In Section 16.4, we gave Green's theorem,

$$\oint_C P\,dx + Q\,dy = \iint_R \left(\frac{\partial Q}{\partial x} - \frac{\partial P}{\partial y}\right) dA, \qquad (1)$$

in a vector form that is equivalent to a two-dimensional version of the divergence theorem. Another vector form of Green's theorem involves the curl of a vector field. Recall from Section 16.1 that if $\mathbf{F} = P\mathbf{i} + Q\mathbf{j} + R\mathbf{k}$ is a vector field, then curl $\mathbf{F}$ is the vector field given by

$$\text{curl } \mathbf{F} = \nabla \times \mathbf{F} = \begin{vmatrix} \mathbf{i} & \mathbf{j} & \mathbf{k} \\ \dfrac{\partial}{\partial x} & \dfrac{\partial}{\partial y} & \dfrac{\partial}{\partial z} \\ P & Q & R \end{vmatrix}$$

$$= \left(\frac{\partial R}{\partial y} - \frac{\partial Q}{\partial z}\right)\mathbf{i} + \left(\frac{\partial P}{\partial z} - \frac{\partial R}{\partial x}\right)\mathbf{j} + \left(\frac{\partial Q}{\partial x} - \frac{\partial P}{\partial y}\right)\mathbf{k}. \qquad (2)$$

The $\mathbf{k}$-component of $\nabla \times \mathbf{F}$ is the integrand of the double integral of Eq. (1). We know from Section 16.2 that we can write the line integral of Eq. (1) as

$$\oint_C \mathbf{F} \cdot \mathbf{T}\,ds,$$

where $\mathbf{T}$ is the positive-directed unit tangent vector to C. Consequently, we can rewrite Green's theorem in the form

$$\oint_C \mathbf{F} \cdot \mathbf{T}\,ds = \iint_R (\text{curl } \mathbf{F}) \cdot \mathbf{k}\,dA. \qquad (3)$$

Stokes' theorem is the generalization of Eq. (3) that we get by replacing the plane region R with a floppy two-dimensional version: an oriented bounded surface S in three-dimensional space with boundary C that consists of one or more simple closed curves in space.

An *oriented* surface is a surface together with a chosen continuous unit normal vector field $\mathbf{n}$. The positive orientation of the boundary C of an oriented surface S corresponds to the unit tangent vector $\mathbf{T}$ such that $\mathbf{n} \times \mathbf{T}$ always points *into* S (Fig. 16.7.1.). Check that for a plane region with unit normal vector $\mathbf{k}$, the positive orientation of its outer boundary is counterclockwise.

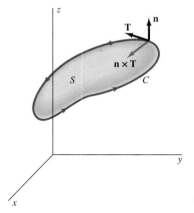

Fig. 16.7.1 Vectors, surface, and boundary curve mentioned in the statement of Stokes' theorem

Stokes' Theorem

Let S be an oriented, bounded, and piecewise smooth surface in space with positively oriented boundary C. Suppose that the components of the vector field $\mathbf{F}$ have continuous first-order partial derivatives in a space region that contains S. Then

$$\oint_C \mathbf{F} \cdot \mathbf{T}\,ds = \iint_S (\text{curl } \mathbf{F}) \cdot \mathbf{n}\,dS. \qquad (4)$$

Thus Stokes' theorem says that *the line integral around the boundary curve of the tangential component of* $\mathbf{F}$ *equals the surface integral of the normal component of* curl $\mathbf{F}$. Compare Eqs. (3) and (4).

The result first appeared publicly as a problem posed by George Stokes (1819–1903) on a prize examination for Cambridge University students in 1854. It had been stated in an 1850 letter to Stokes from the physicist William Thomson (Lord Kelvin, 1824–1907).

In terms of the components of $\mathbf{F} = P\mathbf{i} + Q\mathbf{j} + R\mathbf{k}$ and those of curl $\mathbf{F}$, we can recast Stokes' theorem—with the aid of Eq. (14) of Section 16.5—in its scalar form:

$$\oint_C P\,dx + Q\,dy + R\,dz$$
$$= \iint_S \left(\frac{\partial R}{\partial y} - \frac{\partial Q}{\partial z}\right) dy\,dz + \left(\frac{\partial P}{\partial z} - \frac{\partial R}{\partial x}\right) dz\,dx + \left(\frac{\partial Q}{\partial x} - \frac{\partial P}{\partial y}\right) dx\,dy. \tag{5}$$

Here, as usual, the parametrization of S must correspond to the given unit normal vector $\mathbf{n}$.

To prove Stokes' theorem, we need only establish the equation

$$\oint_C P\,dx = \iint_S \left(\frac{\partial P}{\partial z}\,dz\,dx - \frac{\partial P}{\partial y}\,dx\,dy\right) \tag{6}$$

and the corresponding two equations that are the Q and R "components" of Eq. (5). Equation (5) itself then follows by adding the three results.

Partial Proof   Suppose first that S is the graph of a function $z = f(x, y)$, (x, y) in D, where S has an upper unit normal vector and D is a region in the xy-plane bounded by the simple closed curve J (Fig. 16.7.2). Then

$$\oint_C P\,dx = \oint_J P(x, y, f(x, y))\,dx$$
$$= \oint_J p(x, y)\,dx \qquad [\text{where } p(x, y) \equiv P(x, y, f(x, y))]$$
$$= -\iint_D \frac{\partial p}{\partial y}\,dx\,dy \qquad (\text{by Green's theorem}).$$

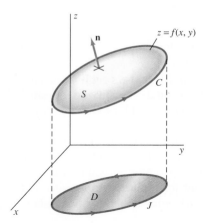

Fig. 16.7.2   The surface S

We now use the chain rule to compute $\partial p/\partial y$ and find that

$$\oint_C P\,dx = -\iint_D \left(\frac{\partial P}{\partial y} + \frac{\partial P}{\partial z}\frac{\partial z}{\partial y}\right) dx\,dy. \tag{7}$$

Next, we use Eq. (15) of Section 16.5:

$$\iint_S P\,dy\,dz + Q\,dz\,dx + R\,dx\,dy = \iint_D \left(-P\frac{\partial z}{\partial x} - Q\frac{\partial z}{\partial y} + R\right) dx\,dy.$$

In this equation we replace P by 0, Q by $\partial P/\partial z$, and R by $-\partial P/\partial y$. This gives

$$\iint_S \left(\frac{\partial P}{\partial z}\,dz\,dx - \frac{\partial P}{\partial y}\,dx\,dy\right) = \iint_D \left(-\frac{\partial P}{\partial z}\frac{\partial z}{\partial y} - \frac{\partial P}{\partial y}\right) dx\,dy. \tag{8}$$

Finally, we compare Eqs. (7) and (8) and see that we have established Eq. (6). If we can write the surface S in the forms $y = g(x, z)$ and

$x = h(y, z)$, then we can derive the Q and R "components" of Eq. (5) in essentially the same way. This proves Stokes' theorem for the special case of a surface S that can be represented as a graph in all three coordinate directions. Stokes' theorem may then be extended to a more general oriented surface by the now-familiar method of subdividing it into simpler surfaces, to each of which the above proof is applicable. ❑

EXAMPLE 1 Apply Stokes' theorem to evaluate

$$\oint_C \mathbf{F} \cdot \mathbf{T} \, ds,$$

where C is the ellipse in which the plane $z = y + 3$ intersects the cylinder $x^2 + y^2 = 1$. Orient the ellipse counterclockwise as viewed from above and take $\mathbf{F}(x, y, z) = 3z\mathbf{i} + 5x\mathbf{j} - 2y\mathbf{k}$.

Solution The plane, cylinder, and ellipse appear in Fig. 16.7.3. The given orientation of C corresponds to the upward unit normal vector $\mathbf{n} = (-\mathbf{j} + \mathbf{k})/\sqrt{2}$ to the elliptical region S in the plane $z = y + 3$ bounded by C. Now

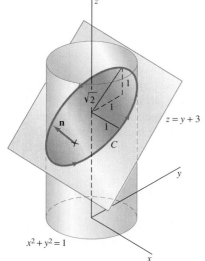

$$\text{curl } \mathbf{F} = \begin{vmatrix} \mathbf{i} & \mathbf{j} & \mathbf{k} \\ \dfrac{\partial}{\partial x} & \dfrac{\partial}{\partial y} & \dfrac{\partial}{\partial z} \\ 3z & 5x & -2y \end{vmatrix} = -2\mathbf{i} + 3\mathbf{j} + 5\mathbf{k},$$

Fig. 16.7.3 The ellipse of Example 1

so

$$(\text{curl } \mathbf{F}) \cdot \mathbf{n} = (-2\mathbf{i} + 3\mathbf{j} + 5\mathbf{k}) \cdot \frac{1}{\sqrt{2}}(-\mathbf{j} + \mathbf{k}) = \frac{-3 + 5}{\sqrt{2}} = \sqrt{2}.$$

Hence by Stokes' theorem,

$$\oint_C \mathbf{F} \cdot \mathbf{T} \, ds = \iint_S (\text{curl } \mathbf{F}) \cdot \mathbf{n} \, dS = \iint_S \sqrt{2} \, dS = \sqrt{2} \, \text{area}(S) = 2\pi,$$

because we can see from Fig. 16.7.3 that S is an ellipse with semiaxes 1 and $\sqrt{2}$. Thus its area is $\pi\sqrt{2}$.

EXAMPLE 2 Apply Stokes' theorem to evaluate

$$\iint_S (\nabla \times \mathbf{F}) \cdot \mathbf{n} \, dS,$$

where $\mathbf{F} = 3z\mathbf{i} + 5x\mathbf{j} - 2y\mathbf{k}$ and S is the part of the parabolic surface $z = x^2 + y^2$ that lies below the plane $z = 4$ and whose orientation is given by the upper unit normal vector (Fig. 16.7.4).

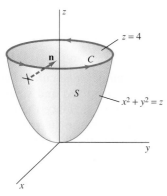

Fig. 16.7.4 The parabolic surface of Example 2

Solution We parametrize the boundary circle C of S by $x = 2 \cos t$, $y = 2 \sin t, z = 4$ for $0 \leqq t \leqq 2\pi$. Then $dx = -2 \sin t \, dt, dy = 2 \cos t \, dt$, and $dz = 0$. So Stokes' theorem yields

$$\iint_S (\nabla \times \mathbf{F}) \cdot \mathbf{n} \, dS = \oint_C \mathbf{F} \cdot \mathbf{T} \, ds = \oint_C 3z \, dx + 5x \, dy - 2y \, dz$$

$$= \int_0^{2\pi} 3 \cdot 4 \cdot (-2 \sin t \, dt) + 5 \cdot (2 \cos t)(2 \cos t \, dt) + 2 \cdot (2 \sin t) \cdot 0$$

$$= \int_0^{2\pi} (-24 \sin t + 20 \cos^2 t) \, dt = \int_0^{2\pi} (-24 \sin t + 10 + 10 \cos 2t) \, dt$$

$$= \left[24 \cos t + 10t + 5 \sin 2t \right]_0^{2\pi} = 20\pi.$$

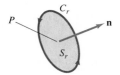

Fig. 16.7.5 A physical interpretation of the curl of a vector field

Just as the divergence theorem yields a physical interpretation of div $\mathbf{F}$ [Eq. (10) of Section 16.6], Stokes' theorem yields a physical interpretation of curl $\mathbf{F}$. Let S_r be a circular disk of radius r, centered at the point P in space and perpendicular to the unit vector $\mathbf{n}$. Let C_r be the boundary circle of S_r (Fig. 16.7.5). Then Stokes' theorem and the average value property of double integrals together give

$$\oint_{C_r} \mathbf{F} \cdot \mathbf{T} \, ds = \iint_{S_r} (\text{curl } \mathbf{F}) \cdot \mathbf{n} \, dS = \pi r^2 \cdot \{(\text{curl } \mathbf{F}) \cdot \mathbf{n}\}(P^*)$$

for some point P^* of S_r, where $\{(\text{curl } \mathbf{F}) \cdot \mathbf{n}\}(P^*)$ denotes the value of $(\text{curl } \mathbf{F}) \cdot \mathbf{n}$ at the point P^*. We divide this equality by πr^2 and then take the limit as $r \to 0$. This gives

$$\{(\text{curl } \mathbf{F}) \cdot \mathbf{n}\}(P) = \lim_{r \to 0} \frac{1}{\pi r^2} \oint_{C_r} \mathbf{F} \cdot \mathbf{T} \, ds. \tag{9}$$

Equation (9) has a natural physical meaning. Suppose that $\mathbf{F} = \rho \mathbf{v}$, where $\mathbf{v}$ is the velocity vector field of a steady-state fluid flow with constant density ρ. Then the value of the integral

$$\Gamma(C) = \oint_C \mathbf{F} \cdot \mathbf{T} \, ds \tag{10}$$

measures the rate of flow of fluid mass *around* the curve C and is therefore called the **circulation** of $\mathbf{F}$ around C. We see from Eq. (9) that

$$\{(\text{curl } \mathbf{F}) \cdot \mathbf{n}\}(P) \approx \frac{\Gamma(C_r)}{\pi r^2}$$

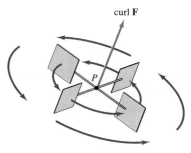

curl F

Fig. 16.7.6 The paddle-wheel interpretation of curl $\mathbf{F}$

if C_r is a circle of very small radius r centered at P and perpendicular to $\mathbf{n}$. If $\{\text{curl } \mathbf{F}\}(P) \neq \mathbf{0}$, it follows that $\Gamma(C_r)$ is greatest (for r fixed and small) when the unit vector $\mathbf{n}$ points in the direction of $\{\text{curl } \mathbf{F}\}(P)$. Hence the line through P determined by $\{\text{curl } \mathbf{F}\}(P)$ is the axis about which the fluid near P is revolving the most rapidly. A tiny paddle wheel placed in the fluid at P (see Fig. 16.7.6) would rotate the fastest if its axis lay along this line. It follows from Miscellaneous Problem 32 at the end of this chapter that $|\text{curl } \mathbf{F}| = 2\rho\omega$ in the case of a fluid revolving steadily about a fixed axis with constant angular speed ω (in radians per second). Thus $\{\text{curl } \mathbf{F}\}(P)$ indicates both the direction *and* rate of rotation of the fluid near P. Because of this interpretation, some older books use the notation "rot **F**" for the curl, an abbreviation that we are happy has disappeared from general use.

If curl **F** = **0** everywhere, then the fluid flow and the vector field **F** are said to be **irrotational**. An infinitesimal straw placed in an irrotational fluid flow would be translated parallel to itself without rotating. A vector field **F** defined on a simply connected region D is irrotational if and only if it is conservative, which in turn is true if and only if the line integral

$$\int_C \mathbf{F} \cdot \mathbf{T} \, ds$$

is independent of the path in D. (The region D is said to be **simply connected** if every simple closed curve in D can be continuously shrunk to a point while staying inside D. The interior of a torus is an example of a space region that is *not* simply connected. It is true, though not obvious, that any piecewise smooth simple closed curve in a simply connected region D is the boundary of a piecewise smooth oriented surface in D.)

Theorem 1   Conservative and Irrotational Fields
Let **F** be a vector field with continuous first-order partial derivatives in a simply connected region D in space. Then the vector field **F** is irrotational if and only if it is conservative; that is, $\nabla \times \mathbf{F} = \mathbf{0}$ if and only if $\mathbf{F} = \nabla\phi$ for some scalar function ϕ defined on D.

Partial Proof   A complete proof of the *if* part of Theorem 1 is easy; by Example 7 of Section 16.1, $\nabla \times (\nabla\phi) = \mathbf{0}$ for any twice-differentiable scalar function ϕ.

Here is a description of how we might show the *only if* part of the proof of Theorem 1. Assume that **F** is irrotational. Let $P_0(x_0, y_0, z_0)$ be a fixed point of D. Given an arbitrary point $P(x, y, z)$ of D, we would like to define

$$\phi(x, y, z) = \int_{C_1} \mathbf{F} \cdot \mathbf{T} \, ds, \tag{11}$$

where C_1 is a path in D from P_0 to P. But we must show that any *other* path C_2 from P_0 to P would give the *same* value for $\phi(x, y, z)$.

We may assume, as suggested by Fig. 16.7.7, that C_1 and C_2 intersect only at their endpoints. Then the simple closed curve $C = C_1 \cup (-C_2)$ bounds an oriented surface S in D, and Stokes' theorem gives

$$\int_{C_1} \mathbf{F} \cdot \mathbf{T} \, ds - \int_{C_2} \mathbf{F} \cdot \mathbf{T} \, ds = \oint_C \mathbf{F} \cdot \mathbf{T} \, ds = \iint_S (\nabla \times \mathbf{F}) \cdot \mathbf{n} \, dS = 0$$

because of the hypothesis that $\nabla \times \mathbf{F} \equiv \mathbf{0}$. This shows that the line integral $\int_C \mathbf{F} \cdot \mathbf{T} \, ds$ is *independent of path*, just as desired. In Problem 21 we ask you to complete this proof by showing that the function ϕ of Eq. (11) is the one whose existence is claimed in Theorem 1. That is, $\mathbf{F} = \nabla\phi$.   ◻

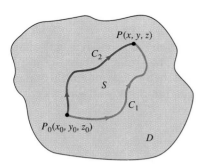

Fig. 16.7.7   Two paths from P_0 to P in the simply connected space region D

EXAMPLE 3   Show that the vector field $\mathbf{F} = 3x^2\mathbf{i} + 5z^2\mathbf{j} + 10yz\mathbf{k}$ is irrotational. Then find a potential function $\phi(x, y, z)$ such that $\nabla\phi = \mathbf{F}$.

Solution To show that **F** is irrotational, we calculate

$$\nabla \times \mathbf{F} = \begin{vmatrix} \mathbf{i} & \mathbf{j} & \mathbf{k} \\ \dfrac{\partial}{\partial x} & \dfrac{\partial}{\partial y} & \dfrac{\partial}{\partial z} \\ 3x^2 & 5z^2 & 10yz \end{vmatrix} = (10z - 10z)\mathbf{i} = \mathbf{0}.$$

Hence Theorem 1 implies that **F** has a potential function ϕ. We can apply Eq. (11) to find ϕ explicitly. If C_1 is the straight line segment from $(0, 0, 0)$ to (x_1, y_1, z_1) that is parametrized by $x = x_1 t$, $y = y_1 t$, $z = z_1 t$ for $0 \leqq t \leqq 1$, then Eq. (11) yields

$$\phi(x_1, y_1, z_1) = \int_{C_1} \mathbf{F} \cdot \mathbf{T} \, ds = \int_{(0,0,0)}^{(x_1, y_1, z_1)} 3x^2 \, dx + 5z^2 \, dy + 10yz \, dz$$

$$= \int_0^1 (3x_1^2 t^2)(x_1 \, dt) + (5z_1^2 t^2)(y_1 \, dt) + (10y_1 t z_1 t)(z_1 \, dt)$$

$$= \int_0^1 (3x_1^3 t^2 + 15 y_1 z_1^2 t^2) \, dt = \left[x_1^3 t^3 + 5 y_1 z_1^2 t^3 \right]_0^1,$$

and thus

$$\phi(x_1, y_1, z_1) = x_1^3 + 5 y_1 z_1^2.$$

We may now drop the subscripts because (x_1, y_1, z_1) is an arbitrary point of space, and therefore we obtain the scalar potential function $\phi(x, y, z) = x^3 + 5yz^2$. As a check, we note that $\phi_x = 3x^2$, $\phi_y = 5z^2$, and $\phi_z = 10yz$, so $\nabla \phi = \mathbf{F}$ as desired.

APPLICATION Suppose that **v** is the velocity field of a steady fluid flow that is both irrotational and incompressible—the density ρ of the fluid is constant. Suppose that S is any closed surface that bounds a region T. Then, because of conservation of mass, the flux of **v** across S must be zero; the mass of fluid within S remains constant. Hence the divergence theorem gives

$$\iiint_T \text{div } \mathbf{v} \, dV = \iint_S \mathbf{v} \cdot \mathbf{n} \, dS = 0.$$

Because this holds for *any* region T, it follows from the usual average value property argument that div $\mathbf{v} = 0$ everywhere. The scalar function ϕ provided by Theorem 1, for which $\mathbf{v} = \nabla \phi$, is called the **velocity potential** of the fluid flow. We substitute $\mathbf{v} = \nabla \phi$ into the equation div $\mathbf{v} = 0$ and thereby obtain

$$\text{div}(\nabla \phi) = \frac{\partial^2 \phi}{\partial x^2} + \frac{\partial^2 \phi}{\partial y^2} + \frac{\partial^2 \phi}{\partial z^2} = 0. \tag{12}$$

Thus the velocity potential ϕ of an irrotational and incompressible fluid flow satisfies *Laplace's equation*.

Laplace's equation appears in numerous other applications. For example, consider a heated body whose temperature function $u = u(x, y, z)$ is independent of time t. Then $\partial u / \partial t \equiv 0$ in the heat equation, Eq. (15) of Section 16.6, shows that the "steady-state temperature" function $u(x, y, z)$ satisfies Laplace's equation

$$\frac{\partial^2 u}{\partial x^2} + \frac{\partial^2 u}{\partial y^2} + \frac{\partial^2 u}{\partial z^2} = 0. \qquad (13)$$

These brief remarks should indicate how the mathematics of this chapter forms the starting point for investigations in a number of areas, including acoustics, aerodynamics, electromagnetism, meteorology, and oceanography. Indeed, the entire subject of vector analysis stems historically from its scientific applications rather than from abstract mathematical considerations. The modern form of the subject is due primarily to J. Willard Gibbs (1839–1903), the first great American physicist, and the English electrical engineer Oliver Heaviside (1850–1925).

16.7 Problems

In Problems 1 through 5, use Stokes' theorem for the evaluation of

$$\iint_S (\text{curl } \mathbf{F}) \cdot \mathbf{n} \ dS.$$

1. $\mathbf{F} = 3y\mathbf{i} - 2x\mathbf{j} + xyz\mathbf{k}$; S is the hemispherical surface $z = \sqrt{4 - x^2 - y^2}$ with upper unit normal vector.

2. $\mathbf{F} = 2y\mathbf{i} + 3x\mathbf{j} + e^z\mathbf{k}$; S is the part of the paraboloid $z = x^2 + y^2$ below the plane $z = 4$ with upper unit normal vector.

3. $\mathbf{F} = \langle xy, -2, \arctan x^2 \rangle$; S is the part of the paraboloid $z = 9 - x^2 - y^2$ above the xy-plane with upper unit normal vector.

4. $\mathbf{F} = yz\mathbf{i} + xz\mathbf{j} + xy\mathbf{k}$; S is the part of the cylinder $x^2 + y^2 = 1$ between the two places $z = 1$ and $z = 3$ with outer unit normal vector.

5. $\mathbf{F} = \langle yz, -xz, z^3 \rangle$; S is the part of the cone $z = \sqrt{x^2 + y^2}$ between the two planes $z = 1$ and $z = 3$ with upper unit normal vector.

In Problems 6 through 10, use Stokes' theorem to evaluate

$$\oint_C \mathbf{F} \cdot \mathbf{T} \ ds.$$

6. $\mathbf{F} = 3y\mathbf{i} - 2x\mathbf{j} + 4x\mathbf{k}$; C is the circle $x^2 + y^2 = 9$, $z = 4$, oriented counterclockwise as viewed from above.

7. $\mathbf{F} = 2z\mathbf{i} + x\mathbf{j} + 3y\mathbf{k}$; C is the ellipse in which the plane $z = x$ meets the cylinder $x^2 + y^2 = 4$, oriented counterclockwise as viewed from above.

8. $\mathbf{F} = y\mathbf{i} + z\mathbf{j} + x\mathbf{k}$; C is the boundary of the triangle with vertices $(0, 0, 0)$, $(2, 0, 0)$, and $(0, 2, 2)$, oriented counterclockwise as viewed from above.

9. $\mathbf{F} = \langle y - x, x - z, x - y \rangle$; C is the boundary of the part of the plane $x + 2y + z = 2$ that lies in the first octant, oriented counterclockwise as viewed from above.

10. $\mathbf{F} = y^2\mathbf{i} + z^2\mathbf{j} + x^2\mathbf{k}$; C is the intersection of the plane $z = y$ and the cylinder $x^2 + y^2 = 2y$, oriented counterclockwise as viewed from above.

In Problems 11 through 14, first show that the given vector field $\mathbf{F}$ is irrotational; then apply the method of Example 3 to find a potential function $\phi = \phi(x, y, z)$ for $\mathbf{F}$.

11. $\mathbf{F} = (3y - 2z)\mathbf{i} + (3x + z)\mathbf{j} + (y - 2x)\mathbf{k}$

12. $\mathbf{F} = (3y^3 - 10xz^2)\mathbf{i} + 9xy^2\mathbf{j} - 10x^2z\mathbf{k}$

13. $\mathbf{F} = (3e^z - 5y \sin x)\mathbf{i} + (5 \cos x)\mathbf{j} + (17 + 3xe^z)\mathbf{k}$

14. $\mathbf{F} = r^3\mathbf{r}$, where $\mathbf{r} = x\mathbf{i} + y\mathbf{j} + z\mathbf{k}$ and $r = |\mathbf{r}|$

15. Suppose that $\mathbf{r} = x\mathbf{i} + y\mathbf{j} + z\mathbf{k}$ and that $\mathbf{a}$ is a constant vector. Show that (a) $\nabla \cdot (\mathbf{a} \times \mathbf{r}) = 0$; (b) $\nabla \times (\mathbf{a} \times \mathbf{r}) = 2\mathbf{a}$; (c) $\nabla \cdot [(\mathbf{r} \cdot \mathbf{r})\mathbf{a}] = 2\mathbf{r} \cdot \mathbf{a}$; (d) $\nabla \times [(\mathbf{r} \cdot \mathbf{r})\mathbf{a}] = 2(\mathbf{r} \times \mathbf{a})$.

16. Prove that

$$\iint_S (\text{curl } \mathbf{F}) \cdot \mathbf{n} \ dS$$

has the same value for all oriented surfaces S that have the same oriented boundary curve C.

17. Suppose that S is a closed surface. Prove in two different ways that

$$\iint_S (\text{curl } \mathbf{F}) \cdot \mathbf{n} \ dS = 0:$$

(a) by using the divergence theorem, with T the region bounded outside by S; and (b) by using Stokes' theorem, with the aid of a simple closed curve C on S.

Line integrals, surface integrals, and triple integrals of vector-valued functions are defined by componentwise integration. Such integrals appear in Problems 18 through 20.

18. Suppose that C and S are as described in the statement of Stokes' theorem and that ϕ is a scalar function. Prove that

$$\oint_C \phi \mathbf{T} \, ds = \iint_S \mathbf{n} \times \nabla\phi \, dS.$$

[*Suggestion:* Apply Stokes' theorem with $\mathbf{F} = \phi\mathbf{a}$, where $\mathbf{a}$ is an arbitrary constant vector.]

19. Suppose that $\mathbf{a}$ and $\mathbf{r}$ are as in Problem 15. Prove that

$$\oint_C (\mathbf{a} \times \mathbf{r}) \cdot \mathbf{T} \, ds = 2\mathbf{a} \cdot \iint_S \mathbf{n} \, dS.$$

20. Suppose that S is a closed surface that bounds the region T. Prove that

$$\iint_S \mathbf{n} \times \mathbf{F} \, dS = \iiint_T \nabla \times \mathbf{F} \, dV.$$

[*Suggestion:* Apply the divergence theorem to $\mathbf{F} \times \mathbf{a}$, where $\mathbf{a}$ is an arbitrary constant vector.]

REMARK *The formulas of Problem 20, the divergence theorem, and Problem 20 of Section 16.6 all fit the pattern*

$$\iint_S \mathbf{n} * (\ ) \, dS = \iiint_T \nabla * (\ ) \, dV,$$

*where * denotes either ordinary multiplication, the dot product, or the vector product, and either a scalar function or a vector-valued function is placed within the parentheses, as appropriate.*

21. Suppose that the line integral $\int_C \mathbf{F} \cdot \mathbf{T} \, ds$ is independent of path. If

$$\phi(x, y, z) = \int_{P_0}^{P} \mathbf{F} \cdot \mathbf{T} \, ds$$

as in Eq. (11), show that $\nabla\phi = \mathbf{F}$. [*Suggestion:* If L is the line segment from (x, y, z) to $(x + \Delta x, y, z)$, then

$$\phi(x + \Delta x, y, z) - \phi(x, y, z) = \int_L \mathbf{F} \cdot \mathbf{T} \, ds = \int_x^{x+\Delta x} P \, dx.]$$

22. Let T be the submerged body of Problem 21 of Section 16.6, with centroid

$$\mathbf{r}_0 = \frac{1}{V} \iiint_T \mathbf{r} \, dV.$$

The torque about $\mathbf{r}_0$ of Archimedes' buoyant force $\mathbf{B} = -W\mathbf{k}$ is given by

$$\mathbf{L} = \iint_S (\mathbf{r} - \mathbf{r}_0) \times (-\rho g z \mathbf{n}) \, dS.$$

(Why?) Apply the result of Problem 20 of this section to prove that $\mathbf{L} = \mathbf{0}$. It follows that $\mathbf{B}$ acts along the vertical line through the centroid $\mathbf{r}_0$ of the submerged body. (Why?)

Chapter 16 Review: DEFINITIONS, CONCEPTS, RESULTS

Use the following list as a guide to concepts that you may need to review.

1. Definition and evaluation of the line integral

$$\int_C f(x, y, z) \, ds$$

2. Definition and evaluation of the line integral

$$\int_C P \, dx + Q \, dy + R \, dz$$

3. Relationship between the two types of line integrals; the line integral of the tangential component of a vector field

4. Line integrals and independence of path

5. Green's theorem

6. Flux and the vector form of Green's theorem

7. The divergence of a vector field

8. Definition and evaluation of the surface integral

$$\iint_S f(x, y, z) \, dS$$

9. Definition and evaluation of the surface integral

$$\iint_S P \, dy \, dz + Q \, dz \, dx + R \, dx \, dy$$

10. Relationship between the two types of surface integrals; the flux of a vector field across a surface

11. The divergence theorem in vector and in scalar notation

12. The curl of a vector field

13. Stokes' theorem in vector and in scalar notation

14. The circulation of a vector field around a simple closed curve

15. Physical interpretation of the divergence and the curl of a vector field

Chapter 16 Miscellaneous Problems

In Problems 1 through 5, evaluate the given line integral.

1. $\int_C (x^2 + y^2)\, ds$, where C is the straight line segment from $(0, 0)$ to $(3, 4)$

2. $\int_C y^2\, dx + x^2\, dy$, where C is the graph of $y = x^2$ from $(-1, 1)$ to $(1, 1)$

3. $\int_C \mathbf{F} \cdot \mathbf{T}\, ds$, where $\mathbf{F} = x\mathbf{i} + y\mathbf{j} + z\mathbf{k}$ and C is the curve $x = e^{2t}$, $y = e^t$, $z = e^{-t}$, $0 \le t \le \ln 2$

4. $\int_C xyz\, ds$, where C is the path from $(1, 1, 2)$ to $(2, 3, 6)$ consisting of three straight line segments, the first parallel to the x-axis, the second parallel to the y-axis, and the third parallel to the z-axis

5. $\int_C \sqrt{z}\, dx + \sqrt{x}\, dy + y^2\, dz$, where C is the curve $x = t$, $y = t^{3/2}$, $z = t^2$, $0 \le t \le 4$

6. Apply Theorem 1 of Section 16.3 to show that the line integral $\int_C y^2\, dx + 2xy\, dy + z\, dz$ is independent of the path C from A to B.

7. Apply Theorem 1 of Section 16.3 to show that the line integral $\int_C x^2 y\, dx + xy^2\, dy$ is not independent of the path C from $(0, 0)$ to $(1, 1)$.

8. A wire shaped like the circle $x^2 + y^2 = a^2$, $z = 0$ has constant density and total mass M. Find its moment of inertia about (a) the z-axis; (b) the x-axis.

9. A wire shaped like the parabola $y = \frac{1}{2}x^2$, $0 \le x \le 2$, has density function $\rho = x$. Find its mass and moment of inertia about the y-axis.

10. Find the work done by the force field $\mathbf{F} = z\mathbf{i} - x\mathbf{j} + y\mathbf{k}$ in moving a particle from $(1, 1, 1)$ to $(2, 4, 8)$ along the curve $y = x^2$, $z = x^3$.

11. Apply Green's theorem to evaluate the line integral

$$\oint_C x^2 y\, dx + xy^2\, dy,$$

where C is the boundary of the region between the two curves $y = x^2$ and $y = 8 - x^2$.

12. Evaluate the line integral

$$\oint_C x^2\, dy,$$

where C is the cardioid $r = 1 + \cos\theta$, by first applying Green's theorem and then changing to polar coordinates.

13. Let C_1 be the circle $x^2 + y^2 = 1$ and C_2 the circle $(x - 1)^2 + y^2 = 9$. Show that if $\mathbf{F} = x^2 y\mathbf{i} - xy^2\mathbf{j}$, then

$$\oint_{C_1} \mathbf{F} \cdot \mathbf{n}\, ds = \oint_{C_2} \mathbf{F} \cdot \mathbf{n}\, ds.$$

14. (a) Let C be the straight line segment from (x_1, y_1) to (x_2, y_2). Show that

$$\frac{1}{2}\int_C -y\, dx + x\, dy = \frac{1}{2}(x_1 y_2 - x_2 y_1).$$

(b) Suppose that the vertices of a polygon are (x_1, y_1), (x_2, y_2), ..., (x_n, y_n), named in counterclockwise order around the polygon. Apply part (a) to show that the area of the polygon is

$$A = \frac{1}{2}\sum_{i=1}^{n}(x_i y_{i+1} - x_{i+1} y_i),$$

where $x_{n+1} = x_1$ and $y_{n+1} = y_1$.

15. Suppose that the line integral $\int_C P\, dx + Q\, dy$ is independent of the path in the plane region D. Prove that

$$\oint_C P\, dx + Q\, dy = 0$$

for every piecewise smooth simple closed curve C in D.

16. Use Green's theorem to prove that

$$\oint_C P\, dx + Q\, dy = 0$$

for every piecewise smooth simple closed curve C in the plane region D if and only if $\partial P/\partial y = \partial Q/\partial x$ at each point of D.

17. Evaluate the surface integral

$$\iint_S (x^2 + y^2 + 2z)\, dS,$$

where S is the part of the paraboloid $z = 2 - x^2 - y^2$ above the xy-plane.

18. Suppose that $\mathbf{F} = (x^2 + y^2 + z^2)(x\mathbf{i} + y\mathbf{j} + z\mathbf{k})$ and that S is the spherical surface $x^2 + y^2 + z^2 = a^2$. Evaluate

$$\iint_S \mathbf{F} \cdot \mathbf{n}\, dS$$

without actually performing an antidifferentiation.

19. Let T be the solid bounded by the paraboloids

$$z = x^2 + 2y^2 \quad \text{and} \quad z = 12 - 2x^2 - y^2,$$

and suppose that $\mathbf{F} = x\mathbf{i} + y\mathbf{j} + z\mathbf{k}$. Find by evaluation of surface integrals the outward flux of $\mathbf{F}$ across the boundary of T.

20. Give a reasonable definition—in terms of a surface integral—of the average distance of the point P from points of the surface S. Then show that the average distance of a fixed point of a spherical surface of radius a from all points of the surface is $4a/3$.

21. Suppose that the surface S is the graph of the equation $x = g(y, z)$ for (y, z) in the region D of the yz-plane. Prove that

$$\iint_S P\,dy\,dz + Q\,dz\,dx + R\,dx\,dy$$
$$= \iint_D \left(P - Q\frac{\partial x}{\partial y} - R\frac{\partial x}{\partial z}\right) dy\,dz.$$

22. Suppose that the surface S is the graph of the equation $y = g(x, z)$ for (x, z) in the region D of the xz-plane. Prove that

$$\iint_S f(x,\,y,\,z)\,dS = \iint_D f(x,\,g(x,\,z),\,z)\,\sec\beta\,dx\,dz,$$

where $\sec\beta = \sqrt{1 + (\partial y/\partial x)^2 + (\partial y/\partial z)^2}$.

23. Let T be a region in space with volume V, boundary surface S, and centroid $(\bar{x}, \bar{y}, \bar{z})$. Use the divergence theorem to show that

$$\bar{z} = \frac{1}{2V}\iint_S z^2\,dx\,dy.$$

24. Apply the result of Problem 23 to find the centroid of the solid hemisphere $x^2 + y^2 + z^2 \leqq a^2$, $z \geqq 0$.

Problems 25 through 27 outline the derivation of the heat equation for a body with temperature $u = u(x, y, z, t)$ at the point (x, y, z) at time t. Denote by K its heat conductivity and by c its heat capacity, both assumed constant, and let $k = K/c$. Let B be a small solid ball within the body, and let S denote the boundary sphere of B.

25. Deduce from the divergence theorem and Eq. (18) of Section 16.5 that the rate of heat flow across S into B is

$$R = \iiint_B k\,\nabla^2 u\,dV.$$

26. The meaning of heat capacity is that, if Δu is small, then $(c\,\Delta u)\,\Delta V$ calories of heat are required to raise the temperature of the volume ΔV by Δu degrees. It follows that the rate at which the volume ΔV is absorbing heat is $c(\partial u/\partial t)\,\Delta V$. (Why?) Conclude that the rate of heat flow into B is

$$R = \iiint_B c\frac{\partial u}{\partial t}\,dV.$$

27. Equate the results of Problems 25 and 26, apply the average value property of triple integrals, and then take the limit as the radius of the ball B approaches zero. You should thereby obtain the heat equation

$$\frac{\partial u}{\partial t} = k\,\nabla^2 u.$$

28. For a *steady-state* temperature function (one that is independent of time t), the heat equation reduces to Laplace's equation,

$$\nabla^2 u = \frac{\partial^2 u}{\partial x^2} + \frac{\partial^2 u}{\partial y^2} + \frac{\partial^2 u}{\partial z^2} = 0.$$

(a) Suppose that u_1 and u_2 are two solutions of Laplace's equation in the region T and that u_1 and u_2 agree on its boundary surface S. Apply Problem 17 of Section 16.6 to the function $f = u_1 - u_2$ to conclude that $\nabla f = \mathbf{0}$ at each point of T. (b) From the facts that $\nabla f = \mathbf{0}$ in T and $f \equiv 0$ on S, conclude that $f \equiv 0$, so $u_1 \equiv u_2$. Thus the steady-state temperatures within a region are *determined* by the boundary-value temperatures.

29. Suppose that $\mathbf{r} = x\mathbf{i} + y\mathbf{j} + z\mathbf{k}$ and that $\phi(r)$ is a scalar function of $r = |\mathbf{r}|$. Compute (a) $\nabla\phi(r)$; (b) $\mathrm{div}[\phi(r)\mathbf{r}]$; (c) $\mathrm{curl}[\phi(r)\mathbf{r}]$.

30. Let S be the upper half of the torus obtained by revolving around the z-axis the circle $(y - a)^2 + z^2 = b^2$ in the yz-plane, with upper unit normal vector. Describe how to subdivide S to establish Stokes' theorem for it. How are the two boundary circles oriented?

31. Explain why the method of subdivision is not sufficient to establish Stokes' theorem for the Möbius strip of Fig. 16.5.3.

32. (a) Suppose that a fluid or a rigid body is rotating with angular speed ω radians per second about the line through the origin determined by the unit vector $\mathbf{u}$. Show that the velocity of the point with position vector $\mathbf{r}$ is $\mathbf{v} = \boldsymbol{\omega} \times \mathbf{r}$, where $\boldsymbol{\omega} = \omega\mathbf{u}$ is the angular velocity vector. Note that $|\mathbf{v}| = \omega|\mathbf{r}|\sin\theta$, where θ is the angle between $\mathbf{r}$ and $\boldsymbol{\omega}$. (b) Use the fact that $\mathbf{v} = \boldsymbol{\omega} \times \mathbf{r}$, established in part (a), to show that $\mathrm{curl}\,\mathbf{v} = 2\boldsymbol{\omega}$.

33. Consider an incompressible fluid flowing in space (no sources or sinks) with variable density $\rho(x, y, z, t)$ and velocity field $\mathbf{v}(x, y, z, t)$. Let B be a small ball with radius r and spherical surface S centered at the point (x_0, y_0, z_0). Then the amount of fluid within S at time t is

$$Q(t) = \iiint_B \rho\,dV,$$

and differentiation under the integral sign yields

$$Q'(t) = \iiint_B \frac{\partial\rho}{\partial t}\,dV.$$

(a) Consider fluid flow across S to get

$$Q'(t) = -\iint_S \rho\mathbf{v}\cdot\mathbf{n}\,dS,$$

where $\mathbf{n}$ is the outer unit normal vector to S. Now apply the divergence theorem to convert this into a volume integral. (b) Equate your two volume integrals for $Q'(t)$, apply the mean value theorem for integrals, and finally take limits as $r \to 0$ to obtain the **continuity equation**

$$\frac{\partial\rho}{\partial t} + \nabla\cdot(\rho\mathbf{v}) = 0.$$

Appendices

Appendix A

Review of Trigonometry

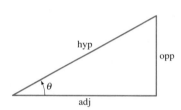

Fig. A.1 The sides and angle θ of a right triangle

In elementary trigonometry, the six basic trigonometric functions of an acute angle θ in a right triangle are defined as ratios between pairs of sides of the triangle. As in Fig. A.1, where "adj" stands for "adjacent," "opp" for "opposite," and "hyp" for "hypotenuse,"

$$\cos \theta = \frac{\text{adj}}{\text{hyp}}, \qquad \sin \theta = \frac{\text{opp}}{\text{hyp}}, \qquad \tan \theta = \frac{\text{opp}}{\text{adj}},$$

$$\sec \theta = \frac{\text{hyp}}{\text{adj}}, \qquad \csc \theta = \frac{\text{hyp}}{\text{opp}}, \qquad \cot \theta = \frac{\text{adj}}{\text{opp}}. \tag{1}$$

We generalize these definitions to *directed* angles of arbitrary size in the following way. Suppose that the initial side of the angle θ is the positive x-axis, so its vertex is at the origin. The angle is **directed** if a direction of rotation from its initial side to its terminal side is specified. We call θ a **positive angle** if this rotation is counterclockwise and a **negative angle** if it is clockwise.

Let $P(x, y)$ be the point at which the terminal side of θ intersects the *unit circle* $x^2 + y^2 = 1$. Then we define

$$\cos \theta = x, \qquad \sin \theta = y, \qquad \tan \theta = \frac{y}{x},$$

$$\sec \theta = \frac{1}{x}, \qquad \csc \theta = \frac{1}{y}, \qquad \cot \theta = \frac{x}{y}. \tag{2}$$

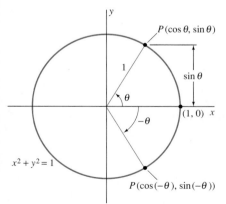

Fig. A.2 Using the unit circle to define the trigonometric functions

We assume that $x \neq 0$ in the case of $\tan \theta$ and $\sec \theta$ and that $y \neq 0$ in the case of $\cot \theta$ and $\csc \theta$. If the angle θ is positive and acute, then it is clear from Fig. A.2 that the definitions in Eqs. (2) agree with the right triangle definitions in Eqs. (1) in terms of the coordinates of P. A glance at the figure also shows which of the functions are positive for angles in each of the four quadrants. Figure A.3 summarizes this information.

Here we discuss primarily the two most basic trigonometric functions, the sine and the cosine. From Eqs. (2) we see immediately that the other four trigonometric functions are defined in terms of $\sin \theta$ and $\cos \theta$ by

$$\tan \theta = \frac{\sin \theta}{\cos \theta}, \qquad \sec \theta = \frac{1}{\cos \theta},$$

$$\cot \theta = \frac{\cos \theta}{\sin \theta}, \qquad \csc \theta = \frac{1}{\sin \theta}. \tag{3}$$

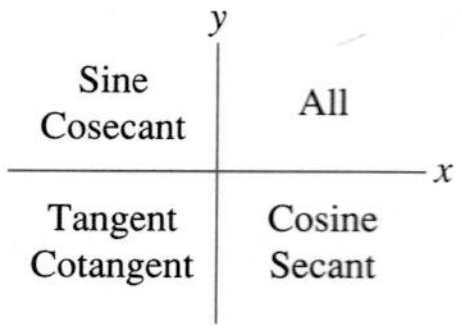

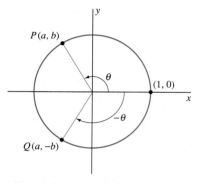

Sine Cosecant	All
Tangent Cotangent	Cosine Secant

Positive in quadrants shown

Fig. A.3 The signs of the trigonometric functions

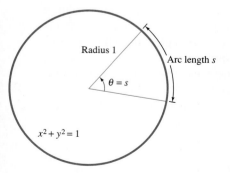

Fig. A.4 The effect of replacing θ by $-\theta$ in sine and cosine functions

Next, we compare the angles θ and $-\theta$ in Fig. A.4. We see that

$$\cos(-\theta) = \cos\theta \quad \text{and} \quad \sin(-\theta) = -\sin\theta. \tag{4}$$

Because $x = \cos\theta$ and $y = \sin\theta$ in Eqs. (2), the equation $x^2 + y^2 = 1$ of the unit circle translates immediately into the **fundamental identity of trigonometry,**

$$\cos^2\theta + \sin^2\theta = 1. \tag{5}$$

Dividing each term of this fundamental identity by $\cos^2\theta$ gives the identity

$$1 + \tan^2\theta = \sec^2\theta. \tag{5'}$$

Similarly, dividing each term in Eq. (5) by $\sin^2\theta$ yields the identity

$$1 + \cot^2\theta = \csc^2\theta. \tag{5''}$$

(See Problem 9 of this appendix.)

In Problems 15 and 16 we outline derivations of the **addition formulas**

$$\sin(\alpha + \beta) = \sin\alpha\cos\beta + \cos\alpha\sin\beta, \tag{6}$$

$$\cos(\alpha + \beta) = \cos\alpha\cos\beta - \sin\alpha\sin\beta. \tag{7}$$

With $\alpha = \theta = \beta$ in Eqs. (6) and (7), we get the **double-angle formulas**

$$\sin 2\theta = 2\sin\theta\cos\theta, \tag{8}$$

$$\cos 2\theta = \cos^2\theta - \sin^2\theta \tag{9}$$

$$= 2\cos^2\theta - 1 \tag{9a}$$

$$= 1 - 2\sin^2\theta, \tag{9b}$$

where Eqs. (9a) and (9b) are obtained from Eq. (9) by use of the fundamental identity in Eq. (5).

If we solve Eq. (9a) for $\cos^2\theta$ and Eq. (9b) for $\sin^2\theta$, we get the **half-angle formulas**

$$\cos^2\theta = \tfrac{1}{2}(1 + \cos 2\theta), \tag{10}$$

$$\sin^2\theta = \tfrac{1}{2}(1 - \cos 2\theta). \tag{11}$$

Equations (10) and (11) are especially important in integral calculus.

RADIAN MEASURE

In elementary mathematics, angles frequently are measured in *degrees,* with $360°$ in one complete revolution. In calculus it is more convenient—and is often essential—to measure angles in *radians.* The **radian measure** of an angle is the length of the arc it subtends in (that is, the arc it cuts out of) the unit circle when the vertex of the angle is at the center of the circle (Fig. A.5).

Recall that the area A and circumference C of a circle of radius r are given by the formulas

$$A = \pi r^2 \quad \text{and} \quad C = 2\pi r,$$

where the irrational number π is approximately 3.14159. Because the circumference of the unit circle is 2π and its central angle is $360°$, it follows that

$$2\pi\,\text{rad} = 360°; \qquad 180° = \pi\,\text{rad} \approx 3.14159\,\text{rad}. \tag{12}$$

Fig. A.5 The radian measure of an angle

A-2

Radians	Degrees
0	0
$\pi/6$	30
$\pi/4$	45
$\pi/3$	60
$\pi/2$	90
$2\pi/3$	120
$3\pi/4$	135
$5\pi/6$	150
π	180
$3\pi/2$	270
2π	360
4π	720

Fig. A.6 Some radian-degree conversions

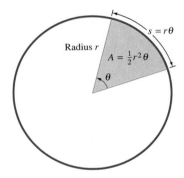

Fig. A.7 The area of a sector and arc length of a circle

Using Eq. (12) we can easily convert back and forth between radians and degrees:

$$1 \text{ rad} = \frac{180°}{\pi} \approx 57° \, 17' \, 44.8'', \tag{12a}$$

$$1° = \frac{\pi}{180} \text{ raq} \approx 0.01745 \text{ rad}. \tag{12b}$$

Figure A.6 shows radian-degree conversioins for some common angles.

Now consider an angle of θ radians at the center of a circle of radius r (Fig. A.7). Denote by s the length of the arc subtended by θ; denote by A the area of the sector of the circle bounded by this angle. Then the proportions

$$\frac{s}{2\pi r} = \frac{A}{\pi r^2} = \frac{\theta}{2\pi}$$

give the formulas

$$s = r\theta \qquad (\theta \text{ in radians}) \tag{13}$$

and

$$A = \tfrac{1}{2} r^2 \theta \qquad (\theta \text{ in radians}). \tag{14}$$

The definitions in Eqs. (2) refer to trigonometric functions of *angles* rather than trigonometric functions of *numbers*. Suppose that t is a real number. Then the number $\sin t$ is, *by definition*, the sine of an angle of t radians—recall that a positive angle is directed counterclockwise from the positive x-axis, whereas a negative angle is directed clockwise. Briefly, $\sin t$ is the sine of an angle of t *radians*. The other trigonometric functions of the number t have similar definitions. Hence, when we write $\sin t$, $\cos t$, and so on, with t a real number, it is *always* in reference to an angle of t radians.

When we need to refer to the sine of an angle of t *degrees*, we will henceforth write $\sin t°$. The point is that $\sin t$ and $\sin t°$ are quite different functions of the variable t. For example, you would get

$$\sin 1° \approx 0.0175 \quad \text{and} \quad \sin 30° = 0.5000$$

on a calculator set in degree mode. But in radian mode, a calculator would give

$$\sin 1 \approx 0.8415 \quad \text{and} \quad \sin 30 \approx -0.9880.$$

The relationship between the functions $\sin t$ and $\sin t°$ is

$$\sin t° = \sin\left(\frac{\pi t}{180}\right). \tag{15}$$

The distinction extends even to programming languages. In FORTRAN, the function `SIN` is the radian sine function, and you must write $\sin t°$ in the form `SIND(T)`. In BASIC you must write `SIN(PI*T/180)` to get the correct value of the sine of an angle of t degrees.

An angle of 2π rad corresponds to one revolution around the unit circle. This implies that the sine and cosine functions have **period** 2π, meaning that

$$\sin(t + 2\pi) = \sin t,$$
$$\cos(t + 2\pi) = \cos t. \tag{16}$$

It follows from Eqs. (16) that

Fig. A.8 Periodicity of the sine and cosine functions

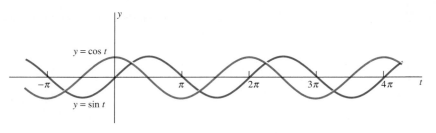

$$\sin(t + 2n\pi) = \sin t \quad \text{and} \quad \cos(t + 2n\pi) = \cos t \qquad (17)$$

for any integer n. This periodicity of the sine and cosine functions is evident in their graphs (Fig. A.8). From Eqs. (3), the other four trigonometric functions also must be periodic, as their graphs in Figs. A.9 and A.10 show.

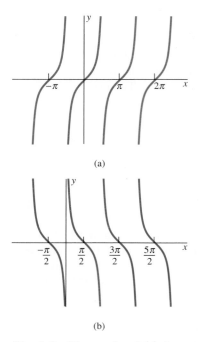

(a)

(b)

Fig. A.9 The graphs of (a) the tangent function and (b) the cotangent function

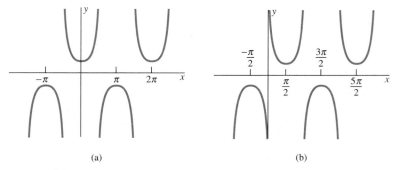

(a)

(b)

Fig. A.10 The graphs of (a) the secant function and (b) the cosecant function

We see from Eqs. (2) that

$$\sin 0 = 0, \qquad \sin \frac{\pi}{2} = 1, \qquad \sin \pi = 0,$$

$$\cos 0 = 1, \qquad \cos \frac{\pi}{2} = 0, \qquad \cos \pi = -1. \qquad (18)$$

The trigonometric functions of $\pi/6$, $\pi/4$, and $\pi/3$ (the radian equivalents of 30°, 45°, and 60°, respectively) are easy to read from the well-known triangles of Fig. A.11. For instance,

$$\sin \frac{\pi}{6} = \cos \frac{\pi}{3} = \frac{1}{2} = \frac{\sqrt{1}}{2},$$

$$\sin \frac{\pi}{4} = \cos \frac{\pi}{4} = \frac{1}{\sqrt{2}} = \frac{\sqrt{2}}{2}, \quad \text{and} \qquad (19)$$

$$\sin \frac{\pi}{3} = \cos \frac{\pi}{6} = \frac{\sqrt{3}}{2}.$$

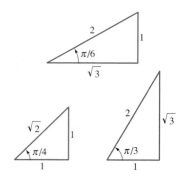

Fig. A.11 Familiar right triangles

To find the values of trigonometric functions of angles larger than $\pi/2$, we can use their periodicity and the identities

$$\sin(\pi \pm \theta) = \mp\sin \theta,$$

$$\cos(\pi \pm \theta) = -\cos \theta, \quad \text{and} \qquad (20)$$

$$\tan(\pi \pm \theta) = \pm\tan \theta$$

(Problem 14) as well as similar identities for the cosecant, secant, and cotangent functions.

EXAMPLE 1

$$\sin \frac{5\pi}{4} = \sin\left(\pi + \frac{\pi}{4}\right) = -\sin \frac{\pi}{4} = -\frac{\sqrt{2}}{2};$$

$$\cos \frac{2\pi}{3} = \cos\left(\pi - \frac{\pi}{3}\right) = -\cos \frac{\pi}{3} = -\frac{1}{2};$$

$$\tan \frac{3\pi}{4} = \tan\left(\pi - \frac{\pi}{4}\right) = -\tan \frac{\pi}{4} = -1;$$

$$\sin \frac{7\pi}{6} = \sin\left(\pi + \frac{\pi}{6}\right) = -\sin \frac{\pi}{6} = -\frac{1}{2};$$

$$\cos \frac{5\pi}{3} = \cos\left(2\pi - \frac{\pi}{3}\right) = \cos\left(-\frac{\pi}{3}\right) = \cos \frac{\pi}{3} = \frac{1}{2};$$

$$\sin \frac{17\pi}{6} = \sin\left(2\pi + \frac{5\pi}{6}\right) = \sin \frac{5\pi}{6}$$

$$= \sin\left(\pi - \frac{\pi}{6}\right) = \sin \frac{\pi}{6} = \frac{1}{2}.$$

EXAMPLE 2   Find the solutions (if any) of the equation

$$\sin^2 x - 3 \cos^2 x + 2 = 0$$

that lie in the interval $[0, \pi]$.

Solution   Using the fundamental identity in Eq. (5), we substitute $\cos^2 x = 1 - \sin^2 x$ into the given equation to obtain

$$\sin^2 x - 3(1 - \sin^2 x) + 2 = 0;$$

$$4 \sin^2 x - 1 = 0;$$

$$\sin x = \pm\tfrac{1}{2}.$$

Because $\sin x \geqq 0$ for x in $[0, \pi]$, $\sin x = -\tfrac{1}{2}$ is impossible. But $\sin x = \tfrac{1}{2}$ for $x = \pi/6$ and for $x = \pi - \pi/6 = 5\pi/6$. These are the solutions of the given equation in $[0, \pi]$.

Appendix A Problems

ANSWERS TO APPENDIX PROBLEMS APPEAR AT THE END OF THE ANSWERS TO ODD-NUMBERED PROBLEMS.

Express in radian measure the angles in Problems 1 through 5.

1. $40°$ **2.** $-270°$ **3.** $315°$ **4.** $210°$ **5.** $-150°$

In Problems 6 through 10, express in degrees the angles given in radian measure.

6. $\dfrac{\pi}{10}$ **7.** $\dfrac{2\pi}{5}$ **8.** 3π **9.** $\dfrac{15\pi}{4}$ **10.** $\dfrac{23\pi}{60}$

In Problems 11 through 14, evaluate the six trigonometric functions of x at the given values.

11. $x = -\dfrac{\pi}{3}$ **12.** $x = \dfrac{3\pi}{4}$

13. $x = \dfrac{7\pi}{6}$ **14.** $x = \dfrac{5\pi}{3}$

Find all solutions x of each equation in Problems 15 through 23.

15. $\sin x = 0$ **16.** $\sin x = 1$ **17.** $\sin x = -1$

18. $\cos x = 0$ **19.** $\cos x = 1$ **20.** $\cos x = -1$

21. $\tan x = 0$ **22.** $\tan x = 1$ **23.** $\tan x = -1$

24. Suppose that $\tan x = \frac{3}{4}$ and that $\sin x < 0$. Find the values of the other five trigonometric functions of x.

25. Suppose that $\csc x = -\frac{5}{3}$ and that $\cos x > 0$. Find the values of the other five trigonometric functions of x.

Deduce the identities in Problems 26 and 27 from the fundamental identity

$$\cos^2 \theta + \sin^2 \theta = 1$$

and from the definitions of the other four trigonometric functions.

26. $1 + \tan^2 \theta = \sec^2 \theta$ **27.** $1 + \cot^2 \theta = \csc^2 \theta$

28. Deduce from the addition formulas for the sine and cosine the addition formula for the tangent:

$$\tan(x + y) = \frac{\tan x + \tan y}{1 - \tan x \tan y}.$$

In Problems 29 through 36, use the method of Example 1 to find the indicated values.

29. $\sin \dfrac{5\pi}{6}$ **30.** $\cos \dfrac{7\pi}{6}$ **31.** $\sin \dfrac{11\pi}{6}$

32. $\cos \dfrac{19\pi}{6}$ **33.** $\sin \dfrac{2\pi}{3}$ **34.** $\cos \dfrac{4\pi}{3}$

35. $\sin \dfrac{5\pi}{3}$ **36.** $\cos \dfrac{10\pi}{3}$

37. Apply the addition formulas for the sine, cosine, and tangent functions (the latter from Problem 28) to show that if $0 < \theta < \pi/2$, then

(a) $\cos\left(\dfrac{\pi}{2} - \theta\right) = \sin \theta$;

(b) $\sin\left(\dfrac{\pi}{2} - \theta\right) = \cos \theta$;

(c) $\cot\left(\dfrac{\pi}{2} - \theta\right) = \tan \theta$.

The prefix *co-* is an abbreviation for the adjective *complementary*, which describes two angles whose sum is $\pi/2$. For example, $\pi/6$ and $\pi/3$ are complementary angles, so (a) implies that $\cos \pi/6 = \sin \pi/3$.

Suppose that $0 < \theta < \pi/2$. Derive the identities in Problems 38 through 40.

38. $\sin(\pi \pm \theta) = \mp \sin \theta$

39. $\cos(\pi \pm \theta) = -\cos \theta$

40. $\tan(\pi \pm \theta) = \pm \tan \theta$

41. The points $A(\cos \theta, -\sin \theta)$, $B(1, 0)$, $C(\cos \phi, \sin \phi)$, and $D(\cos(\theta + \phi), \sin(\theta + \phi))$ are shown in Fig. A.12; all are points on the unit circle. Deduce from the fact that the line segments AC and BD have the same length (because they are subtended by the same angle $\theta + \phi$) that

$$\cos(\theta + \phi) = \cos \theta \cos \phi - \sin \theta \sin \phi.$$

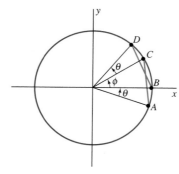

Fig. A.12 Deriving the cosine addition formula (Problem 41)

42. (a) Use the triangles shown in Fig. A.13 to deduce that

$$\sin\left(\theta + \frac{\pi}{2}\right) = \cos \theta \quad \text{and} \quad \cos\left(\theta + \frac{\pi}{2}\right) = -\sin \theta.$$

(b) Use the results of Problem 41 and part (a) to derive the addition formula for the sine function.

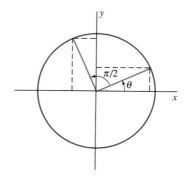

Fig. A.13 Deriving the identities of Problem 42

In Problems 43 through 48, find all solutions of the given equation that lie in the interval $[0, \pi]$.

43. $3 \sin^2 x - \cos^2 x = 2$ **44.** $\sin^2 x = \cos^2 x$

45. $2 \cos^2 x + 3 \sin^2 x = 3$

46. $2 \sin^2 x + \cos x = 2$

47. $8 \sin^2 x \cos^2 x = 1$

48. $\cos 2\theta - 3 \cos \theta = -2$

A-6

Appendix B
Proofs of the Limit Laws

Recall the definition of the limit:

$$\lim_{x \to a} F(x) = L$$

provided that, given $\epsilon > 0$, there exists a number $\delta > 0$ such that

$$0 < |x - a| < \delta \quad \text{implies that} \quad |F(x) - L| < \epsilon. \qquad (1)$$

Note that the number ϵ comes *first*. *Then* a value of $\delta > 0$ must be found so that the implication in (1) holds. To prove that $F(x) \to L$ as $x \to a$, you must, in effect, be able to stop the next person you see and ask him or her to pick a positive number ϵ at random. Then you must *always* be ready to respond with a positive number δ. This number δ must have the property that the implication in (1) holds for your number δ and the given number ϵ. The *only* restriction on x is that

$$0 < |x - a| < \delta,$$

as given in (1).

To do all this, you will ordinarily need to give an explicit method—a recipe or formula—for producing a value of δ that works for each value of ϵ. As Examples 1 through 3 show, the method will depend on the particular function F under study as well as the values of a and L.

EXAMPLE 1 Prove that $\lim_{x \to 3} (2x - 1) = 5$.

Solution Given $\epsilon > 0$, we must find $\delta > 0$ such that

$$|(2x - 1) - 5| < \epsilon \quad \text{if} \quad 0 < |x - 3| < \delta.$$

Now

$$|(2x - 1) - 5| = |2x - 6| = 2|x - 3|,$$

so

$$0 < |x - 3| < \frac{\epsilon}{2} \quad \text{implies that} \quad |(2x - 1) - 5| < 2 \cdot \frac{\epsilon}{2} = \epsilon.$$

Hence, given $\epsilon > 0$, it suffices to choose $\delta = \epsilon/2$. This illustrates the observation that the required number δ is generally a function of the given number ϵ.

EXAMPLE 2 Prove that $\lim_{x \to 2} (3x^2 + 5) = 17$.

Solution Given $\epsilon > 0$, we must find $\delta > 0$ such that

$$0 < |x - 2| < \delta \quad \text{implies that} \quad |(3x^2 + 5) - 17| < \epsilon.$$

Now

$$|(3x^2 + 5) - 17| = |3x^2 - 12| = 3 \cdot |x + 2| \cdot |x - 2|.$$

Our problem, therefore, is to show that $|x + 2| \cdot |x - 2|$ can be made as small as we please by choosing $x - 2$ sufficiently small. The idea is that $|x + 2|$ cannot be too large if $|x - 2|$ is fairly small. For example, if $|x - 2| < 1$, then

$$|x + 2| = |(x - 2) + 4| \leqq |x - 2| + 4 < 5.$$

Therefore,

$$0 < |x - 2| < 1 \quad \text{implies that} \quad |(3x^2 + 5) - 17| < 15 \cdot |x - 2|.$$

Consequently, let us choose δ to be the minimum of the two numbers 1 and $\epsilon/15$. Then

$$0 < |x - 2| < \delta \quad \text{implies that} \quad |(3x^2 + 5) - 17| < 15 \cdot \frac{\epsilon}{15} = \epsilon,$$

as desired.

EXAMPLE 3   Prove that

$$\lim_{x \to a} \frac{1}{x} = \frac{1}{a} \quad \text{if} \quad a \neq 0.$$

Solution   For simplicity, we will consider only the case in which $a > 0$ (the case $a < 0$ is similar).

Suppose that $\epsilon > 0$ is given. We must find a number δ such that

$$0 < |x - a| < \delta \quad \text{implies that} \quad \left| \frac{1}{x} - \frac{1}{a} \right| < \epsilon.$$

Now

$$\left| \frac{1}{x} - \frac{1}{a} \right| = \left| \frac{a - x}{ax} \right| = \frac{|x - a|}{a|x|}.$$

The idea is that $1/|x|$ cannot be too large if $|x - a|$ is fairly small. For example, if $|x - a| < a/2$, then $a/2 < x < 3a/2$. Therefore,

$$|x| > \frac{a}{2}, \quad \text{so} \quad \frac{1}{|x|} < \frac{2}{a}.$$

In this case it would follow that

$$\left| \frac{1}{x} - \frac{1}{a} \right| < \frac{2}{a^2} \cdot |x - a|$$

if $|x - a| < a/2$. Thus, if we choose δ to be the minimum of the two numbers $a/2$ and $a^2\epsilon/2$, then

$$0 < |x - a| < \delta \quad \text{implies that} \quad \left| \frac{1}{x} - \frac{1}{a} \right| < \frac{2}{a^2} \cdot \frac{a^2\epsilon}{2} = \epsilon.$$

Therefore,

$$\lim_{x \to a} \frac{1}{x} = \frac{1}{a} \quad \text{if} \quad a \neq 0,$$

as desired.

We are now ready to give proofs of the limit laws stated in Section 2.2.

Constant Law

If $f(x) \equiv C$, a constant, then

$$\lim_{x \to a} f(x) = \lim_{x \to a} C = C.$$

Proof Because $|C - C| = 0$, we merely choose $\delta = 1$, regardless of the previously given value of $\epsilon > 0$. Then, if $0 < |x - a| < \delta$, it is automatic that $|C - C| < \epsilon$. ❏

Addition Law

If $\lim\limits_{x \to a} F(x) = L$ and $\lim\limits_{x \to a} G(x) = M$, then

$$\lim_{x \to a} [F(x) + G(x)] = L + M.$$

Proof Let $\epsilon > 0$ be given. Because L is the limit of $F(x)$ as $x \to a$, there exists a number $\delta_1 > 0$ such that

$$0 < |x - a| < \delta_1 \quad \text{implies that} \quad |F(x) - L| < \frac{\epsilon}{2}.$$

Because M is the limit of $G(x)$ as $x \to a$, there exists a number $\delta_2 > 0$ such that

$$0 < |x - a| < \delta_2 \quad \text{implies that} \quad |G(x) - M| < \frac{\epsilon}{2}.$$

Let $\delta = \min\{\delta_1, \delta_2\}$. Then

$$0 < |x - a| < \delta \quad \text{implies that} \quad |(F(x) - G(x)) - (L + M)|$$
$$\leqq |F(x) - L| + |G(x) - M| < \frac{\epsilon}{2} + \frac{\epsilon}{2} = \epsilon.$$

Therefore,

$$\lim_{x \to a}[F(x) + G(x)] = L + M,$$

as desired. ❏

Product Law

If $\lim\limits_{x \to a} F(x) = L$ and $\lim\limits_{x \to a} G(x) = M$, then

$$\lim_{x \to a} [F(x) \cdot G(x)] = L \cdot M.$$

Proof Given $\epsilon > 0$, we must find a number $\delta > 0$ such that

$$0 < |x - a| < \delta \quad \text{implies that} \quad |F(x) \cdot G(x) - L \cdot M| < \epsilon.$$

But first, the triangle inequality gives the result

$$|F(x) \cdot G(x) - L \cdot M| = |F(x) \cdot G(x) - L \cdot G(x) + L \cdot G(x) - L \cdot M|$$
$$\leqq |G(x)| \cdot |F(x) - L| + |L| \cdot |G(x) - M|. \quad (2)$$

Because $\lim\limits_{x \to a} F(x) = L$, there exists $\delta_1 > 0$ such that

$$0 < |x - a| < \delta_1 \quad \text{implies that} \quad |F(x) - L| < \frac{\epsilon}{2(|M| + 1)}. \quad (3)$$

And because $\lim_{x \to a} G(x) = M$, there exists $\delta_2 > 0$ such that

$$0 < |x - a| < \delta_2 \quad \text{implies that} \quad |G(x) - M| < \frac{\epsilon}{2(|L| + 1)}. \quad (4)$$

Moreover, there is a *third* number $\delta_3 > 0$ such that

$$0 < |x - a| < \delta_3 \quad \text{implies that} \quad |G(x) - M| < 1,$$

which in turn implies that

$$|G(x)| < |M| + 1. \quad (5)$$

We now choose $\delta = \min\{\delta_1, \delta_2, \delta_3\}$. Then we substitute (3), (4), and (5) into (2) and, finally, see that $0 < |x - a| < \delta$ implies

$$|F(x) \cdot G(x) - L \cdot M| < (|M| + 1) \cdot \frac{\epsilon}{2(|M| + 1)} + |L| \cdot \frac{\epsilon}{2(|L| + 1)}$$

$$< \frac{\epsilon}{2} + \frac{\epsilon}{2} = \epsilon,$$

as desired. The use of $|M| + 1$ and $|L| + 1$ in the denominators avoids the technical difficulty that arises should either L or M be zero. ❏

Substitution Law

If $\lim_{x \to a} g(x) = L$ and $\lim_{x \to L} f(x) = f(L)$, then

$$\lim_{x \to a} f(g(x)) = f(L).$$

Proof Let $\epsilon > 0$ be given. We must find a number $\delta > 0$ such that

$$0 < |x - a| < \delta \quad \text{implies that} \quad |f(g(x)) - f(L)| < \epsilon.$$

Because $\lim_{y \to L} f(y) = f(L)$, there exists $\delta_1 > 0$ such that

$$0 < |y - L| < \delta_1 \quad \text{implies that} \quad |f(y) - f(L)| < \epsilon. \quad (6)$$

Also, because $\lim_{x \to a} g(x) = L$, we can find $\delta > 0$ such that

$$0 < |x - a| < \delta \quad \text{implies that} \quad |g(x) - L| < \delta_1,$$

that is, such that

$$|y - L| < \delta_1,$$

where $y = g(x)$. From (6) we see that

$$0 < |x - a| < \delta \text{ implies that } |f(g(x)) - f(L)| = |f(y) - f(L)| < \epsilon,$$

as desired. ❏

Reciprocal Law

If $\lim_{x \to a} g(x) = L$ and $L \neq 0$, then

$$\lim_{x \to a} \frac{1}{g(x)} = \frac{1}{L}.$$

Proof   Let $f(x) = 1/x$. Then, as we saw in Example 3,

$$\lim_{x \to a} f(x) = \lim_{x \to a} \frac{1}{x} = \frac{1}{L} = f(L).$$

Hence the substitution law gives the result

$$\lim_{x \to a} \frac{1}{g(x)} = \lim_{x \to a} f(g(x)) = f(L) = \frac{1}{L},$$

as desired.   ❏

Quotient Law

If $\lim\limits_{x \to a} F(x) = L$ and $\lim\limits_{x \to a} G(x) = M \neq 0$, then

$$\lim_{x \to a} \frac{F(x)}{G(x)} = \frac{L}{M}.$$

Proof   It follows immediately from the product and reciprocal laws that

$$\lim_{x \to a} \frac{F(x)}{G(x)} = \lim_{x \to a} F(x) \cdot \frac{1}{G(x)} = \left(\lim_{x \to a} F(x) \right)\left(\lim_{x \to a} \frac{1}{G(x)} \right) = L \cdot \frac{1}{M} = \frac{L}{M},$$

as desired.   ❏

Squeeze Law

Suppose that $f(x) \leqq g(x) \leqq h(x)$ in some deleted neighborhood of a and that

$$\lim_{x \to a} f(x) = L = \lim_{x \to a} h(x).$$

Then

$$\lim_{x \to a} g(x) = L.$$

Proof   Given $\epsilon > 0$, we choose $\delta_1 > 0$ and $\delta_2 > 0$ such that

$$0 < |x - a| < \delta_1 \quad \text{implies that} \quad |f(x) - L| < \epsilon$$

and

$$0 < |x - a| < \delta_2 \quad \text{implies that} \quad |h(x) - L| < \epsilon.$$

Let $\delta = \min\{\delta_1, \delta_2\}$. Then $\delta > 0$. Moreover, if $0 < |x - a| < \delta$, then both $f(x)$ and $h(x)$ are points of the open interval $(L - \epsilon, L + \epsilon)$. So

$$L - \epsilon < f(x) \leqq g(x) \leqq h(x) < L + \epsilon.$$

Thus

$$0 < |x - a| < \delta \quad \text{implies that} \quad |g(x) - L| < \epsilon,$$

as desired.   ❏

Appendix B Problems

In Problems 1 through 10, apply the definition of the limit to establish the given equality.

1. $\lim_{x \to a} x = a$

2. $\lim_{x \to 2} 3x = 6$

3. $\lim_{x \to 2} (x + 3) = 5$

4. $\lim_{x \to -3} (2x + 1) = -5$

5. $\lim_{x \to 1} x^2 = 1$

6. $\lim_{x \to a} x^2 = a^2$

7. $\lim_{x \to -1} (2x^2 - 1) = 1$

8. $\lim_{x \to a} \dfrac{1}{x^2} = \dfrac{1}{a^2}$ if $a \neq 0$

9. $\lim_{x \to a} \dfrac{1}{x^2 + 1} = \dfrac{1}{a^2 + 1}$

10. $\lim_{x \to a} \dfrac{1}{\sqrt{x}} = \dfrac{1}{\sqrt{a}}$ if $a > 0$

11. Suppose that $\lim_{x \to a} f(x) = L$ and that $\lim_{x \to a} f(x) = M$. Apply the definition of the limit to prove that $L = M$. Thus a limit of a function is unique if it exists.

12. Suppose that C is a constant and that $\lim_{x \to a} f(x) = L$. Apply the definition of the limit to prove that

$$\lim_{x \to a} C \cdot f(x) = C \cdot L.$$

13. Suppose that $L \neq 0$ and that $\lim_{x \to a} f(x) = L$. Use the method of Example 3 and the definition of the limit to

show directly that

$$\lim_{x \to a} \frac{1}{f(x)} = \frac{1}{L}.$$

14. Use the algebraic identity

$$x^n - a^n = (x - a)(x^{n-1} + x^{n-2}a + \cdots + xa^{n-2} + a^{n-1})$$

to show directly from the definition of the limit that $\lim_{x \to a} x^n = a^n$ if n is a positive integer.

15. Apply the identity

$$|\sqrt{x} - \sqrt{a}| = \frac{|x - a|}{\sqrt{x} + \sqrt{a}}$$

to show directly from the definition of the limit that $\lim_{x \to a} \sqrt{x} = \sqrt{a}$ if $a > 0$.

16. Suppose that $\lim_{x \to a} f(x) = f(a) > 0$. Prove that there exists a neighborhood of a on which $f(x) > 0$; that is, prove that there exists $\delta > 0$ such that

$$|x - a| < \delta \quad \text{implies that} \quad f(x) > 0.$$

Appendix C
The Completeness of the Real Number System

Here we present a self-contained treatment of those consequences of the completeness of the real number system that are relevant to this text. Our principal objective is to prove the intermediate value theorem and the maximum value theorem. We begin with the least upper bound property of the real numbers, which we take to be an axiom.

Definition *Upper Bound and Lower Bound*

The set S of real numbers is said to be **bounded above** if there is a number b such that $x \leqq b$ for every number x in S, and the number b is called an **upper bound** for S. Similarly, if there is a number a such that $x \geqq a$ for every number x in S, then S is said to be **bounded below,** and a is called a **lower bound** for S.

Definition *Least Upper Bound and Greatest Lower Bound*

The number λ is said to be a **least upper bound** for the set S of real numbers provided that

1. λ is an upper bound for S, and

2. If b is an upper bound for S, then $\lambda \leqq b$.

Similarly, the number γ is said to be a **greatest lower bound** for S if γ is a lower bound for S and $\gamma \geqq a$ for every lower bound a of S.

EXERCISE   Prove that if a set S has a least upper bound λ, then it is unique. That is, prove that if λ and μ are both least upper bounds for S, then $\lambda = \mu$.

It is easy to show that the greatest lower bound γ of a set S, if any, is also unique. At this point you should construct examples to illustrate that a set with a least upper bound λ may or may not contain λ and that a similar statement is true of the set's greatest lower bound.

We now state the *completeness axiom* of the real number system.

Least Upper Bound Axiom

If the nonempty set S of real numbers has an upper bound, then it has a least upper bound.

By working with the set T consisting of the numbers $-x$, where x is in S, it is not difficult to show the following consequence of the least upper bound axiom: If the nonempty set S of real numbers is bounded below, then S has a greatest lower bound. Because of this symmetry, we need only one axiom, not two; results for least upper bounds also hold for greatest lower bounds, provided that some attention is paid to the direction of the inequalities.

The restriction that S be nonempty is annoying but necessary. If S is the "empty" set of real numbers, then 15 is an upper bound for S, but S has no least upper bound because $14, 13, 12, \ldots, 0, -1, -2, \ldots$ are also upper bounds for S.

Definition   *Increasing, Decreasing, and Monotonic Sequences*

The infinite sequence $x_1, x_2, x_3, \ldots, x_k, \ldots$ is said to be **nondecreasing** if $x_n \leqq x_{n+1}$ for every $n \geqq 1$. This sequence is said to be **nonincreasing** if $x_n \geqq x_{n+1}$ for every $n \geqq 1$. If the sequence $\{x_n\}$ is either nonincreasing or nondecreasing, then it is said to be **monotonic.**

Theorem 1 gives the **bounded monotonic sequence property** of the set of real numbers. (Recall that a set S of real numbers is said to be **bounded** if it is contained in an interval of the form $[a, b]$.)

Theorem 1   *Bounded Monotonic Sequences*

Every bounded monotonic sequence of real numbers converges.

Proof   Suppose that the sequence

$$S = \{x_n\} = \{x_1, x_2, x_3, \ldots, x_k, \ldots\}$$

is bounded and nondecreasing. By the least upper bound axiom, S has a least upper bound λ. We claim that λ is the limit of the sequence $\{x_n\}$. Consider an open interval centered at λ—that is, an interval of the form $I = (\lambda - \epsilon, \lambda + \epsilon)$, where $\epsilon > 0$. Some terms of the sequence must lie within I, or else $\lambda - \epsilon$ would be an upper bound for S that is less than its least upper bound λ. But if x_N is within I, then—because we are dealing with a nondecreasing sequence—x_k must also lie in I for all $k \geqq N$. Because ϵ is an arbitrary positive number, λ is by definition (Problem 39 of Section 11.2) the limit of the sequence $\{x_n\}$. That is, a bounded nondecreasing sequence converges. A similar proof can be constructed for nonincreasing sequences by working with the greatest lower bound. ❑

Therefore, the least upper bound axiom implies the bounded monotonic sequence property of the real numbers. With just a little effort, you can prove that the two are logically equivalent: If you take the bounded monotonic sequence property as an axiom, then the least upper bound property follows as a theorem. The **nested interval property** of Theorem 2 is also equivalent to the least upper bound property, but we shall prove only that it follows from the least upper bound property, because we have chosen the latter as the fundamental completeness axiom for the real number system.

Theorem 2 *Nested Interval Property of the Real Numbers*

Suppose that $I_1, I_2, I_3, \ldots, I_n, \ldots$ is a sequence of closed intervals (so I_n is of the form $[a_n, b_n]$ for each positive integer n) such that

1. I_n contains I_{n+1} for each $n \geqq 1$, and
2. $\lim\limits_{n \to \infty} (b_n - a_n) = 0$.

Then there exists exactly one real number c such that c belongs to I_n for each n. Thus

$$\{c\} = I_1 \cap I_2 \cap I_3 \cap \ldots.$$

Proof   It is clear from hypothesis (2) of Theorem 2 that there is at most one such number c. The sequence $\{a_n\}$ of the left-hand endpoints of the intervals is a bounded (by b_1) nondecreasing sequence and thus has a limit a by the bounded monotonic sequence property. Similarly, the sequence $\{b_n\}$ has a limit b. Because $a_n \leqq b_n$ for all n, it follows easily that $a \leqq b$. It is clear that $a_n \leqq a \leqq b_n$ for all $n \geqq 1$, so a belongs to every interval I_n; so does b, by a similar argument. But then property (2) of Theorem 2 implies that $a = b$, and clearly this common value—call it c—is the number satisfying the conclusion of Theorem 2. ❑

We can now use these results to prove several important theorems used in the text.

Theorem 3 *Intermediate Value Property of Continuous Functions*

If the function f is continuous on the interval $[a, b]$ and $f(a) < K < f(b)$, then $K = f(c)$ for some number c in (a, b).

Proof   Let $I_1 = [a, b]$. Suppose that I_n has been defined for $n \geqq 1$. We describe (inductively) how to define I_{n+1}, and this shows in particular how to define I_2, I_3, and so forth. Let a_n be the left-hand endpoint of I_n, b_n be its right-hand endpoint, and m_n be its midpoint. If $f(m_n) > K$, then $f(a_n) < K < f(m_n)$; in this case, let $a_{n+1} = a_n$, $b_{n+1} = m_n$, and $I_{n+1} = [a_{n+1}, b_{n+1}]$. If $f(m_n) < K$, then let $a_{n+1} = m_n$ and $b_{n+1} = b_n$. Thus at each stage we bisect I_n and let I_{n+1} be the half of I_n on which f takes on values both above and below K. Note that if $f(m_n)$ is ever actually equal to K, we simply let $c = m_n$ and stop.

It is easy to show that the sequence $\{I_n\}$ of intervals satisfies the hypotheses of Theorem 2. Let c be the (unique) real number common to all the intervals I_n. We will show that $f(c) = K$, and this will conclude the proof.

The sequence $\{b_n\}$ has limit c, so by the continuity of f, the sequence $\{f(b_n)\}$ has limit $f(c)$. But $f(b_n) > K$ for all n, so the limit of $\{f(b_n)\}$ can be no less than K; that is, $f(c) \geqq K$. By considering the sequence $\{a_n\}$, it follows that $f(c) \leqq K$. Therefore, $f(c) = K$. ❏

Lemma 1

If f is continuous on the closed interval $[a, b]$, then f is bounded there.

Proof Suppose by way of contradiction that f is not bounded on $I_1 = [a, b]$. Bisect I_1 and let I_2 be either half on which f is unbounded—if f is unbounded on both halves, then let I_2 be the left half of I_1. In general, let I_{n+1} be a half of I_n on which f is unbounded.

Again it is easy to show that the sequence $\{I_n\}$ of closed intervals satisfies the hypotheses of Theorem 2. Let c be the number common to them all. Because f is continuous, there is a number $\epsilon > 0$ such that f is bounded on the interval $(c - \epsilon, c + \epsilon)$. But for sufficiently large values of n, I_n is a subset of $(c - \epsilon, c + \epsilon)$. This contradiction shows that f must be bounded on $[a, b]$. ❏

Theorem 4   *Maximum Value Property of Continuous Functions*

If the function f is continuous on the closed and bounded interval $[a, b]$, then there exists a number c in $[a, b]$ such that $f(x) \leqq f(c)$ for all x in $[a, b]$.

Proof Consider the set $S = \{f(x) \mid a \leqq x \leqq b\}$. By Lemma 1, this set is bounded; let λ be its least upper bound. Our goal is to show that λ is a value $f(c)$ of f.

With $I_1 = [a, b]$, bisect I_1 as before. Note that λ is the least upper bound of the values of f on at least one of the two halves of I_1; let I_2 be that half. Having defined I_n, let I_{n+1} be the half of I_n on which λ is the least upper bound of the values of f. Let c be the number common to all these intervals. It then follows from the continuity of f, much as in the proof of Theorem 3, that $f(c) = \lambda$. And it is clear that $f(x) \leqq \lambda$ for all x in $[a, b]$. ❏

The technique we are using in these proofs is called the *method of bisection*. We now use it once again to establish the *Bolzano-Weierstrass property* of the real number system.

> **Definition** *Limit Point*
> Let S be a set of real numbers. The number p is said to be a **limit point** of S if every open interval containing p also contains points of S other than p.

> **Theorem 5** *Bolzano-Weierstrass Theorem*
> Every bounded infinite set of real numbers has a limit point.

Proof   Let I_0 be a closed interval containing the bounded infinite set S of real numbers. Let I_1 be one of the closed half-intervals of I_0 that contains infinitely many points of S. If I_n has been chosen, let I_{n+1} be one of the closed half-intervals of I_n containing infinitely many points of S. An application of Theorem 2 yields a number p common to all the intervals I_n. If J is an open interval containing p, then J contains I_n for some sufficiently large value of n and thus contains infinitely many points of S. Therefore, p is a limit point of S. ❑

Our final goal is in sight: We can now prove that a sequence of real numbers converges if and only if it is a Cauchy sequence.

> **Definition** *Cauchy Sequence*
> The sequence $\{a_n\}_1^\infty$ is said to be a **Cauchy sequence** if, for every $\epsilon > 0$, there exists an integer N such that
> $$|a_m - a_n| < \epsilon$$
> for all $m, n \geqq N$.

> **Lemma 2** *Convergent Subsequences*
> Every bounded sequence of real numbers has a convergent subsequence.

Proof   If $\{a_n\}$ has only a finite number of values, then the conclusion of Lemma 2 follows easily. We therefore focus our attention on the case in which $\{a_n\}$ is an infinite set. It is easy to show that this set is also bounded, and thus we may apply the Bolzano-Weierstrass theorem to obtain a limit point p of $\{a_n\}$. For each integer $k \geqq 1$, let $a_{n(k)}$ be a term of the sequence $\{a_n\}$ such that

1. $n(k + 1) > n(k)$ for all $k \geqq 1$, and

2. $\left| a_{n(k)} - p \right| < \dfrac{1}{k}$.

It is then easy to show that $\{a_{n(k)}\}$ is a convergent (to p) subsequence of $\{a_n\}$. ❑

> **Theorem 6** *Convergence of Cauchy Sequences*
> A sequence of real numbers converges if and only if it is a Cauchy sequence.

Proof   It follows immediately from the triangle inequality that every convergent sequence is a Cauchy sequence. Thus suppose that the sequence $\{a_n\}$ is a Cauchy sequence.

Choose N such that

$$|a_m - a_n| < 1$$

if $m, n \geq N$. It follows that if $n \geq N$, then a_n lies in the closed interval $[a_N - 1, a_N + 1]$. This implies that the sequence $\{a_n\}$ is bounded, and thus by Lemma 2 it has a convergent subsequence $\{a_{n(k)}\}$. Let p be the limit of this subsequence.

We claim that $\{a_n\}$ itself converges to p. Given $\epsilon > 0$, choose M such that

$$|a_m - a_n| < \frac{\epsilon}{2}$$

if $m, n \geq N$. Next choose K such that $n(K) \geq M$ and

$$|a_{n(K)} - p| < \frac{\epsilon}{2}.$$

Then if $n \geq M$,

$$|a_n - p| \leq |a_n - a_{n(K)}| + |a_{n(K)} - p| < \epsilon.$$

Therefore, $\{a_n\}$ converges to p by definition. $\square$

Appendix D
Proof of the Chain Rule

To prove the chain rule, we need to show that if f is differentiable at a and g is differentiable at $f(a)$, then

$$\lim_{h \to 0} \frac{g(f(a + h)) - g(f(a))}{h} = g'(f(a)) \cdot f'(a). \tag{1}$$

If the quantities h and

$$k(h) = f(a + h) - f(a) \tag{2}$$

are nonzero, then we can write the difference quotient on the left-hand side of Eq. (1) as

$$\frac{g(f(a + h)) - g(f(a))}{h} = \frac{g(f(a) + k(h)) - g(f(a))}{k(h)} \cdot \frac{k(h)}{h}. \tag{3}$$

To investigate the first factor on the right-hand side of Eq. (3), we define a new function ϕ as follows:

$$\phi(k) = \begin{cases} \dfrac{g(f(a) + k) - g(f(a))}{k} & \text{if } k \neq 0; \\ g'(f(a)) & \text{if } k = 0. \end{cases} \tag{4}$$

By the definition of the derivative of g, we see from Eq. (4) that ϕ is continuous at $k = 0$; that is,

$$\lim_{k \to 0} \phi(k) = g'(f(a)). \tag{5}$$

Next,

$$\lim_{h \to 0} k(h) = \lim_{h \to 0} [f(a + h) - f(a)] = 0 \tag{6}$$

because f is continuous at $x = a$, and $\phi(0) = g'(f(a))$. It therefore follows from Eq. (5) that

$$\lim_{h \to 0} \phi(k(h)) = g'(f(a)). \tag{7}$$

We are now ready to assemble all this information. By Eq. (3), if $h \neq 0$, then

$$\frac{g(f(a + h)) - g(f(a))}{h} = \phi(k(h)) \cdot \frac{f(a + h) - f(a)}{h} \tag{8}$$

even if $k(h) = 0$, because in this case both sides of Eq. (8) are zero. Hence the product rule for limits yields

$$\lim_{h \to 0} \frac{g(f(a + h)) - g(f(a))}{h} = \lim_{h \to 0} \phi(k(h)) \cdot \frac{f(a + h) - f(a)}{h}$$

$$= g'(f(a)) \cdot f'(a),$$

a consequence of Eq. (7) and the definition of the derivative of the function f. We have therefore established the chain rule in the form of Eq. (1). ❑

Appendix E
Existence of the Integral

When the basic computational algorithms of the calculus were discovered by Newton and Leibniz in the latter half of the seventeenth century, the logical rigor that had been a feature of the Greek method of exhaustion was largely abandoned. When computing the area A under the curve $y = f(x)$, for example, Newton took it as intuitively obvious that the area function existed, and he proceeded to compute it as the antiderivative of the height function $f(x)$. Leibniz regarded A as an infinite sum of infinitesimal area elements, each of the form $dA = f(x)\, dx$, but in practice computed the area

$$A = \int_a^b f(x)\, dx$$

by antidifferentiation just as Newton did—that is, by computing

$$A = \left[D^{-1} f(x) \right]_a^b.$$

The question of the *existence* of the area function—one of the conditions that a function f must satisfy in order for its integral to exist—did not at first seem to be of much importance. Eighteenth-century mathematicians were mainly occupied (and satisfied) with the impressive applications of calculus to the solution of real-world problems and did not concentrate on the logical foundations of the subject.

The first attempt at a precise definition of the integral and a proof of its existence for continuous functions was that of the French mathematician

Augustin Louis Cauchy (1789–1857). Curiously enough, Cauchy was trained as an engineer, and much of his research in mathematics was in fields that we today regard as applications-oriented: hydrodynamics, waves in elastic media, vibrations of elastic membranes, polarization of light, and the like. But he was a prolific researcher, and his writings cover the entire spectrum of mathematics, with occasional essays into almost unrelated fields.

Around 1824, Cauchy defined the integral of a continuous function in a way that is familiar to us, as a limit of left-endpoint approximations:

$$\int_a^b f(x)\,dx = \lim_{\Delta x \to 0} \sum_{i=1}^n f(x_{i-1})\,\Delta x.$$

This is a much more complicated sort of limit than the ones we discussed in Chapter 2. Cauchy was not entirely clear about the nature of the limit process involved in this equation, nor was he clear about the precise role that the hypothesis of the continuity of f played in proving that the limit exists.

A complete definition of the integral, as we gave in Section 5.4, was finally produced in the 1850s by the German mathematician Georg Bernhard Riemann. Riemann was a student of Gauss; he met Gauss upon his arrival at Göttingen, Germany, for the purpose of studying theology, when he was about 20 years old and Gauss was about 70. Riemann soon decided to study mathematics and became known as one of the truly great mathematicians of the nineteenth century. Like Cauchy, he was particularly interested in applications of mathematics to the real world; his research particularly emphasized electricity, heat, light, acoustics, fluid dynamics, and—as you might infer from the fact that Wilhelm Weber was a major influence on Riemann's education—magnetism. Riemann also made significant contributions to mathematics itself, particularly in the field of complex analysis. A major conjecture of his, involving the zeta function.

$$\zeta(s) = \sum_{n=1}^{\infty} \frac{1}{n^s}, \tag{1}$$

remains unsolved to this day and has important consequences in the theory of the distribution of prime numbers because

$$\zeta(k) = \prod \left(1 - \frac{1}{p^k} \right)^{-1},$$

where the product $\prod$ is taken over all primes p. [The zeta function is defined in Eq. (1) for complex numbers s to the right of the vertical line at $x = 1$ and is extended to other complex numbers by the requirement that it be differentiable.] Riemann died of tuberculosis shortly before his fortieth birthday.

Here we give a proof of the existence of the integral of a continuous function. We will follow Riemann's approach. Specifically, suppose that the function f is continuous on the closed and bounded interval $[a, b]$. We will prove that the definite integral

$$\int_a^b f(x)\,dx$$

exists. That is, we will demonstrate the existence of a number I that satisfies the following condition: For every $\epsilon > 0$ there exists $\delta > 0$ such that, for *every* Riemann sum R associated with *any* partition P with $|P| < \delta$,

$$|I - R| < \epsilon.$$

(Recall that the mesh $|P|$ of the partition P is the length of the longest subinterval in the partition.) In other words, every Riemann sum associated with every sufficiently "fine" partition is close to the number I. If this happens, then the definite integral

$$\int_a^b f(x)\, dx$$

is said to **exist,** and I is its **value.**

Now we begin the proof. Suppose throughout that f is a function continuous on the closed interval $[a, b]$. Given $\epsilon > 0$, we need to show the existence of a number $\delta > 0$ such that

$$\left| I - \sum_{i=1}^{n} f(x_i{}^*)\, \Delta x_i \right| < \epsilon \tag{2}$$

for every Riemann sum associated with any partition P of $[a, b]$ with $|P| < \delta$.

Given a partition P of $[a, b]$ into n subintervals that are *not necessarily of equal length*, let p_i be a point in the subinterval $[x_{i-1}, x_i]$ at which f attains its minimum value $f(p_i)$. Similarly, let $f(q_i)$ be its maximum value there. These numbers exist for $i = 1, 2, 3, \ldots, n$ because of the maximum value property of continuous functions (Theorem 4 of Appendix C).

In what follows we will denote the resulting lower and upper Riemann sums associated with P by

$$L(P) = \sum_{i=1}^{n} f(p_i)\, \Delta x_i \tag{3a}$$

and

$$U(P) = \sum_{i=1}^{n} f(q_i)\, \Delta x_i, \tag{3b}$$

respectively. Then Lemma 1 is obvious.

Lemma 1

For any partition P of $[a, b]$, $L(P) \leqq U(P)$.

Now we need a definition. The partition P' is called a *refinement* of the partition P if each subinterval of P' is contained in some subinterval of P. That is, P' is obtained from P by adding more points of subdivision to P.

Lemma 2

Suppose that P' is a refinement of P. Then

$$L(P) \leqq L(P') \leqq U(P') \leqq U(P). \tag{4}$$

Proof The inequality $L(P') \leqq U(P')$ is a consequence of Lemma 1. We will show that $L(P) \leqq L(P')$; the proof that $U(P') \leqq U(P)$ is similar.

The refinement P' is obtained from P by adding one or more points of subdivision to P. So all we need show is that the Riemann sum $L(P)$ cannot be decreased by adding a single point of subdivision. Thus we will suppose that the partition P' is obtained from P by dividing the kth subinterval $[x_{k-1}, x_k]$ of P into two subintervals $[x_{k-1}, z]$ and $[z, x_k]$ by means of the new point z.

The only resulting effect on the corresponding Riemann sum is to replace the term

$$f(p_k) \cdot (x_k - x_{k-1})$$

in $L(P)$ by the two-term sum

$$f(u) \cdot (z - x_{k-1}) + f(v) \cdot (x_k - z),$$

where $f(u)$ is the minimum of f on $[x_{k-1}, z]$ and $f(v)$ is the minimum of f on $[z, x_k]$. But

$$f(p_k) \leqq f(u) \quad \text{and} \quad f(p_k) \leqq f(v).$$

Hence

$$f(u) \cdot (z - x_{k-1}) + f(v) \cdot (x_k - z) \geqq f(p_k) \cdot (z - x_{k-1}) + f(p_k) \cdot (x_k - z)$$
$$= f(p_k) \cdot (z - x_{k-1} + x_k - z)$$
$$= f(p_k) \cdot (x_k - x_{k-1}).$$

So the replacement of $f(p_k) \cdot (x_k - x_{k-1})$ cannot decrease the sum $L(P)$ in question, and therefore $L(P) \leqq L(P')$. Because this is all we needed to show, we have completed the proof of Lemma 2. ❏

To prove that all the Riemann sums for sufficiently fine partitions are close to some number I, we must first give a construction of I. This is accomplished through Lemma 3.

Lemma 3

Let P_n denote the regular partition of $[a, b]$ into 2^n subintervals of equal length. Then the (sequential) limit

$$I = \lim_{n \to \infty} L(P_n) \tag{5}$$

exists.

Proof We begin with the observation that each partition P_{n+1} is a refinement of P_n, so (by Lemma 2)

$$L(P_1) \leqq L(P_2) \leqq \cdots \leqq L(P_n) \leqq \cdots .$$

Therefore, $\{L(P_n)\}$ is a nondecreasing sequence of real numbers. Moreover,

$$L(P_n) = \sum_{i=1}^{2^n} f(p_i)\, \Delta x_i \leqq M \sum_{i=1}^{2^n} \Delta x_i = M(b - a),$$

where M is the maximum value of f on $[a, b]$.

Theorem 1 of Appendix C guarantees that a bounded monotonic sequence of real numbers must converge. Thus the number

$$I = \lim_{n \to \infty} L(P_n)$$

exists. This establishes Eq. (5), and the proof of Lemma 3 is complete. ❑

It is proved in advanced calculus that if f is continuous on $[a, b]$, then—for every number $\epsilon > 0$—there exists a number $\delta > 0$ such that

$$|f(u) - f(v)| < \epsilon$$

for every two points u and v of $[a, b]$ such that

$$|u - v| < \delta.$$

This property of a function is called **uniform continuity** of f on the interval $[a, b]$. Thus the theorem from advanced calculus that we need to use states that every continuous function on a closed and bounded interval is uniformly continuous there.

NOTE The fact that f is continuous on $[a, b]$ means that for each number u in the interval and each $\epsilon > 0$, there exists $\delta > 0$ such that if v is a number in the interval with $|u - v| < \delta$, then $|f(u) - f(v)| < \epsilon$. But *uniform* continuity is a more stringent condition. It means that given $\epsilon > 0$, you can find not only a value δ_1 that "works" for u_1, a value δ_2 that works for u_2, and so on, but more: You can find a universal value of δ that works for *all* values of u in the interval. This should not be obvious when you notice the possibility that $\delta_1 = 1$, $\delta_2 = \frac{1}{2}$, $\delta_3 = \frac{1}{3}$, and so on. In any case, it is clear that uniform continuity of f on an interval implies its continuity there.

Remember that throughout we have a continuous function f defined on the closed interval $[a, b]$.

Lemma 4

Suppose that $\epsilon > 0$ is given. Then there exists a number $\delta > 0$ such that if P is a partition of $[a, b]$ with $|P| < \delta$ and P' is a refinement of P, then

$$|R(P) - R(P')| < \frac{\epsilon}{3} \qquad (6)$$

for any two Riemann sums $R(P)$ associated with P and $R(P')$ associated with P'.

Proof Because f must be uniformly continuous on $[a, b]$, there exists a number $\delta > 0$ such that if

$$|u - v| < \delta, \quad \text{then } |f(u) - f(v)| < \frac{\epsilon}{3(b - a)}.$$

Suppose now that P is a partition of $[a, b]$ with $|P| < \delta$. Then

$$|U(P) - L(P)| = \sum_{i=1}^{n} |f(q_i) - f(p_i)| \, \Delta x_i < \frac{\epsilon}{3(b - a)} \sum_{i=1}^{n} \Delta x_i = \frac{\epsilon}{3}.$$

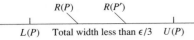

Fig. E.1 Part of the proof of Lemma 4

This is valid because $|p_i - q_i| < \delta$, for both p_i and q_i belong to the same subinterval $[x_{i-1}, x_i]$ of P, and $|P| < \delta$.

Now, as shown in Fig. E.1, we know that $L(P)$ and $U(P)$ differ by less than $\epsilon/3$. We know also that

$$L(P) \leqq R(P) \leqq U(P)$$

for every Riemann sum $R(P)$ associated with P. But

$$L(P) \leqq L(P') \leqq U(P') \leqq U(P)$$

by Lemma 2, because P' is a refinement of P; moreover,

$$L(P') \leqq R(P') \leqq U(P')$$

for every Riemann sum $R(P')$ associated with P'.

As Fig. E.1 shows, both the numbers $R(P)$ and $R(P')$ belong to the interval $[L(P), U(P)]$ of length less than $\epsilon/3$, so Eq. (6) follows, as desired. This concludes the proof of Lemma 4. $\square$

Theorem 1 *Existence of the Integral*

If f is continuous on the closed and bounded interval $[a, b]$, then the integral

$$\int_a^b f(x)\, dx$$

exists.

Proof   Suppose that $\epsilon > 0$ is given. We must show the existence of a number $\delta > 0$ such that, for every partition P of $[a, b]$ with $|P| < \delta$, we have

$$|I - R(P)| < \epsilon,$$

where I is the number given in Lemma 3 and $R(P)$ is an arbitrary Riemann sum for f associated with P.

We choose the number δ provided by Lemma 4 such that

$$|R(P) - R(P')| < \frac{\epsilon}{3}$$

if $|P| < \delta$ and P' is a refinement of P.

By Lemma 3, we can choose an integer N so large that

$$|P_N| < \delta \quad \text{and} \quad |L(P_N) - I| < \frac{\epsilon}{3}. \tag{7}$$

Given an arbitrary partition P such that $|P| < \delta$, let P' be a common refinement of both P and P_N. You can obtain such a partition P', for example, by using all the points of subdivision of both P and P_N to form the subintervals of $[a, b]$ that constitute P'.

Because P' is a refinement of both P and P_N and both the latter partitions have mesh less than δ, Lemma 4 implies that

$$|R(P) - R(P')| < \frac{\epsilon}{3} \quad \text{and} \quad |L(P_N) - R(P')| < \frac{\epsilon}{3}. \tag{8}$$

Here $R(P)$ and $R(P')$ are (arbitrary) Riemann sums associated with P and P', respectively.

Given an arbitrary Riemann sum $R(P)$ associated with the partition P with mesh less than δ, we see that

$$|I - R(P)| = |I - L(P_N) + L(P_N) - R(P') + R(P') - R(P)|$$
$$\leq |I - L(P_N)| + |L(P_N) - R(P')| + |R(P') - R(P)|.$$

In the last sum, both of the last two terms are less than $\epsilon/3$ by virtue of the inequalities in (8). We also know, by (7), that the first term is less than $\epsilon/3$. Consequently,

$$|I - R(P)| < \epsilon.$$

This establishes Theorem 1. ❑

We close with an example that shows that some hypothesis of continuity is required for integrability.

EXAMPLE 1   Suppose that f is defined for $0 \leq x \leq 1$ as follows:

$$f(x) = \begin{cases} 1 & \text{if } x \text{ is irrational;} \\ 0 & \text{if } x \text{ is rational.} \end{cases}$$

Then f is not continuous anywhere. (Why?) Given a partition P of $[0, 1]$, let p_i be a rational point and q_i an irrational point of the ith subinterval of P for $i = 1, 2, 3, \ldots, n$. As before, f attains its minimum value 0 at each p_i and its maximum value 1 at each q_i. Also

$$L(P) = \sum_{i=1}^{n} f(p_i)\,\Delta x_i = 0, \quad \text{whereas} \quad U(P) = \sum_{i=1}^{n} f(q_i)\,\Delta x_i = 1.$$

Thus if we choose $\epsilon = \frac{1}{2}$, then there is *no* number I that can lie within ϵ of both $L(P)$ and $U(P)$, no matter how small the mesh of P. It follows that f is *not* Riemann integrable on $[0, 1]$.

Appendix F
Approximations and Riemann Sums

Several times in Chapter 6 our attempt to compute some quantity Q led to the following situation. Beginning with a regular partition of an appropriate interval $[a, b]$ into n subintervals, each of length Δx, we find an approximation A_n to Q of the form

$$A_n = \sum_{i=1}^{n} g(u_i)h(v_i)\,\Delta x, \tag{1}$$

where u_i and v_i are two (generally different) points of the ith subinterval $[x_{i-1}, x_i]$. For example, in our discussion of surface area of revolution that precedes Eq. (8) of Section 6.4, we found the approximation

$$\sum_{i=1}^{n} 2\pi f(u_i)\sqrt{1 + [f'(v_i)]^2}\,\Delta x \tag{2}$$

to the area of the surface generated by revolving the curve $y = f(x)$, $a \leq x \leq b$, around the x-axis. (In Section 6.4 we wrote x_i^{**} for u_i and x_i^{*} for

v_i.) Note that the expression in (2) is the same as the right-hand side of Eq. (1); take $g(x) = 2\pi f(x)$ and $h(x) = \sqrt{1 + [f'(x)]^2}$.

In such a situation we observe that if u_i and v_i were the *same* point x_i^* of $[x_{i-1}, x_i]$ for each i ($i = 1, 2, 3, \ldots, n$), then the approximation in Eq. (1) would be a Riemann sum for the function $g(x)h(x)$ on $[a, b]$. This leads us to suspect that

$$\lim_{\Delta x \to 0} \sum_{i=1}^{n} g(u_i)h(v_i) \, \Delta x = \int_a^b g(x)h(x) \, dx. \tag{3}$$

In Section 6.4, we assumed the validity of Eq. (3) and concluded from the approximation in (2) that the surface area of revolution ought to be defined to be

$$A = \lim_{\Delta x \to 0} \sum_{i=1}^{n} 2\pi f(u_i)\sqrt{1 + [f'(v_i)]^2} \, \Delta x = \int_a^b 2\pi f(x)\sqrt{1 + [f'(x)]^2} \, dx.$$

Theorem 1 guarantees that Eq. (3) holds under mild restrictions on the functions g and h.

Theorem 1   A Generalization of Riemann Sums

Suppose that h and g' are continuous on $[a, b]$. Then

$$\lim_{\Delta x \to 0} \sum_{i=1}^{n} g(u_i)h(v_i) \, \Delta x = \int_a^b g(x)h(x) \, dx, \tag{3}$$

where u_i and v_i are arbitrary points of the ith subinterval of a regular partition of $[a, b]$ into n subintervals, each of length Δx.

Proof   Let M_1 and M_2 denote the maximum values on $[a, b]$ of $|g'(x)|$ and $|h(x)|$, respectively. Note that

$$\sum_{i=1}^{n} g(u_i)h(v_i) \, \Delta x = R_n + S_n, \quad \text{where} \quad R_n = \sum_{i=1}^{n} g(v_i)h(v_i) \, \Delta x$$

is a Riemann sum approaching $\int_a^b g(x)h(x) \, dx$ as $\Delta x \to 0$, and

$$S_n = \sum_{i=1}^{n} [g(u_i) - g(v_i)] h(v_i) \, \Delta x.$$

To prove Eq. (3) it is sufficient to show that $S_n \to 0$ as $\Delta x \to 0$. The mean value theorem gives

$$|g(u_i) - g(v_i)| = |g'(\overline{x}_i)| \cdot |u_i - v_i| \qquad [\overline{x}_i \text{ in } (u_i, v_i)]$$

$$\leqq M_1 \, \Delta x,$$

because both u_i and v_i are points of the interval $[x_{i-1}, x_i]$ of length Δx. Then

$$|S_n| \leqq \sum_{i=1}^{n} |g(u_i) - g(v_i)| \cdot |h(v_i)| \, \Delta x \leqq \sum_{i=1}^{n} (M_1 \, \Delta x) \cdot (M_2 \, \Delta x)$$

$$= (M_1 M_2 \, \Delta x) \sum_{i=1}^{n} \Delta x = M_1 M_2 (b - a) \, \Delta x,$$

from which it follows that $S_n \to 0$ as $\Delta x \to 0$, as desired.   ☐

As an application of Theorem 1, let us give a rigorous derivation of Eq. (2) of Section 6.3,

$$V = \int_a^b 2\pi x f(x) \, dx, \tag{4}$$

for the volume of the solid generated by revolving around the y-axis the region lying below $y = f(x)$, $a \leq x \leq b$. Beginning with the usual regular partition of $[a, b]$, let $f(x_i^\flat)$ and $f(x_i^\sharp)$ denote the minimum and maximum values of f on the ith subinterval $[x_{i-1}, x_i]$. Denote by x_i^* the midpoint of this subinterval. From Fig. F.1, we see that the part of the solid generated by revolving the region below $y = f(x)$, $x_{i-1} \leq x \leq x_i$, contains a cylindrical shell with average radius x_i^*, thickness Δx, and height $f(x_i^\flat)$ and is contained in another cylindrical shell with the same average radius and thickness but with height $f(x_i^\sharp)$. Hence the volume ΔV_i of this part of the solid satisfies the inequalities

$$2\pi x_i^* f(x_i^\flat) \, \Delta x \leq \Delta V_i \leq 2\pi x_i^* f(x_i^\sharp) \, \Delta x.$$

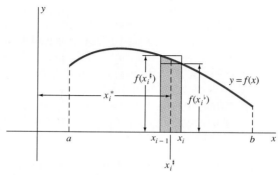

Fig. F.1 A careful estimate of the volume of a solid of revolution around the y-axis

We add these inequalities for $i = 1, 2, 3, \ldots, n$ and find that

$$\sum_{i=1}^n 2\pi x_i^* f(x_i^\flat) \, \Delta x \leq V \leq \sum_{i=1}^n 2\pi x_i^* f(x_i^\sharp) \, \Delta x.$$

Because Theorem 1 implies that both of the last two sums approach $\int_a^b f(x) \, dx$, the squeeze law of limits now implies Eq. (4).

We will occasionally need a generalization of Theorem 1 that involves the notion of a continuous function $F(x, y)$ of two variables. We say that F is *continuous* at the point (x_0, y_0) provided that the value $F(x, y)$ can be made arbitrarily close to $F(x_0, y_0)$ merely be choosing the point (x, y) sufficiently close to (x_0, y_0). We discuss continuity of functions of two variables in Chapter 14. Here it will suffice to accept the following facts: If $g(x)$ and $h(y)$ are continuous functions of the single variables x and y, respectively, then simple combinations such as

$$g(x) \pm h(y), \qquad g(x)h(y), \quad \text{and} \quad \sqrt{[g(x)]^2 + [h(y)]^2}$$

are continuous functions of the two variables x and y.

Now consider a regular partition of $[a, b]$ into n subintervals, each of length Δx, and let u_i and v_i denote arbitrary points of the ith subinterval $[x_{i-1}, x_i]$. Theorem 2—we omit the proof—tells us how to find the limit as $\Delta x \to 0$ of a sum such as

$$\sum_{i=1}^{n} F(u_i, v_i) \, \Delta x.$$

Theorem 2   A *Further Generalization*

Let $F(x, y)$ be continuous for x and y both in the interval $[a, b]$. Then, in the notation of the preceding paragraph,

$$\lim_{\Delta x \to 0} \sum_{i=1}^{n} F(u_i, v_i) \, \Delta x = \int_{a}^{b} F(x, x) \, dx. \qquad (5)$$

Theorem 1 is the special case $F(x, y) = g(x)h(y)$ of Theorem 2. Moreover, the integrand $F(x, x)$ on the right in Eq. (5) is merely an ordinary function of the single variable x. As a formal matter, the integral corresponding to the sum in Eq. (5) is obtained by replacing the summation symbol with an integral sign, changing both u_i and v_i to x, replacing Δx by dx, and inserting the correct limits of integration. For example, if the interval $[a, b]$ is $[0, 4]$, then

$$\lim_{\Delta x \to 0} \sum_{i=1}^{n} \sqrt{9u_i^2 + v_i^4} \, \Delta x = \int_{0}^{4} \sqrt{9x^2 + x^4} \, dx$$

$$= \int_{0}^{4} x(9 + x^2)^{1/2} \, dx = \left[\frac{1}{3}(9 + x^2)^{3/2} \right]_{0}^{4}$$

$$= \tfrac{1}{3}[(25)^{3/2} - (9)^{3/2}] = \tfrac{98}{3}.$$

Appendix F Problems

In Problems 1 through 7, u_i and v_i are arbitrary points of the ith subinterval of a regular partition of $[a, b]$ into n subintervals, each of length Δx. Express the given limit as an integral from a to b, then compute the value of this integral.

1. $\displaystyle \lim_{\Delta x \to 0} \sum_{i=1}^{n} u_i v_i \, \Delta x; \quad a = 0, b = 1$

2. $\displaystyle \lim_{\Delta x \to 0} \sum_{j=1}^{n} (3u_j + 5v_j) \, \Delta x; \quad a = -1, b = 3$

3. $\displaystyle \lim_{\Delta x \to 0} \sum_{i=1}^{n} u_i \sqrt{4 - v_i^2} \, \Delta x; \quad a = 0, b = 2$

4. $\displaystyle \lim_{\Delta x \to 0} \sum_{i=1}^{n} \frac{u_i \, \Delta x}{\sqrt{16 + v_i^2}}; \quad a = 0, b = 3$

5. $\displaystyle \lim_{\Delta x \to 0} \sum_{i=1}^{n} \sin u_i \cos v_i \, \Delta x; \quad a = 0, b = \frac{\pi}{2}$

6. $\displaystyle \lim_{\Delta x \to 0} \sum_{i=1}^{n} \sqrt{\sin^2 u_i + \cos^2 v_i} \, \Delta x; \quad a = 0, b = \pi$

7. $\displaystyle \lim_{\Delta x \to 0} \sum_{k=1}^{n} \sqrt{u_k^4 + v_k^7} \, \Delta x; \quad a = 0, b = 2$

8. Explain how Theorem 1 applies to show that Eq. (8) of Section 6.4 follows from the discussion that precedes it in that section.

9. Use Theorem 1 to derive Eq. (10) of Section 6.4.

Appendix G

L'Hôpital's Rule and Cauchy's Mean Value Theorem

Here we give a proof of l'Hôpital's rule,

$$\lim_{x \to a} \frac{f(x)}{g(x)} = \lim_{x \to a} \frac{f'(x)}{g'(x)}, \tag{1}$$

under the hypotheses of Theorem 1 in Section 8.3. The proof is based on a generalization of the mean value theorem due to the French mathematician Augustin Louis Cauchy. Cauchy used this generalization in the early nineteenth century to give rigorous proofs of several calculus results not previously established firmly.

Cauchy's Mean Value Theorem

Suppose that the functions f and g are continuous on the closed and bounded interval $[a, b]$ and differentiable on (a, b). Then there exists a number c in (a, b) such that

$$[f(b) - f(a)]g'(c) = [g(b) - g(a)]f'(c). \tag{2}$$

REMARK 1 To see that this theorem is indeed a generalization of the (ordinary) mean value theorem, we take $g(x) = x$. Then $g'(x) \equiv 1$, and the conclusion in Eq. (2) reduces to the fact that

$$f(b) - f(a) = (b - a)f'(c)$$

for some number c in (a, b).

REMARK 2 Equation (2) has a geometric interpretation like that of the ordinary mean value theorem. Let us think of the equations $x = g(t), y = f(t)$ as describing the motion of a point $P(x, y)$ moving along a curve C in the xy-plane as t increases from a to b (Fig. G.1). That is, $P(x, y) = P(g(t), f(t))$ is the location of the point P at time t. Under the assumption that $g(b) \neq g(a)$, the slope of the line L connecting the endpoints of the curve C is

$$m = \frac{f(b) - f(a)}{g(b) - g(a)}. \tag{3}$$

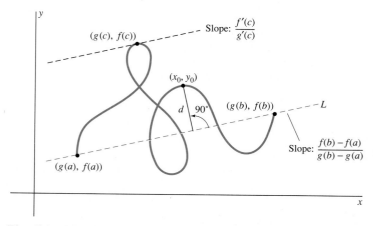

Fig. G.1 The idea of Cauchy's mean value theorem

A-28

But if $g'(c) \neq 0$, then the chain rule gives

$$\frac{dy}{dx} = \frac{dy/dt}{dx/dt} = \frac{f'(c)}{g'(c)} \tag{4}$$

for the slope of the line tangent to the curve C at the point $(g(c),\ f(c))$. But if $g(b) \neq g(a)$ and $g'(c) \neq 0$, then Eq. (2) may be written as

$$\frac{f(b) - f(a)}{g(b) - g(a)} = \frac{f'(c)}{g'(c)}, \tag{5}$$

so the two slopes in Eqs. (3) and (4) are equal. Thus Cauchy's mean value theorem implies that (under our assumptions) there is a point on the curve C where the tangent line is *parallel* to the line joining the endpoints of C. This is exactly what the (ordinary) mean value theorem says for an explicitly defined curve $y = f(x)$. This geometric interpretation motivates the following proof of Cauchy's mean value theorem.

Proof   The line L through the endpoints in Fig. G.1 has point-slope equation

$$y - f(a) = \frac{f(b) - f(a)}{g(b) - g(a)}\, [x - g(a)],$$

which can be rewritten in the form $Ax + By + C = 0$ with

$$A = f(b) - f(a), \qquad B = -[g(b) - g(a)], \quad \text{and}$$
$$C = f(a)\,[g(b) - g(a)] - g(a)\,[f(b) - f(a)]. \tag{6}$$

According to Miscellaneous Problem 71 at the end of Chapter 3, the (perpendicular) distance from the point $(x_0,\ y_0)$ to the line L is

$$d = \frac{|\,Ax_0 + By_0 + C\,|}{\sqrt{A^2 + B^2}}.$$

Figure G.1 suggests that the point $(g(c),\ f(c))$ will maximize this distance d for points on the curve C.

We are motivated, therefore, to define the auxiliary function

$$\phi(t) = Ag(t) + Bf(t) + C, \tag{7}$$

with the constants A, B, and C as defined in Eq. (6). Thus $\phi(t)$ is essentially a constant multiple of the distance from $(g(t),\ f(t))$ to the line L in Fig. G.1.

Now $\phi(a) = 0 = \phi(b)$ (why?), so Rolle's theorem (Section 4.3) implies the existence of a number c in $(a,\ b)$ such that

$$\phi'(c) = Ag'(c) + Bf'(c) = 0. \tag{8}$$

We substitute the values of A and B from Eq. (6) into (8) and obtain the equation

$$[f(b) - f(a)]g'(c) - [g(b) - g(a)]f'(c) = 0.$$

This is the same as Eq. (2) in the conclusion of Cauchy's mean value theorem, and the proof is complete.   ❏

NOTE Whereas the assumptions that $g(b) \neq g(a)$ and $g'(c) \neq 0$ were needed for our geometric interpretation of the theorem, they were not used in its proof—only in the motivation for the method of proof.

PROOF OF L'HÔPITAL'S RULE

Suppose that $f(x)/g(x)$ has the indeterminate form $0/0$ at $x = a$. We may invoke continuity of f and g to allow the assumption that $f(a) = 0 = g(a)$. That is, we simply define $f(a)$ and $g(a)$ to be zero in case their values at $x = a$ are not originally given.

Now we restrict our attention to values of x in a fixed deleted neighborhood of a on which both f and g are differentiable. Choose one such value of x, and hold it temporarily constant. Then apply Cauchy's mean value theorem on the interval $[a, x]$. (If $x < a$, use the interval $[x, a]$.) We find that there is a number z between a and x that behaves as c does in Eq. (2). Hence, by virtue of Eq. (2), we obtain the equation

$$\frac{f(x)}{g(x)} = \frac{f(x) - f(a)}{g(x) - g(a)} = \frac{f'(z)}{g'(z)}.$$

Now z depends on x, but z is trapped between x and a, so z is forced to approach a as $x \to a$. We conclude that

$$\lim_{x \to a} \frac{f(x)}{g(x)} = \lim_{z \to a} \frac{f'(z)}{g'(z)} = \lim_{x \to a} \frac{f'(x)}{g'(x)},$$

under the assumption that the right-hand limit exists. Thus we have verified l'Hôpital's rule in the form of Eq. (1). ❏

Appendix H
Proof of Taylor's Formula

Several different proofs of Taylor's formula (Theorem 2 of Section 11.4) are known, but none of them seems very well motivated—each requires some "trick" to begin the proof. The trick we employ here (suggested by C. R. MacCluer) is to begin by introducing an auxiliary function $F(x)$, defined as follows:

$$F(x) = f(b) - f(x) - f'(x)(b - x) - \frac{f''(x)}{2!}(b - x)^2$$

$$- \cdots - \frac{f^{(n)}(x)}{n!}(b - x)^n - K(b - x)^{n+1}, \qquad (1)$$

where the *constant* K is chosen so that $F(a) = 0$. To see that there *is* such a value of K, we could substitute $x = a$ on the right and $F(x) = F(a) = 0$ on the left in Eq. (1) and then solve routinely for K, but we have no need to do this explicitly.

Equation (1) makes it quite obvious that $F(b) = 0$ as well. Therefore, Rolle's theorem (Section 4.3) implies that

$$F'(z) = 0 \qquad (2)$$

for some point z of the open interval (a, b) (under the assumption that $a < b$).

To see what Eq. (2) means, we differentiate both sides of Eq. (1) and find that

$$F'(x) = -f'(x) + [f'(x) - f''(x)(b - x)]$$
$$+ \left[f''(x)(b - x) - \frac{1}{2!}f^{(3)}(x)(b - x)^2 \right]$$
$$+ \left[\frac{1}{2!}f^{(3)}(x)(b - x)^2 - \frac{1}{3!}f^{(4)}(x)(b - x)^3 \right]$$
$$+ \cdots + \left[\frac{1}{(n - 1)!}f^{(n)}(x)(b - x)^{n-1} - \frac{1}{n!}f^{(n+1)}(x)(b - x)^n \right]$$
$$+ (n + 1)K(b - x)^n.$$

Upon careful inspection of this result, we see that all terms except the final two cancel in pairs. Thus the sum "telescopes" to give

$$F'(x) = (n + 1)K(b - x)^n - \frac{f^{(n+1)}(x)}{n!}(b - x)^n. \tag{3}$$

Hence Eq. (2) means that

$$(n + 1)K(b - z)^n - \frac{f^{(n+1)}(z)}{n!}(b - z)^n = 0.$$

Consequently we can cancel $(b - z)^n$ and solve for

$$K = \frac{f^{(n+1)}(z)}{(n + 1)!}. \tag{4}$$

Finally, we return to Eq. (1) and substitute $x = a$, $F(x) = 0$, and the value of K given in Eq. (4). The result is the equation

$$0 = f(b) - f(a) - f'(a)(b - a) - \frac{f''(a)}{2!}(b - a)^2$$
$$- \cdots - \frac{f^{(n)}(a)}{n!}(b - a)^n - \frac{f^{(n+1)}(z)}{(n + 1)!}(b - a)^{n+1},$$

which is equivalent to the desired Taylor's formula, Eq. (11) of Section 11.4. ❏

Appendix I
Units of Measurement and Conversion Factors

MKS SCIENTIFIC UNITS

❑ *Length* in meters (m), *mass* in kilograms (kg), *time* in seconds (s)

❑ *Force* in newtons (N); a force of 1 N imparts an acceleration of 1 m/s² to a mass of 1 kg.

❑ *Work* in joules (J); 1 J is the work done by a force of 1 N acting through a distance of 1 m.

❑ *Power* in watts (W); 1 W is 1 J/s.

BRITISH ENGINEERING UNITS (fps)

❑ *Length* in feet (ft), *force* in pounds (lb), *time* in seconds (s)

❑ *Mass* in slugs; 1 lb of force imparts an acceleration of 1 ft/s² to a mass of

1 slug. A mass of m slugs at the surface of the earth has a *weight* of $w = mg$ pounds (lb), where $g \approx 32.17$ ft/s^2.

❑ *Work* in ft·lb, *power* in ft·lb/s.

CONVERSION FACTORS

$$1 \text{ in.} = 2.54 \text{ cm} = 0.0254 \text{ m}, \quad 1 \text{ m} \approx 3.2808 \text{ ft}$$

$$1 \text{ mi} = 5280 \text{ ft}; \quad 60 \text{ mi/h} = 88 \text{ ft/s}$$

$$1 \text{ lb} \approx 4.4482 \text{ N}; \quad 1 \text{ slug} \approx 14.594 \text{ kg}$$

$$1 \text{ hp} = 550 \text{ ft·lb/s} \approx 745.7 \text{ W}$$

❑ *Gravitational acceleration:* $g \approx 32.17$ ft/s^2 ≈ 9.807 m/s^2

❑ *Atmospheric pressure:* 1 atm is the pressure exerted by a column of mercury 76 cm high; 1 atm ≈ 14.70 lb/in.^{2} $\approx 1.013 \times 10^5$ N/m^2

❑ *Heat energy:* 1 Btu ≈ 778 ft·lb ≈ 252 cal, 1 cal ≈ 4.184 J

Appendix J
Formulas from Algebra, Geometry, and Trigonometry

LAWS OF EXPONENTS

$$a^m a^n = a^{m+n}, \quad (a^m)^n = a^{mn}, \quad (ab)^n = a^n b^n, \quad a^{m/n} = \sqrt[n]{a^m};$$

in particular,

$$a^{1/2} = \sqrt{a}.$$

If $a \neq 0$, then

$$a^{m-n} = \frac{a^m}{a^n}, \qquad a^{-n} = \frac{1}{a^n}, \quad \text{and} \quad a^0 = 1.$$

QUADRATIC FORMULA

The quadratic equation

$$ax^2 + bx + c = 0 \quad (a \neq 0)$$

has solutions

$$x = \frac{-b \pm \sqrt{b^2 - 4ac}}{2a}.$$

FACTORING

$$a^2 - b^2 = (a - b)(a + b)$$

$$a^3 - b^3 = (a - b)(a^2 + ab + b^2)$$

$$a^4 - b^4 = (a - b)(a^3 + a^2 b + ab^2 + b^3)$$

$$= (a - b)(a + b)(a^2 + b^2)$$

$$a^5 - b^5 = (a - b)(a^4 + a^3 b + a^2 b^2 + ab^3 + b^4)$$

A-32

(The pattern continues.)

$$a^3 + b^3 = (a + b)(a^2 - ab + b^2)$$

$$a^5 + b^5 = (a + b)(a^4 - a^3b + a^2b^2 - ab^3 + b^4)$$

(The pattern continues for odd exponents.)

BINOMIAL FORMULA

$$(a + b)^n = a^n + na^{n-1}b + \frac{n(n - 1)}{1 \cdot 2}a^{n-2}b^2$$

$$+ \frac{n(n - 1)(n - 2)}{1 \cdot 2 \cdot 3}a^{n-3}b^3 + \cdots + nab^{n-1} + b^n$$

if n is a positive integer.

AREA AND VOLUME

In Fig. J.1, the symbols have the following meanings.

A:	area	b:	length of base	r:	radius
B:	area of base	C:	circumference	V:	volume
h:	height	ℓ:	length	w:	width

Rectangle: $A = bh$

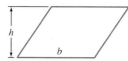
Parallelogram: $A = bh$

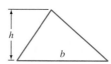
Triangle: $A = \frac{1}{2}bh$

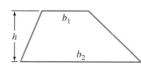
Trapezoid: $A = \frac{1}{2}(b_1 + b_2)h$

Circle: $C = 2\pi r$ and $A = \pi r^2$

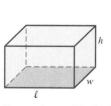
Rectangular parallelepiped:
$V = \ell w h$

Pyramid:
$V = \frac{1}{3}Bh$

Right circular cone:
$V = \frac{1}{3}\pi r^2 h = \frac{1}{3}Bh$

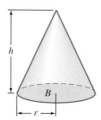
Right circular cylinder:
$V = \pi r^2 h = Bh$

Sphere:
$V = \frac{4}{3}\pi r^3$ and $A = 4\pi r^2$

Fig. J.1 The basic geometric shapes

PYTHAGOREAN THEOREM

In a right triangle with legs a and b and hypotenuse c,

$$a^2 + b^2 = c^2.$$

FORMULAS FROM TRIGONOMETRY

$$\sin(-\theta) = -\sin\theta, \qquad \cos(-\theta) = \cos\theta$$

$$\sin^2\theta + \cos^2\theta = 1$$

$$\sin 2\theta = 2\sin\theta\cos\theta$$

$$\cos 2\theta = \cos^2\theta - \sin^2\theta$$

$$\sin(\alpha + \beta) = \sin\alpha\cos\beta + \cos\alpha\sin\beta$$

$$\cos(\alpha + \beta) = \cos\alpha\cos\beta - \sin\alpha\sin\beta$$

$$\tan(\alpha + \beta) = \frac{\tan\alpha + \tan\beta}{1 - \tan\alpha\tan\beta}$$

$$\sin^2\frac{\theta}{2} = \frac{1}{2}(1 - \cos\theta)$$

$$\cos^2\frac{\theta}{2} = \frac{1}{2}(1 + \cos\theta)$$

For an arbitrary triangle (Fig J.2):

Law of cosines: $\qquad c^2 = a^2 + b^2 - 2ab\cos C.$

Law of sines: $\qquad \dfrac{\sin A}{a} = \dfrac{\sin B}{b} = \dfrac{\sin C}{c}.$

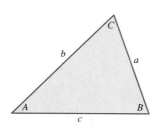

Fig. J.2 An arbitrary triangle

Appendix K
The Greek Alphabet

A	α	alpha	I	ι	iota	P	ρ	rho
B	β	beta	K	κ	kappa	Σ	σ	sigma
Γ	γ	gamma	Λ	λ	lambda	T	τ	tau
Δ	δ	delta	M	μ	mu	Υ	υ	upsilon
E	ϵ	epsilon	N	ν	nu	Φ	ϕ	phi
Z	ζ	zeta	Ξ	ξ	xi	X	χ	chi
H	η	eta	O	o	omicron	Ψ	ψ	psi
Θ	θ	theta	Π	π	pi	Ω	ω	omega

Answers to Odd-Numbered Problems

Section 1.1 (page 11)

1. 14 **3.** 0.5 **5.** 25 **7.** 27 **9.** $\frac{22}{7} - \pi$

11. (a) $-\dfrac{1}{a}$; (b) a; (c) $\dfrac{1}{\sqrt{a}}$; (d) $\dfrac{1}{a^2}$

13. (a) $\dfrac{1}{a^2 + 5}$; (b) $\dfrac{a^2}{1 + 5a^2}$; (c) $\dfrac{1}{a + 5}$; (d) $\dfrac{1}{a^4 + 5}$

15. $\frac{1}{3}$ **17.** ± 3 **19.** 100 **21.** $3h$

23. $2ah + h^2$ **25.** $-\dfrac{h}{a(a + h)}$ **27.** $\{-1, 0, 1\}$

29. $\{-1, 1\}$ **31.** R **33.** R **35.** $[\frac{5}{3}, \infty)$

37. $t \leqq \frac{1}{2}$ **39.** All real numbers other than 3

41. R **43.** $[0, 16]$

45. All real numbers other than 0

47. $C(A) = 2\sqrt{\pi A}, A \geqq 0$ (or $A > 0$)

49. $C(F) = \frac{5}{9}(F - 32), F > -459.67$

51. $A(x) = x\sqrt{16 - x^2}, 0 \leqq x \leqq 4$ (or $0 < x < 4$)

53. $C(x) = 3x^2 + \dfrac{1296}{x}, x > 0$

55. $A(r) = 2\pi r^2 + \dfrac{2000}{r}, r > 0$

57. $V(x) = x(50 - 2x)^2, 0 \leqq x \leqq 25$ (or $0 < x < 25$)

59. Drill 10 new wells **61.** 0.38 **63.** 1.24

65. 0.72 **67.** 3.21 **69.** 1.62

Section 1.2 (page 22)

1. AB and BC have slope 1.

3. AB has slope -2, but BC has slope $-\frac{4}{3}$.

5. AB and CD have slope $-\frac{1}{2}$; BC and DA have slope 2.

7. AB has slope 2, and AC has slope $-\frac{1}{2}$.

9. $m = \frac{2}{3}, b = 0$ **11.** $m = 2, b = 3$

13. $m = -\frac{2}{5}, b = \frac{3}{5}$ **15.** $y = -5$

17. $y - 3 = 2(x - 5)$ **19.** $y - 2 = 4 - x$

21. $y - 5 = -2(x - 1)$ **23.** $x + 2y = 13$

25. $\frac{4}{13}\sqrt{26} \approx 1.568929$

33. $K = \dfrac{125F + 57,461}{225}; F = -459.688$ when $K = 0$.

35. 1136 gal/week **37.** $x = -2.75, y = 3.5$

39. $x = \frac{37}{6}, y = -\frac{1}{2}$ **41.** $x = \frac{22}{5}, y = -\frac{1}{5}$

43. $x = -\frac{7}{4}, y = \frac{33}{8}$ **45.** $x = \frac{119}{12}, y = -\frac{19}{4}$

Section 1.3 (page 30)

1. Center $(2, 0)$, radius 2

3. Center $(-1, -1)$, radius 2

5. Center $(-\frac{1}{2}, \frac{1}{2})$, radius 1

7. Opens upward, vertex at $(3, 0)$

9. Opens upward, vertex at $(-1, 3)$

11. Opens upward, vertex at $(-2, 3)$

13. Circle, center $(3, -4)$, radius 5

15. There are no points on this graph.

17. The graph is the straight line segment joining and including the two points $(-1, 7)$ and $(1, -3)$.

19. Parabola, opening downward, vertex at $(0, 10)$

21. **23.**

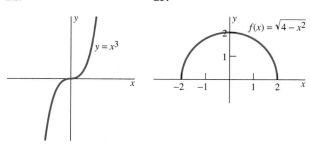

25. The domain of f consists of those numbers x such that $|x| \geqq 3$.

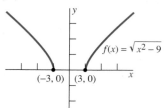

$f(x) = \sqrt{x^2 - 9}$

$(-3, 0)$ $(3, 0)$

27.

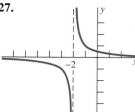

29.

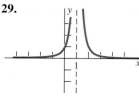

31.

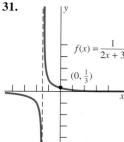

$f(x) = \dfrac{1}{2x + 3}$

$(0, \frac{1}{3})$

33.

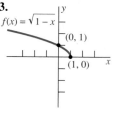

$f(x) = \sqrt{1 - x}$

$(0, 1)$

$(1, 0)$

35.

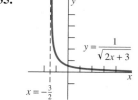

$y = \dfrac{1}{\sqrt{2x + 3}}$

$x = -\dfrac{3}{2}$

37.

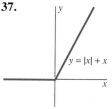

$y = |x| + x$

39.

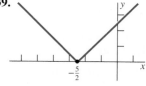

$-\dfrac{5}{2}$

41.

43.

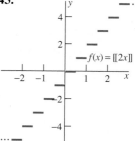

$f(x) = [[2x]]$

45.

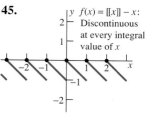

$f(x) = [[x]] - x$:
Discontinuous at every integral value of x

47. $(1.5, 2.5)$ **49.** $(2.25, 1.75)$ **51.** $(2.25, 8.5)$
53. $\left(-\frac{4}{3}, \frac{25}{3}\right)$ **55.** 144 ft **57.** 625

Section 1.4 (page 40)

1. $(f + g)(x) = x^2 + 3x - 2$, domain R;
$(f \cdot g)(x) = x^3 + 3x^2 - x - 3$, domain R; and
$(f/g)(x) = \dfrac{x + 1}{x^2 + 2x - 3}$, domain $x \neq 1, -3$

3. $(f + g)(x) = \sqrt{x} + \sqrt{x - 2}$, domain $x \geqq 2$;
$(f \cdot g)(x) = \sqrt{x^2 - 2x}$, domain $x \geqq 2$;
$(f/g)(x) = \sqrt{\dfrac{x}{x - 2}}$, domain $x > 2$

5. $(f + g)(x) = \sqrt{x^2 + 1} + \dfrac{1}{\sqrt{4 - x^2}}$,
$(f \cdot g)(x) = \dfrac{\sqrt{x^2 + 1}}{\sqrt{4 - x^2}}$, $(f/g)(x) = \sqrt{4 + 3x^2 - x^4}$,
each with domain $-2 < x < 2$

7. $(f + g)(x) = x + \sin x$, domain R;
$(f \cdot g)(x) = x \sin x$, domain R; $(f/g)(x) = \dfrac{x}{\sin x}$,
domain all real numbers not integral multiples of π

9. $(f + g)(x) = \sqrt{x^2 + 1} + \tan x$, $(f \cdot g)(x) = (x^2 + 1)^{1/2} \tan x$, both with domain all real numbers x such that x is not an odd integral multiple of $\pi/2$;
$(f/g)(x) = \dfrac{\sqrt{x^2 + 1}}{\tan x}$, domain all real numbers x such that x is not an integral multiple of $\pi/2$

11. Fig. 1.4.20 **13.** Fig. 1.4.21 **15.** Fig. 1.4.24
17. Fig. 1.4.23 **19.** Fig. 1.4.28 **21.** Fig. 1.4.29
23. Fig. 1.4.30 **25.** 3 **27.** 1 **29.** 0
31. 1 **33.** 5 **35.** 3

Chapter 1 Miscellaneous Problems (page 47)

1. $x \geqq 4$ **3.** $x \neq \pm 3$ **5.** $x \geqq 0$ **7.** $x \leqq \frac{2}{3}$
9. R **11.** $4 \leqq p \leqq 8$ **13.** $2 < I < 4$
15. $V(S) = (S/6)^{3/2}$, $0 < S < \infty$
17. $A = A(P) = \dfrac{\sqrt{3}}{36} P^2$, $0 < P < \infty$
19. $y - 5 = 2(x + 3)$ **21.** $2y = x - 10$
23. $x + 2y = 11$ **25.**

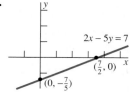

$2x - 5y = 7$

$\left(\frac{7}{2}, 0\right)$

$(0, -\frac{7}{5})$

27.

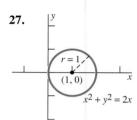

$r = 1$, $(1, 0)$

$x^2 + y^2 = 2x$

29.

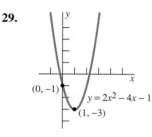

$(0, -1)$

$y = 2x^2 - 4x - 1$

$(1, -3)$

31.

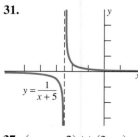

$y = \dfrac{1}{x + 5}$

33.

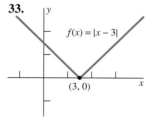

$f(x) = |x - 3|$

$(3, 0)$

37. $(-\infty, -2) \cup (3, \infty)$ **39.** $(-\infty, -2) \cup (4, \infty)$

41. $-1.140, 6.140$ **43.** $1.191, 2.309$

45. $-5.021, 0.896$ **47.** $\left(\frac{5}{2}, \frac{3}{4}\right)$

49. $\left(\frac{7}{4}, -\frac{5}{4}\right)$ **51.** $\left(-\frac{33}{16}, \frac{31}{32}\right)$

53. $x \approx 0.4505$ **55.** 3

57. 3 **59.** 3

Section 2.1 (page 57)

1. 0 **3.** $2x$ **5.** 4 **7.** $4x - 3$ **9.** $4x + 6$

11. $2 - \dfrac{x}{50}$ **13.** $8x$ **15.** $(0, 10)$ **17.** $(1, 0)$

19. $(50, 25)$ **21.** $f'(x) = 3$; $y - 5 = 3(x - 2)$

23. $f'(x) = 4x - 3$; $y - 7 = 5(x - 2)$

25. $f'(x) = 2x - 2$; $y - 1 = 2(x - 2)$

27. $f'(x) = -\dfrac{1}{x^2}$; $x + 4y = 4$

29. $f'(x) = -\dfrac{2}{x^3}$; $x + 4y = 3$

31. $f'(x) = -\dfrac{2}{(x - 1)^2}$; $y - 2 = -2(x - 2)$

33. $\dfrac{dy}{dx} = 2x$; $y - 4 = -4(x + 2)$, $y - 4 = \dfrac{1}{4}(x + 2)$

35. $\dfrac{dy}{dx} = 4x + 3$; $y - 9 = 11(x - 2)$,

$y - 9 = -\frac{1}{11}(x - 2)$

37. $y(3) = 144$ (ft)   **39.** 625 **41.** $y = 12(x - 3)$

43. $(1, 1)$ **45.** 12 **47.** 0.5 **49.** -1

Section 2.2 (page 69)

1. -12 **3.** -1 **5.** 128 **7.** $-\frac{1}{3}$ **9.** 6

11. $16\sqrt{2}$ **13.** 1 **15.** $\frac{1}{4}$ **17.** $-\frac{1}{54}$ **19.** 2

21. 4 **23.** $-\frac{3}{2}$ **25.** -32 **27.** 1 **29.** 0

31. 9 **33.** 2 **35.** 2 **37.** 13

39. Does not exist   **41.** $-\frac{1}{2} x^{-3/2}$

43. $\dfrac{1}{(2x + 1)^2}$ **45.** $\dfrac{x^2 + 2x}{(x + 1)^2}$

49. The limit exists exactly when the real number a is not an integer.

51. $g(x) \to 0$ as $x \to 3$, but $g(3) = 1$.

53. $h\left(\frac{1}{2}\right) = 0$, but $h(1/\pi) < 0$.

Section 2.3 (page 80)

1. 0 **3.** $\frac{1}{2}$ **5.** $-\infty$ (or "does not exist")

7. 5 **9.** Does not exist   **11.** $\frac{1}{3}$ **13.** 0

15. 1 **17.** $\frac{1}{2}$ **19.** 1 **21.** $\frac{1}{3}$

23. $\frac{1}{4}$ **25.** $\frac{1}{4}$ **27.** 0 **29.** 3

31. Does not exist   **33.** 0 **35.** 0

37. $+\infty$ (or "does not exist")   **39.** -1 **41.** 1

43. -1 **45.** 2 **47.** -1

49. $f(x) \to +\infty$ as $x \to 1^+$, $f(x) \to -\infty$ as $x \to 1^-$.

51. $f(x) \to +\infty$ as $x \to -1^-$, $f(x) \to -\infty$ as $x \to -1^+$.

53. $f(x) \to +\infty$ as $x \to -2^-$, $f(x) \to -\infty$ as $x \to -2^+$.

55. $f(x) \to +\infty$ as $x \to 1$.

57. $f(x) \to -\infty$ as $x \to -2^+$, $f(x) \to +\infty$ as $x \to -2^-$, and $f(x) \to -\frac{1}{4}$ as $x \to 2$.

59. If n is an integer, then $f(x) \to 2$ as $x \to n$.

61. If n is an even integer, then the right-hand limit at n is 1 and the left-hand limit at n is -1. If n is an odd integer, then the right-hand limit at n is -1 and the left-hand limit at n is 1.

63. The right-hand limit at the integer n is 1; the left-hand limit at n is zero.

Section 2.4 (page 90)

1. $f(g(x)) = -4x^2 - 12x - 8$; $g(f(x)) = -2x^2 + 5$

3. $f(g(x)) = \sqrt{x^4 + 6x^2 + 6}$, $g(f(x)) = x^2$

5. $f(g(x)) = g(f(x)) = x$

7. $f(g(x)) = \sin x^3$; $g(f(x)) = \sin^3 x$

9. $f(g(x)) = 1 + \tan^2 x$; $g(f(x)) = \tan(1 + x^2)$

Problems 11 through 19 have many correct answers. We give only the most natural.

11. $k = 2$, $g(x) = 2 + 3x$ **13.** $k = \frac{1}{2}$, $g(x) = 2x - x^2$

15. $k = \frac{3}{2}$, $g(x) = 5 - x^2$ **17.** $k = -1$, $g(x) = x + 1$

19. $k = -\frac{1}{2}$, $g(x) = x + 10$

21. R **23.** $(-\infty, -3) \cup (-3, +\infty)$

25. R **27.** $(-\infty, 5) \cup (5, +\infty)$

29. $(-\infty, 2) \cup (2, +\infty)$ **31.** $(-\infty, 1) \cup (1, +\infty)$

33. $(-\infty, 0) \cup (0, 1) \cup (1, +\infty)$ **35.** $(-2, 2)$

37. $(-\infty, 0) \cup (0, +\infty)$

39. At every real number other than the integral multiples of $\pi/2$

41. R

43. Nonremovable discontinuity at $x = -3$

45. Removable discontinuity at $x = 2$; let $f(2) = \frac{1}{4}$
Nonremovable discontinuity at $x = -2$

47. Nonremovable discontinuities at $x = \pm 1$

49. Nonremovable discontinuity at $x = 17$

51. Removable discontinuity at $x = 0$; let $f(0) = 0$

59. Given: $f(x) = x^3 - 4x + 1$. Values of $f(x)$:

x	-3	-2	-1	0	1	2	3
$f(x)$	-14	1	4	1	-2	1	16

63. Discontinuous at $x = 3n$ for each integer n

Chapter 2 Miscellaneous Problems   (page 92)

1. 4   **3.** 0   **5.** $-\frac{5}{3}$ **7.** -2 **9.** 0

11. 4   **13.** 8   **15.** $\frac{1}{6}$ **17.** $-\frac{1}{54}$ **19.** -1

21. 1   **23.** Does not exist   **25.** $+\infty$ (or "does

not exist")   **27.** $+\infty$ **29.** $-\infty$ **31.** 3   **33.** $\frac{3}{2}$

35. 0   **37.** $\frac{9}{4}$ **39.** 2

41. $f'(x) = 4x$; $y - 5 = 4(x - 1)$

43. $f'(x) = 6x + 4$; $y - 2 = 10(x - 1)$

45. $f'(x) = 4x - 3$; $y = x - 1$

47. $4x + 3$ **49.** $\dfrac{1}{(3 - x)^2}$ **51.** $1 + \dfrac{1}{x^2}$

53. $-\dfrac{2}{(x - 1)^2}$ **55.** $a = 3 \pm \sqrt{5}$

57. $g(x) = x - x^2$ or $g(x) = x^2 - x$

59. $g(x) = x + 1$ **61.** $g(x) = \sqrt{x^4 + 1}$

63. Nonremovable discontinuity at $x = -1$; removable discontinuity at $x = 1$; let $f(1) = \frac{1}{2}$.

65. Nonremovable discontinuity at $x = -3$; removable discontinuity at $x = 1$; let $f(1) = \frac{3}{4}$.

Section 3.1   (page 105)

1. $f'(x) = 4$ **3.** $h'(z) = 25 - 2z$

5. $\dfrac{dy}{dx} = 4x + 3$ **7.** $\dfrac{dz}{du} = 10u - 3$

9. $\dfrac{dx}{dy} = -10y + 17$ **11.** $f'(x) = 2$

13. $f'(x) = 2x$ **15.** $f'(x) = \dfrac{-2}{(2x + 1)^2}$

17. $f'(x) = \dfrac{1}{\sqrt{2x + 1}}$ **19.** $f'(x) = \dfrac{1}{(1 - 2x)^2}$

21. $x(0) = 100$

23. $x(2.5) = 99$ **25.** $x(-2) = 120$

27. $y(2) = 64$ (ft)   **29.** $y(3) = 194$ (ft)

31. $\dfrac{dA}{dC} = \dfrac{C}{2\pi}$ **33.** 500 ft; 10 s

35. (a) 2.5 months; (b) 50 chipmunks per month

37. *Very* roughly: $v(20) = 50$ mi/h; $v(40) = 62$ mi/h

41. $V'(30) = -\dfrac{25\pi}{12}$ in.³/s; that is, air is leaking out at about 6.545 in.³/s then.

43. (a) $V'(6) = -144\pi$ cm³/h; (b) -156π cm³/h

45. At $t = 2$ s, $v = 0$ m/s

Section 3.2   (page 115)

1. $f'(x) = 6x - 1$

3. $f'(x) = 2(3x - 2) + 3(2x + 3)$

5. $h'(x) = 3x^2 + 6x + 3$ **7.** $f'(y) = 12y^2 - 1$

9. $g'(x) = \dfrac{1}{(x - 1)^2} - \dfrac{1}{(x + 1)^2}$

11. $h'(x) = -\dfrac{3(2x + 1)}{(x^2 + x + 1)^2}$

13. $g'(t) = (t^2 + 1)(3t^2 + 2t) + (t^3 + t^2 + 1)(2t)$

15. $g'(z) = -\dfrac{1}{2z^2} + \dfrac{2}{3z^3}$

17. $g'(y) = 30y^4 + 48y^3 + 48y^2 - 8y - 6$

19. $g'(t) = \dfrac{3 - t}{(t + 1)^3}$ **21.** $v'(t) = -\dfrac{3}{(t - 1)^4}$

23. $g'(x) = -\dfrac{6x^3 + 15}{(x^3 + 7x - 5)^2}$

25. $g'(x) = \dfrac{4x^3 - 13x^2 + 12x}{(2x - 3)^2}$

27. $\dfrac{dy}{dx} = 3x^2 - 30x^4 - 6x^{-5}$

29. $\dfrac{dy}{dx} = \dfrac{2x^5 + 4x^2 - 15}{x^4}$

31. $\dfrac{dy}{dx} = 3 + \dfrac{1}{2x^3}$ **33.** $\dfrac{dy}{dx} = \dfrac{2x - 1 - 4x^2}{3x^2(x - 1)^2}$

35. $\dfrac{dy}{dx} = \dfrac{x^4 + 31x^2 - 10x - 36}{(x^2 + 9)^2}$

37. $\dfrac{dy}{dx} = \dfrac{30x^5(5x^5 - 8)}{(15x^5 - 4)^2}$ **39.** $\dfrac{dy}{dx} = \dfrac{x^2 + 2x}{(x + 1)^2}$

41. $12x - y = 16$ **43.** $x + y = 3$

45. $5x - y = 10$ **47.** $18x - y = -25$

49. $3x + y = 0$

51. (a) It contracts; (b) -0.06427 cm³/°C

53. $14{,}400\pi \approx 45{,}239$ cm³/cm

55. $y = 3x + 2$

57. Suppose that some line is tangent at both (a, a^2) and (b, b^2). Use the derivative to show that $a = b$.

59. $x = \dfrac{n-1}{n}x_0$

65. $g'(x) = 17(x^3 - 17x + 35)^{16}(3x^2 - 17)$

71. $0, \pm\sqrt{3}$

Section 3.3 (page 124)

1. $\dfrac{dy}{dx} = 15(3x + 4)^4$ **3.** $\dfrac{dy}{dx} = -3(3x - 2)^{-2}$

5. $\dfrac{dy}{dx} = 3(x^2 + 3x + 4)^2(2x + 3)$

7. $\dfrac{dy}{dx} = -4(2 - x)^3(3 + x)^7 + 7(2 - x)^4(3 + x)^6$

9. $\dfrac{dy}{dx} = -\dfrac{6x + 22}{(3x - 4)^4}$

11. $\dfrac{dy}{dx} = 12[1 + (1 + x)^3]^3(1 + x)^2$

13. $\dfrac{dy}{dx} = -\dfrac{6}{x^3}\left(\dfrac{1}{x^2} + 1\right)^2$

15. $\dfrac{dy}{dx} = 48[1 + (4x - 1)^4]^2(4x - 1)^3$

17. $\dfrac{dy}{dx} = \dfrac{12(1 - x^{-4})^2}{x^9} - \dfrac{4(1 - x^{-4})^3}{x^5}$

$\qquad = -\dfrac{4(x^{12} - 6x^8 + 9x^4 - 4)}{x^{17}}$

19. $\dfrac{dy}{dx} = -4x^{-5}(x^{-2} - x^{-8})^3$

$\qquad + 3x^{-4}(8x^{-9} - 2x^{-3})(x^{-2} - x^{-8})^2$

$\qquad = -\dfrac{2(x^3 - 1)^2(x^3 + 1)^2(5x^6 - 14)}{x^{29}}$

21. $u(x) = 2x - x^2, n = 3;$
$\quad f'(x) = 3(2x - x^2)^2(2 - 2x)$

23. $u(x) = 1 - x^2, n = -4; f'(x) = 8x(1 - x^2)^{-5}$

25. $u(x) = \dfrac{x + 1}{x - 1}, n = 7; f'(x) = -\dfrac{14(x + 1)^6}{(x - 1)^8}$

27. $g'(y) = 1 + 10(2y - 3)^4$

29. $F'(s) = 3(s - s^{-2})^2(1 + 2s^{-3})$

31. $f'(u) = 8u(u + 1)^3(u^2 + 1)^3 + 3(u + 1)^2(u^2 + 1)^4$

33. $h'(v) = 2(v - 1)(v^2 - 2v + 2)(v^{-3})(2 - v)^{-3}$

35. $F'(z) = 10(4 - 25z^4)(3 - 4z + 5z^5)^{-11}$

37. $\dfrac{dy}{dx} = 4(x^3)^3 \cdot 3x^2 = 12x^{11}$

39. $\dfrac{dy}{dx} = 2(x^2 - 1)(2x) = 4x^3 - 4x$

41. $\dfrac{dy}{dx} = 4(x + 1)^3 = 4x^3 + 12x^2 + 12x + 4$

43. $\dfrac{dy}{dx} = -\dfrac{2x}{(x^2 + 1)^2}$ **45.** $f'(x) = 3x^2 \cos x^3$

47. $g'(z) = 6(\sin 2z)^2 \cos 2z$ **49.** 40π in.²/s

51. 40 in.²/s **53.** 600 in.³/h **55.** -18

57. $400\pi \approx 1256.64$ cm³/s **59.** 5 cm

61. Total melting time: $2/(2 - 4^{1/3}) \approx 4.85$ h; all melted by about 2:50:50 P.M. that day.

Section 3.4 (page 129)

1. $f'(x) = 10x^{3/2} - x^{-3/2}$ **3.** $f'(x) = (2x + 1)^{-1/2}$

5. $f'(x) = -3x^{-3/2} - \frac{3}{2}x^{1/2}$ **7.** $f'(x) = 3(2x + 3)^{1/2}$

9. $f'(x) = 6x(3 - 2x^2)^{-5/2}$ **11.** $f'(x) = \dfrac{3x^2}{2\sqrt{x^3 + 1}}$

13. $f'(x) = 2x(2x^2 + 1)^{-1/2}$

15. $f'(t) = 3t^2(2t^3)^{-1/2} = \frac{3}{2}\sqrt{2t}$

17. $f'(x) = \frac{3}{2}(2x^2 - x + 7)^{1/2}(4x - 1)$

19. $g'(x) = -\frac{4}{3}(x - 2x^3)^{-7/3}(1 - 6x^2)$

21. $f'(x) = (1 - x^2)^{1/2} - x^2(1 - x^2)^{-1/2}$
$\qquad = (1 - 2x^2)(1 - x^2)^{-1/2}$

23. $f'(t) = \dfrac{1}{2}\left(\dfrac{t^2 + 1}{t^2 - 1}\right)^{-1/2} \cdot \dfrac{(t^2 - 1)(2t) - (2t)(t^2 + 1)}{(t^2 - 1)^2}$
$\qquad = -2t(t^2 + 1)^{-1/2}(t^2 - 1)^{-3/2}$

25. $f'(x) = 3\left(x - \dfrac{1}{x}\right)^2\left(1 + \dfrac{1}{x^2}\right)$

27. $f'(v) = -\dfrac{v + 2}{2v^2\sqrt{v + 1}}$

29. $f'(x) = \frac{1}{3}(1 - x^2)^{-2/3}(-2x)$

31. $f'(x) = (3 - 4x)^{1/2} - 2x(3 - 4x)^{-1/2}$

33. $f'(x) = (-2x)(2x + 4)^{4/3} + \frac{8}{3}(1 - x^2)(2x + 4)^{1/3}$

35. $g'(t) = -2t^{-2}(1 + t^{-1})(3t^2 + 1)^{1/2}$
$\qquad + 3t(1 + t^{-1})^2(3t^2 + 1)^{-1/2} = \dfrac{3t^4 - 3t^2 - 2t - 2}{t^3\sqrt{3t^2 + 1}}$

37. $f'(x) = \dfrac{2(3x + 4)^5 - 15(3x + 4)^4(2x - 1)}{(3x + 4)^{10}}$
$\qquad = \dfrac{23 - 24x}{(3x + 4)^6}$

39. $f'(x) =$
$\qquad \dfrac{(3x + 4)^{1/3}(2x + 1)^{-1/2} - (3x + 4)^{-2/3}(2x + 1)^{1/2}}{(3x + 4)^{2/3}}$
$\qquad = \dfrac{x + 3}{(3x + 4)^{4/3}(2x + 1)^{1/2}}$

41. $h'(y) = \dfrac{(1 + y)^{-1/2} - (1 - y)^{-1/2}}{2y^{5/3}}$

$\qquad - \dfrac{5[(1 + y)^{1/2} + (1 - y)^{1/2}]}{3y^{8/3}}$

$\qquad = \dfrac{(7y - 10)\sqrt{1 + y} - (7y + 10)\sqrt{1 - y}}{6y^{8/3}\sqrt{1 - y^2}}$

43. $g'(t) = \frac{1}{2}[t + (t + t^{1/2})^{1/2}]^{-1/2} \times$
$\quad [1 + \frac{1}{2}(t + t^{1/2})^{-1/2}(1 + \frac{1}{2}t^{-1/2})]$

$\qquad = \dfrac{1 + \dfrac{1 + \dfrac{1}{2\sqrt{t}}}{2\sqrt{t + \sqrt{t}}}}{2\sqrt{t + \sqrt{t + \sqrt{t}}}}$

45. No horizontal tangents; vertical tangent line at $(0, 0)$

47. Horizontal tangent where $x = \frac{1}{3}$ and $y = \frac{2}{9}\sqrt{3}$; vertical tangent line at $(0, 0)$

49. No horizontal or vertical tangents

51. $\pi^2/32 \approx 0.3084$ (s/ft)

53. $(2/\sqrt{5}, 1/\sqrt{5})$ and $(-2/\sqrt{5}, -1/\sqrt{5})$

55. $x + 4y = 18$

57. $3x + 2y = 5$ and $3x - 2y = -5$

59. Equation (3) is an *identity,* and if two functions have identical graphs on an interval, then their derivatives are also identically equal on that interval.

Section 3.5 (page 138)

1. Max.: 2; no min.   **3.** No max.; min.: 0

5. Max.: 2; min.: 0   **7.** Max.: 2; min.: 0

9. Max.: $-\frac{1}{6}$; min.: $-\frac{1}{2}$ **11.** Max.: 7; min.: -8

13. Max.: 3; min.: -5 **15.** Max.: 9; min.: 0

17. Max.: 52; min.: -2 **19.** Max.: 5; min.: 4

21. Max.: 5; min.: 1   **23.** Max.: 9; min.: -16

25. Max.: 10; min.: -22 **27.** Max.: 56; min.: -56

29. Max.: 13; min.: 5   **31.** Max.: 17; min.: 0

33. Max.: $\frac{3}{4}$; min.: 0
35. Max.: $\frac{1}{2}$ (at $x = -1$); min.: $-\frac{1}{6}$ (at $x = 3$)
37. Max.: $f(1/\sqrt{2}) = \frac{1}{2}$; min.: $f(-1/\sqrt{2}) = -\frac{1}{2}$

39. Max.: $f(\frac{3}{2}) = 3 \cdot 2^{-4/3}$; min.: $f(3) = -3$
41. Consider the cases $A = 0$ and $A \neq 0$.
47. (c)   **49.** (d)   **51.** (a)

Section 3.6 (page 149)

1. 25 and 25   **3.** 1250   **5.** 500 in.^{3}

7. 1152   **9.** 250   **11.** 11,250 yd^2 **13.** 128

15. Approximately $3.9665°C$ **17.** 1000 cm^3

19. 0.25 m^3 (all cubes, no open-top boxes)

21. Two equal pieces yield minimum total area 200 in.^{2}; no cut yields one square of maximum area 400 in.^{2}

23. 30,000 m^2 **25.** Approximately 9259.26 in.^{3}

27. Five presses

29. The minimizing value of x is $-2 + \frac{10}{3}\sqrt{6}$ in. To the nearest integer, use $x = 6$ in. of insulation for an annual saving of \$285.

31. Either \$1.10 or \$1.15   **33.** Radius $\frac{2}{3}R$, height $\frac{1}{3}H$

35. Let R denote the radius of the circle, and remember that R is constant.

37. $\dfrac{2000\pi\sqrt{3}}{27}$ **39.** Max.: 4, min.: $\sqrt[3]{16}$ **41.** $\frac{1}{2}\sqrt{3}$

43. Each plank has width $\dfrac{-3\sqrt{2} + \sqrt{34}}{8} \approx 0.198539$,

height $\dfrac{\sqrt{7 - \sqrt{17}}}{2} \approx 0.848071$, and area

$\dfrac{\sqrt{7 - \sqrt{17}}\,(-3\sqrt{2} + \sqrt{34})}{4} \approx 0.673500$.

45. $\frac{2}{3}\sqrt{3} \approx 1.1547$ km from the point nearest the island

47. $\frac{1}{3}\sqrt{3}$ **49.** Actual $x \approx 3.45246$.

51. To minimize the sum, choose the radius of the sphere to be $5[10/(\pi + 6)]^{1/2}$ and the edge length of the cube to be $10[10/(\pi + 6)]^{1/2}$. To maximize the sum, choose the edge length of the cube to be zero.

Section 3.7 (page 161)

1. $f'(x) = 6 \sin x \cos x$ **3.** $f'(x) = \cos x - x \sin x$

5. $f'(x) = \dfrac{x \cos x - \sin x}{x^2}$

7. $f'(x) = \cos^3 x - 2 \sin^2 x \cos x$

9. $g'(t) = 4(1 + \sin t)^3 \cos t$

11. $g'(t) = \dfrac{\sin t - \cos t}{(\sin t + \cos t)^2}$

13. $f'(x) = 2 \sin x + 2x \cos x - 6x \cos x + 3x^2 \sin x$

15. $f'(x) = 3 \cos 2x \cos 3x - 2 \sin 2x \sin 3x$

17. $g'(t) = 3t^2 \sin^2 2t + 4t^3 \sin 2t \cos 2t$

19. $g'(t) = -\frac{5}{2}(\cos 3t + \cos 5t)^{3/2}(3 \sin 3t + 5 \sin 5t)$

21. $\dfrac{dy}{dx} = \dfrac{1}{\sqrt{x}}\sin \sqrt{x} \cos \sqrt{x}$

23. $\dfrac{dy}{dx} = 2x \cos(3x^2 - 1) - 6x^3 \sin(3x^2 - 1)$

25. $\dfrac{dy}{dx} = 2 \cos 2x \cos 3x - 3 \sin 2x \sin 3x$

27. $\dfrac{dy}{dx} = -\dfrac{3 \sin 5x \sin 3x + 5 \cos 5x \cos 3x}{\sin^2 5x}$

29. $\dfrac{dy}{dx} = 4x \sin x^2 \cos x^2$

31. $\dfrac{dy}{dx} = \dfrac{\cos 2\sqrt{x}}{\sqrt{x}}$

33. $\dfrac{dy}{dx} = \sin x^2 + 2x^2 \cos x^2$

35. $\dfrac{dy}{dx} = \frac{1}{2}x^{-1/2} \sin x^{1/2} + \frac{1}{2}\cos x^{1/2}$

37. $\dfrac{dy}{dx} = \frac{1}{2}x^{-1/2}(x - \cos x)^3$
$\qquad\qquad + 3x^{1/2}(x - \cos x)^2(1 + \sin x)$

39. $\dfrac{dy}{dx} = -2x[\sin(\sin x^2)]\cos x^2$

41. $\dfrac{dy}{dx} = 7x^6 \sec^2 x^7$ **43.** $\dfrac{dy}{dx} = 7\sec^2 x \tan^6 x$

45. $\dfrac{dy}{dx} = 5x^7 \sec^2 5x + 7x^6 \tan 5x$

47. $\dfrac{dy}{dx} = \dfrac{\sec\sqrt{x} + \sqrt{x}\sec\sqrt{x}\tan\sqrt{x}}{2\sqrt{x}}$

49. $\dfrac{dy}{dx} = \dfrac{2\cot\dfrac{1}{x^2}\csc\dfrac{1}{x^2}}{x^3}$

51. $\dfrac{dy}{dx} = \dfrac{5\tan 3x \sec 5x \tan 5x - 3\sec 5x \sec^2 3x}{\tan^2 3x}$
$\qquad = 5\cot 3x \sec 5x \tan 5x - 3\csc^2 3x \sec 5x$

53. $\dfrac{dy}{dx} = \sec x \csc x + x\sec x \tan x \csc x$
$\qquad\qquad - x\sec x \csc x \cot x$
$\qquad = x\sec^2 x + \sec x \csc x - x\csc^2 x$

55. $\dfrac{dy}{dx} = [\sec(\sin x)\tan(\sin x)]\cos x$

57. $\dfrac{dy}{dx} = \dfrac{\sec x \cos x - \sin x \sec x \tan x}{\sec^2 x}$
$\qquad = \cos^2 x - \sin^2 x$

59. $\dfrac{dy}{dx} = -\dfrac{5\csc^2 5x}{2\sqrt{1 + \cot 5x}}$ **63.** $\pi/4$

65. $\dfrac{\pi}{18}\sec^2 \dfrac{5\pi}{18} \approx 0.4224$ mi/s (about 1521 mi/h)

67. $\dfrac{2000\pi}{27}$ ft/s (about 158.67 mi/h)   **69.** $\pi/3$

71. $\frac{8}{3}\pi R^3$, twice the volume of the sphere!

73. $\dfrac{3\sqrt{3}}{4}$

75. *Suggestion:* $A(\theta) = \dfrac{s^2(\theta - \sin\theta)}{2\theta^2}$

Section 3.8   (page 173)

1. 128   **3.** 64   **5.** 1   **7.** 16

9. 16   **11.** 4   **13.** 3   **15.** 3

17. $3\ln 2$ **19.** $\ln 2 + \ln 3$

21. $3\ln 2 + 2\ln 3$ **23.** $3\ln 2 - 3\ln 3$

25. $3\ln 3 - 3\ln 2 - \ln 5$ **27.** $2^{81} > 2^{12}$

29. There are three solutions of the equation. The one *not* obvious by inspection is approximately -0.7666647.

31. 6   **33.** -2 **35.** 1, 2   **37.** $81 = 3^4$

39. 0   **41.** $(x + 1)e^x$ **43.** $(x^{1/2} + \frac{1}{2}x^{-1/2})e^x$

45. $x^{-3}(x - 2)e^x$ **47.** $1 + \ln x$

49. $x^{-1/2}(1 + \frac{1}{2}\ln x)$ **51.** $(1 - x)e^{-x}$

53. $3/x$ **57.** $f'(x) = \frac{1}{10}e^{x/10}$

61. $P'(t) = 3^t \ln 3$ **63.** $P'(t) = -(2^{-t}\ln 2)$

65. (a) $P'(0) = \ln 3 \approx 1.09861$ (millions per hour); (b) $P'(4) = 3^4 \cdot \ln 3 \approx 88.9876$ (millions per hour)

Section 3.9   (page 180)

1. $\dfrac{dy}{dx} = \dfrac{x}{y}$ **3.** $\dfrac{dy}{dx} = -\dfrac{16x}{25y}$ **5.** $\dfrac{dy}{dx} = -\sqrt{\dfrac{y}{x}}$

7. $\dfrac{dy}{dx} = -\left(\dfrac{y}{x}\right)^{1/3}$ **9.** $\dfrac{dy}{dx} = \dfrac{3x^2 - 2xy - y^2}{3y^2 + 2xy + x^2}$

11. $\dfrac{dy}{dx} = -\dfrac{x}{y}$; $3x - 4y = 25$

13. $\dfrac{dy}{dx} = \dfrac{1 - 2xy}{x^2}$; $3x + 4y = 10$

15. $\dfrac{dy}{dx} = -\dfrac{2xy + y^2}{2xy + x^2}$; $y = -2$

17. $\dfrac{dy}{dx} = \dfrac{25y - 24x}{24y - 25x}$; $4x = 3y$

19. $\dfrac{dy}{dx} = -\dfrac{y^4}{x^4}$; $x + y = 2$

21. $\dfrac{dy}{dx} = \dfrac{5x^4 y^2 - y^3}{3xy^2 - 2x^5 y}$,   so if $y \neq 0$,
$\dfrac{dy}{dx} = \dfrac{5x^4 y - y^2}{3xy - 2x^5}$; the slope at $(1, 2)$ is $\frac{3}{2}$.

23. $\dfrac{dy}{dx} = \dfrac{y - x^2}{y^2 - x}$, but there are no horizonal tangents (see Problem 63).

25. $\left(2, 2 \pm 2\sqrt{2}\right)$ **27.** $y = 2(x - 3)$, $y = 2(x + 3)$

29. Horizontal tangents at all four points where $|x| = \frac{1}{4}\sqrt{6}$ and $|y| = \frac{1}{4}\sqrt{2}$; vertical tangents at the two points $(-1, 0)$ and $(1, 0)$

31. $\dfrac{4}{5\pi} \approx 0.25645$ ft/s   **33.** $\dfrac{32\pi}{125} \approx 0.80425$ m/h

35. 20 cm²/s   **37.** 0.25 cm/s   **39.** 6 ft/s

41. 384 mi/h

43. (a) About 0.047 ft/min; (b) about 0.083 ft/min

45. $\frac{400}{9} \approx 44.44$ ft/s   **47.** Increasing at 16π cm³/s

49. 6000 mi/h

51. (a) $\frac{11}{15}\sqrt{21} \approx 3.36$ ft/s downward; (b) $\frac{242}{15}\sqrt{119}$ ft/s downward

53. At $t = 12$ min; $32\sqrt{13} \approx 115.38$ mi.

55. $-\dfrac{50}{81\pi} \approx -0.1965$ ft/s

57. $-\dfrac{10}{81\pi} \approx -0.0393$ in./min

59. $300\sqrt{2} \approx 424.26$ mi/h   **61.** $\frac{1}{30}$ ft/s

63. $x^3 + y^3 - 3xy + 1$
$= \frac{1}{2}(x + y + 1)[(x - y)^2 + (x - 1)^2 + (y - 1)^2]$.

Section 3.10   (page 192)

Note: In this section your results may differ from these answers in the last one or two decimal places due to differences in calculators or in methods of solving the equations.

1. 2.2361   **3.** 2.5119   **5.** 0.3028   **7.** −0.7402

9. 0.7391   **11.** 1.2361   **13.** 2.3393   **15.** 2.0288

17. 0.5671   **19.** 0.4429   **21.** (b) 1.25992

23. 0.45018   **27.** 0.755 (0.75487766624669276)

29. −1.8955, 0, and 1.8955

31. $dy/dx = e^x \cos x - e^x \sin x$

33. $dy/dx = -3e^x(2 + 3e^x)[1 + (2 + 3e^x)^{-3/2}]^{-1/3}$

35. $dy/dx = -x^{-1} \sin(1 + \ln x)$ **37.** 0.2261

39. $\alpha_1 \approx 2.029 \approx 1.29\,\dfrac{\pi}{2}$, $\alpha_2 \approx 4.913 \approx 3.13\,\dfrac{\pi}{2}$

(also, $\alpha_3 \approx 5.08\,\dfrac{\pi}{2}$, $\alpha_4 \approx 7.06\,\dfrac{\pi}{2}$)

Chapter 3 Miscellaneous Problems   (page 197)

1. $\dfrac{dy}{dx} = 2x - \dfrac{6}{x^3}$ **3.** $\dfrac{dy}{dx} = \dfrac{1}{2\sqrt{x}} - \dfrac{1}{3x^{4/3}}$

5. $\dfrac{dy}{dx} = 7(x - 1)^6(3x + 2)^9 + 27(x - 1)^7(3x + 2)^8$

7. $\dfrac{dy}{dx} = 4(3x - \frac{1}{2}x^{-2})^3(3 + x^{-3})$

9. $\dfrac{dy}{dx} = -\dfrac{y}{x} = -\dfrac{9}{x^2}$

11. $\dfrac{dy}{dx} = -\frac{3}{2}(x^3 - x)^{-5/2}(3x^2 - 1)$

13. $\dfrac{dy}{dx} = \dfrac{-2(1 + x^2)^3}{(x^4 + 2x^2 + 2)^2} \cdot \dfrac{-2x}{(1 + x^2)^2}$
$= \dfrac{4x(1 + x^2)}{(x^4 + 2x^2 + 2)^2}$

15. $\dfrac{dy}{dx} = \frac{7}{3}[x^{1/2} + (2x)^{1/3}]^{4/3}[\frac{1}{2}x^{-1/2} + \frac{2}{3}(2x)^{-2/3}]$

17. $\dfrac{dy}{dx} = \dfrac{dy}{du} \cdot \dfrac{du}{dx} = \dfrac{(u - 1) - (u + 1)}{(u - 1)^2} \cdot \dfrac{1}{2}(x + 1)^{-1/2}$
$= -\dfrac{1}{\sqrt{x + 1}\,(\sqrt{x + 1} - 1)^2}$

19. $\dfrac{dy}{dx} = \dfrac{1 - 2xy^2}{2x^2y - 1}$

21. $\dfrac{dy}{dx} = \dfrac{1 + \dfrac{2 + \dfrac{\sqrt{3}}{2\sqrt{x}}}{2\sqrt{\sqrt{3x} + 2x}}}{2\sqrt{x + \sqrt{\sqrt{3x} + 2x}}}$

23. $\dfrac{dy}{dx} = -\left(\dfrac{y}{x}\right)^{2/3}$

25. $\dfrac{dy}{dx} = -\dfrac{18(x^3 + 3x^2 + 3x + 3)^2}{(x + 1)^{10}}$

27. $\dfrac{dy}{dx} = \dfrac{(2\cos x + \cos^2 x + 1)\sin x}{2(1 + \cos x)^2\sqrt{\dfrac{\sin^2 x}{1 + \cos x}}}$

29. $\dfrac{dy}{dx} = -\dfrac{3\cos 2x \cos 3x + 4\sin 2x \sin 3x}{2(\sin 3x)^{3/2}}$

31. $\dfrac{dy}{dx} = (\sin^3 2x)(2)(\cos 3x)(-\sin 3x)(3)$
$\quad + (\cos^2 3x)(3\sin^2 2x)(2\cos 2x)$
$\quad = 6\cos 3x \cos 5x \sin^2 2x$

33. $\dfrac{dy}{dx} = 5\left[\sin^4\!\left(x + \dfrac{1}{x}\right)\right]\left[\cos\!\left(x + \dfrac{1}{x}\right)\right]\left(1 - \dfrac{1}{x^2}\right)$

35. $\dfrac{dy}{dx}$
$= [\cos^2(x^4 + 1)^{1/3}][-\sin(x^4 + 1)^{1/3}](x^4 + 1)^{-2/3}(4x^3)$

37. $x = 1$ **39.** $x = 0$ **41.** 0.5 ft/min

43. $\frac{1}{3}$ **45.** $\frac{1}{4}$ **47.** 0

49. $h'(x) = -x(x^2 + 25)^{-3/2}$ **51.** $h'(x) = \frac{5}{3}(x - 1)^{2/3}$

53. $h'(x) = -2x \sin(x^2 + 1)$

55. $\dfrac{dV}{dS} = \frac{1}{4}\sqrt{\dfrac{S}{\pi}}$

57. $\dfrac{2}{\cos^2 50°} \cdot \dfrac{\pi}{36} \approx 0.4224$ mi/s, about 1521 mi/h

59. R^2

61. Minimum area: $(36\pi V^2)^{1/3}$, obtained by making *one* sphere of radius $(3V/4\pi)^{1/3}$; maximum area: $(72\pi V^2)^{1/3}$, obtained by making two equal spheres of radius $\frac{1}{2}(3V/\pi)^{1/3}$

63. $32\pi R^3/81$ **65.** $M/2$ **67.** 36 ft³   **69.** $3\sqrt{3}$

73. 2 mi from the shore point nearest the first town

75. (a) $\dfrac{m^2v^2}{64(m^2 + 1)}$; (b) when $m = 1$, thus when $\alpha = \pi/4$

77. 2.6458   **79.** 2.3714   **81.** -0.3473

83. 0.7402   **85.** -0.7391 **87.** -1.2361

89. Approximately 1.54785 ft

91. $-2.7225, 0.8013, 2.3100$ **97.** 4 in.²/s

99. $-50/(9\pi) \approx -1.7684$ ft/min   **101.** 1 in./min

Section 4.2   (page 208)

1. $(6x + 8x^{-3})\, dx$ **3.** $[1 + \frac{3}{2}x^2(4 - x^3)^{-1/2}]\, dx$

5. $[6x(x - 3)^{3/2} + \frac{9}{2}x^2(x - 3)^{1/2}]\, dx$

7. $[(x^2 + 25)^{1/4} + \frac{1}{2}x^2(x^2 + 25)^{-3/4}]\, dx$

9. $-\frac{1}{2}x^{-1/2}\sin x^{1/2}\, dx$

11. $(2\cos^2 2x - 2\sin^2 2x)\, dx$

13. $(\frac{2}{3}x^{-1}\cos 2x - \frac{1}{3}x^{-2}\sin 2x)\, dx$

15. $(\sin x + x\cos x)(1 - x\sin x)^{-2}\, dx$

17. $f(x) \approx 1 + x$ **19.** $f(x) \approx 1 + 2x$

21. $f(x) \approx 1 - 3x$ **23.** $f(x) \approx x$

25. $3 - \frac{2}{27} \approx 2.926$ **27.** $2 - \frac{1}{32} \approx 1.969$

29. $\frac{95}{1536} \approx 0.06185$

31. $\dfrac{1 + \dfrac{\pi}{90}}{\sqrt{2}} \approx 0.7318$ **33.** $\sin\dfrac{\pi}{2} - \dfrac{\pi}{90}\cos\dfrac{\pi}{2} = 1.000$

35. $\dfrac{dy}{dx} = -\dfrac{x}{y}$ **37.** $\dfrac{dy}{dx} = \dfrac{y - x^2}{y^2 - x}$

41. -4 in.²   **43.** $-405\pi/2$ cm³

45. 10 ft   **47.** 6 W

49. $25\pi \approx 78.54$ in.³   **51.** $4\pi \approx 12.57$ m²

Section 4.3   (page 218)

1. Increasing for $x < 0$, decreasing for $x > 0$; (c)

3. Decreasing for $x < -2$, increasing for $x > -2$; (f)

5. Increasing for $x < -1$ and for $x > 2$, decreasing on $(-1, 2)$; (d)

7. $f(x) = 2x^2 + 5$ **9.** $f(x) = 2 - \dfrac{1}{x}$

11. Increasing on $\boldsymbol{R}$

13. Increasing for $x < 0$, decreasing for $x > 0$

15. Increasing for $x < \frac{3}{2}$, decreasing for $x > \frac{3}{2}$

17. Increasing on $(-1, 0)$ and for $x > 1$, decreasing for $x < -1$ and on $(0, 1)$

19. Increasing on $(-2, 0)$ and for $x > 1$, decreasing for $x < -2$ and on $(0, 1)$

21. Increasing for $x < 2$, decreasing for $x > 2$

23. Increasing for $x < -\sqrt{3}$, for $-\sqrt{3} < x < 1$, and for $x > 3$; decreasing for $1 < x < \sqrt{3}$ and for $\sqrt{3} < x < 3$

25. $f(0) = 0 = f(2)$, $f'(x) = 2x - 2$; $c = 1$

27. $f(-1) = 0 = f(1)$, $f'(x) = -\dfrac{4x}{(1 + x^2)^2}$; $c = 0$

29. $f'(0)$ does not exist.

31. $f(0) \neq f(1)$ **33.** $c = -\frac{1}{2}$ **35.** $c = \frac{35}{27}$

37. The average slope is $\frac{1}{3}$, but $|f'(x)| = 1$ where $f'(x)$ exists.

39. The average slope is 1, but $f'(x) = 0$ wherever it exists.

41. If $g(x) = x^5 + 2x - 3$, then $g'(x) > 0$ for all x in $[0, 1]$ and $g(1) = 0$. So $x = 1$ is the only root of the equation in the given interval.

43. If $g(x) = x^4 - 3x - 20$, then $g(2) = -10$ and $g(3) = 52$. If x is in $[2, 3]$, then

$$g'(x) = 4x^3 - 3 \geqq 4 \cdot 2^3 - 3 = 29 > 0,$$

so g is an increasing function on $[2, 3]$. Hence $g(x)$ can have at most one zero in $[2, 3]$. It has at least one solution because $g(2) < 0 < g(3)$ and g is continuous.

45. Note that $f'(x) = \frac{3}{2}\left(-1 + \sqrt{x + 1}\right)$.

47. Assume that $f'(x)$ has the form

$$a_0 + a_1 x + \cdots + a_{n-1}x^{n-1}.$$

Construct a polynomial such that $p'(x) = f'(x)$. Conclude that $f(x) = p(x) + C$ on $[a, b]$.

Section 4.4   (page 228)

1. Global min. at $x = 2$

3. Local max. at $x = 0$, local min. at $x = 2$

5. No extremum at $x = 1$

7. Local min. at $x = -2$, local max. at $x = 5$

9. Global min. at $x = \pm 1$, local max. at $x = 0$

11. Local max. at $x = -1$, local min. at $x = 1$

13. Local min. at $x = 1$

15. Global min. at $(0, 0)$, local max. at $(2, 4e^{-2})$

17. Global max. at $(\pi/2, 1)$

19. Global max. at $(\pi/2, 1)$, global min. at $(-\pi/2, -1)$

21. Global min. at $(0, 0)$

23. Global max. at (π, π), global min. at $(-\pi, -\pi)$

25. Global max. at $(e^{1/2}, 1/(2e^2))$ **27.** -10 and 10

29. $(1, 1)$ **31.** 9 in. wide, 18 in. long, 6 in. high

33. Radius $5\pi^{-1/3}$ cm, height $10\pi^{-1/3}$ cm

37. Base 5 in. by 5 in., height 2.5 in.

39. Radius $(25/\pi)^{1/3} \approx 1.9965$ in., height 4 times that radius

41. $(\frac{1}{2}\sqrt{6}, \frac{3}{2})$ and $(-\frac{1}{2}\sqrt{6}, \frac{3}{2})$; $(0, 0)$ is *not* the nearest point.

43. 8 cm

45. $L = \sqrt{20 + 12\sqrt[3]{4} + 24\sqrt[3]{2}} \approx 8.324$ m

49. Height $(6V)^{1/3}$, base edge $(\frac{9}{2}V^2)^{1/6}$

Section 4.5 (page 237)

1. (c)   **3.** (d)

5. Parabola opening upward; global min. at $(1, 2)$

7. Increasing if $|x| > 2$, decreasing on $(-2, 2)$; local max. at $(-2, 16)$, local min. at $(2, -16)$

9. Increasing for $x < 1$ and for $x > 3$, decreasing on $(1, 3)$; local max. at $(1, 4)$, local min. at $(3, 0)$

11. Increasing for all x, no extrema

13. Decreasing for $x < -2$ and on $(-0.5, 1)$, increasing on $(-2, -0.5)$ and for $x > 1$; global min. at $(-2, 0)$ and $(1, 0)$, local max. at $(-0.5, 5.0625)$.

15. Increasing for $0 < x < 1$, decreasing for $x > 1$; global max. at $(1, 2)$, no graph for $x < 0$

17. Increasing for $|x| > 1$, decreasing for $|x| < 1$, but with a horizontal tangent at $(0, 0)$; local max. at $(-1, 2)$, local min. at $(1, -2)$

19. Decreasing for $x < -2$ and on $(0, 2)$, increasing for $x > 2$ and on $(-2, 0)$; global min. at $(-2, -9)$ and at $(2, -9)$, local max. at $(0, 7)$

21. Decreasing for $x < \frac{3}{4}$, increasing for $x > \frac{3}{4}$

23. Decreasing on $(-2, 1)$, increasing for $x < -2$ and for $x > 1$; local max. at $(-2, 20)$, local min. at $(1, -7)$

25. Increasing for $x < 0.6$ and for $x > 0.8$, decreasing on $(0.6, 0.8)$; local max. at $(0.6, 16.2)$, local min. at $(0.8, 16.0)$

27. Increasing for $x > 2$ and on $(-1, 0)$, decreasing for $x < -1$ and on $(0, 2)$; local max. at $(0, 8)$, local min. at $(-1, 3)$, global min. at $(2, -24)$

29. Increasing for $|x| > 2$, decreasing for $|x| < 2$; local max. at $(-2, 64)$, local min. at $(2, -64)$

31. Increasing everywhere, no extrema; graph passes through origin; minimum slope $\frac{9}{2}$ occurs at $x = -\frac{1}{2}$

33. Increasing for $x < -\sqrt{2}$ and on $(0, \sqrt{2})$, decreasing for $x > \sqrt{2}$ and on $(-\sqrt{2}, 0)$; global max. 16 occurs where $x = \pm\sqrt{2}$, local min. at $(0, 0)$

35. Increasing for $x < 1$, decreasing for $x > 1$; global max. at $(1, 3)$, vertical tangent at $(0, 0)$

37. Decreasing on $(0.6, 1)$, increasing for $x < 0.6$ and for $x > 1$; local min. and a cusp at $(1, 0)$, local max. at $(0.6, 0.3527)$ (ordinate approximate)

39.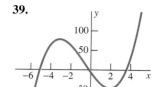

41.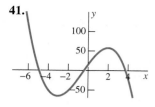

43.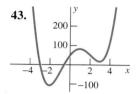

45. (b) $x^3 - 3x + 3 \approx (x + 2.1038)(x^2 - 2.1038x + 1.42599)$; (c) $x \approx 1.0519 \pm 0.5652i$

Section 4.6 (page 250)

1. $8x^3 - 9x^2 + 6$, $24x^2 - 18x$, $48x - 18$

3. $-8(2x - 1)^{-3}$, $48(2x - 1)^{-4}$, $-384(2x - 1)^{-5}$

5. $4(3t - 2)^{1/3}$, $4(3t - 2)^{-2/3}$, $-8(3t - 2)^{-5/3}$

7. $(y + 1)^{-2}$, $-2(y + 1)^{-3}$, $6(y + 1)^{-4}$

9. $g'(t) = t + 2t \ln t$, $g''(t) = 3 + 2 \ln t$, $g'''(t) = 2/t$

11. $3 \cos 3x$, $-9 \sin 3x$, $-27 \cos 3x$

13. $\cos^2 x - \sin^2 x$, $-4 \sin x \cos x$, $4 \sin^2 x - 4 \cos^2 x$

15. $\dfrac{x \cos x - \sin x}{x^2}$, $\dfrac{2 \sin x - 2x \cos x - x^2 \sin x}{x^3}$, $\dfrac{3x^2 \sin x - x^3 \cos x + 6x \cos x - 6 \sin x}{x^4}$

17. $-\dfrac{2x + y}{x + 2y}$, $-\dfrac{18}{(x + 2y)^3}$

19. $-\dfrac{1 + 2x}{3y^2}$, $\dfrac{2y^3 - 42}{9y^5}$

21. $\dfrac{y}{\cos y - x}$, $\dfrac{2y \cos y - 2xy + y^2 \sin y}{(\cos y - x)^3}$

23. $(-3, 81)$, $(5, -175)$, $(1, -47)$

25. $(\frac{9}{2}, -\frac{941}{2})$, $(-\frac{7}{2}, \frac{1107}{2})$, $(\frac{1}{2}, \frac{83}{2})$

27. $(3\sqrt{3}, -492)$, $(-3\sqrt{3}, -492)$, $(0, 237)$, $(3, -168)$, $(-3, -168)$

29. $(\frac{16}{3}, -\frac{181,144}{81})$, $(0, 1000)$, $(4, -1048)$

31. Local min. at $(2, -1)$; no inflection points

33. Local max. at $(-1, 3)$, local min. at $(1, -1)$; inflection point at $(0, 1)$

35. No local extrema; inflection point at $(0, 0)$

37. No local extrema; inflection point at $(0, 0)$

39. Local max. at $(\frac{1}{2}, \frac{1}{16})$, local min. at $(0, 0)$ and $(1, 0)$; the abscissas of the inflection points are the roots of $6x^2 - 6x + 1 = 0$, and the inflection points are at (approximately) $(0.79, 0.03)$ and $(1.21, 0.03)$.

41. Global max. at $(\pi/2, 1)$, global min. at $(3\pi/2, -1)$; inflection point at $(\pi, 0)$

43. Inflection point at $(0, 0)$

45. Global max. at $(0, 1)$ and $(\pi, 1)$, global min. at $(\pi/2, 0)$; inflection points at $(-\pi/4, \frac{1}{2})$, $(\pi/4, \frac{1}{2})$, $(3\pi/4, \frac{1}{2})$, and $(5\pi/4, \frac{1}{2})$

47. Local max. at $(1, e^{-1})$, inflection point at $(2, 2e^{-2})$

49. Local max. at (e, e^{-1}), inflection point at $(e^{3/2}, \frac{3}{2}e^{-3/2})$

63. Increasing for $x < -1$ and for $x > 2$, decreasing on $(-1, 2)$, local max. at $(-1, 10)$, local min. at $(2, -17)$, inflection point at $(\frac{1}{2}, -\frac{7}{2})$

65. Increasing for $x < -2$ and on $(0, 2)$, decreasing for $x > 2$ and on $(-2, 0)$, global max. at $(\pm 2, 22)$, local min. at $(0, 6)$, inflection points where $x^2 = \frac{4}{3}$ (and $y = \frac{134}{9}$)

67. Decreasing for $x < -1$ and on $(0, 2)$, increasing for $x > 2$ and on $(-1, 0)$, local min. at $(-1, -6)$, local max. at $(0, -1)$, global min. at $(2, -33)$, inflection points with approximate coordinates $(1.22, -19.36)$ and $(-0.55, -3.68)$

69. Local max. at $(\frac{3}{7}, \frac{6912}{823,543}) \approx (0.43, 0.0084)$, local min. at $(1, 0)$, inflection points at $(0, 0)$ and at the two solutions of $7x^2 - 6x + 1 = 0$: approximately $(0.22, 0.0042)$ and $(0.63, 0.0047)$. See graph.

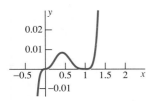

71. The graph is increasing everywhere, with a vertical tangent and inflection point at $(0, 1)$; the other intercept is $(-1, 0)$, and there are no extrema.

73. Global min. at $(0, 0)$, increasing for all $x > 0$, inflection point at $(1, 4)$, concave up for $x > 1$, vertical tangent at the origin

75. Increasing for $x < 1$, decreasing for $x > 1$, vertical tangent and inflection point at $(0, 0)$, another inflection point at $(-2, -7.56)$ (ordinate approximate), global max. at $(1, 3)$

77. (c)   **79.** (b)   **81.** (d)

89. $a = 3pV^2 \approx 3,583,858.8$, $b = \dfrac{V}{3} = 42.7$,

$R = \dfrac{8pV}{3T} \approx 81.80421$

1. 1   **3.** 3   **5.** 2   **7.** 1   **9.** 4

11. 0   **13.** 2

15. $+\infty$ (or "does not exist")   **17.** (g)

19. (a)   **21.** (f)   **23.** (j)   **25.** (l)   **27.** (k)

29. No critical points or inflection points, vertical asymptote $x = 3$, horizontal asymptote $y = 0$, sole intercept at $(0, -\frac{2}{3})$

31. No critical points or inflection points, vertical asymptote $x = -2$, horizontal asymptote $y = 0$, sole intercept at $(0, -\frac{3}{4})$

33. No critical points or inflection points, vertical asymptote $x = \frac{3}{2}$, horizontal asymptote $y = 0$

35. Global min. at $(0, 0)$, inflection points where $3x^2 = 1$ (and $y = \frac{1}{4}$), horizontal asymptote $y = 1$

37. Local max. at $(0, -\frac{1}{9})$, no inflection points, vertical asymptotes $x = \pm 3$, horizontal asymptote $y = 0$

39. Local max. at $(-\frac{1}{2}, -\frac{4}{25})$, horizontal asymptote $y = 0$, vertical asymptotes $x = -3$ and $x = 2$, no inflection points

41. Local min. at $(1, 2)$, local max. at $(-1, -2)$, no inflection points, vertical asymptote $x = 0$; the line $y = x$ is also an asymptote.

43. Local min. at $(2, 4)$, local max. at $(0, 0)$, no inflection points, asymptotes $x = 1$ and $y = x + 1$

45.   **47.**

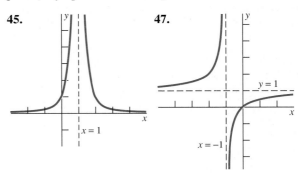

49.   **51.**

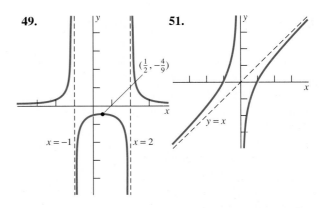

53.

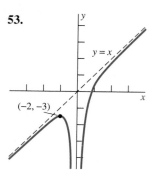

$y = x$

$(-2, -3)$

55. Local min. at $(1, 3)$, inflection point at $(\sqrt[3]{-2}, 0)$, vertical asymptote $x = 0$

Chapter 4 Miscellaneous Problems (page 263)

1. $dy = 3(4x - x^2)^{1/2}(2 - x)\,dx$

3. $dy = -2(x - 1)^{-2}\,dx$

5. $dy = (2x \cos x^{1/2} - \tfrac{1}{2}x^{3/2} \sin x^{1/2})\,dx$

7. $\frac{12,801}{160} = 80.00625$ (true value: about $80.00624\,975588$)

9. $\frac{128,192}{125} = 1025.536$ (true value: about 1025.537037)

11. $\frac{601}{60} \approx 10.016667$ (true value: about 10.016639)

13. 132.5 (true value: about 132.574507)

15. 2.03125 (true value: about $2.03054\,3185$)

17. 7.5 (in.³)   **19.** $10\pi \approx 31.416$ (cm³)

21. $\pi/96 \approx 0.0327$ s   **23.** $c = \sqrt{3}$ **25.** $c = 1$

27. $c = \sqrt[4]{2.2}$

29. Decreasing for $x < 3$, increasing for $x > 3$, global min. at $(3, -5)$; concave upward everywhere

31. Increasing everywhere, no extrema

33. Increasing for $x < \tfrac{1}{4}$, decreasing for $x > \tfrac{1}{4}$, vertical tangent at $(0, 0)$, global max. at $x = \tfrac{1}{4}$

35. $3x^2 - 2, 6x, 6$

37. $-(t^{-2}) + 2(2t + 1)^{-2}, 2t^{-3} - 8(2t + 1)^{-3},$ $-6t^{-4} + 48(2t + 1)^{-4}$

39. $3t^{1/2} - 4t^{1/3}, \tfrac{3}{2}t^{-1/2} - \tfrac{4}{3}t^{-2/3}, -\tfrac{3}{4}t^{-3/2} + \tfrac{8}{9}t^{-5/3}$

41. $-4(t - 2)^{-2}, 8(t - 2)^{-3}, -24(t - 2)^{-4}$

43. $-\tfrac{4}{3}(5 - 4x)^{-2/3}, -\tfrac{32}{9}(5 - 4x)^{-5/3},$ $-\tfrac{640}{27}(5 - 4x)^{-8/3}$

45. $\dfrac{dy}{dx} = -\left(\dfrac{y}{x}\right)^{2/3}, \dfrac{d^2y}{dx^2} = \dfrac{2y^{1/3}}{3x^{5/3}}$

47. $\dfrac{dy}{dx} = \dfrac{1}{2\sqrt{x}(5y^4 - 4)},$

$\dfrac{d^2y}{dx^2} = -\dfrac{20y^3\sqrt{x} + (5y^4 - 4)^2}{4x\sqrt{x}(5y^4 - 4)^3}$

49. $\dfrac{dy}{dx} = \dfrac{5y - 2x}{2y - 5x}$

51. $\dfrac{dy}{dx} = \dfrac{2xy}{3y^2 - x^2 - 1}$

53. Global min. at $(2, -48)$, x-intercepts 0 and (approximately) 3.1748, no inflection points or asymptotes; concave upward everywhere

55. Local max. at $(0, 0)$, global min. where $x^2 = \tfrac{4}{3}$ (and $y = -\tfrac{32}{27}$), inflection points where $x^2 = \tfrac{4}{5}$ (and $y = -\tfrac{96}{125}$), no asymptotes

57. Global max. at $(3, 3)$, inflection points where $x = 6$ and $(4, 0)$ (and a vertical tangent at the latter point), no asymptotes

59. Local max. at $(0, -\tfrac{1}{4})$, no inflection points, horizontal asymptote $y = 1$, vertical asymptotes $x = \pm 2$

61. The inflection point has abscissa the only real solution of $x^3 + 6x^2 + 4 = 0$, approximately -6.10724.

63.   **65.**

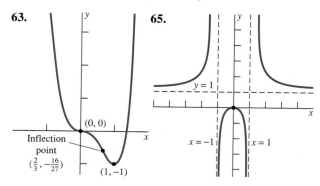

(0, 0)

Inflection point $\left(\tfrac{2}{3}, -\tfrac{16}{27}\right)$ $(1, -1)$

$y = 1$ $x = -1$ $x = 1$

67.   **69.**

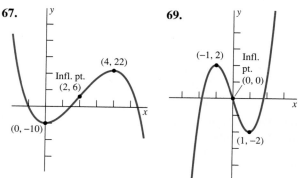

$(4, 22)$

Infl. pt. $(2, 6)$

$(0, -10)$

$(-1, 2)$ Infl. pt. $(0, 0)$

$(1, -2)$

71.

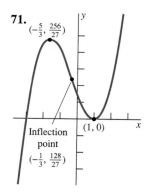

$\left(-\tfrac{5}{3}, \tfrac{256}{27}\right)$

$(1, 0)$

Inflection point $\left(-\tfrac{1}{3}, \tfrac{128}{27}\right)$

73. Maximum value 1 at $x = -1$

75. Base 15×30, height 10

77. Base 5×10, height 8

79. $100 \cdot (\frac{2}{9})^{2/5} \approx 54.79$ mi/h

81. Two horizontal tangents, where $x \approx 0.42$, $y \approx \pm 0.62$; intercepts at $(0, 0)$, $(1, 0)$, and $(2, 0)$; vertical tangent at each intercept; inflection points corresponding to the only positive solution of $3x^2(x - 2)^2 = 4$—that is, $x \approx 2.46789$, $y \approx \pm 1.30191$; no asymptotes

83. 240 ft   **85.** $2\sqrt{2A(n + 2)}$ ft

87. There is neither a maximum nor a minimum.

89. 288 in.^{2} **91.** 270 cm^2

93. In both cases, $m = 1$ and $b = -\frac{2}{3}$.

Section 5.2 (page 278)

1. $x^3 + x^2 + x + C$ **3.** $x - \frac{2}{3}x^3 + \frac{3}{4}x^4 + C$

5. $-\frac{3}{2}x^{-2} + \frac{4}{5}x^{5/2} - x + C$ **7.** $t^{3/2} + 7t + C$

9. $\frac{3}{5}x^{5/3} - 16x^{-1/4} + C$ **11.** $x^4 - 2x^2 + 6x + C$

13. $49e^{x/7} + C$ **15.** $\frac{1}{5}(x + 1)^5 + C$

17. $-\frac{1}{6}(x - 10)^{-6} + C$

19. $\frac{2}{3}x^{3/2} - \frac{4}{5}x^{5/2} + \frac{2}{7}x^{7/2} + C$

21. $\frac{2}{21}x^3 - \frac{3}{14}x^2 - \frac{5}{7}x^{-1} + C$ **23.** $\frac{1}{54}(9t + 11)^6 + C$

25. $\frac{1}{2}e^{2x} - \frac{1}{2}e^{-2x} + C$

27. $\frac{1}{2}\sin 10x + 2\cos 5x + C$

29. $\dfrac{3\sin \pi t}{\pi} + \dfrac{\sin 3\pi t}{3\pi} + C$

33. $\frac{1}{2}(x - \sin x \cos x) + C$; $\frac{1}{2}(x + \sin x \cos x) + C$

35. $y = f(x) = x^2 + x + 3$

37. $y = f(x) = \frac{2}{3}(x^{3/2} - 8)$

39. $y = f(x) = 2\sqrt{x + 2} - 5$

41. $y = f(x) = \frac{3}{4}x^4 - 2x^{-1} + \frac{9}{4}$

43. $y = f(x) = \frac{1}{4}(x - 1)^4 + \frac{7}{4}$

45. $y = f(x) = 2\sqrt{x - 13} - 2$

47. 144 ft; 6 s   **49.** 144 ft   **51.** 5 s; 112 ft/s

53. $\sqrt{60} \approx 7.75$ s; $32\sqrt{60} \approx 247.87$ ft/s

55. 120 ft/s   **57.** 5 s; -160 ft/s   **59.** 400 ft

61. $\frac{1}{4}(-5 + 2\sqrt{145}) \approx 4.77$ s; $16\sqrt{145} \approx 192.6655$ ft/s

63. $\frac{544}{3}$ ft/s   **65.** 22 ft/s^2

67. Approximately 886.154 ft

69. $5\sqrt{210} \approx 72.5$ mi/h

Section 5.3 (page 290)

1. $\displaystyle\sum_{i=1}^{5} i^2$ **3.** $\displaystyle\sum_{n=1}^{5} \frac{1}{n}$ **5.** $\displaystyle\sum_{j=1}^{6} \frac{1}{2^j}$ **7.** $\displaystyle\sum_{k=1}^{5} \left(\frac{2}{3}\right)^k$

9. 190   **11.** 1165   **13.** 224   **15.** 350

17. 338,350   **19.** $\frac{1}{3}$ **21.** n^2 **23.** $\frac{2}{5}, \frac{3}{5}$

25. $\frac{33}{2}, \frac{39}{2}$ **27.** $\frac{6}{25}, \frac{11}{25}$ **29.** $\frac{378}{25}, \frac{513}{25}$ **31.** $\frac{81}{400}, \frac{121}{400}$

35. $\dfrac{n(n + 1)}{2n^2} \to \dfrac{1}{2}$ as $n \to \infty$

37. $\dfrac{81n^2(n + 1)^2}{4n^4} \to \dfrac{81}{4}$ as $n \to \infty$

39. $5n \cdot \dfrac{1}{n} - \dfrac{3n(n + 1)}{2n^2} \to \dfrac{7}{2}$ as $n \to \infty$

Section 5.4 (page 298)

1. $\displaystyle\int_1^3 (2x - 1)\, dx$ **3.** $\displaystyle\int_0^{10} (x^2 + 4)\, dx$

5. $\displaystyle\int_4^9 \sqrt{x}\, dx$ **7.** $\displaystyle\int_3^8 \frac{1}{\sqrt{1 + x}}\, dx$

9. $\displaystyle\int_0^{1/2} \sin 2\pi x\, dx$ **11.** 0.44   **13.** 1.45

15. 19.5   **17.** 58.8   **19.** $-\pi/6$ **21.** 0.24

23. $\frac{137}{60} \approx 2.28333$ **25.** 16.5   **27.** 26.4   **29.** $\pi/6$

31. 0.33   **33.** $\frac{6086}{3465} \approx 1.75462$ **35.** 18   **37.** 40.575

39. 0   **41.** 1.83753   **43.** $\frac{8}{3}$ **45.** 12   **47.** 30

Section 5.5 (page 307)

1. $\frac{55}{12}$ **3.** $\frac{49}{60}$ **5.** $-\frac{1}{20}$ **7.** $\frac{1}{4}$ **9.** $\frac{16}{3}$ **11.** 24

13. 0   **15.** $\frac{32}{3}$ **17.** 0   **19.** $\frac{93}{5}$ **21.** 0

23. $\frac{28}{3}$ **25.** $\ln 2$ **27.** $\frac{1}{2}e^2 - 2e + \frac{5}{2}$ **29.** $\frac{1}{4}$

31. $\frac{2}{5}$ **33.** $-\frac{1}{3}$ **35.** $4/\pi$ **37.** 0   **39.** $\frac{1}{2}$

41. $\frac{2}{3}$ **43.** $25\pi/4 \approx 19.635$

47. $1000 + \displaystyle\int_0^{30} V'(t)\, dt = 160$ (gal)

49. Let $Q = \left(\dfrac{1}{1.2} + \dfrac{1}{1.4} + \dfrac{1}{1.6} + \dfrac{1}{1.8}\right)(0.2) \approx 0.5456$

and $I = \displaystyle\int_1^2 \frac{1}{x}\, dx$. Then $Q + 0.1 \leq I \leq Q + 0.2$.

Hence $0.64 < I < 0.75$.

Section 5.6 (page 316)

1. $\frac{16}{5}$ **3.** $\frac{26}{3}$ **5.** 0   **7.** $\frac{125}{4}$ **9.** $\frac{14}{9}$ **11.** 0

13. 4   **15.** $\frac{1}{3}$ **17.** $-\frac{22}{81} \approx -0.271605$ **19.** 0

21. $\frac{35}{24} \approx 1.458333$ **23.** $e + e^{-1} - 2$ **25.** 4

27. $\ln 2$ **29.** $\frac{31}{20}$ **31.** $\frac{81}{4} + \frac{81}{4} = \frac{81}{2}$

33. Average height: $\frac{800}{3} \approx 266.67$ ft; average velocity: -80 ft/s

35. $\frac{5000}{3} \approx 1666.67$ (L)   **37.** $\frac{200}{3}$ **39.** $\pi/3$

41. $f'(x) = (x^2 + 1)^{17}$ **43.** $h'(z) = (z - 1)^{1/3}$

45. $f'(x) = -(e^x - e^{-x})$ **47.** $G'(x) = (x + 4)^{1/2}$

49. $G'(x) = (x^3 + 1)^{1/2}$ **51.** $f'(x) = 3\sin 9x^2$

53. $f'(x) = 2x \sin x^2$ **55.** $f'(x) = \dfrac{2x}{x^2 + 1}$

57. $y(x) = \displaystyle\int_1^x \dfrac{1}{t}\, dt$ **59.** $y(x) = 10 + \displaystyle\int_5^x \sqrt{1 + t^2}\, dt$

61. The integral does not exist.

Section 5.7 (page 323)

1. $\frac{1}{7}(x + 1)^7 + C$ **3.** $-\frac{1}{24}(4 - 3x)^8 + C$

5. $\frac{2}{7}\sqrt{7x + 5} + C$ **7.** $-\dfrac{1}{\pi}\cos(\pi x + 1) + C$

9. $\frac{1}{2}\sec 2\theta + C$ **11.** $\frac{1}{3}(x^2 - 1)^{3/2} + C$

13. $-\frac{1}{9}(2 - 3x^2)^{3/2} + C$ **15.** $\frac{1}{6}(x^4 + 1)^{3/2} + C$

17. $\frac{1}{6}\sin 2x^3 + C$ **19.** $-\frac{1}{2}e^{-x^2} + C$

21. $-\frac{1}{4}\cos^4 x + C$ **23.** $\frac{1}{4}\tan^4 \theta + C$

25. $2 \sin \sqrt{x} + C$

27. $\frac{1}{10}(x + 1)^{10} + C = \frac{1}{10}(x^2 + 2x + 1)^5 + C$

29. $\frac{1}{2}\ln(x^2 + 4x + 5) + C$

31. $\frac{5}{72}$ **33.** $\frac{98}{3}$ **35.** $\frac{1192}{15}$ **37.** $\frac{15}{128}$ **39.** $\frac{62}{15}$

41. $e - 1$ **43.** $2e^2 - 2e$ **45.** $\frac{1}{2}x - \frac{1}{4}\sin 2x + C$

47. $\dfrac{\pi}{2}$ **49.** $-x + \tan x + C$

Section 5.8 (page 332)

1. $\frac{2}{3}$ **3.** $\frac{16}{3}$ **5.** $+\frac{1}{4}$ **7.** $\frac{1}{4}$ **9.** $+\frac{16}{3}$ **11.** $\frac{1}{4}$

13. $e + e^{-1} - 2$ **15.** $\frac{1}{2}(1 - e^{-4})$

17. $\frac{32}{3}$ **19.** $\frac{128}{3}\sqrt{2}$ **21.** $\frac{500}{3}$ **23.** $\frac{64}{3}$ **25.** $\frac{500}{3}$

27. $\frac{4}{3}$ **29.** $\frac{1}{12}$ **31.** $\frac{16}{3}$ **33.** $\frac{16}{3}$ **35.** $\frac{27}{2}$

37. $\frac{1}{2}(e^2 - e^{-2})$ **39.** $\frac{1}{2}(e - 1)$ **41.** $\sqrt{2} - 1$

47. $\frac{253}{6}$ **49.** $f(x) = 5x^4$ **51.** $f(x) = \frac{1}{2}x$

Section 5.9 (page 345)

1. $T_4 = 8$; true value: 8   **3.** $T_5 \approx 0.65$; true value: $\frac{2}{3}$

5. $T_3 = 0.98$; true value: 1   **7.** $M_4 = 8$; true value: 8

9. $M_5 \approx 0.67$; true value: $\frac{2}{3}$

11. $M_3 \approx 1.01$; true value: 1

13. $T_4 = 8.75$; $S_4 \approx 8.6667$; true value: $\frac{26}{3}$

15. $T_4 \approx 0.0973$; $S_4 \approx 0.0940$; true value: 0.09375

17. $T_6 \approx 3.2599$; $S_6 \approx 3.2411$; true value: approximately 3.2413

19. $T_8 \approx 5.0140$, $S_8 \approx 5.0197$

21. (a) 3.0200; (b) 3.0717   **23.** (a) 2519; (b) 2521

25. (a) 2.09; (b) 2.15   **27.** 19

Chapter 5 Miscellaneous Problems (page 349)

1. $\frac{1}{3}x^3 + 2x^{-1} - \frac{5}{2}x^{-2} + C$

3. $-\frac{1}{30}(1 - 3x)^{10} + C$ **5.** $\frac{3}{16}(4x + 9)^{4/3} + C$

7. $\frac{1}{24}(x^4 + 1)^6 + C$ **9.** $-\frac{3}{8}(1 - x^2)^{4/3} + C$

11. $\frac{7}{5}\sin 5x + \frac{5}{7}\cos 7x + C$

13. $\frac{1}{6}(1 + x^4)^{3/2} + C$ **15.** $C - \dfrac{2}{1 + \sqrt{x}}$

17. $\frac{1}{12}\sin 4x^3 + C$ **19.** $\frac{1}{30}(x^2 + 1)^{15} + C$

21. $\frac{2}{5}(4 - x)^{5/2} - \frac{8}{3}(4 - x)^{3/2} + C$

23. $\sqrt{1 + x^4} + C$ **25.** $y(x) = x^3 + x^2 + 5$

27. $y(x) = \frac{1}{12}(2x + 1)^6 + \frac{23}{12}$ **29.** $y(x) = \frac{3}{2}x^{2/3} - \frac{1}{2}$

31. 6 s; 396 ft   **33.** 120 ft/s

35. Impact time: $t = 2\sqrt{10} \approx 6.32$ s; impact speed: $20\sqrt{10} \approx 63.25$ ft/s

37. 176 ft   **39.** 1700   **41.** 2845

43. $2(\sqrt{2} - 1) \approx 0.82843$

45. $\frac{2}{3}\pi (2\sqrt{2} - 1) \approx 3.82945$

47. Show that every Riemann sum is equal to $c(b - a)$.

51. $\dfrac{2x\sqrt{2x}}{3} + \dfrac{2}{\sqrt{3x}} + C$ **53.** $-2x^{-1} - \frac{1}{4}x^2 + C$

55. $\frac{2}{3}\sin x^{3/2} + C$ **57.** $\cos(t^{-1}) + C$

59. $-\frac{3}{8}(1 + u^{4/3})^{-2} + C$ **61.** $\frac{38}{3}$

63. $\frac{1}{3}x^{-3}(4x^2 - 1)^{3/2} + C$ (use $u = 1/x$)

65. $\frac{1}{30}$ **67.** $\frac{44}{15}$ **69.** $\frac{125}{6}$

71. Semicircle, center $(1, 0)$, radius 1: area $\pi/2$

73. $f(x) = \sqrt{4x^2 - 1}$

75. Use $n \geq 9$. $L_{10} \approx 1.12767$, $R_{10} \approx 1.16909$. The integrand is increasing on $[0, 1]$, so the average of these approximations yields 1.1483 ± 0.05.

77. $M_5 \approx 0.28667$, $T_5 \approx 0.28971$. They bound the true value of the integral because the second derivative of the integrand is positive on $[1, 2]$.

Section 6.1 (page 359)

1. 1   **3.** $2/\pi$ **5.** $\frac{98}{3}$ **7.** 4   **9.** -1

11. $\displaystyle\int_1^4 2\pi x f(x)\, dx$ **13.** $\displaystyle\int_0^{10} \sqrt{1 + [f(x)]^2}\, dx$

15. 1000   **17.** $\frac{500}{3}$ **19.** -320; 320

21. -50; 106.25   **23.** 65; 97   **25.** 1; 1

27. Both answers: $10(1 - e^{-100}) \approx 10$

31. $625\pi/2$ **33.** 550 gal

35. 385,000   **37.** 109.5 in.   **39.** $\frac{3}{4}$

41. $f(x) = 400\pi x^2(1 + x)$; $700\pi/3$ lb

Section 6.2 (page 368)

1. $\pi/5$ **3.** 8π **5.** $\pi^2/2$ **7.** $3\pi/10$

9. $2\pi \ln 2$ **11.** $16\pi/15$ **13.** $\pi/2$ **15.** 8π

17. $121\pi/210$ **19.** 8π **21.** $49\pi/30$

23. $17\pi/10$ **25.** 9π **27.** $\frac{4}{3}\pi a^2 b$ **29.** $\frac{16}{3}a^3$
31. $\frac{4}{3}a^3\sqrt{3}$ **37.** $\frac{16}{3}a^3$ **43.** (a) \$3000; (b) \$3000

Section 6.3 (page 377)

1. 8π **3.** $625\pi/2$ **5.** 16π **7.** π **9.** $6\pi/5$
11. $256\pi/15$ **13.** $4\pi/15$ **15.** $11\pi/15$
17. $56\pi/5$ **19.** $8\pi/3$ **21.** $2\pi/15$ **23.** $\pi/2$
25. $16\pi/3$ **27.** $2\pi \ln 2$ **31.** $\frac{4}{3}\pi a^2 b$
33. $V = 2\pi^2 a^2 b$ **35.** $V = 2\pi^2 a^3$ **37.** (a) $V = \frac{1}{6}\pi h^3$

Section 6.4 (page 386)

Note: In 1 through 19, the integrand is given, followed by the interval of integration.

1. $\sqrt{1 + 4x^2}$; $[0, 1]$
3. $\sqrt{1 + 36(x^4 - 2x^3 + x^2)}$; $[0, 2]$
5. $\sqrt{1 + 4x^2}$; $[0, 100]$ **7.** $\sqrt{1 + 16y^6}$; $[-1, 2]$
9. $\dfrac{\sqrt{x^4 + 1}}{x^2}$; $[1, 2]$ **11.** $2\pi x^2\sqrt{1 + 4x^2}$; $[0, 4]$
13. $2\pi(x - x^2)\sqrt{2 - 4x + 4x^2}$; $[0, 1]$
15. $2\pi(2 - x)\sqrt{1 + 4x^2}$; $[0, 1]$
17. $\pi\sqrt{4x + 1}$; $[1, 4]$
19. $\pi(x + 1)\sqrt{4 + 9x}$; $[1, 4]$ **21.** $\frac{22}{3}$ **23.** $\frac{14}{3}$
25. $\frac{123}{32} = 3.84375$
27. $\frac{1}{27}(104\sqrt{13} - 125) \approx 9.25842$
29. $\frac{1}{6}\pi(5\sqrt{5} - 1) \approx 5.3304$ **31.** $\frac{339}{16}\pi \approx 66.5625$
33. $\frac{1}{9}\pi(82\sqrt{82} - 1) \approx 258.8468$ **35.** 4π
37. 3.8194 (true value: approximately 3.820197789)
41. Avoid the problem when $x = 0$ as follows:

$$L = 8\int_{1/(2\sqrt{2})}^{1} x^{-1/3}\,dx = 6.$$

Section 6.5 (page 394)

1. $y(x) = (\frac{1}{2}x^2 + C)^2$ **3.** $3y^{-2} + 2x^3 = C$
5. $y(x) = 1 + (\frac{1}{2}x^2 + C)^2$
7. $3y + 2y^{3/2} = 3x + 2x^{3/2} + C$
9. $y^3 + y = x - x^{-1} + C$ **11.** $y(x) = (1 - x)^{-1}$
13. $y(x) = (x + 1)^{1/4}$ **15.** $y(x) = (\frac{1}{2} - \frac{1}{3}x^{3/2})^{-2}$
17. $x^2 + y^2 = 169$ **19.** $y(x) = (1 + x - x^3)^{-1}$
21. 20 weeks **23.** (a) 169,000; (b) early in 2011
25. $P(t) \to +\infty$ as $t \to 6^-$. **27.** 3 h
29. 1 h 18 min 40 s after the plug is pulled
31. 1:20 P.M. **33.** About 6 min 2.9 s
35. $f(x) = (\pi/86,400)^2 x^4$; the radius of the hole should be $\dfrac{1}{240\sqrt{3}}$ ft, about 0.02887 in.

Section 6.6 (page 403)

1. 30 **3.** 9 **5.** 0 **7.** 15 ft·lb
9. 2.816×10^9 ft·lb (with $R = 4000$ mi, $g = 32$ ft/s²)
11. $13,000\pi \approx 40,841$ ft·lb
13. $125,000\pi/3 \approx 130,900$ ft·lb
15. $156,000\pi \approx 490,088$ ft·lb
17. $4,160,000\pi \approx 13,069,025$ ft·lb
19. 8750 ft·lb **21.** 11,250 ft·lb
23. $25,000 \cdot [1 - (0.1)^{0.4}]$ in.·lb ≈ 1254 ft·lb
25. 16π ft·lb **27.** $1,382,400\pi$ ft·lb
29. About 690.53 ft·lb **31.** 249.6 lb
33. 748.8 lb **35.** 19,500 lb
37. $700\rho/3 \approx 14,560$ lb **39.** About 32,574 tons

Chapter 6 Miscellaneous Problems (page 406)

1. $-\frac{3}{2}$; $\frac{31}{6}$ **3.** 1; 3 **5.** $\frac{14}{3}$ **7.** $2\pi/15$ **9.** 12 in.
11. $41\pi/105$ **13.** $10.625\pi \approx 33.379$ g
19. $f(x) = \sqrt{1 + 3x}$ **21.** $\dfrac{24 - 2\pi^2}{3\pi} \approx 0.4521$
23. $\frac{10}{3}$ **25.** $\frac{63}{8}$ **27.** $52\pi/5$
31. $y(x) = x^2 + \sin x$ **33.** $y(x) = (C - x)^{-1} - 1$
35. $y(x) = (1 - x^3)^{-1}$ **37.** $y(x) = (4 - 3x^{-1})^{1/3}$
39. $y(x) = (1 - \sin x)^{-1}$
41. $(2y^{1/2} - 1)x = (Cx + 2x^{1/2} - 1)y$ **43.** 1 ft
45. $W = 4\pi R^4 \rho$ **47.** 10,454,400 ft·lb
49. 36,400 tons
51. There is no maximum volume; $c = \frac{1}{3}\sqrt{5}$ minimizes the volume V.

Section 7.2 (page 418)

1. $f'(x) = \dfrac{3}{3x - 1}$ **3.** $f'(x) = \dfrac{1}{1 + 2x}$
5. $f'(x) = \dfrac{3x^2 - 1}{3x^3 - 3x}$ **7.** $f'(x) = -\dfrac{1}{x}\sin(\ln x)$
9. $f'(x) = -\dfrac{1}{x(\ln x)^2}$ **11.** $f'(x) = \dfrac{1}{x} + \dfrac{x}{x^2 + 1}$
13. $f'(x) = -\tan x$
15. $f'(t) = 2t\ln(\cos t) - t^2\tan t$
17. $g'(t) = (\ln t)^2 + 2\ln t$
19. $f'(x) = 6(2x + 1)^{-1} + 8x(x^2 - 4)^{-1}$
21. $f'(x) = -x(4 - x^2)^{-1} - x(9 + x^2)^{-1}$
23. $f'(x) = \dfrac{1}{x + 1} - \dfrac{1}{x - 1}$
25. $g'(t) = 2t^{-1} - 2t(t^2 + 1)^{-1}$
27. $f'(x) = -x^{-1} + \cot x$

29. $\dfrac{y \ln y}{y - x}$

31. $\dfrac{y}{-x + \cot y}$

33. $\frac{1}{2} \ln |2x - 1| + C$

35. $\frac{1}{6} \ln(1 + 3x^2) + C$

37. $\frac{1}{4} \ln |2x^2 + 4x + 1| + C$

39. $\frac{1}{3}(\ln x)^3 + C$

41. $\ln |x + 1| + C$

43. $\ln(x^2 + x + 1) + C$

45. $\frac{1}{2}(\ln x)^2 + C$

47. $\frac{1}{2} \ln(1 - \cos 2x) + C$

49. $\frac{1}{3} \ln |x^3 - 3x^2 + 1| + C$

51. 0 **53.** 0 **55.** 0

59. $m \approx -0.2479, k \approx 291.7616$

65. $y \to 0$ as $x \to 0^+$; $dy/dx \to 0$ as $x \to 0^+$. The point $(0, 0)$ is not on the graph. Intercept at $(1, 0)$; global min. where $x = e^{-1/2}$—the coordinates are approximately $(0.61, -0.18)$; inflection point where $x = e^{-3/2}$—the coordinates are approximately $(0.22, -0.07)$.

67. $y \to -\infty$ as $x \to 0^+$; $y \to 0$ as $x \to +\infty$. Global max. at $(e^2, 2/e)$; only intercept is $(1, 0)$; inflection point where $x = e^{8/3}$. The x-axis is a horizontal asymptote, and the y-axis is a vertical asymptote.

71. Midpoint estimate: approximately 872.47; trapezoidal estimate: approximately 872.60; true value of integral: approximately 872.5174.

Section 7.3 (page 425)

1. $f'(x) = 2e^{2x}$

3. $f'(x) = 2xe^{(x^2)} = 2xe^{x^2} = 2x \exp(x^2)$

5. $f(x) = -2x^{-3} \exp(x^{-2})$

7. $g'(t) = \left(1 + \frac{1}{2}\sqrt{t}\right)e^{\sqrt{t}}$

9. $g'(t) = (1 + 2t - t^2)e^{-t}$

11. $g'(t) = -e^{\cos t} \sin t$

13. $f'(x) = -e^{-x} \sin(1 - e^{-x})$

15. $f'(x) = \dfrac{1 - e^{-x}}{x + e^{-x}}$

17. $f'(x) = e^{-2x}(3 \cos 3x - 2 \sin 3x)$

19. $g'(t) = 15(e^t - t^{-1})(e^t - \ln t)^4$

21. $f'(x) = -(5 + 12x)e^{-4x}$

23. $g'(t) = (e^{-t} + te^{-t} - 1)t^{-2}$

25. $f'(x) = (x - 2)e^{-x}$

27. $f'(x) = e^x \exp(e^x) = e^x e^{e^x}$

29. $f'(x) = 2e^x \cos 2e^x$

31. $\dfrac{e^y}{1 - xe^y} = \dfrac{e^y}{1 - y}$

33. $\dfrac{e^x - ye^{xy}}{xe^{xy} - e^y}$

35. $\dfrac{e^{x-y} - y}{e^{x-y} + x} = \dfrac{xy - y}{xy + x}$

37. $-\frac{1}{2}e^{1-2x} + C$

39. $\frac{1}{9} \exp(3x^3 - 1) + C$

41. $\frac{1}{2} \ln(1 + e^{2x}) + C$

43. $\frac{1}{2} \exp(1 - \cos 2x) + C$

45. $\frac{1}{2} \ln(x^2 + e^{2x}) + C$

47. $-\exp(-\frac{1}{2}t^2) + C$

49. $2e^{\sqrt{x}} + C$

51. $\ln(1 + e^x) + C$

53. $-\frac{2}{3} \exp(-x^{3/2}) + C$

55. e^2 **57.** e **59.** $+\infty$ **61.** $+\infty$

63. Global min. and intercept at $(0, 0)$, local max. at $(2, 4e^{-2})$; inflection points where $x = 2 \pm \sqrt{2}$. The x-axis is an asymptote.

65. Global max. at $(0, 1)$, the only intercept; inflection points where $x = \pm\sqrt{2}$. The x-axis is the only asymptote.

67. $\frac{1}{2}\pi(e^2 - 1) \approx 10.0359$

69. $\frac{1}{2}(e - e^{-1}) \approx 1.1752$

71. The solution is approximately 1.278. Note that if $f(x) = e^{-x} - x + 1$, then $f'(x) < 0$ for all x.

73. $f'(x) = 0$ when $x = 0$ and when $x = n$; f is increasing on $(0, n)$ and decreasing for $x > n$. Thus $x = n$ yields the absolute max. value of $f(x)$ for $x \geq 0$. The x-axis is a horizontal asymptote, and there are inflection points where $x = n \pm \sqrt{n}$.

77. $y(x) = e^{-2x} + 4e^x$

Section 7.4 (page 432)

1. $f'(x) = 10^x \ln 10$

3. $f'(x) = 3^x 4^{-x} \ln 3 - 3^x 4^{-x} \ln 4 = \left(\frac{3}{4}\right)^x \ln \left(\frac{3}{4}\right)$

5. $f'(x) = -(7^{\cos x})(\ln 7)(\sin x)$

7. $f'(x) = 2^{x\sqrt{x}} \left(\frac{3}{2} \ln 2\right)\sqrt{x}$

9. $f'(x) = x^{-1} 2^{\ln x} \ln 2$

11. $f'(x) = 17^x \ln 17$

13. $f'(x) = -x^{-2} 10^{1/x} \ln 10$

15. $f'(x) = (2^{2^x} \ln 2)(2^x \ln 2)$

17. $f'(x) = \dfrac{1}{\ln 3} \cdot \dfrac{x}{x^2 + 4}$

19. $f'(x) = \dfrac{\ln 2}{\ln 3} = \log_3 2$

21. $f'(x) = \dfrac{1}{x(\ln 2)(\ln x)}$

23. $f'(x) = \dfrac{\exp(\log_{10} x)}{x \ln 10}$

25. $\dfrac{3^{2x}}{2 \ln 3} + C$

27. $\dfrac{2 \cdot 2^{\sqrt{x}}}{\ln 2} + C$

29. $\dfrac{7^{x^3 + 1}}{3 \ln 7} + C$

31. $\dfrac{(\ln x)^2}{2 \ln 2} + C$

33. $[x(x^2 - 4)^{-1} + (4x + 2)^{-1}](x^2 - 4)^{1/2}(2x + 1)^{1/4}$

35. $2^x \ln 2$

37. $\dfrac{(x^{\ln x})(2 \ln x)}{x}$

39. $\dfrac{y}{3}\left(\dfrac{1}{x + 1} + \dfrac{1}{x + 2} - \dfrac{2x}{x^2 + 1} - \dfrac{2x}{x^2 + 2}\right)$

41. $(\ln x)^{\sqrt{x}}[\frac{1}{2}x^{-1/2} \ln(\ln x) + x^{-1/2}(\ln x)^{-1}]$

43. $\left(\dfrac{3x}{1 + x^2} - \dfrac{4x^2}{1 + x^3}\right)(1 + x^2)^{3/2}(1 + x^3)^{-4/3}$

45. $\left[\dfrac{2x^3}{x^2 + 1} + 2x \ln(x^2 + 1)\right](x^2 + 1)^{(x^2)}$

47. $\frac{1}{4}x^{-1/2}(2 + \ln x)(\sqrt{x})^{\sqrt{x}}$ **49.** e^x

51. $x^{\exp(x)} e^x(x^{-1} + \ln x)$

57. Note that $\ln \dfrac{x^x}{e^x} = x \ln \dfrac{x}{e}$.

59. $\dfrac{dy}{dx} = -\dfrac{\ln 2}{x(\ln x)^2}$

Section 7.5   (page 441)

1. \$119.35; \$396.24   **3.** About 3.8685 h

5. About 686 yr old

7. (a) 9.308%; (b) 9.381%; (c) 9.409%; (d) 9.416%; (e) 9.417%

9. \$44.52   **11.** After an additional 32.26 days

13. About 35 yr   **15.** About 4.2521×10^9 yr old

17. 2.40942 min

19. (a) 20.486 in.; 9.604 in. (b) 3.4524 mi, about 18,230 ft

Section 7.6   (page 447)

1. $y(x) = -1 + 2e^x$ **3.** $y(x) = \frac{1}{2}(e^{2x} + 3)$

5. $x(t) = 1 - e^{2t}$ **7.** $x(t) = 27e^{5t} - 2$

9. $v(t) = 10(1 - e^{-10t})$ **11.** 4,870,238

15. About 46 days after the rumor starts

19. $\dfrac{400}{\ln 2} \approx 577$ ft   **23.** (b) \$1,308,283

25. (a) $x(t) = 100{,}000 - 80{,}000e^{-kt}$, where $k = \frac{1}{14}\ln 2$; (b) on March 29; (c) everyone gets the flu.

Chapter 7 Miscellaneous Problems   (page 449)

1. $f'(x) = \dfrac{1}{2x}$ **3.** $f'(x) = \dfrac{1 - e^x}{x - e^x}$

5. $f'(x) \equiv \ln 2$ **7.** $f'(x) = (2 + 3x^2)e^{-1/x^2}$

9. $f'(x) = \dfrac{1 + \ln \ln x}{x}$ **11.** $f'(x) = x^{-1} 2^{\ln x} \ln 2$

13. $f'(x) = -\dfrac{2}{(x - 1)^2} \exp\!\left(\dfrac{x + 1}{x - 1}\right)$

15. $f'(x) = \dfrac{3}{2}\left(\dfrac{1}{x - 1} + \dfrac{8x}{3 - 4x^2}\right)$

17. $f'(x) = \dfrac{(\sin x \cos x) \exp(\sqrt{1 + \sin^2 x})}{\sqrt{1 + \sin^2 x}}$

19. $f'(x) = \cot x + \ln 3$ **21.** $f'(x) = \dfrac{x^{1/x}(1 - \ln x)}{x^2}$

23. $f'(x) = \left(\dfrac{1 + \ln \ln x}{x}\right)(\ln x)^{\ln x}$

25. $-\frac{1}{2}\ln|1 - 2x| + C$

27. $\frac{1}{2}\ln|1 + 6x - x^2| + C$

29. $-\ln(2 + \cos x) + C$ **31.** $\dfrac{2 \cdot 10^{\sqrt{x}}}{\ln 10} + C$

33. $\frac{2}{3}(1 + e^x)^{3/2} + C$ **35.** $\dfrac{6^x}{\ln 6} + C$

37. $x(t) = t^2 + 17$ **39.** $x(t) = 1 + e^t$

41. $x(t) = \frac{1}{3}(2 + 7e^{3t})$ **43.** $x(t) = \sqrt{2}\,e^{\sin t}$

45. Horizontal asymptote: the x-axis; global max. where $x = \frac{1}{2}$; inflection point where $x = (1 + \sqrt{2})/2$—approximately $(1.21, 0.33)$; global min. and intercept at $(0, 0)$, with a vertical tangent there as well

47. Global min. at $(4, 2 - \ln 4)$, inflection point at $(16, 1.23)$ (ordinate approximate). The y-axis is a vertical asymptote; there is no horizontal asymptote. (The graph continues to rise for large increasing x.)

49. Inflection point at $(0.5, e^{-2})$. The horizontal line $y = 1$ and the x-axis are asymptotes. The point $(0, 0)$ is *not* on the graph. As $x \to 0^+$, $y \to 0$; as $x \to 0^-$, $y \to +\infty$. As $|x| \to +\infty$, $y \to 1$.

51. Sell immediately!

53. (b) The minimizing value is about 10.516. But because the batch size must be an integer, 11 (rather than 10) minimizes $f(x)$. (c) \$977.85

57. 20 weeks

59. (a) \$925.20; (b) \$1262.88

61. About 22.567 h after the power failure; that is, at about 9:34 P.M. the following evening

63. (b) $v(10) = 176(1 - e^{-1}) \approx 111.2532$ ft/s, about 75.85 mi/h. The limiting velocity is $a/\rho = 176$ ft/s, exactly 120 mi/h.

65. (a) The minimum of $f(x) - g(x)$ occurs when $x = 4$ and is $2(1 - \ln 2) > 0$. Hence $f(x) > g(x)$ for all $x > 0$. (b) The (large) solution of $g(x) = h(x)$ is (by Newton's method) approximately 93.354460835. (c) $p = e$

Section 8.2   (page 461)

1. (a) $\pi/6$, (b) $-\pi/6$, (c) $\pi/4$, (d) $-\pi/3$

3. (a) 0, (b) $\pi/4$, (c) $-\pi/4$, (d) $\pi/3$

5. $f'(x) = 100x^{99}(1 - x^{-200})^{-1/2}$

7. $f'(x) = \dfrac{1}{x|\ln x|\sqrt{(\ln x)^2 - 1}}$

9. $f'(x) = \dfrac{\sec^2 x}{\sqrt{1 - \tan^2 x}}$

11. $f'(x) = \dfrac{e^x}{\sqrt{1 - e^{2x}}}$

13. $f'(x) = -\dfrac{2}{\sqrt{1 - x^2}}$

15. $f'(x) = -\dfrac{2}{x\sqrt{x^4 - 1}}$

17. $f'(x) = -\dfrac{1}{(1 + x^2)(\arctan x)^2}$

19. $f'(x) = \dfrac{1}{x[1 + (\ln x)^2]}$

21. $f'(x) = \dfrac{2e^x}{1 + e^{2x}}$

23. $f'(x) = \dfrac{\cos(\arctan x)}{1 + x^2}$

25. $f'(x) = \dfrac{1 - 4x \arctan x}{(1 + x^2)^3}$

27. $\dfrac{dy}{dx} = -\dfrac{1 + y^2}{1 + x^2}; x + y = 2$

29. $\dfrac{dy}{dx} = -\dfrac{\sqrt{1 - y^2}\,\arcsin y}{\sqrt{1 - x^2}\,\arcsin x}; x + y = \sqrt{2}$

31. $\pi/4$ **33.** $\pi/12$ **35.** $\pi/12$

37. $\frac{1}{2}\arcsin 2x + C$ **39.** $\frac{1}{5}\operatorname{arcsec}|x/5| + C$

41. $\arctan(e^x) + C$ **43.** $\frac{1}{15}\operatorname{arcsec}|x^3/5| + C$

45. Both $\arcsin(2x - 1) + C$ and $2 \arcsin x^{1/2} + C$ are correct.

47. $\frac{1}{50}\arctan(x^{50}) + C$ **49.** $\arctan(\ln x) + C$

51. $\pi/4$ **53.** $\pi/2$ **55.** $\pi/12$ **59.** 8 m

63. $\pi/2$ **65.** (b) $A = 1 - \frac{1}{3}\pi, B = 1 + \frac{2}{3}\pi$

Section 8.3 (page 466)

1. $\frac{1}{2}$ **3.** $\frac{2}{5}$ **5.** 0 **7.** 0 **9.** $\frac{1}{2}$

11. 2 **13.** 0 **15.** 1 **17.** 1 **19.** $\frac{3}{5}$

21. $\frac{3}{2}$ **23.** $\frac{1}{3}$ **25.** $\dfrac{\ln 2}{\ln 3}$ **27.** $\frac{1}{2}$ **29.** 1

31. $\frac{1}{3}$ **33.** $-\frac{1}{2}$ **35.** 1 **37.** $\frac{1}{4}$ **39.** $\frac{2}{3}$

41. 6 **43.** $\frac{4}{3}$ **45.** $\frac{2}{3}$ **47.** 0

Section 8.4 (page 471)

1. 1 **3.** $\frac{3}{8}$ **5.** $\frac{1}{4}$ **7.** 1 **9.** 0

11. -1 **13.** $-\infty$ **15.** $-\infty$ **17.** $-\frac{1}{2}$ **19.** 0

21. 1 **23.** 1 **25.** $e^{-1/6}$ **27.** $e^{-1/2}$ **29.** 1

31. e^{-1} **33.** $-\infty$

Section 8.5 (page 480)

1. $f'(x) = 3 \sinh(3x - 2)$

3. $f'(x) = 2x \tanh(1/x) - \operatorname{sech}^2(1/x)$

5. $f'(x) = -12 \coth^2 4x \operatorname{csch}^2 4x$

7. $f'(x) - e^{\operatorname{csch} x} \operatorname{csch} x \coth x$

9. $f'(x) = (\cosh x) \cos(\sinh x)$

11. $f'(x) = 4x^3 \cosh x^4$ **13.** $f'(x) = -\dfrac{1 + \operatorname{sech}^2 x}{(x + \tanh x)^2}$

15. $\frac{1}{2}\cosh x^2 + C$ **17.** $x - \frac{1}{3}\tanh 3x + C$

19. $\frac{1}{6}\sinh^3 2x + C$ **21.** $-\frac{1}{2}\operatorname{sech}^2 x + C$

23. $-\frac{1}{2}\operatorname{csch}^2 x + C$ **25.** $\ln(1 + \cosh x) + C$

27. $\frac{1}{4}\tanh x + C$ **29.** $f'(x) = 2(4x^2 + 1)^{-1/2}$

31. $f'(x) = \frac{1}{2}x^{-1/2}(1 - x)^{-1}$

33. $f'(x) = (x^2 - 1)^{-1/2}$

35. $f'(x) = \frac{3}{2}(\sinh^{-1} x)^{1/2}(x^2 + 1)^{-1/2}$

37. $f'(x) = (1 - x^2)^{-1}(\tanh^{-1} x)^{-1}$

39. $\operatorname{arcsinh}(x/3) + C$ **41.** $\frac{1}{4}\ln \frac{9}{5} \approx 0.14695$

43. $-\frac{1}{2}\operatorname{sech}^{-1}|3x/2| + C$ **45.** $\sinh^{-1}(e^x) + C$

47. $-\operatorname{sech}^{-1}(e^x) + C$ **53.** $\sinh a$

57. $\ln(1 + \sqrt{2}) \approx 0.881373587$

Chapter 8 Miscellaneous Problems (page 482)

1. $f'(x) = 3(1 - 9x^2)^{-1/2}$

3. $g'(t) = 2t^{-1}(t^4 - 1)^{-1/2}$

5. $f'(x) = -(\sin x)(1 - \cos^2 x)^{-1/2} = -\dfrac{\sin x}{|\sin x|}$

7. $g'(t) = 10(100t^2 - 1)^{-1/2}$

9. $f'(x) = -2x^{-1}(x^4 - 1)^{-1/2}$

11. $f'(x) = \frac{1}{2}x^{-1/2}(1 - x)^{-1/2}$

13. $f'(x) = \dfrac{2x}{x^4 + 2x^2 + 2}$

15. $f'(x) = e^x \sinh e^x + e^{2x} \cosh e^x$

17. $f'(x) \equiv 0$

19. $f'(x) = \dfrac{x}{|x|\sqrt{x^2 + 1}}$

21. $\frac{1}{2}\arcsin 2x + C$

23. $\arcsin(x/2) + C$

25. $\sin^{-1}(e^x) + C$

27. $\frac{1}{2}\arcsin(2x/3) + C$

29. $\frac{1}{3}\arctan(x^3) + C$

31. $\sec^{-1}|2x| + C$

33. $\operatorname{arcsec}(e^x) + C$

35. $2 \cosh \sqrt{x} + C$

37. $\frac{1}{2}(\arctan x)^2 + C$

39. $\frac{1}{2}\sinh^{-1}(2x/3) + C$

41. $\frac{1}{4}$ **43.** $\frac{1}{2}$ **45.** $-\frac{1}{6}$ **47.** 1 **49.** $-\infty$

51. $+\infty$ **53.** e^2 **55.** $-\dfrac{e}{2}$ **57.** $\dfrac{\pi^2}{6}$

61. $x \approx 4.730041$

Section 9.2 (page 487)

1. $-\frac{1}{15}(2 - 3x)^5 + C$ **3.** $\frac{1}{9}(2x^3 - 4)^{3/2} + C$

5. $\frac{9}{8}(2x^2 + 3)^{2/3} + C$ **7.** $-2 \csc \sqrt{y} + C$

9. $\frac{1}{6}(1 + \sin \theta)^6 + C$ **11.** $e^{-\cot x} + C$

13. $\frac{1}{11}(\ln t)^{11} + C$ **15.** $\frac{1}{3} \arcsin 3t + C$

17. $\frac{1}{2} \arctan (e^{2x}) + C$ **19.** $\frac{3}{2} \arcsin (x^2) + C$

21. $\frac{1}{15} \tan^5 3x + C$ **23.** $\tan^{-1}(\sin \theta) + C$

25. $\frac{2}{5}(1 + \sqrt{x})^5 + C$ **27.** $\ln |\arctan t| + C$

29. $\sec^{-1} e^x + C$

31. $\frac{2}{7}(x - 2)^{7/2} + \frac{8}{5}(x - 2)^{5/2} + \frac{8}{3}(x - 2)^{3/2} + C$

33. $\frac{1}{3}(2x + 3)^{1/2}(x - 3) + C$

35. $\frac{3}{10}(x + 1)^{2/3}(2x - 3) + C$

37. $\dfrac{1}{60} \ln \left| \dfrac{3x + 10}{3x - 10} \right| + C$

39. $\frac{1}{2}x(4 + 9x^2)^{1/2} + \frac{2}{3} \ln |3x + (4 + 9x^2)^{1/2}| + C$

41. $\frac{1}{32}x(16x^2 + 9)^{1/2} - \frac{9}{128} \ln |4x + (16x^2 + 9)^{1/2}| + C$

43. $\frac{1}{128}x(32x^2 - 25)(25 - 16x^2)^{1/2} + \frac{625}{512} \arcsin \frac{4}{5}x + C$

45. The substitution $u = e^x$ leads to an integral in the form of Formula (44) in the Table of Integrals inside the back cover. The answer is

$$\tfrac{1}{2}e^x(9 + e^{2x})^{1/2} + \tfrac{9}{2} \ln[e^x + (9 + e^{2x})^{1/2}] + C.$$

47. With $u = x^2$ and (47) in the Table of Integrals, $\frac{1}{2}((x^4 - 1)^{1/2} - \operatorname{arcsec} x^2) + C$

49. With $u = \ln x$ and (48) in the Table of Integrals, $\frac{1}{8}\{(\ln x)[2(\ln x)^2 + 1][(\ln x)^2 + 1]^{1/2}$
$- \ln |(\ln x) + [(\ln x)^2 + 1]^{1/2}|\} + C$

53. $\sin^{-1}(x - 1) + C$

Section 9.3   (page 495)

1. $\frac{1}{4}(2x - \sin 2x \ \cos 2x) + C$

3. $2 \tan \dfrac{x}{2} + C$

5. $\frac{1}{3} \ln |\sec 3x| + C$

7. $\frac{1}{3} \ln |\sec 3x + \tan 3x| + C$

9. $\frac{1}{2}(x - \sin x \cos x) + C$

11. $\frac{1}{3} \cos^3 x - \cos x + C$

13. $\frac{1}{3} \sin^3 \theta - \frac{1}{5} \sin^5 \theta + C$

15. $\frac{1}{5} \sin^5 x - \frac{2}{3} \sin^3 x + \sin x + C$

17. $\frac{2}{5}(\cos x)^{5/2} - 2(\cos x)^{1/2} + C$

19. $-\frac{1}{14} \cos^7 2z + \frac{1}{5} \cos^5 2z - \frac{1}{6} \cos^3 2z + C$

21. $\frac{1}{4}(\sec 4x + \cos 4x) + C$

23. $\frac{1}{3} \tan^3 t + \tan t + C$

25. $-\frac{1}{4} \csc^2 2x - \frac{1}{2} \ln |\sin 2x| + C$

27. $\frac{1}{12} \tan^6 2x + C$

29. $-\frac{1}{10} \cot^5 2t - \frac{1}{3} \cot^3 2t - \frac{1}{2} \cot 2t + C$

31. $\frac{1}{4} \cos^4 \theta - \frac{1}{2} \cos^2 \theta + C_1 = \frac{1}{4} \sin^4 \theta + C_2$

33. $\frac{2}{3}(\sec t)^{3/2} + 2(\sec t)^{-1/2} + C$

35. $\frac{1}{3} \sin^3 \theta + C$

37. $\frac{1}{5} \sin 5t - \frac{1}{15} \sin^3 5t + C$

39. $t + \frac{1}{3} \cot 3t - \frac{1}{9} \cot^3 3t + C$

41. $-\frac{1}{5} \cos^{5/2} 2t + \frac{2}{9} \cos^{9/2} 2t - \frac{1}{13} \cos^{13/2} 2t + C$

43. $\frac{1}{2} \sin^2 x - \cos x + C$

45. $-\cot x - \frac{1}{3} \cot^3 x - \frac{1}{2} \csc^2 x + C$

49. $\frac{1}{4} \cos 2x - \frac{1}{16} \cos 8x + C$

51. $\frac{1}{6} \sin 3x + \frac{1}{10} \sin 5x + C$

57. $\pi/4$

59. $2 \ln 2 \approx 1.386294436112$

Section 9.4   (page 501)

1. $\frac{1}{2} xe^{2x} - \frac{1}{4} e^{2x} + C$

3. $-t \cos t + \sin t + C$

5. $\frac{1}{3} x \sin 3x + \frac{1}{9} \cos 3x + C$

7. $\frac{1}{4} x^4 \ln x - \frac{1}{16} x^4 + C$

9. $x \arctan x - \frac{1}{2} \ln(1 + x^2) + C$

11. $\frac{2}{3} y^{3/2} \ln y - \frac{4}{9} y^{3/2} + C$

13. $t(\ln t)^2 - 2t \ln t + 2t + C$

15. $\frac{2}{3} x(x + 3)^{3/2} - \frac{4}{15}(x + 3)^{5/2} + C$

17. $\frac{2}{9} x^3(x^3 + 1)^{3/2} - \frac{4}{45}(x^3 + 1)^{5/2} + C$

19. $-\frac{1}{2}(\csc \theta \cot \theta + \ln |\csc \theta + \cot \theta|) + C$

21. $\frac{1}{3} x^3 \arctan x - \frac{1}{6} x^2 + \frac{1}{6} \ln(1 + x^2) + C$

23. $x \operatorname{arcsec} x^{1/2} - (x - 1)^{1/2} + C$

25. $(x + 1) \arctan x^{1/2} - x^{1/2} + C$

27. $-x \cot x + \ln |\sin x| + C$

29. $\frac{1}{2} x^2 \sin x^2 + \frac{1}{2} \cos x^2 + C$

31. $-2x^{-1/2}(2 + \ln x) + C$

33. $x \sinh x - \cosh x + C$

35. $x^2 \cosh x - 2x \sinh x + 2 \cosh x + C$

37. $\pi(e - 2) \approx 2.25655$

39. $\frac{1}{2}(x - 1)e^x \sin x + \frac{1}{2} xe^x \cos x + C$

47. $6 - 2e \approx 0.563436$

49. $6 - 2e$

57. $\frac{1}{80}(2\pi^6 - 10\pi^4 + 15\pi^2) \approx 13.709144$

Section 9.5   (page 510)

1. $\frac{1}{2} x^2 - x + \ln |x + 1| + C$

3. $\dfrac{1}{3} \ln \left| \dfrac{x - 3}{x} \right| + C$

5. $\dfrac{1}{5} \ln \left| \dfrac{x - 2}{x + 3} \right| + C$

7. $\frac{1}{4} \ln |x| - \frac{1}{8} \ln(x^2 + 4) + C$

9. $\frac{1}{3} x^3 - 4x + 8 \arctan \frac{1}{2} x + C$

11. $x - 2 \ln |x + 1| + C$

13. $x + (x + 1)^{-1} + C$

15. $\dfrac{1}{4} \ln \left| \dfrac{x - 2}{x + 2} \right| + C$

17. $\dfrac{3}{2} \ln |2x - 1| - \ln |x + 3| + C$

19. $\ln |x| + 2(x + 1)^{-1} + C$

21. $\dfrac{3}{2} \ln |x^2 - 4| + \dfrac{1}{2} \ln |x^2 - 1| + C$

23. $\ln |x + 2| + 4(x + 2)^{-1} - 2(x + 2)^{-2} + C$

25. $\dfrac{1}{2} \ln \left(\dfrac{x^2}{x^2 + 1} \right) + C$

27. $\dfrac{1}{2} \ln \left(\dfrac{x^2}{x^2 + 4} \right) + \dfrac{1}{2} \arctan \dfrac{x}{2} + C$

29. $-\dfrac{1}{2} \ln |x + 1| + \dfrac{1}{4} \ln(x^2 + 1) + \dfrac{1}{2} \arctan x + C$

31. $\arctan \dfrac{1}{2} x - \dfrac{3}{2} \sqrt{2} \arctan x \sqrt{2} + C$

33. $\dfrac{1}{\sqrt{2}} \arctan \dfrac{x}{\sqrt{2}} + \dfrac{1}{2} \ln(x^2 + 3) + C$

35. $x + \dfrac{1}{2} \ln |x - 1| - 5(2x - 2)^{-1}$
$\qquad + \dfrac{3}{4} \ln(x^2 + 1) + 2 \arctan x + C$

37. $\dfrac{1}{4} (1 - 2e^{2t})(e^{2t} - 1)^{-2} + C$

39. $\dfrac{1}{4} \ln |3 + 2 \ln t| + \dfrac{1}{4} (3 + 2 \ln t)^{-1} + C$

41. $\dfrac{4}{15} \pi (52 - 75 \ln 2) \approx 0.0116963\,24$

43. $x(t) = \dfrac{2e^t}{2e^t - 1}$

45. $x(t) = \dfrac{2e^{2t} + 1}{2e^{2t} - 1}$

47. $x(t) = \dfrac{21e^t - 16}{8 - 7e^t}$

49. About 153,700,000

51. (a) 1.37 s; (b) 200 g

53. $P(t) = \dfrac{200}{2 - e^{t/100}}$;

(a) $t = 100 \ln 1.8 \approx 58.8$ (days);
(b) $t = 100 \ln 2 \approx 69.3$ (days)

Section 9.6   (page 516)

1. $\arcsin \dfrac{1}{4} x + C$

3. $-\dfrac{\sqrt{4 - x^2}}{4x} + C$

5. $8 \arcsin \dfrac{x}{4} - \dfrac{x \sqrt{16 - x^2}}{2} + C$

7. $\dfrac{x}{9 \sqrt{9 - 16x^2}} + C$

9. $\ln |x + \sqrt{x^2 - 1}| - \dfrac{\sqrt{x^2 - 1}}{x} + C$

11. $\dfrac{1}{80} [(9 + 4x^2)^{5/2} - 15(9 + 4x^2)^{3/2}] + C$

13. $\sqrt{1 - 4x^2} - \ln \left| \dfrac{1 + \sqrt{1 - 4x^2}}{2x} \right| + C$

15. $\dfrac{1}{2} \ln |2x + \sqrt{9 + 4x^2}| + C$

17. $\dfrac{25}{2} \arcsin \dfrac{1}{5} x - \dfrac{1}{2} x \sqrt{25 - x^2} + C$

19. $\dfrac{1}{2} x \sqrt{x^2 + 1} - \dfrac{1}{2} \ln |x + \sqrt{1 + x^2}| + C$

21. $\dfrac{1}{18} x \sqrt{4 + 9x^2} - \dfrac{2}{27} \ln |3x + \sqrt{4 + 9x^2}| + C$

23. $x(1 + x^2)^{-1/2} + C$

25. $\dfrac{1}{512} \left[3 \ln \left| \dfrac{x + 2}{x - 2} \right| - \dfrac{12x}{x^2 - 4} + \dfrac{32x}{(x^2 - 4)^2} \right] + C$

27. $\dfrac{1}{2} x \sqrt{9 + 16x^2} + \dfrac{9}{8} \ln |4x + \sqrt{9 + 16x^2}| + C$

29. $\sqrt{x^2 - 25} - 5 \operatorname{arcsec} \dfrac{1}{5} x + C$

31. $\dfrac{1}{8} x(2x^2 - 1)(x^2 - 1)^{1/2} - \dfrac{1}{8} \ln |x + (x^2 - 1)^{1/2}| + C$

33. $-x(4x^2 - 1)^{-1/2} + C$

35. $-x^{-1}(x^2 - 5)^{1/2} + \ln |x + (x^2 - 5)^{1/2}| + C$

37. $\operatorname{arcsinh} \dfrac{1}{5} x + C$

39. $\operatorname{arccosh} \dfrac{1}{2} x - x^{-1}(x^2 - 4)^{1/2} + C$

41. $\dfrac{1}{8} x(1 + 2x^2)(1 + x^2)^{1/2} - \dfrac{1}{8} \operatorname{arcsinh} x + C$

43. $\dfrac{1}{32} \pi [18\sqrt{5} - \ln(2 + \sqrt{5})] \approx 3.8097$

45. $\sqrt{5} - \sqrt{2} + \ln \left(\dfrac{2 + 2\sqrt{2}}{1 + \sqrt{5}} \right) \approx 1.222016$

49. $2\pi [\sqrt{2} + \ln(1 + \sqrt{2})] \approx 14.4236$

53. $\$6\frac{2}{3}$ million

Section 9.7   (page 522)

1. $\arctan(x + 2) + C$

3. $11 \arctan(x + 2) - \dfrac{3}{2} \ln(x^2 + 4x + 5) + C$

5. $\arcsin \dfrac{1}{2} (x + 1) + C$

7. $-2 \arcsin \dfrac{1}{2} (x + 1) - \dfrac{1}{2} (x + 1)(3 - 2x - x^2)^{1/2}$
$\qquad - \dfrac{1}{3} (3 - 2x - x^2)^{3/2} + C$

9. $\dfrac{5}{16} \ln |2x + 3| + \dfrac{7}{16} \ln |2x - 1| + C$

11. $\dfrac{1}{3} \arctan \dfrac{1}{3} (x + 2) + C$

13. $\dfrac{1}{4} \ln \left| \dfrac{1 + x}{3 - x} \right| + C$

15. $\ln(x^2 + 2x + 2) - 7 \arctan(x + 1) + C$

17. $\dfrac{2}{9} \arcsin(x - \dfrac{2}{3}) - \dfrac{1}{9} (5 + 12x - 9x^2)^{1/2} + C$

19. $\dfrac{75}{4} \arcsin \dfrac{2}{5} (x - 2) + \dfrac{3}{2} (x - 2)(9 + 16x - 4x^2)^{1/2}$
$\qquad + \dfrac{1}{6} (9 + 16x - 4x^2)^{3/2} + C$

21. $\dfrac{1}{9} (7x - 12)(6x - x^2)^{-1/2} + C$

23. $-(16x^2 + 48x + 52)^{-1} + C$

25. $\dfrac{3}{2} \ln(x^2 + x + 1) - \dfrac{5}{3} \sqrt{3} \arctan(\dfrac{1}{3} \sqrt{3} [2x + 1]) + C$

27. $\dfrac{1}{32} \ln \left| \dfrac{x + 2}{x - 2} \right| - \dfrac{x}{8(x^2 - 4)} + C$

29. $\ln |x| - \dfrac{2}{3} \sqrt{3} \arctan(\dfrac{1}{3} \sqrt{3} [2x + 1]) + C$

31. $-\dfrac{5}{4} (x - 1)^{-1} - \dfrac{1}{4} \ln |x - 1| - \dfrac{5}{4} (x + 1)^{-1}$
$\qquad + \dfrac{1}{4} \ln |x + 1| + C$

33. $-\dfrac{1}{4} (x + 7)(x^2 + 2x + 5)^{-1} - \dfrac{1}{8} \arctan(\dfrac{1}{2} [x + 1]) + C$

37. About 3.69 mi

39. $\ln|x - 1| - \frac{1}{2}\ln(x^2 + 2x + 2) + \arctan(x + 1) + C$

41. $\frac{1}{2}x^2 + \ln|x - 1| + \frac{1}{2}\ln(x^2 + x + 1)$
$+ \frac{1}{3}\sqrt{3}\arctan(\frac{1}{3}\sqrt{3}[2x + 1]) + C$

43. $\frac{1}{2}\ln(x^4 + x^2 + 1) - \frac{2\sqrt{3}}{3}\arctan\left(\frac{\sqrt{3}}{2x^2 + 1}\right) + C$

Section 9.8 (page 529)

1. 1   **3.** $+\infty$ **5.** $+\infty$ **7.** 1

9. $+\infty$ **11.** $-\frac{1}{2}$ **13.** $\frac{9}{2}$ **15.** $+\infty$

17. Does not exist   **19.** $2(e - 1)$

21. $+\infty$ **23.** $\frac{1}{4}$

Chapter 9 Miscellaneous Problems (page 533)

Note: Different techniques of integration may produce answers that appear to differ from these. If both are correct, they must differ by only a constant.

1. $2\arctan\sqrt{x} + C$ **3.** $\ln|\sec x| + C$

5. $\frac{1}{2}\sec^2\theta + C$

7. $x\tan x - \frac{1}{2}x^2 + \ln|\cos x| + C$

9. $\frac{2}{15}(2 - x^3)^{5/2} - \frac{4}{9}(2 - x^3)^{3/2} + C$

11. $\frac{1}{2}x(25 + x^2)^{1/2} - \frac{25}{2}\ln|x + (25 + x^2)^{1/2}| + C$

13. $\frac{2}{3}\sqrt{3}\arctan(\frac{1}{3}\sqrt{3}[2x - 1]) + C$

15. $\frac{103}{87}\sqrt{29}\arctan(\frac{1}{29}\sqrt{29}[3x - 2])$
$+ \frac{5}{6}\ln(9x^2 - 12x + 33) + C$

17. $\frac{2}{9}(1 + x^3)^{3/2} + C$ **19.** $\arcsin(\frac{1}{2}\sin x) + C$

21. $-\ln|\ln\cos x| + C$

23. $(x + 1)\ln(x + 1) - x + C$

25. $\frac{1}{2}x\sqrt{x^2 + 9} + \frac{9}{2}\ln|x + \sqrt{x^2 + 9}| + C$

27. $\frac{1}{2}(x - 1)(2x - x^2)^{1/2} + \frac{1}{2}\arcsin(x - 1) + C$

29. $\frac{1}{3}x^3 + 2x - \sqrt{2}\ln\left|\frac{x + \sqrt{2}}{x - \sqrt{2}}\right| + C$

31. $\frac{1}{2}(x^2 + x)(x^2 + 2x + 2)^{-1} - \frac{1}{2}\arctan(x + 1) + C$

33. $\frac{\sin 2\theta}{2(1 + \cos 2\theta)} + C$ **35.** $\frac{1}{5}\sec^5 x - \frac{1}{3}\sec^3 x + C$

37. $\frac{1}{8}x^2[4(\ln x)^3 - 6(\ln x)^2 + 6(\ln x) - 3] + C$

39. $\frac{1}{2}e^x(1 + e^{2x})^{1/2} + \frac{1}{2}\ln[e^x + (1 + e^{2x})^{1/2}] + C$

41. $\frac{1}{54}\arcsec|\frac{1}{3}x| + \frac{1}{18}x^{-2}(x^2 - 9)^{1/2} + C$

43. $\ln|x| + \frac{1}{2}\arctan 2x + C$

45. $\frac{1}{2}(\sec x \tan x - \ln|\sec x + \tan x|) + C$

47. $\ln|x + 1| - \frac{2}{3}x^{-3} + C$

49. $\ln|x - 1| + \ln(x^2 + x + 1) + (x - 1)^{-1}$
$- 2(x^2 + x + 1)^{-1} + C$

51. $x[(\ln x)^6 - 6(\ln x)^5 + 30(\ln x)^4 - 120(\ln x)^3$
$+ 360(\ln x)^2 - 720\ln x + 720] + C$

53. $\frac{1}{3}(\arcsin x)^3 + C$ **55.** $\frac{1}{2}\sec^2 z + \ln|\cos z| + C$

57. $\frac{1}{2}\arctan(\exp(x^2)) + C$

59. $-\frac{1}{2}(x^2 + 1)\exp(-x^2) + C$

61. $-\frac{1}{x}\arcsin x - \ln\left|\frac{1 + \sqrt{1 - x^2}}{x}\right| + C$

63. $\frac{1}{8}\arcsin x + \frac{1}{8}x(2x^2 - 1)(1 - x^2)^{1/2} + C$

65. $\frac{1}{4}\ln|2x + 1| + \frac{5}{4}(2x + 1)^{-1} + C$

67. $\frac{1}{2}\ln|e^{2x} - 1| + C$

69. $2\ln|x + 1| + 3(x + 1)^{-1} - \frac{5}{3}(x + 1)^{-3} + C$

71. $\frac{1}{2}\ln(x^2 + 1) + \arctan x - \frac{1}{2}(x^2 + 1)^{-1} + C$

73. $\frac{1}{45}(x^3 + 1)^{3/2}(6x^3 + 4) + C$

75. $\frac{2}{3}(1 + \sin x)^{3/2} + C$

77. $\frac{1}{2}\ln|\sec x + \tan x| + C$

79. $-2(1 - \sin t)^{1/2} + C$

81. $-2x + \sqrt{3}\arctan(\frac{1}{3}\sqrt{3}[2x + 1])$
$+ \frac{1}{2}(2x + 1)\ln(x^2 + x + 1) + C$

83. $-x^{-1}\arctan x + \ln|x(1 + x^2)^{-1/2}| + C$

85. $\frac{1}{2}\ln(x^2 + 1) + \frac{1}{2}(x^2 + 1)^{-1} + C$

87. $\frac{1}{2}(x - 6)(x^2 + 4)^{-1/2} + C$

89. $\frac{1}{3}(1 + \sin^2 x)^{3/2} + C$

91. $\frac{1}{2}e^x(x\sin x - x\cos x + \cos x) + C$

93. $-\frac{1}{2}(x - 1)^{-2}\arctan x + \frac{1}{8}\ln(x^2 + 1)$
$- \frac{1}{4}\ln|x - 1| - \frac{1}{4}(x - 1)^{-1} + C$

95. $\frac{11}{9}\arcsin\frac{1}{2}(3x - 1) - \frac{2}{9}(3 + 6x - 9x^2)^{1/2} + C$

97. $\frac{1}{3}x^3 + x^2 + 3x + 4\ln|x - 1| - (x - 1)^{-1} + C$

99. $x\arcsec x^{1/2} - (x - 1)^{1/2} + C$

101. $\frac{1}{4}\pi(e^2 - e^{-2} + 4)$

103. (a) $A_t = \pi\left(\sqrt{2} - e^{-t}(1 + e^{-2t})^{1/2}\right.$
$\left. + \ln\left[\frac{1 + \sqrt{2}}{e^{-t} + (1 + e^{-2t})^{1/2}}\right]\right)$;
(b) $\pi[\sqrt{2} + \ln(1 + \sqrt{2})] \approx 7.2118$

105. $\frac{\pi\sqrt{2}}{2}\left[2\sqrt{14} - \sqrt{2} + \ln\left(\frac{1 + \sqrt{2}}{2\sqrt{2} + \sqrt{7}}\right)\right]$
≈ 11.66353

109. $\frac{5}{4}\pi \approx 3.92699$

111. The value of the integral is $\frac{1}{630}$.

113. $\frac{5\sqrt{6} - 3\sqrt{2}}{2} + \frac{1}{2}\ln\left(\frac{1 + \sqrt{2}}{\sqrt{3} + \sqrt{2}}\right) \approx 3.869983$

115. The substitution is $u = e^x$.
(a) $\frac{2}{3}\sqrt{3}\arctan(\frac{1}{3}\sqrt{3}[1 + 2e^x]) + C$

119. $\frac{1}{4}\sqrt{2}\ln\left|\frac{1 + \tan\theta - \sqrt{2}\tan\theta}{1 + \tan\theta + \sqrt{2}\tan\theta}\right|$
$- \frac{1}{2}\sqrt{2}\arctan\sqrt{2}\cot\theta + C$

121. $\frac{2}{25,515}(3x - 2)^{3/2}(945x^3 + 540x^2 + 288x + 128) + C$

123. $\frac{3}{4}(x^2 - 1)^{-1/3}(x^2 - 3) + C$

125. $\frac{2}{9}(x^3 + 1)^{1/2}(x^3 - 2) + C$

127. $2 \arctan\sqrt{\frac{1 + x}{1 - x}} - (1 - x)\sqrt{\frac{1 + x}{1 - x}} + C$

129. $3(x + 1)^{1/3} - \sqrt{3}\arctan\frac{1 + 2(x + 1)^{1/3}}{\sqrt{3}}$
$\qquad + \ln|(x + 1)^{1/3} - 1|$
$\qquad - \frac{1}{2}\ln|(x + 1)^{2/3} + (x + 1)^{1/3} + 1| + C$

131. $\sqrt{1 + e^{2x}} + \frac{1}{2}\ln\left(\frac{\sqrt{1 + e^{2x}} - 1}{\sqrt{1 + e^{2x}} + 1}\right) + C$

133. $\frac{8}{15}$

135. $\tan\frac{\theta}{2} + C = \frac{1 - \cos\theta}{\sin\theta} + C = \frac{\sin\theta}{1 + \cos\theta} + C$

137. $-\frac{2\sin\theta}{1 + \sin\theta - \cos\theta} + C$

139. $\frac{\sqrt{2}}{2}\ln\left|\frac{1 - \cos\theta + (\sqrt{2} - 1)\sin\theta}{1 - \cos\theta - (\sqrt{2} + 1)\sin\theta}\right| + C$

141. $\ln\frac{1 - \cos\theta}{1 - \cos\theta + \sin^2\theta} + C$

Section 10.1 (page 542)

1. $x + 2y + 3 = 0$ **3.** $4y + 25 = 3x$

5. $x + y = 1$ **7.** Center $(-1, 0)$, radius $\sqrt{5}$

9. Center $(2, -3)$, radius 4

11. Center $(0.5, 0)$, radius 1

13. Center $(0.5, -1.5)$, radius 3

15. Center $\left(-\frac{1}{3}, \frac{4}{3}\right)$, radius 2   **17.** The point $(3, 2)$

19. No points   **21.** $(x + 1)^2 + (y + 2)^2 = 34$

23. $(x - 6)^2 + (y - 6)^2 = 0.8$

25. $2x + y = 13$

27. $(x - 6)^2 + (y - 11)^2 = 18$

29. $(x/5)^2 + (y/3)^2 = 1$

31. $y - 7 + 4\sqrt{3} = (4 - 2\sqrt{3})(x - 2 + \sqrt{3})$,
$\qquad y - 7 - 4\sqrt{3} = (4 + 2\sqrt{3})(x - 2 - \sqrt{3})$

33. $y - 1 = 4(x - 4)$ and $y + 1 = 4(x + 4)$

35. $a = |h^2 - p^2|^{1/2}$, $b = a(e^2 - 1)^{1/2}$, $h = \dfrac{p(e^2 + 1)}{1 - e^2}$

Section 10.2 (page 548)

1. (a) $\left(\frac{1}{2}\sqrt{2}, \frac{1}{2}\sqrt{2}\right)$ (b) $(1, -\sqrt{3})$
     (c) $\left(\frac{1}{2}, -\frac{1}{2}\sqrt{3}\right)$ (d) $(0, -3)$
     (e) $(\sqrt{2}, \sqrt{2})$ (f) $(\sqrt{3}, -1)$
     (g) $(-\sqrt{3}, 1)$

3. $r\cos\theta = 4$ **5.** $\theta = \arctan\frac{1}{3}$

7. $r^2\cos\theta\sin\theta = 1$ **9.** $r = \tan\theta\sec\theta$

11. $x^2 + y^2 = 9$ **13.** $x^2 + 5x + y^2 = 0$

15. $(x^2 + y^2)^3 = 4y^4$ **17.** $x = 3$

19. $x = 2$; $r = 2\sec\theta$

21. $x + y = 1$; $r = \dfrac{1}{\cos\theta + \sin\theta}$

23. $y = x + 2$; $r = \dfrac{2}{\sin\theta - \cos\theta}$

25. $x^2 + y^2 + 8y = 0$; $r = -8\sin\theta$

27. $x^2 + y^2 = 2x + 2y$; $r = 2(\cos\theta + \sin\theta)$

29. Fig. 10.2.21   **31.** Fig. 10.2.15

33. Fig. 10.2.19   **35.** Fig. 10.2.18

37. Fig. 10.2.16

39. Symmetric about the x-axis

41. Symmetric about the x-axis

43. Symmetric about the x-axis

45. Symmetric about the origin

47. Symmetric about both axes and the origin

49. Symmetric about the x-axis

51. Symmetric about the y-axis

53. No points of intersection

55. $(0, 0)$, $\left(\frac{1}{2}, \pi/6\right)$, $\left(\frac{1}{2}, 5\pi/6\right)$, $(1, \pi/2)$

57. The pole, the point $(r, \theta) = (2, \pi)$, and the two points $r = 2(\sqrt{2} - 1)$, $|\theta| = \arccos(3 - 2\sqrt{2})$

59. (a) $r\cos(\theta - \alpha) = p$

Section 10.3 (page 554)

1. π **3.** $3\pi/2$ **5.** $9\pi/2$

7. 4π **9.** $19\pi/2$

11. $\pi/2$ (one of *four* loops)

13. $\pi/4$ (one of *eight* loops)

15. 2 (one of *two* loops)   **17.** 4 (one of *two* loops)

19. $\frac{1}{6}(2\pi + 3\sqrt{3})$ **21.** $\frac{1}{24}(5\pi - 6\sqrt{3})$

23. $\frac{1}{6}(39\sqrt{3} - 10\pi)$ **25.** $\frac{1}{2}(2 - \sqrt{2})$

27. $\frac{1}{6}(20\pi + 21\sqrt{3})$ **29.** $\frac{1}{2}(2 + \pi)$ **31.** $\pi/2$

Section 10.4 (page 559)

1. $y^2 = 12x$ **3.** $(x - 2)^2 = -8(y - 3)$

5. $(y - 3)^2 = -8(x - 2)$

7. $x^2 = -6(y + \frac{3}{2})$ **9.** $x^2 = 4(y + 1)$

11. $y^2 = 12x$; vertex $(0, 0)$, axis the x-axis

13. $y^2 = -6x$; vertex $(0\ 0)$, axis the x-axis

15. $x^2 - 4x - 4y = 0$; vertex $(2, -1)$, axis the line $x = 2$

17. $4y = -12 - (2x + 1)^2$; vertex $(-0.5, -3)$, axis the line $x = -0.5$

23. About 0.693 days; that is, about 16 h 38 min

27. $\alpha = \frac{1}{2} \arcsin(0.49) \approx 0.256045$ (14° 40′ 13″),
$\alpha = \frac{1}{2}(\pi - \arcsin(0.49)) \approx 1.314751$ (75° 19′ 47″)

29. *Suggestion:* $x^2 - 2xy + y^2 - 2ax - 2ay + a^2 = 0$

Section 10.5 (page 565)

1. $\left(\dfrac{x}{4}\right)^2 + \left(\dfrac{y}{5}\right)^2 = 1$ **3.** $\left(\dfrac{x}{15}\right)^2 + \left(\dfrac{y}{17}\right)^2 = 1$

5. $\dfrac{x^2}{16} + \dfrac{y^2}{7} = 1$ **7.** $\dfrac{x^2}{100} + \dfrac{y^2}{75} = 1$

9. $\dfrac{x^2}{16} + \dfrac{y^2}{12} = 1$

11. $\dfrac{(x-2)^2}{16} + \dfrac{(y-3)^2}{4} = 1$

13. $\dfrac{(x-1)^2}{25} + \dfrac{(y-1)^2}{16} = 1$

15. $\dfrac{(x-1)^2}{81} + \dfrac{(y-2)^2}{72} = 1$

17. Center $(0, 0)$, foci $(\pm 2\sqrt{5}, 0)$, major axis 12, minor axis 8

19. Center $(0, 4)$, foci $(0, 4 \pm \sqrt{5})$, major axis 6, minor axis 4

21. About 3466.36 AU—that is, about 3.22×10^{11} mi, or about 20 light-days

27. $\dfrac{(x-1)^2}{4} + \dfrac{y^2}{16/3} = 1$

Section 10.6 (page 571)

1. $\dfrac{x^2}{1} - \dfrac{y^2}{15} = 1$ **3.** $\dfrac{x^2}{16} - \dfrac{y^2}{9} = 1$

5. $\dfrac{y^2}{25} - \dfrac{x^2}{25} = 1$ **7.** $\dfrac{y^2}{9} - \dfrac{x^2}{27} = 1$

9. $\dfrac{x^2}{4} - \dfrac{y^2}{12} = 1$ **11.** $\dfrac{(x-2)^2}{9} - \dfrac{(y-2)^2}{27} = 1$

13. $\dfrac{(y+2)^2}{9} - \dfrac{(x-1)^2}{4} = 1$

15. Center $(1, 2)$, foci $(1 \pm \sqrt{2}, 2)$, asymptotes $y - 2 = \pm(x - 1)$

17. Center $(0, 3)$, foci $(0, 3 \pm 2\sqrt{3})$, asymptotes $y = 3 \pm x\sqrt{3}$

19. Center $(-1, 1)$, foci $(-1 \pm \sqrt{13}, 1)$, asymptotes $y = \frac{1}{2}(3x + 5)$, $y = -\frac{1}{2}(3x + 1)$

21. There are no points on the graph if $c > 15$.

25. $16x^2 + 50xy + 16y^2 = 369$

27. About 16.42 mi north of B and 8.66 mi west of B; that is, about 18.56 mi from B at a bearing of 27°48′ west of north.

Section 10.7 (page 575)

1. $2(x')^2 + (y')^2 = 4$: ellipse; origin $(2, 3)$

3. $9(x')^2 - 16(y')^2 = 144$: hyperbola; origin $(1, -1)$

5. The single point $(2, 3)$

7. Ellipse; 45°, $4(x')^2 + 2(y')^2 = 1$

9. The two parallel lines (a "degenerate parabola") $(x')^2 = 4$; $\tan^{-1}(\frac{1}{2}) \approx 26.57°$

11. Hyperbola; $\tan^{-1}(\frac{1}{3}) \approx 18.43°$, $(x')^2 - (y')^2 = 1$

13. Ellipse; $\tan^{-1}(\frac{2}{3}) \approx 33.69°$, $2(x')^2 + (y')^2 = 2$

15. Ellipse; $\tan^{-1}(\frac{4}{3}) \approx 53.13°$, $4(x')^2 + (y')^2 = 4$

17. Ellipse; $\tan^{-1}(\frac{1}{4}) \approx 14.04°$, $2(x')^2 + (y')^2 = 4$

19. The two perpendicular lines $y' = x'$ and $y' = -x'$ (a "degenerate hyperbola"); $\tan^{-1}(\frac{5}{12}) \approx 22.62°$

21. Ellipse, $\tan^{-1}(\frac{4}{3}) \approx 53.13°$, $25(x' - 1)^2 + 50(y')^2 = 50$

23. Hyperbola, $\tan^{-1}(\frac{4}{3}) \approx 53.13°$, $2(y' - 1)^2 - (x' - 2)^2 = 1$

25. Hyperbola, $\tan^{-1}(\frac{8}{15}) \approx 28.07°$, $(x' - 1)^2 - (y')^2 = 1$

Chapter 10 Miscellaneous Problems (page 577)

1. Circle, center $(1, 1)$, radius 2

3. Circle, center $(3, -1)$, radius 1

5. Parabola, vertex $(4, -2)$, opening downward

7. Ellipse, center $(2, 0)$, major axis 6, minor axis 4

9. Hyperbola, center $(-1, 1)$, vertical axis, foci at $(-1, 1 \pm \sqrt{3})$

11. There are no points on the graph.

13. Hyperbola, axis inclined at 22.5° from the horizontal

15. Ellipse, center at the origin, major axis $2\sqrt{2}$, minor axis 1, rotated through the angle $\alpha = \pi/4$

17. Parabola, vertex $(0, 0)$, opening to the "northeast," axis at angle $\alpha = \tan^{-1}(\frac{3}{4})$ from the horizontal

19. Circle, center $(1, 0)$, radius 1

21. Straight line $y = x + 1$

23. Horizontal line $y = 3$

25. Two ovals tangent to each other and to the y-axis at $(0, 0)$

27. Apple-shaped curve, symmetric about the y-axis

29. Ellipse, one focus at $(0, 0)$, directrix $x = 4$, eccentricity $e = 0.5$

31. $\frac{1}{2}(\pi - 2)$ **33.** $\frac{1}{6}(39\sqrt{3} - 10\pi) \approx 6.02234$

35. 2   **37.** $5\pi/4$ **39.** $r = 2p\cos(\theta - \alpha)$

41. If $a > b$, the maximum is $2a$ and the minimum is $2b$.

43. $b^2 y = 4hx(b - x)$; alternatively,

$$r = b\sec\theta - \frac{b^2}{4h}\sec\theta\tan\theta$$

45. *Suggestion:* Let θ be the angle that QR makes with the

$$r = b\sec\theta - \frac{b^2}{4h}\sec\theta\tan\theta$$

x-axis.

49. The curve is a hyperbola with one focus at the origin, directrix $x = -\frac{3}{2}$, and eccentricity $e = 2$.

51. $\frac{3}{2}$

Section 11.2 (page 587)

1. $\frac{2}{5}$ **3.** 0   **5.** 1   **7.** Does not converge

9. 0   **11.** 0   **13.** 0   **15.** 1

17. 0   **19.** 0   **21.** 0   **23.** 0

25. e **27.** e^{-2} **29.** 2   **31.** 1

33. Does not converge

35. 0   **41.** (b) 4

Section 11.3 (page 596)

1. $\frac{3}{2}$ **3.** Diverges   **5.** Diverges

7. 6   **9.** Diverges   **11.** Diverges

13. Diverges   **15.** $\dfrac{\sqrt{2}}{\sqrt{2}-1} = 2 + \sqrt{2}$

17. Diverges   **19.** $\frac{1}{12}$ **21.** $\dfrac{e}{\pi - e}$

23. Diverges   **25.** $\frac{65}{12}$ **27.** $\frac{247}{8}$ **29.** $\frac{1}{4}$

31. $\frac{47}{99}$ **33.** $\frac{41}{333}$ **35.** $\frac{314,156}{99,999}$

37. $S_n = \ln(n + 1)$; diverges

39. $S_n = \dfrac{3}{2} - \dfrac{1}{n} - \dfrac{1}{n+1}$; the sum is $\frac{3}{2}$.

45. Computations with S are meaningless because S is not a number.

47. 4.5 s   **49.** (a) $M_n = (0.95)^n M_0$; (b) 0

51. Peter $\frac{4}{7}$, Paul $\frac{2}{7}$, Mary $\frac{1}{7}$ **53.** $\frac{1}{12}$

Section 11.4 (page 610)

1. $e^{-x} = 1 - \dfrac{x}{1!} + \dfrac{x^2}{2!} - \dfrac{x^3}{3!} + \dfrac{x^4}{4!} - \dfrac{x^5}{5!} + \left(\dfrac{e^{-z}}{6!}\right)x^6$ for some z between 0 and x.

3. $\cos x = 1 - \dfrac{x^2}{2!} + \dfrac{x^4}{4!} - \left(\dfrac{\sin z}{5!}\right)x^5$ for some z between 0 and x.

5. $\sqrt{1 + x} = 1 + \dfrac{x}{1!\,2} - \dfrac{x^2}{2!\,4} + \dfrac{3x^3}{3!\,8} - \dfrac{5x^4}{128}(1 + z)^{-7/2}$ for some z between 0 and x.

7. $\tan x = \dfrac{x}{1!} + \dfrac{2x^3}{3!}$

$+ \dfrac{16\sec^4 z\tan z + 8\sec^2 z\tan^3 z}{4!}x^4$ for some z between 0 and x.

9. $\sin^{-1} z = \dfrac{x}{1!} + \dfrac{x^3}{3!}\cdot\dfrac{1 + 2z^2}{(1 - z^2)^{5/2}}$ for some z between 0 and x.

11. $e^x = e + \dfrac{e}{1!}(x - 1) + \dfrac{e}{2!}(x - 1)^2 + \dfrac{e}{3!}(x - 1)^3$

$+ \dfrac{e}{4!}(x - 1)^4 + \dfrac{e^z}{5!}(x - 1)^5$ for some z between 1 and x.

13. $\sin x = \dfrac{1}{2} + \dfrac{\sqrt{3}}{1!\,2}\left(x - \dfrac{\pi}{6}\right) - \dfrac{1}{2!\,2}\left(x - \dfrac{\pi}{6}\right)^2$

$- \dfrac{\sqrt{3}}{3!\,2}\left(x - \dfrac{\pi}{6}\right)^3 + \dfrac{\sin z}{4!\,2}\left(x - \dfrac{\pi}{6}\right)^4$ for some z between $\pi/6$ and x.

15. $\dfrac{1}{(x - 4)^2} = 1 - 2(x - 5) + 3(x - 5)^2 -$

$4(x - 5)^3 + 5(x - 5)^4 - 6(x - 5)^5 +$

$\dfrac{7}{(z - 4)^8}(x - 5)^6$ for some z between 5 and x.

17. $\cos x = -1 + \dfrac{(x - \pi)^2}{2!} - \dfrac{(x - \pi)^4}{4!} -$

$\dfrac{(x - \pi)^5}{5!}\sin z$ for some z between π and x.

19. $x^{3/2} = 1 + \dfrac{3(x - 1)}{2} + \dfrac{3(x - 1)^2}{8} - \dfrac{(x - 1)^3}{16}$

$+ \dfrac{3(x - 1)^4}{128} - \dfrac{3(x - 1)^5}{256z^{7/2}}$ for some z between 1 and x.

21. $e^{-x} = 1 - x + \dfrac{x^2}{2!} - \dfrac{x^3}{3!} + \dfrac{x^4}{4!} - \cdots$

23. $e^{-3x} = 1 - 3x + \dfrac{9x^2}{2} - \dfrac{27x^3}{3!} + \dfrac{81x^4}{4!} - \cdots$

25. $\cos 2x = 1 - \dfrac{4x^2}{2!} + \dfrac{16x^4}{4!} - \dfrac{64x^6}{6!} + \dfrac{256x^8}{8!} - \cdots$

27. $\sin x^2 = x^2 - \dfrac{x^6}{3!} + \dfrac{x^{10}}{5!} - \dfrac{x^{14}}{7!} + \dfrac{x^{18}}{9!} - \cdots$

29. $\ln(1 + x) = x - \frac{1}{2}x^2 + \frac{1}{3}x^3 - \frac{1}{4}x^4 + \frac{1}{5}x^5 - \cdots$

31. $e^{-x} = 1 - x + \dfrac{x^2}{2!} - \dfrac{x^3}{3!} + \dfrac{x^4}{4!} - \cdots$

33. $\ln x = (x - 1) - \frac{1}{2}(x - 1)^2 + \frac{1}{3}(x - 1)^3$
$\quad - \frac{1}{4}(x - 1)^4 + \cdots$

35. $\cos x = \dfrac{\sqrt{2}}{2} - \dfrac{\sqrt{2}}{1!\,2}\left(x - \dfrac{\pi}{2}\right) - \dfrac{\sqrt{2}}{2!\,2}\left(x - \dfrac{\pi}{2}\right)^2$
$\quad + \dfrac{\sqrt{2}}{3!\,2}\left(x - \dfrac{\pi}{2}\right)^3 + \dfrac{\sqrt{2}}{4!\,2}\left(x - \dfrac{\pi}{2}\right)^4 - \dfrac{\sqrt{2}}{5!\,2}\left(x - \dfrac{\pi}{2}\right)^5$
$\quad - \dfrac{\sqrt{2}}{6!\,2}\left(x - \dfrac{\pi}{2}\right)^6 + \dfrac{\sqrt{2}}{7!\,2}\left(x - \dfrac{\pi}{2}\right)^7 + \cdots$

37. $\sinh x = x + \dfrac{x^3}{3!} + \dfrac{x^5}{5!} + \dfrac{x^7}{7!} + \dfrac{x^9}{9!} + \cdots$

39. $\dfrac{1}{x} = 1 - (x - 1) + (x - 1)^2 - (x - 1)^3$
$\quad + (x - 1)^4 - \cdots$

41. $\sin x = \dfrac{\sqrt{2}}{2}\left[1 + \left(x - \dfrac{\pi}{4}\right) - \dfrac{(x - \pi/4)^2}{2!}\right.$
$\quad \left. - \dfrac{(x - \pi/4)^3}{3!} + \dfrac{(x - \pi/4)^4}{4!} + \cdots\right]$

Section 11.5   (page 617)

1. Diverges   **3.** Diverges   **5.** Converges
7. Diverges   **9.** Converges   **11.** Converges
13. Converges   **15.** Converges   **17.** Diverges
19. Converges   **21.** Diverges   **23.** Diverges
25. Converges   **27.** Converges
29. The terms are not nonnegative.
31. The terms are not decreasing.
33. $n = 100$
37. With $n = 6$, $1.0368 < S < 1.0370$. The true value of the sum is not known exactly but is $1.03692\,77551\,4337$ to the accuracy shown.
39. About a million centuries
41. With $n = 10$, $S_{10} \approx 1.08203\,658$ and $3.141566 < \pi < 3.141627$.

Section 11.6   (page 623)

1. Converges   **3.** Diverges   **5.** Converges
7. Diverges   **9.** Converges   **11.** Converges
13. Diverges   **15.** Converges   **17.** Converges
19. Converges   **21.** Converges   **23.** Converges
25. Converges   **27.** Diverges   **29.** Diverges
31. Converges   **33.** Converges   **35.** Diverges

Section 11.7   (page 631)

1. Converges   **3.** Converges   **5.** Converges
7. Converges   **9.** Diverges

11. Converges absolutely
13. Converges conditionally
15. Converges absolutely   **17.** Converges absolutely
19. Diverges   **21.** Converges absolutely
23. Converges conditionally   **25.** Diverges
27. Diverges   **29.** Converges absolutely
31. Converges absolutely
33. $n = 1999$ **35.** $n = 6$
37. The first six terms give the estimate 0.6065.
39. The first three terms give the estimate 0.0953.
41. Using the first ten terms: $S_{10} \approx 0.81796\,22$ and $3.1329 < \pi < 3.1488$.

Section 11.8   (page 641)

1. $[-1, 1)$ **3.** $(-1, 1)$ **5.** $[-1, 1]$
7. $(0.4, 0.8)$ **9.** $[2.5, 3.5]$
11. Converges only for $x = 0$
13. $(-4, 2)$ **15.** $[2, 4]$
17. Converges only for $x = 5$ **19.** $(-1, 1)$
21. $f(x) = x^2 - 3x^3 + \dfrac{9x^4}{2!} - \dfrac{27x^5}{3!} + \dfrac{81x^6}{4!} - \cdots$;
$R = +\infty$.
23. $f(x) = x^2 - \dfrac{x^6}{3!} + \dfrac{x^{10}}{5!} - \dfrac{x^{14}}{7!} + \dfrac{x^{18}}{9!} - \cdots$;
$R = +\infty$
25. $(1 - x)^{1/3} = 1 - \dfrac{x}{3} - \dfrac{2x^2}{2!\,3^2} - \dfrac{2 \cdot 5 x^3}{3!\,3^3} - \dfrac{2 \cdot 5 \cdot 8 x^4}{4!\,3^4}$
$\quad - \cdots$; $R = 1$
27. $f(x) = 1 - 3x + 6x^2 - 10x^3 + 15x^4 - \cdots$;
$R = 1$
29. $f(x) = 1 - \frac{1}{2}x + \frac{1}{3}x^2 - \frac{1}{4}x^3 + \frac{1}{5}x^4 - \cdots$; $R = 1$
31. $f(x) = \dfrac{x^4}{4} - \dfrac{x^{10}}{3!\,10} + \dfrac{x^{16}}{5!\,16} - \dfrac{x^{22}}{7!\,22} + \cdots$
$\quad = \displaystyle\sum_{n=0}^{\infty} \dfrac{(-1)^n}{(2n + 1)!\,(6n + 4)} x^{6n+4}$
33. $f(x) = x - \dfrac{x^4}{4} + \dfrac{x^7}{2!\,7} - \dfrac{x^{10}}{3!\,10} + \dfrac{x^{13}}{4!\,13} - \cdots$
$\quad = \displaystyle\sum_{n=0}^{\infty} \dfrac{(-1)^n}{n!\,(3n + 1)} x^{3n+1}$
35. $f(x) = x - \dfrac{x^3}{2!\,3} + \dfrac{x^5}{3!\,5} - \dfrac{x^7}{4!\,7} + \dfrac{x^9}{5!\,9} - \cdots$
$\quad = \displaystyle\sum_{n=1}^{\infty} \dfrac{(-1)^{n+1}}{n!\,(2n - 1)} x^{2n-1}$
37. Using six terms: $3.14130\,878 < \pi < 3.1416744$

Section 11.9 (page 647)

1. $65^{1/3} = (4 + \frac{1}{64})^{1/3}$. The first four terms of the binomial series give $65^{1/3} \approx 4.020726$; answer: 4.021.

3. Three terms of the usual sine series give 0.479427 with error less than 0.000002; answer: 0.479.

5. Five terms of the usual arctangent series give 0.463684 with error less than 0.000045; answer: 0.464.

7. 0.309   **9.** 0.174   **11.** 0.946   **13.** 0.487

15. 0.0976   **17.** 0.444   **19.** 0.747   **21.** -0.5

23. 0.5   **25.** 0

31. The first five coefficients are 1, 0, $\frac{1}{2}$, 0, and $\frac{5}{24}$.

Chapter 11 Miscellaneous Problems (page 650)

1. 1   **3.** 10   **5.** 0   **7.** 0   **9.** No limit

11. 0   **13.** $+\infty$ (or "no limit")   **15.** 1

17. Converges   **19.** Converges   **21.** Converges

23. Diverges   **25.** Converges   **27.** Converges

29. Diverges   **31.** $(-\infty, +\infty)$

33. $[-2, 4)$ **35.** $[-1, 1]$

37. Converges only for $x = 0$ **39.** $(-\infty, +\infty)$

41. Converges for *no x*   **43.** Converges for all x

51. Seven terms of the binomial series give 1.084.

53. 0.461   **55.** 0.797

Section 12.1 (page 659)

1. $y = 2x - 3$ **3.** $y^2 = x^3$

5. $y = 2x^2 - 5x + 2$ **7.** $y = 4x^2, x > 0$

9. $\left(\frac{x}{5}\right)^2 + \left(\frac{y}{3}\right)^2 = 1$ **11.** $x^2 - y^2 = 1$

13. (a) $y - 5 = \frac{9}{4}(x - 3)$; (b) $\dfrac{d^2y}{dx^2} = \dfrac{9}{16}t^{-1}$,
concave upward at $t = 1$

15. (a) $y = -\frac{1}{2}\pi(x - \frac{1}{2}\pi)$; (b) concave downward

17. (a) $x + y = 3$; (b) concave downward

19. $\psi = \pi/6$ (constant)   **21.** $\psi = \pi/2$

25. $x = \dfrac{p}{m^2}, y = \dfrac{2p}{m}$

Section 12.2 (page 666)

1. $\frac{22}{5}$ **3.** $\frac{4}{3}$ **5.** $\frac{1}{2}(1 + e^\pi) \approx 12.0703$

7. $\frac{358}{35}\pi \approx 32.13400$ **9.** $\frac{16}{15}\pi \approx 3.35103$

11. $\frac{74}{3}$ **13.** $\frac{1}{4}\pi\sqrt{2} \approx 1.11072$

15. $(e^{2\pi} - 1)\sqrt{5} \approx 1195.1597$

17. $\frac{8}{3}\pi(5\sqrt{5} - 2\sqrt{2}) \approx 69.96882$

19. $\frac{2}{27}\pi(13\sqrt{13} - 8) \approx 9.04596$

21. $16\pi^2 \approx 157.91367$ **23.** $5\pi^2 a^3$

25. (a) $A = \pi ab$; (b) $V = \frac{4}{3}\pi ab^2$

27. $\pi\sqrt{1 + 4\pi^2} + \frac{1}{2}\ln(2\pi + \sqrt{1 + 4\pi^2}) \approx 21.25629$

29. $\frac{3}{8}\pi a^2$ **31.** $\frac{12}{5}\pi a^2$ **33.** $\frac{216}{5}\sqrt{3} \approx 74.8246$

35. $\frac{243}{4}\pi\sqrt{3} \approx 330.5649$ **39.** $6\pi^3 a^3$

Section 12.3 (page 675)

1. $\sqrt{5}, 2\sqrt{13}, 4\sqrt{2}, \langle -2, 0 \rangle, \langle 9, -10 \rangle$; no

3. $2\sqrt{2}, 10, \sqrt{5}, \langle -5, -6 \rangle, \langle 0, 2 \rangle$; no

5. $\sqrt{10}, 2\sqrt{29}, \sqrt{65}; 3\mathbf{i} - 2\mathbf{j}; -\mathbf{i} + 19\mathbf{j}$; no

7. 4, 14, $\sqrt{65}$, $4\mathbf{i} - 7\mathbf{j}$, $12\mathbf{i} + 14\mathbf{j}$; yes

9. $\mathbf{u} = \langle -\frac{3}{5}, -\frac{4}{5} \rangle$; $\mathbf{v} = \langle \frac{3}{5}, \frac{4}{5} \rangle$

11. $\mathbf{u} = \frac{8}{17}\mathbf{i} + \frac{15}{17}\mathbf{j}$; $\mathbf{v} = -\frac{8}{17}\mathbf{i} - \frac{15}{17}\mathbf{j}$

13. $-4\mathbf{j}$ **15.** $8\mathbf{i} - 14\mathbf{j}$ **17.** Yes   **19.** No

21. (a) $15\mathbf{i} - 21\mathbf{j}$; (b) $\frac{5}{3}\mathbf{i} - \frac{7}{3}\mathbf{j}$

23. (a) $\frac{35}{58}\mathbf{i}\sqrt{58} - \frac{15}{58}\mathbf{j}\sqrt{58}$; (b) $-\frac{40}{89}\mathbf{i}\sqrt{89} - \frac{25}{89}\mathbf{j}\sqrt{89}$

25. $c = 0$ **33.** $\mathbf{v}_a = (500 + 25\sqrt{2})\mathbf{i} + 25\mathbf{j}\sqrt{2}$

35. $\mathbf{v}_a = -225\mathbf{i}\sqrt{2} + 275\mathbf{j}\sqrt{2}$

41. The question makes no sense.

Section 12.4 (page 682)

1. 0, 0   **3.** $2\mathbf{i} - \mathbf{j}, 4\mathbf{i} + \mathbf{j}$ **5.** $6\pi\mathbf{i}, 12\pi^2\mathbf{j}$

7. $\mathbf{j}, \mathbf{i}$ **9.** $\frac{1}{2}(2 - \sqrt{2})\mathbf{i} + \mathbf{j}\sqrt{2}$ **11.** $\frac{484}{15}\mathbf{i}$

13. 11   **15.** 0   **17.** $\mathbf{i} + 2t\mathbf{j}, t\mathbf{i} + t^2\mathbf{j}$

19. $\frac{1}{2}t^2\mathbf{i} + \frac{1}{3}t^3\mathbf{j}$; $(1 + \frac{1}{6}t^3)\mathbf{i} + \frac{1}{12}t^4\mathbf{j}$

21. $v_0 \approx 411.047$ ft/s

25. (a) 100 ft, $400\sqrt{3}$ ft; (b) 200, 800; (c) 300, $400\sqrt{3}$

27. $140\sqrt{5} \approx 313$ (m/s)

29. Inclination angle
$$\alpha = \arctan\left(\frac{8049 - 280\sqrt{10}}{8000}\right) \approx 0.730293 \text{ (about}$$
$41° 50'34'')$; initial velocity
$$v_0 = \frac{5600}{(20\sqrt{10} - 7)\cos \alpha} \approx 133.64595 \text{ m/s}$$

35. Begin with $\dfrac{d}{dt}(\mathbf{v} \cdot \mathbf{v}) = 0$.

37. A repulsive force acting directly away from the origin, with magnitude proportional to distance from the origin

Section 12.5 (page 689)

1. $\mathbf{v} = a\mathbf{u}_\theta, \mathbf{a} = -a\mathbf{u}_r$

3. $\mathbf{v} = \mathbf{u}_r + t\mathbf{u}_\theta, \mathbf{a} = -t\mathbf{u}_r + 2\mathbf{u}_\theta$

5. $\mathbf{v} = (12\cos 4t)\mathbf{u}_r + (6\sin 4t)\mathbf{u}_\theta$,
$\mathbf{a} = (-60\sin 4t)\mathbf{u}_r + (48\cos 4t)\mathbf{u}_\theta$

9. 36.65 mi/s, 24.13 mi/s

11. 0.672 mi/s, 0.602 mi/s

13. About -795 mi—thus it's not possible.

15. About 1.962 h

Chapter 12 Miscellaneous Problems (page 690)

1. The straight line $y = x + 2$

3. The circle $(x - 2)^2 + (y - 1)^2 = 1$

5. Equation: $y^2 = (x - 1)^3$

7. $y - 2\sqrt{2} = -\frac{4}{3}(x - \frac{3}{2}\sqrt{2})$ **9.** $2\pi y + 4x = \pi^2$

11. 24   **13.** 3π **15.** $\frac{1}{27}(13\sqrt{13} - 8) \approx 1.4397$

17. $\frac{43}{6}$ **19.** $\frac{1}{8}(4\pi - 3\sqrt{3}) \approx 0.92128$

21. $\frac{471,295}{1024}\pi \approx 1445.915$

23. $\frac{1}{2}\pi\sqrt{5}(e^\pi + 1) \approx 84.7919$

25. $x = a\theta - b\sin\theta$, $y = a - b\cos\theta$

27. *Suggestion*: Compute $r^2 = x^2 + y^2$.   **29.** $6\pi^3 a^3$

35. Two solutions: $\alpha \approx 0.033364$ rad (about $1°54'53''$) and $\alpha \approx 1.29116$ rad (about $73°58'40''$)

Section 13.1   (page 700)

1. (a) $\langle 5, 8, -11 \rangle$; (b) $\langle 2, 23, 0 \rangle$; (c) 4; (d) $\sqrt{51}$; (e) $\frac{1}{15}\sqrt{5}\langle 2, 5, -4 \rangle$

3. (a) $2\mathbf{i} + 3\mathbf{j} + \mathbf{k}$; (b) $3\mathbf{i} - \mathbf{j} + 7\mathbf{k}$; (c) 0; (d) $\sqrt{5}$; (e) $\frac{1}{3}\sqrt{3}(\mathbf{i} + \mathbf{j} + \mathbf{k})$

5. (a) $4\mathbf{i} - \mathbf{j} - 3\mathbf{k}$; (b) $6\mathbf{i} - 7\mathbf{j} + 12\mathbf{k}$; (c) -1; (d) $\sqrt{17}$; (e) $\frac{1}{5}\sqrt{5}(2\mathbf{i} - \mathbf{j})$

7. $\theta = \cos^{-1}(-\frac{13}{50}\sqrt{10}) \approx 2.536$

9. $\theta = \cos^{-1}\left(-\dfrac{34}{\sqrt{3154}}\right) \approx 2.221$

11. $\text{comp}_b\, \mathbf{a} = \frac{2}{7}\sqrt{14}$, $\text{comp}_a\, \mathbf{b} = \frac{4}{15}\sqrt{5}$

13. $\text{comp}_b\, \mathbf{a} = 0$, $\text{comp}_a\, \mathbf{b} = 0$

15. $\text{comp}_b\, \mathbf{a} = -\frac{1}{10}\sqrt{10}$, $\text{comp}_a\, \mathbf{b} = -\frac{1}{5}\sqrt{5}$

17. $(x + 2)^2 + (y - 1)^2 + (z + 5)^2 = 7$

19. $(x - 4)^2 + (y - 5)^2 + (z + 2)^2 = 38$

21. Center $(-2, 3, 0)$, radius $\sqrt{13}$

23. Center $(0, 0, 3)$, radius 5

25. A plane perpendicular to the z-axis at $z = 10$

27. All points in the three coordinate planes

29. The point $(1, 0, 0)$

31. $\alpha = \cos^{-1}(\frac{1}{9}\sqrt{6}) \approx 74.21°$, $\beta = \gamma = \cos^{-1}(\frac{5}{18}\sqrt{6}) \approx 47.12°$

33. $\alpha = \cos^{-1}(\frac{3}{10}\sqrt{2}) \approx 64.90°$, $\beta = \cos^{-1}(\frac{2}{5}\sqrt{2}) \approx 55.55°$, $\gamma = \cos^{-1}(\frac{1}{2}\sqrt{2}) = 45°$

35. 48

37. If there's no friction, the work done is *mgh*.

41. $A = \frac{3}{2}\sqrt{69} \approx 12.46$

43. $\cos^{-1}(\frac{1}{3}\sqrt{3}) \approx 0.9553$, about $54.7356°$

49. $2x + 9y - 5z = 23$; the plane through the midpoint of the segment AB perpendicular to AB

51. $60°$

Section 13.2   (page 708)

1. $\langle 0, -14, 7 \rangle$ **3.** $\langle -10, -7, 1 \rangle$

7. $(\mathbf{a} \times \mathbf{b}) \times \mathbf{c} = \langle -1, 1, 0 \rangle$, $\mathbf{a} \times (\mathbf{b} \times \mathbf{c}) = \langle 0, 0, -1 \rangle$

11. $A = \frac{1}{2}\sqrt{2546} \approx 25.229$

13. (a) 55; (b) $\frac{55}{6}$ **15.** 4395.657 (m²)

17. 31,271.643 (ft²)   **21.** (b) $\frac{1}{38}\sqrt{9842} \approx 2.6107$

Section 13.3   (page 715)

1. $x = t$, $y = 2t$, $z = 3t$

3. $x = 4 + 2t$, $y = 13$, $z = -3 - 3t$

5. $x = 3 + 3t$, $y = 5 - 13t$, $z = 7 + 3t$

7. $x + 2y + 3z = 0$

9. $x - z + 8 = 0$

11. $x = 2 + t$, $y = 3 - t$, $z = -4 - 2t$; $x - 2 = -y + 3 = \dfrac{-z - 4}{2}$

13. $x = 1$, $y = 1$, $z = 1 + t$; $x - 1 = 0 = y - 1$, z arbitrary

15. $x = 2 + 2t$, $y = -3 - t$, $z = 4 + 3t$; $\dfrac{x - 2}{2} = -y - 3 = \dfrac{z - 4}{3}$

17. $y = 7$ **19.** $7x + 11y = 114$

21. $3x + 4y - z = 0$ **23.** $2x - y - z = 0$

25. $\theta = \cos^{-1}\left(\dfrac{1}{\sqrt{3}}\right) \approx 54.736°$

27. The planes are parallel: $\theta = 0$.

29. $\dfrac{x - 3}{2} = y - 3 = \dfrac{-z + 1}{5}$

31. $3x + 2y + z = 6$ **33.** $7x - 5y - 2z = 9$

35. $x - 2y + 4z = 3$ **37.** $\frac{10}{3}\sqrt{3}$

41. (b) $\dfrac{133}{\sqrt{501}} \approx 5.942$

Section 13.4   (page 720)

1. $\mathbf{v} = 5\mathbf{k}$, $\mathbf{a} = \mathbf{0}$, $v = 5$

3. $\mathbf{v} = \mathbf{i} + 2t\mathbf{j} + 3t^2\mathbf{k}$, $\mathbf{a} = 2\mathbf{j} + 6t\mathbf{k}$, $v = \sqrt{1 + 4t^2 + 9t^4}$

5. $\mathbf{v} = \mathbf{i} + 3e^t\mathbf{j} + 4e^t\mathbf{k}$, $\mathbf{a} = 3e^t\mathbf{j} + 4e^t\mathbf{k}$, $v = \sqrt{1 + 25e^t}$

7. $\mathbf{v} = (-3 \sin t)\mathbf{i} + (3 \cos t)\mathbf{j} - 4\mathbf{k}$, $\mathbf{a} = (-3 \cos t)\mathbf{i} - (3 \sin t)\mathbf{j}$, $v = 5$

9. $\mathbf{r}(t) = t^2\mathbf{i} + 10t\mathbf{j} - 2t^2\mathbf{k}$

11. $\mathbf{r}(t) = 2\mathbf{i} + t^2\mathbf{j} + (5t - t^3)\mathbf{k}$

13. $\mathbf{r}(t) = (10 + \frac{1}{6}t^3)\mathbf{i} + (10t + \frac{1}{12}t^4)\mathbf{j} + \frac{1}{20}t^5\mathbf{k}$

15. $\mathbf{r}(t) = (1 - t - \cos t)\mathbf{i} + (1 + t - \sin t)\mathbf{j} + 5t\mathbf{k}$

17. $\mathbf{v} = 3\sqrt{2}(\mathbf{i} + \mathbf{j}) + 8\mathbf{k}$, $v = 10$, $\mathbf{a} = 6\sqrt{2}(-\mathbf{i} + \mathbf{j})$

21. *Suggestion*: Compute $\dfrac{d}{dt}(\mathbf{r} \cdot \mathbf{r})$.

23. 100 ft, $\sqrt{6425} \approx 80.16$ ft/s

27. $9 \cdot (2 - \sqrt{3})$ ft

Section 13.5 (page 731)

1. 10π **3.** $19(e - 1) \approx 32.647$

5. $2 + \frac{9}{10} \ln 3 \approx 2.9888$ **7.** 0   **9.** 1

11. $\dfrac{40\sqrt{2}}{41\sqrt{41}} \approx 0.2155$ **13.** At $(-\frac{1}{2} \ln 2, \frac{1}{2}\sqrt{2})$

15. Maximum curvature $\frac{5}{9}$ at $(\pm 5, 0)$, minimum curvature $\frac{3}{25}$ at $(0, \pm 3)$

17. $\mathbf{T} = \frac{1}{10}\sqrt{10}(\mathbf{i} + 3\mathbf{j})$, $\mathbf{N} = \frac{1}{10}\sqrt{10}(3\mathbf{i} - \mathbf{j})$

19. $\mathbf{T} = \frac{1}{57}\sqrt{57}(3\mathbf{i} - 4\mathbf{j}\sqrt{3})$, $\mathbf{N} = \frac{1}{57}\sqrt{57}(4\mathbf{i}\sqrt{3} + 3\mathbf{j})$

21. $\mathbf{T} = -\frac{1}{2}\sqrt{2}(\mathbf{i} + \mathbf{j})$, $\mathbf{N} = \frac{1}{2}\sqrt{2}(\mathbf{i} - \mathbf{j})$

23. $a_T = 18t(9t^2 + 1)^{-1/2}$, $a_N = 6(9t^2 + 1)^{-1/2}$

25. $a_T = t(1 + t^2)^{-1/2}$, $a_N = (2 + t^2)(1 + t^2)^{-1/2}$

27. $1/a$ **29.** $x^2 + (y - \frac{1}{2})^2 = \frac{1}{4}$

31. $(x - \frac{3}{2})^2 + (y - \frac{3}{2})^2 = 2$ **33.** $\frac{1}{2}$

35. $\frac{1}{3}e^{-t}\sqrt{2}$ **37.** $a_T = 0 = a_N$

39. $a_T = (4t + 18t^3)(1 + 4t^2 + 9t^4)^{-1/2}$, $a_N = 2(1 + 9t^2 + 9t^4)^{1/2}(1 + 4t^2 + 9t^4)^{-1/2}$

41. $a_T = t(t^2 + 2)^{-1/2}$, $a_N = (t^4 + 5t^2 + 8)^{1/2}(t^2 + 2)^{-1/2}$

43. $\mathbf{T} = \frac{1}{2}\sqrt{2}\langle 1, \cos t, -\sin t \rangle$, $\mathbf{N} = \langle 0, -\sin t, -\cos t \rangle$; at $(0, 0, 1)$, $\mathbf{T} = \frac{1}{2}\sqrt{2}\langle 1, 1, 0 \rangle$, $\mathbf{N} = \langle 0, 0, -1 \rangle$

45. $\mathbf{T} = \frac{1}{3}\sqrt{3}(\mathbf{i} + \mathbf{j} + \mathbf{k})$, $\mathbf{N} = \frac{1}{2}\sqrt{2}(-\mathbf{i} + \mathbf{j})$

47. $x = 2 + \frac{4}{13}s$, $y = 1 - \frac{12}{13}s$, $z = 3 + \frac{3}{13}s$

49. $x(s) = 3\cos\frac{1}{5}s$, $y(s) = 3\sin\frac{1}{5}s$, $z(s) = \frac{4}{5}s$

51. Begin with $\dfrac{d}{dt}(\mathbf{v} \cdot \mathbf{v})$.

53. $|t|^{-1}$

55. $A = 3$, $B = -8$, $C = 6$, $D = 0$, $E = 0$, $F = 0$

Section 13.6 (page 741)

1. A plane with intercepts $(\frac{20}{3}, 0, 0)$, $(0, 10, 0)$, and $(0, 0, 2)$

3. A vertical circular cylinder with radius 3

5. A vertical cylinder intersecting the xy-plane in the rectangular hyperbola $xy = 4$

7. An elliptical paraboloid opening upward from its vertex at the origin

9. A circular paraboloid opening downward from its vertex at $(0, 0, 4)$

11. A paraboloid opening upward, vertex at the origin, axis the z-axis

13. A cone, vertex the origin, axis the z-axis (both nappes)

15. A parabolic cylinder perpendicular to the xz-plane, its trace there the parabola opening upward with axis the z-axis and vertex at $(x, z) = (0, -2)$

17. An elliptical cylinder perpendicular to the xy-plane, its trace there the ellipse with center $(0, 0)$ and intercepts $(\pm 1, 0)$ and $(0, \pm 2)$

19. An elliptical cone, vertex $(0, 0, 0)$, axis the x-axis

21. A paraboloid, opening downward, vertex at the origin, axis the z-axis

23. A hyperbolic paraboloid, saddle point at the origin, meeting the xz-plane in a parabola with vertex the origin and opening downward, meeting the xy-plane in a parabola with vertex the origin and opening upward, meeting each plane parallel to the yz-plane in a hyperbola with directrices parallel to the y-axis

25. A hyperboloid of one sheet, axis the z-axis, trace in the xy-plane the circle with center $(0, 0)$ and radius 3, traces in parallel planes larger circles, and traces in planes parallel to the z-axis hyperbolas

27. An elliptic paraboloid, axis the y-axis, vertex at the origin

29. A hyperboloid of two sheets, axis the y-axis

31. A paraboloid, axis the x-axis, vertex at the origin, equation $x = 2(y^2 + z^2)$

33. Hyperboloid of one sheet, equation $x^2 + y^2 - z^2 = 1$

35. A paraboloid, vertex at the origin, axis the x-axis, equation $y^2 + z^2 = 4x$

37. The surface resembles a rug covering a turtle: highest point $(0, 0, 1)$; $z \to 0$ from above as $|x|$ or $|y|$ (or both) increase without bound, equation $z = \exp(-x^2 - y^2)$.

39. A circular cone with axis of symmetry the z-axis

41. Ellipses with semiaxes 2 and 1

43. Circles   **45.** Parabolas opening downward

47. Parabolas opening upward if $k > 0$, downward if $k < 0$

51. The projection of the intersection has equation $x^2 + y^2 = 2y$; it is the circle with center $(0, 1)$ and radius 1.

53. Equation: $5x^2 + 8xy + 8y^2 - 8x - 8y = 0$. Because the discriminant is negative, it is an ellipse. In a uv-plane rotated approximately $55°16'41''$ from the xy-plane, the ellipse has center $(u, v) = (0.517, -0.453)$, minor axis 0.352 in the u-direction, and major axis 0.774 in the v-direction.

Section 13.7 (page 747)

1. $(0, 0, 5)_{cyl}$, $(5, 0, 0)_{sph}$

3. $\left(\sqrt{2}, \dfrac{\pi}{4}, 0\right)_{cyl}$, $\left(\sqrt{2}, \dfrac{\pi}{2}, \dfrac{\pi}{4}\right)_{sph}$

5. $\left(\sqrt{2}, \dfrac{\pi}{4}, 1\right)_{cyl}$, $\left(\sqrt{3}, \tan^{-1}\sqrt{2}, \dfrac{\pi}{4}\right)_{sph}$

7. $(\sqrt{5}, \tan^{-1}\frac{1}{2}, -2)_{cyl}$,
$\left(3, \dfrac{\pi}{2} + \tan^{-1}(\frac{1}{2}\sqrt{5}), \tan^{-1}\frac{1}{2}\right)_{sph}$

9. $(5, \tan^{-1}\frac{4}{3}, 12)_{cyl}$, $(13, \tan^{-1}\frac{5}{12}, \tan^{-1}\frac{4}{3})_{sph}$

11. A cylinder, radius 5, axis the z-axis

13. The *plane* $y = x$

15. The upper nappe of the cone $x^2 + y^2 = 3z^2$

17. The xy-plane

19. A cylinder, axis the vertical line $x = 0$, $y = 1$; its trace in the xy-plane is the circle $x^2 + (y - 1)^2 = 1$.

21. The vertical plane with trace the line $y = -x$ in the xy-plane

23. The horizontal plane $z = 1$

25. $r^2 + z^2 = 25$, $\rho = 5$

27. $r(\cos\theta + \sin\theta) + z = 1$,
$\rho(\sin\phi\cos\theta + \sin\phi\sin\theta + \cos\phi) = 1$

29. $r^2 + z^2 = r(\cos\theta + \sin\theta) + z$,
$\rho = \sin\phi\cos\theta + \sin\phi\sin\theta + \cos\phi$

31. $z = r^2$

33. (a) $1 \leq r^2 \leq 4 - z^2$; (b) $\csc\phi \leq \rho \leq 2$
(and, as a consequence, $\dfrac{\pi}{6} \leq \phi \leq \dfrac{5\pi}{6}$)

35. About 3821 mi $\approx$ 6149 km

37. Just under 50 km $\approx$ 31 mi

39. $0 \leq \rho \leq H\sec\phi$, $0 \leq \phi \leq \arctan\left(\dfrac{R}{H}\right)$, θ arbitrary

Chapter 13 Miscellaneous Problems (page 750)

7. $x = 1 + 2t$, $y = -1 + 3t$, $z = 2 - 3t$;
$\dfrac{x - 1}{2} = \dfrac{y + 1}{3} = \dfrac{2 - z}{3}$

9. $-13x + 22y + 6z = -23$

11. $x - y + 2z = 3$

15. 3

17. $\frac{1}{9}$; $a_T = 2$, $a_N = 1$

21. $3x - 3y + z = 1$

27. $\rho = 2\cos\phi$

29. $\rho^2 = 2\cos 2\phi$; shaped like an hourglass with rounded ends

37. The curvature is zero when x is an integral multiple of π and reaches the maximum value 1 when x is an odd integral multiple of $\pi/2$.

39. $\mathbf{N} = -\dfrac{2}{\sqrt{\pi^2 + 4}}\mathbf{i} - \dfrac{\pi}{\sqrt{\pi^2 + 4}}\mathbf{j}$,

$\mathbf{T} = -\dfrac{\pi}{\sqrt{\pi^2 + 4}}\mathbf{i} + \dfrac{2}{\sqrt{\pi^2 + 4}}\mathbf{j}$

43. $A = \frac{15}{8}$, $B = -\frac{5}{4}$, $C = \frac{3}{8}$

Section 14.2 (page 761)

1. All (x, y)

3. Except on the line $x = y$

5. Except on the coordinate axes $x = 0$ and $y = 0$

7. Except on the lines $y = \pm x$

9. Except at the origin $(0, 0, 0)$

11. The horizontal plane with equation $z = 10$

13. A plane that makes a $45°$ angle with the xy-plane, intersecting it in the line $x + y = 0$

15. A circular paraboloid opening upward from its vertex at the origin

17. The upper hemispherical surface of radius 2 centered at the origin

19. A circular cone opening downward from its vertex at $(0, 0, 10)$

21. Straight lines of slope 1

23. Ellipses centered at $(0, 0)$, each with major axis twice the minor axis and lying on the x-axis

25. Vertical (y-direction) translates of the curve $y = x^3$

27. Circles centered at the point $(2, 0)$

29. Circles centered at the origin

31. Circular paraboloids opening upward, each with its vertex on the z-axis

33. Spheres centered at $(2, -1, 3)$

35. Elliptical cylinders, each with axis the vertical line through $(2, 1, 0)$ and with the length of the x-semiaxis twice that of the y-semiaxis

37.

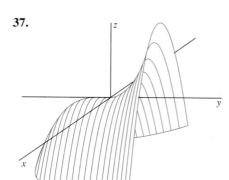

39.

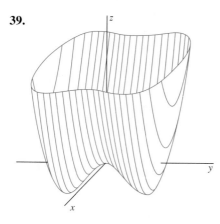

41. Corresponding figures are 14.2.27 and 14.2.35; 14.2.28 and 14.2.36; 14.2.29 and 14.2.33; 14.2.30 and 14.2.34; 14.2.31 and 14.2.38; 14.2.32 and 14.2.37.

Section 14.3 (page 767)

1. 7   **3.** e **5.** $\frac{5}{3}$ **7.** 0

9. 1   **11.** $-\frac{3}{2}$ **13.** 1   **15.** -4

17. y, x **19.** $y^2, 2xy$

29. Note that $f(x, y) = (x - \frac{1}{2})^2 + (y + 1)^2 - \frac{1}{4} \geqq -\frac{1}{4}$ for all x.

Section 14.4 (page 774)

1. $\dfrac{\partial f}{\partial x} = 4x^3 - 3x^2y + 2xy^2 - y^3,$

$\dfrac{\partial f}{\partial y} = -x^3 + 2x^2y - 3xy^2 + 4y^3$

3. $\dfrac{\partial f}{\partial x} = e^x(\cos y - \sin y), \dfrac{\partial f}{\partial y} = -e^x(\cos y + \sin y)$

5. $\dfrac{\partial f}{\partial x} = -\dfrac{2y}{(x - y)^2}, \dfrac{\partial f}{\partial y} = \dfrac{2x}{(x - y)^2}$

7. $\dfrac{\partial f}{\partial x} = \dfrac{2x}{x^2 + y^2}, \dfrac{\partial f}{\partial y} = \dfrac{2y}{x^2 + y^2}$

9. $\dfrac{\partial f}{\partial x} = yx^{y-1}, \dfrac{\partial f}{\partial y} = x^y \ln x$

11. $\dfrac{\partial f}{\partial x} = 2xy^3z^4, \dfrac{\partial f}{\partial y} = 3x^2y^2z^4, \dfrac{\partial f}{\partial z} = 4x^2y^3z^3$

13. $\dfrac{\partial f}{\partial x} = yze^{xyz}, \dfrac{\partial f}{\partial y} = xze^{xyz}, \dfrac{\partial f}{\partial z} = xye^{xyz}$

15. $\dfrac{\partial f}{\partial x} = 2xe^y \ln z, \dfrac{\partial f}{\partial y} = x^2e^y \ln z, \dfrac{\partial f}{\partial z} = \dfrac{x^2e^y}{z}$

17. $\dfrac{\partial f}{\partial r} = \dfrac{4rs^2}{(r^2 + s^2)^2}, \dfrac{\partial f}{\partial s} = -\dfrac{4r^2s}{(r^2 + s^2)^2}$

19. $\dfrac{\partial f}{\partial u} = e^v + we^u, \dfrac{\partial f}{\partial v} = e^w + ue^v, \dfrac{\partial f}{\partial w} = e^u + ve^w$

21. $z_{xy} = z_{yx} = -4$

23. $z_{xy} = z_{yx} = -4xy \exp(-y^2)$

25. $z_{xy} = z_{yx} = \dfrac{1}{(x + y)^2}$

27. $z_{xy} = z_{yx} = 3e^{-3x} \sin y$

29. $z_{xy} = z_{yx} = -4xy^{-3} \sinh y^{-2}$

31. $6x + 8y - z = 25$ **33.** $z \equiv -1$

35. $27x - 12y - z = 38$ **37.** $x - y + z = 1$

39. $10x - 16y - z = 9$

43. $f_{xyz}(x, y, z) = (x^2y^2z^2 + 3xyz + 1)e^{xyz}$

51. $(10, -7, -58)$

53. (a) A decrease of about 2750 cm³; (b) an increase of about 82.5 cm³

Section 14.5 (page 784)

1. None   **3.** $(0, 0, 5)$ **5.** $(3, -1, -5)$

7. $(-2, 0, -4)$ **9.** $(-2, 0, -7)$ and $(-2, 1, -9)$

11. $(0, 0, 0)$, $(1, 0, 2/e)$, $(-1, 0, 2/e)$, $(0, 1, 3/e)$, and $(0, -1, 3/e)$

13. Min. value: $f(1, 1) = 1$

15. Max. value: $f(1, \pm1) = 2$

17. Min. value: $f(4, -2) = f(-4, 2) = -16$

19. Max. value: $f(1, -2) = e^5$

21. Max. value: $f(1, 1) = 3$, min. value: $f(-1, -1) = -3$

23. Max value: $f(0, 2) = 4$, min. value: $f(1, 0) = -1$

25. Let $t = 1/\sqrt{2}$. Max. value: $f(t, t) = f(-t, -t) = 1$, min. value: $f(t, -t) = f(-t, t) = -1$

27. $10 \times 10 \times 10$ **29.** $10 \times 10 \times 10$ cm

31. $\frac{4}{3}$; $x = 2, y = 1, z = \frac{2}{3}$

33. Height 10 ft, front and back 40 ft wide, sides 20 ft deep

35. Base 2 m $\times$ 2 m, height 3 m

37. 11,664 in.³   **39.** $\frac{1}{2}$

41. Max. area: 900 (one square), min. area: 300 (three equal squares)

45. Local min. at $(1, 1)$, saddle point at $(-\frac{1}{3}, -\frac{1}{3})$

47. Global max. 1 at $(\pm 1, 1)$, global min. -1 at $(\pm 1, -1)$, local max. at *every* point $(0, y)$ for which $-1 \leq y < 0$, local min. at every point $(0, y)$ for which $0 < y \leq 1$, a sort of half-saddle at $(0, 0)$

49. 1 m³   **51.** $abc/27$

53. Base 20×20 in., height 60 in.   **55.** $(\frac{2}{3}, \frac{1}{3})$

57. The base b of each triangular end should equal its height; the depth of the A-frame should be $b\sqrt{2}$.

Section 14.6 (page 793)

1. $dw = (6x + 4y)\, dx + (4x - 6y^2)\, dy$

3. $dw = (x\, dx + y\, dy)/\sqrt{1 + x^2 + y^2}$

5. $dw = (-y\, dx + x\, dy)/(x^2 + y^2)$

7. $dw = (2x\, dx + 2y\, dy + 2z\, dz)/(x^2 + y^2 + z^2)$

9. $dw = (\tan yz)\, dx + (xz \sec^2 yz)\, dy + (xy \sec^2 yz)\, dz$

11. $dw = -e^{-xyz}(yz\, dx + xz\, dy + xy\, dz)$

13. $dw = \exp(-v^2)(2u\, du - 2u^2 v\, dv)$

15. $dw = (x\, dx + y\, dy + z\, dz)/\sqrt{x^2 + y^2 + z^2}$

17. $\Delta f \approx 0.014$ (true value: about 0.01422975)

19. $\Delta f \approx -0.0007$ **21.** $\Delta f \approx \frac{53}{1300} \approx 0.04077$

23. $\Delta f \approx 0.06$ **25.** 191.1   **27.** 1.4

29. $x \approx 1.95$ **31.** 8.18 in.³   **33.** 0.022 acres

35. The period increases by about 0.0278 s.

37. About 303.8 ft

Section 14.7 (page 800)

1. $-(2t + 1) \exp(-t^2 - t)$ **3.** $6t^5 \cos t^6$

5. $\dfrac{\partial w}{\partial s} = \dfrac{\partial w}{\partial t} = \dfrac{2}{s + t}$ **7.** $\dfrac{\partial w}{\partial s} = 0$, $\dfrac{\partial w}{\partial t} = 5e^t$

9. $\partial r/\partial x = (y + z) \exp(yz + xy + xz)$,
$\partial r/\partial y = (x + z) \exp(yz + xy + xz)$,
$\partial r/\partial z = (x + y) \exp(yz + xy + xz)$

11. $\dfrac{\partial z}{\partial x} = -\left(\dfrac{z}{x}\right)^{1/3}$, $\dfrac{\partial z}{\partial y} = -\left(\dfrac{z}{y}\right)^{1/3}$

13. $\dfrac{\partial z}{\partial x} = -\dfrac{yz(e^{xy} + e^{xz}) + (xy + 1)e^{xy}}{e^{xy} + xye^{yz}}$,

$\dfrac{\partial z}{\partial y} = -\dfrac{x(x + z)\, e^{xy} + e^{xz}}{xye^{xz} + e^{xy}}$

15. $\dfrac{\partial z}{\partial x} = -\dfrac{c^2 x}{a^2 z}$, $\dfrac{\partial z}{\partial y} = -\dfrac{c^2 y}{b^2 z}$

17. $\dfrac{\partial w}{\partial x} = \dfrac{(2x^2 - y^2)y^{1/2}}{2x^{1/2}(x^2 - y^2)^{3/4}}$,

$\dfrac{\partial w}{\partial y} = \dfrac{(x^2 - 2y^2)x^{1/2}}{2y^{1/2}(x^2 - y^2)^{3/4}}$

19. $\dfrac{\partial w}{\partial x} = (y^3 - 3x^2 y)(x^2 + y^2)^{-3} - y$,

$\dfrac{\partial w}{\partial y} = (x^3 - 3xy^2)(x^2 + y^2)^{-3} - x$

21. $x + y - 2z = 7$

23. $5x + 5y + 11z = 31$

27. $\partial w/\partial x = f'(u)(\partial u/\partial x) = f'(u)$, and so on.

29. Show that $w_u = w_x + w_y$. Then note that

$$w_{uv} = \frac{\partial}{\partial v}w_u = \frac{\partial w_u}{\partial x} \cdot \frac{\partial x}{\partial v} + \frac{\partial w_u}{\partial y} \cdot \frac{\partial y}{\partial v}.$$

Section 14.8 (page 809)

1. $\langle 3, -7 \rangle$ **3.** $\langle 0, 0 \rangle$ **5.** $\langle 0, 6, -4 \rangle$

7. $\langle 1, 1, 1 \rangle$ **9.** $\langle 2, -\frac{3}{2}, -2 \rangle$ **11.** $8\sqrt{2}$

13. $\frac{12}{13}\sqrt{13}$ **15.** $-\frac{13}{20}$ **17.** $-\frac{1}{6}$ **19.** $-6\sqrt{2}$

21. Max.: $\sqrt{170}$; direction: $\langle 7, 11 \rangle$

23. Max.: $14\sqrt{2}$; direction: $\langle 3, 5, -8 \rangle$

25. Max.: $2\sqrt{14}$; direction: $\langle 1, 2, 3 \rangle$

27. $29(x - 2) - 4(y + 3) = 0$

29. $x + y + z = 1$

35. (a) $\frac{34}{3}$°C/ft; (b) 13°C/ft and $\langle 4, 3, 12 \rangle$

37. (a) $z = \frac{3}{10}x + \frac{1}{5}y - \frac{2}{5}$ (b) 0.44 (true value: 0.448)

39. $x - 2y + z + 10 = 0$

43. Compass heading about 36.87°; climbing at 45°

45. Compass heading about 203.2°; climbing at about 75.29°

Section 14.9 (page 818)

1. Max. 4 at $(\pm 2, 0)$, min. -4 at $(0, \pm 2)$

3. Max. 3 at $(\frac{3}{2}\sqrt{2}, \sqrt{2})$ and $(-\frac{3}{2}\sqrt{2}, -\sqrt{2})$, min. -3 at $(-\frac{3}{2}\sqrt{2}, \sqrt{2})$ and $(\frac{3}{2}\sqrt{2}, -\sqrt{2})$

5. Min. $\frac{18}{7}$ at $(\frac{9}{7}, \frac{6}{7}, \frac{3}{7})$, no max.

7. Max. 7 at $(\frac{36}{7}, \frac{9}{7}, \frac{4}{7})$, min. -7 at $(-\frac{36}{7}, -\frac{9}{7}, -\frac{4}{7})$

9. Min. $\frac{25}{3}$ at $(-\frac{5}{3}, \frac{1}{3}, \frac{7}{3})$. There is no max. because, in effect, we seek the extrema of the square of the distance between the origin and a point (x, y, z) on an unbounded straight line.

21. $(2, -2, 1)$ and $(-2, 2, 1)$

25. $(2, 3)$ and $(-2, -3)$

27. Highest: $(\frac{2}{5}\sqrt{5}, \frac{1}{5}\sqrt{5}, \sqrt{5} - 4)$; lowest: $(-\frac{2}{5}\sqrt{5}, -\frac{1}{5}\sqrt{5}, -\sqrt{5} - 4)$

29. Farthest: $x = -\frac{1}{20}(15 + 9\sqrt{5})$, $y = 2x$, $z = \frac{1}{4}(9 + 3\sqrt{5})$; nearest: $x = -\frac{1}{20}(15 - 9\sqrt{5})$, $y = 2x$, $z = \frac{1}{4}(9 - 3\sqrt{5})$

33. $\dfrac{3 - 2\sqrt{2}}{4} P^2$

Section 14.10 (page 827)

1. Min. $(-1, 2, -1)$, no other extrema

3. Saddle point $(-\frac{1}{2}, -\frac{1}{2}, \frac{29}{4})$, no extrema

5. Min. $(-3, 4, -9)$, no other extrema

7. Saddle point $(0, 0, 3)$, local max. $(-1, -1, 4)$

9. No extrema

11. Saddle point $(0, 0, 0)$, local min. $(-1, -1, -2)$ and $(1, 1, -2)$

13. Saddle point $(-1, 1, 5)$, local min. $(3, -3, -27)$

15. Saddle point $(0, -2, 32)$, local min. $(-5, 3, -93)$

17. Saddle point $(0, 0, 0)$, local max. $(1, 2, 2)$ and $(-1, -2, 2)$

19. Saddle point $(-1, 0, 17)$, local min. $(2, 0, -10)$

21. Saddle point $(0, 0, 0)$, local (actually, global) max. $\left(\dfrac{\sqrt{2}}{2}, \dfrac{\sqrt{2}}{2}, \dfrac{1}{2e}\right)$ and $\left(-\dfrac{\sqrt{2}}{2}, -\dfrac{\sqrt{2}}{2}, \dfrac{1}{2e}\right)$, local (actually, global) min. $\left(-\dfrac{\sqrt{2}}{2}, \dfrac{\sqrt{2}}{2}, -\dfrac{1}{2e}\right)$ and $\left(\dfrac{\sqrt{2}}{2}, -\dfrac{\sqrt{2}}{2}, -\dfrac{1}{2e}\right)$

23. Local (actually, global) min.

25. Local (actually, global) max.

27. Min. value 3 at $(1, 1)$ and at $(-1, -1)$

29. See Problem 41 of Section 14.5 and its answer

31. The critical points are of the form (m, n), where both m and n are either even integers or odd integers. The critical point (m, n) is a saddle point if both m and n are even, but a local maximum if both m and n are of the form $4k + 1$ or of the form $4k + 3$. It is a local minimum in the remaining cases.

35. Local min. at $(-1.8794, 0)$ and $(1.5321, 0)$, saddle point at $(0.3473, 0)$

37. Local min. at $(-1.8794, 1.8794)$ and $(1.5321, -1.5321)$, saddle point at $(0.3473, -0.3473)$

39. Local min. at $(3.6247, 3.9842)$ and $(3.6247, -3.9842)$, saddle point at $(0, 0)$

Chapter 14 Miscellaneous Problems (page 830)

3. On the line $y = x$, $g(x, y) \equiv \frac{1}{2}$, except that $g(0, 0) = 0$.

5. $f(x, y) = x^2 y^3 + e^x \sin y + y + C$

7. All points of the form $(a, b, \frac{1}{2})$ (so $a^2 + b^2 = \frac{1}{2}$) together with $(0, 0, 0)$

9. The normal to the cone at $(a, b, \sqrt{a^2 + b^2})$ meets the z-axis at $(0, 0, 2\sqrt{a^2 + b^2})$.

15. Base $2\sqrt[3]{3} \times 2\sqrt[3]{3}$ ft, height $5\sqrt[3]{3}$ ft

17. $200 \pm 2\ \Omega$

19. 3%

21. $(\pm4, 0, 0)$, $(0, \pm2, 0)$, and $(0, 0, \pm\frac{4}{3})$

25. Parallel to the vector $\langle 4, -3 \rangle$; that is, at an approximate bearing of either $126.87°$ or $306.87°$

27. 1

31. Semiaxes 1 and 2

33. There is no such triangle of minimum perimeter, unless we consider as a triangle the figure with all sides of length zero—a single point on the circumference of the circle. The triangle of *maximum* perimeter is equilateral, with perimeter $3\sqrt{3}$.

35. Closest: $(\frac{1}{3}\sqrt{6}, \frac{1}{6}\sqrt{6})$, farthest: $(-\frac{1}{3}\sqrt{6}, -\frac{1}{6}\sqrt{6})$

39. Max. 1, min. $-\frac{1}{2}$

41. Local min. -1 at $(1, 1)$ and at $(-1, -1)$; horizontal tangent plane (but no extrema) at $(0, 0, 0)$, $(\sqrt{3}, 0, 0)$, and $(-\sqrt{3}, 0, 0)$

43. Local min. -8 at $(2, 2)$, horizontal tangent plane at $(0, 0, 0)$

45. Local max. $\frac{1}{432}$ at $(\frac{1}{2}, \frac{1}{3})$. All points on the intervals $(-\infty, 0)$ and $(1, +\infty)$ on the x-axis are local min. (value 0), and all points on the interval $(0, 1)$ on the x-axis are local max. (value 0); saddle point at $(0, 1)$. Horizontal tangent plane at the origin; it really isn't a saddle point.

47. Saddle point at $(0, 0)$; each point on the hyperbola $xy = \ln 2$ yields a global min.

49. No extrema; saddle points at $(1, 1)$ and $(-1, -1)$

Section 15.1 (page 839)

1. 80   **3.** -78 **5.** $\frac{513}{4}$

7. $-\frac{9}{2}$ **9.** 1   **11.** $\frac{1}{2}(e - 1)$

13. $2(e - 1)$ **15.** $2\pi + \frac{1}{4}\pi^4$ **17.** 1

19. $2\ln 2$ **21.** -32 **23.** $\frac{4}{15}(9\sqrt{3} - 8\sqrt{2} + 1)$

Section 15.2 (page 845)

1. $\frac{5}{6}$ **3.** $\frac{1}{2}$ **5.** $\frac{1}{12}$ **7.** $\frac{1}{20}$

9. $-\frac{1}{18}$ **11.** $\frac{1}{2}(e - 2)$ **13.** $\frac{61}{3}$

15. $\displaystyle\int_0^4 \int_{-\sqrt{y}}^{\sqrt{y}} x^2 y\, dx\, dy = \frac{512}{21}$

17. $\displaystyle\int_0^1 \int_{-\sqrt{y}}^{\sqrt{y}} x\, dx\, dy + \int_1^9 \int_{(y-3)/2}^{\sqrt{y}} x\, dx\, dy = \frac{32}{3}$

19. $\int_0^4 \int_{2-\sqrt{4-y}}^{y/2} 1\, dx\, dy = \frac{4}{3}$ **21.** $\int_0^\pi \int_0^y \frac{\sin y}{y}\, dx\, dy = 2$

23. $\int_0^1 \int_0^x \frac{1}{1+x^4}\, dy\, dx = \frac{\pi}{8}$

Section 15.3 (page 850)

1. $\frac{1}{6}$ **3.** $\frac{32}{3}$ **5.** $\frac{5}{6}$ **7.** $\frac{32}{3}$

9. $2\ln 2$ **11.** 2 **13.** $2e$ **15.** $\frac{1}{3}$

17. $\frac{41}{60}$ **19.** $\frac{4}{15}$ **21.** $\frac{10}{3}$ **23.** 19

25. $\frac{4}{3}$ **27.** $\frac{1}{6}abc$ **29.** $\frac{2}{3}$

33. $\frac{625}{2}\pi \approx 981.748$

35. $\frac{1}{6}(2\pi + 3\sqrt{3})R^3 \approx (1.913)R^3$ **37.** $\frac{256}{15}$

Section 15.4 (page 858)

3. $\frac{3}{2}\pi$ **5.** $\frac{1}{6}(4\pi - 3\sqrt{3}) \approx 1.22837$

7. $\frac{1}{2}(2\pi - 3\sqrt{3}) \approx 0.5435$ **9.** $\frac{16}{3}\pi$

11. $\frac{23}{8}\pi$ **13.** $\frac{1}{4}\pi\ln 2$ **15.** $\frac{16}{5}\pi$

17. $\frac{1}{4}\pi(1 - \cos 1) \approx 0.36105$

19. 2π **21.** 4π **27.** 2π

29. $\frac{1}{3}\pi a^3(2 - \sqrt{2}) \approx (0.6134)a^3$

31. $\frac{1}{4}\pi$ **35.** $2\pi^2 a^2 b$

Section 15.5 (page 868)

1. $(2, 3)$ **3.** $(1, 1)$ **5.** $(\frac{4}{3}, \frac{2}{3})$

7. $(\frac{3}{2}, \frac{6}{5})$ **9.** $(0, -\frac{8}{5})$ **11.** $\frac{1}{24}, (\frac{2}{5}, \frac{2}{5})$

13. $\frac{256}{15}, (0, \frac{16}{7})$ **15.** $\frac{1}{12}, (\frac{9}{14}, \frac{9}{14})$ **17.** $\frac{1}{3}, (0, \frac{22}{35})$

19. $2, (\frac{1}{2}\pi, \frac{1}{8}\pi)$ **21.** $a^3, (\frac{7}{12}a, \frac{7}{12}a)$ **23.** $\frac{128}{5}, (0, \frac{20}{7})$

25. $\pi; \bar{x} = \frac{\pi^2 - 4}{\pi} \approx 1.87, \bar{y} = \frac{\pi}{8} \approx 0.39$

27. $\frac{1}{3}\pi a^3; \bar{x} = 0, \bar{y} = \frac{3a}{2\pi}$

29. $\frac{2}{3}\pi + \frac{1}{4}\sqrt{3}; \bar{x} = 0, \bar{y} = \frac{36\pi + 33\sqrt{3}}{32\pi + 12\sqrt{3}} \approx 1.4034$

31. $\frac{2\pi a^{n+4}}{n + 4}$ **33.** $\frac{3}{2}\pi k$ **35.** $\frac{1}{9}$

37. $\hat{x} = \frac{2}{21}\sqrt{105}, \hat{y} = \frac{4}{3}\sqrt{5}$

39. $\hat{x} = \hat{y} = \frac{1}{10}a\sqrt{30}$

41. $(4r/3\pi, 4r/3\pi)$ **43.** $(2r/\pi, 2r/\pi)$

51. (a) $\bar{x} = 0, \bar{y} = \frac{4a^2 + 3\pi ab + 6b^2}{12b + 3\pi a}$;

(b) $\frac{1}{3}\pi a(4a^2 + 2\pi ab + 6b^2)$

53. $(1, \frac{1}{4})$ **55.** $\frac{484}{3}k$

Section 15.6 (page 876)

1. 18 **3.** 128 **5.** $\frac{1}{60}$ **7.** 0 **9.** 12

11. $V = \int_0^3 \int_0^{3-(2x/3)} \int_0^{6-2x-2y} 1\, dz\, dy\, dx = 6$

13. $\frac{128}{5}$ **15.** $\frac{332}{105}$ **17.** $\frac{256}{15}$

19. $\frac{11}{30}$ **21.** $(0, \frac{20}{7}, \frac{20}{7})$ **23.** $(0, \frac{8}{7}, \frac{12}{7})$

25. $\bar{x} = 0, \bar{y} = \frac{44 - 9\pi}{72 - 9\pi}, \bar{z} = \frac{9\pi - 16}{72 - 9\pi}$

27. $\frac{8}{7}$ **29.** $\frac{1}{30}$ **33.** $\frac{2}{3}a^5$

35. $\frac{38}{45}ka^7$ **37.** $\frac{1}{3}k$ **39.** $(\frac{9}{64}\pi, \frac{9}{64}\pi, \frac{3}{8})$

41. 24π **43.** $\frac{1}{6}\pi$

Section 15.7 (page 884)

1. 8π **5.** $\frac{4}{3}\pi(8 - 3\sqrt{3})$ **7.** $\frac{1}{2}\pi a^2 h^2$

9. $\frac{1}{4}\pi a^4 h^2$ **11.** $\frac{81}{2}\pi, (0, 0, 3)$ **13.** 24π

15. $\frac{1}{6}\pi(8\sqrt{2} - 7)$ **17.** $\frac{1}{12}\pi\delta a^2 h(3a^2 + 4h^2)$

19. $\frac{1}{3}\pi$ **21.** $(0, 0, 3a/8)$ **23.** $\frac{1}{3}\pi$

25. $\frac{1}{3}\pi a^3(2 - \sqrt{2}); \bar{x} = 0 = \bar{y}, \bar{z} = \frac{3}{16}(2 + \sqrt{2})a \approx (0.6402)a$

27. $\frac{7}{5}ma^2$

29. The surface obtained by rotating the circle in the xz-plane with center $(a, 0)$ and radius a around the z-axis—a doughnut with an infinitesimal hole; $2\pi^2 a^3$

31. $\frac{2}{15}(128 - 51\sqrt{3})\pi a^5$ **33.** $\frac{37}{48}\pi a^4, \bar{z} = \frac{105}{74}a$

Section 15.8 (page 891)

1. $6\pi\sqrt{11}$ **3.** $\frac{1}{6}\pi(17\sqrt{17} - 1)$

5. $3\sqrt{2} + \frac{1}{2}\ln(3 + 2\sqrt{2}) \approx 5.124$ **7.** $3\sqrt{14}$

9. $\frac{2}{3}\pi(2\sqrt{2} - 1) \approx 3.829$ **11.** $\frac{1}{6}\pi(65\sqrt{65} - 1)$

15. $8a^2$

Section 15.9 (page 899)

1. $x = \frac{1}{2}(u + v), y = \frac{1}{2}(u - v); J = -\frac{1}{2}$

3. $x = \sqrt{u/v}, y = \sqrt{uv}; J = (2v)^{-1}$

5. $x = \frac{1}{2}(u + v), y = \frac{1}{2}\sqrt{u - v}; J = -\frac{1}{4}(u - v)^{-1/2}$

7. $\frac{3}{5}$ **9.** $\ln 2$ **11.** $\frac{1}{8}(2 - \sqrt{2})$

13. $\frac{39}{2}\pi$ **15.** 8

17. S is the region $3u^2 + v^2 \leqq 3$; the value of the integral is $\frac{2\pi\sqrt{3}(e^3 - 1)}{3e^3}$.

Chapter 15 Miscellaneous Problems (page 901)

1. $\frac{1}{3}(2 - \sqrt{2})$ **3.** $\frac{e - 1}{2e}$ **5.** $\frac{1}{4}(e^4 - 1)$

7. $\frac{4}{3}$ **9.** $9\pi, \bar{z} = \frac{9}{16}$ **11.** 4π

13. 4π **15.** $\frac{1}{16}(\pi - 2)$ **17.** $\frac{128}{15}, (\frac{32}{7}, 0)$

19. $k\pi$, $(1, 0)$ **21.** $\bar{y} = 4b/3\pi$ **23.** $(0, \frac{8}{5})$

25. $\frac{10}{3}\pi(\sqrt{5} - 2) \approx 2.4721$

27. $\frac{3}{10}Ma^2$ **29.** $\frac{1}{5}M(b^2 + c^2)$

31. $\frac{128}{225}\delta(15\pi - 26) \approx (12.017)\delta$, where δ is the (constant) density

33. $\frac{8}{3}\pi$ **41.** $\frac{18}{7}$

43. $\frac{1}{6}\pi(37\sqrt{37} - 17\sqrt{17}) \approx 81.1418$

47. $4\sqrt{2}$ **48.** Approximately 3.49608

51. 3δ **53.** $\frac{8}{15}\pi abc$

Section 16.1 (page 909)

1.

3.

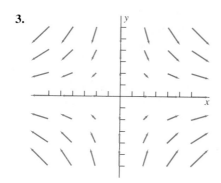

5.

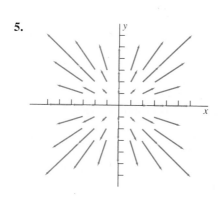

7.

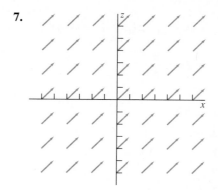

9.

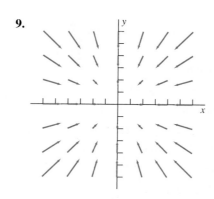

11. 3, **0**

13. 0, **0**

15. $x^2 + y^2 + z^2$, $\langle -2yz, -2xz, -2xy \rangle$

17. 0, $\langle 2y - 2z, 2z - 2x, 2x - 2y \rangle$

19. 3, $\langle x \cos xy - x \cos xz, y \cos yz - y \cos xy, z \cos xz - z \cos yz \rangle$

Section 16.2 (page 918)

1. $\frac{310}{3}$, $\frac{248}{3}$, 62 **3.** $3\sqrt{2}$, 3, 3

5. $\frac{49}{24}$, $\frac{3}{2}$, $\frac{4}{3}$ **7.** $\frac{6}{5}$

9. 315 **11.** $\frac{19}{60}$

13. $\pi + 2\pi^2$ **15.** 28

17. $\frac{1}{6}(14\sqrt{14} - 1) \approx 8.563867$

19. $(0, 2a/\pi)$

21. $10k\pi$, $(0, 0, 4\pi)$

23. $\frac{1}{2}ka^3$, $(\frac{2}{3}a, \frac{2}{3}a)$, $I_x = I_y = \frac{1}{4}ka^5$

25. (a) $\frac{1}{2}k \ln 2$; (b) $-\frac{1}{2}k \ln 2$

Section 16.3 (page 925)

1. $f(x, y) = x^2 + 3xy + y^2$

3. $f(x, y) = x^3 + 2xy^2 + 2y^3$

A-68

5. $f(x, y) = \frac{1}{4}x^4 + \frac{1}{3}y^3 + y \ln x$

7. $f(x, y) = \sin x + x \ln y + e^y$

9. $f(x, y) = x^3y^3 + xy^4 + \frac{1}{5}y^5$

11. $f(x, y) = x^2y^{-1} + y^2x^{-3} + 2y^{1/2}$

17. 6

19. $1/e$

21. $-\pi$

23. $f(x, y, z) = xyz$

25. $f(x, y, z) = xy \cos z - yze^x$

27. $\mathbf{F}$ is not conservative on any region containing $(0, 0)$.

29. $Q_z = 0 \neq 2y = R_y$

Section 16.4   (page 933)

1. 0   **3.** 3   **5.** $\frac{3}{10}$ **7.** 2   **9.** 0

11. $\frac{16}{105}$ **13.** πa^2 **15.** $\frac{3}{8}\pi$ **25.** $\frac{3}{2}$ **31.** 2π

Section 16.5 (page 941)

1. $\frac{1}{120}\sqrt{3}$ **3.** $\frac{1}{60}\pi(1 + 391\sqrt{17}) \approx 84.4635$

5. $\frac{16}{3}\pi$ **7.** 24π **9.** 0

11. $(\frac{1}{2}a, \frac{1}{2}a, \frac{1}{2}a)$

13. $\bar{z} = \dfrac{1 + (24a^4 + 2a^2 - 1)\sqrt{1 + 4a^2}}{10[(1 + 4a^2)^{3/2} - 1]}$,

$I_z = \frac{1}{60}\pi\delta\left[1 + (24a^4 + 2a^2 - 1)\sqrt{1 + 4a^2}\right]$

15. $\bar{x} = \dfrac{4}{3\pi - 6} \approx 1.16796$, $\bar{y} = 0$,

$\bar{z} = \dfrac{\pi}{2\pi - 4} \approx 1.13797$

17. Net flux: 0

19. Net flux: 1458π

Section 16.6   (page 949)

1. Both values: 4π **3.** Both values: 24

5. Both values: $\frac{1}{2}$ **7.** $\frac{2385}{2}\pi$ **9.** $\frac{1}{4}$

11. $\frac{703,125}{4}\pi$ **13.** 16π

23. $\frac{1}{48}\pi(482,620 + 29,403 \ln 11) \approx 36,201.967$

Section 16.7   (page 957)

1. -20π **3.** 0   **5.** -52π **7.** -8π **9.** -2

11. $\phi(x, y, z) = yz - 2xz + 3xy$

13. $\phi(x, y, z) = 3xe^z + 5y \cos x + 17$

Chapter 16 Miscellaneous Problems   (page 959)

1. $\frac{125}{3}$

3. $\frac{69}{8}$ (Use the fact that the integral of $\mathbf{F} \cdot \mathbf{T}$ is independent of the path.)

5. $\frac{2148}{5}$

9. $\frac{1}{3}(5\sqrt{5} - 1) \approx 3.3934$; $I_y = \frac{1}{15}(2 + 50\sqrt{5})$
≈ 7.5869

11. $\frac{2816}{7}$

17. $\frac{371}{30}\pi$

19. 72π

29. (a) $\phi'(r)(\mathbf{r}/r)$; (b) $3\phi(r) + r\phi'(r)$; (c) $\mathbf{0}$

Appendix A

1. $2\pi/9$ **3.** $7\pi/4$ **5.** $-5\pi/6$ **7.** $72°$

9. $675°$

11. $\sin x = -\frac{1}{2}\sqrt{3}$, $\cos x = \frac{1}{2}$, $\tan x = -\sqrt{3}$,
$\csc x = -2/\sqrt{3}$, $\sec x = 2$, $\cot x = -1/\sqrt{3}$

13. $\sin x = -\frac{1}{2}$, $\cos x = -\frac{1}{2}\sqrt{3}$, $\cot x = \sqrt{3}$,
$\csc x = -2$, $\sec x = -2/\sqrt{3}$, $\tan x = 1/\sqrt{3}$

15. $x = n\pi$ (n any integer)

17. $x = \frac{3}{2}\pi + 2n\pi$ (n any integer)

19. $x = 2n\pi$ (n any integer)

21. $x = n\pi$ (n any integer)

23. $x = \frac{3}{4}\pi + n\pi$ (n any integer)

25. $\sin x = -\frac{3}{5}$, $\cos x = \frac{4}{5}$, $\tan x = -\frac{3}{4}$,
$\sec x = \frac{5}{4}$, $\cot x = -\frac{4}{3}$

29. $\frac{1}{2}$ **31.** $-\frac{1}{2}$ **33.** $\frac{1}{2}\sqrt{3}$

35. $-\frac{1}{2}\sqrt{3}$ **43.** $\pi/3, 2\pi/3$ **45.** $\pi/2$

47. $\pi/8, 3\pi/8, 5\pi/8, 7\pi/8$

Appendix F

1. $\displaystyle\int_0^1 x^2 \, dx = \frac{1}{3}$

3. $\displaystyle\int_0^2 x\sqrt{4 - x^2} \, dx = \frac{8}{3}$

5. $\displaystyle\int_0^{\pi/2} \sin x \cos x \, dx = \frac{1}{2}$

7. $\displaystyle\int_0^2 \sqrt{x^4 + x^7} \, dx = \frac{52}{9}$

References for Further Study

References 2, 3, 7, and 10 may be consulted for historical topics pertinent to calculus. Reference 14 provides a more theoretical treatment of single-variable calculus topics than ours. References 4, 5, 8, and 15 include advanced topics in multivariable calculus. Reference 11 is a standard work on infinite series. References 1, 9, and 13 are differential equations textbooks. Reference 6 discusses topics in calculus together with computing and programming in BASIC. Those who would like to pursue the topic of fractals should look at Reference 12.

1. BOYCE, WILLIAM E., and RICHARD C. DiPRIMA, *Elementary Differential Equations* (5th ed.). New York: John Wiley, 1991.

2. BOYER, CARL B., *A History of Mathematics* (2nd ed.). New York: John Wiley, 1991.

3. BOYER, CARL B., *The History of the Calculus and Its Conceptual Development*. New York: Dover Publications, 1959.

4. BUCK, R. CREIGHTON, *Advanced Calculus* (3rd ed.). New York: McGraw-Hill, 1977.

5. COURANT, RICHARD, and FRITZ JOHN, *Introduction to Calculus and Analysis*. Vols. I and II. New York: Springer-Verlag, 1989.

6. EDWARDS, C. H., JR., *Calculus and the Personal Computer*. Englewood Cliffs, N.J.: Prentice Hall, 1986.

7. EDWARDS, C. H., JR., *The Historical Development of the Calculus*. New York: Springer-Verlag, 1982.

8. EDWARDS, C. H., JR., *Advanced Calculus of Several Variables*. New York: Academic Press, 1973.

9. EDWARDS, C. H., JR., and DAVID E. PENNEY, *Elementary Differential Equations with Boundary Value Problems* (3rd ed.). Englewood Cliffs, N.J.: Prentice Hall, 1993.

10. KLINE, MORRIS, *Mathematical Thought from Ancient to Modern Times*. Vols. I, II, and III. New York: Oxford University Press, 1990.

11. KNOPP, KONRAD, *Theory and Application of Infinite Series* (2nd ed.). New York: Hafner Press, 1990.

12. PEITGEN, H. O. and P. H. RICHTER, *The Beauty of Fractals*. New York: Springer-Verlag, 1986.

13. SIMMONS, GEORGE F., *Differential Equations with Applications and Historical Notes*. New York: McGraw-Hill, 1972.

14. SPIVAK, MICHAEL E., *Calculus* (2nd ed.). Berkeley: Publish or Perish, 1980.

15. TAYLOR, ANGUS E., and W. ROBERT MANN, *Advanced Calculus* (3rd ed.). New York: John Wiley, 1983.

Index

Boldface type indicates page on which a term is defined.

FORMS INVOLVING $\sqrt{u^2 \pm a^2}$

45 $\displaystyle\int \frac{du}{\sqrt{u^2 \pm a^2}} = \ln\left|u + \sqrt{u^2 \pm a^2}\right| + C$

44 $\displaystyle\int \sqrt{u^2 \pm a^2}\, du = \frac{u}{2}\sqrt{u^2 \pm a^2} \pm \frac{a^2}{2}\ln\left|u + \sqrt{u^2 \pm a^2}\right| + C$

46 $\displaystyle\int \frac{\sqrt{u^2 + a^2}}{u}\, du = \sqrt{u^2 + a^2} - a\ln\left(\frac{a + \sqrt{u^2 + a^2}}{u}\right) + C$

47 $\displaystyle\int \frac{\sqrt{u^2 - a^2}}{u}\, du = \sqrt{u^2 - a^2} - a\sec^{-1}\frac{u}{a} + C$

48 $\displaystyle\int u^2\sqrt{u^2 \pm a^2}\, du = \frac{u}{8}(2u^2 \pm a^2)\sqrt{u^2 \pm a^2} - \frac{a^4}{8}\ln\left|u + \sqrt{u^2 \pm a^2}\right| + C$

49 $\displaystyle\int \frac{u^2\, du}{\sqrt{u^2 \pm a^2}} = \frac{u}{2}\sqrt{u^2 \pm a^2} \mp \frac{a^2}{2}\ln\left|u + \sqrt{u^2 \pm a^2}\right| + C$

50 $\displaystyle\int \frac{du}{u^2\sqrt{u^2 \pm a^2}} = \mp \frac{\sqrt{u^2 \pm a^2}}{a^2 u} + C$

51 $\displaystyle\int \frac{\sqrt{u^2 \pm a^2}}{u^2}\, du = -\frac{\sqrt{u^2 \pm a^2}}{u} + \ln\left|u + \sqrt{u^2 \pm a^2}\right| + C$

52 $\displaystyle\int \frac{du}{(u^2 \pm a^2)^{3/2}} = \pm \frac{u}{a^2\sqrt{u^2 \pm a^2}} + C$

53 $\displaystyle\int (u^2 \pm a^2)^{3/2}\, du = \frac{u}{8}(2u^2 \pm 5a^2)\sqrt{u^2 \pm a^2} + \frac{3a^4}{8}\ln\left|u + \sqrt{u^2 \pm a^2}\right| + C$

FORMS INVOLVING $\sqrt{a^2 - u^2}$

54 $\displaystyle\int \sqrt{a^2 - u^2}\, du = \frac{u}{2}\sqrt{a^2 - u^2} + \frac{a^2}{2}\sin^{-1}\frac{u}{a} + C$

55 $\displaystyle\int \frac{\sqrt{a^2 - u^2}}{u}\, du = \sqrt{a^2 - u^2} - a\ln\left|\frac{a + \sqrt{a^2 - u^2}}{u}\right| + C$

56 $\displaystyle\int \frac{u^2\, du}{\sqrt{a^2 - u^2}} = -\frac{u}{2}\sqrt{a^2 - u^2} + \frac{a^2}{2}\sin^{-1}\frac{u}{a} + C$

57 $\displaystyle\int u^2\sqrt{a^2 - u^2}\, du = \frac{u}{8}(2u^2 - a^2)\sqrt{a^2 - u^2} + \frac{a^4}{8}\sin^{-1}\frac{u}{a} + C$

58 $\displaystyle\int \frac{du}{u^2\sqrt{a^2 - u^2}} = -\frac{\sqrt{a^2 - u^2}}{a^2 u} + C$

59 $\displaystyle\int \frac{\sqrt{a^2 - u^2}}{u^2}\, du = -\frac{\sqrt{a^2 - u^2}}{u} - \sin^{-1}\frac{u}{a} + C$

60 $\displaystyle\int \frac{du}{u\sqrt{a^2 - u^2}} = -\frac{1}{a}\ln\left|\frac{a + \sqrt{a^2 - u^2}}{u}\right| + C$

61 $\displaystyle\int \frac{du}{(a^2 - u^2)^{3/2}} = \frac{u}{a^2\sqrt{a^2 - u^2}} + C$

62 $\displaystyle\int (a^2 - u^2)^{3/2}\, du = \frac{u}{8}(5a^2 - 2u^2)\sqrt{a^2 - u^2} + \frac{3a^4}{8}\sin^{-1}\frac{u}{a} + C$

EXPONENTIAL AND LOGARITHMIC FORMS

63 $\displaystyle\int u e^u\, du = (u - 1)e^u + C$

64 $\displaystyle\int u^n e^u\, du = u^n e^u - n\int u^{n-1} e^u\, du$

65 $\displaystyle\int \ln u\, du = u\ln u - u + C$

66 $\displaystyle\int u^n \ln u\, du = \frac{u^{n+1}}{n+1}\ln u - \frac{u^{n+1}}{(n+1)^2} + C$

67 $\displaystyle\int e^{au}\sin bu\, du = \frac{e^{au}}{a^2 + b^2}(a\sin bu - b\cos bu) + C$

68 $\displaystyle\int e^{au}\cos bu\, du = \frac{e^{au}}{a^2 + b^2}(a\cos bu + b\sin bu) + C$

INVERSE TRIGONOMETRIC FORMS

69 $\displaystyle\int \sin^{-1} u\, du = u\sin^{-1} u + \sqrt{1 - u^2} + C$

70 $\displaystyle\int \tan^{-1} u\, du = u\tan^{-1} u - \frac{1}{2}\ln(1 + u^2) + C$

71 $\displaystyle\int \sec^{-1} u\, du = u\sec^{-1} u - \ln\left|u + \sqrt{u^2 - 1}\right| + C$

72 $\displaystyle\int u\sin^{-1} u\, du = \frac{1}{4}(2u^2 - 1)\sin^{-1} u + \frac{u}{4}\sqrt{1 - u^2} + C$

73 $\displaystyle\int u\tan^{-1} u\, du = \frac{1}{2}(u^2 + 1)\tan^{-1} u - \frac{u}{2} + C$

74 $\displaystyle\int u\sec^{-1} u\, du = \frac{u^2}{2}\sec^{-1} u - \frac{1}{2}\sqrt{u^2 - 1} + C$

75 $\displaystyle\int u^n \sin^{-1} u\, du = \frac{u^{n+1}}{n+1}\sin^{-1} u - \frac{1}{n+1}\int \frac{u^{n+1}}{\sqrt{1 - u^2}}\, du \quad \text{if } n \neq -1$

76 $\displaystyle\int u^n \tan^{-1} u\, du = \frac{u^{n+1}}{n+1}\tan^{-1} u - \frac{1}{n+1}\int \frac{u^{n+1}}{1 + u^2}\, du \quad \text{if } n \neq -1$

77 $\displaystyle\int u^n \sec^{-1} u\, du = \frac{u^{n+1}}{n+1}\sec^{-1} u - \frac{1}{n+1}\int \frac{u^n}{\sqrt{u^2 - 1}}\, du \quad \text{if } n \neq -1$

(Table of Integrals continues from previous page)

HYPERBOLIC FORMS

78 $\displaystyle\int \sinh u \, du = \cosh u + C$

79 $\displaystyle\int \cosh u \, du = \sinh u + C$

80 $\displaystyle\int \tanh u \, du = \ln (\cosh u) + C$

81 $\displaystyle\int \coth u \, du = \ln |\sinh u| + C$

82 $\displaystyle\int \operatorname{sech} u \, du = \tan^{-1} |\sinh u| + C$

83 $\displaystyle\int \operatorname{csch} u \, du = \ln \left| \tanh \frac{u}{2} \right| + C$

84 $\displaystyle\int \sinh^2 u \, du = \frac{1}{4} \sinh 2u - \frac{u}{2} + C$

85 $\displaystyle\int \cosh^2 u \, du = \frac{1}{4} \sinh 2u + \frac{u}{2} + C$

86 $\displaystyle\int \tanh^2 u \, du = u - \tanh u + C$

87 $\displaystyle\int \coth^2 u \, du = u - \coth u + C$

88 $\displaystyle\int \operatorname{sech}^2 u \, du = \tanh u + C$

89 $\displaystyle\int \operatorname{csch}^2 u \, du = -\coth u + C$

90 $\displaystyle\int \operatorname{sech} u \tanh u \, du = -\operatorname{sech} u + C$

91 $\displaystyle\int \operatorname{csch} u \coth u \, du = -\operatorname{csch} u + C$

MISCELLANEOUS ALGEBRAIC FORMS

92 $\displaystyle\int u(au + b)^{-1} du = \frac{u}{a} - \frac{b}{a^2} \ln |au + b| + C$

93 $\displaystyle\int u(au + b)^{-2} du = \frac{1}{a^2} \left(\ln |au + b| + \frac{b}{au + b} \right) + C$

94 $\displaystyle\int u(au + b)^{n} du = \frac{(au + b)^{n+1}}{a^2} \left(\frac{au+b}{n+2} - \frac{b}{n+1} \right) + C$ if $n \neq -1, -2$

95 $\displaystyle\int \frac{du}{(a^2 \pm u^2)^n} = \frac{1}{2a^2(n - 1)} \left(\frac{u}{(a^2 \pm u^2)^{n-1}} + (2n - 3) \int \frac{du}{(a^2 \pm u^2)^{n-1}} \right)$ if $n \neq 1$

96 $\displaystyle\int u \sqrt{au + b} \, du = \frac{2}{15a^2} (3au - 2b) (au + b)^{3/2} + C$

97 $\displaystyle\int u^n \sqrt{au + b} \, du = \frac{2}{a(2n + 3)} \left(u^n(au + b)^{3/2} - nb \int u^{n-1} \sqrt{au + b} \, du \right)$

98 $\displaystyle\int \frac{u \, du}{\sqrt{au + b}} = \frac{2}{3a^2} (au - 2b) \sqrt{au + b} + C$

99 $\displaystyle\int \frac{u^n \, du}{\sqrt{au + b}} = \frac{2}{a(2n + 1)} \left(u^n \sqrt{au + b} - nb \int \frac{u^{n-1} \, du}{\sqrt{au + b}} \right)$

100a $\displaystyle\int \frac{du}{u \sqrt{au + b}} = \frac{1}{\sqrt{b}} \ln \left| \frac{\sqrt{au + b} - \sqrt{b}}{\sqrt{au + b} + \sqrt{b}} \right| + C$ if $b > 0$

100b $\displaystyle\int \frac{du}{u \sqrt{au + b}} = \frac{2}{\sqrt{-b}} \tan^{-1} \sqrt{\frac{au + b}{-b}} + C$ if $b < 0$

101 $\displaystyle\int \frac{du}{u^n \sqrt{au + b}} = -\frac{\sqrt{au + b}}{b(n - 1)u^{n-1}} - \frac{(2n - 3)a}{(2n - 2)b} \int \frac{du}{u^{n-1} \sqrt{au + b}}$ if $n \neq 1$

102 $\displaystyle\int \sqrt{2au - u^2} \, du = \frac{u - a}{2} \sqrt{2au - u^2} + \frac{a^2}{2} \sin^{-1} \frac{u - a}{a} + C$

103 $\displaystyle\int \frac{du}{\sqrt{2au - u^2}} = \sin^{-1} \frac{u - a}{a} + C$

104 $\displaystyle\int u^n \sqrt{2au - u^2} \, du = -\frac{u^{n-1}(2au - u^2)^{3/2}}{n + 2} + \frac{(2n + 1)a}{n + 2} \int u^{n-1} \sqrt{2au - u^2} \, du$

105 $\displaystyle\int \frac{u^n \, du}{\sqrt{2au - u^2}} = -\frac{u^{n-1}}{n} \sqrt{2au - u^2} + \frac{(2n - 1)a}{n} \int \frac{u^{n-1} \, du}{\sqrt{2au - u^2}}$

106 $\displaystyle\int \frac{\sqrt{2au - u^2}}{u} \, du = \sqrt{2au - u^2} + a \sin^{-1} \frac{u - a}{a} + C$

107 $\displaystyle\int \frac{\sqrt{2au - u^2}}{u^n} \, du = \frac{(2au - u^2)^{3/2}}{(3 - 2n)au^n} + \frac{n - 3}{(2n - 3)a} \int \frac{\sqrt{2au - u^2}}{u^{n-1}} \, du$

108 $\displaystyle\int \frac{du}{u^n \sqrt{2au - u^2}} = \frac{\sqrt{2au - u^2}}{a(1 - 2n)u^n} + \frac{n - 1}{(2n - 1)a} \int \frac{du}{u^{n-1} \sqrt{2au - u^2}}$

109 $\displaystyle\int \left(\sqrt{2au - u^2} \right)^n du = \frac{u - a}{n + 1} (2au - u^2)^{n/2} + \frac{na^2}{n + 1} \int \left(\sqrt{2au - u^2} \right)^{n-2} du$

110 $\displaystyle\int \frac{du}{\left(\sqrt{2au - u^2} \right)^n} = \frac{u - a}{(n - 2)a^2.} + \left(\sqrt{2au - u^2} \right)^{2 - n} + \frac{n - 3}{(n - 2)a^2} \int \frac{du}{\left(\sqrt{2au - u^2} \right)^{n-2}}$

DEFINITE INTEGRALS

111 $\displaystyle\int_0^\infty u^n e^{-u} \, du = \Gamma(n + 1) = n! \quad (n \geq 0)$

112 $\displaystyle\int_0^\infty e^{-au^2} \, du = \frac{1}{2} \sqrt{\frac{\pi}{a}} \quad (a > 0)$

113 $\displaystyle\int_0^{\pi/2} \sin^n u \, du = \int_0^{\pi/2} \cos^n u \, du = \begin{cases} \dfrac{1 \cdot 3 \cdot 5 \cdot \cdots \cdot (n - 1)}{2 \cdot 4 \cdot 6 \cdot \cdots \cdot n} \dfrac{\pi}{2} & \text{if } n \text{ is an even integer and } n \geq 2 \\ \dfrac{2 \cdot 4 \cdot 6 \cdot \cdots \cdot (n - 1)}{3 \cdot 5 \cdot 7 \cdot \cdots \cdot n} & \text{if } n \text{ is an odd integer and } n \geq 3 \end{cases}$